Alkaloids: Chemical and Biological Perspectives

ALKALOIDS: CHEMICAL AND BIOLOGICAL PERSPECTIVES

Volume Five

Edited by

S. WILLIAM PELLETIER
Institute for Natural Products Research

and

The Department of Chemistry
University of Georgia, Athens

A Wiley-Interscience Publication

JOHN WILEY & SONS

New York Chichester Brisbane Toronto Singapore

Library of Congress Cataloging in Publication Data:

Alkaloids : chemical and biological perspectives.

   "A Wiley-Interscience publication."
   Includes bibliographies and indexes.
   1. Alkaloids.   I. Pelletier, S. W., 1924–

QD421.A56   1983   574.19′242   82-11071
ISBN 0-471-85372-0 (v. 5)

Printed in the United States of America

10   9   8   7   6   5   4   3   2   1

Dedicated to
the memory of

SIR ROBERT ROBINSON, O.M., F.R.S.

(1886–1975)

A pioneer in developing the theory of alkaloid biogenesis, he brought a new dimension to the study of natural products. He was the driving force behind the revolution in the electronic theory of organic chemical reactions and introduced the idea of the aromatic sextet, which rationalized the study of aromatic substitution reactions. He made major contributions to the art of synthesis, such as the Robinson annelation reaction and the synthesis of tropinones, and elucidated the structures of the anthocyanin pigments. He was a founder of the international organic chemistry journals *Tetrahedron* and *Tetrahedron Letters*.

Contributors

A. Cavé, Laboratoire de Pharmacognosie, E.R.A. 317 C.N.R.S., Faculté de Pharmacie, 92290 Châtenay-Malabry, France

A. D. Elbein, Department of Biochemistry, University of Texas Health Science Center, San Antonio, Texas

E. Gellert, Middle Park, Melbourne, Victoria, Australia

T. Hudlicky, Department of Chemistry, Virginia Polytechnic Institute and State University, Blacksburg, Virginia

L. D. Kwart, Department of Chemistry, Virginia Polytechnic Institute and State University, Blacksburg, Virginia

M. Leboeuf, Laboratoire de Pharmacognosie, E.R.A. 317 C.N.R.S., Faculté de Pharmacie, Châtenay-Malabry, France

R. J. Molyneux, Plant Protection Research Unit, United States Department of Agriculture, Western Regional Research Center, Albany, California

J. W. Reed, Department of Chemistry, Virginia Polytechnic Institute and State University, Blacksburg, Virginia

P. L. Schiff, Jr., Department of Pharmaceutical Sciences, School of Pharmacy, University of Pittsburgh, Pittsburgh, Pennsylvania

P. G. Waterman, Phytochemistry Research Laboratories, Department of Pharmacy, University of Strathclyde, Glasgow, Scotland, United Kingdom

Preface

Volume 5 of *Alkaloids: Chemical and Biological Perspectives* presents timely reviews on five important alkaloid topics.

A biochemist and chemist collaborated to author Chapter 1 on the chemistry and biochemistry of some simple indolizidines and related alkaloids. A discussion of inhibition of glycosidases by indolizidine and pyrrolidine alkaloids and of structure–activity relationships is especially significant.

The structure and synthesis of alkaloids of the phenanthroindolizidine type are reviewed in Chapter 2. Chapter 3 treats the aporphinoid alkaloids of the Annonaceae, one of the largest plant families of the Magnoliales (about 2000 species) that occur throughout the tropics.

Chapter 4 treats the alkaloids of *Thalictrum* species. These alkaloids are phenylalanine- or tyrosine-derived bases that occur as a variety of monomeric or dimeric structural variants. Extracts of *Thalictrum* species have been used in folk medicine for centuries all over the world for treatment of such conditions as jaundice, leprosy, rheumatism, pregnancy-induced vomiting and certain internal and external infections. The chapter is a comprehensive one including descriptions of the techniques of isolation, the elucidation of structures, and the pharmacology of individual alkaloids.

The discovery about 30 years ago of potent antileukemic activity in certain esters of cephalotaxine, such as harringtonine and homoharringtonine, stimulated great interest in the synthesis of alkaloids of the genus *Cephalotaxus*. Chapter 5 provides a concise review of major development in the chemical synthesis of these interesting and potentially useful alkaloids.

Each chapter in this volume has been reviewed by at least one expert in the field. Indexes for both subjects and organisms are provided.

I thank the authors for their work in preparing chapters for this volume and James L. Smith in the Wiley-Interscience Division, who has been a source of advice, encouragement, and help. Special thanks are due Shirley Thomas and her colleagues in the Production Division for their patience, skill, and care in producing such a handsome volume.

The editor invites prospective contributors to write him about topics for review in future volumes of this series.

S. WILLIAM PELLETIER

Athens, Georgia
April 1987

Contents of Previous Volumes

Contents

Alkaloids: Chemical and Biological Perspectives

Chapter One

The Chemistry and Biochemistry of Simple Indolizidine and Related Polyhydroxy Alkaloids

Alan D. Elbein
Department of Biochemistry
University of Texas Health Science Center
San Antonio, Texas 78284

Russell J. Molyneux
Plant Protection Research Unit
United States Department of Agriculture
Western Regional Research Center
Albany, California 94710

CONTENTS

1. INTRODUCTION

Among those bicyclic alkaloids possessing a bridgehead nitrogen atom, the indolizidine alkaloids, having a fused 6/5 ring system, occupy a structurally intermediate position between the pyrrolizidine and quinolizidine classes. In contrast to the latter groups, however, the indolizidines have not generally been identified so clearly as a specific class of alkaloids. This may be due in part to the occurrence of the indolizidine ring system in a number of alkaloids as a structural moiety associated with additional heterocyclic systems in such a manner that these logically take precedence, because of structural or biogenetic considerations. Examples of such alkaloids are the *Aspidosperma* [1] and *Erythrina* [2] classes, the *Amaryllidaceae* alkaloids [3], and certain steroidal alkaloids occurring in the *Solanaceae* [4].

Excellent general reviews of the chemistry of alkaloids in which the indolizidine ring predominates are those of Gellert [5], Saxton [6], and Lamberton [7]. The more complex *Tylophora* (phenanthroindolizidine) [8] and *Elaeocarpus* [9] alkaloids have also been reviewed in detail, as have the gephrotoxins [10], which occur as constituents of Neotropical poison-dart frogs. The purpose of this chapter is not to reiterate these summaries, but rather to discuss in detail the chemistry and biochemistry of those recently isolated simple indolizidine alkaloids, at present few in number, which have generated intense interest among scientists in diverse fields because of their enzyme-inhibitory properties. We hope that such a treatment will stimulate the isolation, structural determination, and synthesis of novel indolizidine alkaloids of this type, which must surely exist in nature; generate interest in the relationship of structure to enzyme-inhibitory activity; and ultimately result in application of the alkaloids or their derivatives for medicinal or other biological purposes.

The parent ring system of the indolizidine alkaloids, 1-azabicyclo[4,3,0]nonane (**1**), also known as δ-coniceine, indolizidine, or octahydropyrrocoline, is not known as a natural product and therefore cannot be defined as a true alkaloid [11]. It has, however, been derived from the poison hemlock (*Conium maculatum* L.) alkaloid (±)-coniine (**2**) by bromination and treatment with sulfuric acid [12] and has been prepared by numerous synthetic procedures [13]. The base is obtained as a colorless oil, b.p. 75°/ 43 mm, and forms picrate (m.p. 228–229°) and hydrobromide (m.p. 196°) salts. The generally accepted numbering system for the indolizidine alkaloids is illustrated in structure **1**.

1   **2**

2. CHEMISTRY

In the discussion of the chemistry of the simple indolizidine alkaloids that follows, the alkaloids are dealt with individually rather than as a group, proceeding in order of increasing molecular weight.

2.1. Swainsonine, $C_8H_{15}NO_3$

2.1.1. Occurrence and Isolation.   Consumption of the western Australian legume *Swainsona canescens* (Benth.) A. Lee and of certain other members of the genus, many of which are commonly known as Darling Pea or Poison Pea, produces locomotor disturbances in cattle, sheep and horses, which are then described as "peastruck" [14]. The disease was shown to be pathologically and biochemically similar to the genetically controlled lysosomal storage disease mannosidosis, which occurs in humans and Angus cattle [15]. Recognition of this similarity stimulated a search for constituents of the plant capable of producing mannosidosis, ultimately leading to the isolation of swainsonine (**3**), a powerful inhibitor of lysosomal α-mannosidase [16].

The well-known "loco" disease of the western United States, produced by consumption of a number of *Astragalus* and *Oxytropis* species (family Leguminosae) commonly known as "locoweeds," has been a serious economic problem for livestock producers for more than 100 years [17]. Numerous attempts to isolate the toxic constituent were unsuccessful, although Fraps and Carlyle [18] obtained from Big Bend locoweed (*A. mollissimus* var. *earlei*), an impure, basic, very hygroscopic material, capable of producing the loco syndrome in cats, which they named "locoine." The early recognition [19] that "loco" and "peastruck" sheep exhibited clinical similarities has recently been supported by the observation that sheep fed *A. lentiginosus, A. pubentissimus,* and *O. sericea* and those fed *Swainsona galegifolia* showed no obvious differences in their microscopic lesions [20]. This finding provided the impetus for the examination and subsequent isolation of swainsonine from spotted locoweed (*A. lentiginosus*) [21].

A third occurrence of swainsonine as a minor metabolite, previously incorrectly identified [22] as 3,4,5-trihydroxypyrindine (**4**), of the fungus *Rhizoctonia leguminicola* has recently been reported [23]. The alkaloid is found in association with the major indolizidine alkaloid slaframine (**37**) (see below). In addition the alkaloid has been isolated from the culture broth of another fungus, *Metarhizium anisopliae* F3622, by Japanese workers [32].

3

4

Swainsonine is very soluble in water and hydroxylic solvents, moderately soluble in polar, nonhydroxylic solvents such as acetone, ethyl acetate, and chloroform, and quite insoluble in nonpolar solvents. Conventional alkaloid isolation and purification techniques, involving partition of extracts into acidic solution, basification, and back-extraction into a water-immisible organic solvent are therefore not applicable. The alkaloid was obtained by extraction of the plant material with ethyl acetate [16] or acetone [21] and purification of the water-soluble portion of the extract by chromatography on an AG50W-X8 (NH_4^+ form) ion-exchange resin. The fraction eluting with dilute ammonium hydroxide was monitored for α-mannosidase inhibitory activity using jack bean α-mannosidase or mouse liver homogenate as an enzyme source and 4-methylumbelliferyl-α-D-mannopyranoside as substrate. Extraction from the residue of the active fraction with ammonia-saturated chloroform and crystallization from the same solvent gave swainsonine as white needles, m.p. 144–145° (dec.) [16], 149–150° (dec.) [24]; $[\alpha]_{589}^{25}$ −84.0°, $[\alpha]_{578}^{25}$ −87.7°, $[\alpha]_{546}^{25}$ −92.2°, and $[\alpha]_{436}^{25}$ −165.4°, ethanol (c, 0.99).

The yield of swainsonine from *Swainsona canescens* was 0.001% and from *Astragalus lentiginosus* was 0.003%. From *Rhizoctonia leguminicola* mycelium a yield of approximately 0.004% was obtained, the alkaloid being purified by formation of the acetate derivative and subsequent hydrolysis rather than by ion-exchange chromatography. Recently swainsonine has been shown to be present in the locoweeds *Astragalus mollissimus, A. wootoni, Oxytropis sericea,* and *O. lambertii,* as well as in the selenium-containing species *A. bisulcatus* and *A. praelongus,* which have not generally been acknowledged to be loco inducers. Estimates of alkaloid content by TLC indicate that higher levels occur in the flowers, pods and seeds than in the leaves and that even 2-year-old, dry, dead stems retain swainsonine [25], a finding that supports field observation that dead locoweeds remain toxic to livestock.

2.1.2. Structure. The structure of swainsonine was determined primarily by spectroscopic methods. Nevertheless the formation of certain derivatives provides confirmatory evidence for the assigned structure. Thus, treatment with acetic anhydride yields a di-*O*-acetate (m.p. 128–129°) at room temperature and a tri-*O*-acetate (oil) at 80°, but no *N*-acetyl derivative, indicating the presence of a tertiary nitrogen atom [16]. With 2,2-dimethoxypropane and *p*-toluenesulfonic acid, an acetonide (m.p. 105–107°) is produced which therefore establishes the existence of a *cis*-diol group [23]. Color tests are somewhat equivocal because although a weak positive reaction is obtained with Dragendorff alkaloid reagent, a definite purple color is produced with ninhydrin, an observation that originally led to the formulation of the fungal metabolite as the pyrindine (4), having a secondary amine group [22].

The 90 MHz ¹H and ¹³C NMR spectra of swainsonine and its acetate derivatives established the absence of unsaturation in the molecule. The off-resonance decoupled ^{13}C spectrum showed the presence of four methine carbons, three of which are bound to oxygen and one to nitrogen, and four methylene carbons, two of which are bound to nitrogen. The fundamental structure of the alkaloid was therefore of the trihydroxypyrrolizidine or trihydroxyindolizidine type. However, the pyrrolizidine alkaloid ring system can be eliminated from consideration because no methylene carbon atoms bound to oxygen are observed.

The indolizidine alkaloid structure is supported by the ^{1}H NMR spectrum. Addition of trifluoroacetic acid to swainsonine in D_2O quaternized the nitrogen atom, inducing large downfield shifts of those protons situated on adjacent carbon atoms. Thus the H(3) and H(5) methylene protons exhibited shifts of 50–90 Hz and the methine proton H(8a) a shift of 109 Hz. Specific couplings were defined by a series of INDOR experiments, the *trans*-diaxial relationship of H(8) and H(8a) being established by the large vicinal coupling constant of 10.0 Hz. Similarly, the *cis* relationship of H(1) to H(8a) was determined by a vicinal coupling constant of 3.3 Hz. Comparison of the spectrum with ^{1}H NMR spectra of various pyrrolizidine alkaloid necine bases provided confirmatory evidence for the assigned stereochemistry.

High-resolution mass spectrometry showed a parent ion at *m/e* 173.1054 (5) and loss of $C_2H_4O_2$ to give a base peak fragment (*m/e* 113.0834) formulated as the ion (6). This fragmentation pattern was analogous with that observed for necine bases such as platynecine (7), in which initial fissions are of bonds β to the nitrogen atom, occurring solely in the ring bearing the 7-hydroxyl group due to promotion by this substituent, yielding the base peak fragment (8) [26]. Other fragmentations were compatible with the formulation of the swainsonine parent ion as (5).

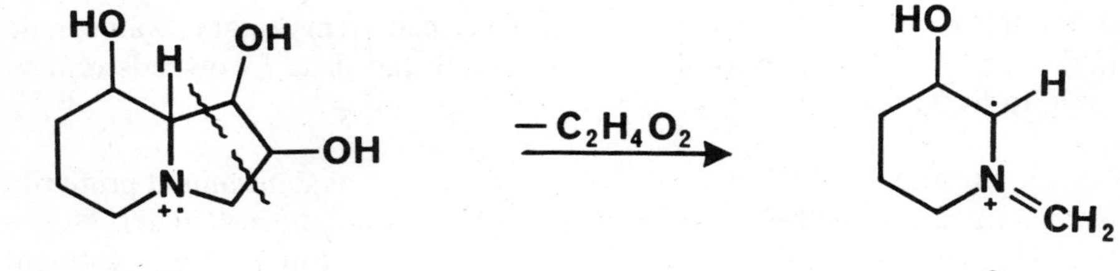

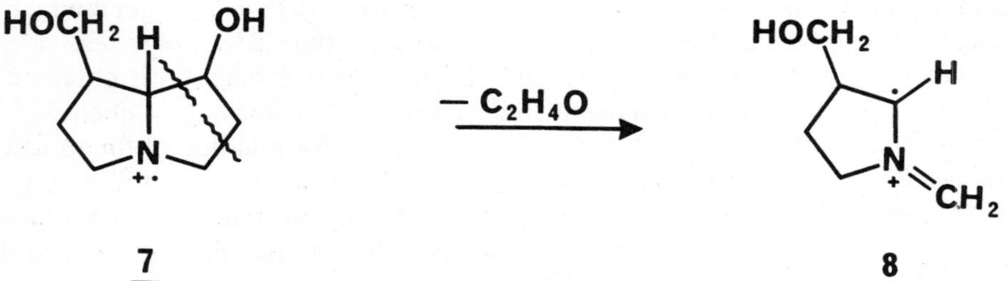

The stereochemistry of the ring junction was established by the presence of strong Bohlmann bands (~2800 cm^{-1}) in the IR spectrum of swainsonine and its triacetate, indicative of two or more α-hydrogen atoms *trans*-diaxial to the nitrogen lone pair, a condition that could only be satisfied by a *trans*-fused system.

An X-ray crystal structure determination of swainsonine diacetate provided supporting evidence for the assigned structure and relative configuration, indicating that the six-membered ring adopts a chair conformation whereas the five-membered ring has a pseudoenvelope conformation [27].

The absolute configuration of swainsonine was determined by the asymmetric induction method of Horeau [28]. The acetonide derivative, when acylated with racemic 2-phenylbutanoic anhydride, yielded an enantiomeric excess of (+)-2-phenylbutanoic acid, indicating an (R) configuration for C(8). The structure of swainsonine is therefore established as (1S,2R,8R,8aR)-1,2,8-trihydroxyindolizidine (2). This assignment has recently been confirmed by stereospecific syntheses (see below).

2.1.3. Synthesis. The intense interest in swainsonine as a biochemical tool for the study of glycoprotein processing stimulated a number of synthetic approaches to this alkaloid, no less than four appearing in print almost simultaneously in early 1984 [29–32] and a fifth being reported [33] but as yet unpublished.

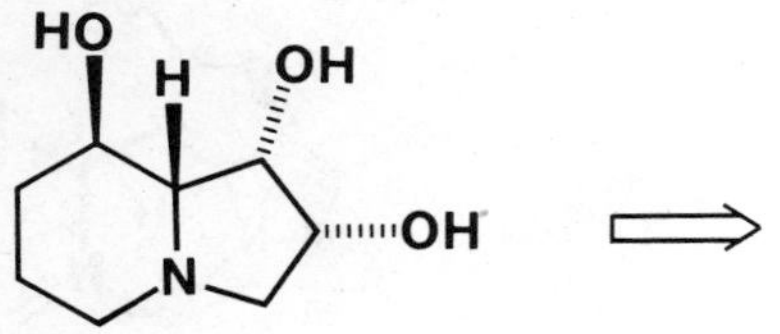

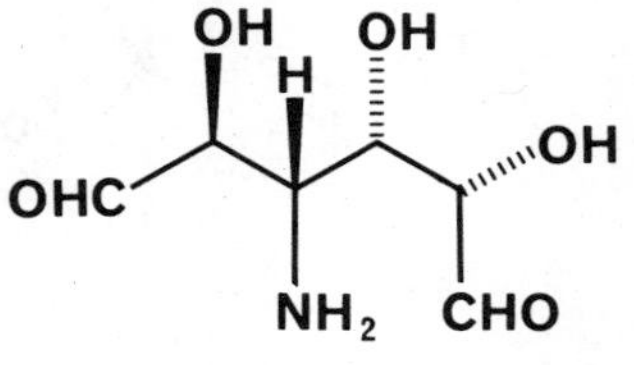

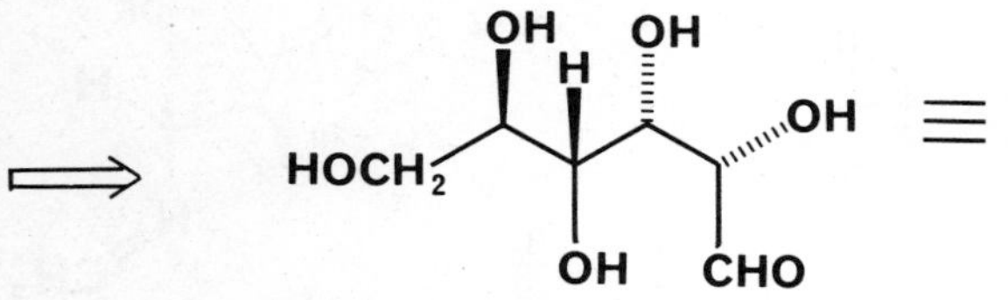

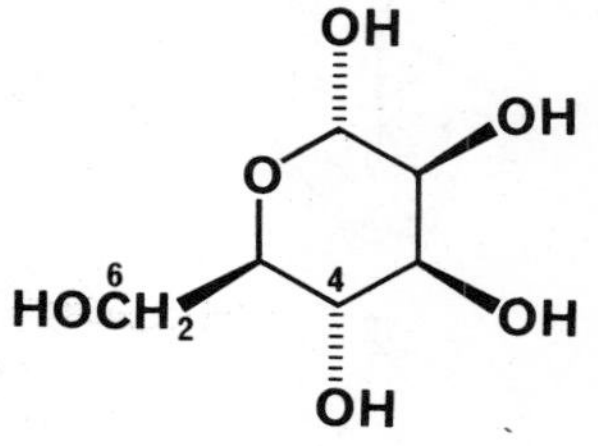

The strategic approach of all the syntheses was based upon the same retrosynthetic reasoning, the four chiral centers in swainsonine (3) being derived from the adjacent chiral centers in either D-glucose [29,30] or D-mannose [31,32], via a suitable form of the aminodialdehyde (9). Thus introduction of a nitrogen function at the 4 position of D-mannose (10) and a two-carbon chain extension at the 6 position would provide a carbon skeleton suitable for generation of the fused heterocyclic rings of swainsonine.

The approaches of the Suami and Richardson groups [29,30] followed an essentially identical route from the pyrrolidine intermediate (17), the synthetic equivalent to the aminodialdehyde (9), differing only in the choice of blocking groups (R = benzyl [29] or R = acetyl [30]). The two routes are outlined in Figure 1. The starting materials methyl 3-acetamido-2,4,6-tri-*O*-

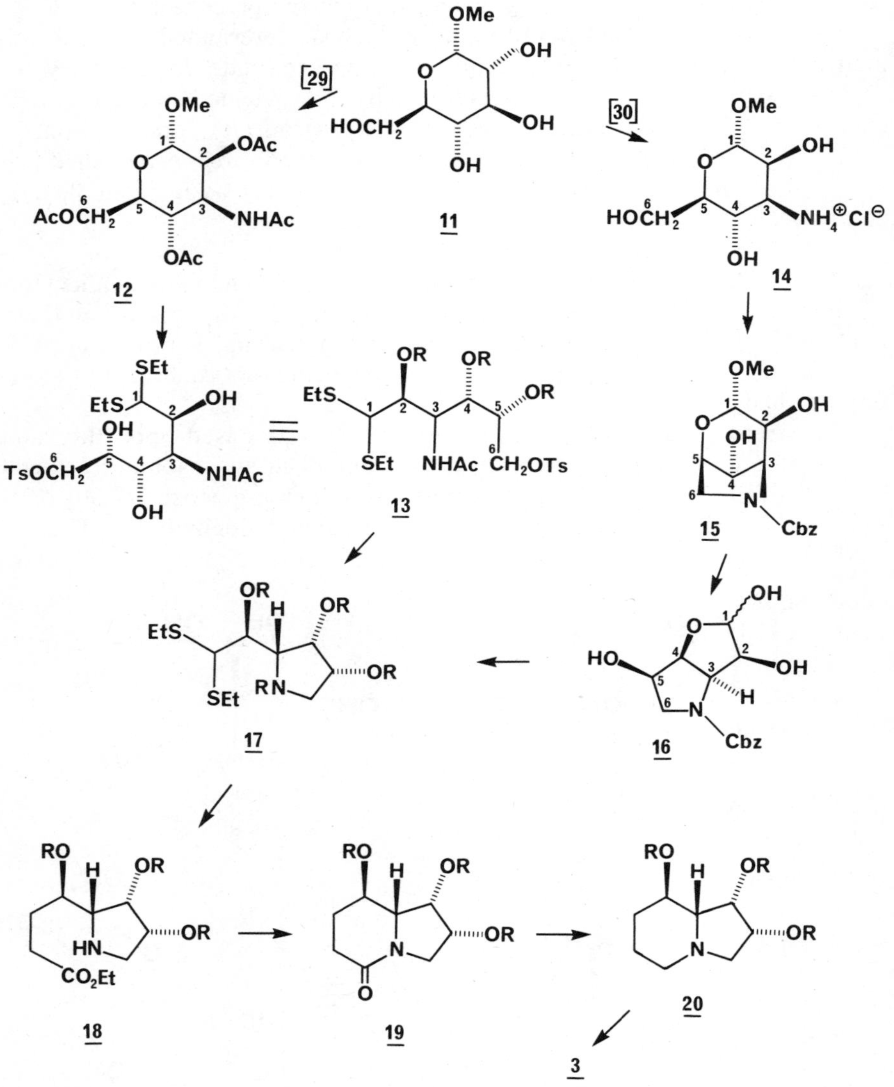

Figure 1. Synthetic routes to swainsonine (3) from D-glucose [29,30].

acetyl-3-deoxy-α-mannopyranoside (12) [29] or methyl 3-amino-3-deoxy-α-D-mannopyranoside hydrochloride (14) [30], both originally derived from methyl α-D-glucopyranoside (11) with inversion of configuration at C(2), were carried through a considerable number of blocking and deblocking steps to the common pyrrolidine intermediate (17). Condensation with ethanethiol of the protected mannose derivative yielded a diethyldithioacetal, the tosylate (13) of which underwent a base-catalyzed nucleophilic displacement to give 17, R = benzyl. Alternatively, the 3,6-imine (15), derived from the mannose derivative 14, underwent acidic hydrolysis to the 3,6-iminohexofuranose (16), which upon reaction with ethanethiol under acidic conditions yielded the corresponding diethyldithioacetal (17, R = H).

Dethioacetylation of the protected pyrrolidine (17) gave the corresponding aldehyde, which underwent condensation with diethyl ethoxycarbonylmethylphosphonate [29] or ethoxycarbonylmethylenetriphenylphosphorane [30] to yield an unsaturated ester as a mixture of stereoisomers. Subsequent hydrogenation of the double bond gave the saturated ester (18). This ester was cyclized to the lactam (19), which was reduced to the tertiary amine (20) and deacetylated or debenzylated to yield authentic swainsonine (3). An overall yield from methyl α-D-glucopyranoside (11) of 0.7% in 18 steps was obtained by Suami's group [29] and 0.7% in 15 steps by Richardson's group [30].

The two syntheses commencing with D-mannose (21) also followed remarkably similar routes to each other, as outlined in Fig. 2, but differed markedly in the overall yield of swainsonine. Thus Fleet's group at Oxford reported an overall yield of 9% in 13 steps [31] whereas Takaya's group achieved a yield of only 0.1% in the same number of steps [32]. In both cases the synthetic equivalent to the aminodialdehyde (9) was the 4-azido-4-deoxy-2,3-O-isopropylidene-α-D-mannopyranoside (22, R = benzyl [31], or R = CH$_3$ [32]). In order to generate this intermediate two inversions at C(4) of the suitably protected sugar were necessary, the second occurring on introduction of the azide functionality using sodium azide in dimethylformamide.

At this point the two routes diverged, a two-carbon chain extension at C(6) via a Wittig reaction giving either the unsaturated aldehyde (23) or the unsaturated ester (24). Treatment of the aldehyde (23) with 10% palladium on charcoal achieves hydrogenation of the double bond, reduction of the azide to an amine group, and subsequent intramolecular reductive amination of the aminoaldehyde to yield the bicyclic secondary amine (25, R = benzyl) in a single step [31]. In contrast the corresponding bicyclic amine (25, R = CH$_3$) was obtained by hydrogenation of the ester (24), cyclization to the lactam, and reduction with borane in tetrahydrofuran [32].

The considerable difference in the overall yield by the two routes resulted primarily from the ease of removal of the benzyl blocking group in the amine (25, R = benzyl) as opposed to the methyl group in the analogous amine (25, R = CH$_3$). Thus, hydrogenolysis of the benzyl group formed a lactol

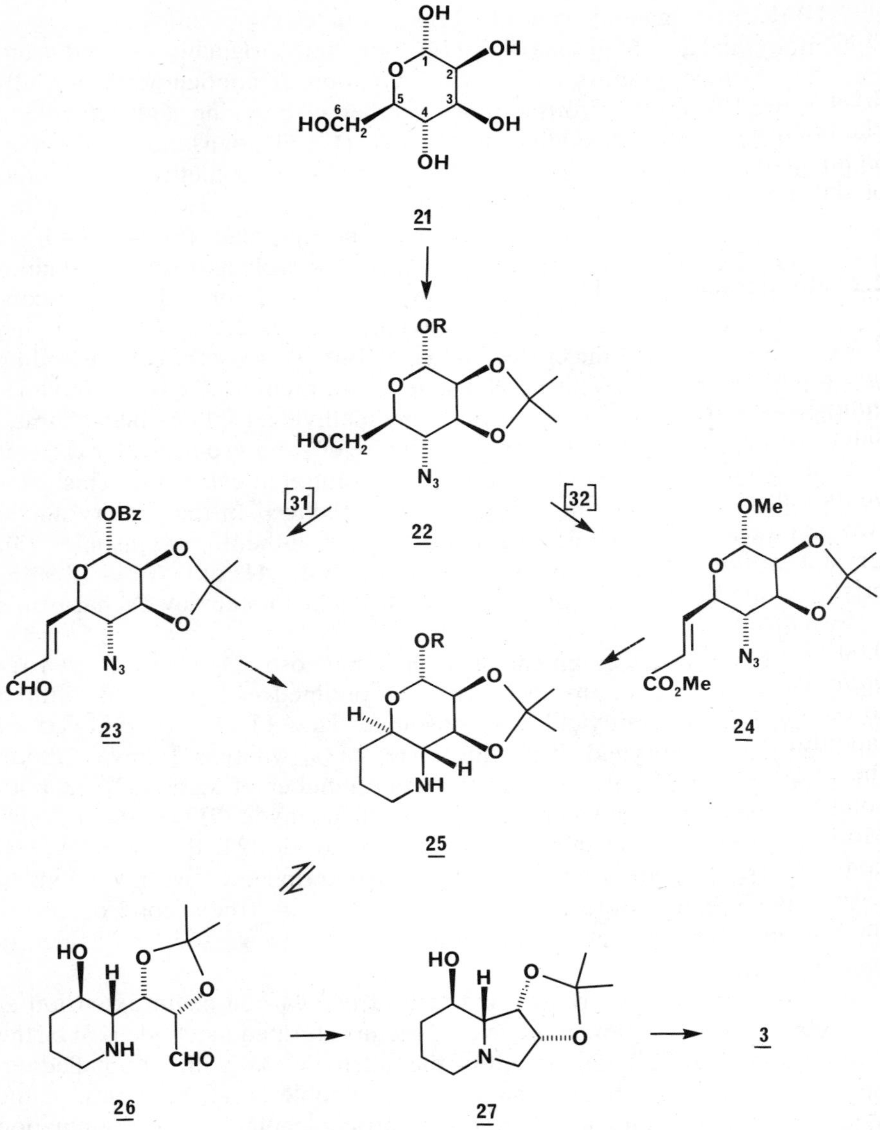

Figure 2.   Synthetic routes to swainsonine (**3**) from D-mannose [31,32].

(**25**, R = H), in equilibrium with the open-chain aminoaldehyde (**26**), which undergoes reductive amination to form swainsonine acetonide (**27**) in high yield. Removal of the isopropylidene protecting group with trifluoroacetic acid gave swainsonine (**3**) [31]. In contrast, demethylation of the amine (**25**, R = CH₃) with boron trichloride gave numerous spots on the TLC and subsequent reduction with sodium cyanoborohydride gave only a 1.8% yield of swainsonine in these final two steps [32].

2.1.4. Biosynthesis. No studies of the biosynthesis of swainsonine in plants have been reported. The biosynthesis of the alkaloid in *Rhizoctonia leguminicola* has been investigated and is discussed in association with that of slaframine (**37**) in Section 2.4.4.

2.2. Swainsonine *N*-Oxide, $C_8H_{15}NO_4$

2.2.1. Occurrence and Isolation.   In addition to swainsonine a second minor alkaloid was detected by Dragendorff's reagent in extracts of *Astragalus lentiginosus*. This compound was isolated in its pure form with considerable difficulty and shown to be swainsonine *N*-oxide (**28**) [21].

Swainsonine *N*-oxide is extremely soluble in water and hydroxylic solvents but in contrast to swainsonine it is insoluble in chloroform, and the two alkaloids can be separated on the basis of this differential solubility. Crystallization from acetone gave pale yellow, extremely hygroscopic prisms, m.p. 161–163° (dec.).

Swainsonine *N*-oxide was isolated from *A. lentiginosus* in a yield of 0.001% and has been detected in a number of other locoweeds, both *Astragalus* and *Oxtropis* species, by TLC [25]. The alkaloid has not been reported to co-occur with swainsonine in *Swainsona canescens,* or as a fungal metabolite. However, it appears unlikely that it is an artifact produced during the extraction process because the proportion of *N*-oxide relative to swainsonine varies widely between different locoweed species and plant parts [25]. Moreover, because *N*-oxides of pyrrolizidine and quinolizidine alkaloids are commonly found, frequently exceeding the proportion of free base alkaloid [34], it seems reasonable to assume that a similar situation exists with the indolizidine alkaloids.

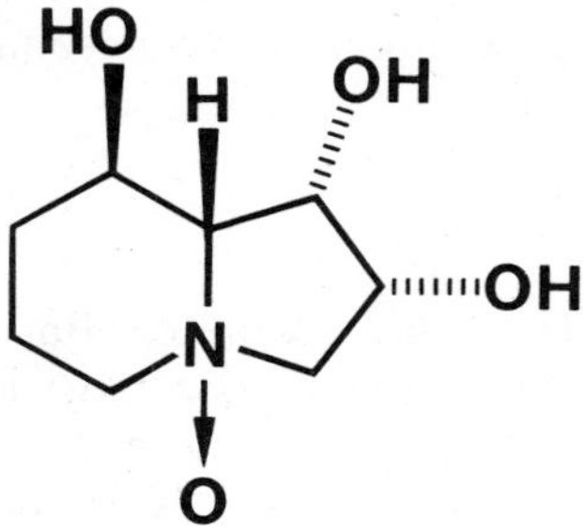

28

2.2.2. Structure. Swainsonine N-oxide underwent reduction with zinc dust in $2N$ hydrochloric acid, or with sodium borohydride, to give a product indistinguishable from swainsonine [24]. Although combustion analysis gave a molecular formula of $C_8H_{15}NO_4$, the EI mass spectrum, obtained at a probe temperature of 180°, was virtually identical with that of swainsonine. However, the isobutane CI mass spectrum determined at 140° exhibited a molecular ion at m/e 190 (100%) and major fragments corresponding to loss of oxygen and sequential loss of three molecules of water. The inability to obtain a molecular ion at higher probe temperatures is probably due to thermal deoxygenation, a phenomenon frequently observed in the mass spectra of N-oxides [35].

The ^{13}C NMR spectrum in D_2O, on comparison with that of swainsonine, provided confirmatory evidence for the N-oxide structure, the methylene carbons bound to nitrogen at C(3) and C(5) showing downfield shifts of 15.7 and 11.2 p.p.m., respectively, and the methine carbon at C(8a) a downfield shift of 4.3 p.p.m.

The stereochemistry of the N-oxide bond relative to the proton H(8a) has not been established. In pyrrolizidine alkaloids the two fused five-membered rings are fixed in a rigid conformation such that the nitrogen lone pair and the corresponding N-oxides necessarily have the same stereochemistry. On the other hand, with quinolizidine alkaloids the two fused six-membered rings are sufficiently flexible that N-oxide formation can occur with either retention or inversion to yield both *cis* and *trans* isomers from a single alkaloid. The situation with the indolizidine alkaloids is equivocal. Methylation of indolizidine gave a mixture of *cis*- and *trans*-fused methiodides [36] but oxidation of swainsonine gave only a single N-oxide [21]. Probably, this product is the *trans*-N-oxide with the same configuration as swainsonine. In this form the hydroxyl groups at C(1) and C(2) would be on the same side of the ring as the oxide group and therefore stabilize the molecule by intramolecular hydrogen bonding.

2.2.3. Synthesis. Oxidation of swainsonine with 30% hydrogen peroxide in ethanol, a general method for formation of N-oxides, yields a single product identical in all respects with natural swainsonine N-oxide [21].

2.3. Castanospermine, $C_{18}H_{15}NO_4$

2.3.1. Occurrence and Isolation. Moreton Bay chestnut or black bean (*Castanospermum australe* A. Cunn.) is a large leguminous tree found in rain forests and along creek banks in eastern Australia from northern New South Wales through Queensland, primarily in coastal regions. It has been introduced as a landscape tree in southern California. Although Aborigines ate the large, chestnut-like seeds after soaking them in water and roasting them, early European settlers found that consumption of raw or roasted seeds produced severe gastrointestinal pain. Recent cases of human poisoning by the raw seeds have been reported [14].

Both cattle and horses ingest the seeds when available, with certain animals apparently developing a preference for them, despite the often fatal gastroenteritis that develops. Cattle appear to be more susceptible to poisoning than horses.

Early reports implied that the toxin was a saponin but recently the indolizidine alkaloid castanospermine (29) has been isolated from *C. australe* seeds [37] and shown to be a potent inhibitor of almond emulsin β-glucosidase and fibroblast lysosomal α- and β-glucosidases [38]. This property may well account for the toxicity and gastrointestinal effect of the seeds in both humans and animals.

Castanospermine is very soluble in water and hydroxylic solvents but insoluble in nonpolar solvents. Extraction of ground *C. australe* seeds with 75% ethanol [37], water [38], or methanol [24] and purification of the water-soluble portion of the extract by ion-exchange chromatography, in a similar manner to that used for the isolation of swainsonine, gave a fraction containing basic materials. Crystallization from aqueous ethanol or methanol gave the alkaloid as white prisms, m.p. 212–215° (dec.) [37]; $[\alpha]^{24}_{589}$ + 79.2°, $[\alpha]^{24}_{578}$ + 82.6°, $[\alpha]^{24}_{546}$ + 93.8°, $[\alpha]^{24}_{436}$ + 59.0°, methanol (c, 0.62) [24].

Castanospermine was isolated from immature *C. australe* seeds in 0.06% yield [37] and from mature seeds in 0.3% yield [38]. The alkaloid has also been detected in the leaves and bark of the tree [24].

2.3.2. Structure. Spectroscopic methods were employed primarily in the determination of the structure of castanospermine. However, treatment with acetic anhydride in pyridine gave a tetra-*O*-acetate (m.p. 110–112°) and a methiodide (oil) was formed with iodomethane, indicating that the compound was a tetrahydroxylated tertiary base alkaloid. In spite of this evidence characteristic alkaloid colors failed to form with Dragendorff's reagent or iodoplatinate, but a yellow-brown color, changing to purple over several days and therefore more characteristic of nonprotein amino acids, was produced with ninhydrin [37]. The molecule was degraded by periodic acid, indicating the presence of vicinal hydroxyl groups. Reaction of castanospermine with

29

30% hydrogen peroxide in ethanol gave a single N-oxide, m.p. 202–203° (dec.), but this compound was not detectable in plant extracts [39].

The ^{13}C NMR spectrum of castanospermine indicated the presence of three methylene and five methine carbon atoms, all of the latter being bound either to oxygen or nitrogen, with no unsaturation present in the molecule. The 360 MHz ^{1}H NMR spectrum clearly showed signals for each of the 15 protons present. Extensive double resonance experiments established the coupling constants and defined the structure of castanospermine as the tetrahydroxyindolizidine alkaloid (29).

The high-resolution EI mass spectrum of the alkaloid showed a molecular ion at m/e 189.1002, corresponding to the molecular formula of $C_8H_{15}NO_4$ established by elemental analysis, and exhibited a major fragment at m/e 145.0737. This ion presumably has a structure analogous to the ion (6) observed in the mass spectrum of swainsonine.

An X-ray crystal structure determination of castanospermine confirmed the structural assignment and showed that the six-membered ring adopts a chair conformation whereas the five-membered ring has a pseudoenvelope conformation, as observed in swainsonine. The X ray defined the relative configuration and an arbitrary choice was made of the enantiomer in which the bridgehead proton was β. A subsequent synthesis of castanospermine has shown that this choice was incorrect and that the structure is therefore (1S,6S,7R,8R,8aR)-1,6,7,8-tetrahydroxyindolizidine (29). It is interesting to note that although swainsonine (3) and castanospermine (29) show close structural similarities as highly hydroxylated indolizidine alkaloids, they have opposite absolute configurations at the bridgehead carbon C(8a). Whether or not this is a reflection of differing biosynthetic pathways remains to be established.

2.3.3. Synthesis.   The total synthesis of castanospermine from D-glucose has been achieved via the enantiospecific route outlined in Fig. 3 [40], thus establishing the absolute configuration. Tri-O-benzyl-D-glucopyranoside (30) was condensed with benzylamine to give an anomeric mixture, which on reduction with lithium aluminum hydride provided the diol (31). Trifluoracetylation of the amine group and a series of protection and deprotection steps of the diol moiety provided the epoxide (32, R = COCF$_3$) with inverted configuration at C(5). The epoxide function provides a suitable leaving group for cyclization of the aminoepoxide (32, R = H), which occurs spontaneously on reductive cleavage of the amide (32, R = COCF$_3$) with sodium borohydride to form the piperidinol (33).

The alcohol function of the piperidine (33) was oxidized to an aldehyde, and condensed with lithio t-butylacetate to yield a 1:1 mixture of diastereoisomers (34). These were separated and the less polar diastereomer hydrogenolyzed and cyclized with acid to the lactam (35). Reduction of this lactam with diisobutylaluminum hydride gave (+)-castanospermine (29) identical with the natural product. The more polar diastereoisomer of (34)

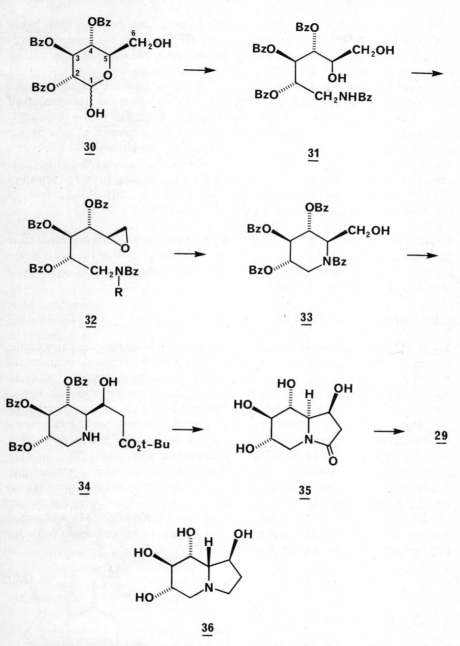

Figure 3. Synthesis of castanospermine (**29**) from D-glucose [40].

was converted by an identical route into 1-epicastanospermine (**36**), $[\alpha]_{589}$ $+6°$, H_2O (c, 0.45). The synthesis of castanospermine by this route involves 15 steps from D-glucose, and although the overall yield was not reported the sequence proceeds to the intermediate (**34**) in ~10% yield.

A number of benefits accrue from the synthetic route. Debenzylation of the piperidinol (**33**) provides a total synthesis of (+)-deoxynojirimycin, itself an inhibitor of a number of glucosidases. The availability of synthetic 1-epicastanospermine (**36**) will provide a means for determining whether or not epimerization of the 1-hydroxyl group alters the glucosidase-inhibitory activity relative to castanospermine. In addition the lactam (**35**), its epimer, and the corresponding pentahydroxyindolizidines arising from reduction of the keto group would provide excellent models for a study of the effects of structure and stereochemistry on glycosidase-inhibitory activity.

2.3.4. Biosynthesis. Investigations of the biosynthesis of castanospermine have not been reported. However, it has been suggested that the trihydroxy-pipecolic acid available synthetically from the piperidinol (**33**) might be a biosynthetic intermediate [40].

2.4. Slaframine, $C_{10}H_{18}N_2O_2$

2.4.1. Occurrence and Isolation. A toxicity problem affecting dairy cattle, horses, and sheep was frequently reported during the 1950s throughout the Midwest. The disease was characterized by excessive salivation (''slobber-ing''), lacrimation, diarrhea, feed refusal, and anorexia, and was subse-quently shown to be associated with consumption of second-cutting red clo-ver and occasionally other legume forages infected with *Rhizoctonia leguminicola*. This fungus is the cause of ''black patch,'' a disease respon-sible for considerable losses of the red clover seed crop [41].

A culture of *R. leguminicola* grown on nontoxic red clover extract pro-duced a mycelium that caused profuse salivation when fed to guinea pigs or infused directly into the rumen of dairy cows [42]. However, the culture filtrate was nontoxic. The indolizidine alkaloid slaframine (**37**) was subse-quently isolated from the mycelium and shown to be responsible for the physiological effects in animals [43,44].

37   **38**

Slaframine is soluble in water and hydroxylic solvents but insoluble in chloroform at acidic or neutral pH. However, at pH 10 the alkaloid will partition from water into chloroform, and this property may be used to advantage for its purification. Extraction of *R. leguminicola* mycelium with ethanol and treatment of the water-soluble portion of the extract with lead diacetate precipitated impurities. Additional contaminants were removed by extraction with chloroform at acidic pH and partitioning of the alkaloid into chloroform at alkaline pH. After repeated partitioning cycles, slaframine was isolated as the dipicrate. The free base has not been isolated in crystalline form but is obtained as a colorless oil that undergoes rapid aerial oxidation. The yield of slaframine dipicrate was ~3.3 g/L of fungal culture.

2.4.2. Structure.   The structure of slaframine was originally reported as 1-acetoxy-8-aminoindolizidine (**38**) [45,46] primarily on the basis of degradative and MS evidence but was subsequently revised to 1-acetoxy-6-aminoindolizidine (**37**) through the use of ^{1}H NMR spin-decoupling experiments [47].

Slaframine dipicrate, m.p. 183–184°, had the elemental analysis $C_{22}H_{24}N_8O_{17}$ and the free base therefore contained two nitrogen atoms, one of which was shown to be primary by Van Slyke analysis. A purple color formed with ninhydrin supported this conclusion and a positive Dragendorff's reaction indicated the presence of a tertiary nitrogen atom. However, more recent studies with swainsonine (**3**) and castanospermine (**29**) have shown that such color tests are not unequivocal because both these alkaloids give a positive color test with ninhydrin and only a weak reaction with Dragendorff's reagent.

Treatment of slaframine with weak base gave a crystalline deacetyl derivative which lacked physiological activity. When this compound was treated with acetic anhydride an *N*-acetyl derivative, m.p. 140–142°, was formed that also lost an acetyl group on alkaline hydrolysis. This hydrolysis product was oxidized with chromic acid to a ketone, proving that slaframine possesses an *O*-acetyl group. Von Braun cleavage of the pyrrolidine ring with cyanogen bromide gave a product formulated as the cyanopiperidine (**39**), which formed the piperidine (**40**, $R_1 = R_2 = H$) via conversion to the

39   **40**

iodide and reduction with lithium aluminum hydride. Methylation of the latter gave the di-*N*-methyl derivative (**40**, R_1 = Me, R_2 = H), and acetylation gave the triacetate (**40**, R_1 = R_2 = Ac).

The ^{1}H NMR spectrum of slaframine hydrochloride in D_2O showed a secondary *O*-acetyl group and the hydrochloride of the deacetyl derivative exhibited a carbinol proton. Spin-decoupling experiments carried out at 100 MHz on *N*-acetylslaframine hydrochloride established that H(8a) was coupled to a single proton at C(1) and two protons at C(8), rather than the —CHNAc proton as required by the original 8-amino structure (**38**). The latter proton was shown to be coupled to both C(5) protons and was therefore located at C(6), having a half-bandwidth consistent with an equatorial configuration. A *cis* configuration for H(1) and H(8a) was established by direct comparison of the NMR spectrum of *N*-acetylslaframine with the spectra of isomeric l-acetoxyindolizidines of known relative stereochemistry [48].

The mass spectra of slaframine and its various salts were virtually identical due to dissociation in the ion source. The molecular ion at m/e 198 was consistent with the elemental analysis of the dipicrate. In contrast to swainsonine and castanospermine, slaframine undergoes fragmentation in the piperidine ring rather than the pyrrolidine ring, giving rise to major fragments at m/e 155.095 (**41**) and m/e 142.087 (**42**).

The absolute configuration of slaframine was determined by Horeau's method. Treatment of *N*-acetyl-*O*-deacetylslaframine with racemic 2-phenylbutanoic anhydride gave an enantiomeric excess of (−)-2-phenylbutanoic acid, indicating an (*S*)-configuration for C(1). The structure of slaframine is therefore established as (1*S*,6*S*,8a*S*)-1-acetoxy-6-aminoindolizidine (**37**).

2.4.3. Synthesis.   Three total syntheses of racemic slaframine have been reported. The two earlier routes [49,50] relied upon Dieckmann cyclization, a classical approach to the formation of the indolizidine ring system, whereas the most recent approach [51] used an intramolecular imino Diels–Alder reaction.

The nonstereoselective route devised by Rinehart's group [49], outlined in Figure 4, commenced with 2-bromo-5-nitropyridine, which was converted

41   **42**

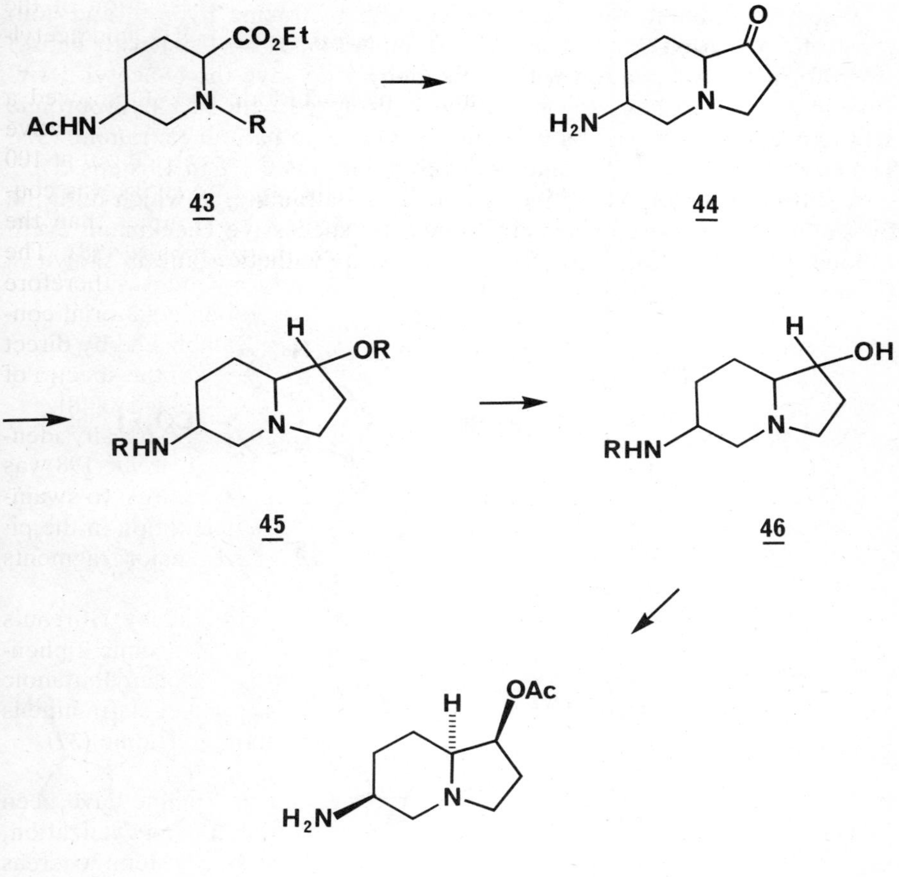

Figure 4. Synthesis of slaframine (**37**) by a nonstereoselective route [49].

by conventional means into ethyl 5-acetamidopipecolate (**43**, R = H). Condensation with ethyl acrylate gave the diester (**43**, R = —CH$_2$CH$_2$CO$_2$Et) which underwent Dieckmann cyclization on treatment with potassium *t*-butoxide in toluene at 0°. Hydrolysis and decarboxylation in refluxing 8*N* hydrochloric acid gave the ketoindolizidine (**44**). Sodium borohydride reduction gave a mixture of stereoisomeric alcohols (**45**, R = H) which were acetylated. The four stereoisomeric 1-acetoxy-6-acetamidoindolizidines (**45**, R = Ac) were separated by chromatography on alumina in chloroform. The third isomer to elute was identical with *N*-acetylslaframine (**46**, R = Ac) by IR and NMR spectra, TLC, and GLC. The stereochemistry of the remaining three isomers was established by their NMR spectra.

N-acetylslaframine was deacetylated with hydrazine hydrate and converted to *N*-carbobenzoxydeacetylslaframine (**46**, R = Cbz) with benzyl chloroformate. Acetylation with acetic anhydride gave the *O*-acetyl derivative and the carbobenzoxy group was hydrolyzed with 30% hydrobromic acid in acetic acid to yield a product identical with natural slaframine (**37**). The overall yield from 2-bromo-5-nitropyridine was 0.1% in 13 steps.

A subsequent stereoselective synthesis of slaframine, in which both the five- and six-membered rings are formed by successive Dieckmann cyclizations, has been described [50]. The general synthetic route is shown in Figure 5.

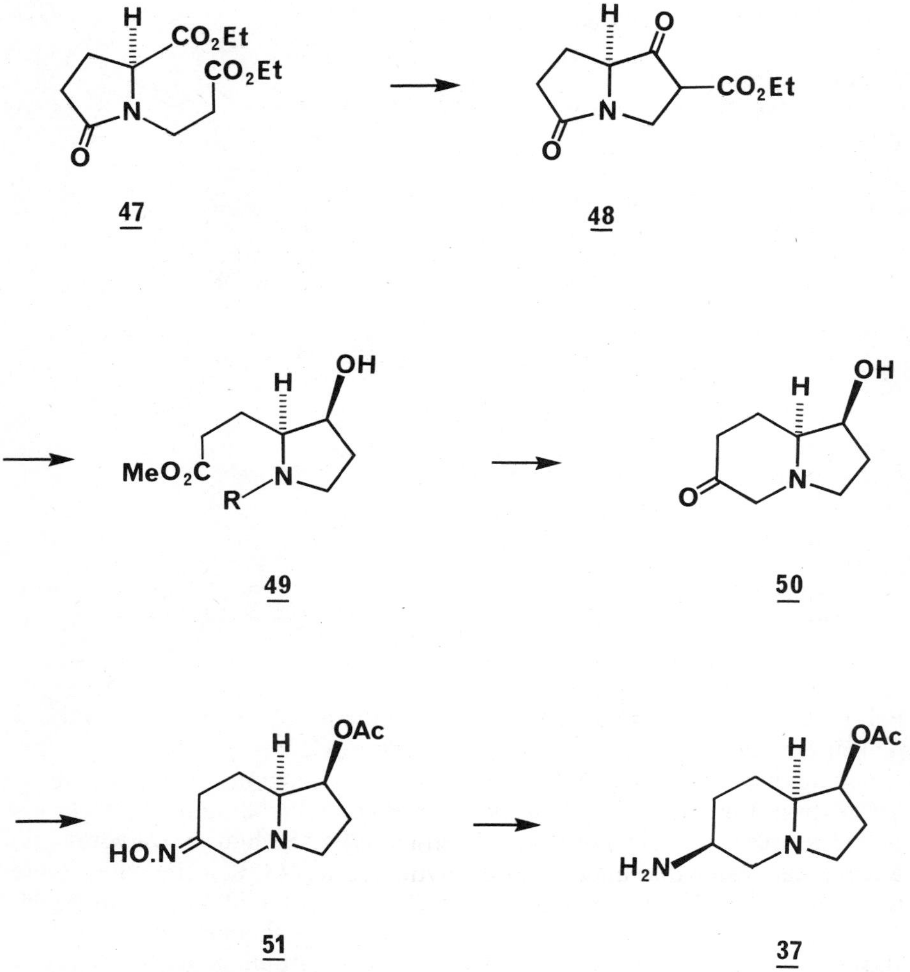

Figure 5. Synthesis of slaframine (**37**) by Dieckmann cyclization route [50].

Ethyl N-(β-carbethoxyethyl)-5-oxopyrrolidine-2-carboxylate (**47**), prepared by condensation of glutamic acid and acrylonitrile, underwent Dieckmann cyclization when treated with sodium ethoxide to yield the pyrrolizidine (**48**). Decarboxylation and concurrent hydrolysis of the lactam ring gave an oxopyrrolidine acid, which was reduced and esterified to yield the hydroxypyrrolidine ester (**49**, R = H) when hydrogenated in methanol. Alkylation with methyl bromoacetate gave the hydroxypyrrolidine diester (**49**, R = CH_2CO_2Me).

Treatment of the diester with sodium hydride gave an unstable Dieckmann cyclization product that was decarboxylated to give the hydroxyketoindolizidine (**50**). Acetylation of the hydroxy group and treatment with hydroxylamine hydrochloride gave a mixture of *syn*- and *anti*-oximes (**51**). Hydrogenation of the mixed oximes gave DL-slaframine (**37**), characterized as the dihydrochloride, dipicrate, and N-acetyl derivative. The stereoselective nature of the synthesis depends on the assumption that the two hydrogenation steps (**48** to **49** and **51** to **37**) occur by insertion of hydrogen from the least-hindered side of the molecule, so that the hydrogen atoms H(1), H(6), and H(8a) all lie on the same side of the two rings. Although this assumption is reasonable it is by no means certain and the synthesis therefore cannot be regarded as unequivocal in its stereochemistry. The overall yield of DL-slaframine from L-(+)-glutamic acid, achieved in 11 steps, was of the order of 0.3%.

A more recent stereospecific synthesis of racemic slaframine, together with 1-epislaframine, has been described [51], and is outlined in Fig. 6. The diene aldehyde (**52**) was treated with the carbanion of bis(trimethylsilyl)acetamide to yield the β-hydroxyamide (**53**). Protection of the hydroxy group as the *t*-butyldimethylsilyl ether, followed by treatment with cesium carbonate and paraformaldehyde and immediate acetylation, provided the acetate (**54**, R = *t*-BuMe$_2$Si). Thermolysis of the latter gave an intermediate acylimine (**55**, R = *t*-BuMe$_2$Si) which underwent intramolecular imino Diels–Alder cyclization to a separable mixture of bicyclic lactams (**56** and **57**, R = *t*-BuMe$_2$Si). The epimer (**57**, R = H) could be converted to the desired stereoisomer (**56**, R = H) by oxidation of the alcohol function to a keto group and subsequent stereospecific reduction with 9-borobicyclo[3.3.1]nonane.

The lactam ester [**56**] was hydrogenated and hydrolyzed to the corresponding acid which was converted to the carbamate lactam (**58**, R = *t*-BuMe$_2$Si) via a Curtius rearrangement. Reduction of the keto function with diborane, hydrolysis of the *t*-BuMe$_2$Si protecting group, and acetylation gave a carbamate ester that underwent catalytic hydrogenolysis to yield racemic slaframine (**37**). The overall yield was 1.7% in 12 steps from the diene aldehyde (**52**).

Similar transformations of the lactam (**57**) gave 1-epislaframine (**59**). However, no information was reported on the biological activity of this epimer.

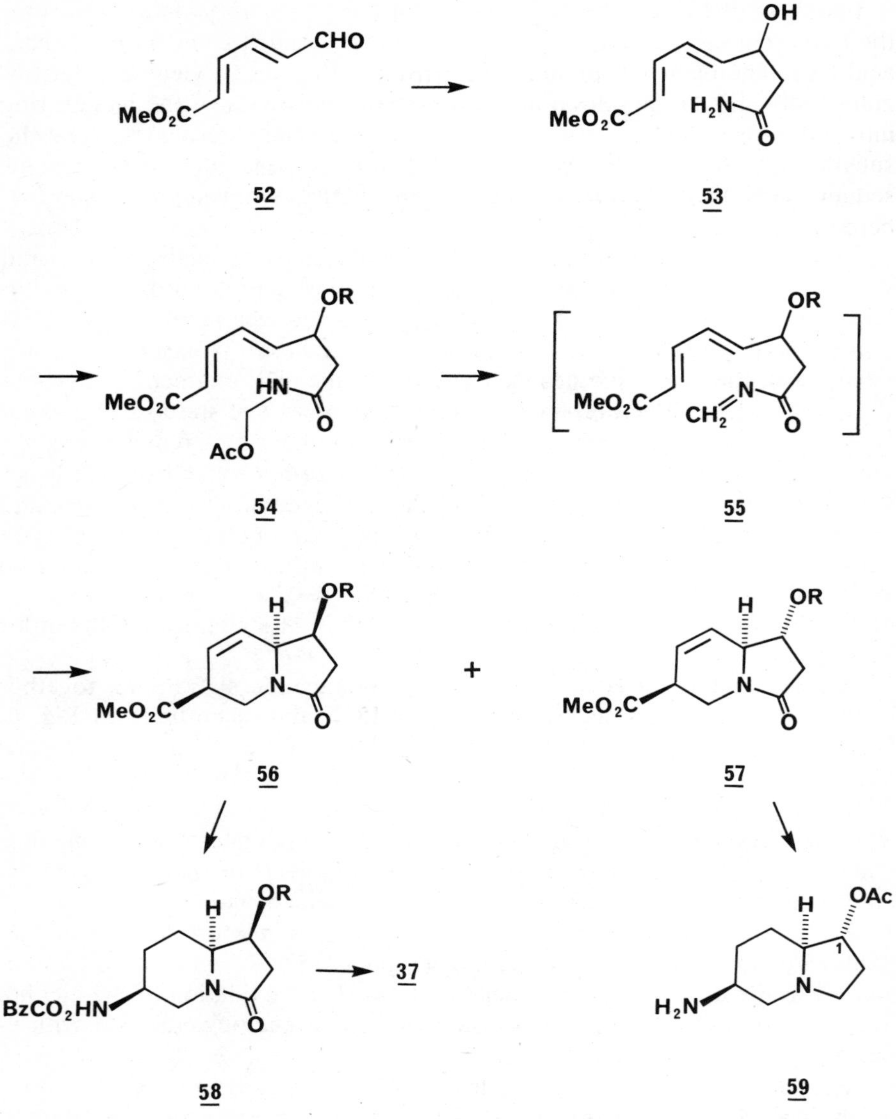

Figure 6. Synthesis of slaframine (**37**) by intramolecular imino Diels–Alder cyclization route [51].

2.4.4. Biosynthesis. The biosynthesis of slaframine (**37**) and swainsonine (**3**) produced by the fungus *Rhizoctonia leguminicola* has been investigated [52–56]. The overall biosynthetic route established for these two alkaloids is shown in Fig. 7. However, it should not be assumed that the same pathway operates for the formation of swainsonine in plants.

Initial work by Guengerich and Broquist [52,53] established that six of the carbon atoms in slaframine (37) were derived from lysine via pipecolic acid (60). The bicyclic precursors 1-ketoindolizidine (61), 1-hydroxyindolizidine (63), and 1,6-dihydroxyindolizidine were also shown to be convertible into slaframine by *Rhizoctonia leguminicola*. The derivation of the 1-acetyl substituent from acetate via acetyl coenzyme A was demonstrated [52]. Subsequently the remaining two carbon atoms, C(2) and C(3), of the five-membered ring were also shown to be acetate-derived, being introduced via ma-

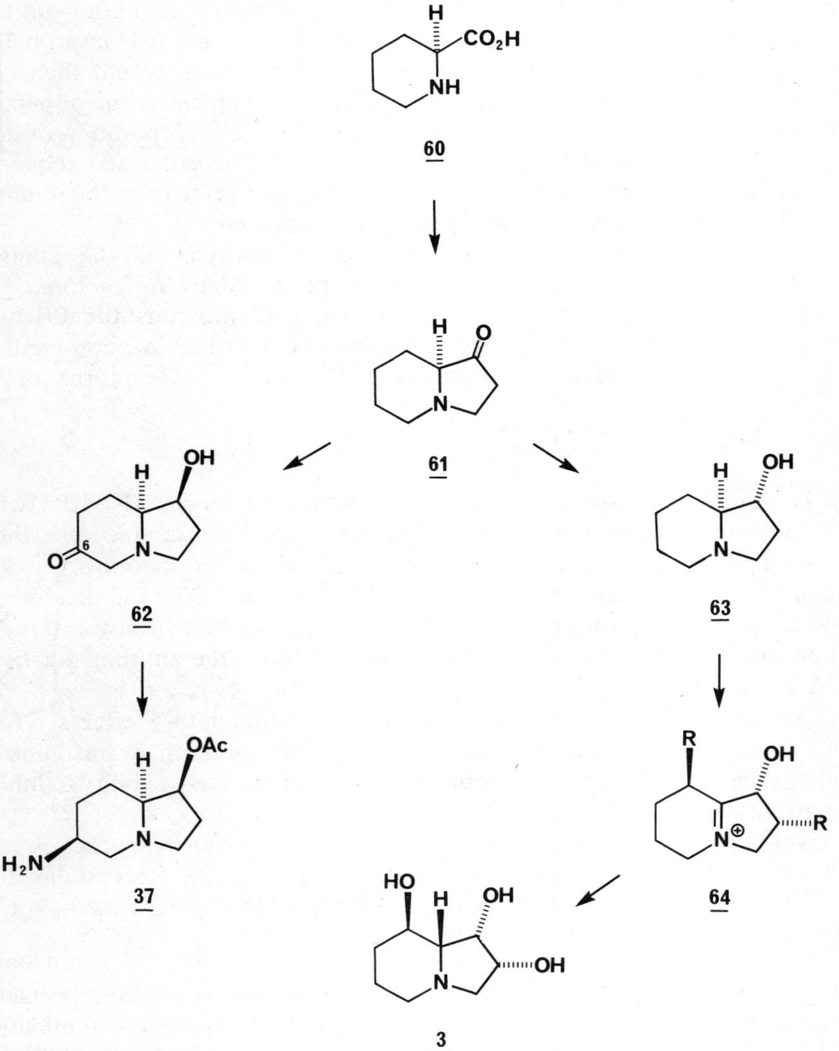

Figure 7. Biosynthesis of slaframine (37) and swainsonine (3).

lonate [54]. The same precursors were shown to give rise to swainsonine (3), although at the time it was thought to be the pyrindine alkaloid (4).

More recently it has been found that perdeuteriopipecolic acid is incorporated into both slaframine and swainsonine with loss of two deuterium atoms. In slaframine both deuterium atoms are lost from the C(6) position, indicating that the 6-ketoindolizidine (62) is a probable intermediate. In contrast, the deuterium atoms in swainsonine are absent from C(8) and C(8a). The introduction of the hydroxyl group at C(8) accounts for the first of these missing atoms. The loss of the C(8a) deuterium atom and the fact that (S)-pipecolic acid (60) yields swainsonine (3) with an inversion of configuration at this center indicated that oxidation to an iminium ion (64) is probable. Reduction from the least hindered side of the molecule would then give swainsonine with an (R) configuration. The involvement of nicotinamide coenzyme in the reduction step was indicated by the incorporation of deuterium from 1,1-dideuterioethanol via acetic acid. The order and sequence of introduction of the C(2) and C(8) hydroxyl groups relative to the iminium ion (64, R = H or OH) have not yet been established.

A point of some interest in the biosynthetic pathway is that the requisite precursors for slaframine and swainsonine produced by *Rhizoctonia leguminicola,* namely, pipecolic acid and malonic acid, are constituents of red clover, the host plant of the fungus. Sterile red clover hay infusion medium is required for maintenance and growth of *R. leguminicola* cultures [52].

2.5. Cyclizidine, $C_{17}H_{25}NO_3$

2.5.1. Occurrence and Isolation.   A *Streptomyces* species (NCIB 11649), previously undescribed, was recently isolated from a soil sample in northern England. When grown under aerobic conditions in the laboratory it was found to have antibiotic activity against *Botrytis allii*. The indolizidine alkaloid cyclizidine (Antibiotic M14791) (65) was isolated from the fermentation broth but was found not to be responsible for the antifungal activity [57].

The alkaloid exhibited nonselective immunostimulatory effects. However, in spite of its similarity in oxygenation pattern to both swainsonine (3) and castanospermine (29), no reports have appeared as to its ability to inhibit glycosidases.

Cyclizidine was isolated from whole fermentation broth by extraction with ethyl acetate and crystallization from the same solvent. Recrystallization from ether gave colorless needles, m.p. 184°; $[\alpha]_{589}^{23.5}$ −46.3°, methanol (c, 2.0). The yield of alkaloid was 0.12 g/L.

2.5.2. Structure.   The structure of cyclizidine was determined primarily by spectroscopic and X-ray crystallographic methods. However, acetylation with acetic anhydride–pyridine gave a monoacetate, m.p. 63–65°, with the elemental analysis $C_{19}H_{27}NO_4$, the IR spectrum of which showed the pres-

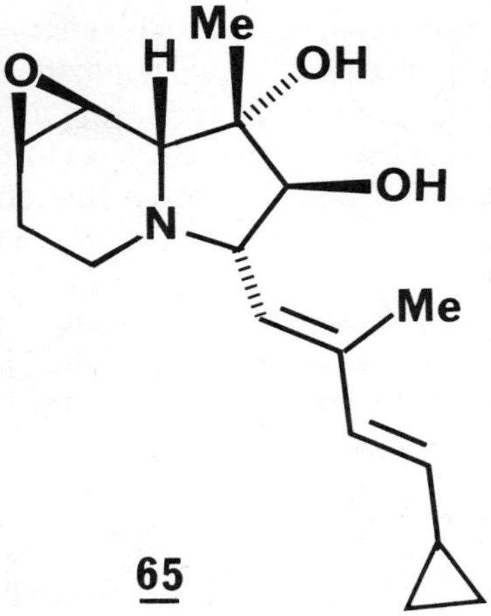

65

ence of a tertiary, unacetylated hydroxyl group. The UV spectrum exhibited a chromophore characteristic of a conjugated triene.

The 400 MHz ^{1}H NMR spectrum of the alkaloid established the presence of a cyclopropyl moiety, the methine proton showing coupling to the protons of a *trans* double bond, with additional long-range coupling to another olefinic proton. The latter was coupled to a vinylic methyl group. The structure of the unusual side chain was thus determined and confirmed by the ^{13}C NMR spectrum. The presence of an epoxide function was also established by ^{1}H and ^{13}C NMR spectroscopy.

The mass spectrum confirmed the molecular formula of $C_{17}H_{25}NO_3$ (M^+ 291.1837) and showed a base peak at *m/e* 161. Although the fragmentation pattern was not delineated, a fragment at *m/e* 217 could well be accounted for by the ion (**66**), analogous to the ion (**6**, R = H or OH) formed by swainsonine and castanospermine.

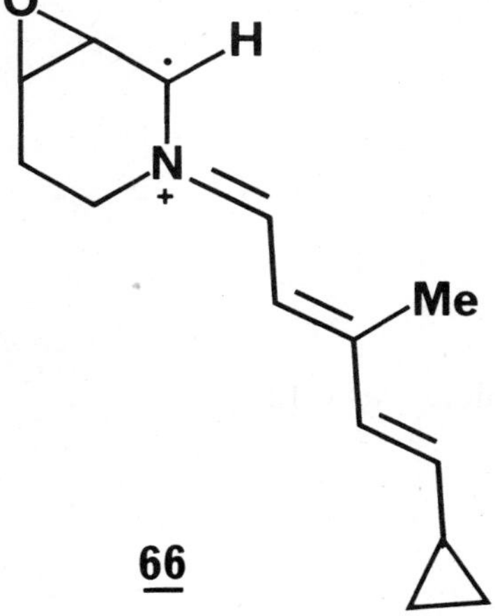

66

The total structure and relative stereochemistry of cyclizidine was determined by X-ray crystallography. The indolizidine ring system was thus established, with the five-membered ring having an envelope conformation and the six-membered ring a twist-boat conformation. This is in contrast to swainsonine and castanospermine, in which the larger ring adopts a chair conformation.

2.5.3. Synthesis.   No synthesis has yet been reported of this alkaloid, which presents a considerable synthetic challenge in terms of number and variety of asymmetric groups located on the indolizidine ring system.

2.6. Related Polyhydroxy Alkaloids

A few polyhydroxy alkaloids are known that possess similarities in structure and biochemical activity to the simple indolizidine alkaloids. These alkaloids are nitrogen analogs of pyranose or furanose sugars but structural analogies may also be noted if they are visualized as indolizidine alkaloids from which portions of either the five- or six-membered rings have been cleaved. Included in this group are nojirimycin (**67**) and deoxynojirimycin (**68**), which have been known for some time and studied quite extensively as glycosidase inhibitors, and the more recently discovered deoxymannojirimycin (**69**) and 2,5-dihydroxymethyl-3,4-dihydroxypyrrolidine (**70**). They are included for brief discussion in this context because of the information they provide as to the structural requirements for glycosidase inhibition.

Nojirimycin (**67**), $C_6H_{13}NO_5$, m.p. 126–130° (dec.), $[\alpha]_{589}^{24}$ +100°, H_2O, occurs in the fermentation broth of several strains of *Streptomyces,* including *S. nojiriensis,* from which the name of the alkaloid is derived [58]. The compound is quite unstable under neutral or acidic conditions but may be isolated as the stable bisulfite adduct, from which it can be recovered in the pure state by alkaline hydrolysis followed by ion-exchange chromatography and crystallization at low temperature.

67   **68**

The structure of nojirimycin was determined by chemical methods in combination with MS and NMR [58]. A particularly significant transformation was its reduction, either catalytically or with sodium borohydride, to deoxynojirimycin (68), the structure of which was established by 220 MHz ^{1}H NMR spectroscopy. Whereas deoxynojirimycin yields a penta-O,N-acetate when treated with acetic anhydride in pyridine, nojirimycin yields a quite different product, 2-acetoxy-5-hydroxypyridine. Acid-catalyzed dehydration gives a related product, 5-hydroxy-2-pyridinemethanol. The D configuration for nojirimycin was established by ORD and the alkaloid is therefore 5-amino-5-deoxy-α-D-glucopyranose.

The structure was confirmed by a nine-step synthesis from D-glucose in 19% overall yield [58].

Deoxynojirimycin (68), $C_6H_{13}NO_4$, m.p. 206°, α_{589}^{22} +44.7°, H_2O(c, 1.0), previously obtained as a reduction product of nojirimycin [58], was subsequently isolated from a number of *Bacillus* species [59] and has also been obtained in 0.067% yield from mulberry plants [60]. The structure was determined during the course of structural elucidation of nojirimycin, primarily by high-resolution NMR spectroscopy, as 1,5-dideoxy-1,5-imino-D-glucitol (68) [58].

Several syntheses of deoxynojirimycin have been reported, the earlier of which were multistep routes involving the use of protective groups [58,61,62]. More recently a four-step chemical–microbiological synthesis, utilizing the microorganism *Gluconobacter oxydans* to produce the *N*-benzoyloxycarbonyl derivative of 6-amino-6-deoxy-L-sorbose, which undergoes intramolecular reductive amination on catalytic hydrogenation, has been used to prepare large quantities of the alkaloid [63]. The synthesis of deoxynojirimycin from D-glucose via an intermediate obtained in the synthesis of castanospermine has also been described [40] (see Section 2.3.3).

Deoxymannojirimycin (69), $C_6H_{13}NO_4$, m.p. 185–187°, $[\alpha]_{578}^{20}$ −39°, H_2O [64], the mannose analog of deoxynojirimycin, has been isolated in 0.24% yield as the hydrochloride from seeds of the legume *Lonchocarpus sericeus* [65]. The structure was elucidated from the EI and CI mass spectra in combination with ^{1}H and ^{13}C NMR spectroscopy. The 100 MHz ^{1}H NMR spectrum exhibited a well-resolved eight-proton system that allowed assignment of all resonances and complete determination of the relative configuration.

The absolute configuration was established by application of the benzoate chirality method [66]. The primary hydroxyl and imino groups were protected by formation of a cyclic carbonate which gave a tribenzoate on treatment with benzoyl chloride in pyridine. The intense negatively split Cotton effect extrema of this derivative established that the vicinal secondary hydroxyl groups were arranged in a counterclockwise manner on the ring and therefore indicated a D configuration. The structure was thus established as 1,5-dideoxy-1,5-imino-D-mannitol (69).

The alkaloid was synthesized [64] in six steps in approximately 4% yield from D-mannose via benzyl-2,3,5,6-di-O-isopropylidene-α-D-mannofuranoside, by a route similar in principle to that developed by Inouye et al. [58] for the synthesis of nojirimycin and deoxynojirimycin. 5-Amino-5-deoxy-D-mannopyranose (i.e., mannojirimycin) was also obtained during the course of this synthesis, although in contrast to nojirimycin it has not been reported to occur as a natural product.

Complementary enantiospecific syntheses of mannojirimycin from D-mannose and D-glucose have been reported by Fleet et al. [67]. The synthesis from D-mannose was achieved in 23% overall yield in nine steps by a sequence similar to that used for the synthesis of swainsonine [31]. The amino function was introduced between C(1) and C(5) of D-mannose and the route involved a double inversion at C(5). On the other hand, with D-glucose as starting material, a single inversion at C(2) was involved subsequent to introduction of the amino function between C(2) and C(6). The latter synthesis proceeded from diacetone glucose in nine steps in 26% overall yield.

2,5-Dihydroxymethyl-3,4-dihydroxypyrrolidine (**70**), $C_6H_{13}NO_4$, $[\alpha]_{589}^{20}$ +56.4°, H_2O(c, 7.0), has been isolated in 0.1% yield from fresh leaves of *Derris elliptica* [68], and more recently from seeds of *Lonchocarpus costaricensis* [69], both of which are members of the Leguminosae. The compound was rapidly degraded on treatment with periodic acid, indicating the presence of vicinal hydroxyl groups. The ^1H NMR spectrum of the hydrochloride derivative showed that the compound contained four hydroxyl groups and eight nonexchangeable hydrogen atoms. The ^{13}C NMR spectrum of the free base showed only three resonances, corresponding to one methylene and two methine groups, all of which are shifted to low field by attachment to either —OH or —NH groups, indicating a remarkable symmetry in the molecule. Mass spectral data supported the imino–alcohol structure (**70**), and this conclusion was confirmed by the 300 MHz NMR spectrum.

The optical activity of the compound eliminated the *meso* forms from consideration and the structure is therefore (2R,5R)-dihydroxymethyl-3R,4R-dihydroxypyrrolidine or its enantiomer. Synthesis of the alkaloid has not so far been reported.

69   **70**

3. BIOCHEMISTRY OF INHIBITION BY INDOLIZIDINE ALKALOIDS

3.1. General Comments

3.1.1. Glycosidases.   Glycosidases are enzymes that catalyze the hydrolysis of glycosidic bonds in simple glycosides, in oligosaccharides and polysaccharides, and in complex carbohydrates such as glycolipids and glycoproteins. Such reactions result in the release of monosaccharides or oligosaccharides that are of lower molecular weight than the initial starting material. These enzymes can be broadly classified as exoglycosidases that attack the glycosidic linkage at the nonreducing terminus of a saccharide chain and liberate monosaccharide units, and endoglycosidases that act on glycosidic linkages within saccharide chains and give rise to smaller oligosaccharides. A large number of different glycosidases are known that have specificity for the sugar residue and for the anomeric configuration [70].

Glycosidases are extremely widespread in nature and appear to be present in all organisms. This great diversity is probably a reflection of their vital catabolic role. Thus extracellular glycosidases, such as those present in the intestinal tract of animals or those secreted by bacteria and fungi, degrade large molecules to prepare them for uptake by the organism [71]. There are also a number of intracellular glycosidases that are typically located in the lysosomes of eukaryotic cells. These enzymes also degrade complex carbohydrates within the cell and are involved in the turnover of various cellular components [72].

Glycosidases may also have other critical functions in metabolism. For example, in the growth of plants, fungi, and bacteria, new cell wall material for expansion of the existing wall must be inserted at various growing points, and glycosidases are believed to be involved in loosening of the old cell wall [73]. In the biosynthesis of the oligosaccharide portion of the N-linked glycoproteins, various glycosidases play key roles in trimming specific sugars to give the different types of oligosaccharide structures (see below for details of these reactions).

3.1.2. Glycoprotein Biosynthesis and Processing.   The N-linked or asparagine-linked glycoproteins are commonly found in eukaryotic cells, both as cell surface or membrane proteins and also as secreted proteins [74]. The oligosaccharide portion of these glycoproteins may be either of the high-mannose, the complex, or the hybrid structure, as shown in Fig. 8. All these structures contain the same basic core region, which is a pentasaccharide composed of a branched trimannose region (enclosed in the box) linked to an N,N'-diacetylchitobiose. This chitobiose is, in turn, attached to the amide nitrogen of an asparagine residue on the protein. In the high-mannose structures shown in Figure 8, this pentasaccharide is further substituted by six α-linked mannose units, but some of the high-mannose types may have fewer

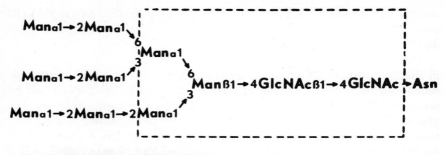

I. HIGH-MANNOSE TYPE

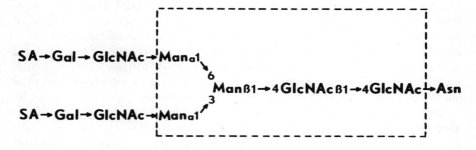

II. COMPLEX TYPE

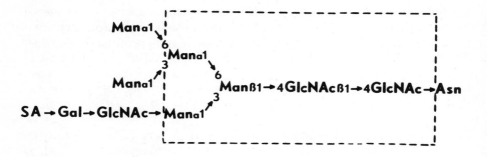

III. HYBRID TYPE

Figure 8.   Structure of the oligosaccharides of *N*-linked glycoproteins.

than nine mannose units. On the other hand, in the complex structures, the pentasaccharide is elongated by the trisaccharide, sialic acid–galactose–GlcNAc. The complex oligosaccharides may have two (biantennary), three (triantennary), or four (tetraantennary) of these trisaccharide units. Finally, the hybrid chains, which have only recently been found in animal cells, are combinations of the high-mannose and complex types and may have several mannose units on the $\alpha1,6$-branch, as well as one or two trisaccharide units on the $\alpha1,3$-branch [75].

The biosynthesis of the core region of each of the oligosaccharides seen in Figure 8 involves a common series of reactions that utilize lipid-linked saccharide intermediates. These reactions are shown in Fig. 9. The pathway is commonly referred to as the dolichol pathway (or cycle), because dolichyl phosphate serves as a carrier of the sugars [76–78]. The sequence of reactions is initiated by the transfer of GlcNAc-1-P, to dolichyl phosphate to form the initial lipid intermediate, GlcNAc–pyrophosphoryl–dolichol [79–81]. A second GlcNAc is then added, also from UDP–GlcNAc, to form N,N'-diacetylchitobiosyl-pyrophosphoryldolichol [82,83]. This lipid-linked saccharide is then the acceptor of mannose and glucose residues to give the final lipid-linked oligosaccharide, that is, $Glc_3Man_9(GlcNAc)_2$-pyrophosphoryldolichol [84–86]. This large oligosaccharide is then transferred from its lipid carrier to asparagine residues on the protein, while the polypeptide is being synthesized on membrane-bound polysomes [87–89]. The enzymes involved in the "dolichol cycle" are all membrane-bound glycosyltransferases and these reactions occur in the endoplasmic reticulum [90–92].

Once the oligosaccharide has been transferred to the protein, the oligosaccharide undergoes a series of processing reactions that are catalyzed by a number of specific and unusual glycosidases. The initial processing reactions begin in the endoplasmic reticulum and continue as the protein is

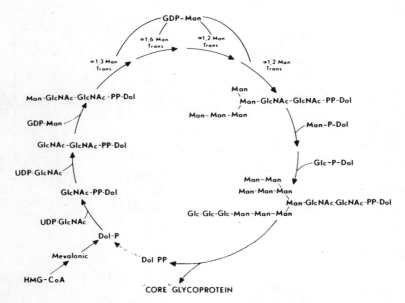

Figure 9. Reaction involved in the formation of lipid-linked oligosaccharides. This pathway leads to formation of $Glc_3Man_9NAc_2$–PP–dolichol.

transported through the Golgi apparatus to its final destination [93,94]. These processing reactions involved in glycoprotein biosynthesis are outlined in Fig. 10. Very soon after attachment of the oligosaccharide to the protein, the glucose residues are removed from the glycoprotein by two different membrane-bound glucosidases [95–97]. Glucosidase I removes the terminal $\alpha 1,2$-linked glucose [98,99], and glucosidase II removes the next two $\alpha 1,3$-

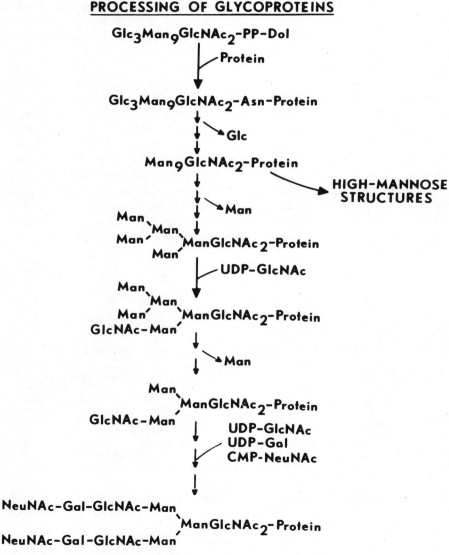

PROCESSING OF GLYCOPROTEINS

Figure 10. Reactions involved in the processing of the oligosaccharide chains of the *N*-linked glycoproteins.

linked glucose residues [100,101]. Several of the alkaloids discussed below are relatively specific inhibitors of these glucosidases. Removal of all three glucose residues gives rise to a glycoprotein having a $Man_9(GlcNAc)_2$ structure. This glycoprotein may become the high-mannose type of structure found in various cells, or it may be further processed to form the precursor to the complex types of structures. These additional processing reactions occur in the endoplasmic reticulum and in the Golgi apparatus and involve the removal of all four $\alpha1,2$-linked mannose units. Thus, specific $\alpha1,2$-mannosidases (mannosidases IA/B) [16] have been isolated and purified from the Golgi apparatus of hepatocytes [102–105], and recently another $\alpha1,2$-mannosidase has been found in the endoplasmic reticulum [106]. These enzymes give rise to a $Man_5(GlcNAc)_2$–protein. A GlcNAc is then added from UDP–GlcNAc to this $Man_5(GlcNAc)_2$–protein, and this GlcNAc apparently is the signal for another mannosidase, called mannosidase II, to remove the $\alpha1,3$- and $\alpha1,6$-linked mannose residues [107–109]. Several of the alkaloids discussed in this chapter are specific inhibitors of these mannosidases. After the removal of the final two mannose units to give a $GlcNAc–Man_3–$$(GlcNAc)_2$–protein, the remaining sugars of the complex chains, that is, GlcNAc, galactose, and sialic acid, are added by the sequential transfer from their sugar nucleotides to form the completed glycoproteins [110,111]. However, all the details of these additions are not known. Nor is it known why some oligosaccharides remain as high-mannose structures, whereas others become complex types, or why some chains are biantennary whereas others are tri- or tetraantennary.

3.2. Swainsonine

3.2.1. Inhibition of α-Mannosidase. The first studies showing a biological role for the indolizidine alkaloids were those done with swainsonine. Early observations had indicated that animals that ingested *Swainsona canescens,* one of the plants that produces swainsonine, exhibited symptoms like those individuals with the hereditary storage disease, mannosidosis ([15]; see below). Because this disease entails an absence of the lysosomal α-mannosidase, Dorling et al. [112] examined the effects of this alkaloid on the activity of the α-mannosidases from various tissues.

Mammalian tissues contain three structurally and genetically distinct types of α-mannosidase with different pH optima and subcellular locations [113]. These have been designated acidic or lysosomal, intermediate or Golgi, and neutral or cytosolic. The acidic α-mannosidase has a pH optimum of 4.0–4.5 and is localized and active in the lysosomes [114]. This enzyme cleaves terminal mannose residues, regardless of whether they are present in $\alpha1,2$, $\alpha1,3$, or $\alpha1,6$ linkages. Thus this enzyme functions in the degradation of the oligosaccharide portion of various glycoproteins. There are at least two different Golgi α-mannosidases, both having pH optima around 5.5–6.0

[102,115]. As discussed previously and shown in Fig. 10, these enzymes are involved in the processing of the oligosaccharide chains of glycoproteins. There are several enzymes, referred to as mannosidase IA/IB, that have specificity for α1,2-mannosyl linkages [102–105]. The other Golgi mannosidase that participates in glycoprotein processing is called mannosidase II [107–109,161]. This enzyme cleaves the α1,3- and α1,6-linked mannosyl residues from the GlcNAc–Man$_5$(GlcNAc)$_2$–protein. Recently, another α-mannosidase was found in the endoplasmic reticulum [106]. Although this enzyme has not been purified as yet, it does cleave α1,2-mannosyl residues, and has a pH optimum of about 6.5. Its precise role is not known, but it is believed to be involved in processing probably at a very early stage of glycoprotein trimming. Finally, an α-mannosidase with a pH optimum of 6.5 is located in the cytoplasm. The function of this enzyme is not known, but it may arise by proteolysis of the endoplasmic reticulum α-mannosidase.

Dorling et al. [112] found that swainsonine, at a concentration of 20 μM, completely inhibited the acidic α-mannosidase (pH 4.0) from all tissues examined, as well as the acidic α-mannosidase from the liver of the lamprey eel and from the seeds of jack bean. On the other hand, α-glucosidase, β-galactosidase, β-hexosaminidase, and β-glucuronidase from mouse liver were not affected by 10 times the swainsonine concentration needed for α-mannosidase inhibition. These workers point out that the pK_a of swainsonine is 7.4, and thus at pH 4.0, the alkaloid would be fully ionized. Using molecular models, they observed that the relative positions of the cationic centers and the three hydroxyl groups are similar to that of the hypothetical mannosyl ion which may serve as an intermediate. This arrangement may account for the apparent specificity of swainsonine for α-mannosidase. Dorling and colleagues present evidence to indicate that swainsonine is a reversible active site-directed inhibitor of lysosomal α-mannosidase.

In another study, swainsonine was tested at a variety of concentrations against a number of commercially available glycosidases in order to determine its specificity. Even at 1 μg/mL (about 50 times the amount needed for 50% inhibition of α-mannosidase), swainsonine had no inhibitory effect on β-mannosidase, α- or β-glucosidase, α- or β-galactosidase, β-N-acetylhexosaminidase, or α-L-fucosidase [116]. A number of kinetic studies were done to determine the type of inhibition by swainsonine on the jack bean α-mannosidase. Lineweaver–Burk plots of substrate concentration (i.e., paranitrophenyl-α-D-mannopyranoside) versus velocity in the presence of various amounts of swainsonine showed considerable curvature at high substrate concentrations, suggesting that swainsonine may be a competitive inhibitor that binds tightly to the enzyme and is only slowly displaced. Periodate oxidation of swainsonine completely destroyed its inhibitory activity.

Studies on the effects of swainsonine on the processing of the N-linked oligosaccharides have shown that this alkaloid alters the normal processing pathway and prevents the formation of complex chains [117,118] (see below for details). As discussed previously, there are several different mannosidases involved in the processing reactions. Mannosidases IA and IB are $\alpha1,2$-specific and act on oligosaccharides from $Man_9(GlcNAc)_2$, whereas mannosidase II cleaves $\alpha1,3$- and $\alpha1,6$-mannosyl residues from $Man_5(GlcNAc)_2$ structures in completing the mannosyl trimming reactions. Swainsonine did not inhibit Golgi mannosidase IA and IB, whereas mannosidase II was very strongly inhibited [119]. In addition, Tulsiani et al. examined the effect of swainsonine on the processing of mannose-labeled oligosaccharides ranging in size from $[^3H]Man_9GlcNAc$ to $[^3H]Man_5GlcNAc$ by rat liver Golgi-rich fractions. Swainsonine had no effect on the yield of $[^3H]Man_5GlcNAc$ or of free $[^3H]$mannose produced by these enzyme preparations, a further indication that the $\alpha1,2$-mannosidases were not inhibited. The processing experiments were also done in the presence of UDP–GlcNAc, to promote the addition of GlcNAc to the $Man_5GlcNAc$ species. In the absence of swainsonine, the Golgi preparations converted the oligosaccharides mainly to $GlcNAc–Man_3GlcNAc$ and $GlcNAc_2–Man_3GlcNAc$ species, whereas in the presence of the alkaloid the product was $GlcNAc–Man_5GlcNAc$. Thus the processing mannosidase II is inhibited by swainsonine [119]. Characterization of the oligosaccharide products produced in cultured mammalian cells in the presence of swainsonine indicates that this site of inhibition also occurs *in vivo* (see below for details).

3.2.2. Inhibition of Glycoprotein Processing.

Because swainsonine was found to be a potent inhibitor of α-mannosidase, it was tested as an inhibitor of the biosynthesis of N-linked glycoproteins. The initial studies were done in cell culture to determine whether this alkaloid would permeate cells and, if so, whether it would affect the processing pathway. MDCK or Chinese hamster ovary cells, when grown in the presence of 2-$[^3H]$mannose or 6-$[^3H]$glucosamine, incorporate these radioisotopes into N-linked glycoproteins that have both high-mannose and complex types of oligosaccharides. Although the amount of each type of oligosaccharide varies with the cell type and with the conditions of growth, animal cells generally have more of the complex types of oligosaccharides at their cell surface. One can easily distinguish the complex structures from the hybrid and high-mannose types by digesting the glycoproteins with the proteolytic enzyme, pronase, to produce glycopeptides, and then examining the size or elution position of the glycopeptides by gel filtration on columns of Biogel P-4 [117]. These glycopeptides can be further examined by treating them with the enzyme, endo-β-N-acetylglucosaminidase H, which cleaves high-mannose and hybrid structures between the two internal N-acetylglucosamine residues. However, this enzyme will not attack complex types of oligosaccharides [120]. Thus after digestion with this enzyme, the high-mannose and hybrid chains show an altered and slower migration on the Biogel columns because they have lost the GlcNAc–peptide portion of the molecule. However, the complex chains continue to migrate in the same position, before or after treatment with the endoglucosaminidase H.

When MDCK or Chinese hamster ovary cells were incubated for several hours in the presence of swainsonine and then labeled with [³H]mannose or [³H]glucosamine, there was a dramatic decrease in the amount of label in the complex types of glycopeptides, and a substantial increase in the amount of label in the high-mannose (or hybrid) types of oligosaccharides. These changes were monitored by examining the amount of radioactivity that became susceptible to digestion by endoglucosaminidase H with increasing concentrations of swainsonine. These experiments did demonstrate that swainsonine was inhibiting one or more steps in the processing pathway, but the specific site of action was not determined [117].

The effect of swainsonine on the biosynthesis and processing of several viral glycoproteins was also studied. Influenza virus is an enveloped virus that has a major coat protein called the hemagglutinin. This protein is an N-linked glycoprotein that has seven or eight oligosaccharide chains, of which five or six are of the complex type and two or three of the high-mannose type [121]. Another enveloped virus is vesicular stomatitis virus, which produces the N-linked glycoprotein called the G protein. This glycoprotein contains two N-linked oligosaccharides, both of which are of the complex type [122]. Influenza virus was grown in MDCK cells in the presence of swainsonine, and the viral glycoprotein was labeled by the addition of 2-[³H]mannose or 6-[³H]glucosamine. At concentrations of swainsonine of 1 μg/mL, more than 90% of the viral glycopeptides were susceptible to the action of endoglucosaminidase H, whereas in the normal virus, more than 70% of the glycopeptides are resistant to the action of this enzyme [118]. Essentially the same results were obtained when VSV was grown in baby hamster kidney cells in the presence of swainsonine. In this case, a complete change in oligosaccharide structure from endoglucosaminidase H-resistant to endoglucosaminidase H-sensitive occurred at a swainsonine concentration of 50 ng/mL of culture media [123]. The major oligosaccharide found in the VSV glycoproteins produced in the presence of swainsonine was characterized by a variety of enzymatic and chemical procedures and shown to be a hybrid structure like III in Fig. 8. With the VSV system, swainsonine had to be added to the infected cells within the first four or five hours of infection in order to cause the observed changes from complex to hybrid structures. Presumably, this is the time when most of the enveloped glycoproteins are synthesized and processed. Interestingly, the effect of swainsonine was reversible. That is, it could be removed by replacing the growth media with fresh media and the cells would revert to the production of the normal complex oligosaccharides, as long as the initial swainsonine concentration was not too high, and as long as it was removed early enough in the infection. In these experiments with virus-infected cells, swainsonine did not appear to affect the production of virus; that is, the same number of viral particles were observed in the medium of swainsonine-grown and control cells. Also, the virus produced in the presence of swainsonine appeared to retain its infectivity. These studies indicate that the presence of complex types of oligosaccharides on the enveloped glycoproteins are not necessary for the production and release of virus from the cells, nor for the infectivity of the virus on other cultured cells.

Swainsonine also caused fibroblasts to produce hybrid types of oligosaccharides. In these experiments, in the presence of swainsonine, at 10 μg/mL, most of the complex glycoproteins were replaced by hybrid types. The principal oligosaccharide was demonstrated to have the same structure as III in Fig. 8 [124]. This same hybrid [^{3}H]oligosaccharide was produced by rat liver Golgi fractions incubated in the presence of [^{3}H]Man$_5$GlcNAc, UDP–GlcNAc, UDP–galactose, CMP–NeuAc, and swainsonine. Swainsonine was also utilized to study the processing of the asparagine-linked carbohydrate chains of α1-antitrypsin in rat hepatocytes [125]. In normal hepatocytes, newly synthesized α1-antitrypsin was found in the cells as a glycoprotein with an apparent molecular weight of 49,000, having oligosaccharide chains of the high-mannose type. However, in the medium, a secreted form of α1-antitrypsin was identified with an apparent molecular weight of 54,000 and oligosaccharides of the complex type. Because pulse chain studies indicated a precursor–product relationship, it seems likely that the high-mannose type is processed to complex oligosaccharides during transport of α1-antitrypsin to the outside. When the hepatocytes were treated with swainsonine, the intracellular form of the α1-antitrypsin was indistinguishable from that of control cells and had an apparent molecular weight of 49,000. However, the α1-antitrypsin secreted from the cells was different from that of control cells. Thus it had a lower molecular weight (51,000), a higher content (i.e., incorporation) of [^{3}H]mannose but only half as much [^{3}H]galactose, and the same amount of [^{3}H]fucose. Furthermore, this 51,000 MW form of α1-antitrypsin was susceptible to the action of endoglucosaminidase H, whereas the 54,000 MW glycoprotein from control cells was resistant to this enzyme. Thus although swainsonine altered the carbohydrate structure of the α1-antitrypsin, this change did not impair the secretion of the molecule [125].

Fibronectin is an interesting and important glycoprotein that is secreted by fibroblasts and is necessary for adhesion of some cells to the substratum. This protein contains about 4–10% carbohydrate, consisting principally of complex oligosaccharides having the biantennary structure [126]. Fibronectin was secreted into the medium of swainsonine-treated fibroblasts and found to contain endoglucosaminidase H-sensitive structures, rather than the normal endoglucosaminidase H-resistant oligosaccharides. The structure of the swainsonine-modified, endoglucosaminidase H-sensitive oligosaccharide was analyzed using various glycosidases in conjunction with gel filtration. The structure was deduced to be Galβ–GlcNAcβ–Manα–[Manα(Manα–)Manα–]Manβ–GlcNAcβ–(± Fuc-)GlcNAc. Other studies showed that the synthesis and secretion of fibronectin were not affected by its change in glycosylation. The results also imply that swainsonine inhibits Golgi mannosidase II in intact cells, and that removal of α1,3- and α1,6-

linked mannoses by that enzyme is not a prerequisite for the addition of galactose to the outer, nonreducing GlcNAc, or the fucose to the inner, reducing GlcNAc. Taken as a whole, these studies indicate that significant changes in oligosaccharide structure have little effect upon the routing and cellular release of a typical N-asparagine-linked glycoprotein [127].

3.2.3. Effect of Swainsonine on Cells.

Because swainsonine appeared to be a valuable tool to alter oligosaccharide structures *in vivo*, it was important to determine whether it inhibited cell growth, or was cytotoxic. Swainsonine, at concentrations as high as 1 µg/mL, did not affect the growth of MDCK, Chinese hamster ovary, SV 101, B-16 melanoma, or Intestine 407 cells, as determined by changes in cell numbers over a 5-day period. There was also no apparent change in cell size or cell shape in these cells grown in the presence of swainsonine. The alkaloid did not appear to be cytotoxic, nor to cause alterations in cell morphology as evidenced by comparisons of normal cell sections to those of sections from cells grown in the alkaloid for up to 5 days [128].

As a measure of changes in cell surface oligosaccharides, the binding of various lectins to normal and swainsonine-grown cells was compared. For example, Concanavalin A binds to α-linked mannose and glucose units on high-mannose or hybrid chains, whereas wheat germ lectin binds to GlcNAc and sialic acid units on complex chains. When various animal cells were grown in the presence of swainsonine for 3–4 days, these cells were able to bind almost twice as much [³H]Concanavalin A as were control cells, whereas the binding of wheat germ agglutinin was significantly decreased in the alkaloid-grown cells. Such results are consistent with the notion that the alkaloid causes an increase in high-mannose or hybrid structures at the cell surface and a decrease in complex structures. In addition, the attachment of E. coli B-886, a bacterium that binds to animal cells by virtue of high-mannose (and hybrid) chains, was also increased two-fold by growing cells in the presence of the alkaloid. Swainsonine did not inhibit protein synthesis as measured by the incorporation of [³H]leucine or [³H]proline into trichloroacetic acid-insoluble material. However, the incorporation of GlcNAc, galactose, and fucose into glycoprotein was decreased as much as 50% by swainsonine, whereas mannose incorporation was significantly increased. These results are consistent with the inhibition of processing, leading to the formation of hybrid structures [128].

The interaction of swainsonine with cultured fibroblasts was examined [129]. Swainsonine was rapidly internalized by the cells and the amount of swainsonine taken up into the cells depended on the length of time in contact and the concentration of swainsonine in the medium. However, at 37°, a plateau of internalized swainsonine occurred after 2 hr at extracellular concentrations of alkaloid of 100 µM or higher. At lower concentrations of swainsonine, the rate of uptake was found to be temperature-dependent, increasing greatly at 20°. The rapidity and temperature sensitivity of the uptake, together with the observation that mannose or mannose-6-phosphate did not prevent the association, suggests that swainsonine enters the cells by permeation rather than by endocytosis. When swainsonine is withdrawn from the culture medium, there is a decrease of cell-associated swainsonine with time of incubation [129].

Swainsonine induces the accumulation of mannose-rich oligosaccharides in human fibroblasts. The composition of the storage products shows that the alkaloid completely inhibits lysosomal α-mannosidase and alters the processing of glycoproteins by inhibiting Golgi mannosidase II. Comparison of the storage products in genetic and swainsonine-induced mannosidosis suggests that human fibroblasts contain a lysosomal α-mannosidase that is unaffected in genetic mannosidosis [130].

However, swainsonine may have other effects in cells besides inhibition of mannosidases and accumulation of storage products. For example, this alkaloid may inhibit the receptor-mediated uptake as well as the degradation of certain ligands. Thus rat pulmonary macrophages were incubated in the presence of a radio-labeled mannosyl-oligosaccharide, and receptor-mediated endocytosis and degradation of this material by the cells was followed in the presence and absence of swainsonine. At 1 μg/mL of swainsonine, both the internalization and degradation of the radio-labeled ligand were inhibited whereas at 0.1 μg/mL only the degradation was inhibited [131]. The inhibition of uptake at higher concentrations of swainsonine may be due to the alkaloid inhibiting the formation of receptor–ligand complex, to inhibition of the internalization of the complex, or to inhibition of dissociation of the endocytosed ligand–receptor complex. However, the inhibition of degradation of the mannosyl-oligosaccharide seen at lower swainsonine concentrations is undoubtedly due to inhibition of lysosomal α-mannosidase. Swainsonine was also found to inhibit the degradation of other endocytosed glycoproteins. Thus the degradation of [^{125}I]asialofetuin by rat liver cells was inhibited 50% at a swainsonine concentration of 6×10^{-7} M. In the presence of inhibitor, there was an increased cellular accumulation of the glycoproteins that corresponded quantitatively to the decreased degradation, indicating that uptake was the same in the presence and absence of inhibitor. The excess asialofetuin that accumulated within the cells in the presence of swainsonine was found entirely within the lysosomal peak on Percoll density gradient centrifugation, and this labeled material migrated like the original asialofetuin upon SDS gel electrophoresis. Thus the inhibition of breakdown occurs prior to any conversion of the protein to intermediates of detectably smaller size. On the other hand, the presence of swainsonine did not inhibit the degradation of nonglycoproteins by the rat liver cells. In contrast, chloroquine and 3-methyladenine inhibited both the degradation of glycoproteins and nonglycoproteins by these cells [132].

3.2.4. Effect of Swainsonine in Animals. Ingestion by grazing livestock of certain leguminous plants of the genera *Astragalus* and *Oxytropis,* as well as the Australian genus *Swainsona,* induces a chronic neurological disease that resembles mannosidosis [133]. Livestock must consume locoweed for an extended period before symptoms of poisoning become evident. If locoweed grazing is discontinued before an animal becomes too emaciated, the animal will recover, but will continue to show neurological signs when stressed [134]. Locoweeds can cause abortion and birth defects if ingested (and a toxin is secreted in the milk of affected animals) [135]. Hence early detection of the disease is essential. In order to explore methods for early and rapid detection, the urine of animals fed locoweed or swainsonine were examined by TLC and by high-performance liquid chromatography (HPLC) to determine the oligosaccharide profiles.

Rats, sheep, and guinea pigs treated with swainsonine excrete "high-mannose" oligosaccharides in the urine. The major oligosaccharide in the urine of rats and guinea pigs is a $Man_5GlcNAc$, whereas sheep excrete a mixture of oligosaccharides of composition of $Man_{2-5}(GlcNAc)_2$ and $Man_{3-5}GlcNAc$. The presence of these oligosaccharides suggests that Golgi mannosidase II as well as lysosomal α-mannosidase is inhibited by swainsonine, resulting in the storage of abnormally processed asparagine-linked glycans from glycoproteins. Although the defect in glycoprotein processing appeared to have little effect on the health of the intoxicated animal, the accompanying lysosomal storage produces a diseased state [136]. In sheep, changes in the pattern of urinary oligosaccharides could be correlated with the onset of visible symptoms, which occurred approximately 5 weeks after the typical urine sugars were first detected [137]. In locoweed-fed sheep, certain oligosaccharides appeared early in the urine and then rapidly declined, whereas others appeared at later times [138]. It seems likely that rapid screening by TLC and HPLC as well as a knowledge of the particular oligosaccharide structure that would be expected in urine should make early detection quite feasible.

Swainsonine was administered to rats by including it in the drinking water, and its effect on tissue enzyme levels was examined. The activity of Golgi mannosidase II was markedly decreased to 22% of control values, but changes in other Golgi enzymes did not occur. However, unexpected and unusual changes occurred in the lysosomal enzymes. In liver, the acid mannosidase increased markedly instead of decreasing, as would be expected from a mannosidase inhibitor, and one which causes a mannosidosis-like condition. Also the principal change in brain was an increase in lysosomal mannosidase levels. In plasma, most lysosomal enzyme levels increased. These results indicate that pathological effects of swainsonine are not solely attributable to its inhibition of lysosomal α-mannosidase, and may also be caused by the inhibition by this alkaloid of glycoprotein processing [139]. Using the pig as an experimental model, the effects of swainsonine and locoweed on the animals were compared directly to determine whether swainsonine was indeed responsible for the symptoms of locoism [140]. Both treatments increased lysosomal acid glycosidase activities in most tissues, decreased liver Golgi mannosidase II levels, increased plasma hydrolase levels, and greatly increased the tissue oligosaccharide levels, especially of $Man_5(GlcNAc)_2$ and $Man_4(GlcNAc)_2$. These results support the idea that swainsonine is the agent in locoweed responsible for the changes in enzyme levels and in the accumulation of oligosaccharides. In addition, the behavior of the animals was the same regardless of whether they were on locoweed or on swainsonine [140].

3.3. Castanospermine

3.3.1. Inhibition of Glycosidases.

Based on the specific inhibition of α-mannosidase by swainsonine, other indolizidine alkaloids were sought with the idea that they would have different stereochemistry and therefore might inhibit different glycosidases. Thus, castanospermine was tested against a variety of commercially available glycosidases to see whether it would inhibit any of these enzymes. Interestingly enough, castanospermine proved to be a potent inhibitor of almond emulsin β-glucosidase, showing almost complete inhibition at 50 μg/mL. However, even at levels of 250 μg/mL, this alkaloid did not inhibit yeast α-glucosidase, α- or β-galactosidase, α-mannosidase, β-glucuronidase, β-N-acetylglucosaminidase, or α-L-fucosidase. Some inhibition of β-xylosidase was also observed but the inhibitor concentration needed for this enzyme was considerably higher than that required for β-glucosidase. The inhibition of β-glucosidase was seen throughout the time course, and the inhibition with respect to p-nitrophenyl-β-glucopyranoside as substrate was of the mixed type. Castanospermine also inhibited β-glucosidase and β-glucocerebrosidase in fibroblast extracts, but with this enzyme preparation, castanospermine also inhibited the lysosomal α-glucosidase. Apparently this lysosomal α-glucosidase is sensitive to the alkaloid whereas the yeast α-glucosidase is not [38].

The mechanism of inhibition of castanospermine on fungal amyloglucosidase (an exo-1,4-α-glucosidase) and almond emulsin β-glucosidase was examined [39]. Castanospermine proved to be a competitive inhibitor of amyloglucosidase, at both pH 4.5 and 6.0, when assayed with the p-nitrophenyl-α-glucoside as substrate. It was also a competitive inhibitor of almond emulsin β-glucosidase at pH 6.5, but previous studies indicated that inhibition was of the mixed type at pH 4.5. In these studies, the pH of the incubation mixture had a marked effect on the inhibition. Thus, in all cases, castanospermine was a much better inhibitor at pH 6.0–6.5 than it was at lower pH values. The pK_a for castanospermine in water has been found to be 6.09, indicating that the alkaloid is probably more active in the unprotonated form. This was also suggested by the finding that the N-oxide of castanospermine, although still a competitive inhibitor of amyloglucosidase, was 50–100 times less active than was castanospermine, and its activity was not markedly altered by changes in the pH of the incubation mixtures [39].

3.3.2. Inhibition of Glycoprotein Processing. Because castanospermine inhibits α- and β-glucosidase activity and several of the glycosidases involved in glycoprotein processing are glucosidases, this alkaloid was tested as an inhibitor of the trimming of the oligosaccharide chains of N-linked glycoproteins [141]. When influenza virus was raised in the presence of castanospermine, at 10 μg/ml or higher, 80–90% of the viral glycopeptides became susceptible to the action of endoglucosaminidase H, whereas in the normal virus 70% of the glycopeptides are resistant to this enzyme. The major oligosaccharide released by endoglucosaminidase H from castanospermine-grown virus migrated like a Hexose$_{10}$GlcNAc upon gel filtration on a calibrated column of Biogel P-4. This oligosaccharide was characterized as a Glc$_3$Man$_7$GlcNAc on the basis of various enzymatic treatments, as well as by methylation analysis of the [2-^{3}H]mannose-labeled or [6-^{3}H]galactose-labeled oligosaccharide. The presence of three glucose residues in the oligosaccharide was also confirmed by periodate oxidation studies of the [^{3}H]glucose-labeled oligosaccharide. Castanospermine did not inhibit the incorporation of [^{3}H]leucine or [^{14}C]alanine into protein in MDCK cells, even at levels as high as 50 μg/mL. In addition, castanospermine did not inhibit the production of influenza virus, nor did it alter the infectivity of virus raised in its presence. The castanospermine did, however, alter the surfaces of cells, because MDCK cells grown for several days in the presence of 10 μg/ml of this alkaloid were able to bind almost twice as much [^{3}H]Concanavalin A as were control cells [141]. These studies indicated that castanospermine inhibited glucosidase I, the first processing glycosidase that removes the outermost glucose residue. Cell-free studies with a rat liver particulate enzyme that has all the processing glycosidase activities showed that castanospermine did inhibit glucosidase I.

Castanospermine also inhibited glycoprotein processing in suspension-cultured soybean cells [142]. Soybean cells were pulse labeled with [2-^{3}H]mannose and chased for varying periods of time in unlabeled medium. In normal cells, the initial glycopeptides (i.e., at 0 min of chase) contained oligosaccharides having Glc$_3$Man$_9$(GlcNAc)$_2$ structures, and these were trimmed with time to Man$_9$(GlcNAc)$_2$ down to Man$_7$(GlcNAc)$_2$. In the presence of castanospermine, no trimming of glucose residues occurred, although some mannose units were apparently still removed. Thus the major oligosaccharide in the glycopeptides of castanospermine-incubated cells after a 90-min chase was the Glc$_3$Man$_7$(GlcNAc)$_2$ structure. Thus in plants castanospermine also appears to inhibit glucosidase I [142].

The influenza virus hemagglutinin contains sulfate in its oligosaccharide chains and this sulfate is apparently added during the processing reactions. Castanospermine and swainsonine were used to determine the point at which sulfate addition occurs. When influenza virus was grown in the presence of castanospermine, the major oligosaccharide was a Glc$_3$Man$_7$(GlcNAc)$_2$, and this oligosaccharide did not contain sulfate. However, when the virus was grown in the presence of swainsonine and labeled with ^{35}SO$_4$ and

[2-^{3}H]mannose, the major structure was a hybrid oligosaccharide, that is, Gal–Glc*N*AcMan$_5$(Glc*N*Ac)$_2$, and this oligosaccharide did contain sulfate. This hybrid oligosaccharide was susceptible to digestion by endoglucosaminidase H, and the sulfate was found to be located on the innermost Glc*N*Ac residue [143]. These data indicate that trimming of the 1,3 branch is necessary for sulfate addition, but trimming of mannose units from the 1,6 branch is not essential for addition of sulfate.

3.3.3. Effect on Animals.

The seeds produced by *Castanospermum australe* have been reported to be toxic to animals and to cause gastrointestinal symptoms. In order to determine whether this toxicity was due to castanospermine, and to determine what effect this alkaloid would have on glucosidase activity in vivo, rats were injected intraperitoneally with various amounts of castanospermine. Animals were injected daily over a period of 3–4 days, and at different times animals were sacrificed and various tissues were examined for the activities of different glycosidases. In animals given 0.5 mg/g body weight of the alkaloid over a 3-day period (one injection daily) the α-glucosidase activity of liver was decreased to 40% of control values whereas activity of brain was only 30% of that of control animals and activity of kidney also 30–40% of control values. Both the acidic (pH 4.5) and the neutral (pH 6.0) α-glucosidase activities were depressed in the various tissues examined. On the other hand, the activity of liver β-*N*-acetylhexosaminidase was increased with castanospermine treatment, whereas β-galactosidase activity was unchanged. α-Mannosidase activity in liver was also decreased in treated animals, but much larger doses of alkaloid were required to inhibit this enzyme than to depress the α-glucosidase. It is not known from these preliminary results whether the castanospermine actually inhibits the formation (and/or translocation) of the α-glucosidase, or whether it simply inhibits and ties up the existing enzyme. Because the acidic α-glucosidase is involved in glycogen metabolism, the livers of treated animals were examined by electron microscopy to determine whether changes in these tissues had occurred. In normal animals, the glycogen in the liver is distributed throughout the cytoplasm and is seen in various individual and aggregated particles. However, in the treated animals, there was very little glycogen in the cytoplasm, but instead, the hepatocytes contained a number of vesicles that were filled with a dense, granular material. That this material was indeed glycogen was indicated by its disappearance from the vesicles when the liver slices were treated with α-amylase [144]. These preliminary results suggest that castanospermine will be a useful tool for studies on glycogen metabolism and may provide a model system for Pompe's disease.

High doses of castanospermine (2 mg/g body weight) caused a severe diarrhea in experimental animals. This was found to be caused by castanospermine inhibition of intestinal maltase and sucrase in these animals. The normal rat diet is high in sucrose and starch, and because these carbohydrates cannot be utilized in the presence of the alkaloid, they provide a rich medium for intestinal bacteria. These problems could be overcome in the above studies by the replacement in the rat diet of sucrose and starch with glucose. Animals on this new, high-glucose diet did not show gastrointestinal symptoms when given castanospermine, and did not have diarrhea.

4. BIOCHEMISTRY OF INHIBITION BY PYRROLIDINE ALKALOIDS

4.1. 2,5-Dihydroxymethyl-3,4-dihydroxypyrrolidine

4.1.1. Inhibition of Glycosidases.   2,5-Dihydroxymethyl-3,4-dihydroxy-pyrrolidine (DMDP) was tested as an inhibitor of various glycosidases using the appropriate p-nitrophenyl glycoside as substrate. DMDP was a potent inhibitor of almond emulsin β-glucosidase ($K_i = 7.3 \times 10^{-6}\,M$), yeast α-glucosidase ($K_i = 6.5 \times 10^{-6}\,M$), and insect trehalase ($K_i = 5.5 \times 10^{-5}\,M$). The kinetics of inhibition of these enzymes by the DMDP were not presented so it is not known whether the inhibitory actions are competitive or how they compare with each other. DMDP did not inhibit several other glycosidases including α-mannosidase, α- or β-galactosidase, or β-glucuronidase, even at concentrations of $10^{-2}\,M$ [65].

4.1.2. Inhibition of Glycoprotein Processing.   DMDP was also tested as an inhibitor of the processing of the influenza viral hemagglutinin. At 250 μg/mL of DMDP, more than 80% of the [³H]mannose-labeled glycopeptides became susceptible to the action of endoglucosaminidase H, whereas in the normal virus, more than 70% of the glycopeptides are resistant to this enzyme. The major oligosaccharide produced in the presence of DMDP and released by endoglucosaminidase H sized like a Hexose$_{11}$GlcNAc on a calibrated column of Biogel P-4. This oligosaccharide was only slightly susceptible to α-mannosidase digestion, indicating that it contained blocking glucose units. The oligosaccharide was also synthesized in the presence of [³H]galactose to label the glucose residues, and the purified Hexose$_{11}$GlcNAc was subjected to methylation. Three methylated glucose derivatives were obtained that corresponded to terminal (2,3,4,6-tetramethylglucose), to 2-linked (3,4,6-trimethylglucose), and to 3-linked (2,4,6-trimethylglucose) glucose. Thus this oligosaccharide must be a Glc$_3$Man$_8$GlcNAc, and DMDP must inhibit glucosidase I [145]. These results are of considerable interest for they indicate that a pyranose ring structure is not necessary for inhibition of glycosidases. It will be of some interest to compare the effects of an indolizidine alkaloid and a pyrrolidine alkaloid on the same enzyme in terms of the kinetic properties and type of inhibition. Such studies could lead to a considerable insight into the mechanism of inhibition and the type of chemical structure that is the best inhibitor. DMDP did not inhibit the incorporation of [³H]leucine into protein in MDCK cells at levels as high as 250 μg/mL. It also did not inhibit the release of virus particles as measured by the hemagglutination assay [145]. Thus DMDP and other pyrrolidine alkaloids should be interesting compounds to study glycoprotein processing.

4.2. 1,4-Dideoxy-1,4-Imino-D-Mannitol

4.2.1. Inhibition of Glycoprotein Processing.
1,4-Dideoxy-1,4-imino-D-mannitol was synthesized from benzyl-α-D-mannopyranoside. This compound was found to be a potent inhibitor of jack bean α-mannosidase, causing 50% inhibition at a concentration of 5×10^{-7} M. This inhibition was of the competitive type and showed a K_i of 7.6×10^{-7} M. On the other hand, the deoxyaminomannofuranose was a weak inhibitor of the hydrolysis of p-nitrophenyl-α-D- and β-D-glucopyranosides by yeast α-glucosidase (50% inhibition of enzyme at 5×10^{-4} M) and almond emulsin β-glucosidase (50% inhibition at 4.5×10^{-4} M). These results and those with DMDP demonstrate that simple nitrogen analogs of furanose sugars are a class of compounds that may allow the design of specific glucosidase inhibitors [146]. In addition, the synthesis of polyhydroxylated piperidines and pyrrolidines may provide a general and predictive method for controlling glycosidase, glycosyl transferase, and other enzyme-mediated reactions that involve carbohydrate substrates. Such compounds would also be valuable tools for probing the mechanism of action.

The 1,4-Dideoxy-1,4-imino-D-mannitol also proved to be an inhibitor of glycoprotein processing [147]. These studies were done using the influenza virus hemagglutinin as the model system, and using essentially the same protocol as described above with the castanospermine studies. In the presence of the iminomannitol, the major oligosaccharide was a $Man_9(GlcNAc)_2$. However, this inhibitor was much less effective than swainsonine, because only about 70% of the viral glycopeptides became susceptible to the action of endoglucosaminidase H, even at 250 μg/mL of inhibitor. It appears most likely that this 1,4-iminomannitol inhibits mannosidase I, but this compound has not yet been tested directly on the isolated enzyme.

5. BIOCHEMISTRY OF INHIBITION BY PIPERIDINE ALKALOIDS

5.1. Nojirimycin and Deoxynojirimycin

Nojirimycin is an antibiotic substance obtained from the culture broth of certain species of *Streptomyces* [148]. Its chemical structure has been determined as 5-amino-5-deoxy-D-glucopyranose. Thus, nojirimycin is essentially a glucose analog in which the ring oxygen has been substituted with a nitrogen [58]. Nojirimycin was initially found to inhibit several microbial glucosidases [149] and also intestinal sucrase [150]. This antibiotic was also tested with the purified lysosomal α-glucosidase from human liver, and found to be three orders of magnitude more potent an inhibitor than the classical inhibitor, turanose [151]. In a liver homogenate, nojirimycin still inhibited the acidic α-glucosidase but it did not inhibit β-glucosidase. The inhibition was of a noncompetitive or competitive nature depending on the substrate being used [151].

1-Deoxynojirimycin is the reduced form of nojirimycin and is produced by *Bacillus* species. This compound is a potent inhibitor of yeast and pancreatic glucosidases [59]. Deoxynojirimycin was also found to be a potent inhibitor of both glucosidase I and glucosidase II from *Saccharomyces cerevesiae,* as well as of the calf pancreas microsomal glucosidases that remove all three glucose residues [152]. The effect of deoxynojirimycin on intact cells was examined using confluent IEC-6 intestinal epithelial cells, after labeling for 24 hr with [2-^{3}H]mannose. The labeled glycopeptides were separated on Biogel P-6 before and after digestion with endoglucosaminidase H. Deoxynojirimycin (5 mM) greatly decreased the proportion of radioactivity present in complex chains and rendered the high-mannose oligosaccharides less susceptible to the action of α-mannosidase [152]. Both nojirimycin and deoxynojirimycin were potent inhibitors of calf liver glucosidase I with 50% inhibition at concentrations of 0.16 mM and 2 μM, respectively. Hydrolysis of Glc$_2$Man$_9$GlcNAc and Glc$_1$Man$_9$GlcNAc was inhibited to a somewhat lower extent, suggesting that the two structures are acted upon by a different enzyme than that which attacks Glc$_3$Man$_9$GlcNAc [153]. Deoxynojirimycin and its N-methyl derivative were also specific inhibitors of the trimming of the outermost glucose residues of the N-linked glycoproteins of fowl plaque virus [154].

In normal hepatocytes, α1-proteinase inhibitor is present in the cells as a 49,000 MW protein having oligosaccharide chains of the Man$_9$(GlcNAc)$_2$ and Man$_8$(GlcNAc)$_2$ structures. In the presence of 1-deoxynojirimycin (5 mM), hepatocytes accumulated α1-proteinase inhibitor as a 51,000 MW protein having oligosaccharides of the high-mannose type that also contained glucose, as shown by their susceptibility to α-glucosidase. The largest oligosaccharide species was a Glc$_3$Man$_9$GlcNAc. Thus conversion to complex chains was inhibited by this drug. Furthermore, at high concentrations of deoxynojirimycin, glycosylation was inhibited, for some of the glycoproteins contained only two oligosaccharide chains instead of the normal three chains. The secretion of the α1-proteinase inhibitor was inhibited about 50% at 5 mM deoxynojirimycin, whereas the secretion of albumin was unaffected. The oligosaccharides of the secreted proteinase inhibitor were characterized by a variety of methods, and deoxynojirimycin was found to block processing incompletely. Thus even in the presence of the inhibitor, α1-proteinase inhibitor molecules were produced that carried one or two complex oligosaccharides. Only those α1-proteinase inhibitor molecules processed to complex oligosaccharides in one or two of their oligosaccharide chains were nearly exclusively secreted [155].

In a pulse chase experiment, 1-deoxynojirimycin greatly reduced the rate of secretion of α1-antitrypsin and α1-antichymotrypsin by human hepatoma HepG2 cells but had marginal effects on the secretion of the glycoprotein C-3 and transferrin, as well as of albumin. Equilibrium density gradient centrifugation of extracts from inhibited cells indicated that the α1-antitrypsin and α1-antichymotrypsin had accumulated in the rough endoplasmic reticulum. The oligosaccharides on the cell-associated α1-antitrypsin and α1-antichymotrypsin, synthesized in the presence of deoxynojirimycin, remained sensitive to endoglucosaminidase H and were thought to have the $Glc_{1-3}Man_9GlcNAc$ structure. Swainsonine did not affect the rate of secretion of these two glycoproteins, although it did alter the oligosaccharide chains to hybrid structures. On the other hand, tunicamycin did have a similar effect on secretion to that of deoxynojirimycin [156]. The authors suggest that movement of α1-antitrypsin and α1-antichymotrypsin from the rough endoplasmic reticulum to the Golgi requires that the N-linked oligosaccharide be processed to at least the $Man_9(GlcNAc)_2$ form. Possibly this oligosaccharide forms part of the recognition site of a transport receptor for certain secretory proteins. A somewhat similar result was found with deoxynojirimycin and castanospermine on the formation of VSV. The data suggest that conditions that prevent the removal of glucose can block the growth of virus and that the effect is due specifically to an effect on the ability of the virus glycoprotein G to mature to a correct functional conformation.

5.2. Deoxymannojirimycin

Deoxymannojirimycin, the mannose analog of deoxynojirimycin, was synthesized chemically and tested as a processing inhibitor. The experimental system used in these studies was a hybridoma cell system that produces IgM and IgD. After preincubation with the inhibitor, the cells were pulse labeled for 10 min with $[^{35}S]$methionine and then chased for various times in the presence of unlabeled methionine plus inhibitor. The IgM and IgD present within the cells and that secreted into the medium were examined by SDS gel electrophoresis and the proteins from control cells were compared with those of cells incubated in inhibitor. The deoxymannojirimycin appeared to inhibit mannosidase I, blocking the conversion of high-mannose chains to complex oligosaccharides. Thus in the presence of this inhibitor, the major oligosaccharide released by Endo H corresponded with the $Man_9GlcNAc$ standard. The deoxymannojirimycin did not inhibit the secretion of IgM and IgD although these glycoproteins had an altered oligosaccharide chain [157]. These reuslts, coupled with studies with other inhibitors, suggest that glycoproteins with high-mannose or hybrid oligosaccharides are still secreted at the normal rate (as compared with the same glycoproteins with its normal complex chain), whereas those glycoproteins produced in the presence of inhibitors that prevent the removal of glucose residues may be altered in terms of movement within the cell, or secretion. Thus the presence of glucose on the oligosaccharides may cause the interaction (or prevent the interaction) with a receptor in the endoplasmic reticulum or in the Golgi apparatus that prevents or retards the secretion of the protein.

Deoxymannojirimycin was also utilized as an inhibitor of the G protein of VSV virus and of the hemagglutinin of influenza virus [158,159]. In one of these studies the formation of the human class I histocompatability antigens was also examined [158]. Essentially, deoxymannojirimycin (and also deoxynojirimycin) greatly altered the oligosaccharide structures of membrane and secreted proteins in these various experimental systems. However, neither inhibitor interfered with their surface expression, as determined by a number of assays such as accessibility to proteases and antibodies. Furthermore, these drugs did not inhibit the production of infectious virus particles. Again, these studies inducate that the presence of complex oligosaccharides is not essential for formation and release of mature, infectious virus.

6. STRUCTURE–ACTIVITY RELATIONSHIPS

The number of simple indolizidine and related polyhydroxy alkaloids capable of inhibiting glycosidases so far described is quite restricted. The information available is therefore too limited to reach conclusions as to the structural requirements for inhibitory activity. It may, however, be worthwhile considering the common structural features possessed by those inhibitors presently known in order to suggest candidate compounds for evaluation, whether isolated as natural products or prepared synthetically.

The following three structural features are common to the glycosidase inhibitors discussed in earlier sections of this review:

1.   A secondary or tertiary nitrogen atom situated in a five-membered (pyrrolidine), six-membered (piperidine), or fused five/six-membered (indolizidine) ring.
2.   At least three hydroxyl groups located in a β-position relative to the nitrogen atom.
3.   Fixed stereochemical relationships between the hydroxyl groups, which may account for the specificity of the inhibitory activity to particular enzymes.

Dorling et al. [112] have proposed that the inhibition of α-mannosidase by swainsonine is due to the structural similarity of the protonated alkaloid to the mannosyl cation, which is a suggested intermediate in enzymatic hydrolysis of mannopyranosides. The potent inhibition of α-mannosidase by

71

72

the recently synthesized 1,4-dideoxy-1,4-imino-D-mannitol (**71**), which possesses the same configuration at the hydroxyl groups as swainsonine [146], lends support to this hypothesis. Moreover, this polyhydroxypyrrolidine, as yet unknown as a natural product, conforms to the three criteria listed above. On the other hand the swainsonine analog, ($\pm$)-(1α,2α,8aα)-indolizidine-1,2-diol (**72**), recently synthesized by Colegate et al. [160], possesses only two hydroxyl groups β to the nitrogen atom and was found to be a very weak inhibitor of α-mannosidase.

Novel hydroxy indolizidine, pyrrolidine, and piperidine alkaloids that may be isolated should be considered as candidates for evaluation of glycosidase inhibitory activity in relation to the above structural criteria. The frequent association of such known alkaloids with the Leguminosae suggests that this plant family would be the most fruitful source for new members of the class. Probably such compounds or their synthetic analogs could be tailored for inhibition of specific carbohydrase-mediated reactions by appropriate selection of ring size and hydroxyl group stereochemistry.

REFERENCES

1. G. A. Cordell in *The Alkaloids: Chemistry and Physiology,* Vol. 17, R. H. F. Manske and R. G. A. Rodrigo, Eds., Academic, New York, 1979, p. 199.
2. S. F. Dyke and S. N. Quessy in *The Alkaloids: Chemistry and Physiology,* Vol. 18, R. H. F. Manske and R. G. A. Rodrigo, Eds., Academic, New York, 1981, p. 1.
3. C. Fuganti in *The Alkaloids: Chemistry and Physiology,* Vol. 15, R. H. F. Manske. Ed., Academic, New York, 1975, p. 83.
4. H. Ripperger and K. Schreiber in *The Alkaloids: Chemistry and Physiology,* Vol. 19, R. H. F. Manske and R. G. A. Rodrigo, Eds., Academic, New York, 1981, p. 81.
5. E. Gellert, *J. Nat. Prod.* **45,** 50 (1982).
6. J. E. Saxton in *Specialist Periodical Reports: The Alkaloids,* Vol. 5, J. E. Saxton, Ed., The Chemical Society, London, 1975, p. 87, and preceding Reports.
7. J. A. Lamberton in *Specialist Periodical Reports: The Alkaloids,* Vol. 13, M. F. Grundon, Ed., Royal Society of Chemistry, London, 1983, p. 82, and preceding Reports.
8. I. R. C. Bick and W. Sinchai in *The Alkaloids: Chemistry and Physiology,* Vol. 19, R. H. F. Manske and R. G. A. Rodrigo, Eds., Academic, New York, 1981, p. 193.

9. S. R. Johns and J. A. Lamberton in *The Alkaloids: Chemistry and Physiology,* Vol. 14, R. H. F. Manske, Ed., Academic, New York, 1973, p. 325.

10. J. W. Daly in *Progress in the Chemistry of Organic Natural Products,* Vol. 41, W. Herz, H. Grisebach, and G. W. Kirby, Eds., Springer, New York, 1982, p. 205; J. W. Daly and T. F. Spande in *Alkaloids: Chemical and Biological Perspectives,* Vol. 4, S. W. Pelletier, Ed., Wiley Interscience, Wiley, New York, 1986, p. 95.

11. S. W. Pelletier in *Alkaloids: Chemical and Biological Perspectives,* Vol. 1, S. W. Pelletier, Ed., Wiley, New York, 1983, p. 1.

12. E. Lellmann, *Chem. Ber.* **23,** 2141 (1890).

13. For leading references see N. A. Katri, H. F. Schmitthenner, J. Shringarpure, and S. M. Weinreb, *J. Am. Chem. Soc.* **103,** 6387 (1981).

14. S. L. Everist, *Poisonous Plants of Australia,* Angus and Robertson, Sydney, 1981, p. 481.

15. P. R. Dorling, C. R. Huxtable, and P. Vogel, *Neuropathol. Appl. Neurobiol.* **4,** 285 (1978).

16. S. M. Colegate, P. R. Dorling, and C. R. Huxtable, *Aust. J. Chem.* **32,** 2257 (1979).

17. C. D. Marsh, "The Locoweed Disease of the Plains," *U.S. Dept. Agric. Bur. Animal Ind. Bull. No.* **112** (1909).

18. G. S. Fraps and E. C. Carlyle, *Texas Agric. Exp. Sta. Bull. No.* **537** (1936).

19. J. H. Maiden, "Plants Reputed to be Poisonous to Livestock in Australia," *Dept. Agric. N.S.W., Misc. Publ.* **477,** 10 (1901).

20. W. J. Hartley in *Effects of Poisonous Plants on Livestock,* R. F. Keeler, K. R. van Kampen, and L. F. James, Eds., Academic, New York, 1978, p. 363.

21. R. J. Molyneux and L. F. James, *Science* **216,** 190 (1982).

22. F. P. Guengerich, S. J. DiMari, and H. P. Broquist, *J. Am. Chem. Soc.* **95,** 2055 (1973).

23. M. J. Schneider, F. S. Ungemach, H. P. Broquist, and T. M. Harris, *Tetrahedron* **39,** 29 (1983).

24. R. J. Molyneux, unpublished results.

25. R. J. Molyneux, L. F. James, and K. E. Panter in *Plant Toxicology,* A. A. Seawright, M. P. Hegarty, R. F. Keeler, and L. F. James, Eds., Queensland Poisonous Plants Committee, Brisbane, 1985, p. 266.

26. A. J. Aasen, C. C. J. Culvenor, and L. W. Smith, *J. Org. Chem.* **34,** 4137 (1969).

27. B. W. Skelton and A. H. White, *Aust. J. Chem.* **33,** 435 (1980).

28. A. Horeau, *Tetrahedron Lett.* 506 (1961).

29. T. Suami, K. Tadano, and Y. Iimura, *Chem. Lett.* 513 (1984).

30. M. H. Ali, L. Hough, and A. C. Richardson, *J. Chem. Soc., Chem. Commun.* 447 (1984).

31. G. W. J. Fleet, M. J. Gough, and P. W. Smith, *Tetrahedron Lett.* **25,** 1853 (1984).

32. N. Yasuda, H. Tsutsumi, and T. Takaya, *Chem. Lett.* 1201 (1984).

33. K. B. Sharpless, Abstract, 103rd Annual Meeting of the Pharmaceutical Society of Japan, Tokyo, 1983, p. 67.

34. R. J. Molyneux, A. E. Johnson, J. N. Roitman, and M. E. Benson, *J. Agric. Food Chem.* **27,** 494 (1979).

35. J. D. Phillipson and S. S. Handa, *Lloydia* **41,** 385 (1978).

36. W. L. Meyer and N. Sapianchay, *J. Am. Chem. Soc.* **86,** 3343 (1964).

37. L. D. Hohenschutz, E. A. Bell, P. J. Jewess, D. P. Leworthy, R. J. Pryce, E. Arnold, and J. Clardy, *Phytochemistry* **20,** 811 (1981).

38. R. Saul, J. P. Chambers, R. J. Molyneux, and A. D. Elbein, *Arch. Biochem. Biophys.* **221,** 593 (1983).

39. R. Saul, R. J. Molyneux, and A. D. Elbein, *Arch. Biochem. Biophys.* **230,** 668 (1984).

40. R. C. Bernotas and B. Ganem, *Tetrahedron Lett.* **25**, 165 (1984).

41. For comprehensive reviews of the *Rhizoctonia* toxin problem see H. P. Broquist and J. J. Snyder in *Microbiol. Toxins,* Vol. VII, *Algal and Fungal Toxins,* S. Kadis, A. Ciegler, and S. J. Ajl, Eds., Academic, New York, 1971, p. 319; and S. D. Aust in *Mycotoxins,* I. F. H. Purchase, Eds., Elsevier, Amsterdam, 1974, p. 97.

42. E. B. Smalley, R. E. Nichols, M. H. Crump, and J. N. Henring, *Phytopathology* **52**, 753 (1962).

43. S. D. Aust and H. P. Broquist, *Nature (London),* 204 (1965).

44. D. P. Rainey, E. B. Smalley, M. H. Crump, and F. M. Strong, *Nature* (London), 203 (1965).

45. S. D. Aust, H. P. Broquist, and K. L. Rinehart, Jr., *J. Am. Chem. Soc.* **88**, 2879 (1966).

46. B. J. Whitlock, D. P. Rainey, N. V. Riggs, and F. M. Strong, *Tetrahedron Lett.* 3819, (1966).

47. R. A. Gardiner, K. L. Rinehart, Jr., J. J. Snyder, and H. P. Broquist, *J. Am. Chem. Soc.,* **90**, 5639 (1968).

48. H. S. Aaron, C. P. Radar, and G. E. Wicks, Jr., *J. Org. Chem.* **31**, 3502 (1966).

49. D. Cartwright, R. A. Gardiner, and K. L. Rinehart, Jr., *J. Am. Chem. Soc.* **92**, 7615 (1970).

50. W. J. Gensler and M. W. Hu, *J. Org. Chem.* **38**, 3848 (1973).

51. R. A. Gobao, M. L. Bremmer, and S. M. Weinreb, *J. Am. Chem. Soc.* **104**, 7065 (1982).

52. F. P. Guengerich, J. J. Snyder, and H. P. Broquist, *Biochemistry* **12**, 4264 (1973).

53. F. P. Guengerich and H. P. Broquist, *Biochemistry* **12**, 4270 (1973).

54. E. C. Clevenstine, H. P. Broquist, and T. M. Harris, *Biochemistry* **18**, 3658 (1979).

55. E. C. Clevenstine, P. Walter, T. M. Harris, and H. P. Broquist, *Biochemistry* **18**, 3663 (1979).

56. M. J. Schneider, F. S. Ungemach, H. P. Broquist, and T. M. Harris, *J. Am. Chem. Soc.* **104**, 6863 (1982).

57. A. A. Freer, D. Gardner, D. Greatbanks, J. P. Poyser, and G. A. Sim, *J. Chem. Soc. Chem. Commun.* 1160, (1982).

58. S. Inouye, T. Tsuruoka, T. Ito, and T. Niida, *Tetrahedron* **23**, 2125 (1968).

59. D. D. Schmidt, W. Frommer, L. Muller, and E. Truscheit, *Naturwissenschaften* **66**, 584 (1979).

60. H. Murai, K. Ohata, H. Enomoto, Y. Yoshikuni, T. Kono, and M. Yagi, German patent No. 2656602; *Chem. Abstr.* **87**, 14127 (1977).

61. H. Paulsen, I. Sangster, and K. Heyns, *Chem. Ber.* **100**, 802 (1967).

62. H. Saeki and E. Ohki, *Chem. Pharm. Bull.* **16**, 2477 (1968).

63. G. Kinast and M. Schedel, *Angew. Chem. Intl. Ed. Engl.* **20**, 805 (1981).

64. G. Legler and E. Julich, *Carbohydrate Res.* **128**, 61 (1984).

65. L. E. Fellows, E. A. Bell, D. G. Lynn, F. Pilkiewicz, I. Miura, and K. Nakanishi, *J. Chem. Soc. Chem. Commun.* 977, (1979).

66. N. Harada and K. Nakanishi, *Acc. Chem. Res.* **5**, 257 (1972).

67. G. W. J. Fleet, M. J. Gough, and T. K. M. Shing, *Tetrahedron Lett.* **25**, 4029 (1984).

68. A. Welter, J. Jadot, G. Dardenne, M. Marlier, and J. Casimir, *Phytochemistry* **15**, 747 (1976).

69. R. J. Nash, S. V. Evans, L. E. Fellows, and E. A. Bell, in *Plant Toxicology,* A. A. Seawright, M. P. Hegarty, R. F. Keeler, and L. F. James, Eds., Queensland Poisonous Plants Committee, Brisbane, 1985, p. 309.

70. V. Ginsburg and E. F. Neufeld, *Methods in Enzymology,* Vol. 8, Academic, New York, 1966.

71. W. W. Shreeve, *Physiological Chemistry of Carbohydrates in Mammals,* Saunders, Philadelphia, 1974.

72. E. F. Neufeld, T. W. Lim, and L. J. Shapiro, *Ann. Rev. Biochem.* **44,** 357 (1975).

73. R. Cleland, *Ann. Rev. Plant Physiol.* **22,** 197 (1971).

74. R. Kornfeld and S. Kornfeld, *Ann. Rev. Biochem.* **45,** 217 (1976).

75. P. V. Wagh and O. P. Bahl, *CRC Crit. Rev. Biochem.* **16,** 307 (1981).

76. A. D. Elbein, *Ann. Rev. Plant Physiol.* **30,** 239 (1979).

77. D. S. Struck and W. J. Lennarz, in *The Biochemistry of Glycoproteins and Proteoglycans,* W. J. Lennarz, Ed., Plenum, New York, 1980, p. 35.

78. R. J. Staneloni and L. F. Leloir, *CRC Crit. Rev. Biochem.* **12,** 289 (1982).

79. L. F. Leloir, R. J. Staneloni, H. Carminatti, and H. Behrens, *Biochem. Biophys. Res. Commun.* **52,** 1285 (1973).

80. A. Heifetz and A. D. Elbein, *J. Biol. Chem.* **252,** 3057 (1977).

81. R. K. Keller, D. Y. Boon, and F. C. Crum, *Biochemistry* **18,** 3946 (1979).

82. J. A. Levy, H. Carminatti, A. I. Cantarella, N. H. Behrens, and L. F. Leloir, *Biochem. Biophys. Res. Commun.* **60,** 118 (1974).

83. L. Lehle and W. Tanner, *FEBS Lett* **71,** 167 (1976).

84. E. Ki, I. Tabas, and S. Kornfeld, *J. Biol. Chem.* **253,** 7762 (1978).

85. T. Jui, B. Stetson, S. Turco, S. C. Hubbard, and P. W. Robbins, *J. Biol. Chem.* **254,** 4554 (1979).

86. H. Hori, P. James, Jr., and A. D. Elbein, *Arch. Biochem. Biophys.* **215,** 12 (1982).

87. M. Kiely, G. S. McKnight, and R. Schimke, *J. Biol. Chem.* **251,** 5490 (1976).

88. V. R. Lingappa, J. R. Lingappa, R. Prasad, K. Ebner, and G. Blobel, *Proc. Natl. Acad. Sci. U.S.A.* **75,** 2338 (1978).

89. J. E. Rothman and H. I. Lodish, *Nature (London)* **269,** 775 (1977).

90. V. Czicki and W. J. Lennarz, *J. Biol. Chem.* **252,** 7901 (1977).

91. I. Vargas and H. Carminatti, *Mol. Cell. Biochem.* **16,** 171 (1977).

92. A. Parodi and J. Martin-Barrientos, *Biochim. Biophys. Acta* **500,** 80 (1977).

93. S. C. Hubbard and R. C. Ivatt, *Ann. Rev. Biochem.* **50,** 555 (1981).

94. L. A. Hunt, J. R. Etchison, and D. F. Summers, *Proc. Natl. Acad. Sci. U.S.A.* **75,** 754 (1978).

95. R. A. Ugalde, R. J. Staneloni, and L. F. Leloir, *FEBS Lett.* **91,** 209 (1978).

96. W. W. Chen and W. J. Lennarz, *J. Biol. Chem.* **253,** 5780 (1978).

97. J. M. Michael and S. Kornfeld, *Arch. Biochem. Biophys.* **199,** 249 (1980).

98. J. J. Elting, W. W. Chen, and W. J. Lennarz, *J. Biol. Chem.* **255,** 2325 (1980).

99. R. D. Kilker, Jr., B. Saunier, J. S. Tkacz, and A. Herscovics, *J. Biol. Chem.* **256,** 5299 (1982).

100. L. S. Grinna and P. W. Robbins, *J. Biol. Chem.* **255,** 2255 (1980).

101. P. M. Burns and O. Touster, *J. Biol. Chem.* **257,** 9991 (1982).

102. D. J. Opheim and O. Touster, *J. Biol. Chem.* **253,** 1017 (1978).

103. D. P. R. Tulsiani, D. J. Opheim, and O. Touster, *J. Biol. Chem.* **252,** 3227 (1977).

104. I. Tabas and S. Kornfeld, *J. Biol. Chem.* **254,** 11655 (1979).

105. W. T. Forsee and J. Schutzbach, *J. Biol. Chem.* **256,** 6577 (1981).

106. J. Bischoff and R. Kornfeld, *J. Biol. Chem.* **258,** 7907 (1983).

107. I. Tabas and S. Kornfeld, *J. Biol. Chem.* **253**, 7779 (1978).

108. S. Narasimhan, P. Stanley, and H. Schachter, *J. Biol. Chem.* **252**, 3926 (1977).

109. N. Harpaz and H. Schachter, *J. Biol. Chem.* **255**, 4885 (1980).

110. H. Schachter and S. Roseman, in *The Biochemistry of Glycoproteins and Proteoglycans,* W. J. Lennarz, Ed., Plenum, New York, 1980, p. 85.

111. T. A. Beyer, J. E. Sadler, J. I. Rearick, J. C. Paulson, and R. L. Hill, *Adv. Enzymol.* **52**, 23 (1981).

112. P. R. Dorling, C. R. Huxtable, and S. M. Colegate, *Biochem. J.* **191**, 649 (1980).

113. B. Winchester, *Biochem. Soc. Trans.* **12**, 522 (1984).

114. N. C. Phillips, D. Robinson, and B. G. Winchester, *Clin. Chim. Acta* **55**, 11 (1974).

115. S. Hirani and B. Winchester, *Biochem. J.* **179**, 583 (1979).

116. M. S. Kang and A. D. Elbein, *Plant Physiol.* **71**, 551 (1983).

117. A. D. Elbein, R. Solf, P. R. Dorling, and K. Vosbeck, *Proc. Natl. Acad. Sci. U.S.A.* **78**, 7393 (1981).

118. A. D. Elbein, P. R. Dorling, K. Vosbeck, and M. Horisberger, *J. Biol. Chem.* **257**, 1573 (1982).

119. D. R. P. Tulsiani, T. M. Harris, and O. Touster, *J. Biol. Chem.* **257**, 7936 (1982).

120. A. L. Tarentino and F. Maley, *J. Biol. Chem.* **249**, 811 (1974).

121. A. Matsumoto, H. Yoshima, and A. Kobata, *Biochemistry* **22**, 188 (1983).

122. C. L. Reading, E. E. Penhoet, and C. E. Ballou, *J. Biol. Chem.* **253**, 5600 (1978).

123. M. S. Kang and A. D. Elbein, *J. Virol.* **46**, 60 (1983).

124. D. R. Tulsiani and O. Touster, *J. Biol. Chem.* **258**, 7578 (1983).

125. V. Gross, T-A. Trans-Thi, K. Vosbeck, and P. C. Heinrich, *J. Biol. Chem.* **258**, 4032 (1983).

126. M. Fukuda and S. Hakomori, *J. Biol. Chem.* **254**, 5442 (1979).

127. R. G. Arumugham and M. L. Tanzer, *J. Biol. Chem.* **258**, 11883 (1983).

128. A. D. Elbein, Y. T. Pan, R. Solf, and K. Vosbeck, *J. Cellular Physiol.* **115**, 265 (1983).

129. K. Chotai, C. Jennings, B. Winchester, and P. Dorling, *J. Cell. Biochem.* **21**, 107 (1983).

130. I. Cenci Di Bello, P. Dorling, and B. Winchester, *Biochem. J.* **215**, 693 (1983).

131. R. G. Arumughan and M. L. Tanzer, *Biochem. Biophys. Res. Commun.* **116**, 922 (1983).

132. J. R. Winkler and H. L. Segal, *J. Biol. Chem.* **259**, 1958 (1984).

133. C. R. Huxtable and P. R. Dorling, *Am. J. Pathol.* **107**, 124 (1982).

134. K. R. Van Kampen, R. W. Knees, and L. F. James, in *Effects of Poisonous Plants on Livestock,* R. F. Keeler, K. R. Van Kampen, and L. F. James, Eds., Academic, New York, 1978, p. 465.

135. L. F. James and W. J. Hartley, *Am. J. Ves. Res.* **38**, 1263 (1977).

136. D. Abraham, W. F. Blakemore, R. D. Jolly, R. Sidebotham, and B. Winchester, *Biochem. J.* **215**, 573 (1983).

137. S. Sadeh, C. D. Warren, P. F. Daniel, B. Bugge, L. F. James, and R. W. Jeanloz, *FEBS Lett.* **163**, 104 (1983).

138. C. D. Warren, S. Sadeh, P. F. Daniel, B. Bugge, L. F. James, and R. W. Jeanloz, *FEBS Lett.* **163**, 99 (1983).

139. D. P. R. Tulsiani and O. Touster, *Arch. Biochem. Biophys.* **181**, 216 (1983).

140. D. P. R. Tulsiani, H. P. Broquist, L. F. James, and O. Touster, *Arch. Biochem. Biophys.* **232**, 76 (1984).

141. Y. T. Pan, H. Hori, R. Saul, B. A. Sanford, R. J. Molyneux, and A. D. Elbein, *Biochemistry* **22**, 3975 (1983).

142. H. Hori, Y. T. Pan, R. J. Molyneux, and A. D. Elbein, *Arch. Biochem. Biophys.* **228**, 525 (1984).

143. R. Merkle, A. D. Elbein, and A. Heifetz, *J. Biol. Chem.*, **260**, 1083 (1985).

144. R. Saul, J. Ghidoni, R. J. Molyneux, and A. D. Elbein, *Proc. Natl. Acad. Sci. U.S.A.* **82**, 93 (1985).

145. A. D. Elbein, M. Mitchell, B. A. Sanford, L. E. Fellows, and S. V. Evans, *J. Biol. Chem.* **259**, 12409 (1984).

146. G. W. Fleet, P. W. Smith, S. V. Evans, and L. E. Fellows, *J. Chem. Soc. Chem. Commun.* 1240 (1984).

147. G. Palmartczyk, M. Mitchell, G. W. Fleet, P. W. Smith and A. D. Elbein, *Arch. Biochem. Biophys.* **243**, 35 (1985).

148. H. Ishida, K. Kumagai, T. Niida, T. Tsuruoka, and H. Yumoto, *J. Antibiotics Ser. A* **20**, 66 (1967).

149. T. Niwa, S. Inouye, T. Tsuruoka, Y. Koaze, and T. Niida, *Agric. Biol. Chem.* **34**, 966 (1970).

150. G. Hanozet, H. Pircher, P. Vanni, B. Oesch, and G. Semenza, *J. Biol. Chem.* **256**, 3703 (1981).

151. J. P. Chambers, A. D. Elbein, and J. C. Williams, *Biochem. Biophys. Res. Commun.* **107**, 1490 (1982).

152. B. Saunier, R. P. Kolker, Jr., J. S. Tkacz, A. Quaroni, and A. Herscovics, *J. Biol. Chem.* **257**, 14155 (1982).

153. H. Hettkamp, E. Bause, and G. L. Legler, *Biosci. Rep.* **2**, 899 (1982).

154. P. A. Romero, R. Datema, and R. T. Schwarz, *Virology* **130**, 238 (1983).

155. V. Gross, T. Geiger, T.-A. Trans-Thi, F. Gauthier, and P. C. Heinrich, *Eur. J. Biochem.* **129**, 317 (1982).

156. H. F. Lodish and N. Kong, *J. Cell Biol.* **58**, 1726 (1984).

157. U. Fuhrmann, E. Bause, G. Legler, and H. Ploegh, *Nature (London)* **307**, 755 (1984).

158. B. Burke, K. Martin, E. Bause, G. Legler, N. Peyrieras, and H. Ploegh, *Embo J.* **3**, 551 (1984).

159. A. D. Elbein, G. Legler, A. Tlusty, W. McDowell, and R. Schwarz, *Arch. Biochem. Biophys.* **235**, in Press (1984).

160. S. M. Colegate, P. R. Dorling, and C. R. Huxtable, *Aust. J. Chem.* **37**, 1503 (1984).

161. D. R. P. Tulsiani, S. C. Hubbard, P. W. Robbins, and O. Touster, *J. Biol. Chem.* **257**, 3660 (1982).

Chapter Two

Structure and Synthesis of Phenanthroindolizidine Alkaloids and Some Related Compounds

Emery Gellert
Department of Chemistry
University of Wollongong
P.O. Box 1144
Wollongong, New South Wales 2500, Australia

CONTENTS

1. INTRODUCTION

Alkaloids containing the phenanthroindolizidine (including secophenan-throindolizidine) ring system have been isolated [1] from several species of five genera (*Antitoxicum*, *Cyanchum*, *Pergularia*, *Tylophora*, and *Vincetoxicum*) of the Asclepiadaceae plant family and from two species (*septica* and *hispida*) of the genus *Ficus* of the Moraceae family. The majority of phenanthroindolizidine alkaloids are present in, and were isolated from, plant species of the genus *Tylophora*, of which *T. indica* (Burn) Merrill, previously known (and sometimes still described) as *T. asthmatica* Wight and Arn, proved to be the richest source. Therefore it became the most often investigated species.

The physiological activity attributed to whole (fresh or dried) or powdered plants, decoctions, extracts, mixtures, or individual compounds was reported in appropriate medicinal and pharmacological publications [2,3], for example, pharmacopoeias. The pharmaceutical use of the drugs produced resulted in intensive chemical studies of not only the isolation of the anticipated active components but also the structural elucidation of the alkaloids isolated and characterized. Although several minor alkaloids (and also others available in somewhat larger quantities) showed interesting biological properties, the most important and significant pharmaceutical activity was exhibited by the alkaloid tylocrebrine, **1** [4]. Tylocrebrine was first isolated

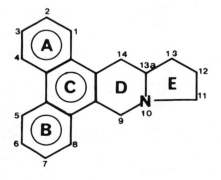

PHENANTHROINDOLIZIDINE

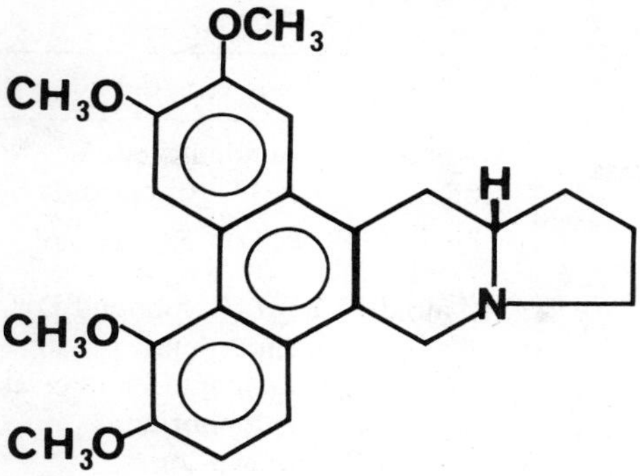

1 TYLOCREBRINE

from *T. crebriflora* S.T. Blake [5,6] and possesses the absolute configuration shown [7] in structure **1**.

2. ISOLATION OF THE ALKALOIDS

2.1. From Plants of the Family Asclepiadaceae

2.1.1. Species of the Genus *Tylophora*

Table 2.1. Alkaloids of *T. indica*

Authors	Alkaloids Isolated	References
Hooper	Unidentified base	[8]
Arata and Gelzer	Presence of alkaloids	[9]
Ratnagiriswaran and Venkatachalam	Two alkaloids: tylophorine, **2**, and tylophorinine, **3**	[10]
Chopra	Presence of alkaloids	[11]
Chopra et al.	(−)-Tylophorine, **2**, and another alkaloid (which could be identical with septicine)	[12,13]
Govindachari et al.	Tylophorine, **2** Tylophorinine, **3**	[14]
Rao	Compound A (identical with tylophorine, **2**) Compound B (identical with tylophorinine, **3**) Compound C (identified later as desmethyltylophorinine, **4**) Compound D (identified later as Alkaloid A, **5**)	[15]

Table 2.1. (*continued*)

Authors	Alkaloids Isolated	References
Rao et al.	Compound E (unidentified) Compound "sixth" (traces only) Tylophorine, 2 Tylophorinine, 3 Alkaloid A, 5 (cf. Compound D) Alkaloid B, 6 (not identical with Compound C) identified later as a desmethyltylophorinine Alkaloid C (either 3- or 6-desmethyltylophorinine but not identical with either Alkaloid B or with Compound C) identified later as tylophorinidine, 7, that is, a 6-desmethyltylophorinine	[16]
Mulchandani et al.	Tylophorinidine, 7 Tylophorinicine	[17]
Govindachari et al.	Tylophorinidine, 7 Isotylocrebrine, 8 (+)-Septicine, 9	[18]
Govindachari et al.	Three quaternary alkaloids which may be artifacts due to isolation procedures used: Dehydrotylophorine, 10 Anhydrodehydrotylophorinine, 11 Anhydrodehydrotylophorinidine, 12	[19]
I.R.C. Bick et al.	Dehydroantofine, 13	[20,21]

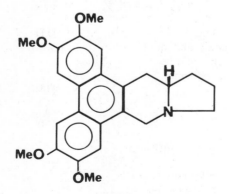

2 TYLOPHORINE

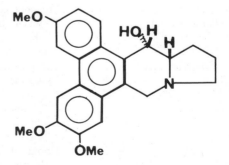

3 TYLOPHORININE

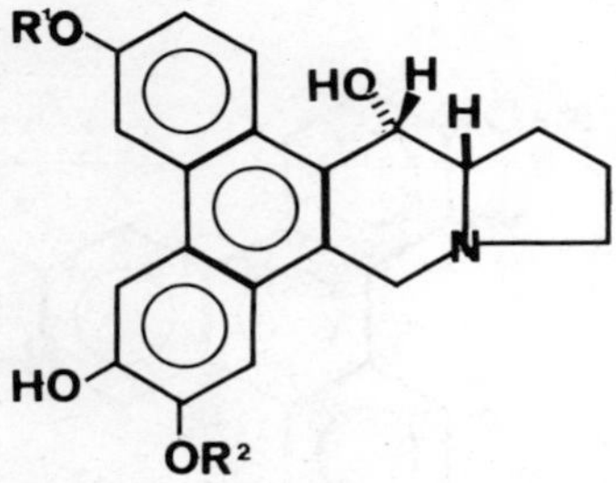

4 A DESMETHYLTYLO-PHORININE

$R^1 + R^2 = H + CH_3$

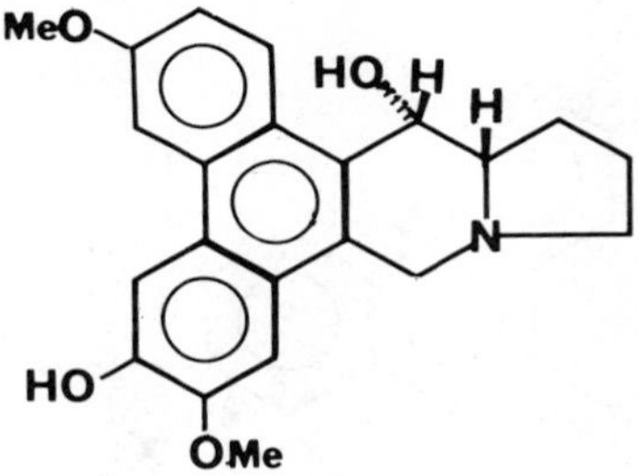

7 TYLOPHORINIDINE

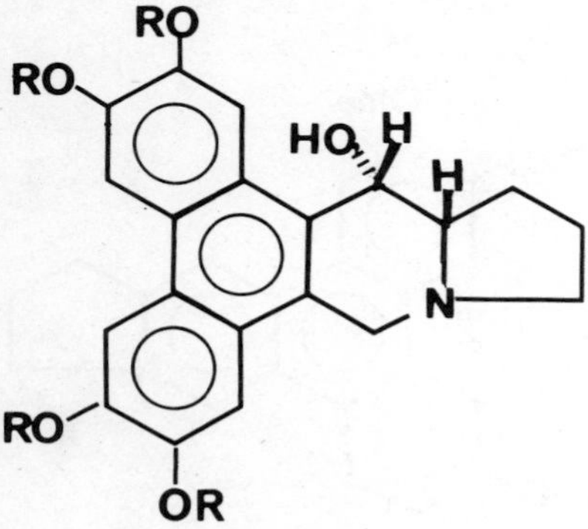

5 Rao's ALKALOID A
from T. indica

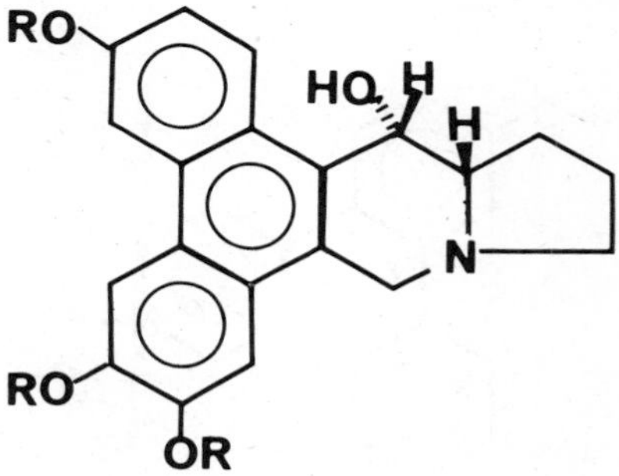

6 Rao's ALKALOID B
from T. indica

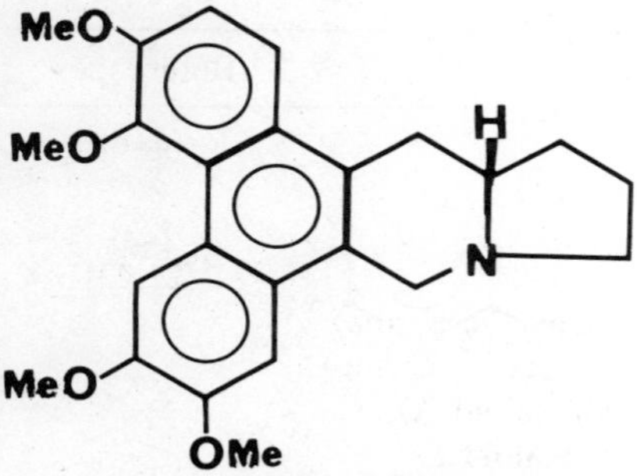

8 ISOTYLOCREBRINE

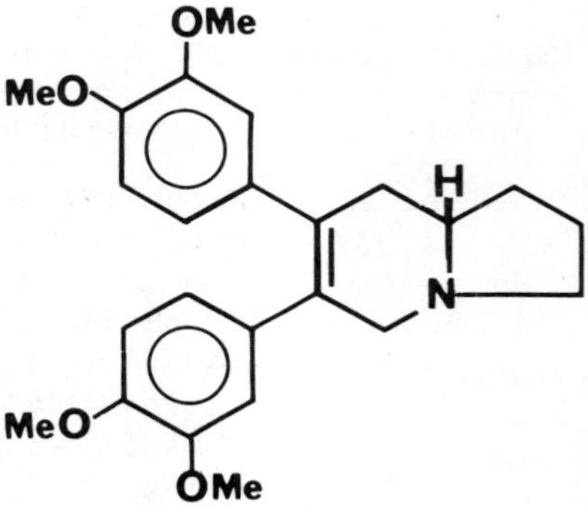

9 SEPTICINE

59

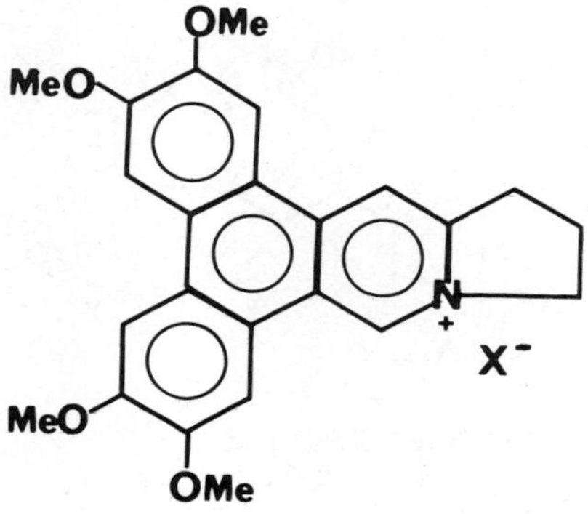

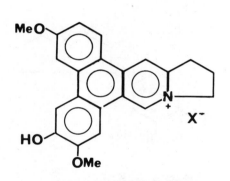

10 DEHYDROTYLOPHORINE

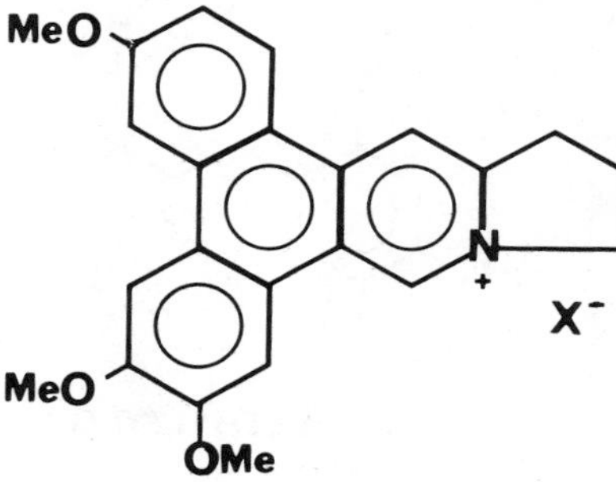

11 ANHYDRODEHYDRO-TYLOPHORININE

12 ANHYDRODEHYDRO-TYLOPHORINIDINE

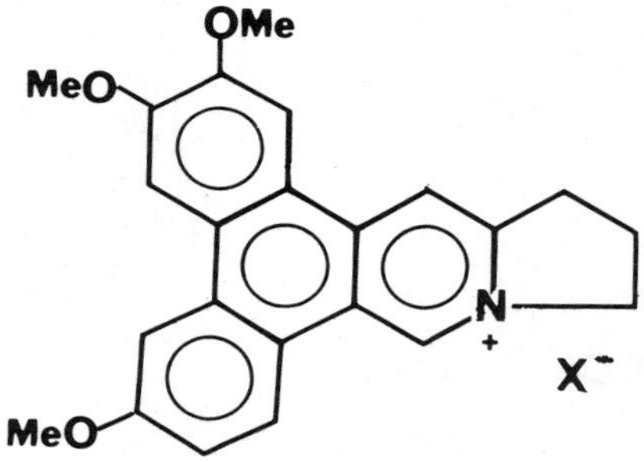

13 DEHYDROANTOFINE

Table 2.2. Alkaloids of *T. crebriflora*

Author	Alkaloids Isolated	References
Gellert et al.	Tylocrebrine, **1**	[4–6]
	Tylophorine, **2**	
Rao	Tylocrebrine, **1**	[23]
Rao et al.	Tylophorine, **2**	[22,23]
	Alkaloid A, **14** (14-hydroxyisotylocrebrine)	
	Alkaloid B, **15** (4-desmethylisotylocrebrine)	
	Alkaloid C, **16** (4-desmethylalkaloid A)	
	Alkaloid D, **17** (14-hydroxyalkaloid E)	
	Alkaloid E, **18** (4-methoxytylophorine)	

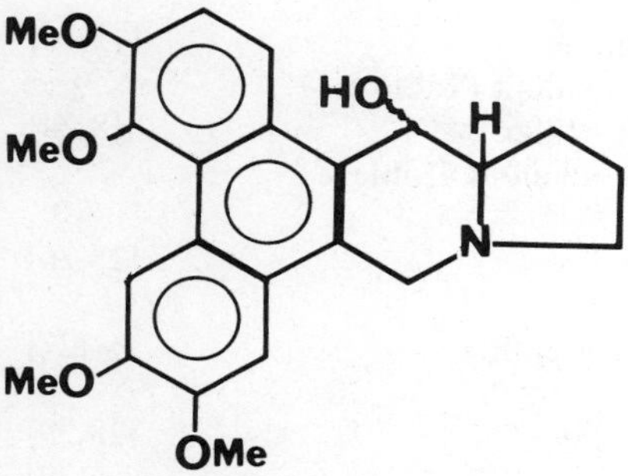

14 Rao's ALKALOID 'A'

14-HYDROXYISOTYLO-
CREBRINE

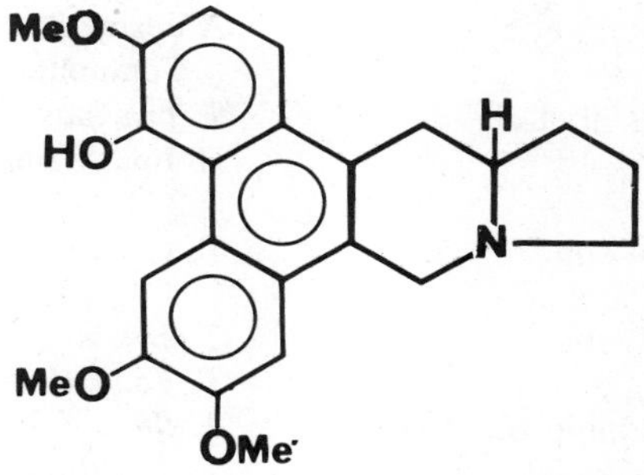

15 Rao's ALKALOID 'B'

4-DESMETHYLISOTYLO-
CREBRINE

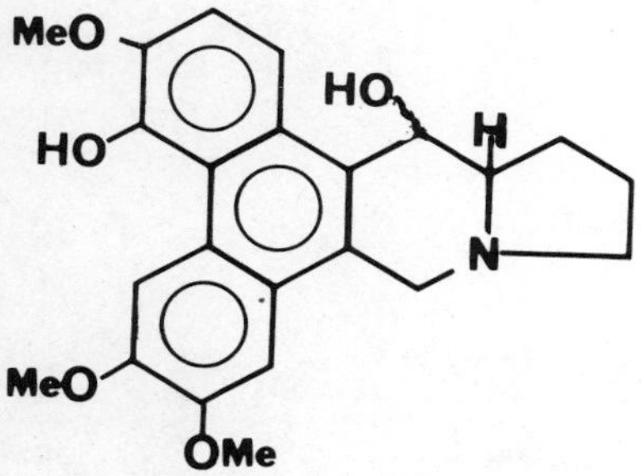

16 Rao's ALKALOID 'C'

4-DESMETHYLALKAL. 'A'

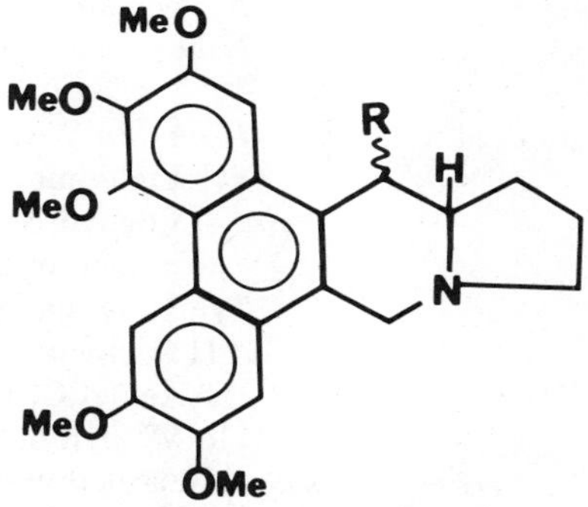

17 Rao's ALKALOID 'D' R=OH

18 Rao's ALKALOID 'E' R=H

Table 2.3. Other Alkaloid-containing *Tylophora* Species

Author	*Tylophora* Species and Any Alkaloids Isolated and Characterized	References
Rao,	*T. dalzellii* Hook f.	
Rao, et al.	A desmethyltylophorinine	[15,16]
	Compound E of the patent (Table 2.1)	
Brill et al	*T. brevipes* (Turcz.) F-Villar	[8,24]
	Hooper's unidentified alkaloid (Table 2.1)	
Webb,	*T. sylvatica* Decne	[25,26]
Patel et al.		
Webb	*T. erecta* F. Muell. ex Benth.	[27]
	T. paniculata R. Br.	
Phillipson et al.	*T. flava* Trimen	[28,29]
Karnick	*T. cordifolia* Thw.	[29]
	T. capparidifolia Wight et Arn	
	T. exelis Coleb	
	T. fasciculata Ham.	
	T. govani Done.	
	T. iphisia Done.	
	T. longifolia Wight	
	T. macrantha Hook f.	
	T. pauciflora Wight et Arn	
	T. rotundifolia Ham.	
	T. tenerrima Wight	
	T. tennis Blume	
	T. zeylamia Done.	
Karnick	*T. hirsuta* Wight	[29]
Fong et al.		[30]
Bhutani et al.	*T. hirsuta* Wight	[174,175]
	Tylohirsutinine, **18a**	
	(13-dehydro derivative of isotylocrebrine, **8**)	
	Tylohirsutinidine, **18b**	
	(13-dehydro derivative of Rao's Alkaloid C, **16**)	
	13a-Methyltylohirsutinidine, **18c**	
	(13a-methyl derivative of either **16** or its enantiomer)	
	13a-Methyltylohirsutine **18d**	
	(13a-methyl derivative of isotylocrebrine, **8**)	
	13a-Hydroxysepticine, **18e**	
	Unidentified base of M^+ 393	
	Unidentified base of M^+ 381	
Govindachari et al.	*T. mollissima* Wight	[29,31]
	Tylophorine, **2**	
	Tylophorinine, **3**	

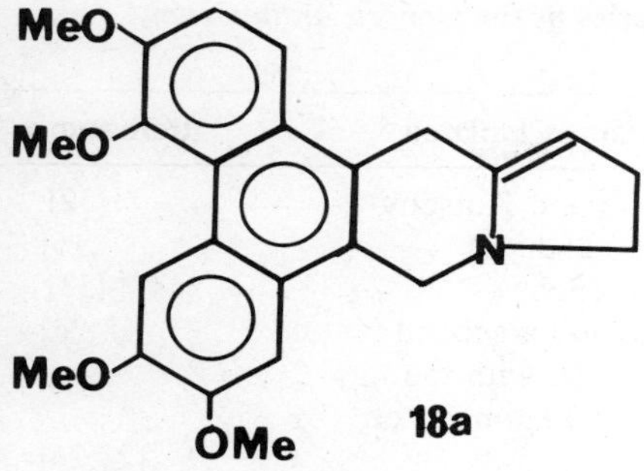

18a

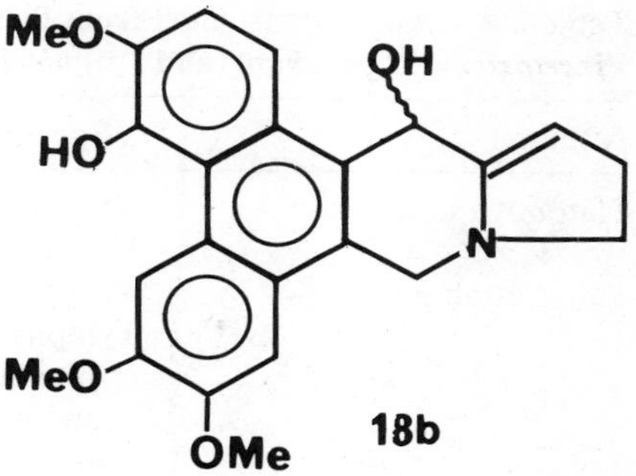

18b

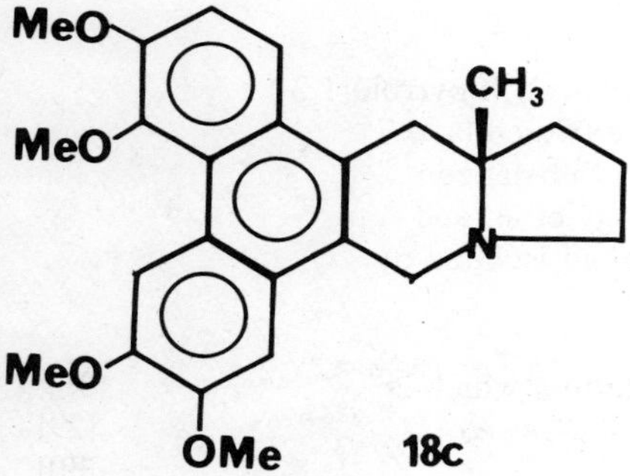

18c

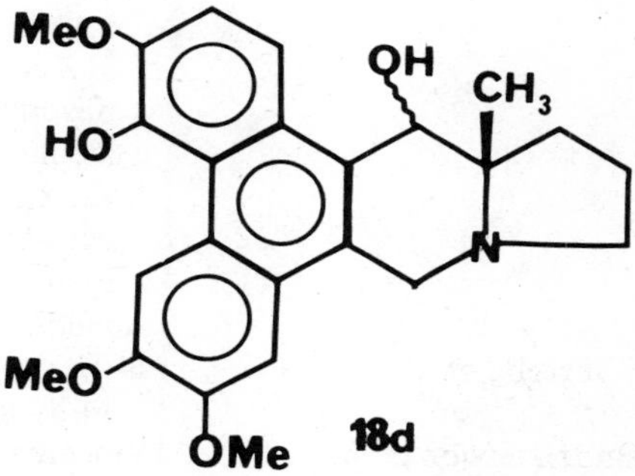

18d

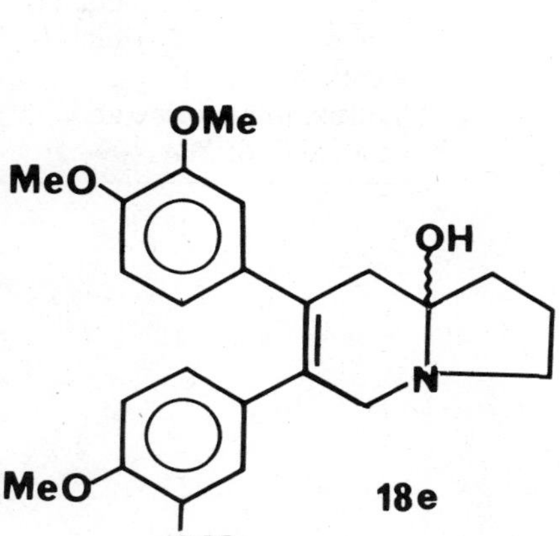

18e

Table 2.4. **Alkaloids Isolated from Plant Species of the Genera** *Antitoxicum,* *Vincetoxicum, Cyanchum,* **and** *Pergularia*

Author	Species, Alkaloids Isolated	References
Platonova et al.	*A. funebre* Boiss and Kotschy Uncharacterized alkaloid	[32]
Pailer et al.	*V. officinale* Moench[a] Tylophorine, **2**, and a second alkaloid (identical with the one isolated by Platonova et al.)	[33]
Haznagy et al.	*C. vincetoxicum* (L.) Pers.[a]	[34,35]
Ferenczy et al.	Three alkaloids, one of them named Substance C-1	[36]
Wiegrebe et al.	*C. vincetoxicum* (L.) Pers.[a] Tylophorine, **2** Alkaloid A, which is (−)-antofine, **19** 2,3,6-Trimethoxy-9,11,12,13,13a,14-hexahydrodibenzo[f,h]pyrrolo[1,2-b]isoquinoline (which is identical with Substance C-1 of Haznagy et al. and with the alkaloid isolated by Platonova et al.) Alkaloid C, **20**	[37]
Wiegrebe et al.	14-Hydroxyantofine (which is 14-hydroxy-alkaloid A), **21**	[38]
Budzikiewicz et al.	Vincetene, **22**	[39]
Mulchandani et al.	*P. pallida* Wight and Arn Tylophorine, **2** Tylophorinine, **3** Pergularinine, **23** (−)-Desoxypergularinine, **24** Tylophoricinine, **25**, which is identical with 14-hydroxytolophorine and alkaloid D of MW 409	[40–43]

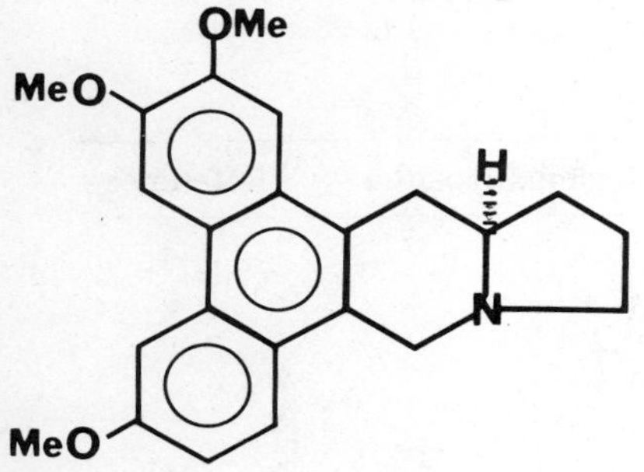

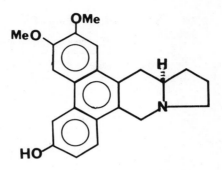

19 ANTOFINE

20 ALKALOID 'C'
of WIEGREBE

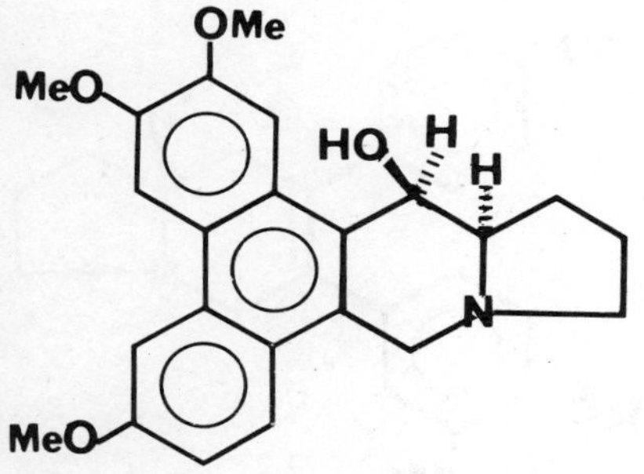

21 14-HYDROXY-
ANTOFINE

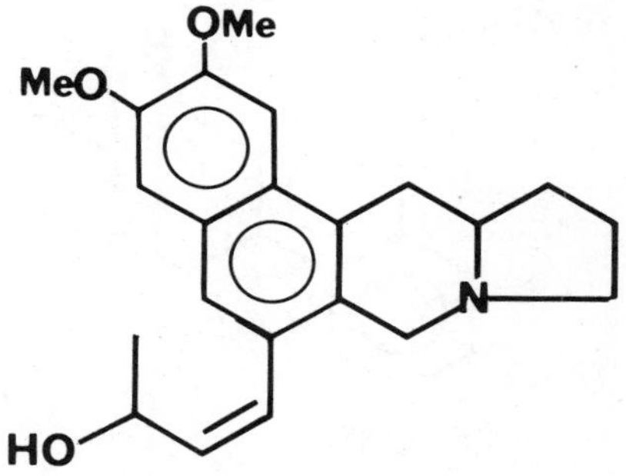

22 VINCETENE

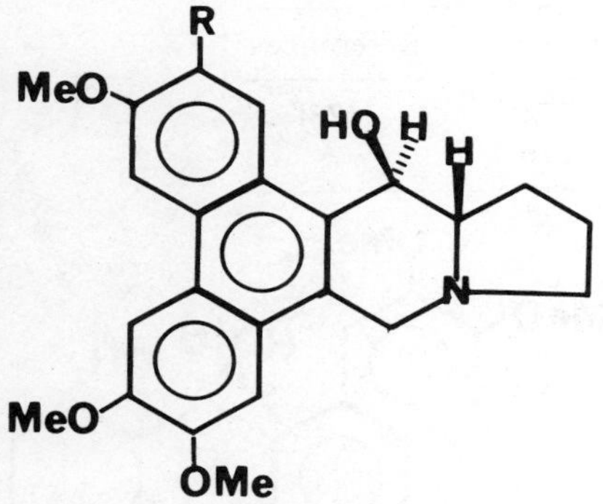

23 PERGULARININE R=H
25 TYLOPHORICININE R=OMe

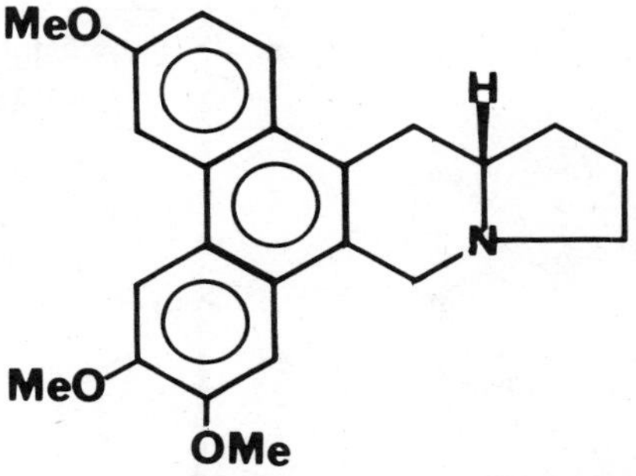

24 (−)-DESOXYPERGULARININE

2.2. From Plants of the Family Moraceae

2.2.1. Species of the Genus *Ficus*

Table 2.5. Species of the Genus *Ficus*

Author	Species, Alkaloids Isolated	References
Russell	*F. septica* Tylocrebrine, **1** Tylophorine, **2** Antofine, **19** (−)-Septicine, **9**	[44]
Herbert et al.	(±)-Septicine, **9**	[45]
Venkatachalam et al.	*F. hispida* (−)-Desoxypergularinine, **24** *O*-Methyltylophorinidine, **26** Hispidine, **27**	[46]

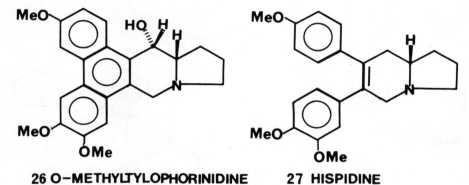

26 O−METHYLTYLOPHORINIDINE   **27 HISPIDINE**

2.3. From a Shrub of the Family Acanthaceae

Table 2.6. The Species *Hypoestes verticillaris*

Author	Alkaloids Isolated	References
Pettit et al.	Hypoestatin 1, **27a** Hypoestatin 2, **27b**	[176]

27a   **27b**

3. ELUCIDATION OF THE STRUCTURE OF THE ALKALOIDS

3.1. Chemical Methods

The structure of tylophorine (2) was established by Govindachari et al. [14]. The alkaloid contains four methoxyl but no *N*-methyl or *C*-methyl groups. It has no readily reducible double bonds, no carbonyl groups, and no active hydrogen atoms. Consequently the nitrogen atom is tertiary so that a methiodide (28a, $X^- = I^-$) is formed readily. This methiodide is stable toward alkali, for it yields when warmed with alkali the racemic methohydroxide 28b ($X^- = {^-}OH$), called the isomethohydroxide, instead of undergoing the expected Hofmann degradation process. Such behavior had already been reported to occur in the case of substituted quinolizidine alkaloids, for ex-

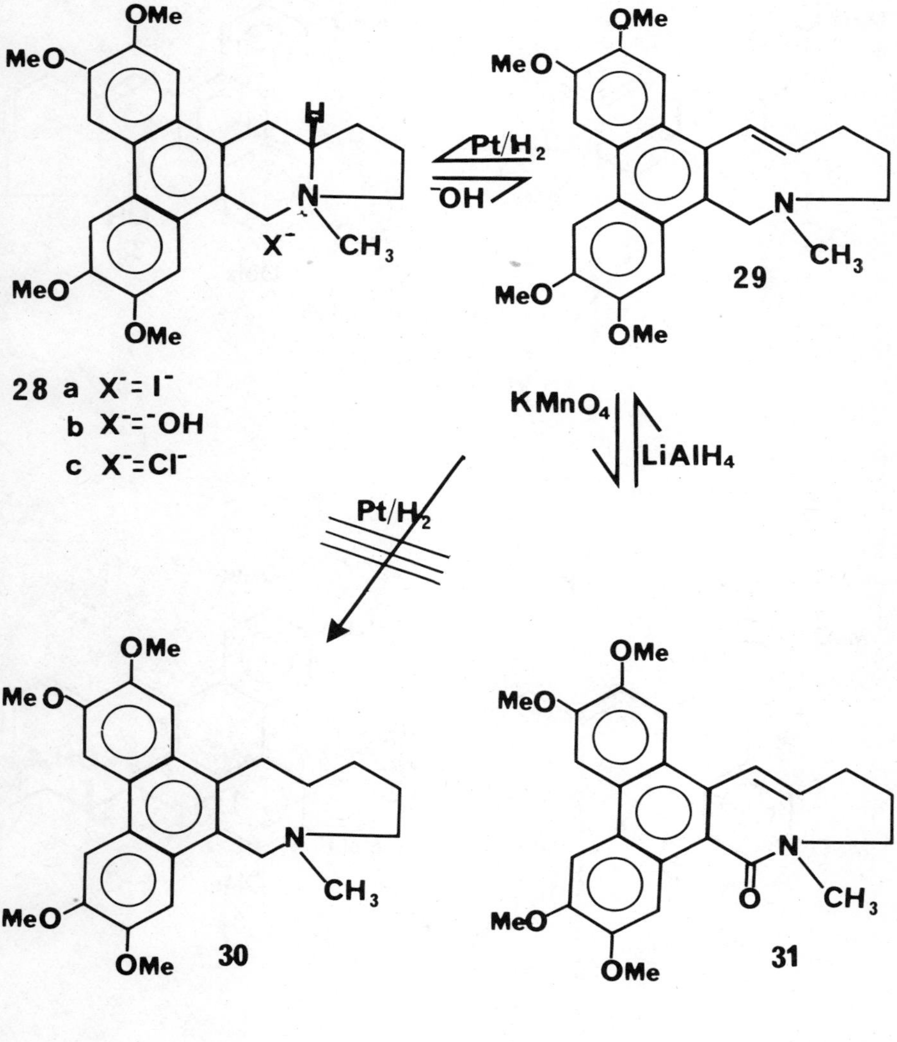

28 a X⁻= I⁻
   b X⁻=⁻OH
   c X⁻=Cl⁻

ample, canadine [47] and cryptopleurine [48–50]. Hofmann degradation could be achieved, however, under more drastic conditions, establishing that the nitrogen atom is, also in this case, common to two rings. Methine-I (**29**), however, did not give a dihydro product (**30**) when catalytic hydrogenation was attempted, but yielded the unchanged racemic quaternary compound **28**, as was the case with the above quinolizidine alkaloids [47–50]. Oxidation of methine-I (**29**) with potassium permanganate produced the unsaturated amide (**31**), which could be reduced back to methine-I by lithium aluminum hydride. Hofmann degradation of methine-I (**29**) yielded methine-II (**32**) and a nitrogen-free product (**33**) in which the dimethylamino group is replaced by a hydroxyl group. Hofmann degradation of methine-II (**32**) gave methine-

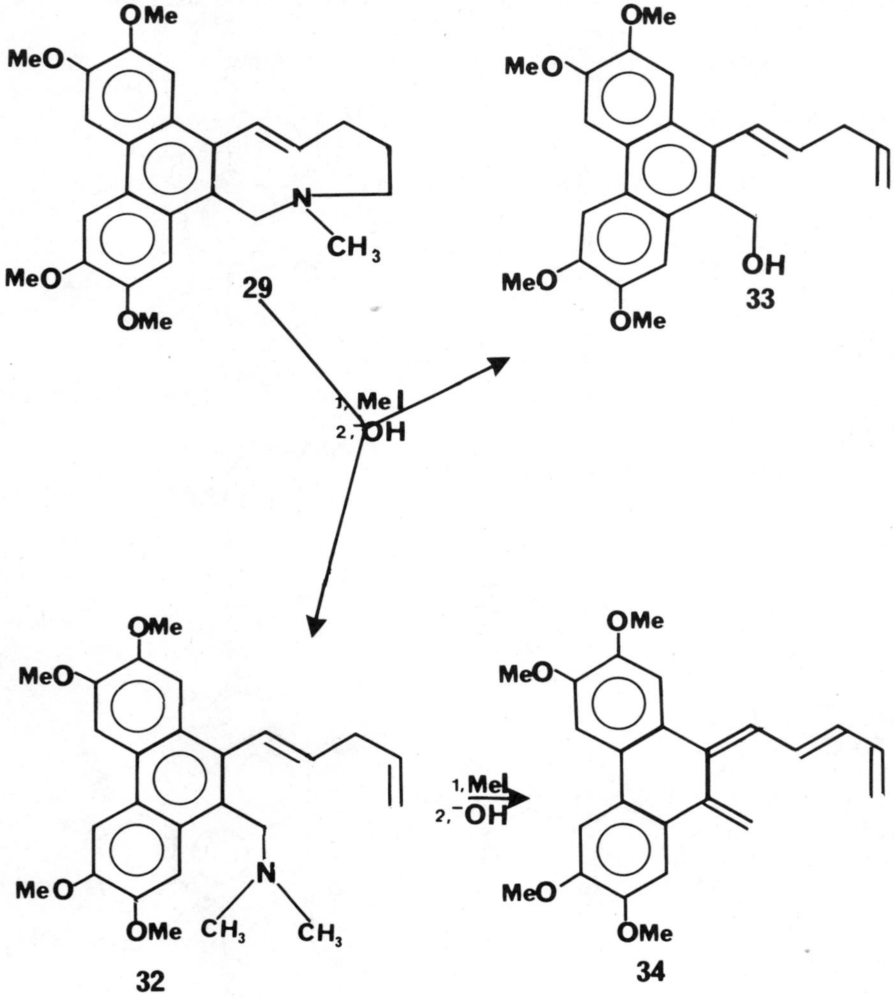

III (34), the final unsaturated product of the Hofmann degradation of tylophorine.

Emde degradation of tylophorine methochloride (28c) gave isodihydro-homotylophorine (35), a pyrrolidine derivative that can be dehydrogenated with the help of palladium–charcoal to the pyrrole derivative (36).

Hofmann degradation of isodihydrohomotylophorine (35) gave a methine (37), which is of course not identical with the Emde degradation product (38) obtained from tylophorine–methine-I (29) [93]. Because both 37 and 38 can be hyrogenated catalytically to 39, the two compounds must represent [51] the *cis* and *trans* isomers of the same product.

Von Braun degradation of tylophorine (2) gave the bromocyanamide (40), which is reduced by sodium borohydride to a hydroxycyanamide (41). The latter regenerates tylophorine when hydrolyzed with acid, a reaction characteristic for 1,4- and 1,5-amino alcohols.

Oxidation in acetone of 37, the *trans*-methine I of the Emde product, yielded two acids [52], namely, 2,3,6,7-tetramethoxy-9-methylphenan-threne-10-carboxylic acid (42a) and 2,3,6,7-tetramethoxyphenanthrene-9,10-dicarboxylic acid (43a), isolated by chromatography of their methyl esters (42b and 43b, respectively). Hydrolysis of 42b followed by decarboxylation gave 2,3,6,7-tetramethoxy-9-methylphenanthrene 42c, whereas hydrolysis of 43b yielded the dicarboxylic acid 43a, which was identified through its anhydride (43c) and imide (43d). All these compounds could also be obtained by permanganate oxidation, either in pyridine or in aqueous pyridine solution, of tylophorine methohydroxide (28b). Oxidation of tylophorine methiodide (28a) with permanganate at 100°, however, gave *meta*-hemipinic acid (44). Finally, tylophorine gave, on zinc dust distillation [53], 9,10-dimethylphenanthrene, thereby establishing unequivocally that the indolizidine ring is fused to the phenanthrene ring in the 9,10 position. The absolute configuration, as shown in formula 2, was established by Govindachari et al [54].

Tylocrebrine [6] (1) is isomeric with tylophorine (2) and contains four methoxy but no N-methyl or C-methyl groups. As in the case of tylophorine it has no readily reducible double bonds or carbonyl groups or any active hydrogen atoms. Its UV spectrum is different from that of tylophorine but it is practically superimposable on that of 3,4,6,7-tetramethoxy-9-methyl-phenanthrene, as opposed to that of 42c. This indicates that the difference between the two alkaloids is likely to be in the methoxy substitution pattern on the phenanthrene moiety. This substitution pattern allows tylocrebrine to be represented by either structure 1 or structure 8. The decision in favor of 1 was made by synthesizing both structures and comparing these with the authentic material [6].

Tylophorinine (3) differs from tylocrebrine and tylophorine by containing only three (not four) methoxyl groups in the 2,3,6-positions of the phenanthrene ring system, a pattern similar to that in the phenanthroquinolizidine alkaloid cryptopleurine [48–50]. However, 3 also possesses an alcoholic hy-

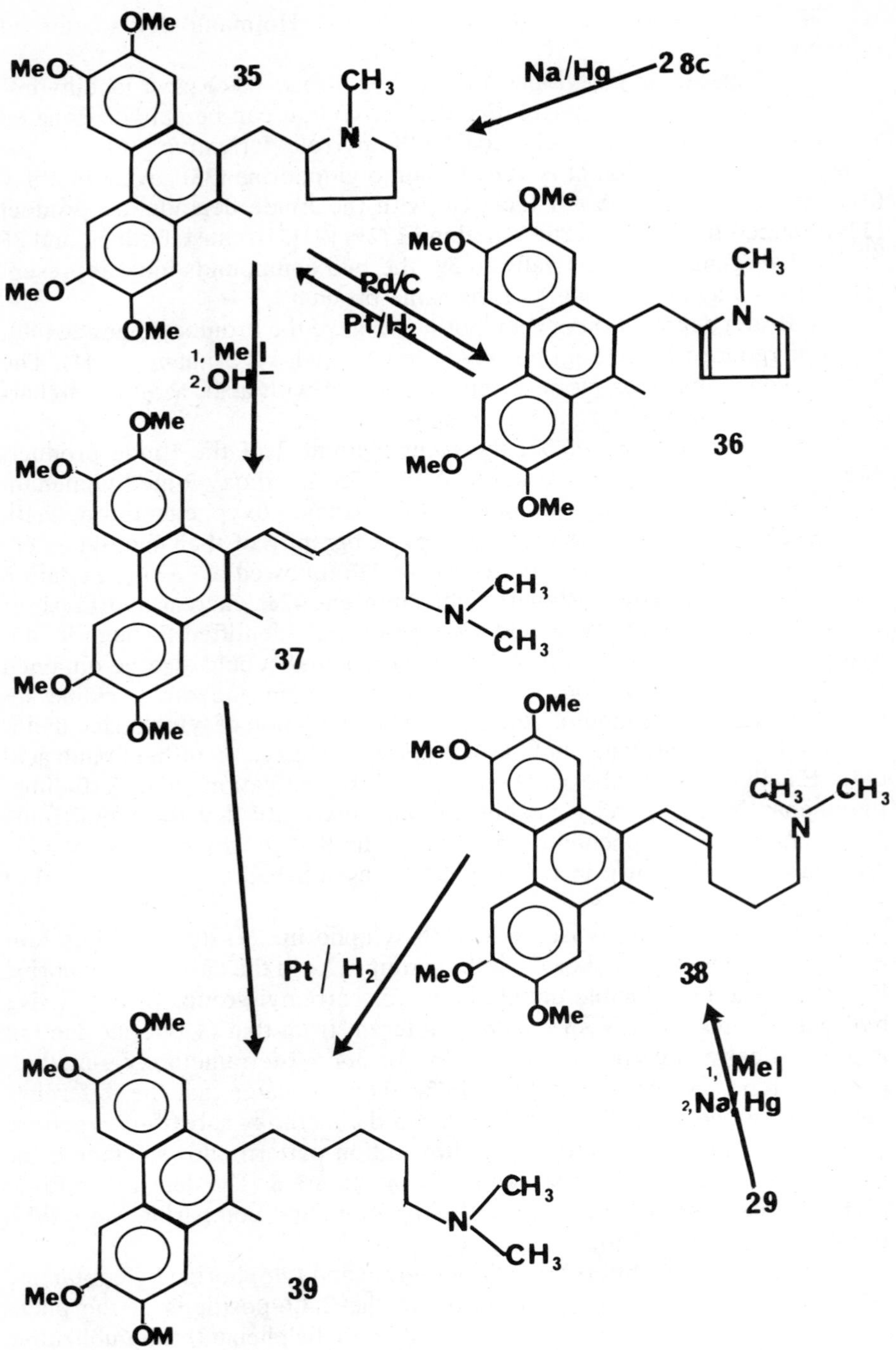

OMe
MeO
35
CH3
N
Na/Hg
28c
Pd/C
Pt/H2
MeO
OMe
1, MeI
2, OH⁻
36
37
CH3
N
CH3
38
Pt / H2
1, MeI
2, Na/Hg
29
39
OM

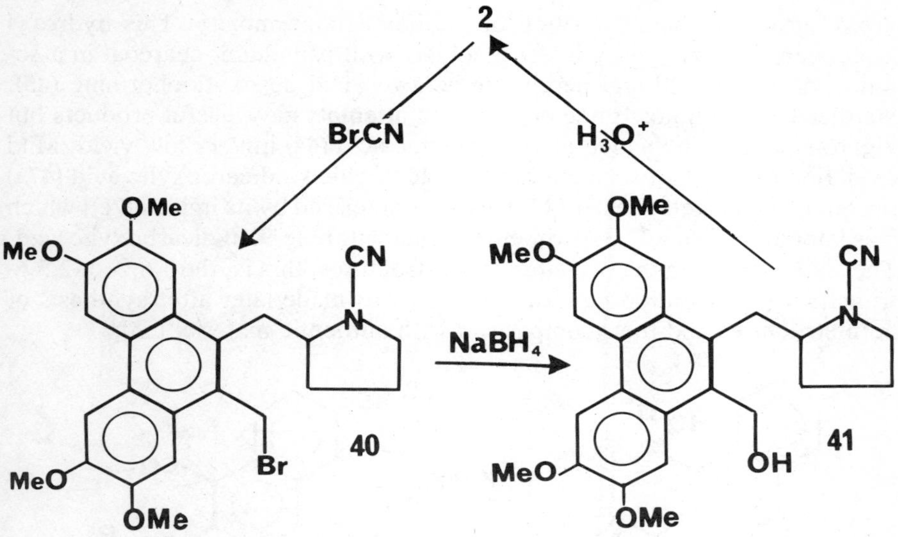
2
BrCN
H3O+
OMe
MeO
CN
N
NaBH4
Br
40
MeO
OMe
OMe
MeO
CN
N
OH
41
MeO
OMe

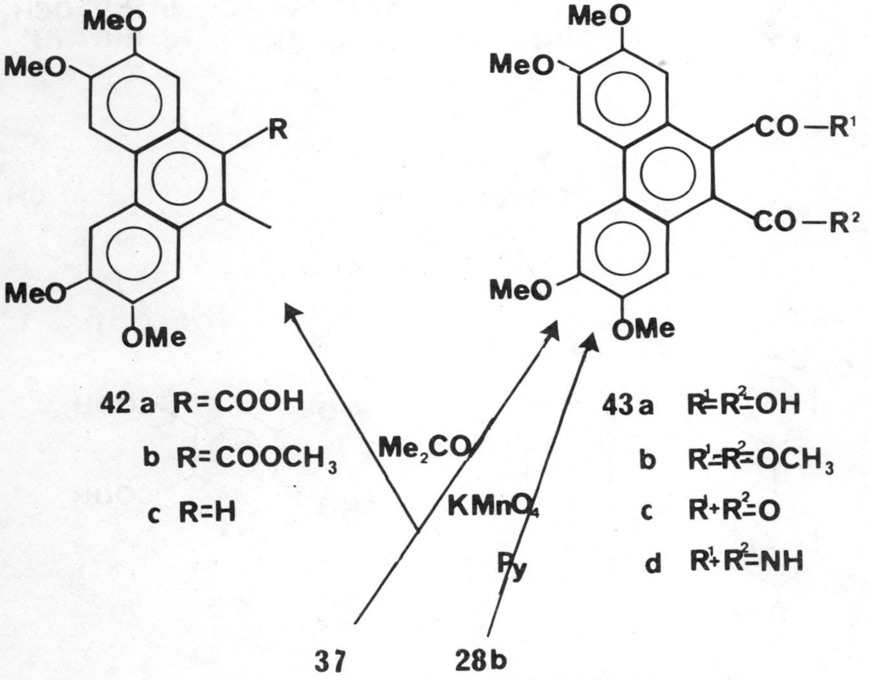
MeO
MeO
R
MeO
OMe
MeO
MeO
CO—R1
CO—R2
MeO
OMe
42 a R=COOH
b R=COOCH3
c R=H
Me2CO
KMnO4
Py
43 a R1=R2=OH
b R1=R2=OCH3
c R1+R2=O
d R1+R2=NH
37
28b

28a $\xrightarrow[100^\circ]{KMnO_4}$ 44

droxyl group in the 14 position of the indolizidine moiety. This hydroxyl group can be removed by hydrogenolysis with palladium–charcoal in a solution of acetic acid and perchloric acid to yield desoxytlophorinine (**45**). Neither Hofmann nor Emde degradation attempts gave useful products but vigorous oxidation yielded *meta*-hemipinic acid (**44**) in very low yield. Mild oxidation of tylophorinine methiodide (**46**) yielded a dicarboxylic acid (**47a**) [isolated via its methyl ester (**47b**) and characterized by its imide (**47c**)] which was later identified as 2,3,6-trimethoxyphenanthrene-9,10-dicarboxylic acid. The choice between the two alternative structures, that is, the 3,6,7- or 2,3,6-trimethoxy substitution pattern (**3** or **48**), was made later after synthesis of both structures and their comparison with authentic material [55].

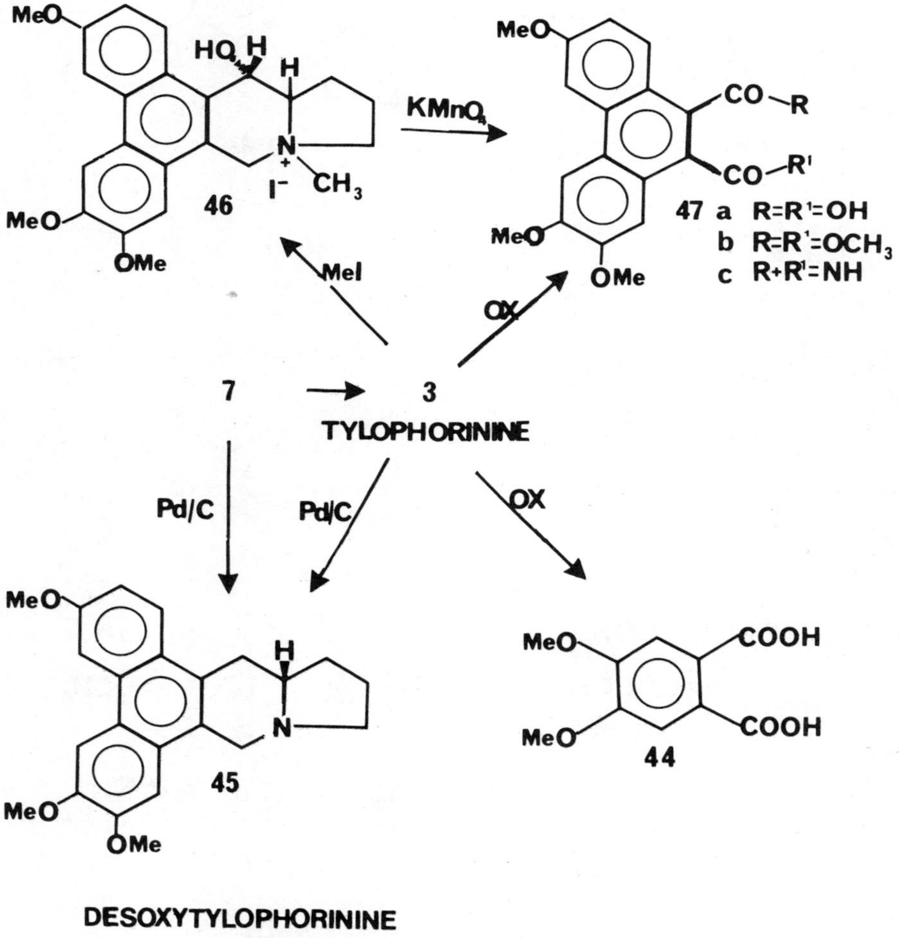

47 a R=R¹=OH
   b R=R¹=OCH₃
   c R+R¹=NH

7 → 3

TYLOPHORININE

45

DESOXYTYLOPHORININE

44

Antofine (**19**) [33–35,37–38] contains three methoxyl but no *N*-methyl or *C*-methyl groups, no active hydrogen atoms, and no readily reducible double bonds or carbonyl groups. Its UV spectrum is practically superimposable on that of tylophorine (**2**). Its structure must therefore be represented either by **19** or by its isomer (**45**). Because **45** was already synthesized [56] and was also prepared from tylophorinine (**3**) by hydrogenolysis (but was shown to be different from antofine), its structure must be represented by **19**, and the structure of 6-desmethylantofine by **20**. Their absolute configuration was established by Wiegrebe et al. as (*R*) [57]. This was later confirmed by Govindachari et al. [18].

Tylophorinidine (**7**) was first considered incorrectly to possess the 2,3,5,7-tetraoxygenated structure [17]. This was quickly changed [58] to the methylated 3,6,7-trimethoxy-14-hydroxy enantiomer (**23**), then to the correct enantiomer (**7**) [59]. The accuracy of this assignment was confirmed by *O*-methylation of **7** to **3**, and also by showing that both **3** and the methylated **7** yield desoxytylophorinine (**45**) on hydrogenolysis. Interestingly, structure **26**, which was initially and incorrectly assigned to tylophorinidine, was later isolated from *F. hispida* [46].

The structures of Rao's alkaloids from *T. indica* [15–18] have been only partially elucidated, with the exception of tylophorinidine (**7**). In the structures of the other alkaloids the positions of the free phenolic hydroxyl groups and the methoxyl groups are still uncertain. It is known only that Compound C (**4**) has 3,6,7-trioxygenation whereas Alkaloids A and B (**5** and **6**) have 2,3,6,7-tetraoxygenation patterns on the phenanthrene moiety. The structures of the three quaternary alkaloids (even though they may turn out to be artifacts), on the other hand, are well established. They were successfully converted, by catalytic hydrogenation, to the corresponding tertiary alkaloids [19]. Thus the reineckate of **10** gave tylophorine (**2**), that of **11** yielded (±)-desoxytylophorinine (**45**), and that of **12** produced (±)-desoxytylophorinidine (**49**).

48

49 DESOXYTYLOPHORINIDINE

Of Rao's alkaloids isolated from *T. crebriflora* [22,23] Alkaloid A exhibited the same methoxylation pattern as tylocrebrine (**1**), but hydrogenolysis yielded not tylocrebrine but isotylocrebrine (**8**) with the "alternative" oxygenation pattern. Alkaloid A is therefore 14-hydroxyisotylocrebrine (**14**) in which the stereochemistry at C(13a) is not yet known. Alkaloid B contains only three methoxyl and one phenolic hydroxyl groups. Because the single phenolic group readily methylates to isotylocrebrine, (**8**) it must be 4-desmethylisotylocrebrine (**15**). Alkaloid C possesses a phenolic hydroxyl group at C(4). Because Alkaloid C methylates to Alkaloid A (**14**), which yields **8** on hydrogenolysis, Alkaloid C is best represented as 4-desmethyl-14-hydroxyisotylocrebrine (**16**). The occurrence of the phenolic hydroxyl group at C(4) in both **15** and **16** was confirmed by NMR spectroscopy and by obtaining, in both cases, a positive Gibbs test, which requires that the position para to the phenolic group be free of substituents. Alkaloids D (**17**) and E (**18**) contain five aromatic methoxyl groups each. In addition Alkaloid D possesses a benzylic hydroxyl group that can be removed by Clemmenson reduction to yield Alkaloid E. The NMR spectra of both alkaloids are very similar to that of tylocrebrine (**1**), which indicates the presence of one of either a 2,3,4,6,7- or 3,4,6,7,8-pentamethoxy substitution pattern in the alkaloids. The question whether the fifth methoxyl group is attached to C(2) or C(8) can be resolved by observing the lack of two ortho protons in the C(2) substituted formula, and considering the downfield shift of one of the aromatic proton signals in alkaloids possessing C(14) hydroxylation, which is assumed responsible for such a shift. Consequently the 2,3,4,6,7-pentasubstituted pattern is the correct one and Alkaloid D must be represented by formula **17** and Alkaloid E by **18**.

(−)-Pergularinine (**23**) [40–43] is a diastereoisomer of *O*-methyltylophorinidine (**26**). Because (−)-desoxypergularinine (**24**) exhibits a negative ORD effect of the same magnitude as tylophorine (**2**), the configuration at C(13a) of both **24** and **2** must be the same. Similarly the same configuration must hold at C(13a) in tylophorinidine (**7**), in pergularinine (**23**), in desoxytylophorinine (**45**), and in desoxytylophoridine (**49**). This then also establishes the difference of pergularinine (**23**) from *O*-methyltylophoridinine (**26**) only in the configuration at C(14). Alkaloid D from *P. pallida* [40–42] was initially characterized only by its molecular ion M$^+$ 409, calculated as $C_{24}H_{27}NO_5$. Later it was renamed tylophoricinine [43] and shown to be identical with 14-hydroxytylophorine (**25**).

The ready oxidation of septicine methiodide to veratric acid (**50**) [18] and the transformation of the alkaloid when irradiated with UV light into a mixture of phenanthroindolizidines containing both **1** and **2** clearly establishes its structure as **9**. The structure of the other secophenanthroindolizidine alkaloid, hispidine [46], was determined by spectroscopic means as **27**. It is traditional to discuss these two secophenanthroindolizidine alkaloids here because they were isolated from the same plant material as the phenanthroindolizidine alkaloids. However, neither phyllostemine nor phyllostem-

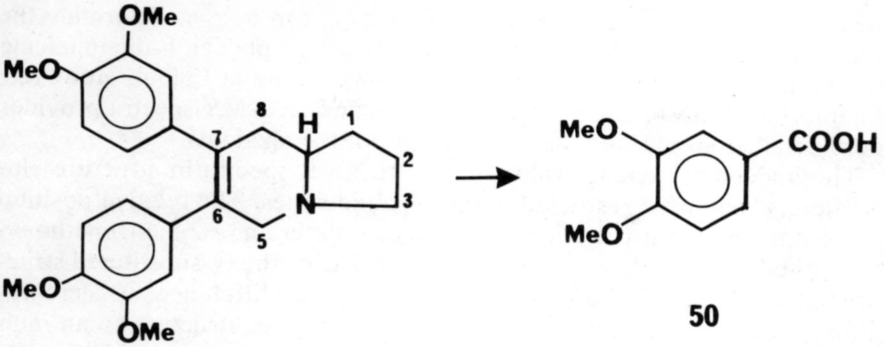

9   SEPTICINE

inine [20] are included in this report although they were isolated, together with antofine and dehydroantofine, from a *Cryptocarya* species of the family Lauraceae.

Finally, there is another alkaloid, vincetene (22) [39], that possesses an incomplete phenanthroindolizidine ring system. Ring B of the above structure is opened to give a tetracyclic, benzopyrroloisoquinoline ring system. The four carbon atoms of the "open ring" remain attached, as a straight alkenyl chain, to position 8 of the isoquinoline ring. It contains one *cis* double bond conjugated with the isoquinoline ring, and an allylic hydroxyl group with a methyl group terminating the chain. Its structure was elucidated mainly by the use of spectroscopic data and by the synthesis of its racemic desoxydihydro derivative. Its absolute configuration, that is, at the bridge-head carbon of the indolizidine ring, is not yet determined, however.

3.2.   Spectroscopic Methods

The use of spectroscopy played, as indicated earlier, an important role in the elucidation of the structure of these alkaloids. Quite early [60] phenanthrene derivatives were recognized to possess UV absorption spectra characteristic of not only the nucleus itself but also of the pattern of (mono- to penta-) substitution of the ring system [6,18,19,52]. Ultraviolet spectra can also help to distinguish between alcoholic and phenolic hydroxyl groups [60–62] by showing whether any new double bond is or is not formed by dehydration of the alcohol, that is, whether it would or would not extend the chromophoric system by exhibiting a bathochromic shift. The question whether the hydroxyl groups are phenolic or alcoholic can also be established by evaluating the absorption bands of their esters. The carbonyl group absorbs at different wave numbers in phenolic and alcoholic esters, with ~30 cm^{-1} difference in the values obtained. The Bohlmann bands [63–66], if present in the IR spectrum around 2800 cm^{-1}, can help to determine the orientation of the hydroxyl group on C(14) of the phenanthroindolizidine ring system relative to the bridgehead hydrogen atom at C(13a). However, the information available from mass spectra and ^{1}H NMR spectra provides more detailed insight into the full structure of the alkaloids.

The undeniable general value of the ^{1}H NMR spectra in structure elucidation is somewhat restricted in this alkaloid series. The relative position of the nitrogen atom present in the molecule differs in a 2,3,6-trimethoxy-substituted structure from the one with a 3,6,7-trimethoxy-substituted structure, although the ^{1}H NMR spectra show very little difference. This creates difficulties when it is necessary to assign the correct structure to an individual alkaloid, for example, when the choice is between tylocrebrine (1) and isotylocrebrine (8) or antofine (19) and desoxypergularinine (24). The problem has to be solved in another way, namely, (a) by synthesizing both compounds and comparing the products with the authentic alkaloid, as was done in case of the former pair (1 and 8) (b) by using the nuclear Overhauser effect, as was the case with the latter pair (19 and 24) [20], or (c) by introduction at the 9 position of the phenanthroindolizidine structure of a magnetic anisotropic substituent. This would allow exploration of the substituent's effect on the proton (if present) in positions 7 and 8 of the phenanthrene ring [37], for the protons in positions 1 and 2 will not be influenced by this protonless substituent. A cyano group, for instance, can be readily introduced into this position by mercuric acetate oxidation (in EDTA) of, for example, antofine [32,37] to its immonium salt-I (51) (and in addition also to a dehydro derivative of 51, 52), which can react with cyanide by nucleophilic attack to yield, in this case, the 9-cyano derivative (53) of antofine.

The presence of a hydroxyl group at C(14) (either as a free hydroxyl group or as its acetate) has a deshielding effect [6,23,38] on the aromatic protons (if any) on C(1) or C(2) when compared with an oxygen-free C(14), that is, after hydrogenolysis of the benzylic hydroxyl group. Such considerations helped in determining the position of the free hydroxyl group in Rao's Alkaloids B and C (15 and 16) [22,23], where the phenolic hydroxy group was assigned either to C(4) or C(5) of the phenanthrene moiety. In order to choose between the two possible positions [i.e., to select C(4) as the correct one] an additional, positive Gibbs test was required. In case of the pentamethoxy-substituted alkaloids no Gibbs test was necessary; NMR measurements were sufficient.

The configuration of the bridgehead, that is, C(13a), can be determined by interconversion of alkaloids into ones possessing known configurations or with the help of ORD spectra [31,54,61,68].

Mass spectral fragmentation of the phenanthroindolizidine alkaloids proceeds, predominantly, by two different pathways depending primarily on whether or not C(14) is oxygenated. If C(14) is not oxygenated fragmentation occurs via the retro-Diels–Alder reaction [17,26,67]. An intense ion, m/z 69

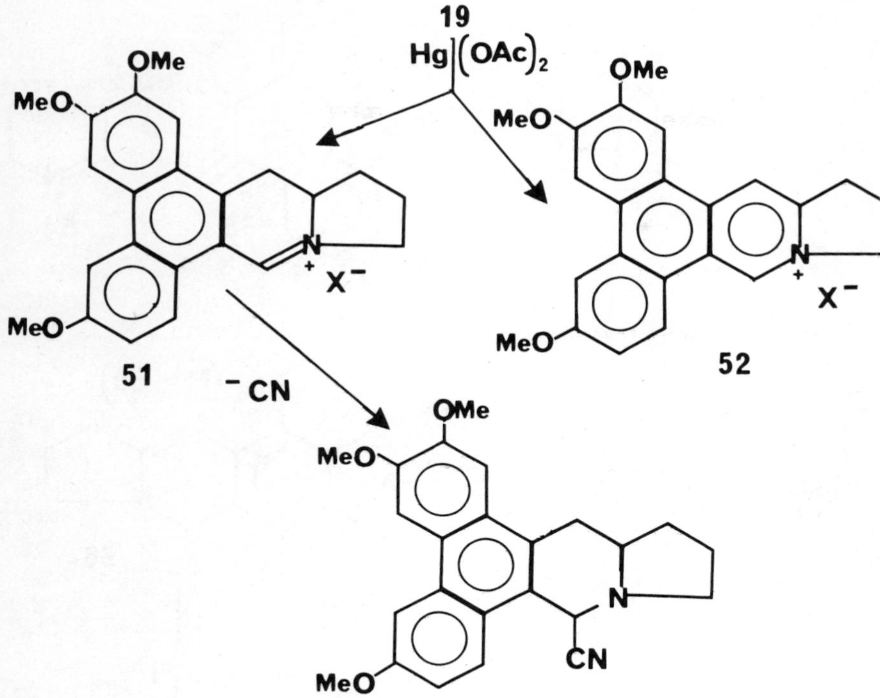

51

52

53

(usually the base peak), representing the 1,2-dehydropyrrolidine ion (**54**), is produced together with an appropriately substituted 9,10-dimethylenephenanthraquinone (**55**). The presence of **54** in the spectrum indicates that the pyrrolidine portion of the phenanthroindolizidine is unsubstituted. If, however, C(14) has a free hydroxyl group or is acetylated then fragmentation via the retro-Diels–Alder reaction occurs only to a small extent (in a similar way, as above) [38]. The main path of fragmentation proceeds via a McLafferty rearrangement initiated by the loss of an acetic acid molecule yielding the intermediate ion **56** (M^+ -60), which loses a hydrogen to yield the stable (substituted) dibenzoisoquinolinium ion (**57**) (M^+ -61), which usually is the base peak. This stable ion can further fragment, for example, by the loss of HCN (m/z 27), to produce smaller ions.

Recently, the structures of five alkaloids isolated from *T. hirsuta* [174,175] were elucidated primarily by the use of spectroscopy. Tylohirsutine (**18a**), $C_{24}H_{25}NO_4$, contains a phenanthroindolizidine skeleton with an extra double bond at C(13) (i.e., not conjugated with the phenanthrene ring) that makes the alkaloid optically inactive. Spectroscopic evidence is in agreement with the dehydroisotylocrebrine structure. Tylohirsutinidine (**18b**), $C_{23}H_{23}NO_5$, also contains a phenanthroindolizidine skeleton with the extra double bond. However, the methoxyl group, which is at C(4) in **18a**, is demethylated to

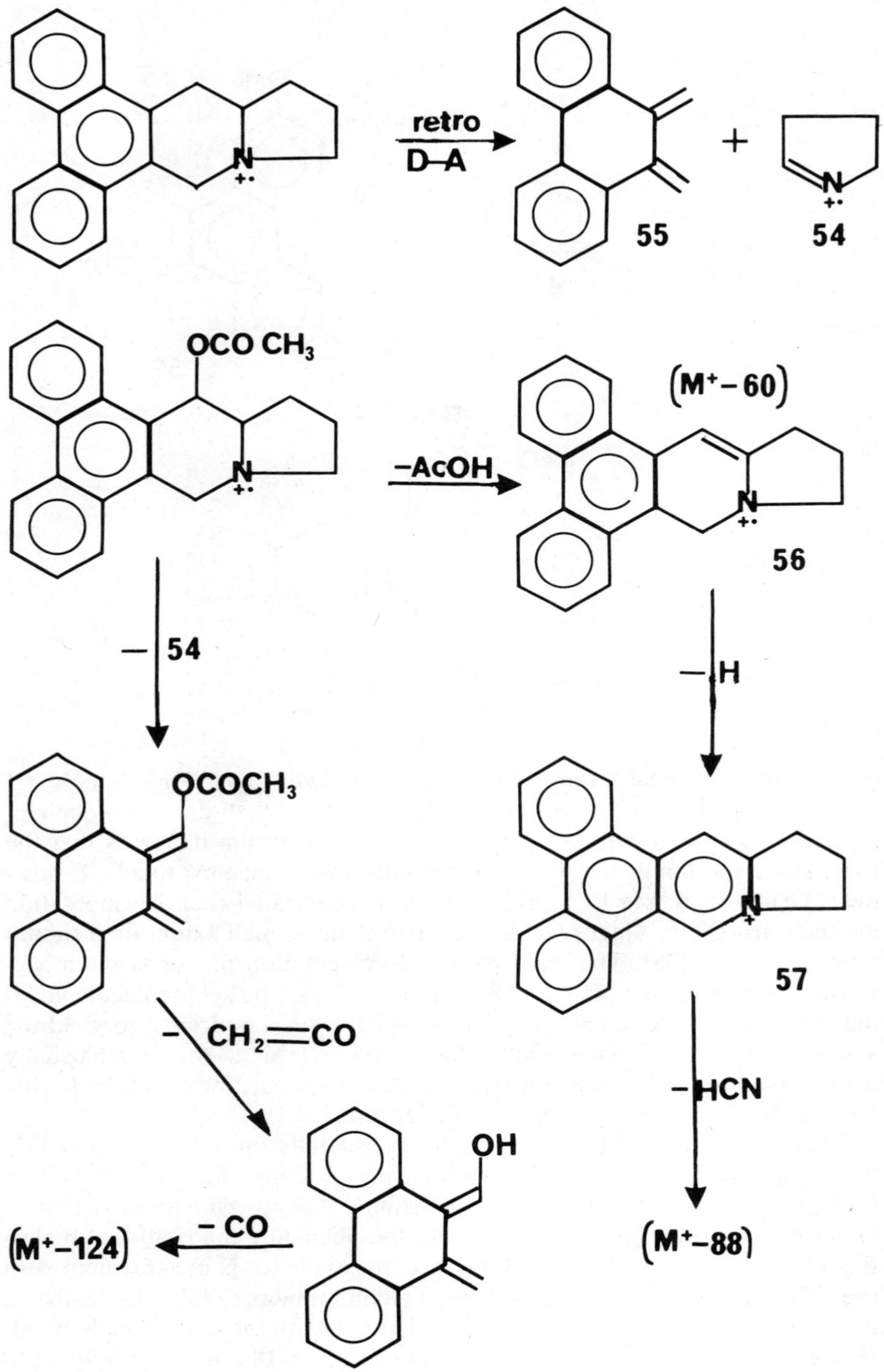

retro D–A

55 + 54

OCOCH₃

(M⁺ – 60)

–AcOH

56

– 54

–H

OCOCH₃

57

– CH₂=CO

–HCN

OH

(M⁺ – 124) ← – CO

(M⁺ – 88)

a phenolic hydroxyl group. Although the presence of an alcoholic hydroxyl group at C(14) provides this alkaloid with a potential asymmetric center, **18b** is racemic. 13a-Methyltylohirsutinidine (**18c**), $C_{24}H_{27}NO_5$, is dextrorotatory. Its phenanthroindolizidine skeleton also possesses a phenolic hydroxyl group at C(4) and an alcoholic hydroxyl group at C(14), making it very similar to that of **18b**. However, it has, instead of the extra double bond, an angular methyl group at C(13a), thereby giving the molecule a second asymmetric center. 13a-Methyltylohirsutine (**18d**) $C_{25}H_{29}NO_4$, is a C(13a)-methyl derivative of isotylocrebrine (**8**); i.e., it has an angular methyl group at C(13a) of the isotylocrebrine molecule (instead of the double bond of **18a**). Spectroscopic data indicated that 13a-hydroxysepticine (**18e**) possesses a structure similar to that of (+)-septicine (**9**). The tertiary nature of the hydroxyl group was assumed because of its inability to form an acetate with acetic acid anhydride and pyridine. Consequently it was assigned to the C(13a) bridgehead carbon atom of the phenanthroindolizidine skeleton, which is C(9) of the indolizidine ring of septicine; that is, the alkaloid **18e** should be named 9-hydroxysepticine. Attempted acetylation, by allowing the alkaloid to stand overnight in acetic anhydride and pyridine at room temperature, resulted in the formation of apparently only one phenanthroindolizidine derivative (**58a**), $C_{24}H_{27}NO_5$, the C(13a)-hydroxy-2 (cf. Russell [44], who found all possible isomers present in the mixture when he cyclized **9** by UV light). Addition of a drop of boron trifluoride diethyletherate to a solution of **18e** and **58a** in dry ether was reported to dehydrate them (in 6 hr below 10°) to **58b** and **58c**, respectively, producing a C(13) double bond, not conjugated with the existing double bond system.

In addition, small amounts of two other phenanthroindolizidine alkaloids were isolated from *Hypoestes verticillaris* [176]: hypoestatin 1 (**27a**) and hypoestatin 2 (**27b**). Their structures were elucidated also on the basis of spectroscopic evidence. The mass spectrum indicated the presence of a 2,3,6-trimethoxyphenanthroindolizidine substitution pattern, and the UV and other spectra showed an alcoholic hydroxyl group at C(14). Comparison of the molecular rotation figures obtained for **27a** and **27b** with those of **23**, **24**, and **26** (but not with those of **19** and **21**) showed that this hydroxyl group is β oriented [42]. Comparison of their circular dichroism spectra with that of **1** elucidated the (*S*) configuration at the methyl-substituted C(13a).

4. BIOLOGICAL ACTIVITY OF THE ALKALOIDS

Extracts of *Tylophora* alkaloids exhibit interesting and potent physiological activities as recorded in, for example, the Bengal [2], Indian [8,10], and Philippine pharmacopoeias [24]. Their emetic and vesicant activity has been mentioned in several publications [6,11,12], which also recorded their lethal toxicity to frogs in contrast to their near inactivity in mice and guinea pigs. In later years the influence of these alkaloids on the circulatory and respiratory systems [69–72] was also explored [12–14], and they were shown to have a depressant effect on the heart muscles and a stimulant effect on the muscles of the blood vessels. The use of these alkaloid extracts was also established in the treatment of allergies, especially of asthma [73–77]. A number of plant extracts and also a few individual alkaloids exhibited promising activity against leukemia [3,4,16,78].

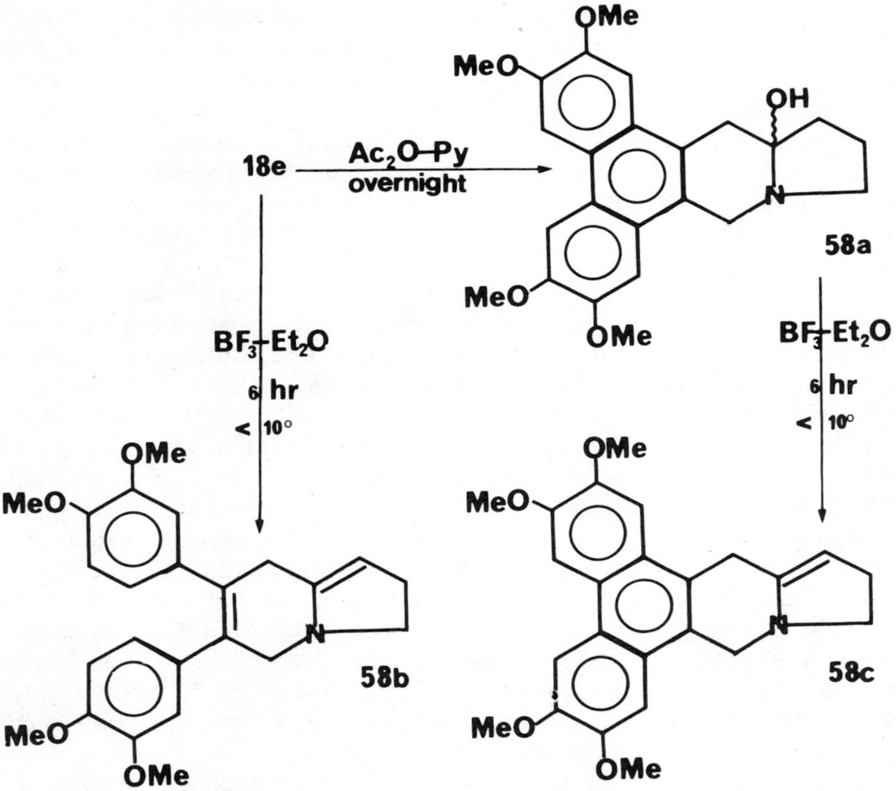

Other extracts possessed strong antibacterial, antifungal, antinflammatory, and immunopathological [79,80] activities. Alkaloids 1 and 2 inhibit the incorporation of [^{14}C]leucine into proteins of Ehrlich tumor cells, but do not affect RNA synthesis at the same low concentration. This indicates that they might compete for protein synthesizing sites in cells [81,82]. They also inhibit, irreversibly, protein synthesis in HeLa cells [83], in yeasts [84,85], and in Chinese hamster ovary cells [86]. The conditions required to produce phenanthroindolizidine alkaloids in tissue culture made from leaf, stem, and root segments of *T. indica* were elucidated [87], though earlier attempts with callus cultures of the same plant species failed to do so [88]. The amebicidal activity of some ipecac alkaloids and their ability to inhibit protein synthesis parallel and correspond with the activity of tylocrebrine (1) [89].

5. SYNTHESIS OF PHENANTHROINDOLIZIDINE ALKALOIDS

Most of the early syntheses of the phenanthroindolizidines relied on the information accumulated from synthetic studies of phenanthroquinolizidines [90–92]. The first phenanthroindolizidine synthesis was, however, established on a different basis.

5.1. The Unsubstituted Phenanthroindolizidine Molecule

5.1.1. Routes 1–3

Route 1.   Govindachari's first synthesis [93] commenced with the condensation of phenanthrene-9-aldehyde with 4-nitrobutyl benzoate to give the nitroalkene (58). This was reduced by lithium aluminum hydride to the amino derivative (59), which could be formylated, and then condensed using phosphorus oxychloride to the immonium salt (60). The immonium salt, which represents a 3,4-dihydroisoquinolinium salt type compound, is reduced readily by sodium borohydride to the unsubstituted phenanthroindolizidine molecule (61).

Route 2.   Chauncy et al. [94] relied on a different approach. Condensation of 9-chloromethylphenanthrene with benzyl L-prolinate gave the N-substituted prolyl ester, which could be hydrolyzed to N-(9-phenanthrylmethyl)proline (62). This was cyclized to the racemic aminoketone (63) by warming it at 92° with polyphosphoric acid, because the usual Friedel–Crafts type of intramolecular cyclization did not succeed. The carbonyl group of the aminoketone (63) was then reduced to the desired methylene group to yield the unsubstituted phenanthroindolizidine (61) either by using the Huang–Minlon variation of the Wolff–Kishner reduction or by employing a more recent reducing agent: sodium dihydrobis(2-methoxyethoxy)aluminate [95].

Route 3.   Takano et al. [96] approached the problem in a completely different way. The reaction between phenanthrene and dichlorocarbene produced 7,7-dichlorodibenzo[a,c]bicyclo[4.1.0] heptane (64), which reacted with potassium t-butoxide in tetrahydrofuran to yield 9-(chloro-2-tetrahydrofurylmethyl)phenanthrene (65). Dehydrohalogenation of 65 with pyridine in dimethylformamide followed by hydrolysis of the product with alcoholic hydrochloric acid rearranged the molecule into a γ-hydroxyketone (66). Warming this ketol derivative with benzylamine readily forms a Schiff base (67) that can be reduced with sodium borohydride to the benzylamino alcohol derivative (68). Stirring 68 with thionyl chloride in chloroform results in ring

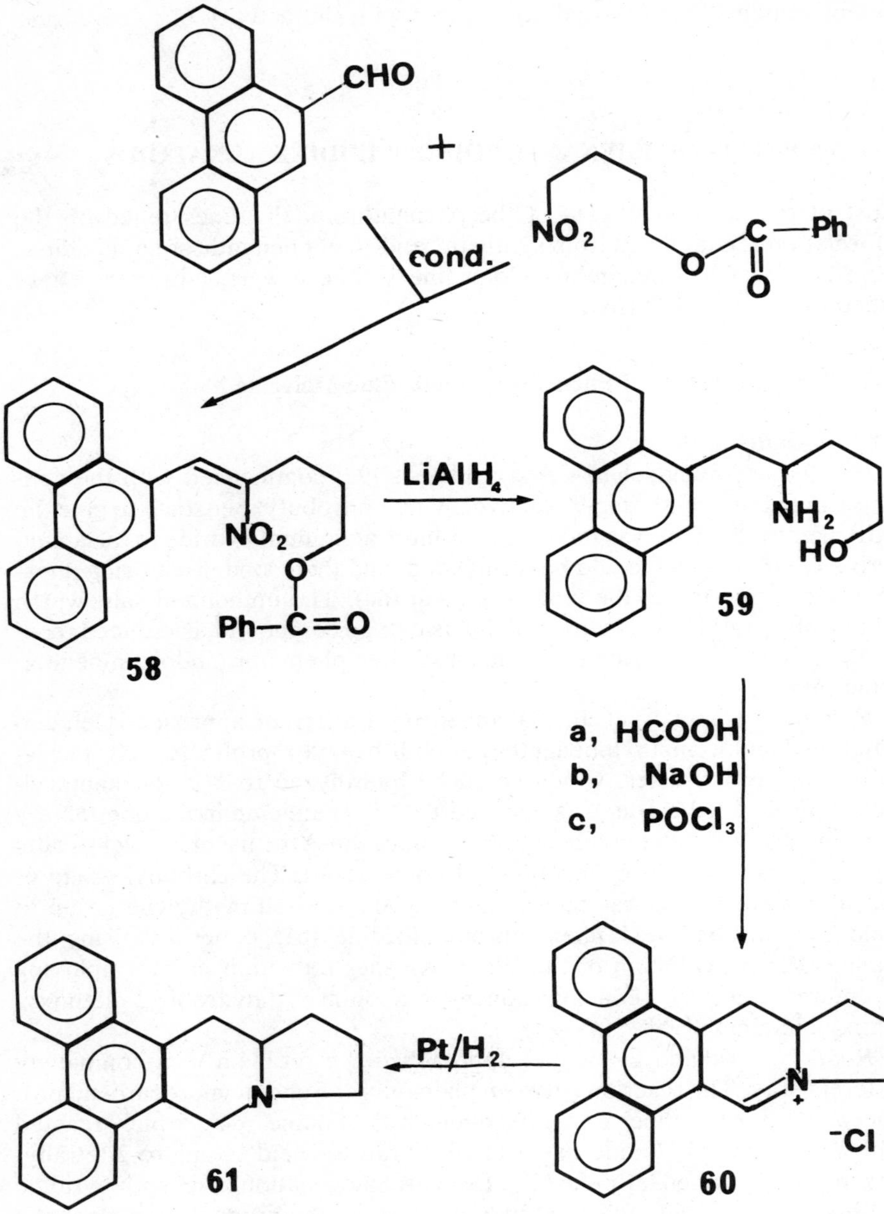

CHO
cond.
NO2
C
Ph
O
O
NO2
O
Ph—C=O
58
LiAlH4
NH2
HO
59
a, HCOOH
b, NaOH
c, POCl3
Pt/H2
N
61
N
Cl
60

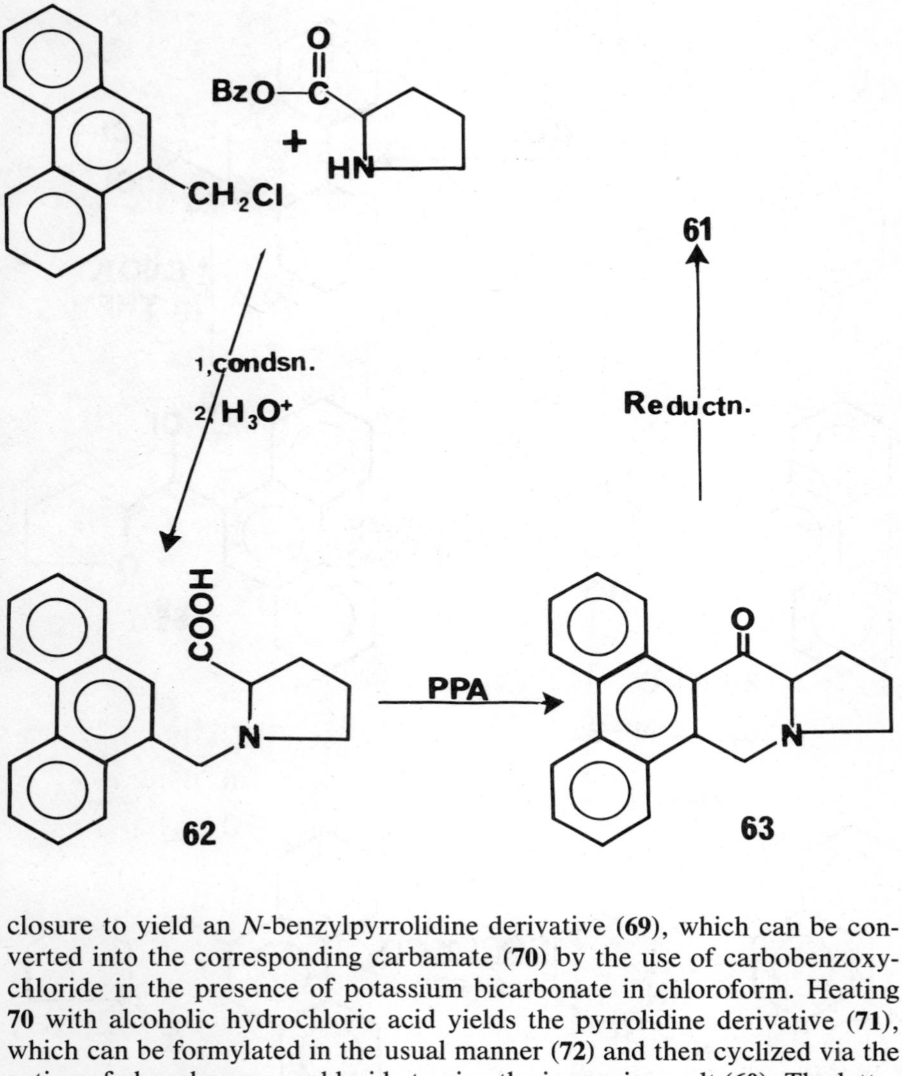

closure to yield an *N*-benzylpyrrolidine derivative (**69**), which can be converted into the corresponding carbamate (**70**) by the use of carbobenzoxychloride in the presence of potassium bicarbonate in chloroform. Heating **70** with alcoholic hydrochloric acid yields the pyrrolidine derivative (**71**), which can be formylated in the usual manner (**72**) and then cyclized via the action of phosphorus oxychloride to give the immonium salt (**60**). The latter was prepared by Govindachari in an earlier experiment but via another sequence of reactions. This salt (**60**) was reduced to **61** in the same manner as reported for route 1, that is, by sodium borohydride.

5.2.   The Actual Alkaloids

Of the three routes employed for the synthesis of the unsubstituted phenanthroindolizidine molecule the first one (cf. route 1) was not used for al-

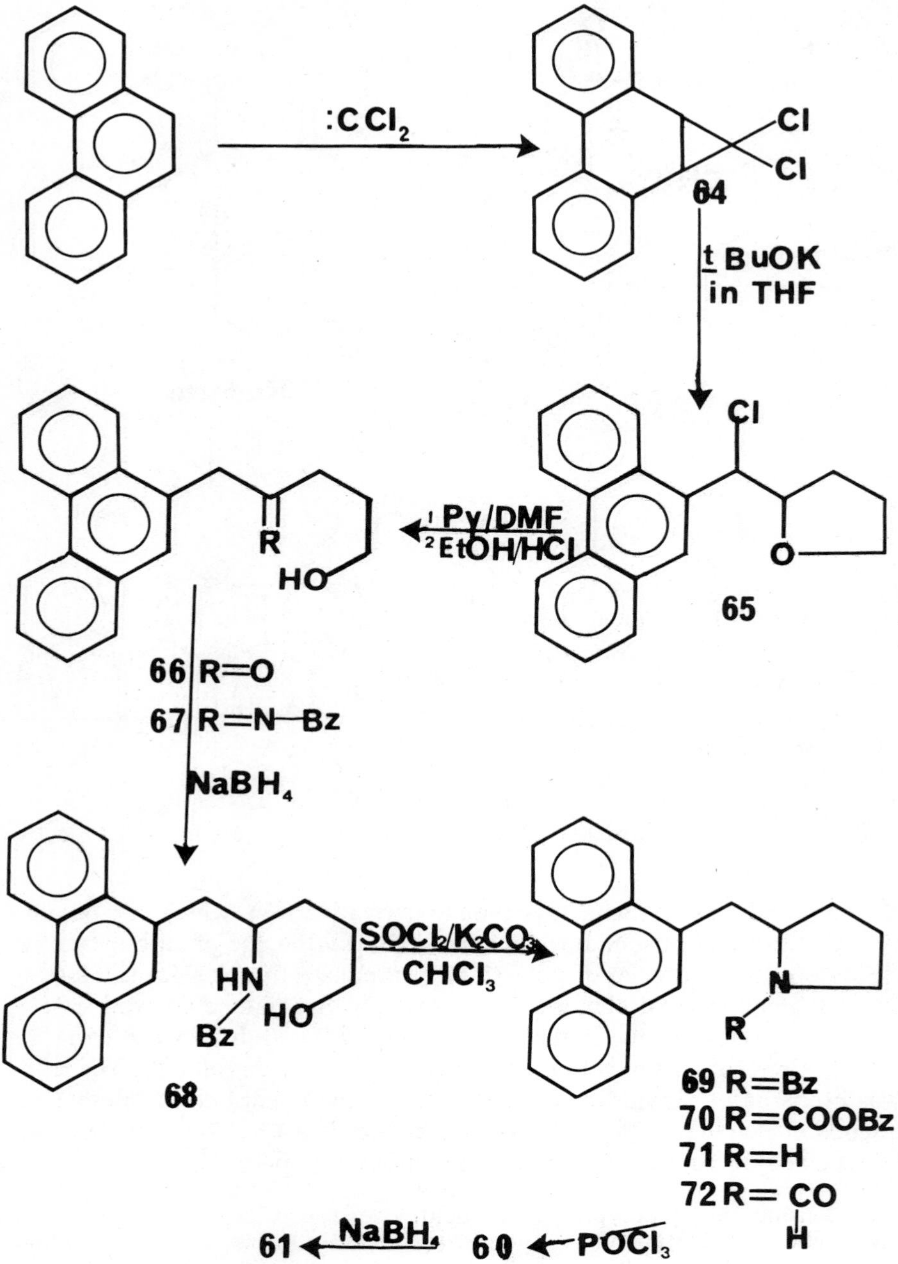
:CCl2
Cl
Cl
64
t BuOK
in THF
Cl
O
65
1 Py/DMF
2 EtOH/HCl
R
HO
66 R=O
67 R=N—Bz
NaBH4
HN
Bz
HO
68
SOCl2/K2CO3
CHCl3
N
R
69 R=Bz
70 R=COOBz
71 R=H
72 R= CO
H
61
NaBH4
60
POCl3

kaloid syntheses. The other two methods (cf. routes 2 and 3) underwent some modifications in order to become usable for the synthesis of the variously substituted alkaloids.

5.2.1. Routes 4–17

Route 4. Pyrrylmagnesium halide was condensed [97,98] with appropriately substituted 9-chloromethylphenanthrenes (in the case of the tylophorine synthesis this is the 2,3,6,7-tetramethoxy substituted one) (73) prepared from phenanthrene-9-carboxylic acids (or their esters) (74), initially by standard methods, then by a more efficient variation [99,100] of the Pschorr synthesis. The carboxylic acids (or their esters) were reduced to the alcohols using either traditional methods or diborane (which requires strictly controlled reaction conditions to avoid reduction of the carbonyl group to the methyl group). Then the hydroxy group was converted to the alkyl halide by halogenation with thionyl chloride. Condensation of 73 with pyrrylmagnesium bromide produced a 2-substituted pyrrole derivative (75), which was reduced to the pyrrolidine derivative (76) by catalytic hydrogenation. Formylation of 76 at the secondary nitrogen atom, followed by cyclization with phosphorus oxychloride to the immonium salt (77), that is, a derivative of 60, gave racemic tylophorine (2) when reduced with sodium borohydride. In addition to (±)-2, (±)-tylocrebrine (1), (±)-tylophorinine (3), (±)-isotylocrebrine (8), (±)-antofine (19), (±)-Wiegrebe's Alkaloid C (20), (±)-pergularinine (23), (±)-desoxypergularinine (24), and (±)-*O*-methyltylophorinidine (26) were also prepared by this method using appropriately substituted starting materials. The racemate of (±)-2 was resolved to give the (+)- and (−)-enantiomers of 2 via diastereoisomeric salt formation with (+)-camphor-10-sulfonic acid.

Route 5. This follows the slightly modified procedure employed by Marchini and Belleau [92] for the synthesis of the phenanthroquinolizidine skeleton. The main difference between the two processes lies in the substitution of a prolyl ester for the ester of 2-pipecolic acid giving an indolizidine instead of the quinolizidine ring system. Benzyl L-prolinate [6,94] (the use of this ester was preferred to that of the very hygroscopic methyl or ethyl esters because of its ready crystallizability) was *N*-alkylated with substituted 9-chloromethylphenanthrenes (cf. 73) in dimethylformamide (instead of the initially used alcohol or dioxane) in the presence of potassium carbonate to yield, e.g., alkyl *N*-(phenanthren-9-ylmethyl)-L-prolinate (78). This amino acid ester was hydrolyzed to the corresponding amino acid derivative (79), which could be cyclized to the appropriately substituted racemic cyclic α-aminoketone (80) by polyphosphoric acid. These rather labile aminoketones were reduced without delay via either the Clemmensen reduction or sodium borohydride reduction of their tosylhydrazone derivatives to give (±)-2, which could be resolved by the same method as in route 4. Compound 80 could also be converted to the corresponding 14-hydroxy derivative (81) by

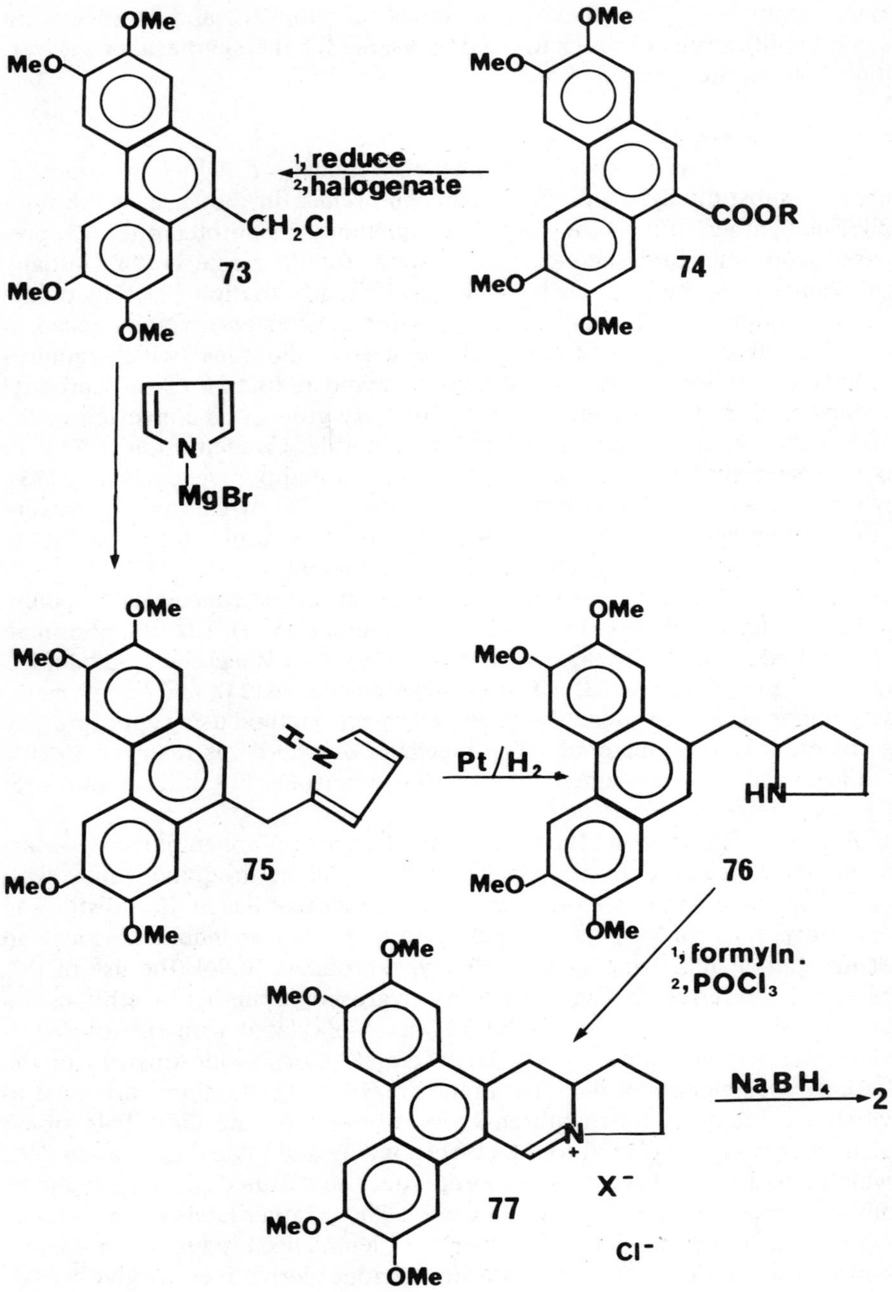

73

74

¹, reduce
², halogenate

CH₂Cl

COOR

N
MgBr

75

Pt/H₂

76

HN

¹, formyln.
², POCl₃

77

NaBH₄ → 2

X⁻

Cl⁻

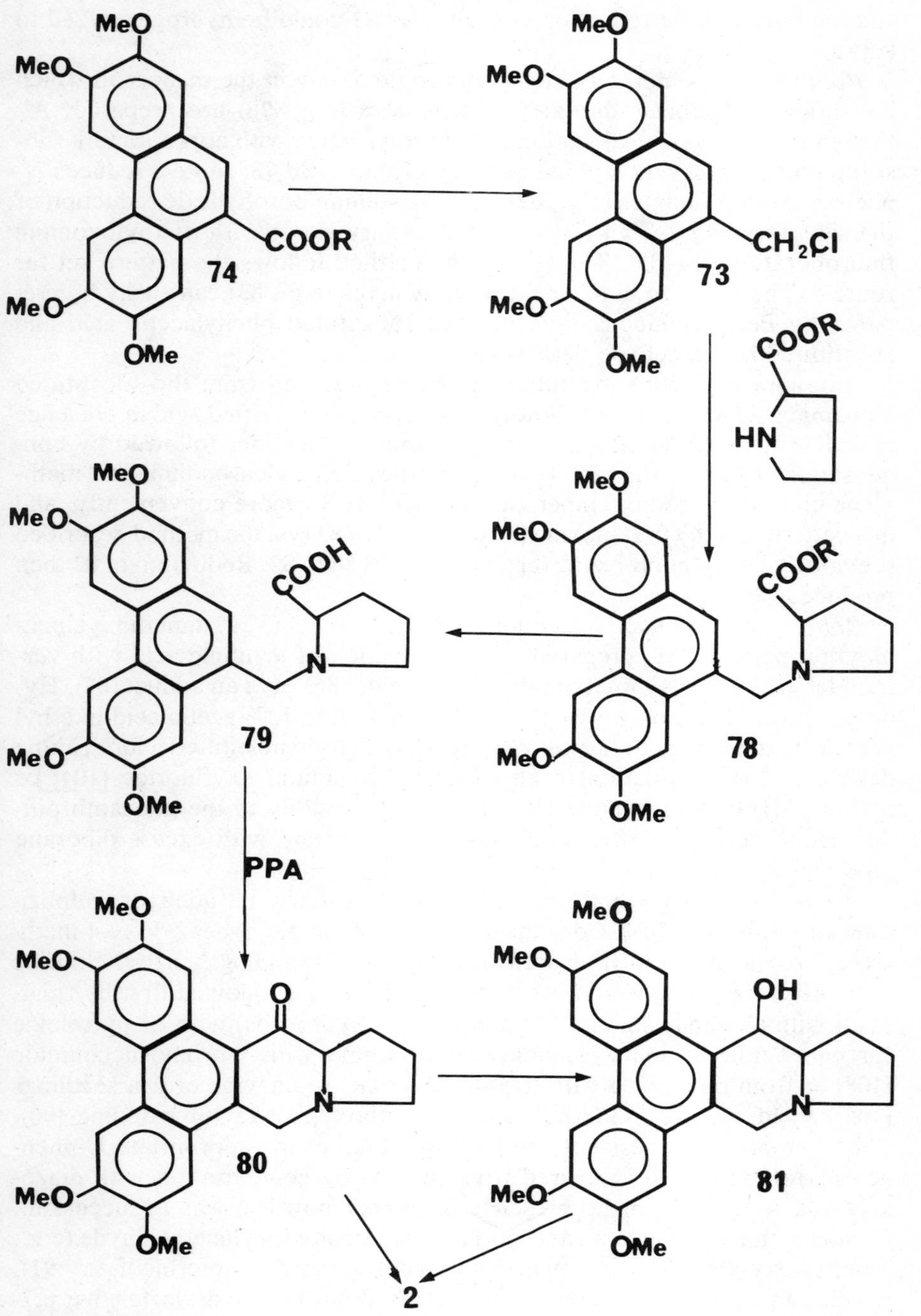
MeO
MeO
COOR
74
MeO
OMe
CH2Cl
73
COOR
HN
COOH
N
79
PPA
COOR
N
78
O
N
80
OH
N
81
2

sodium borohydride reduction. The alcohol **81** could be hydrogenolyzed to (±)-**2**.

Route 6. This method differs from route 5 only in the manner in which the different *N*-phenanthrylmethyl prolinates (e.g., **78**) are prepared. Although route 5 uses *N*-alkylation of the prolyl esters with appropriately substituted phenanthrylmethyl halides (e.g., **73**) to yield **78**, route 6 reduces *N*-phenanthroyl prolinates (e.g., **82**) to **78** by sodium borohydride reduction of the salts formed by *O*-alkylation of the tertiary amide with triethyloxonium fluoroborate [101–105]. The rest of the method follows the pattern set for route 5. The required cinnamic acid derivatives (e.g., **83**) can easily be prepared by Perkin condensation between substituted phenylacetic acid and substituted benzaldehyde derivatives.

Any appropriately substituted **82** can be prepared from the substituted cinnamic acid **83**, either by photolysis [101] of the esterified acid in presence of iodine to yield **74** after chromatographic purification, followed by condensation of **74** with the prolyl ester using dicyclohexylcarbodiimide in methylene chloride at room temperature to yield **82**, or more conveniently, and in a better yield, by first making **84**, the amide of **83** (via the method described previously), then photolyzing (again as above) **84** to **82**. Reduction of **82** then produces **78**.

Route 7. Michael addition [104] of 3-(3′,4′-dimethoxycinnamoyl)propanoate (**85**, prepared by condensation of levulinic acid with veratraldehyde) and 3,4-dimethoxybenzyl cyanide (**86**) gave an adduct (**87**). Hydrogenation of this over 10% platinum–charcoal in 10% acetic acid in ethyl acetate results in double ring closure to a diaryl-substituted indolizidone derivative (**88**). Oxidation of **88** either by vanadium oxyfluoride [104] or thallium(III) trifluoroacetate [105] proceeds smoothly to the phenanthroindolizidone derivative (**89**). Reduction of this lactam with excess diborane produces **2**.

Route 8. This method proceeds through a diaryl tetradehydroindolizidine structure [106,107]. Condensation of, for example, 3-benzyloxy-4-methoxybenzoylacetic acid (prepared from the corresponding benzoyl chloride derivative via condensation with ethyl sodioacetate followed first by treatment with alcoholic sodium acetate, then by hydrolysis) with Δ^1-pyrroleine (prepared either from the α-amino acid, ornithine, with *N*-bromosuccinimide [108] or from putrescine with the diamine oxidase enzyme of pea seedlings [105]) yields at pH 7 2-(3-benzyloxy-4-methoxyphenacyl)pyrrolidine (**90**). The attempt to synthesize **90** by hydrogenation of the appropriate *N*-phenacylpyrrole [109,110] (prepared from pyrrole by condensation with diazoacetophenone [111] in the presence of copper powder) was unsuccessful. Condensation of **90** with an appropriately substituted arylacetaldehyde (e.g., 3-benzyloxy-4-methoxyphenylacetaldehyde) gives an intermediate (**91**), which can be reduced in situ [103] with sodium borohydride to give 6,7-bis(3-benzyloxy-4-methoxyphenyl)-6,7-dehydroindolizidine (**92**). This dehydroindolizidine derivative was then oxidized with potassium ferricyanide in a two-phase system [112] to a tetrasubstituted phenanthroindolizidine. Methylation of its hydrogenolyzed derivative with diazomethane gave 2.

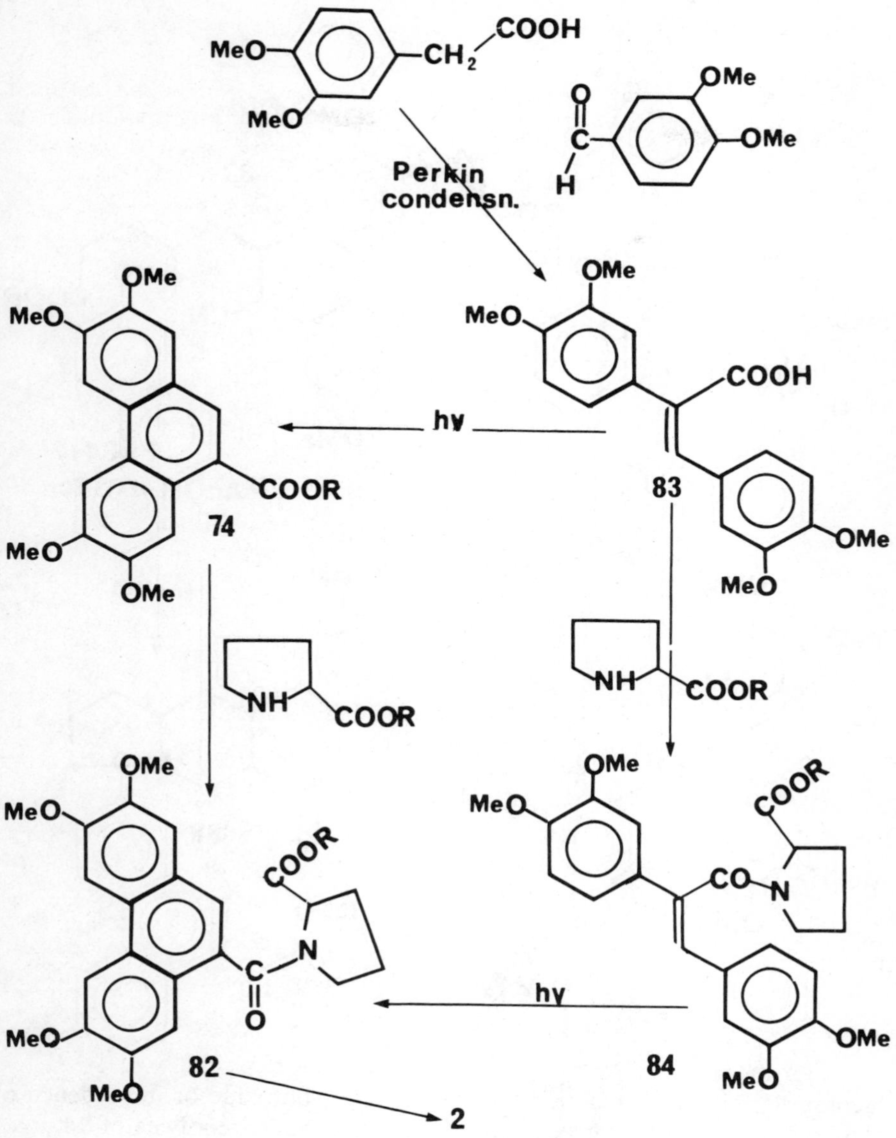

Route 9. This is a variation [113] of the preceding approach (cf. route 8) that commences with compound 90. Trigo et al. [114] treated 90 with 4-benzyloxyphenylacetyl chloride to yield (93), which was converted to the

lactam (94) either by reaction with potassium *t*-butoxide or in presence of alcoholic potassium hydroxide. Acid-catalyzed hydrogenolysis of 94 gives the corresponding dihydroxy compound (95), which can be oxidized by the phenolic oxidation process, here by using either MnO_2/SiO_2 [115] or Cu_2Cl_2/O_2/pyridine [116], to the intermediate dienone (96). (Methylation of 96 with diazomethane gives 97.) Warming 96 with methanolic sulfuric acid at 60° produces (the nonisolated) rearranged compound (98), which gives 99 when methylated with diazomethane. Reduction of 99 with lithium aluminum hydride yields racemic 19. Similarly, warming of 97, as above, gives 100, which yields, when reduced with lithium aluminum hydride, racemic 20.

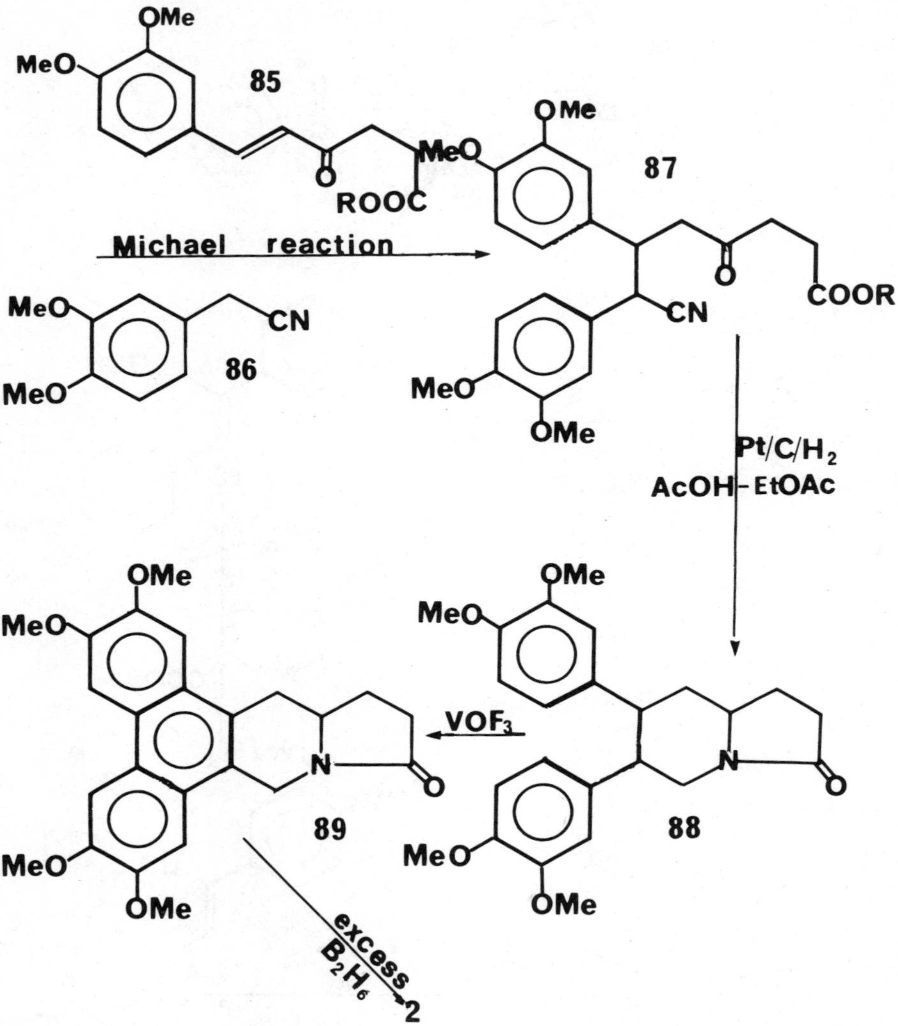

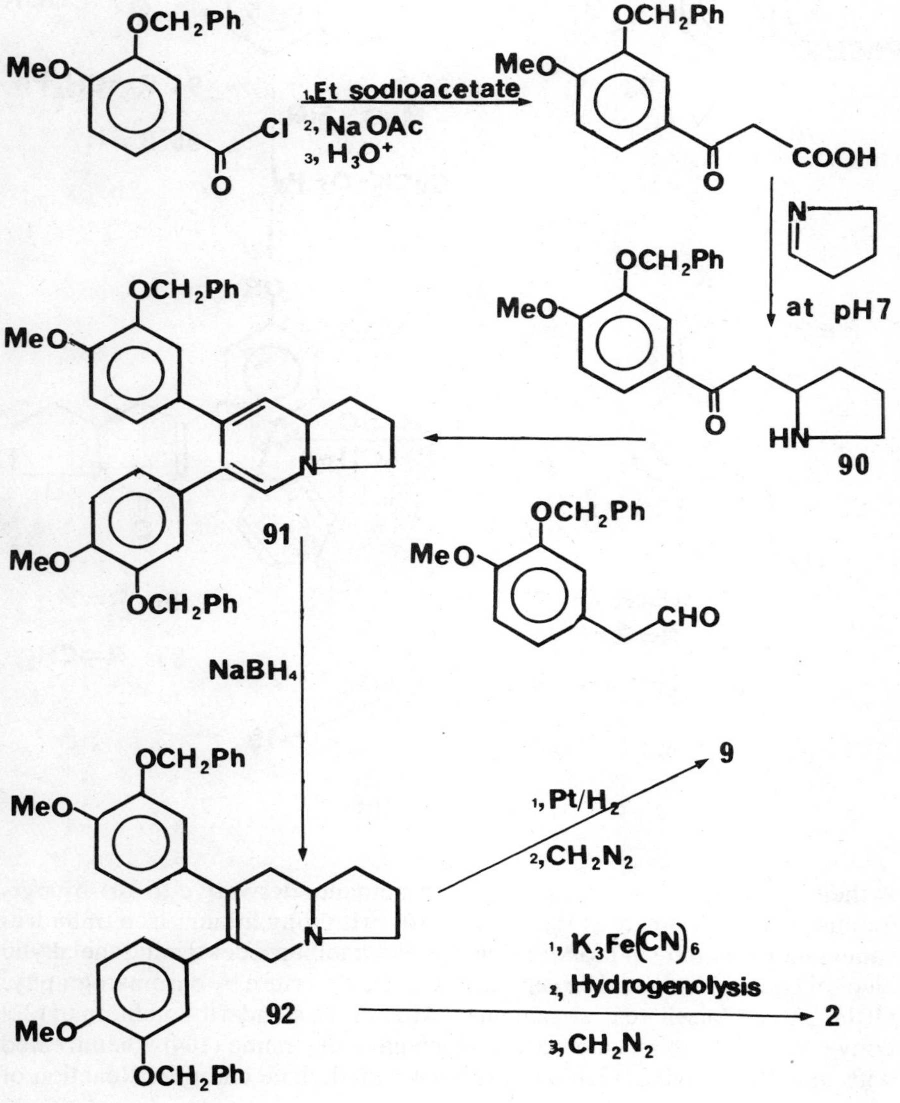

Route 10. This method represents a Diels–Alder cycloaddition reaction of conjugated dienes with imino dienophiles to produce nitrogen-containing natural products [117], indolizidine [118] and phenanthroindolizidine [119] ring systems. The sequence commences with the synthesis of, for example, 2,3,6,7-tetramethoxyphenanthrene-9-aldehyde (**101**) [120], prepared from the appropriately substituted phenanthrene-9-carboxylate (e.g., **74**). The two-step reaction from **74** to **101** starts with the reduction of the ester group by lithium aluminum hydride to the 9-phenanthrylmethanol derivative, which

is then followed by the oxidation of the methanol derivative to **101** by pyridinium chlorochromate [121]. Stirring **101** with vinyllithium in tetrahydrofuran in an ice cold solution for 1 hr yielded a mixture containing the allylic alcohol (**102**) which could be separated from the mixture by chromatography. Ortho ester Claisen rearrangement of **102** under standard conditions [122] converted it first to an ester (**103**), which gave the amide (**104**) when treated with dimethylaluminum amide in refluxing methylene chloride. Reaction of

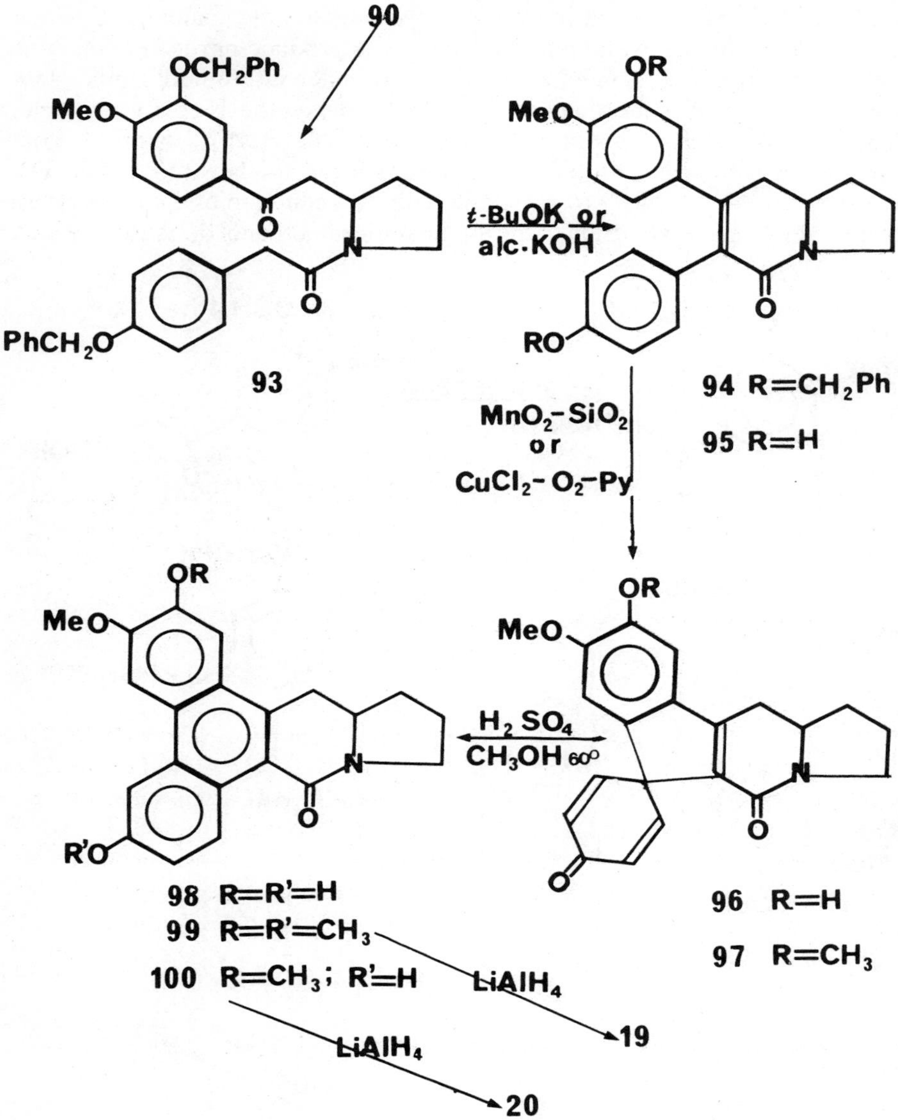

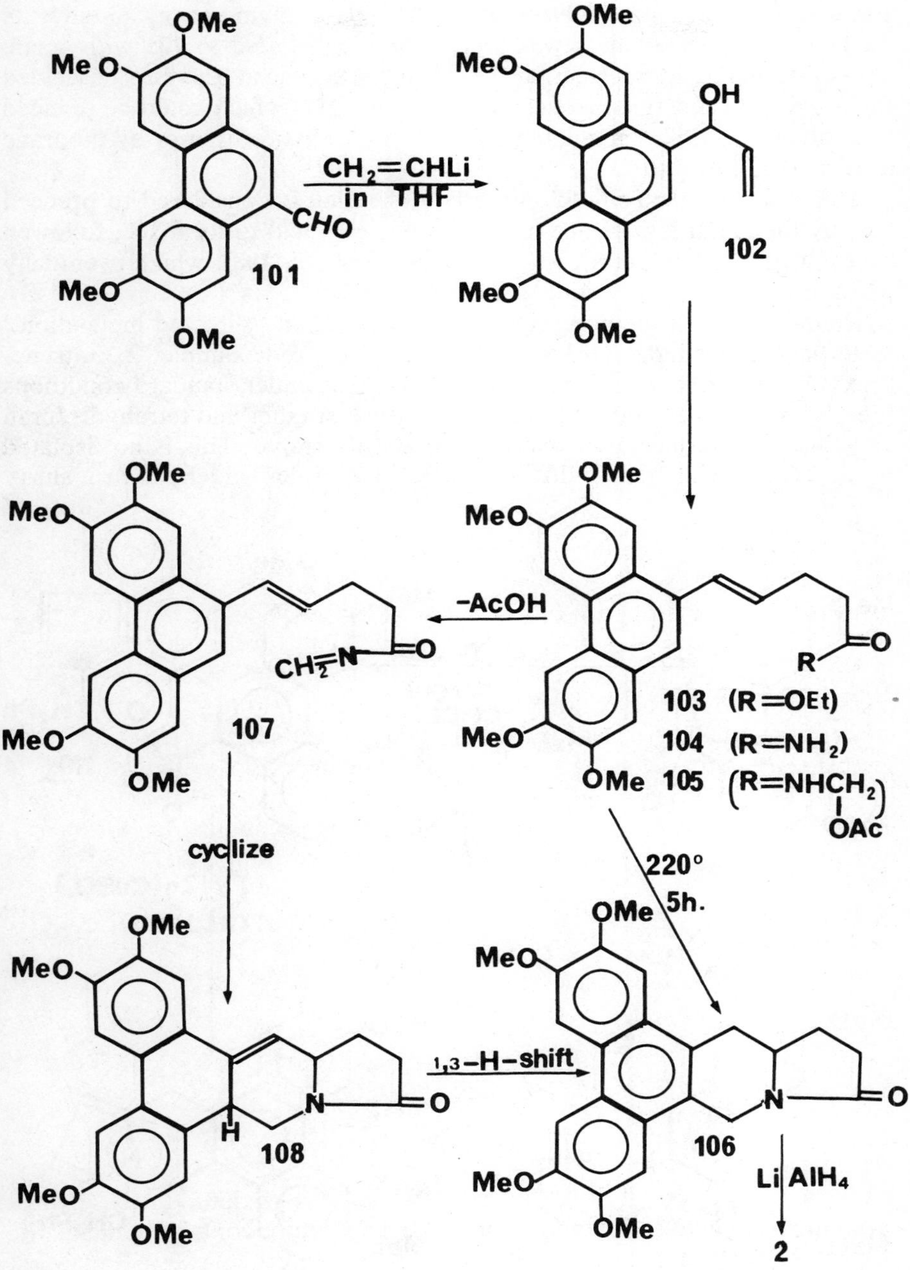

OMe
MeO
CHO
101
MeO
OMe
CH2=CHLi
in THF
OMe
MeO
OH
102
MeO
OMe
OMe
MeO
-AcOH
CH2=N
O
107
MeO
OMe
OMe
MeO
R
O
103 (R=OEt)
104 (R=NH2)
105 (R=NHCH2 OAc)
MeO
OMe
cyclize
220°
5h.
OMe
MeO
N
O
H
108
MeO
OMe
1,3-H-shift
OMe
MeO
N
O
106
MeO
OMe
LiAlH4
2

104 with 37% formaldehyde/sodium hydroxide in glyme gave, initially, its methanol derivative which was immediately acetylated to 105 with acetic anhydride and pyridine. Heating 105 in bromobenzene at 220° for 5 hr yielded the previously described lactam (106) (cf. also 107) which could be reduced to 2 either by lithium aluminum hydride in tetrahydrofuran or by diborane as described in route 7.

The rearrangement of 105 into 106 by heating is considered to proceed first by the initial loss of one molecule of acetic acid to form 107, followed by cyclization of 107 to a pentacyclic intermediate (108), which eventually rearomatizes by a 1,3 H-shift to 106.

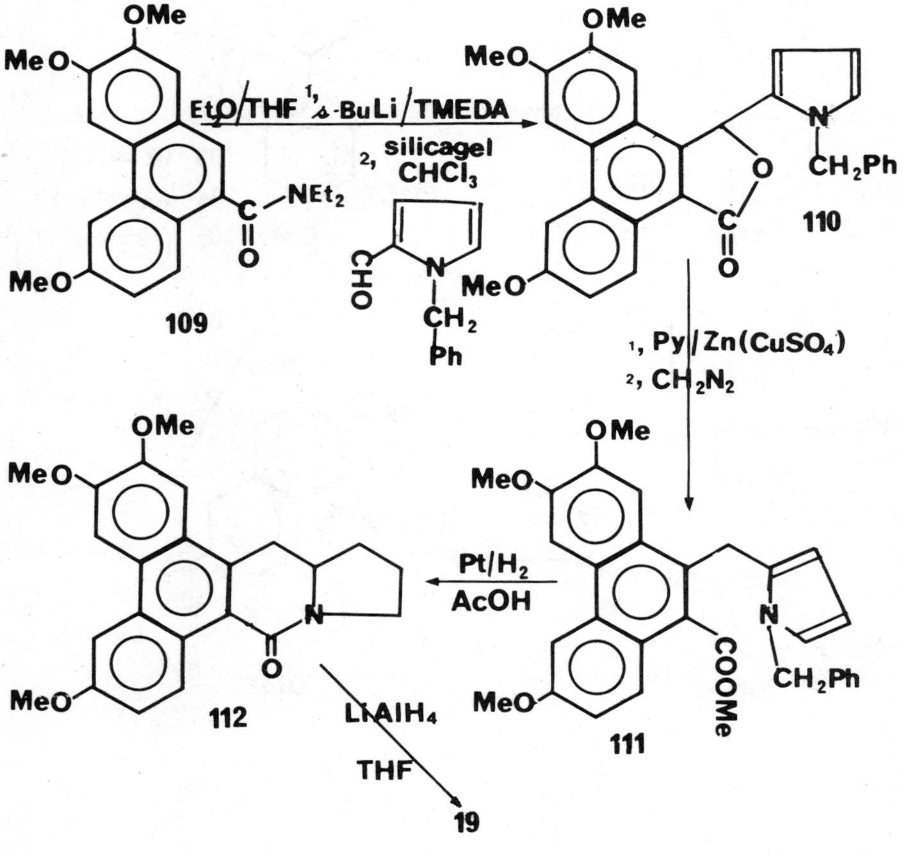

Route 11. This method represents the so-called "directed metalation" of tertiary benzamides [123,124]. Metalation of, for example, 2,3,6-trimethoxyphenanthroic acid diethylamide (109) gives under standard conditions (i.e., with secondary butyllithium in a mixture of ether and tetrahydrofuran at −78°) an intermediate *ortho* aryllithium derivative. This is not isolated but is heated with freshly distilled N-benzylpyrrole-2-aldehyde in a slurry

of silica gel and chloroform to convert it into the *N*-benzylpyrrolophthalide (**110**). Reduction of **110** with zinc (copper sulfate) and potassium hydroxide in pyridine hydrolyzes the phthalide to the intermediate acid, which can be esterified with diazomethane to yield the corresponding substituted methyl benzoate derivative (**111**). Catalytic hydrogenation of **111** in acetic acid converts it to a lactam (**112**), which can readily be reduced to (±)-antofine (**19**) by lithium aluminum hydride. Antofine (**19**) could of course be readily synthesized also via routes 4 and 9.

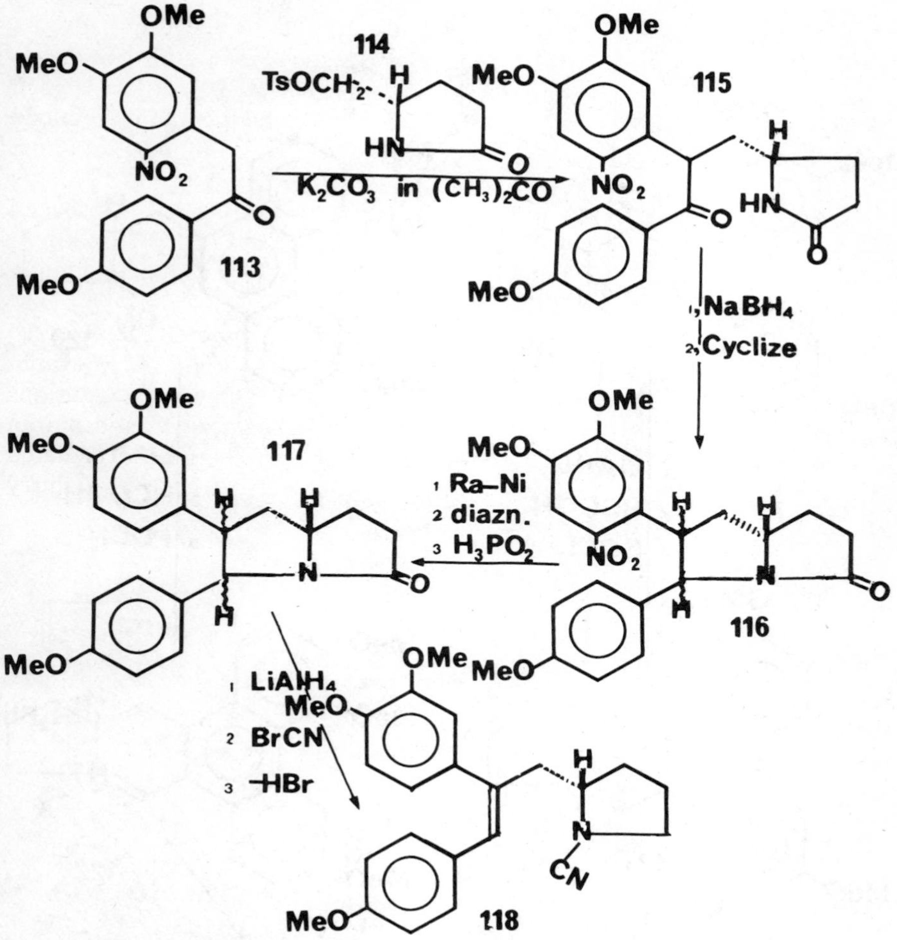

Route 12. Although the synthesis of racemic antofine (**19**) [and of Wiegrebe's Alkaloid *C* (**20**)] has already been accomplished by more than one method, this route represents a stereospecific synthesis [125,126], confirming also the absolute stereochemistry. Alkylation of 4,4′,5′-trimethoxy-2′-nitrodesoxybenzoin (**113**) with (*S*)-(2-pyrrolidon-5-yl)methyl-*p*-toluenesulfonate (**114**) yields the ketopyrrolidone derivative (**115**), which can be re-

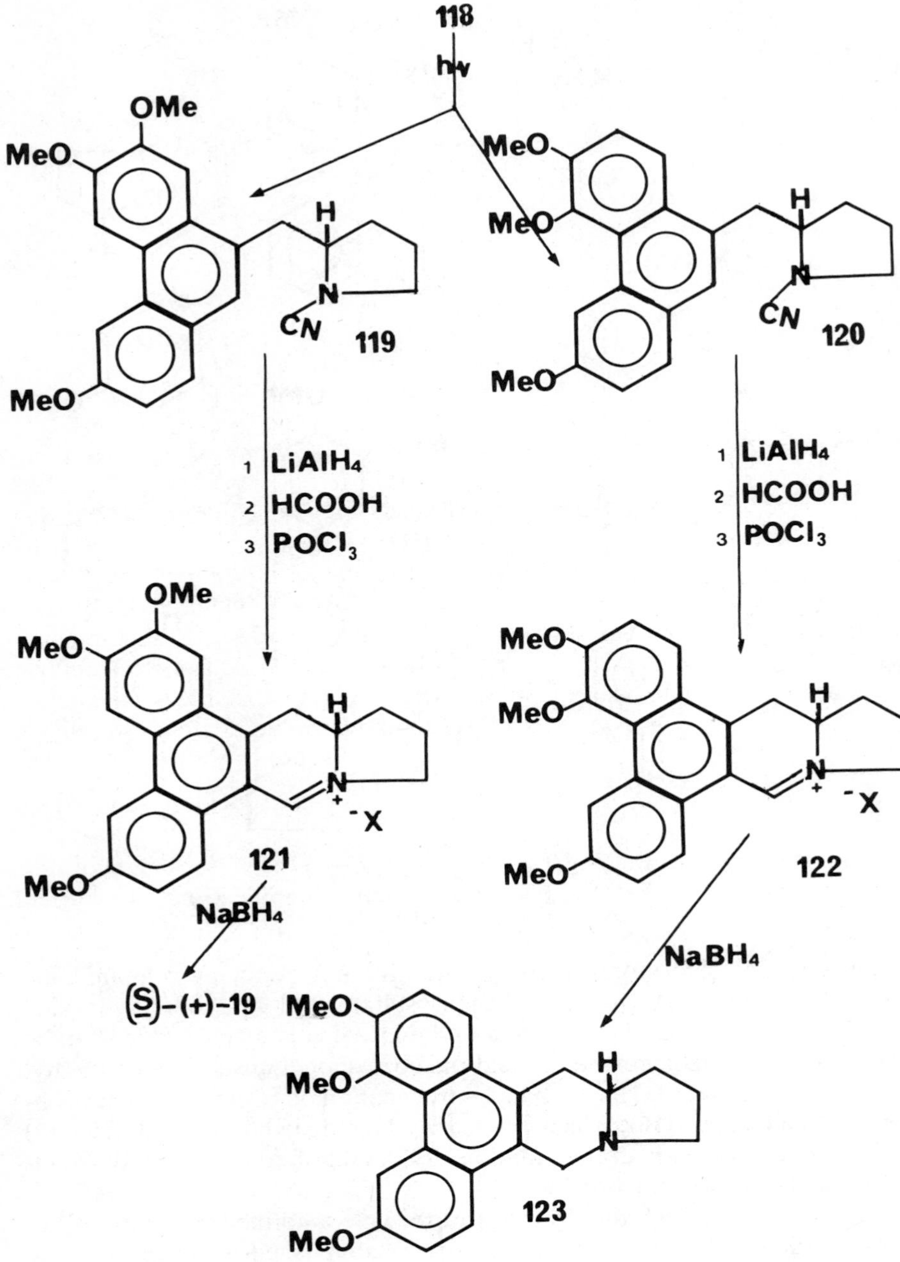
118
hν
OMe
MeO
MeO
H
N
CN
119
MeO
MeO
H
N
CN
120
MeO
1 LiAlH4
2 HCOOH
3 POCl3
1 LiAlH4
2 HCOOH
3 POCl3
OMe
MeO
H
N
+
-X
121
MeO
MeO
MeO
H
N
+
-X
122
MeO
NaBH4
(S)-(+)-19
NaBH4
MeO
MeO
H
N
123
MeO

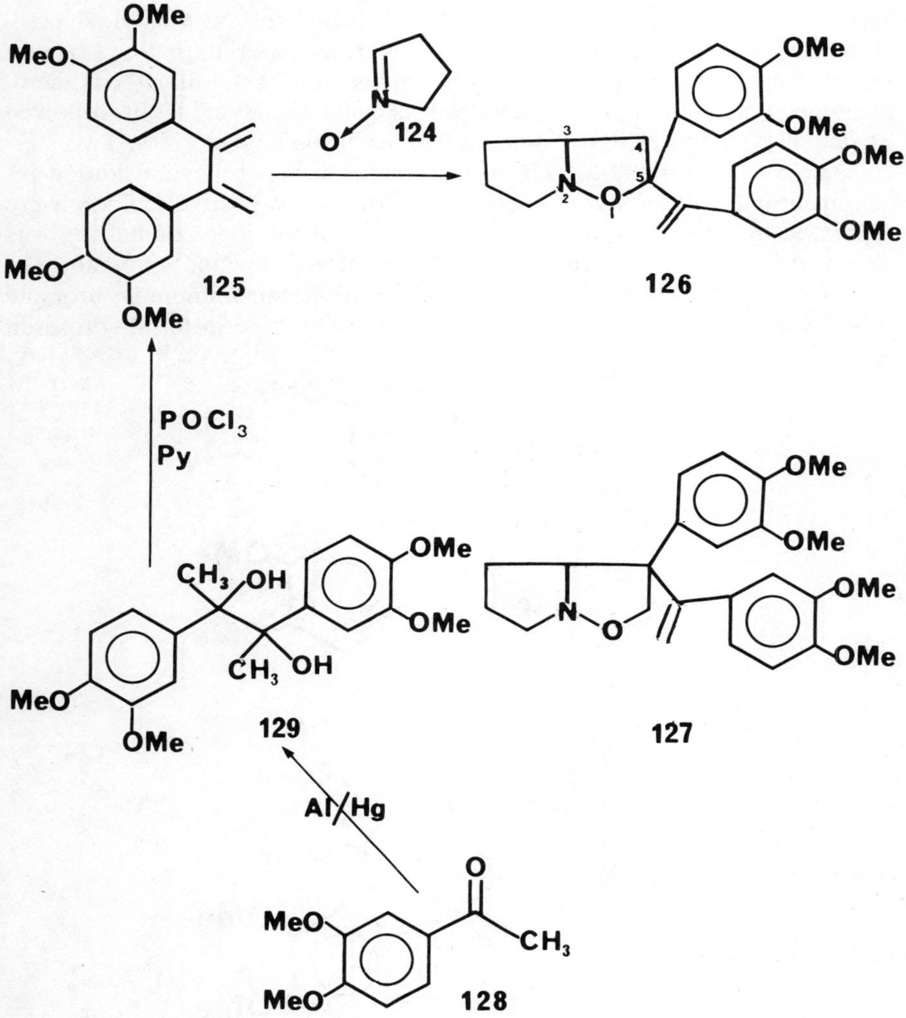

duced with sodium borohydride to the hydroxypyrrolidone derivative. Dehydration of this followed by acid-catalyzed cyclization gives the diarylpyrrolizidone derivative (**116**). The amino group generated by reduction of the nitro group in **116** (by hydrogenation with Raney nickel) can then be replaced by hydrogen via the diazotization process to yield the denitrated product (**117**). The reduction of the pyrrolizidone ring by lithium aluminum hydride, followed by the opening of the pyrrolidine ring with cyanogen bromide with consequent dehydrobromination and double bond formation between the two aromatic rings, gives the *N*-cyanopyrrolidine derivative (**118**). Photolysis of **118** by irradiation with UV light yields two phenanthrene derivatives (**119** and **120**), which differ from each other in their methoxylation pattern owing to cyclization of the stilbene derivative in the ortho and para positions, respectively. Reduction, separately, of **119** and **120** with lithium aluminum hydride followed by formylation and then ring closure with phosphorus oxychloride yields the two immonium salts (**121** and **122**). The salts can be reduced by sodium borohydride (as usual with these types of immonium salts) to give the (*S*)-(+)-isomer of **19** and also the (*S*)-(+)-3,4,6-trimethoxyphenanthroindolizidine (**123**).

Route 13. Synthesis of the racemic septicine (**9**) can also be achieved by a regioselective method [127]. Because the intermediate products were chiral it must be assumed that the penultimate chiral product racemized because **9** was stated to be racemic. 1,3-Cycloaddition of 1-pyrroline 1-oxide (**124**) [128] to 2,3-bis(3′,4′-dimethoxyphenyl)butadiene (**125**) yields preferentially the 5-vinyl isoxazolidine derivative (**126**), not the 4-vinyl isoxazolidine derivative (**127**). (Preparation of the starting material, that is, **125**, was achieved by a pinacol-type condensation reaction of 3,4-dimethoxyacetophenone (**128**) with aluminum amalgam to yield the glycol (**129**), followed by dehydration of **129** in pyridine with phosphorus oxychloride). Two stereoisomers (**130** and **131**) of **126** were formed during the cycloaddition reaction, in a 2:1 ratio, by *exo* and *endo* addition, respectively. They were separated by TLC on alumina. The formation of the indolizidine ring was achieved first by bromination of **130** in chloroform at room temperature to yield, by bromine addition and cyclization, the bromoammonium bromide (**132**). Reduction of **132** with lithium aluminum hydride in tetrahydrofuran

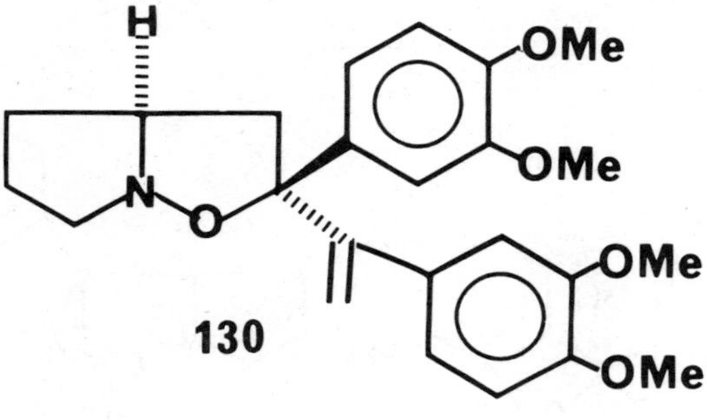

130

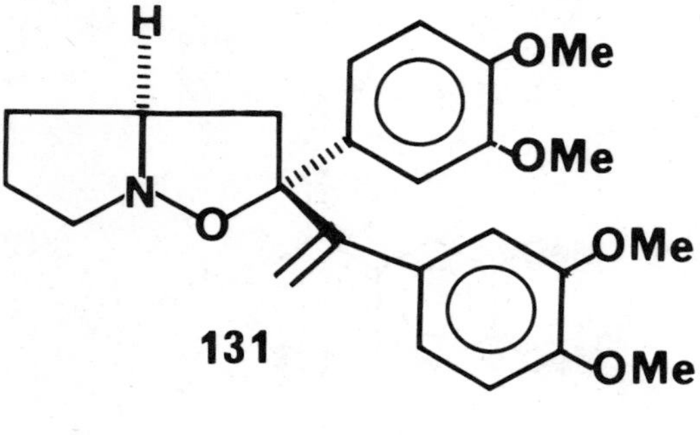

131

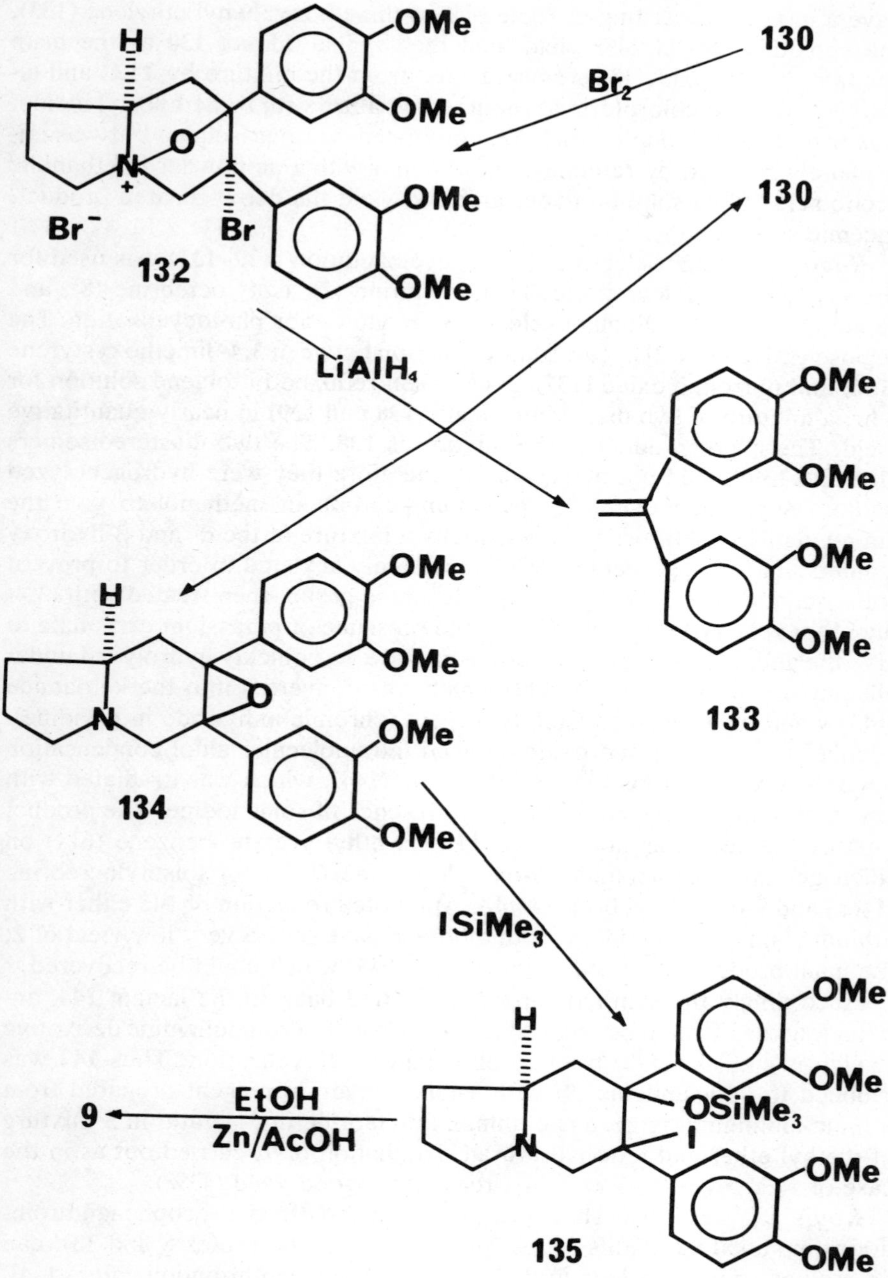

130

Br₂

132

Br⁻   Br

LiAlH₄

130

133

134

ISiMe₃

9

EtOH
Zn/AcOH

135

gave a mixture consisting of some bis(3,4-dimethoxyphenyl)ethylene (133), the epoxide (134) in 24% yield, and the original adduct 130 as the main product. The epoxide 134 was separated from the mixture by TLC and allowed to react in chloroform at room temperature with trimethylsilyl iodide. The trimethylsilyl iodohydrine (135) produced was not isolated but was immediately reduced by refluxing it for 90 min with zinc powder in ethanolic acetic acid (10:1) solution under argon to yield the deoxygenated product, racemic septicine (9).

Route 14. Regioselective [3 + 2] cycloaddition [129–131] was used for the synthesis of tylocrebrine (1), tylophorine (2), isotylocrebrine (8), and septicine (9), via 1,3-dipolar cycloaddition followed by photocyclization. The regioselective [3 + 2] cycloaddition reaction between 3,4-dimethoxystyrene (136) and pyrroline oxide (137) gives, when refluxed in toluene solution for 3 hr, a mixture of two diastereoisomers (138 and 139) in nearly quantitative yield. The major product in the mixture is 138. The two diastereoisomers (138 and 139) could not be separated; therefore they were hydrogenolyzed without separation over 10% palladium–carbon in methanol to give the amino alcohol (140), which is essentially a mixture of the α- and β-hydroxy isomers. This amino alcohol was immediately acylated in order to prevent subsequent oxidation of its susceptible amino group, then treated with (3,4-dimethoxyphenyl)acetyl chloride in the presence of potassium carbonate to give the amido ester (141). The crude mixture was quickly hydrolyzed under alkaline conditions to yield (142) which was converted into the ketoamide (143) when oxidized with Collins reagent (chromic anhydride in pyridine). Alcoholic potassium hydroxide induced intramolecular aldol condensation under reflux conditions to give the lactam (144), which was irradiated with UV light in methylene chloride in the presence of some iodine. The product mixture yielded after preparative TLC in ethyl acetate–benzene (6:1) on silica gel three compounds: 9-oxotylophorine (145), 9-oxoisotylocrebrine (146a) and 9-oxotylocrebrine (146b). Attempted reduction of 145 either with lithium aluminum hydride or with diborane gave only a very low yield of 2; the main product consisted of unchanged 145, which could be recovered.

Accordingly the synthetic approach shifted back to the lactam 144, anticipating that it could be reduced to the diaryldehydroindolizidine derivative (which is similar to 92) prior to photochemical cyclization. Thus 144 was reduced to (±)-septicine (9) with a mixed hydride reagent prepared from lithium aluminum hydride and aluminum chloride in 3:1 ratio in a mixture of diethyl ether and tetrahydrofuran. Irradiation of 9, carried out as in the case of 145, gave (±)-2 now in a reasonably good yield (43%).

Route 15. As it was shown earlier, septicine (9), the secophenanthroindolizidine alkaloid of this series (prepared, e.g., via routes 8 and 13), can be used as an intermediate in the synthesis of phenanthroindolizidines [44]. The conversion can be achieved by irradiation of 9 with UV light. Accordingly any synthesis of 9 can also represent a synthesis of tetramethoxyphenanthroindolizidine alkaloids. The first synthesis of racemic 9 [132] com-

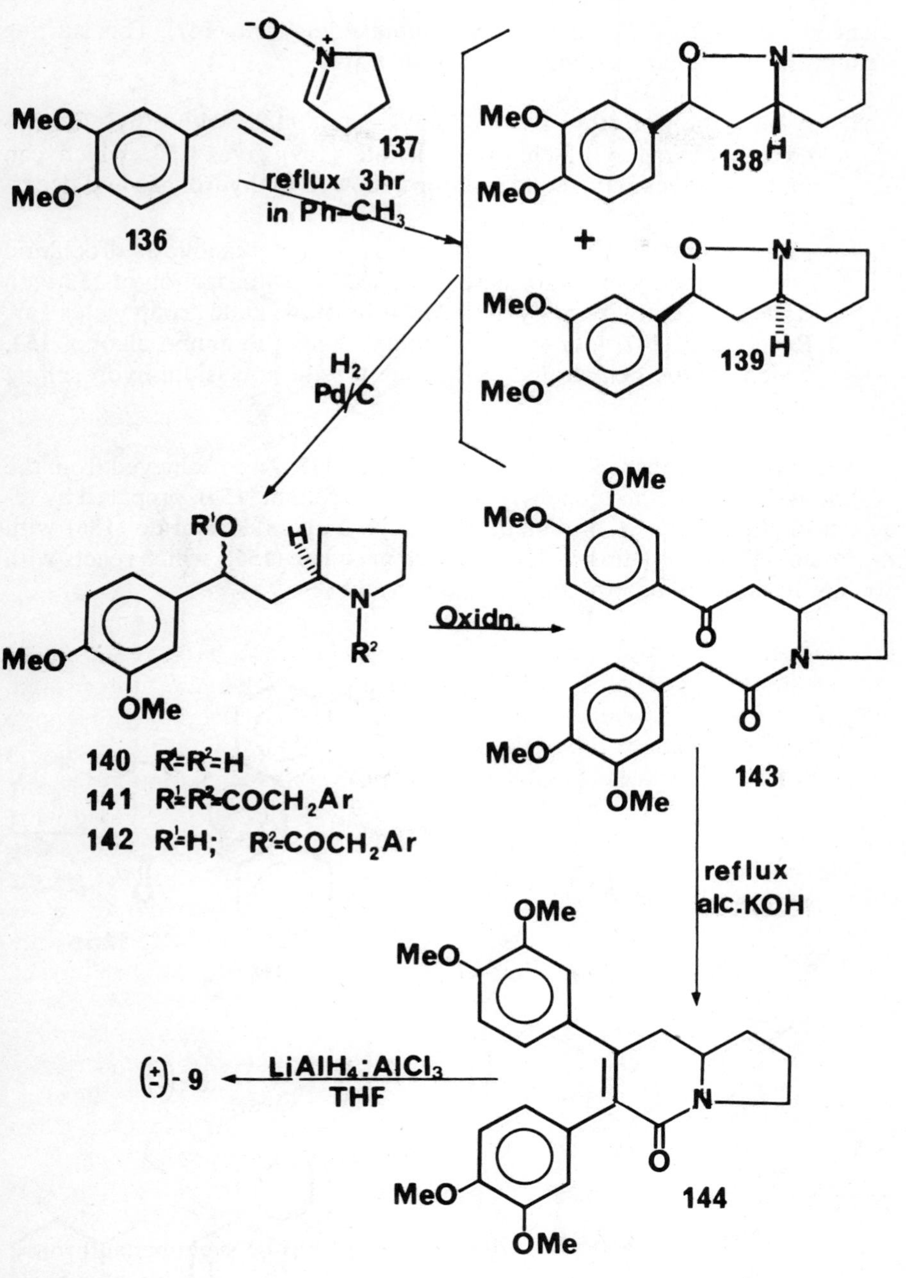
MeO
MeO
136
137
reflux 3hr
in Ph-CH3
138
139
H2
Pd/C
R'O
R2
OMe
140 R1=R2=H
141 R1=R2=COCH2Ar
142 R1=H; R2=COCH2Ar
Oxidn.
143
reflux
alc.KOH
144
(±)-9
LiAlH4:AlCl3
THF

menced with 5-(3′,4′-dimethoxyphenyl)indolizidin-6-one (**147**). This starting material may be prepared in two different ways:

1. *N*-alkylation of ethyl 2-pyrrolidinylacetate (**148**) with ethyl 2-(3′,4′-dimethyoxyphenyl)-3-chloropropionate (**149**) gives **150**, which can undergo Dieckmann condensation followed by hydrolysis and decarboxylation to yield **147**.

2. Ring closure [133] of **151** with the help of methanolic hydrochloric acid and methyl orthoformate gives **152**. Desulfurization of **152** with Raney nickel followed by the hydrolysis of the ketal group yields **147**. Reaction of **147** with veratryl lithium gives the amino alcohol **153**, which can be dehydrated by heating it with potassium hydrosulfate to produce (±)-**9**.

Route 16. Stereospecific synthesis of (−)-**9** [134] was achieved from the *N*-2,3-bis(3′,4′-dimethoxyphenyl)-2-alkenyl-L-prolinol (**154**), prepared by reaction of the 2,3,-bis(3′,4′-dimethoxyphenyl)-2-alkenyl chloride (**155**) with L-prolinol. The mesylation of **154** gives the mesylate (**156**), which reacts with sodium hydride in dimethylformamide to yield (−)-**9**.

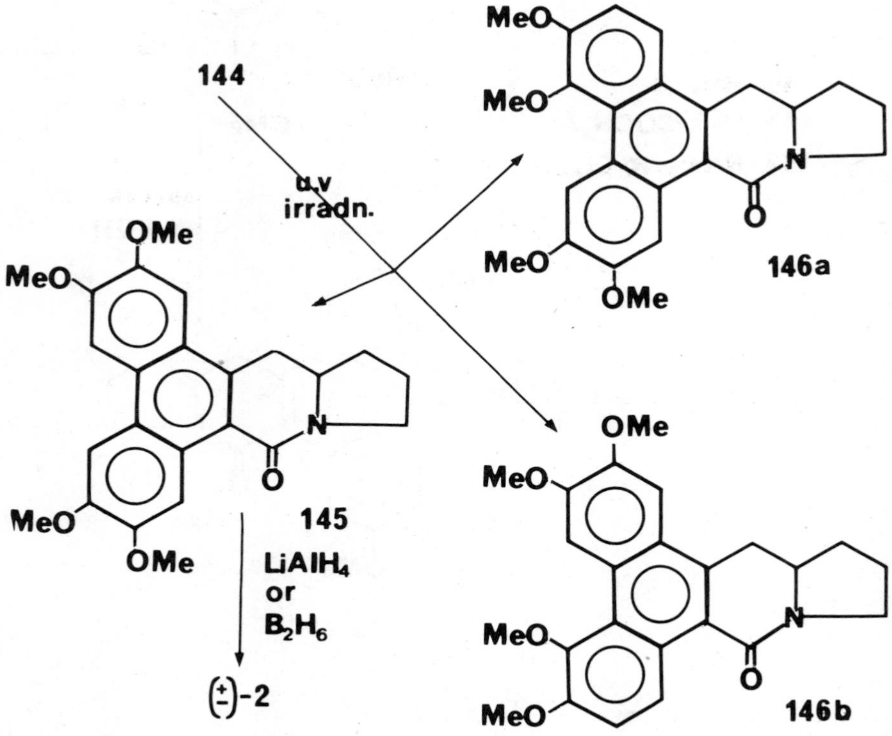

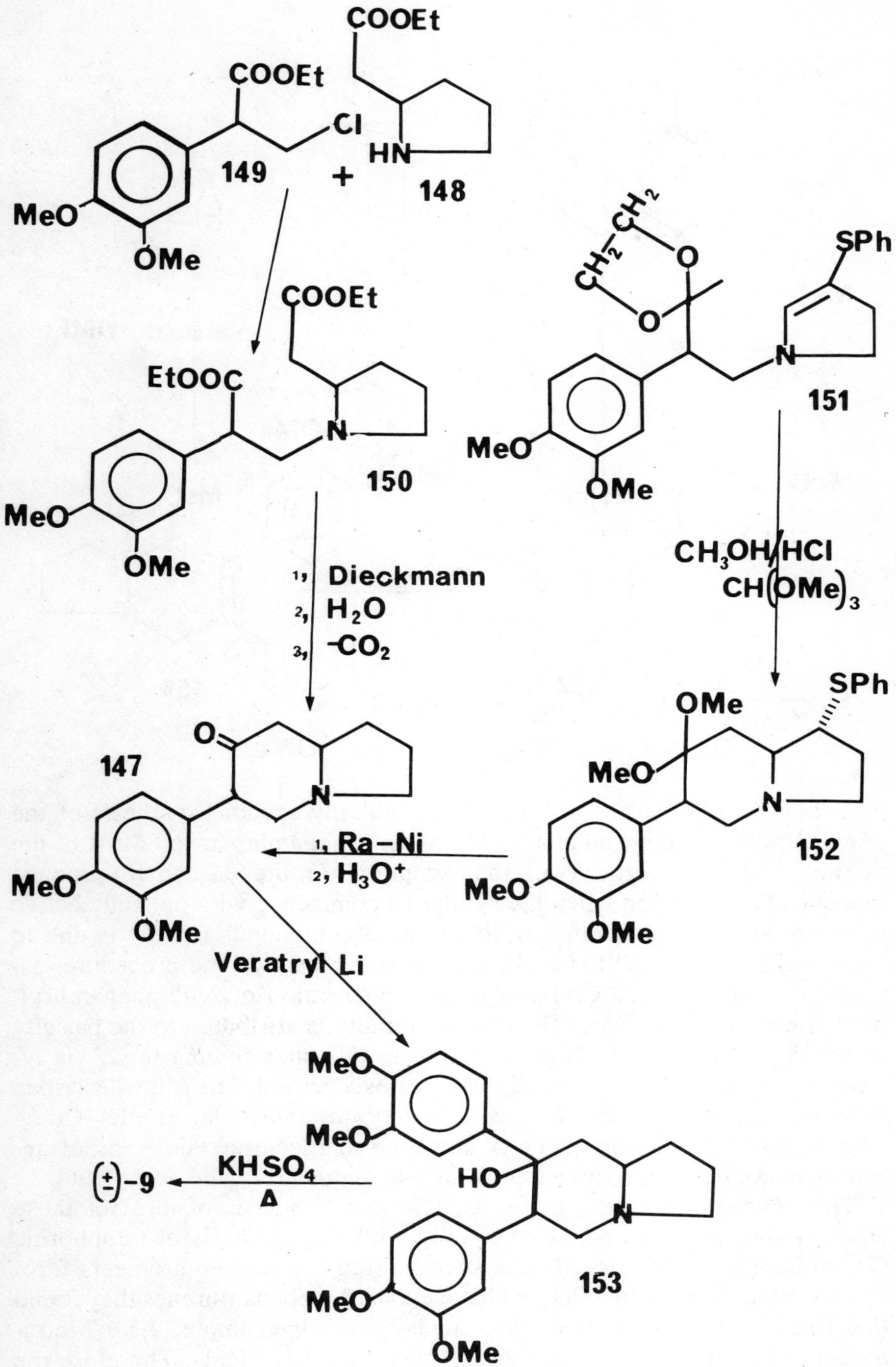

COOEt

COOEt

Cl

HN **148**

149

+

MeO

OMe

COOEt

EtOOC

N

MeO

OMe

150

CH₂—CH₂

O O

SPh

N

MeO

OMe

151

CH₃OH/HCl
CH(OMe)₃

1, Dieckmann
2, H₂O
3, -CO₂

O

147

N

MeO

OMe

OMe

MeO

SPh

N

MeO

OMe

152

1, Ra–Ni
2, H₃O⁺

Veratryl Li

MeO

MeO

HO

N

MeO

OMe

153

(±)-9

KHSO₄
Δ

103

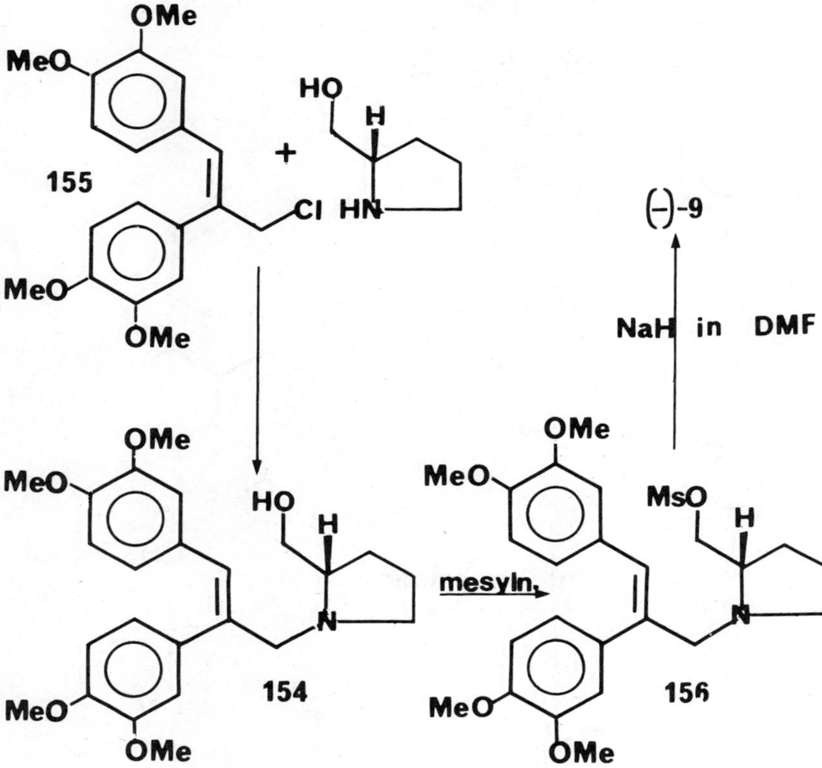

Route 17. This method deals with a chirally specific synthesis of the phenanthroindolizidine alkaloids [135] from (S)-α-amino acids. Most of the methods described earlier (with the exception of routes 12 and 16) produce racemic alkaloids even when the syntheses commence with optically active α-amino acids. Racemization of these chiral compounds usually is due to the conditions required for cyclization at some stage of the procedure, especially during Friedel–Crafts type of condensation of N-(9-phenanthryl-methyl)proline derivatives. The loss of chirality is attributed to the basicity of the compounds, but efforts to eliminate this shortcoming (e.g., via N-oxide formation prior to cyclization) were unsuccessful. The route described here overcomes this problem by carrying out intramolecular Friedel–Crafts ring closure of nonbasic cyclic N-alkylated amidocarboxylic acids of appropriate configuration, prepared from α-aminodicarboxylic acids [136].

This route could, a priori, be used for the synthesis of most of these alkaloids but the cited paper deals only with the synthesis of tylophorine (**2**). Although the authors made a specific study of the requirements for a facile synthesis of methoxylated 9-halogenomethylphenanthrenes they found that the N-alkylation of the amino acids with, for example, 2,3,6,7-tetra-methoxy-9-chloromethylphenanthrene gave very low yields. Therefore the corresponding phenanthrene-9-aldehyde derivatives were used. 2,3,6,7-Tetramethoxy-9-formylphenanthrene (**157**) was condensed with diisopropyl (*S*)-(+)-glutamate (**158**) (used for the purpose of preventing premature pyroglutamate formation) to give the interim Schiff base (**159**). Because **159** has the tendency to be converted to the aminal it was quickly reduced with sodium cyanoborohydride, in the presence of a catalytic amount of glacial acetic acid, to the required diisopropyl (*S*)-(+)-(2,3,6,7-tetramethoxy-9-phenanthrylmethyl)glutamate (**160**). Cyclization [137] of **160** in warm methanol–acetic acid gave the amido ester (**161**), which was first saponified with methanolic potassium hydroxide in dioxane and then acidified with phos-

158
157
159
AcOH
NaCNBH₃
160
R=i-Pr
161 R= i-Pr
162 R= H
ClCOCOCl/CH₂Cl₂/DMF
SnCl₄
163

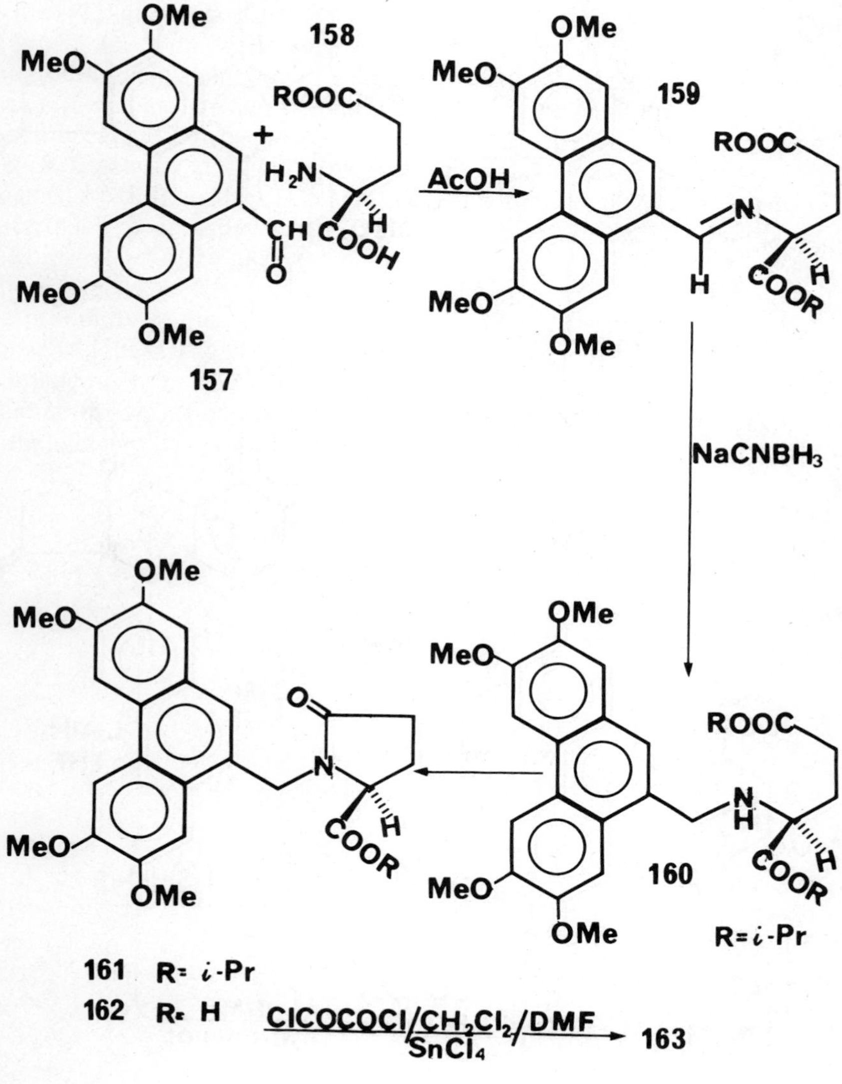

phoric acid to give the amido acid (**162**). Addition of oxalyl chloride to **162** in methylene chloride and dimethylformamide with stirring for 1½ hr at room temperature gave the acid chloride, which was not isolated but was refluxed for 4 hr, in the same reaction mixture, with stannic chloride to yield (*S*)-2,3,6,7-tetramethoxyphenanthro[9,10–b]-11,14-indolizidinedione (**163**). Catalytic hydrogenation of **163** with 10% palladium–carbon gave a mixture of two diastereoisomeric alcohols (**164** and **165**). The use of Pearlman's catalyst gave the β-alcohol (**164**) as the major product and the α-alcohol (**165**) in a

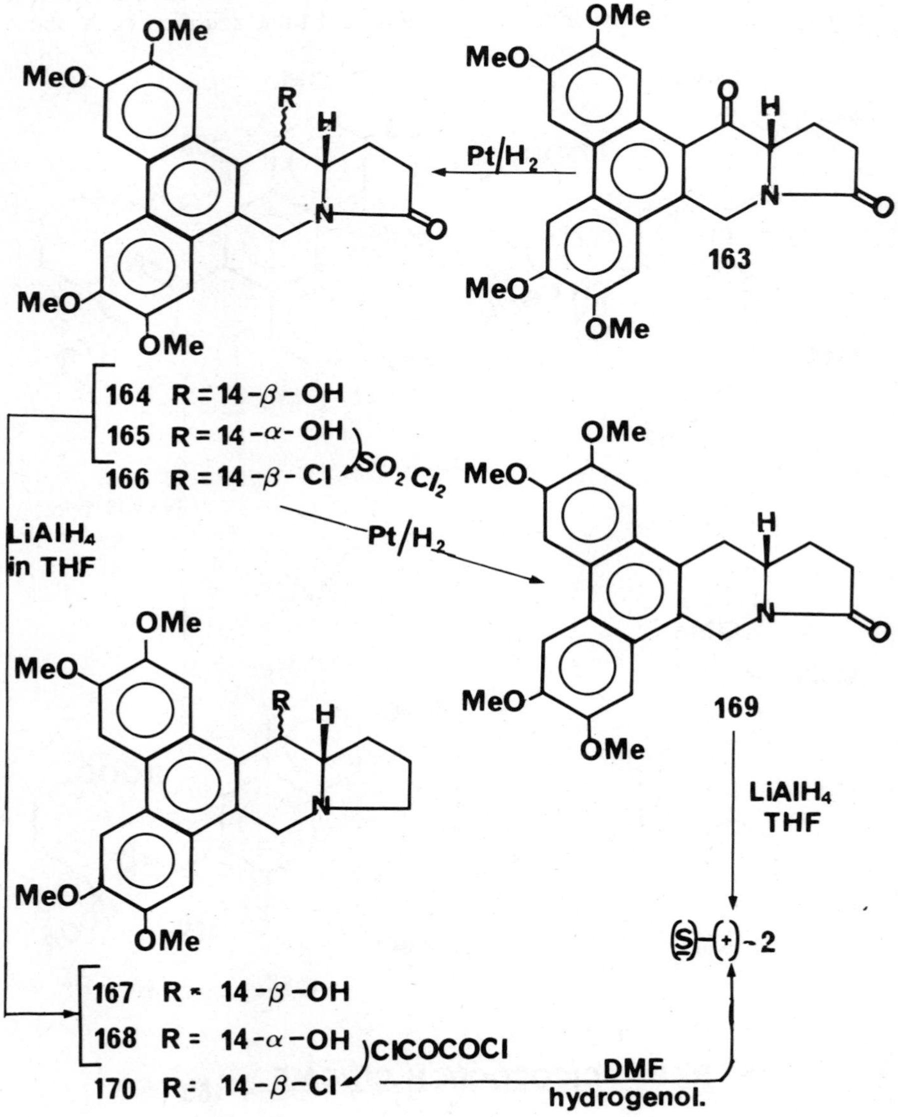

OMe
MeO
R
H
N
MeO
OMe
Pt/H2
OMe
MeO
O
H
N
O
163
MeO
OMe
164 R=14-β-OH
165 R=14-α-OH
166 R=14-β-Cl
SO2Cl2
LiAlH4
in THF
Pt/H2
OMe
MeO
H
N
O
MeO
OMe
169
OMe
MeO
R
H
N
MeO
OMe
LiAlH4
THF
(S)-(+)-2
167 R= 14-β-OH
168 R= 14-α-OH
170 R= 14-β-Cl
ClCOCOCl
DMF
hydrogenol.

trace amount. The NMR coupling constants (6–8 Hz) for the diaxial vicinal protons of the acetate of **164** are consistent with this assignment and also with those of similar compounds [138]. Reduction of **163** with either lithium aluminum hydride or sodium borohydride gave a mixture of two diastereoisomeric alcohols in which **164**, i.e., the β-alcohol predominates. Bulkier hydride reagents yielded the α-alcohol (**165**). For identification and assignment of configuration note that the α-alcohols dehydrate more quickly and are more rapidly converted to the halides than the corresponding β-alcohols. The acetates of the α-alcohols exhibit small coupling constants for the *cis* vicinal hydrogens.

Reduction of **164** and **165** by lithium aluminum hydride in tetrahydrofuran gave the optically active amino alcohols **167** and **168**, respectively, in good yield. Acetylation of these alcohols allowed reconfirmation of the stereochemical assignments made with the amino alcohols **164** and **165**.

As mentioned earlier, the 14β-hydroxyamido alcohol (**164**) cannot easily be converted to the corresponding halide whereas the corresponding 14α-hydroxyamido alcohol (**165**) readily formed a halide with thionyl chloride, identified as the 14β-amido chloride (**166**). The chlorine atom could then be removed by catalytic hydrogenation to yield the optically active amide (**169**), which can be reduced to the corresponding amine, (*S*)-(+)-tylophorine (**2**) by lithium aluminum hydride in boiling tetrahydrofuran. Similarly, chloride formation with oxalyl chloride from the amino alcohols (**167** and **168**) gave the chloroamine (**170**) which undergoes hydrogenolysis in dimethylformamide also to (*S*)-(+)-**2**.

This chirally specific synthesis of tylophorine contradicts some of the assignments of absolute stereochemistry [7] of several phenanthroindolizidine alkaloids based on previously assigned absolute configurations for, for example, tylophorine as (*S*)-(−)-**2** [57] and antofine as (*R*)-(−)-**19** [57].

6. BIOSYNTHESIS OF THE ALKALOIDS

The biogenesis of the phenanthroindolizidine alkaloids is of intrinsic interest because of their physiological activity. Investigation of their natural synthetic path commenced shortly after the structures of some of these alkaloids were elucidated. The initial hypothesis outlining a probable pathway was proposed by Mulchandani et al. [139]. They suggested that the reaction between 3,4-dihydroxybenzoylacetic acid (**171**) and Δ¹-pyrroleine [105] (which was assumed to originate from either the amino acid ornithine or from putrescine) gives 2-(3′,4′-dihydroxyphenacyl)pyrrolidine (**172**). Condensation of **172** with 3,4-dihydroxyphenylpyruvic acid (**173**) yields the intermediate (**174**), which can readily be transformed to tylophorine (**2**) by oxidative coupling. To test the validity of this proposition tyrosine (labeled with ^{14}C at position 2) was fed by the wick technique to $1\frac{1}{2}$-yr-old *T. indica* plants, which were then harvested after 14 days. The position of the ^{14}C label in **2** was

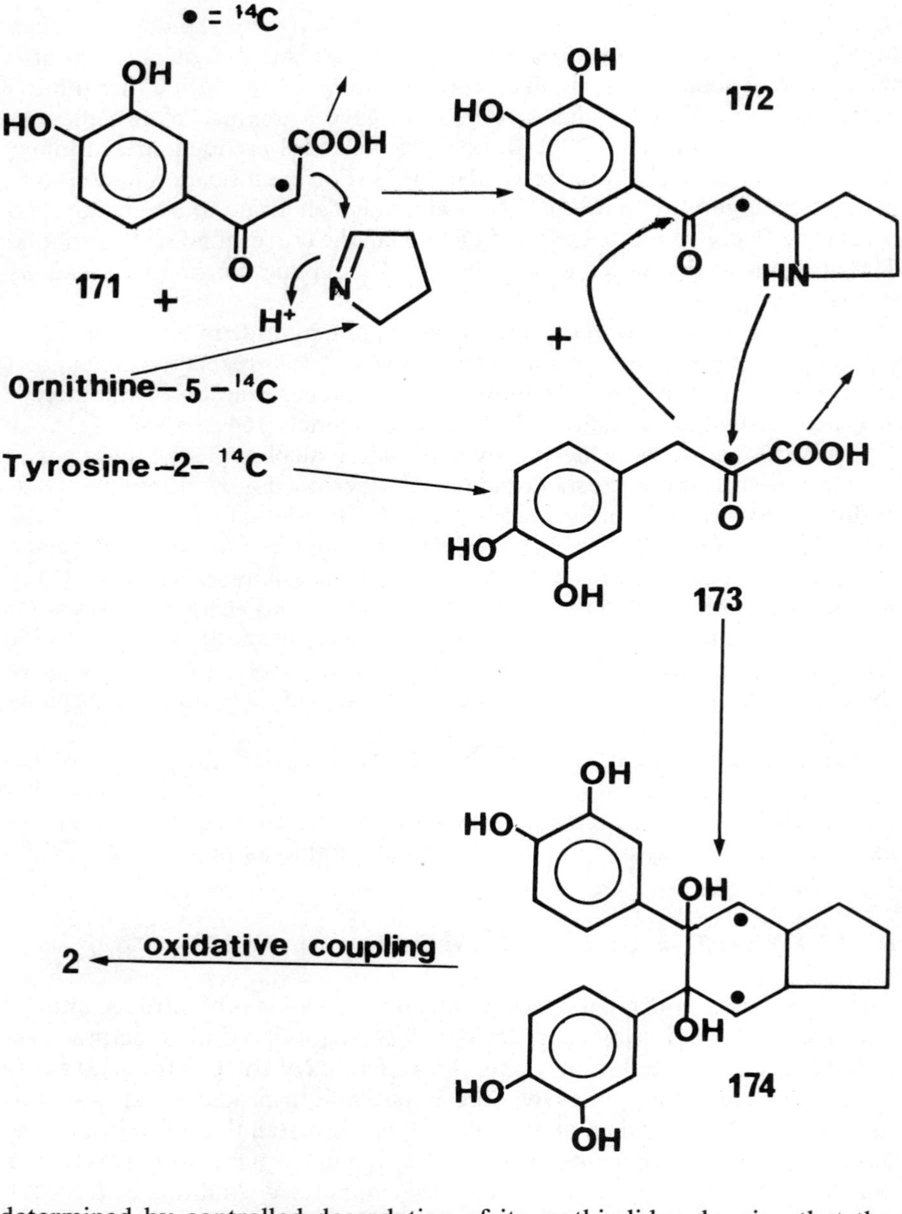

$\bullet = {}^{14}C$

171 + Ornithine–5–^{14}C

Tyrosine–2–^{14}C

172

173

2 ← oxidative coupling

174

determined by controlled degradation of its methiodide, showing that the label is present only at C(9), proving that tyrosine is the precursor of only C(9) and ring B.

Because the origin of ring A remained undetermined by the tyrosine experiments, feeding experiments with other possible ^{14}C-labeled precursors were undertaken [140]. Of these, [1-^{14}C]benzoic acid did not show any in-

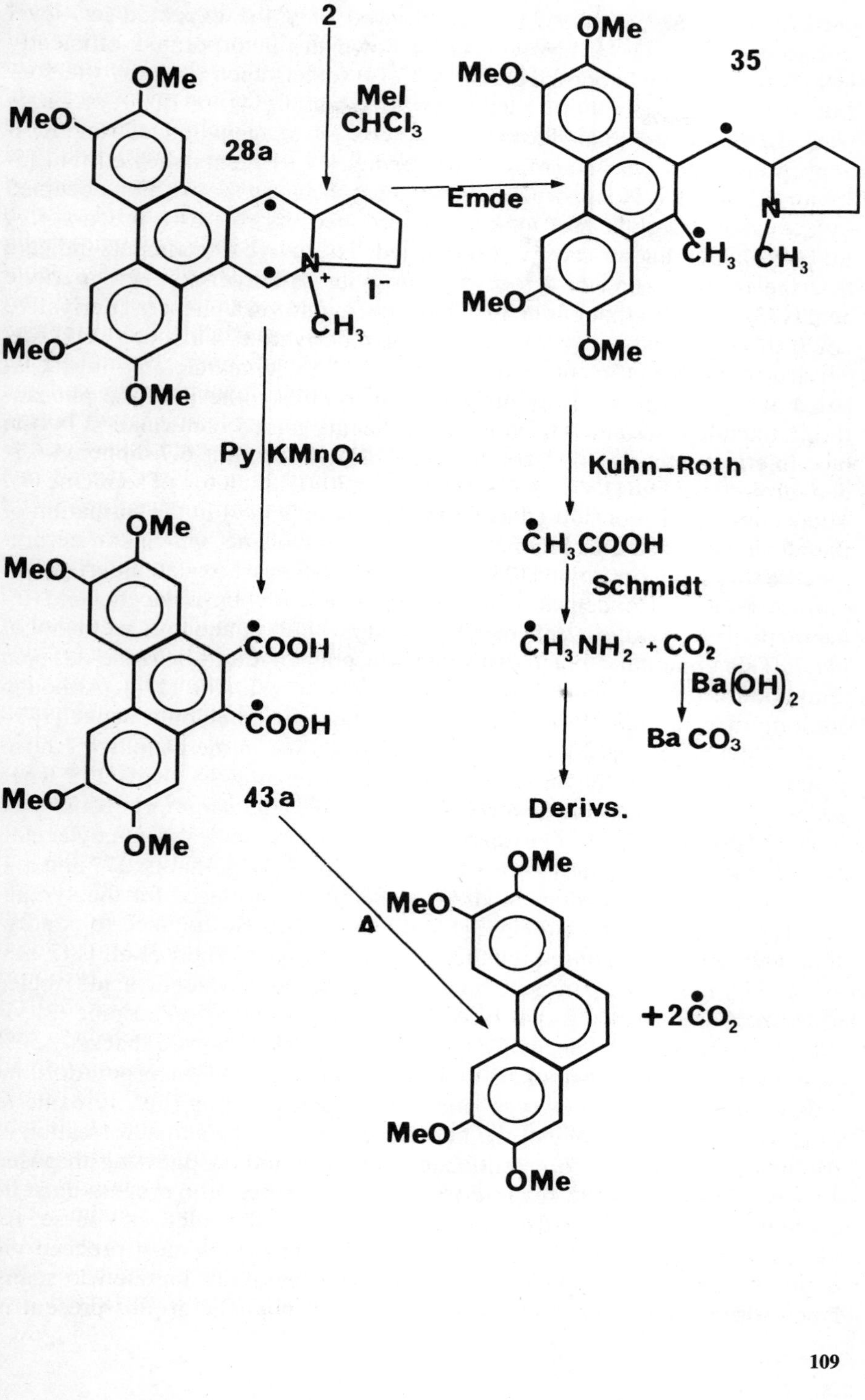
2
MeI
CHCl3
28a
Emde
35
Py KMnO4
Kuhn–Roth
CH3COOH
Schmidt
CH3NH2 + CO2
Ba(OH)2
BaCO3
Derivs.
43a
Δ
+2CO2
OMe
MeO
CH3
I⁻

corporation, whereas [2-^{14}C]acetate showed only the expected low-level overall labeling. [2-^{14}C]Phenylalanine, however, incorporated efficiently into **2**, the ^{14}C label appearing only at C(14) (determined again by the controlled degradation method), establishing that both C(14) and ring A originate from unsubstituted phenylalanine (followed by oxygenation of ring A, if appropriate, at a later stage). Similar feeding experiments showed that [5-^{14}C]ornithine was also incorporated into **2** but, because the position occupied by the labeled carbon has not been determined, its role as a precursor of the pyrrolidine ring was not yet established. The above experiments indicate that the incorporation of tyrosine occurs via 3,4-dihydroxyphenylpyruvic acid (**173**), whereas that of phenylalanine occurring via a phenacyl derivative could therefore proceed via a cinnamic acid derivative. This idea [141] was confirmed by the efficient incorporation of [2-^{14}C]cinnamic acid into **2** labeled at C(14). The pathway proposed for the biosynthesis of the phenanthroindolizidine alkaloids from the amino acids gained confirmation by the biomimetic synthesis of Herbert et al. [103] of the first 6,7-diphenyl-6,7-dehydroindolizidine (**175**), that is, the unsubstituted skeleton of septicine (**9**). This compound is considered as an intermediate [23,44] in the formation of the alkaloids. Their starting material was Δ^1-pyrroleine, which can be prepared either from ornithine [105,108] or from its decarboxylated derivative, putrescine [142]. Condensation of Δ^1-pyrroleine with benzoylacetic acid (cf. synthetic route 8) gives 2-phenacylpyrrolidine (**176**) in aqueous methanol at pH 7. This compound reacts with phenylacetaldehyde in benzene at room temperature (and without catalyst) and yields an enamine (**178**). Although addition of enamines to carbonyl groups does not commonly occur [143–144], **178** cyclizes readily to **179**, then (on standing in methanol for 1 hr at room temperature) dehydrates to **180**, which eventually yields **175** when reduced with sodium borohydride in methanol. The same procedure also yields **9** instead of **175** if instead of benzoylacetic acid and phenylacetaldehyde their appropriately substituted derivatives [145], that is, **177** and 3,4-dimethoxyphenylacetaldehyde [146], respectively, are used for the synthesis. The success of the biomimetic synthesis indicated the need to confirm the intermediates proposed for the actual pathway. Herbert et al. [147,148] showed by feeding experiments in *T. indica* that benzoylacetic acid (labeled at the carbonyl carbon atom) incorporated intact into tylophorinine (**3**) via the corresponding 2-phenacylpyrrolidines. They also showed that only 50% of the ditritiated (in positions 3 and 5) tyrosine achieved incorporation, indicating that the second oxygenation in the benzene ring (i.e., tyrosine to DOPA) must occur at one of the two tritiated sites. The actual formation of the alkaloids from the aforementioned precursors and the pathway proposed suggest strongly that the key intermediate of the biogenetic process must be septicine (**9**, or its variously oxygenated analogs). This idea, of course, requires that the ring closure between the two aryl groups must proceed via oxidative coupling (of phenols), a method successfully imitated in many laboratory synthetic procedures. (Protection of phenolic groups present in

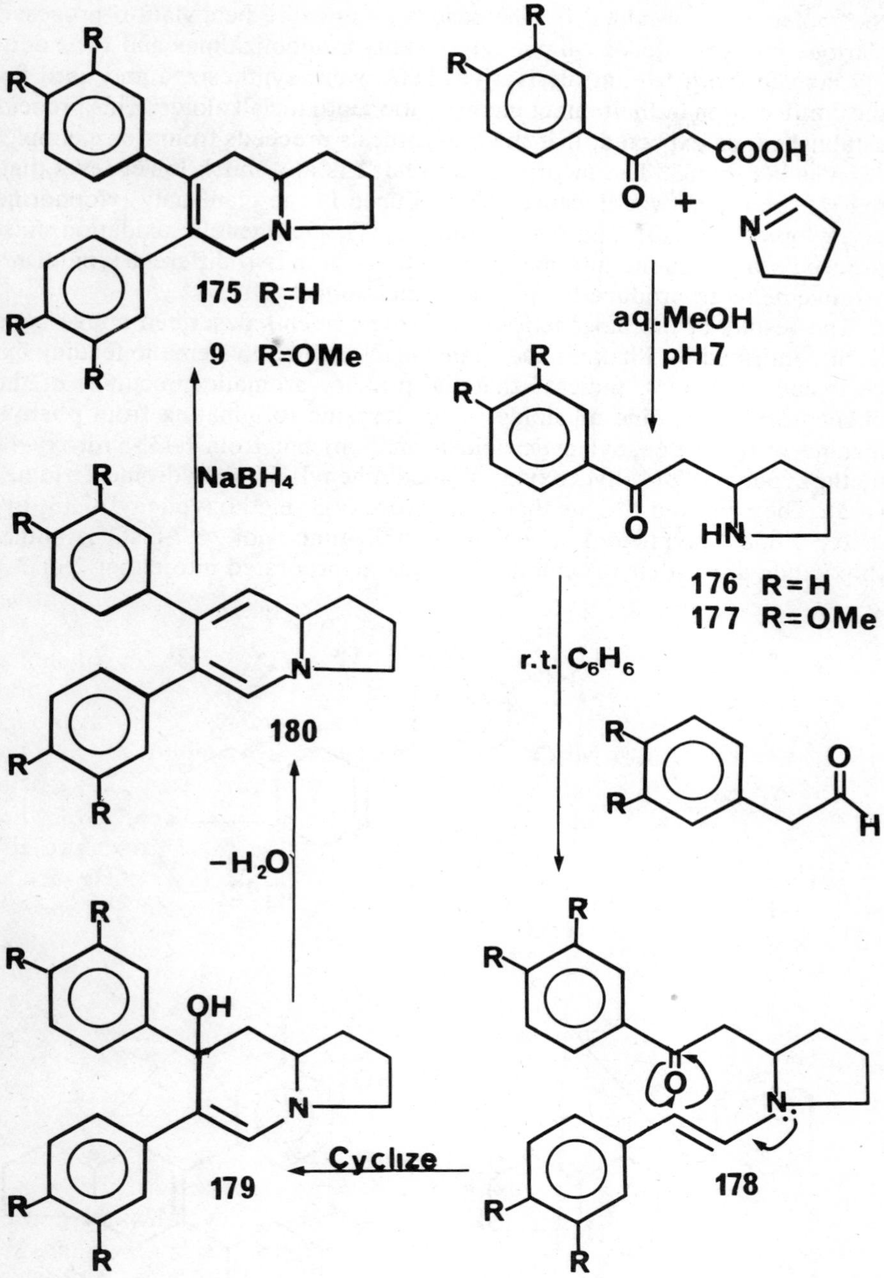

175 R=H

9 R=OMe

176 R=H
177 R=OMe

180

179

178

aq. MeOH
pH 7

NaBH₄

−H₂O

r.t. C₆H₆

Cyclize

the molecule is provided by the readily removable benzylation process.) Variously oxygenated 6,7-diphenyl-6,7-dehydroindolizidines and their doubly labeled (with tritium) derivatives [148] were synthesized and used for the confirmation of their intact incorporation into the alkaloids. This process established, as expected, that the biosynthesis proceeds from, for example, 181 via 182 to 183. The incorporation of 181 is at a much lower level than that of the other two. Because 183 produced in the plant only tylophorine (2), tylophorinine (3), and tylophorinidine (7), the phenolic oxidation must proceed via a dienone intermediate (184) utilizing two different type of rearrangements to produce the alkaloids mentioned.

The results of the biogenetic feeding experiments described above have been confirmed by Bhakuni and Mangla [149]. Their systematic feeding experiments, however, indicate that the primary aromatic precursor of the phenanthroindolizidine alkaloids is not tyrosine (originating from phenylalanine prior to benzoylacetic acid formation) but from 6-(3'-hydroxy-4'-methoxyphenyl)-7-(4'-hydroxy-3'-methoxyphenyl)-6,7-dehydroindolizidine (185). They showed that neither 6-(4'-hydroxy-3'-methoxyphenyl)-7-(3'-hydroxy-4'-methoxyphenyl)-6,7-dehydroindolizidine nor 6,7-bis(3',4'-dimethoxyphenyl)-6,7-dehydroindolizidine was incorporated into either 2 or 3.

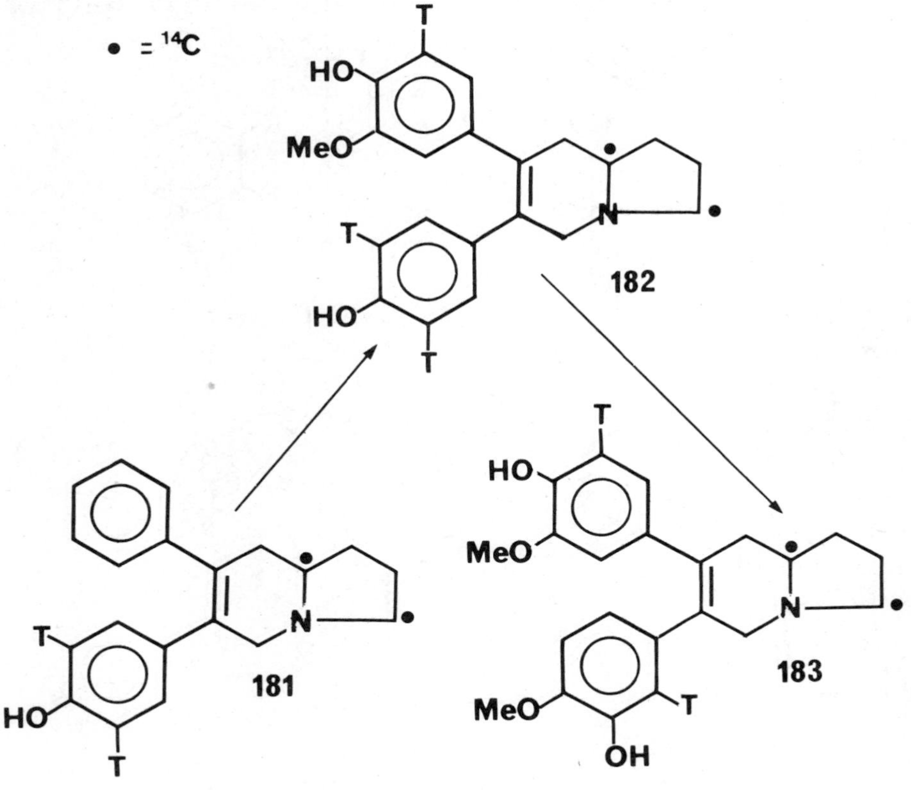

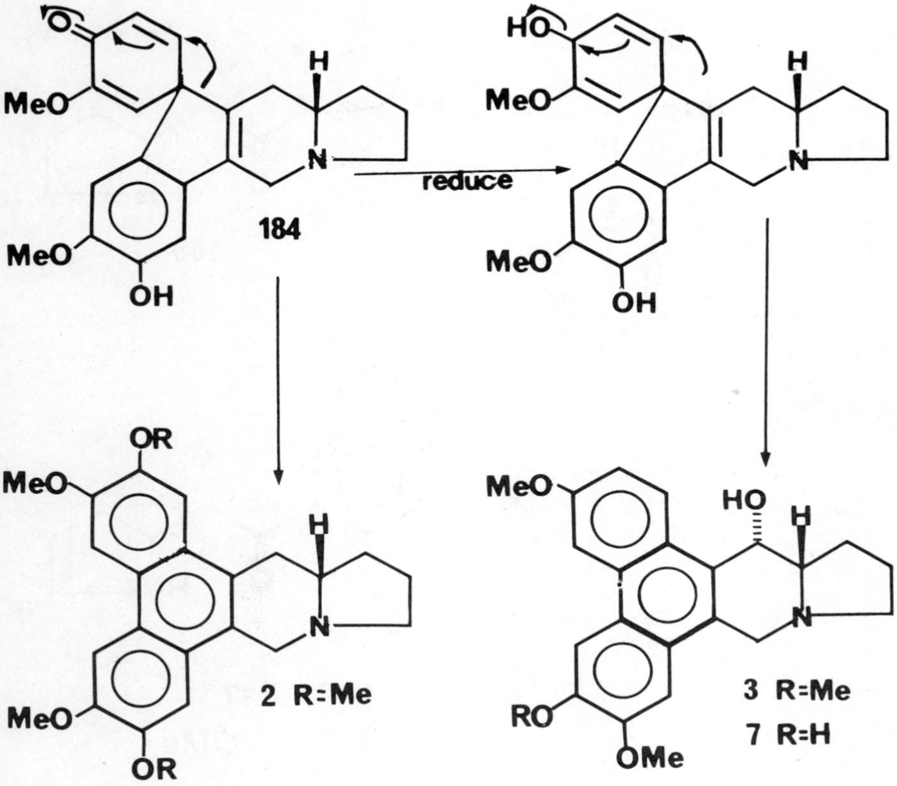

In addition they also established that trioxygenated (instead of tetraoxygenated) derivatives of 6,7-diphenyl-6,7-dehydroindolizidine do not incorporate into **2** or **3**. Accordingly they propose that the biosynthesis of **2** and **3** proceeds from phenylalanine via 3,4-dihydroxyphenylalanine, 6,7-bis(3′,4′-dihydroxyphenyl)-6,7-dehydroindolizidine, 6-(3′,4′-dihydroxyphenyl)-7-(4′-hydroxy-3′-methoxyphenyl)-6,7-dehydroindolizidine to the primary precursor **185**, then to **2** and **3** as proposed earlier.

Additional evidence [151] provided later confirmed that the earlier stages of the proposed mechanism are also correct. 2-Phenacylpyrrolidines were shown to be incorporated intact into the correspondingly substituted 6,7-diaryl-6,7-dehydroindolizidines, which have already been shown to incorporate intact into **2** and **3**. The [14]C- and ^{3}H-labeled derivatives [150] (**186–188**) were fed separately to *T. indica* plants. **186** and **188** were incorporated intact into **3**, whereas incorporation of **187** resulted in the loss of 50% of its tritium label. This result can be easily explained because entry of a second hydroxy group into the benzene ring must take place in one or the other ortho tritiated positions in the same way as was done in the incorporation experiments with tritiated tyrosine [148]. By the use of ^{14}C-labeled ornithine,

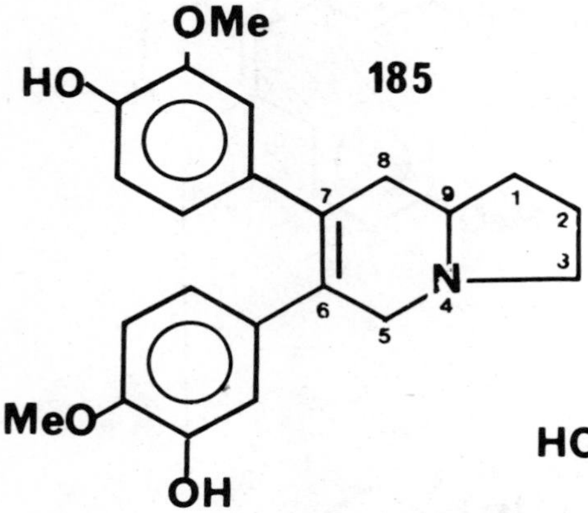

OMe

HO

185

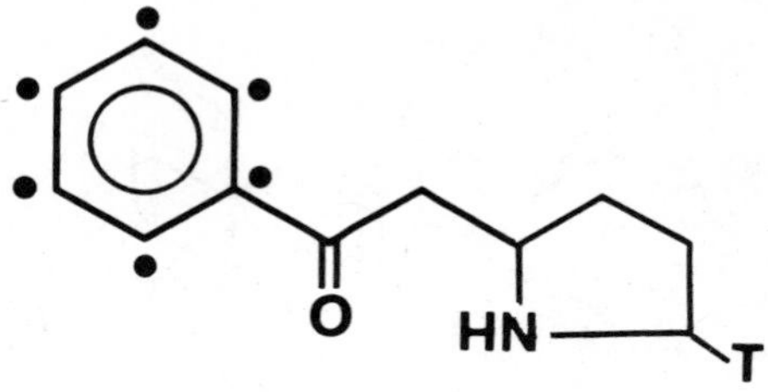

8

7   9   1

2

3

N

6

5   4

MeO

OH

O

HN

186

T

T

HO

R

O   HN

T

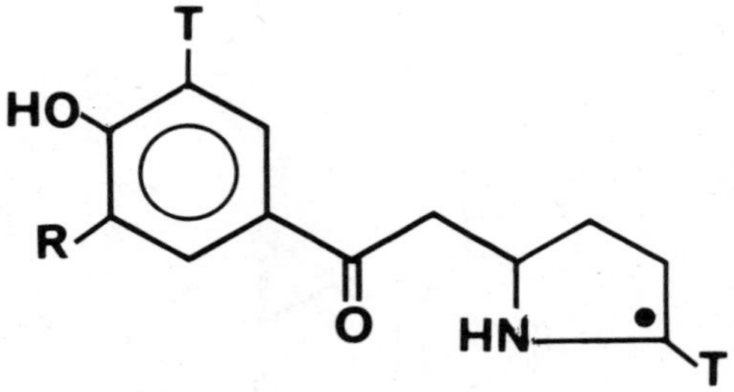

187 R=T
188 R=OMe

C(11) in the phenanthroindolizidine alkaloids was shown to be the same as C(5) of ornithine. It is easy to predict that C(5) of ornithine, after reaction with N-bromosuccinimide, will appear as C(5) both in Δ^1-pyrroleine and in 2-phenacylpyrrolidine (**189**). The position of the label in ornithine, however, had to be proved to be actually at C(5). This was achieved by controlled degradation of the labeled **189**. N-Methylation of **189** to **190** was carried out, not by the usual Eschweiler–Clarke reaction, which gave very poor yields, but via reduction [152] of its imine (formed by treating it with formaldehyde) with sodium cyanoborohydride. Hofmann degradation of the methiodide of **190** gave the methine, which can be hydrogenated, depending on reaction conditions, to either **191** or **192**. Attempted Hofmann degradation of the methiodides of both **191** and **192** was unsuccessful but the N-oxide of **193**, prepared by allowing **192** to stand in methanolic hydrogen peroxide overnight, formed an olefin (**194**), when heated at 170° at 10^{-4} mm Hg. Oxidation of **194** with osmium, tetroxide and sodium metaperiodate gave formaldehyde isolated as its dimedone derivative. This result accounts for the total radioactivity of the starting material, that is, ornithine.

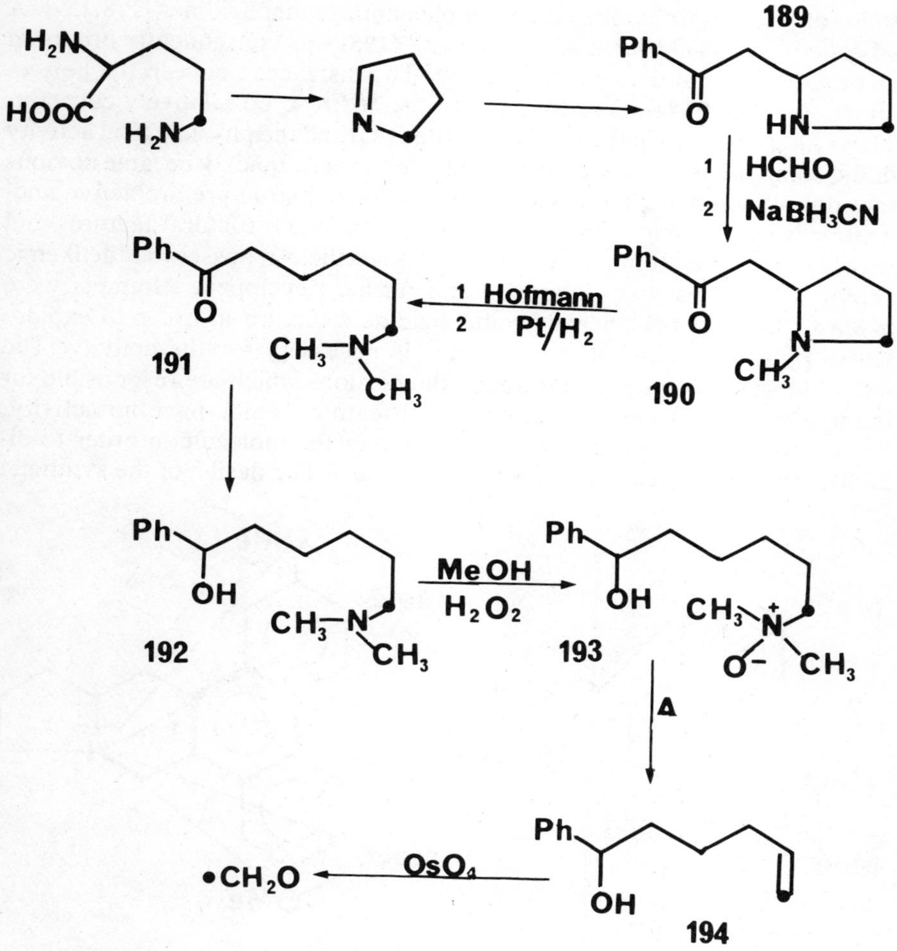

7. CORRELATION OF STRUCTURE WITH PHYSIOLOGICAL ACTIVITY

7.1. Definition and Hypothesis

The known physiological activity of a number of phenanthroindolizidine alkaloids, especially their antineoplastic activity [4,153,154], led to the creation of hypotheses to explain their pharmacological and toxicological activity, followed by attempts to improve the activity of the natural products. For this reason a number of 3,4-diphenylpiperidines [155] and benzoisoquinolines [156–157] were synthesized. Their properties were compared with those of phenanthroindolizidines and phenanthroquinolizidines [158,159]. A so-called "N–O–O triangulation pattern" (**195**) was consequently proposed as necessary for antineoplastic activity. The distances between the hetero-atoms (see Fig. 1) required for activity seemed to keep relatively constant. These interatomic distances were recorded [83] and the physiological activity of the compounds found was evaluated. However, it quickly became obvious that not all compounds possessing the postulated structure are active and, contrarily, that compounds without the postulated structural features and with varying interatomic distances may nevertheless possess antileukemic activity. This idea was elaborated and further developed. Attempts were made to simplify the phenanthroindolizidine structure in order to explore which portion (if any) of the molecule is necessary for the activity. The intention was, of course, (a) to retain the portions which are responsible for the activity, (b) to remove the structural features which have no activity, and (c) to modify some necessary portions of the molecule in order to di-minish (if not fully remove) harmful side effects. The design of the synthetic

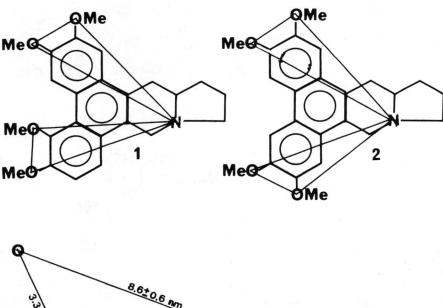

Figure 1

approach so far undertaken incorporates the following approaches:

1. Deletion of rings A and B.
2. Deletion of ring A only.
3. Deletion of ring B only.
4. Deletion of rings A and E.
5. Deletion of rings B and E.

Although the first reported experiments (which included only the first three approaches) dealt only with racemic compounds, the later ones ensured stereospecific syntheses and retained (and elucidated the nature of) the chiral centers at carbons equivalent to C(13a) and C(14) of the alkaloid structure [e.g., at carbons C(7) and C(7a) in **196** type structures and at carbons C(11a) and C(12) in **197** type structures].

7.2. Syntheses of Simplified Analogs

7.2.1. Racemic Compounds. The first limited synthetic approach on the lines put forward in the preceding section was reported only recently [160]. The simplification idea was embraced and carried out on standard routes described for the preparation of pyrroloisoquinoline derivatives. All compounds synthesized were racemates and the limited amount of physiological testing undertaken is incorporated in this report. The authors synthesized three basic types of compounds representing deletion of ring A, the deletion of ring B, and the deletion of both rings A and B from the phenanthroindolizidine structure, but retained ring E in all cases.

7.2.1.1. The General Method of Synthesis Employed. The method of synthesis used by the authors proceeded in a fashion similar to that of Govandachari's [93] first synthesis of the phenanthroindolizidine skeleton. The aromatic aldehydes were condensed with 4-nitrobutanoate to yield 5-aryl-4-nitropent-4-enoates which gave, when reduced with lithium aluminum hydride, the corresponding 4-amino-5-arylpentan-1-ols (**200**). Cyclization of the amino alcohols to the corresponding 2-arylmethylpyrrolidines (**201**) was carried out via the preparation of their sulfate esters with concentrated sulfuric acid followed by treatment with sodium hydroxide [161]. The pyrrolidines (**201**) were then formylated by refluxing them with anhydrous ethyl formate to yield the N-formyl derivatives (**202**) which were cyclodehydrated by the Bischler–Napieralsky method, that is, with phosphorus oxychloride in toluene to the quaternary ammonium salts (**203**), reducible with sodium borohydride to the various isoquinolines.

7.2.1.2. Syntheses with Deletion of Rings A and B or Ring A or Ring B. Three types of pyrroloisoquinolines were prepared by the above method:

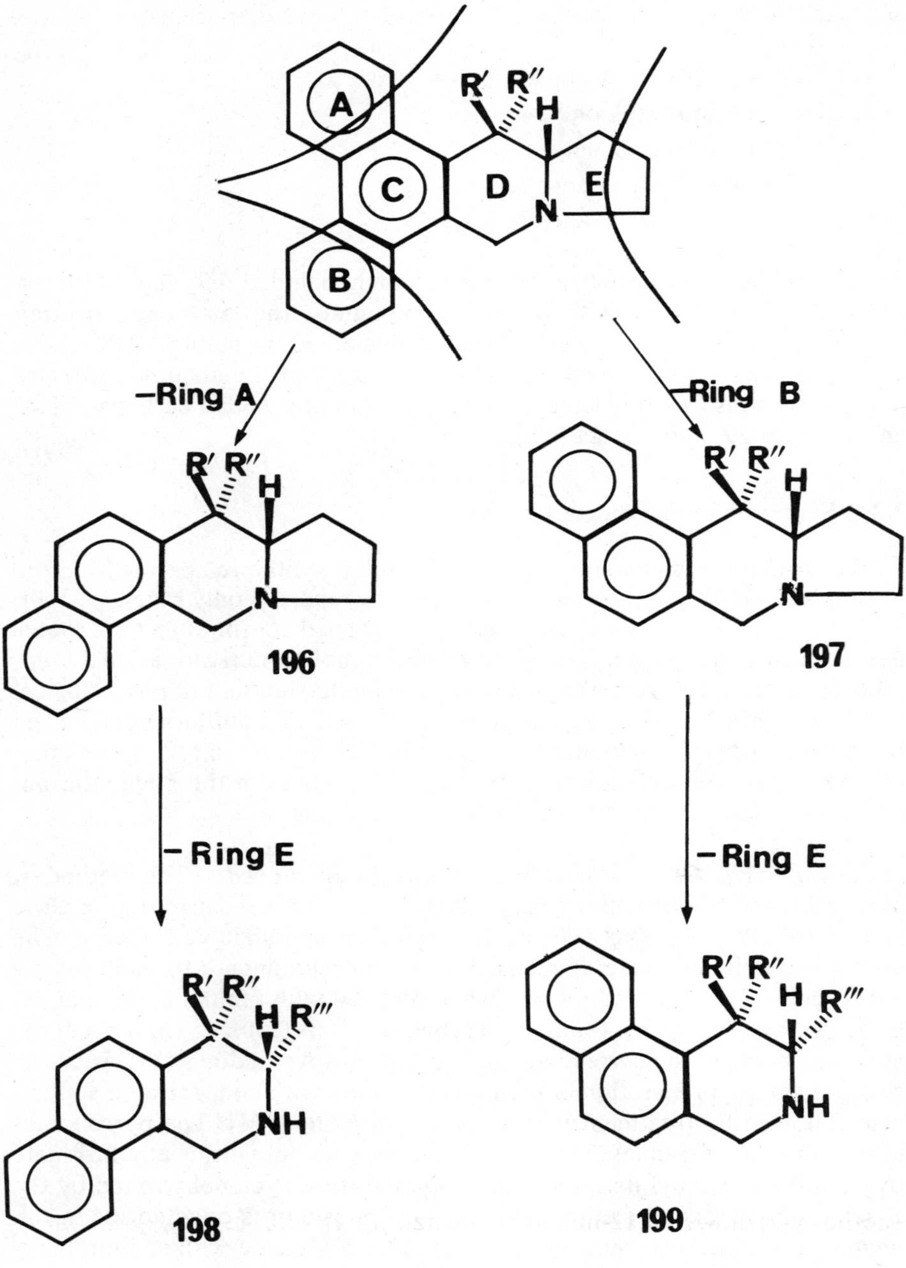
A
B
C
D
E
R′
R″
H
N
−Ring A
−Ring B
196
197
− Ring E
− Ring E
R‴
NH
198
199

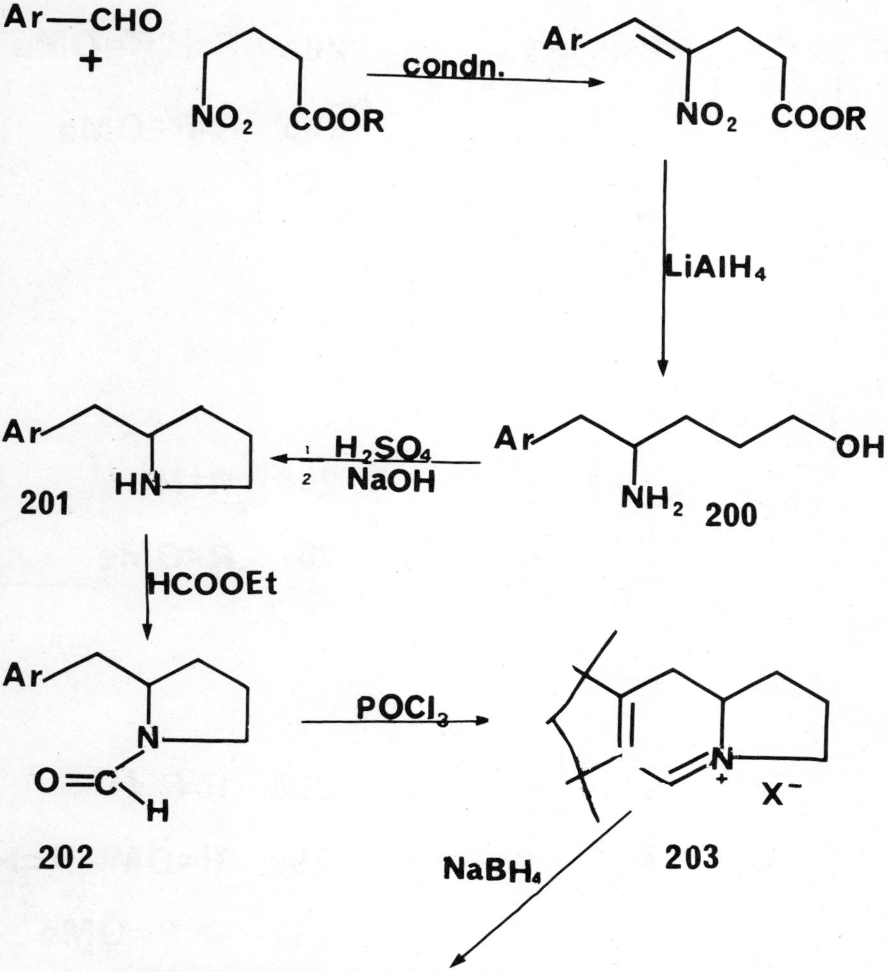

Hexahydropyrroloisoquinolines

8-methoxy-1,2,3,5,10,10a-hexahydropyrrolo[1,2-b]isoquinoline (**204**), 7,8-dimethoxy-1,2,3,5,10,10a-hexahydropyrrolo[1,2-b]isoquinoline (**205**), 7,9,10,11,11a,12-hexahydrobenzo[f]pyrrolo[1,2-b]isoquinoline (**206**), 3-methoxy-7,9,10,11,11a,12-hexahydrobenzo[f]pyrrolo[1,2-b]isoquinoline (**207**), 7,7a,8,9,10,12-hexahydrobenzo[h]pyrrolo[1,2-b]isoquinoline (**208**), 3-methoxy-7,7a,8,9,10,12-hexahydrobenzo[h]pyrrolo[1,2-b]isoquinoline (**209**), and 2,3-dimethoxy-7,7a,8,9,10,12-hexahydrobenzo[h]pyrrolo[1,2-b]isoquinoline (**210**). None of these compounds (**200–210**) exhibited anti-cancer activity in the standard tests run by the National Cancer Institute of NIH, USA. Standard pharmacological tests [162] showed a variety of activities within this group of compounds, for example, depressant, stimulant,

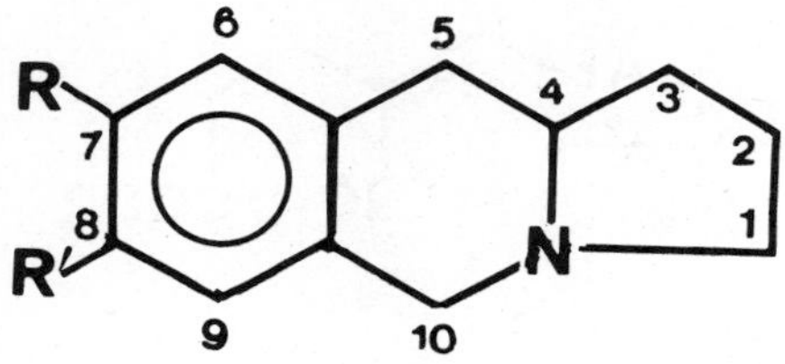

204 R=H,R′=OMe

205 R=R′=OMe

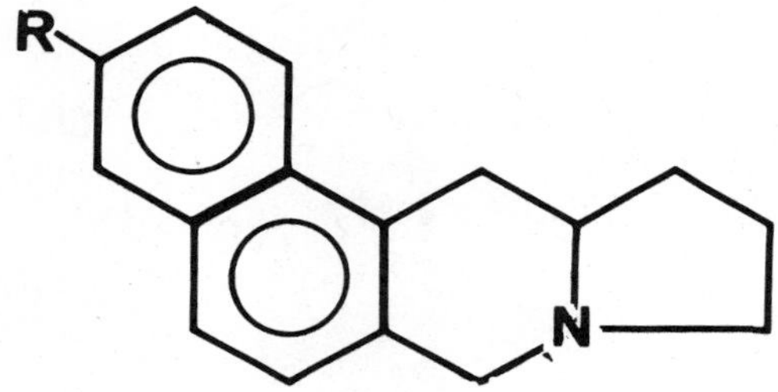

206 R=H

207 R′=OMe

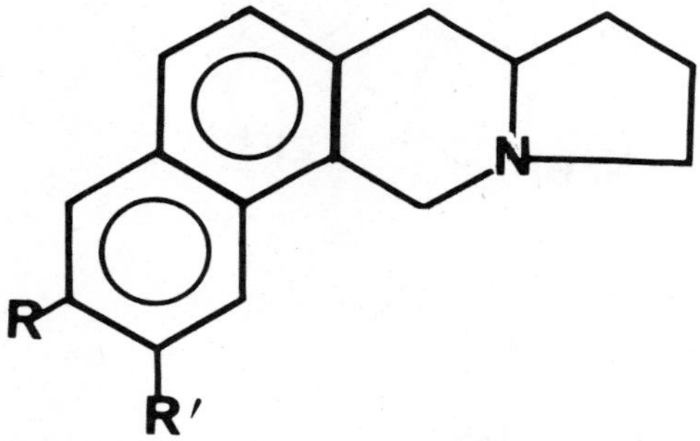

208 R=R′=H

209 R=OMe,R′=H

210 R=R′=OMe

antidepressant, antihistaminic, antiinflammatory, and hypertensive activities, but they have not yet been tested in the clinics.

7.2.2. Chiral Compounds. The following papers [163–164] dealt with stereospecific synthesis of all four types of simplified analog structures represented by formulas **196–199**. Although these syntheses utilize the same Friedel–Crafts type of ring closure techniques developed earlier [94,99] for the preparation of the racemic alkaloids (cf., e.g., synthetic routes 2 and 5), they now employ reaction conditions which produce ring closure with retention of the (S) configuration of the amino acids in the respective tri- and tetra-cyclic α-aminoketones and α-amino alcohols.

Compounds possessing structures of type **198** and **199** were prepared from esters of the natural amino acids by N-alkylation with 1-chloromethylna-phthalene (**211**) [165] and 2-chloromethylnaphthalene (**224**), respectively [166–167]. N-Alkylation of amino acids containing primary amino groups is more complicated than that of L-proline for both mono- and dialkylation can take place, just as they are known to occur also in peptide synthesis during protection of the amino groups by benzyl chloride [168–170]. If, however, the reaction was carried out in refluxing propan-2-ol in presence of anhy-drous sodium carbonate, the amount of the unwanted dinaphthalenylme-thylated product is diminished. Further improvement can be achieved by the addition of the naphthalenylmethyl chloride *slowly* to the amino acid ester. (Any remaining dialkylated product can then be removed by ether extraction from a 1.5 M hydrochloric acid solution.) Saponification followed by treatment with acid gave the N-alkylated amino acid, which did not cy-clize under the conditions used for the preparation of the racemates and their analogs. This behavior was attributed to the basicity of the nitrogen and the removal of this basicity was sought prior to cyclization. Of the two commonly used protecting agents the authors found p-toluenesulfonation the more sat-isfactory method because it provided better protection and gave better yields when tosylation was carried out in acetone at pH 10.4–10.5 (cf. protection by the use of cyclized aminodicarboxylic acid lactams [135–137]). The to-sylated amino acid derivatives did not cyclize (with or without racemization) to the aminoketones. They could, however, be quantitatively converted to their respective acid chlorides with phosphorus pentachloride in benzene at 45–50°. (At higher temperatures they formed intractable tarry polymers [17]). Because the acid chlorides were moisture sensitive they were cyclized im-mediately. The cyclization process, however, was very sensitive and success depended on the reaction conditions selected: solvent, catalyst, reaction time, and temperature played important roles; some conditions could even cause decarboxylation [172]. The most successful cyclization conditions were found to be a solvent of low dielectric constant (e.g., benzene), a mild Friedel–Crafts catalyst (e.g., stannic chloride), and room temperature. The aminoketones were reduced by sodium borohydride or lithium aluminum hydride, in each case to two diastereoisomeric alcohols in about 3:1 or 4:1 ratio to each other.

Similar N-alkylation reactions of L-proline followed by ring closure to the corresponding aminoketones and their reduction to the amino alcohols ex-cluded the possibility of protection by tosylation of the nitrogen of the N-alkylated amino acid. To overcome this difficulty ring closure via the Friedel–Crafts method needed a stronger catalyst (e.g., aluminum chloride) and, most of the times, the reaction required initiation, for example, heating until commencement of hydrochloric acid formation. Consequently the yields of the compounds of type **196** and **197** were lower than those utilizing protected *N*-alkylated primary amino acids, for example, in compounds of type **198** and **199**.

7.2.2.1. Deletion of Rings A and E. Reaction of 1-chloromethylnaphthalene (**211**) with ethyl alaninate hydrochloride (**212**) in refluxing propan-2-ol in the presence of anhydrous sodium carbonate gave ethyl (*S*)-(+)-*N*-(naphthalen-1-ylmethyl)alaninate (**214**). This ester was saponified with alkali, then acidified to the amino acid, which was then tosylated at the nitrogen with *p*-toluenesulfonyl chloride in acetone solution, while the pH was adjusted to pH 10.6 by addition of 1 *N* sodium hydroxide. Acidification of the product gave (*S*)-(+)-*N*-(naphthalen-1-ylmethyl)-*N*-tosylalanine (**216**). Stirring a mixture of **216** and phosphorus pentachloride in dry benzene for 2 hr at room temperature gave the acid chloride, which was not isolated. The benzene solution of the acid chloride was cooled in ice and was kept below 10° while stannic chloride was added slowly to it. After stirring overnight the reaction mixture was diluted with ether and cold 1:1 water–hydrochloric acid added. The organic layer was then washed, dried, and evaporated to dryness. The residue after chromatography on silica gel and crystallization gave (*S*)-(+)-3-methyl-2-tosyl-1,2-dihydrobenz[h]isoquinolin-4(3*H*)-one (**218**). Reduction of **218** with a stirred suspension of lithium aluminum hydride in tetrahydrofuran overnight gave two diastereoisomeric alcohols: the minor isomer was shown to be (3*S*,4*R*)-3-methyl-1,2,3,4-tetrahydrobenz[h]isoquinolin-4-ol (**220**) and the major isomer turned out to be (3*S*,4*S*)-3-methyl-1,2,3,4-tetrahydrobenz[h]isoquinolin-4-ol (**222**). Similarly, reaction of **211** with ethyl 2-aminobutanoate hydrochloride (**213**) gave **215**, which could be tosylated to **217**, then chlorinated with phosphorus pentachloride to the acid chloride and ring-closed with stannic chloride to give (*S*)-(+)-3-ethyl-2-tosyl-1,2-dihydrobenzo[h]isoquinolin-4(3*H*)-one (**219**). Reduction of **219** as above gave the major diastereoisomeric alcohol, (3*S*,4*R*)-3-ethyl-1,2,3,4-tetrahydrobenz[h]isoquinolin-4-ol (**221**), and the minor product, (3*S*,4*S*)-3-ethyl-1,2,3,4-tetrahydrobenz[h]isoquinolin-4-ol (**223**).

7.2.2.2. Deletion of Rings B and E. The starting material here is 2-chloromethylnaphthalene (**224**) instead of the 1-substituted naphthalene. Reaction of **224** with ethyl alaninate hydrochloride (**212**) in refluxing propan-2-ol in the presence of anhydrous sodium carbonate gave ethyl (*S*)-(+)-*N*-(naphthalen-2-ylmethyl)alaninate (**225**). The sequence then proceeded as described in Section 7.2.2.1, yielding the tosylated compound (**227**), then (*S*)-(+)-2-methyl-3-tosyl-3,4-dihydrobenz[f]isoquinolin-1(2*H*)-one (**229**), which was reduced to the minor (1*S*,2*S*)-2-methyl-1,2,3,4-tetrahydrobenz[f]isoquinolin-1-ol (**231**) and the major (1*R*,2*S*)-2-methyl-1,2,3,4-tetrahydrobenz[f]isoquinolin-1-ol (**234**).

Similarly, reaction of **224** with ethyl 2-aminobutanoate hydrochloride (**213**) gave (**226**) and the reaction sequence followed to (**228**) and to (*S*)-(+)-

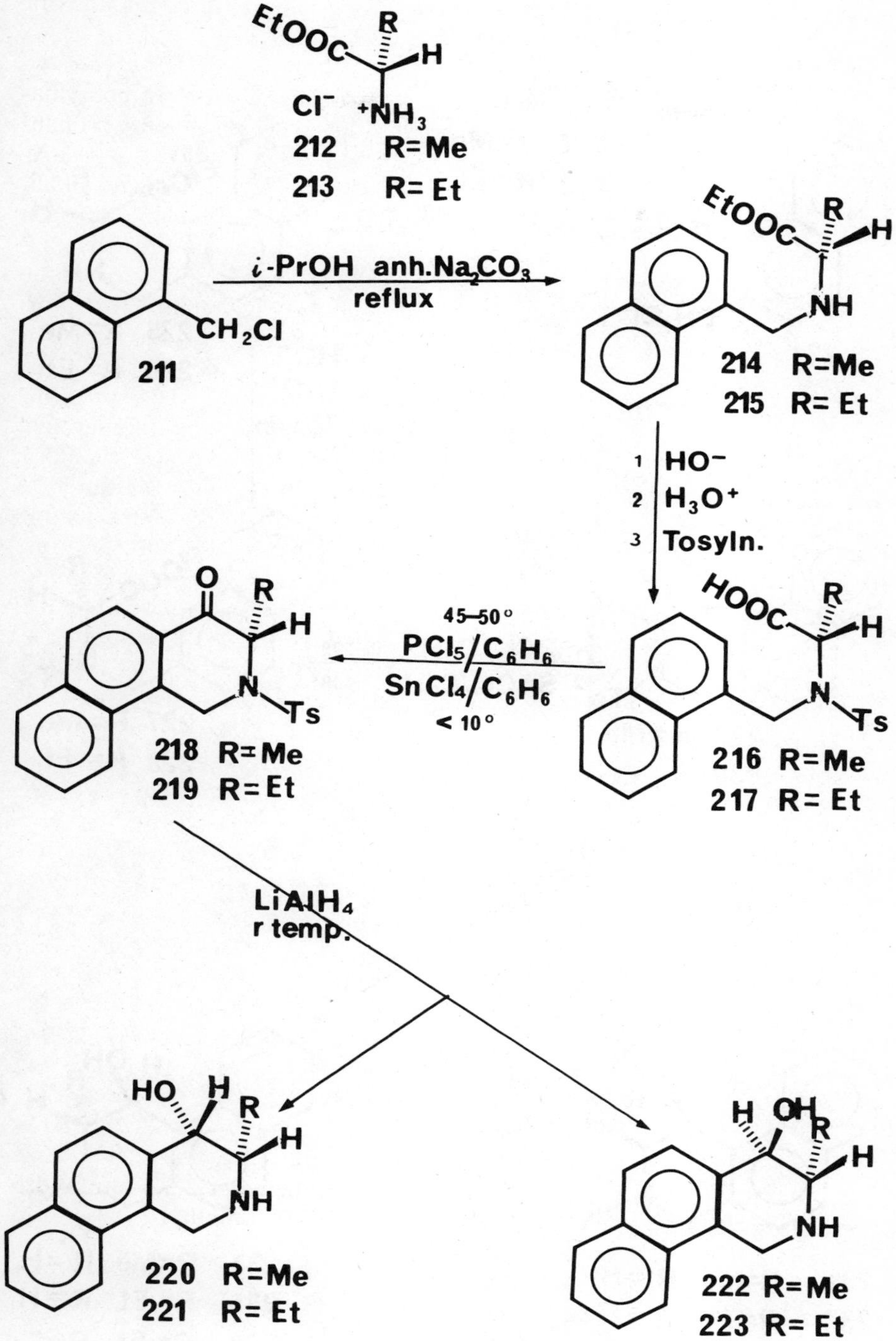
EtOOC
R
H
Cl⁻ ⁺NH₃
212 R=Me
213 R= Et
i-PrOH anh.Na₂CO₃
reflux
CH₂Cl
211
214 R=Me
215 R= Et
1 HO⁻
2 H₃O⁺
3 Tosyln.
HOOC
N
Ts
216 R=Me
217 R= Et
45–50°
PCl₅/C₆H₆
SnCl₄/C₆H₆
< 10°
O
218 R=Me
219 R= Et
LiAlH₄
r temp.
HO
NH
220 R=Me
221 R= Et
OH
222 R=Me
223 R= Et

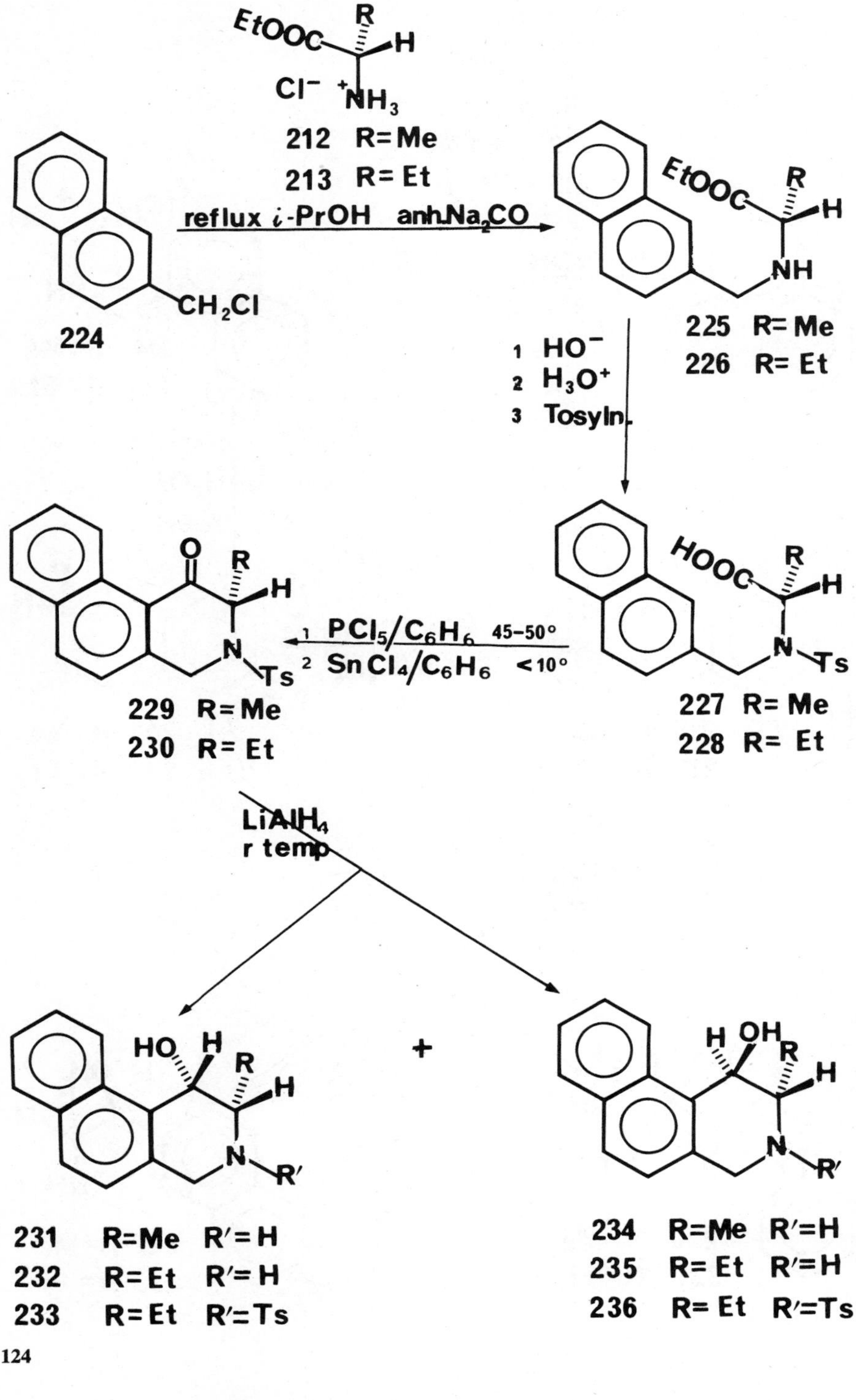
EtOOC
R
H
Cl⁻ ⁺NH₃
212 R=Me
213 R=Et
reflux i-PrOH anh.Na₂CO
CH₂Cl
224
EtOOC
NH
225 R=Me
226 R=Et
1 HO⁻
2 H₃O⁺
3 Tosyln.
HOOC
N
Ts
227 R=Me
228 R=Et
1 PCl₅/C₆H₆ 45–50°
2 SnCl₄/C₆H₆ <10°
O
229 R=Me
230 R=Et
LiAlH₄
r temp
HO
R′
+
H OH
231 R=Me R′=H
232 R=Et R′=H
233 R=Et R′=Ts
234 R=Me R′=H
235 R=Et R′=H
236 R=Et R′=Ts

2-ethyl-3-tosyl-3,4-dihydrobenz[f]isoquinolin-1(2*H*)-one (**230**). Reduction of **230** with lithium aluminum hydride overnight at room temperature, as in all previous experiments, gave the major alcohol, (1*S*,2*R*)-2-ethyl-1,2,3,4-te-trahydrobenz[f]isoquinolin-1-ol (**232**) and the minor alcohol, (1*R*,2*R*)-2-ethyl-1,2,3,4-tetrahydrobenz[f]isoquinolin-1-ol (**235**). However, if the lithium aluminum hydride reduction of **230** was stopped after 30 min only the corresponding *N*-tosyl derivatives (**233** and **236**) were formed. They possess the same absolute configuration as **232** and **235**, respectively. This result establishes that the alcohols are formed by reduction of keto group first, followed by removal of the tosyl group, and therefore their absolute stereochemistry is determined by consideration of Cram's rule [173] applied to their tosylates.

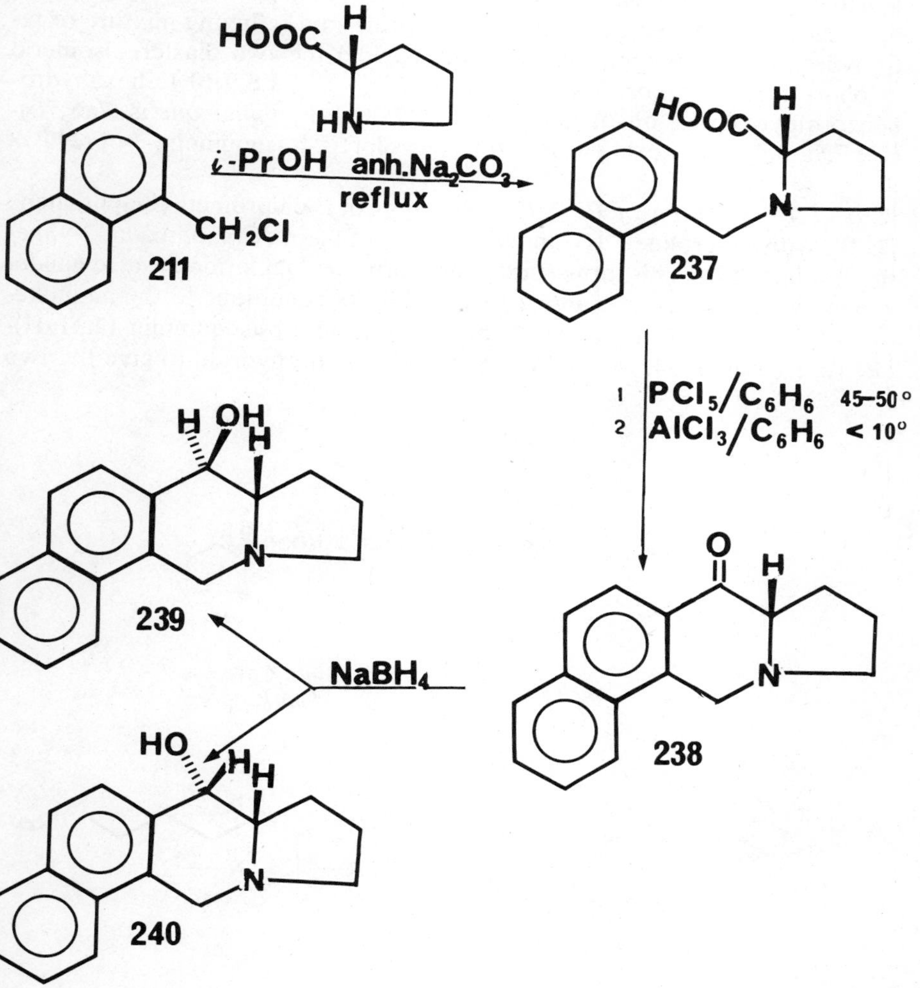

7.2.2.3. Deletion of Ring A. Reaction of 1-chloromethylnaphthalene (211) with L-proline in refluxing propan-2-ol and anhydrous sodium carbonate yields (S)-(−)-N-(naphthalen-1-ylmethyl)proline (237). This forms the acid chloride with phosphorus pentachloride under nitrogen after some initial heating. After azeotropic distillation of the phosphorus compounds, the acid halide was dissolved in dry benzene and cooled, and finely powdered aluminum chloride was added while the mixture was stirred at 10° overnight. The reaction mixture was then diluted with ether, then 1:1 water–hydrochloric acid added. Evaporation of the organic solvents was followed by basification of the aqueous acidic solution to pH ~11. Extraction of the aminoketone with ether from the aqueous basic solution produced (S)-(−)-8,9,10,12-tetrahydrobenzo[h]pyrrolo[1,2-b]isoquinolin-7(7aH)-one (238). Finally 238 was reduced by sodium borohydride in a refluxing mixture of tetrahydrofuran, methanol, and dioxane to yield the two diastereoisomeric alcohols. The *major one* is (7aS,7R)-(−)-7,7a,8,9,10,12-hexahydrobenzo[h]pyrrolo[1,2-b]isoquinolin-7-ol (239) and the *minor one* is (7aS,7S)-(−)-7,7a,8,9,10,12-hexahydrobenzo[h]pyrrolo[1,2-b]isoquinolin-7-ol (240).

7.2.2.4. Deletion of Ring B. Reaction of 2-chloromethylnaphthalene (224) with L-proline as above yields (S)-(−)-N-(naphthalen-2-ylmethyl)proline (241). This forms with phosphorus pentachloride the acid halide, which cyclizes under the above Friedel–Crafts conditions to the aminoketone, (S)-7,9,10,11-tetrahydrobenzo[f]pyrrolo[1,2-b]isoquinolin-12(11aH)-one (242). Again, 242 was reduced by sodium borohydride to give the two

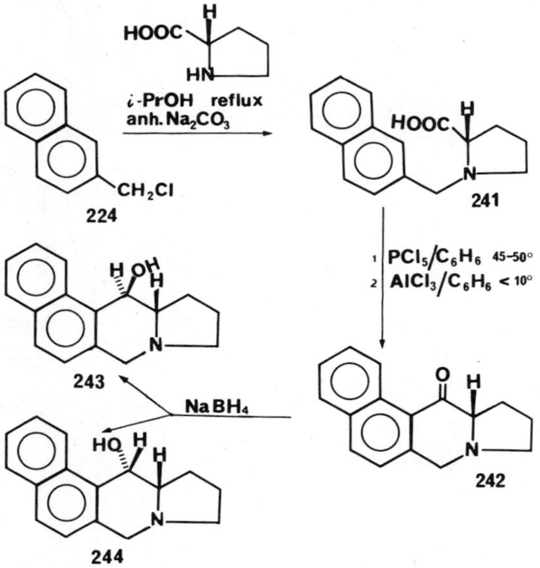

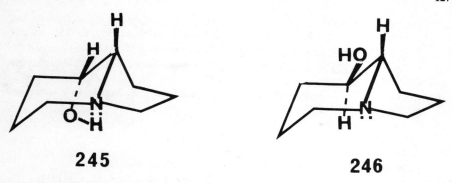

245

246

diastereoisomeric alcohols of which the major one is (11*S*,12*R*)-7,9,10,11,
11a,12-hexahydrobenzo[f]pyrrolo[1,2-b]isoquinolin-12-ol (**243**) and the minor
one, (11*S*,12*S*)-7,9,10,11,11a,12-hexahydrobenzo[f]pyrrolo[1,2-b]isoquin-
olin-12-ol (**244**).

7.2.3. The Absolute Stereochemistry of the Chiral Compounds Synthesized.

The formation of the two diastereoisomeric pairs of alcohols obtained by
the reduction of the respective aminoketones and the proportions in which
they formed are governed by Cram's rule [173]. Both isomers show Bohlmann's bands [64] indicative of *trans* junctions (i.e., the bridgehead hydrogen
is *trans* to the unshared pair of electrons on the nitrogen atom). The two
alcohols therefore differ only in their configuration relative to the bridgehead
carbon atom. Because one of the two diastereoisomeric alcohols showed an
intramolecularly bonded broad OH absorption band at about 3566 cm^{-1}
(**245**), it must possess an axial hydroxyl *trans* to the bridgehead (and is
consequently capable of hydrogen bonding to the lone pair of electrons of
the nitrogen atom). The other isomer exhibits a sharp OH band indicative
of an equatorial hydroxyl group *cis* to the bridgehead hydrogen atom (**246**)
(and is incapable of hydrogen bonding to the lone pair of electrons of the
nitrogen atom); configurations can thus readily be assigned to all four pairs
of diastereoisomeric alcohols so produced. One only has to take into account
the fact that in the case of the tosylated (i.e., protected) aminoketones
Cram's rule must be applied in the presence of the large *N*-tosyl substituent
because the reduction of the carbonyl group is very much faster than the
removal of the tosyl group, as proved by the results of the brief lithium
aluminum hydride reduction. The use of sterically hindered bulky reducing
agents, such as the ones described by Buckley and Rapoport [135–137], have
not yet been attempted.

REFERENCES

1. E. Gellert, *J. Nat. Prod.* **45**, 50 (1982).
2. *Bengal Pharmacopoeia*, 1844.

3. J. L. Hartwell and B. J. Abbot in *Advances in Pharmacology and Chemotherapy*, Vol. 7, R. J. Schnitzer and A. Golden, Eds., Academic, New York, 1969, p. 125.

4. E. Gellert and R. Rudzats, *J. Med. Chem.* **7**, 361 (1964).

5. E. Gellert, *7th IUPAC Int. Symp. Chem. Nat. Prod.*, Melbourne, 1960, Abstract No. CM13/H2, p. 35.

6. E. Gellert, T. R. Govindachari, M. V. Lakshmikantham, I. S. Ragade, R. Rudzats, and N. Viswanathan. *J. Chem. Soc.* **1962**, 1008.

7. E. Gellert, R. Rudzats, J. C. Craig, S. K. Roy, and R. Woodard, *Aust. J. Chem.* **31**, 2095 (1978).

8. D. Hooper, *Pharm. J. (Trans.)* **21**, 617 (1891).

9. D. Arata and C. Gelzer, *Ber. Deutsch. Chem. Gesell.* **24**, 1849 (1891).

10. A. N. Ratnagiriswaran and K. Venkatachalam, *Ind. J. Med. Res.* **22**, 433 (1935).

11. R. N. Chopra, *Indigenous Drugs of India*, Calcutta, 1933.

12. R. N. Chopra, N. N. Ghosh, J. B. Bose, and S. Ghosh, *Arch. Pharm.* **275**, 236 (1937).

13. R. N. Chopra, S. L. Nayar, and I. C. Chopra, *Glossary of Indian Medicinal Plants*, C.S.I.R., New Delhi, India, 1956, p. 250.

14. T. R. Gavindachari, B. R. Pai, and K. Nagarajan, *J. Chem. Soc.* **1954**, 2801.

15. K. V. Rao, U.S. patent No. 3497593.

16. K. V. Rao, R. A. Wilson, and B. Cunnings, *J. Pharm. Sci.* **60**, 1725 (1971).

17. N. B. Mulchandani, S. S. Iyer, and L. P. Badheka, *Chem. Ind. (London)* **1971**, 505.

18. T. R. Govindachari, N. Viswanathan, J. Radhakrishnan, B. R. Pai, S. Natarajan, and P. S. Subramaniam, *Tetrahedron* **29**, 891 (1973).

19. T. R. Govindachari, N. Viswanathan, J. Radhakrishnan, R. Charubala, N. Nityananda Rao, and B. R. Pai, *Indian J. Chem.* **11**, 1215 (1973).

20. I. R. C. Bick, W. Sinchai, T. Sevenet, A. Ranaivo, M. Nieto, and A. Cave, *Planta Med.* **39**, 205 (1980).

21. I. R. C. Bick and W. Sinchai in *The Alkaloids*, Vol. 19, R. Manske, Ed., 1981, p. 193.

22. K. V. Rao, R. Wilson, and B. Cummings, *J. Pharm. Sci.* **59**, 1501 (1970).

23. K. V. Rao, *J. Pharm. Sci.* **59**, 1608 (1970).

24. H. C. Brill and A. H. Wells, *Philipp. J. Sci.* **12**, 167 (1917).

25. L. J. Webb, *Guide to the Medicinal and Poisonous Plants of Queensland*, C.S.I.R. Bulletin No. 232, Melbourne, 1932.

26. M. B. Patel and J. M. Rawson, *Planta Med.* **12**, 33 (1964).

27. L. J. Webb, *Australian Phytochemical Survey*, Part II, C.S.I.R. Bulletin No. 263, Melbourne, 1952.

28. J. D. Phillipson, I. Tezcan, and P. J. Hylands, *Planta Med.* **25**, 301 (1974).

29. C. R. Karnick, *Planta Med.* **27**, 333 (1975).

30. H. H. S. Fong, M. Trojankova, J. Trojanek, and N. R. Farnsworth, *Lloydia* **35**, 117 (1972).

31. T. R. Govindachari and N. Viswanathan, *Heterocycles* **11**, 587 (1978).

32. T. F. Platonova, A. D. Kuzovkov, and P. S. Massagetow, *Zh. Obshkh. Khim.* **28**, 3131 (1958); *Chem. Abstr.* **53**, 7506d (1959).

33. M. Pailer and W. Streicher, *Monatsh. Chem.* **96**, 1094 (1965).

34. A. Haznagy, K. Szendrei, and L. Toth, *Pharmazie* **20**, 541 (1965).

35. A. Haznagy, L. Toth, and K. Szendrei, *Pharmazie* **20**, 649 (1965).

36. L. Ferenczy, J. Zsolt, A. Haznagy, L. Toth, and K. Szendrei, *Acta Microbiol. Acad. Sci. Hung.* **12**, 337 (1965/66).

37. W. Wiegrebe, L. Faber, H. Brockman, Jr., H. Budzikiewicz, and U. Kruger, *J. Liebig's Ann. Chem.* **721**, 154 (1969).

38. W. Wiegrebe, H. Budzikiewicz, and L. Faber, *Arch. Pharm.* **303**, 1009 (1970).

39. H. Budzikiewicz, L. Faber, E-G. Herrmann, F. F. Perrollaz, U. P. Schlunegger, and W. Wiegrebe, *J. Liebig's Ann. Chem.* **1979**, 1212.

40. N. B. Mulchandani and S. R. Venkatachalam, *8th Int. Symp. Chem. Nat. Prod.*, New Delhi, 1972.

41. N. B. Mulchandani and S. R. Venkatachalam, Atomic Energy Commission India, Bhabka Atomic Res. Centre Report BARC-764, 1974, p. 8.

42. N. B. Mulchandani and S. R. Venkatachalam, *Phytochemistry* **15**, 1561 (1976).

43. N. B. Mulchandani and S. R. Venkatachalam, *Phytochemistry* **23**, 1206 (1984).

44. J. H. Russell *Naturwissenschaften* **50**, 443 (1963).

45. R. B. Herbert and C. J. Moody, *Phytochemistry* **11**, 1184 (1972).

46. S. R. Venkatachalam and N. B. Mulchandani, *Naturwissenschaften* **69**, 287 (1982).

47. F. L. Pyman, *J. Chem. Soc.* **103**, 817 (1913).

48. E. Gellert and N. V. Riggs, *Aust. J. Chem.* **7**, 113 (1954).

49. E. Gellert, *Chem. Ind. (London)* 983 (1955).

50. E. Gellert, *Aust. J. Chem.* **9**, 489 (1956).

51. A. T. Blomquist, L. H. Liu, and J. C. Bohrer, *J. Am. Chem. Soc.* **74**, 3643 (1952).

52. T. R. Govindachari, M. V. Lakshmikantham, N. Nagarajan, and B. R. Pai, *Chem. Ind. (London)* **1957**, 1484.

53. T. R. Govindachari, M. V. Lakshmikantham, B. R. Pai, and S. Rajappa, *Tetrahedron* **9**, 53 (1960).

54. T. R. Govindachari, T. G. Rajagopalan, and N. Viswanathan, *J. Chem. Soc., Perkin Trans. 1.* **1974**, 1161.

55. T. R. Govindachari, I. S. Ragade, and N. Viswanathan, *J. Chem. Soc. (London)* **1962**, 1357.

56. T. R. Govindachari, B. R. Pai, S. Prabhakar, and R. S. Savitri, *Tetrahedron* **21**, 2573 (1965).

57. W. Wiegrebe, L. Faber, and Th. Breyhan, *Arch. Pharm.* **304**, 188 (1971).

58. V. Snieckus, in *The Alkaloids*, Vol. 14, R.H.F. Manske, Ed., New York, Academic, 1973, p. 425.

59. V. K. Wadhawan, S. K. Sikka, and N. B. Mulchandani, *Tetrahedron* **29**, 5091 (1973).

60. A. W. Sangster and K. L. Stuart, *Chem. Rev.* **65**, 69 (1965).

61. T. R. Govindachari, B. R. Pai, I. S. Ragade, S. Rajappa, and N. Viswanathan, *Tetrahedron* **14**, 288 (1961).

62. W. Wiegrebe, *Pharm. Zeit.* **117**, 1509 (1972).

63. F. Bohlmann, *Angew Chem.* **69**, 641 (1957).

64. F. Bohlmann, *Chem. Ber.* **91**, 2157 (1958).

65. F. Bohlmann, D. Schumann, and H. Schultz, *Tetrahedron Lett.* 173 (1965).

66. T. A. Crabb, R. F. Newton, and D. Jackson, *Chem. Rev.* **71**, 109 (1971).

67. T. R. Govindachari, *J. Ind. Chem. Soc.* **50**, 1 (1973).

68. T. R. Govindachari, N. Viswanathan, and B. R. Pai, *Indian J. Chem.* **12**, 866 (1974).

69. R. N. Chopra, N. N. De, and M. Chakerburty, *Ind. J. Med. Res.* **23**, 263 (1935).

70. R. Dhananjayan, C. Gopalakrishnan, and L. Kameswaran, *Indian J. Pharmacol.* **7**, 13 (1975).

71. V. Raina and S. Raina, *Biochem. Biophys. Res. Commun.* **94**, 1074 (1980).

72. S. S. Rao, A. W. Bhagwat, and S. S. Gupta, *Indian J. Pharmacol.* **14**, 125 (1982).
73. D. N. Shivpuri, M. P. Menon, and D. Parkash, *J. Assoc. Physicians India* **16**, 9 (1968).
74. D. N. Shivpuri, M. P. Menon, and D. Parkash, *J. Allergy (U.S.A.)* **43**, 145 (1969).
75. D. N. Shivpuri, S. C. Singhal, and D. Parkash, *Ann. Allergy* **30**, 407 (1972).
76. D. N. Shivpuri and M. K. Agarwal, *Ann. Allergy* **31**, 87 (1973).
77. P. S. R. K. Haranath and K. Shyamala, *Indian J. Med. Res.* **63**, 661 (1975).
78. M. P. Chitnis, D. D. D. Khandalekar, and M. K. Adwanka, *Indian J. Med. Res.* **60**, 359 (1972).
79. C. Gopalakrishnan, D. Shankaranarayan, L. Kameswaran, and S. Natarajan, *Indian J. Med. Res.* **69**, 513 (1979).
80. C. Gopalakrishnan, D. Shankaranarayan, and L. Kameswaran, *Indian J. Med. Res.* **71**, 940 (1980).
81. M. R. Atkinson, G. R. Donaldson, and A. W. Murray, *Biochem. Biophys. Res. Commun.* **31**, 104 (1968).
82. M. T. Huang and A. P. Grollman, *Mol. Pharmacol.* **8**, 538 (1972).
83. N. Kumar, Ph.D. thesis, University of Wollongong, N.S.W., Australia, 1983.
84. P. Grant, L. Sanchez, and A. Jimenez, *J. Bacteriol.* **120**, 1308 (1974).
85. H. Dolz, D. Vazquez, and A. Jimenez, *Biochemistry* **21**, 3181 (1982).
86. R. S. Gupta and L. Simonovitch, *Biochemistry* **16**, 3209 (1977).
87. B. D. Benjamin, M. R. Heble, and M. S. Chadha, *Z. Pflanzenphysiol.* **92**, 77 (1979).
88. B. D. Benjamin and N. B. Mulchandani, *Planta Med.* **23**, 394 (1973).
89. N. Entner and A. P. Grollman, *J. Protozool.* **20**, 160 (1973).
90. C. K. Bradsher and H. Berger, *J. Am. Chem. Soc.* **79**, 3287 (1957).
91. C. K. Bradsher and H. Berger, *J. Am. Chem. Soc.* **80**, 930 (1958).
92. P. Marchini and B. Belleau, *Can. J. Chem.* **36**, 581 (1958).
93. T. R. Govindachari, M. V. Lakshmikantham, N. V. Nagarajan, and B. R. Pai, *Tetrahedron* **4**, 311 (1958).
94. B. Chauncy, E. Gellert, and K. N. Trivedi, *Aust. J. Chem.* **22**, 427 (1969).
95. D. O. Shah and K. N. Trivedi, *Indian J. Chem.* **15B**, 599 (1977).
96. S. Takano, K. Yuta, and K. Ogasawara, *Heterocycles* **4**, 947 (1976).
97. T. R. Govindachari, M. V. Lakshmikantham, and S. Rajadurai, *Chem. Ind.,* **1960**, 664.
98. T. R. Govindachari, M. V. Lakshmikantham, and S. Rajadurai, *Tetrahedron* **14**, 284 (1961).
99. B. Chauncy and E. Gellert, *Aust. J. Chem.* **23**, 2503 (1970).
100. B. Chauncy and E. Gellert, *Aust. J. Chem.* **22**, 993 (1969).
101. R. B. Herbert and C. J. Moody, *J. Chem. Soc. Chem. Commun.* **1970**, 121.
102. R. F. Borch, *Tetrahedron Lett.* **1968**, 61.
103. R. H. Herbert, F. B. Jackson, and I. T. Nicholson, *J. Chem. Soc. Chem. Commun.* **1976**, 450.
104. A. J. Liepa and R. E. Summons, *J. Chem. Soc. Chem. Commun.* **1977**, 826.
105. J. E. Cragg, R. B. Herbert, F. B. Jackson, C. J. Moody, and I. T. Nicholson, *J. Chem. Soc. Perkin I.* **1982**, 2477.
106. V. K. Mangla and D. S. Bhakuni, *Indian J. Chem.* **19B**, 748 (1980).
107. V. K. Mangla and D. S. Bhakuni, *Tetrahedron* **36**, 2489 (1980).
108. W. B. Jacoby and J. Fredericks, *J. Biol. Chem.* **234**, 2145 (1959).
109. N. K. Hart, S. R. Johns, and J. A. Lamberton, *Aust. J. Chem.* **25**, 817 (1972).

110. M. Tanaka and I. Iijima, *Tetrahedron* **29**, 1285 (1973).

111. W. Bradley and R. Robinson, *J. Chem. Soc.* **1928**, 1310.

112. D. S. Bhakuni and A. N. Singh, *Tetrahedron* **35**, 2365 (1979).

113. D. S. Bhakuni and P. K. Gupta, *Indian J. Chem.* **21B**, 393 (1982).

114. G. G. Trigo, E. Galvez, and M. M. Sollhuber, *J. Heterocycl. Chem.* **16**, 1625 (1979).

115. J. M. Paton, P. L. Pauson, and T. S. Stevens, *J. Chem. Soc.* **1969**, 1309.

116. T. Kametani, Y. Satoh, M. Takemura, Y. Ohta, and M. Ihara, *Heterocycles* **5**, 175 (1976).

117. S. M. Weinreb and J. I. Levin, *Heterocycles* **12**, 949 (1979).

118. S. M. Weinreb, N. A. Khatri, and J. Shringarpure, *J. Am. Chem. Soc.* **101**, 5073 (1979).

119. H. F. Schmitthenner and S. M. Weinreb, *J. Org. Chem.* **45**, 3372 (1980).

120. N. A. Khatri, H. F. Schmitthenner, J. Shringarpure, and S. M. Weinreb, *J. Am. Chem. Soc.* **103**, 6387 (1981).

121. E. J. Corey and J. W. Suggs, *Tetrahedron Lett.* **1975**, 2647.

122. W. S. Johnson, L. Werthemann, W. R. Bartlett, T. J. Brocksom, T. Li, D. J. Faulkner, and M. R. Peterson, *J. Am. Chem. Soc.* **92**, 741 (1970).

123. M. Iwao, M. Watanabe, S. O. de Silva, and V. Snieckus, *Tetrahedron Lett.* **22**, 2349 (1981).

124. V. Snieckus, *Heterocycles* **14**, 1649 (1980).

125. L. Faber and W. Wiegrebe, *Helv. Chim. Acta* **56**, 2882 (1973).

126. L. Faber and W. Wiegrebe, *Helv. Chim. Acta* **59**, 2201 (1976).

127. T. Iwashita, M. Suzuki, T. Kusumi, and H. Kakisawa, *Chem. Lett.* **1980**, 383.

128. R. Huisgen, R. Grashey, H. Seidl, and H. Hauk, *Chem. Ber.* **101**, 2559 (1968).

129. H. Iida and C. Kibayashi, *Tetrahedron Lett.* **22**, 1913 (1981).

130. H. Iida, M. Tanaka, and C. Kibayashi, *J. Chem. Soc. Chem. Commun.* **1983**, 271.

131. H. Iida, Y. Watanabe, M. Tanaka, and C. Kibayashi, *J. Org. Chem.* **49**, 2412 (1984).

132. T. R. Govindachari and N. Viswanathan, *Tetrahedron* **26**, 715 (1970).

133. R. V. Stevens and Y. Luh, *Tetrahedron Lett.* **1977**, 979.

134. J. H. Russel and H. Hunziker, *Tetrahedron Lett.* **1969**, 4035.

135. T. F. Buckley, III, and H. Rapoport, *J. Org. Chem.* **48**, 4222 (1983).

136. T. F. Buckley, III, and H. Rapoport, *J. Am. Chem. Soc.* **102**, 3056 (1980).

137. T. F. Buckley, III, and H. Rapoport, *J. Am. Chem. Soc.* **103**, 6157 (1981).

138. T. Poettinger and W. Wiegrebe, *Arch. Pharm.* **314**, 240 (1981).

139. N. B. Mulchandani, S. S. Iyer, and L. P. Badheka, *Phytochemistry* **8**, 1931 (1969).

140. N. B. Mulchandani, S. S. Iyer, and L. P. Badheka, *Phytochemistry* **10**, 1047 (1971).

141. N. B. Mulchandani, S. S. Iyer, and L. P. Badheka, *Phytochemistry* **15**, 1697 (1976).

142. A. J. Clarke and P. J. G. Mann, *Biochem. J.* **71**, 596 (1959).

143. P. Houdewind, J. C. L. Armande, and U. K. Pandit, *Tetrahedron Lett.* 1974, p. 591.

144. D. Burrows and E. P. Burrows, *J. Org. Chem.* **28**, 1180 (1963).

145. K. Kratzl and G. E. Micksche, *Monatsh. Chem.* **94**, 434 (1963).

146. Y. Ban and T. Oischi, *Chem. Pharm. Bull. (Japan)* **6**, 574 (1958).

147. R. B. Herbert, F. B. Jackson, and I. T. Nicholson, *J. Chem. Soc. Chem. Commun.* **1976**, 865.

148. R. B. Herbert and F. B. Jackson, *J. Chem. Soc. Chem. Commun.* **1977**, 955.

149. D. S. Bhakuni and V. K. Mangla, *Tetrahedron* **37**, 401 (1981).

150. G. W. Kirby and L. Ogunkoya, *J. Chem. Soc.* **1965**, 6914.

151. R. B. Herbert, F. B. Jackson, and I. T. Nickolson, *J. Chem. Soc. Perkin Trans. I* **1984**, p. 825.

152. R. F. Borch, M. D. Bernstein, and H. D. Durst, *J. Am. Chem. Soc.* **93**, 2897 (1971).

153. K. Y. Zee-Cheng and C. C. Cheng, *J. Med. Chem.* **12**, 157 (1969).

154. K. Y. Zee-Cheng and C. C. Cheng, *J. Pharm. Sci.* **59**, 1630 (1970).

155. E. Gellert, R. Rudzats, R. E. Summons, and B. R. Worth, *Aust. J. Chem.* **24**, 843 (1971).

156. K. Y. Zee-Cheng, W. H. Nyberg, and C. C. Cheng, *J. Heterocycl. Chem.* **9**, 805 (1972).

157. C. S. Menon, K. Y. Zee-Cheng, and C. C. Cheng, *J. Heterocycl. Chem.* **14**, 905 (1977).

158. G. De Martino, M. Scalzo, S. Massa, and R. Giuliano, *Il Farmaco-Ed. Sci.* **28**, 976 (1973).

159. M. Sollhuber, M. T. Grande, G. G. Trigo, D. Vazquez, and A. Jimenez, *Curr. Microbiol.* **4**, 81 (1980).

160. S. P. Gaur, P. C. Jain, and N. Anand, *Indian J. Chem.* **21B**, 46 (1982).

161. S. J. Brois, *J. Org. Chem.* **27**, 3532 (1962).

162. R. C. Srimal and B. N. Dhawan, *Indian J. Pharm.* **34**, 172 (1972).

163. E. Gellert, N. Kumar, and D. Tober, *Aust. J. Chem.* **36**, 157 (1983).

164. E. Gellert and N. Kumar, *Aust. J. Chem.* **37**, 819 (1984).

165. O. Grummitt and A. C. Buck, *J. Am. Chem. Soc.* **65**, 295 (1943).

166. H. Leonard, *J. Am. Chem. Soc.* **71**, 1392 (1949).

167. K. Fries and K. Schimmelschmidt, *Chem. Ber.* **1925**, 2835.

168. L. Velluz, G. Amiard, and R. Heymes, *Bull. Soc. Chim.* **1954**, 1012.

169. L. Velluz, J. Anatel, and G. Amiard, *Bull. Soc. Chim.* **1954**, 1449.

170. L. Velluz, G. Amiard, and R. Heymes, *Bull. Soc. Chim.* **1955**, 201.

171. W. J. Johnson and M. G. Glenn, *J. Am. Chem. Soc.* **71**, 1092 (1948).

172. G. R. Proctor and R. H. Thomson, *J. Chem. Soc.* **1957**, 2302.

173. D. J. Cram and F. A. Elhafez Abd, *J. Am. Chem. Soc.* **74**, 5828 (1952).

174. K. K. Bhutani, M. Ali, and C. K. Atal, *Indian J. Pharm. Sci.* **46**, 47 (1984).

175. K. K. Bhutani, M. Ali, and C. K. Atal, *Phytochemistry* **23**, 1765 (1984).

176. G. R. Pettit, A. Goswami, G. M. Cragg, J. M. Schmidt, and Ji-C. Zou, *J. Nat. Prod.* **47**, 913 (1984).

Chapter Three

The Aporphinoid Alkaloids of the Annonaceae

André Cavé and Michel Leboeuf
Laboratoire de Pharmacognosie
U.A. 496 C.N.R.S.
Faculté de Pharmacie
92290 Châtenay-Malabry, France

Peter G. Waterman
Phytochemistry Research Laboratories
Department of Pharmacy
University of Strathclyde
Glasgow G1 1XW, Scotland, United Kingdom

CONTENTS

---

1. INTRODUCTION

The Annonaceae is one of the largest of the plant families of the Magnoliales (Annonales, Ranales), which is generally considered to make up the most archaic order of extant dicotyledonous plants [1,2]. The family is distributed throughout the tropics, in the Old World occurring predominantly as trees or lianas in humid, lowland, forest regions, but in the Neotropics found most commonly as trees or shrubs growing in drier, more open environments [3]. In all, the family Annonaceae has been estimated to contain in excess of 2000 species and about 120 genera [4].

The phytochemistry of the Annonaceae has recently been the subject of an exhaustive review by Leboeuf et al. [5], and the biogenetic and structural diversity of the secondary metabolites listed therein clearly demonstrate that the family has considerable potential as a source of novel compounds. The review of Leboeuf et al. [5] also demonstrated that, as in many other families of the Magnolialean (Ranalean) alliance [6,7], a central theme in the secondary metabolic profile of the Annonaceae is the widespread production of phenylalanine- and tyrosine-derived alkaloids formed via 1-benzyltetrahydroisoquinoline (1-btiq) intermediates. The purpose of this chapter is to review aspects of the chemistry and distribution of the major group of 1-btiq-derived alkaloids, the aporphinoids,* within the Annonaceae. For the purposes of this discussion we have divided the aporphinoids as follows:

(a)   Proaporphines (**1**). Numerically a minor group, but with an important role as precursors of some of the other aporphinoid types.

(b)   Aporphines (**2**). The major group, which we have, for convenience, divided into four subgroups. The *aporphines sensu stricto* are taken to include all those alkaloids based on **2** that exhibit *O*-substitution only on ring A or on rings A and D. All levels of *N*-substitution are

---

* For the purpose of this chapter the aporphinoids are considered to include all those alkaloids with a true aporphine nucleus (structure **2**) and alkaloids based on that nucleus but with a modified skeleton. A complete listing of aporphinoids, except proaporphines, and their known botanical sources is available in the literature [8–10].

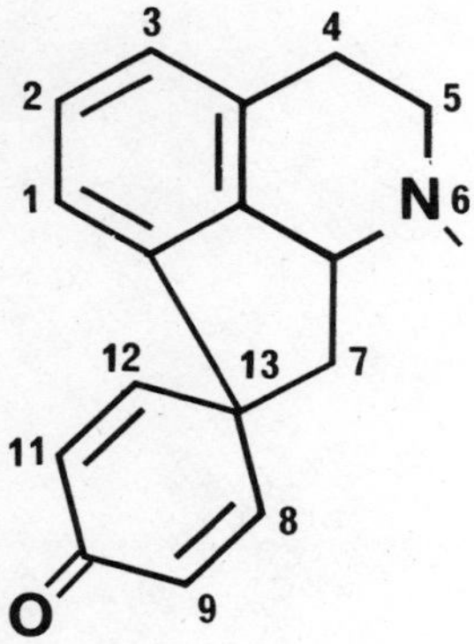

1

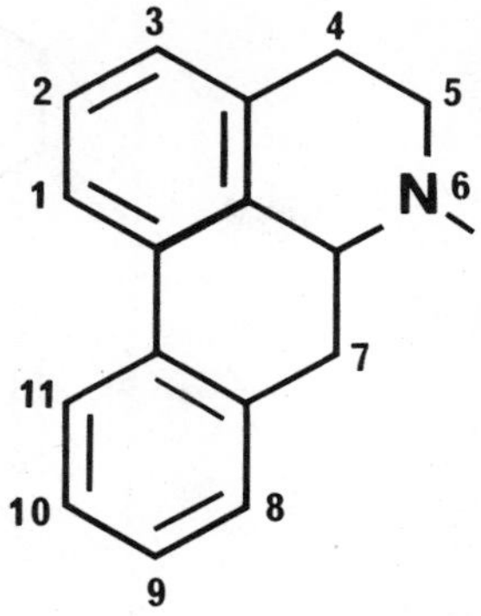

2

included, from the secondary amines (noraporphines) through to *N,N*-dimethyl and *N*-methyl-*N*-oxide quaternary compounds and *N*-formyl and *N*-carbamoyl substituents. The *dehydroaporphines* are a small subgroup distinguished from the aporphines *s.s.* by an additional center of unsaturation between C(6a) and C(7). It should be noted that the term dehydroaporphine is restricted to alkaloids with C(6a)/C(7) unsaturation; for other alkaloids with additional unsaturation the position must be specified, for example, 4,5-dehydroaporphines. *7-Alkylaporphinoids*, in which the alkyl group always originates as a methyl substituent, also invariably contain additional centers of unsaturation, sometimes as a dehydroaporphine or as a 6,6a-dehydroaporphine, or in some cases as a 4,5,6,6a-tetradehydroaporphine with a fully aromatized ring B. The final subgroup is the *7-oxygenated* or, occasionally, *4,7-dioxygenated* aporphines. These intergrade with the preceding subgroup inasmuch as they include some alkaloids with geminal methyl and hydroxyl substituents at C(7). None of this group is a dehydroaporphine.

(c)   Oxoaporphines (**3**). Alkaloids with an oxo substituent at C(7) and a

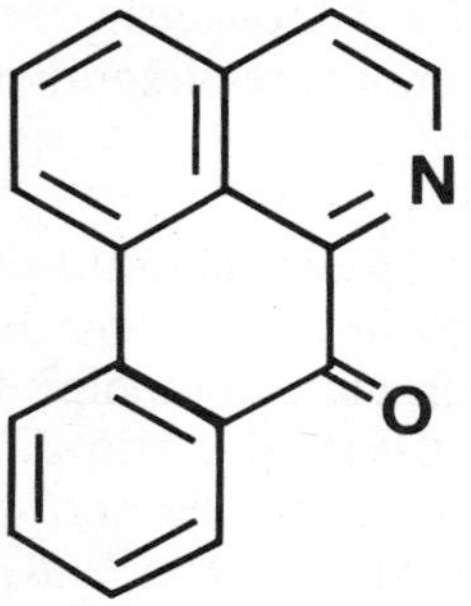

3

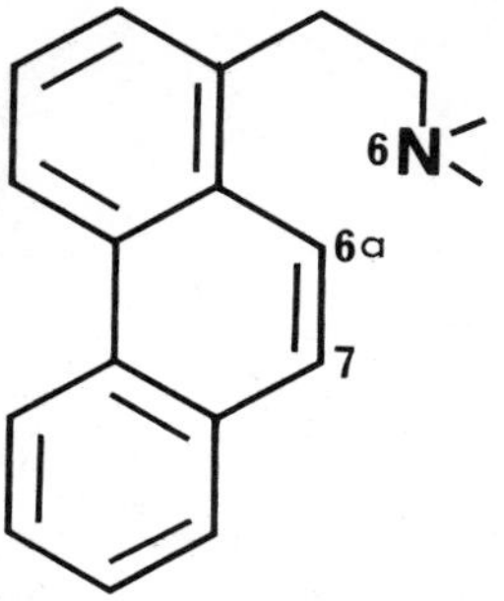

4

fully aromatic aporphine skeleton. Some of the oxoaporphines are among the most widely distributed alkaloids of the Annonaceae.

(d) Ring B Oxoaporphines. A very small group of alkaloids with a 5-oxo or a 4,5-dioxo substitution pattern in ring B.

(e) Dimeric aporphinoids. Recently several interesting dimeric aporphinoids have been characterized from the Annonaceae. The majority are based on an oxoaporphine linked through C(4) to C(7) of a 4,5,6a,7-tetradehydroaporphine. The other dimers are symmetrical, identical monomers linked through C(7).

(f) Phenanthrenes (**4**). Alkaloids of aporphine origin that have undergone fission between N(6) and C(6a) to leave a phenanthrene nucleus with an ethylamine side chain, the nitrogen carrying one or two methyl substituents. It should be noted that for this review the biogenetic numbering system advocated by Leboeuf and Cavé [11] has been adopted because it permits direct comparisons with the aporphine precursors. Technically, numbering of the phenanthrenes should commence at C(3a) on **4**.

(g) Miscellaneous, degraded aporphinoids. Among the most interesting of the alkaloids of the Annonaceae are a small group assumed to originate from aporphinoid precursors, but that have undergone quite extensive degradation.

Although the above make up the majority of all alkaloids isolated from the Annonaceae there are a number of others that are not considered in this review. These include the simple isoquinolines, 1-btiq monomers (**5**), numerous O-linked and some C-linked 1-btiq dimers, the 1-btiq-derived tetrahydroprotoberberines (**6**) and protoberberines (**7**), and the morphinanedienones pallidine (**8**, R = H) and sebiferine (**8**, R = Me). These have been reviewed by Leboeuf et al. [5] and with the exception of the protoberberine–

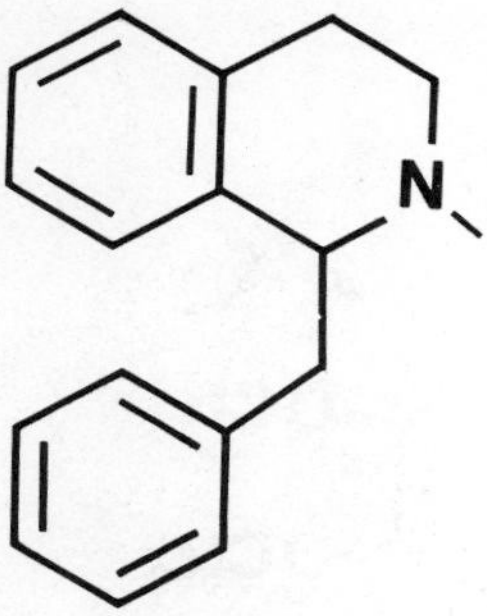

5 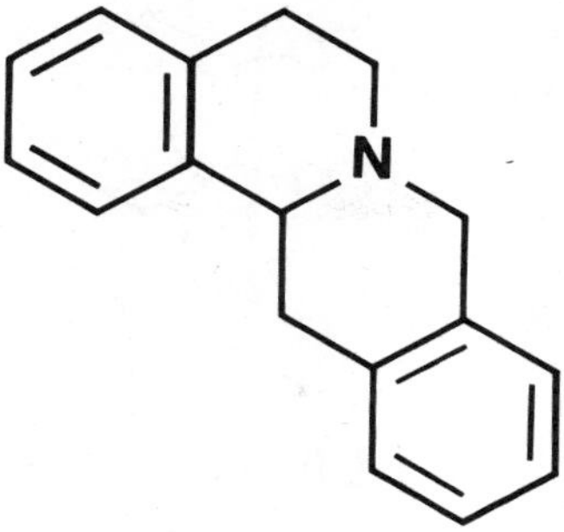 **6**

styrene adduct staudine (**9**) and the recently isolated [12] and highly unusual
C(11)/C(11′) *C*-linked bis-1-btiq type alkaloids such as antioquine (**10**), se-
cantioquine (**11**), and pisopowine (**12**), they have not shown the originality
found among the aporphinoids.

In addition to alkaloids originating from phenylalanine and tyrosine, a
number of tryptophan and indole compounds have also been reported in the
past few years. These include the unique pyrimidine-β-carbolines (**13**, R =
H and R = OMe) from *Annona muricata* [5]; indolosesquiterpenes such as
polyalthenol (**14**) and greenwayodendrin-3-one (**15**) from African species of
Polyalthia or *Greenwayodendron*, which have been reviewed by Waterman
[13] in a previous volume of this series, and a number of *C*-benzylated indole
and dihydroindol-3-one alkaloids from *Uvaria angolensis* [14], for example,

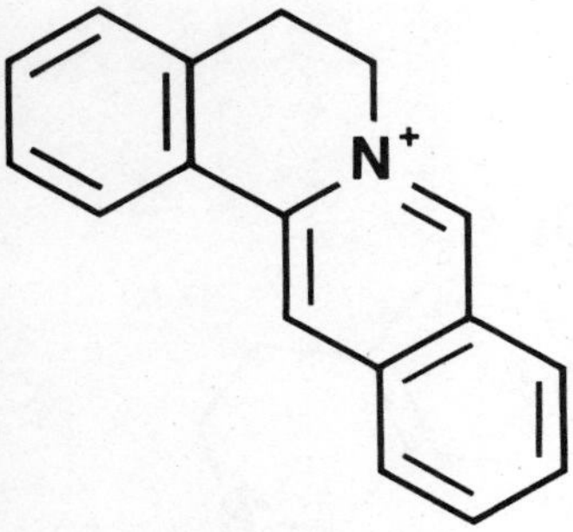

7

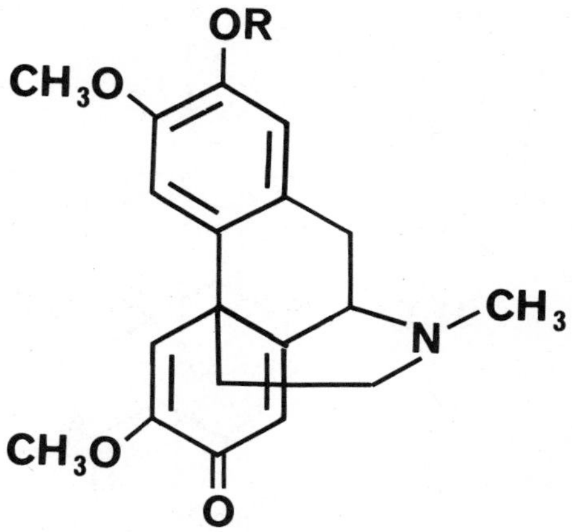

8

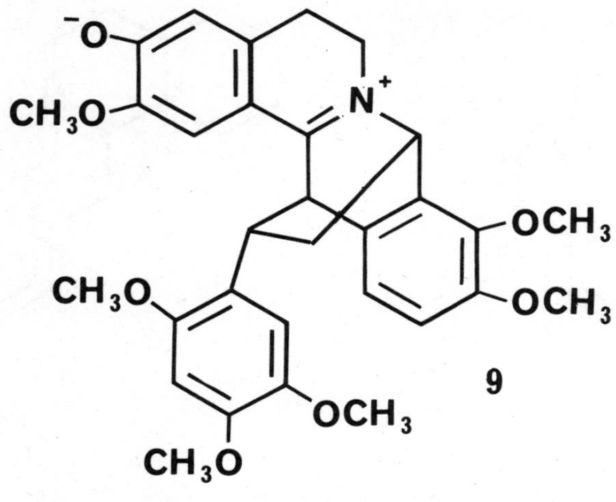

O
CH3O
N+
OCH3
OCH3
CH3O
OCH3
CH3O

9

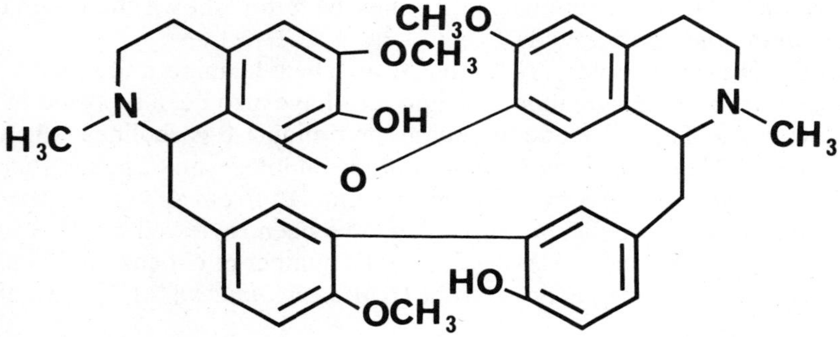

CH3O
OCH3
N
OH
N
H3C
CH3
O
HO
OCH3

10

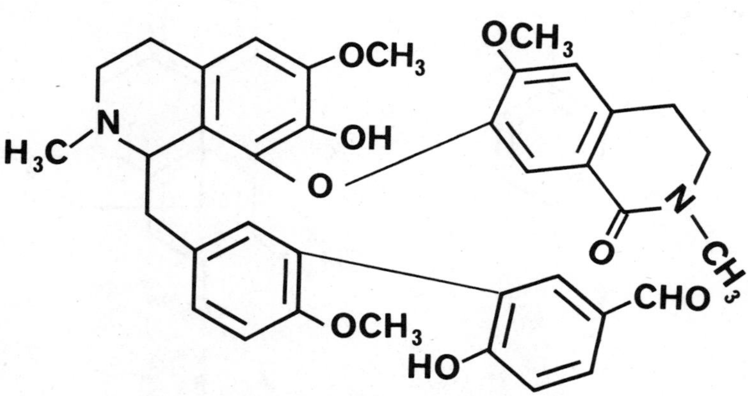

OCH3
OCH3
N
OH
H3C
O
N
O
CH3
CHO
OCH3
HO

11

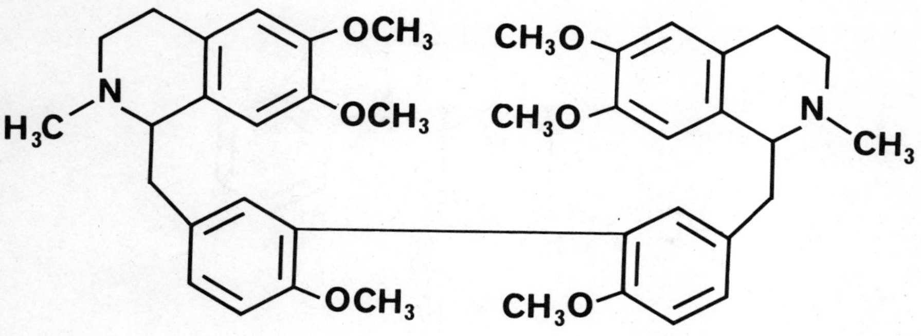

12

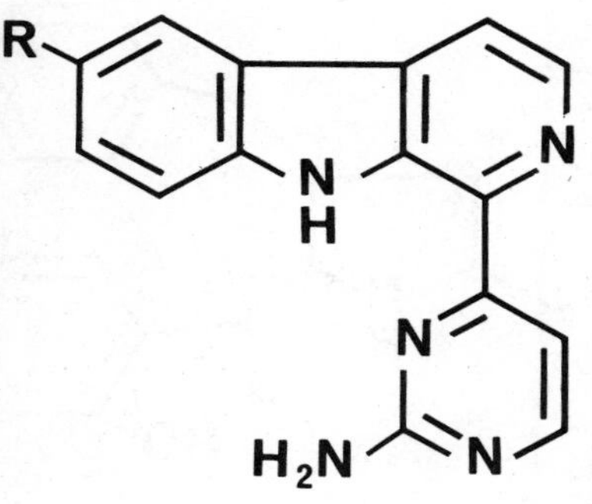

13

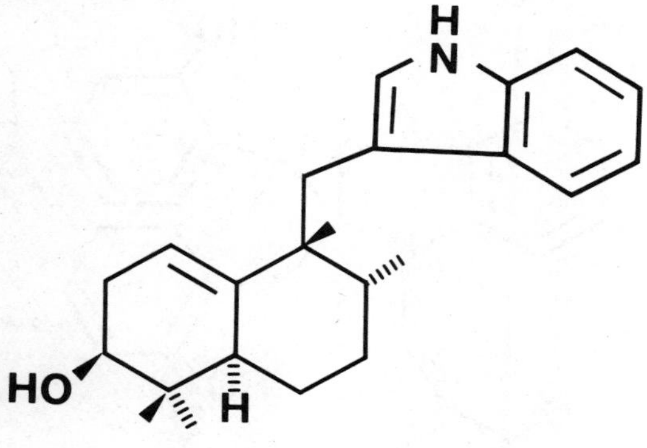

14

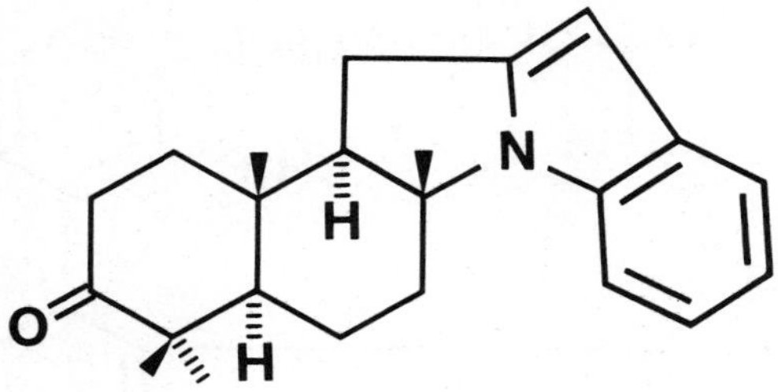

15

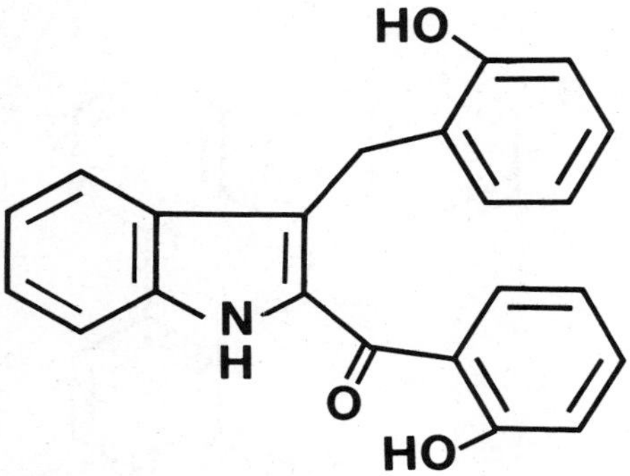

16

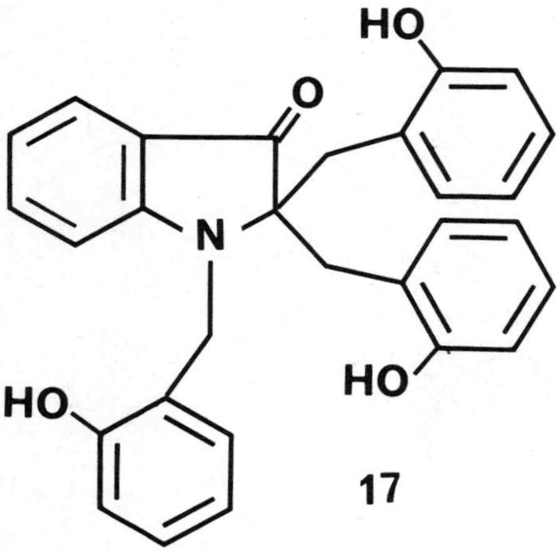

17

uvarindole-A (16) and uvarindole-C (17). Many of these indole derivatives are either nonbasic or only weakly basic and a more extensive search of nonpolar extracts of Annonaceae could well reveal a wider occurrence than present data suggest.

2. BIOSYNTHETIC PATHWAYS TO APORPHINOIDS OF THE ANNONACEAE: AN OVERVIEW

A striking feature of the annonaceous aporphinoids is the wide range of O-substitution patterns that are found. Both C(1) and C(2) are invariably substituted as a result of the comparable substitution of the 1-btiq precursors. Some positions, notably C(3) and C(7), are presumed to obtain O-substitution at a later stage, by oxidation of a preformed aporphinoid nucleus, although this may not be the case with the other occasionally oxygenated positions, C(4) and C(5). The oxygenation pattern found on ring D is considered, on the other hand, to be a function of the actual mechanism by which biosynthesis of the aporphine nucleus is achieved. To date the following ring D patterns have been noted in the Annonaceae:

(a) No substitution.
(b) Monosubstitution, at any one of the four possible positions, although C(8) substitution is rare.
(c) Disubstitution, usually C(9)/C(10) or C(10)/C(11), but examples of C(8)/C(9) and C(9)/C(11) have been recorded.

The trisubstituted ring D, at C(8), C(9), and C(10), found in *Ocotea* species (Lauraceae) has not yet been recorded in the Annonaceae. In this section the possible routes to those patterns of substitution found in the Annonaceae are discussed, with the exception of the 1,2,9,11 pattern (see Section 5.2).

The precursors of the aporphine skeleton are clearly established as 1-btiq alkaloids [15–19]. It is equally certain that the 1-btiq precursors originate in turn from an interaction between dopamine and a deaminated dopamine-derived 6C–2C or 6C–3C aldehyde or acid [15–17]. Recently an enzyme-catalyzing coupling between dopamine and 3,4-dihydroxyphenylacetaldehyde to give the 1-btiq norlaudanosoline (18) has been characterized [20]. For the corresponding trisubstituted 1-btiq coclaurine (19, R = Me), which occurs quite widely in the Annonaceae, replacement of the 3,4-substituted phenylacetaldehyde with the 4-hydroxy compound has been found to lead to the corresponding O-demethylcoclaurine (19, R = H) [21].

Thus the route from amino acid to 1-btiq precursor is well defined and, likewise, the role of 1-btiq as an intermediary in the formation of aporphinoids is well supported by numerous *in vivo* and *in vitro* studies [18,19]. It is also generally agreed that the driving force behind the conversion of the

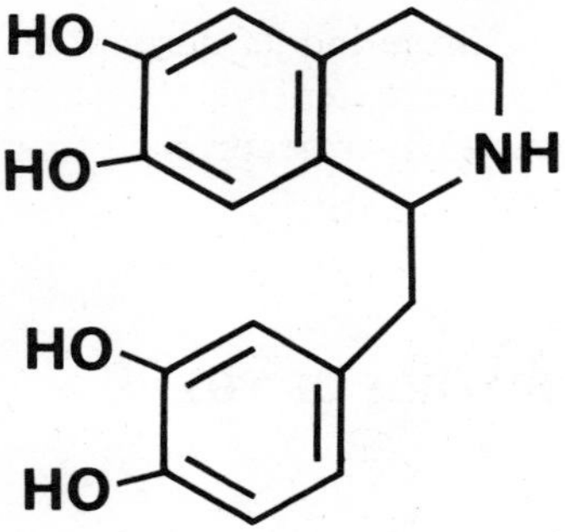

18

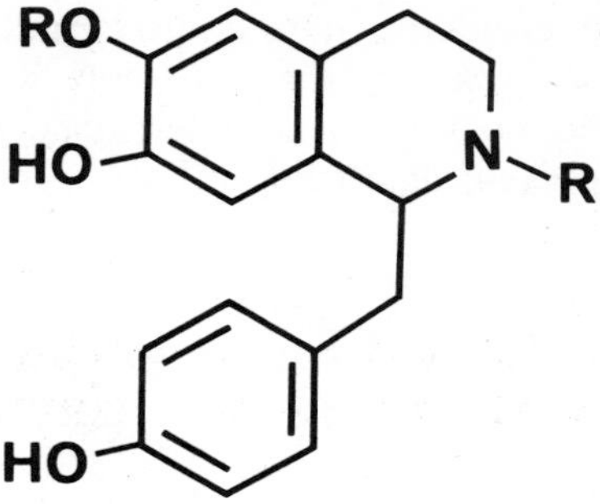

19

tricyclic 1-btiq into the tetracyclic aporphine is an oxidative coupling achieved *via* an intramolecular diradical that arises in a partially methylated 1-btiq [15,18]. Unfortunately, despite this seemingly well-resolved situation there remain a number of contentious points concerning the cyclization mechanism and intermediates, and for many of the substitution patterns of ring D more than one route of biosynthesis appears feasible.

The most simple pathway (Scheme 1) envisages the direct oxidative coupling of reticuline (**20**) in the *bis*-dienone radical form. Hence the two radicals can be *ortho–ortho* to the initiating ring A and D phenol subsituents, in which case they would result in a 1,2,10,11-substitution pattern, or *ortho–para*, with the resulting aporphine having 1,2,9,10 substitution. Third and fourth alternatives arise from a *para* radical in ring A and an *ortho* or *para* radical in ring D, which leads to morphinanedienones with differing substitution patterns. The pathways to the aporphines appear to be favored over those to the morphinanedienones in most taxa, including the Annonaceae.

The pathways of Scheme 1 can certainly be invoked as routes to some 1,2,9,10- and 1,2,10,11-substituted aporphines [19], but do not rationalize

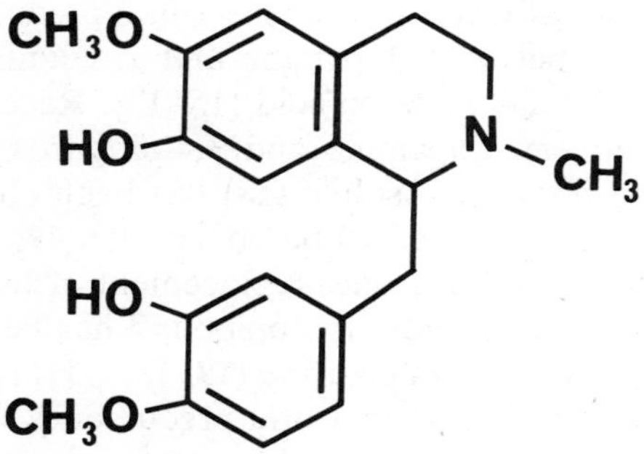

20

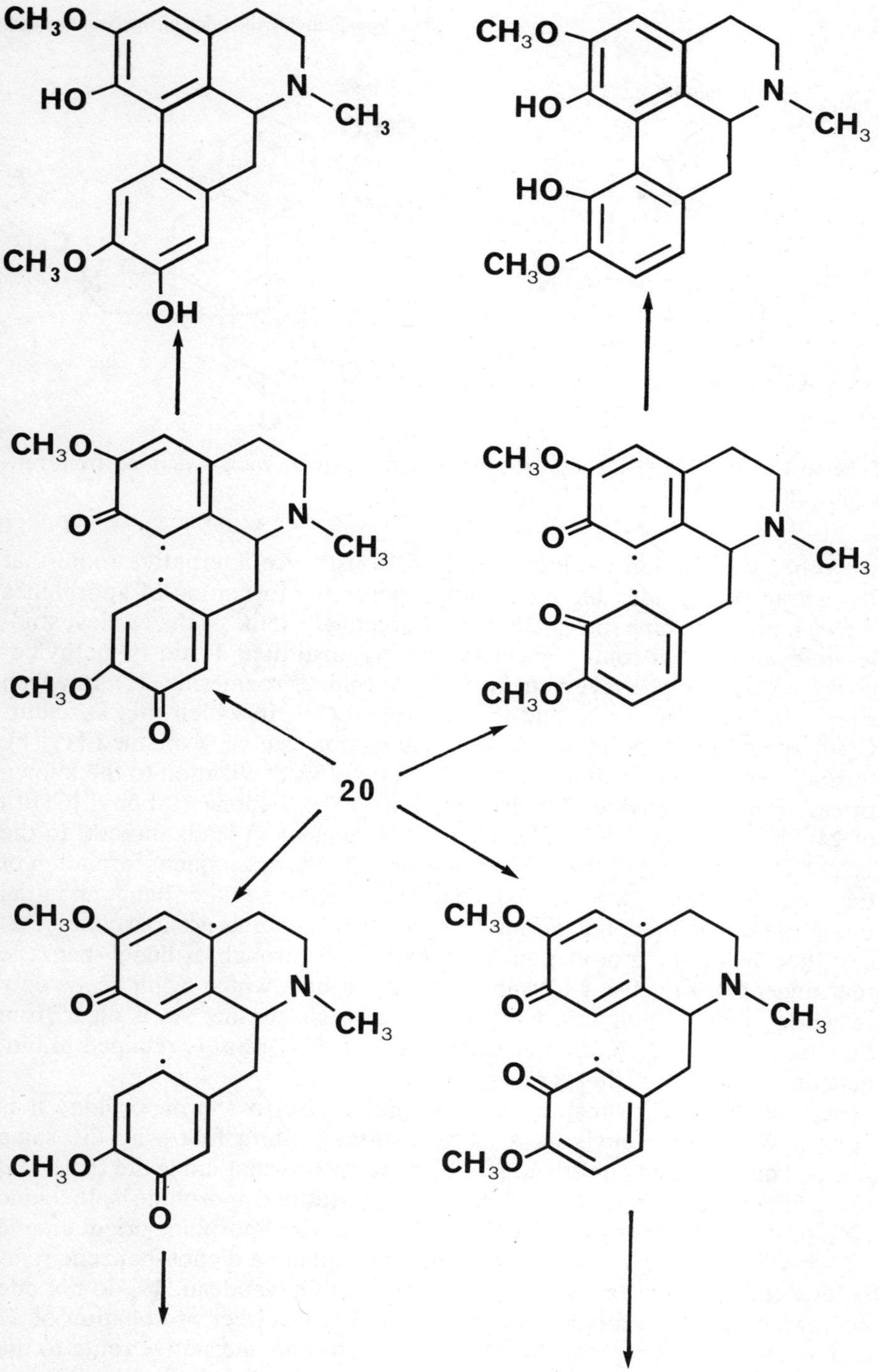

Scheme 1. *(continued on next page)*

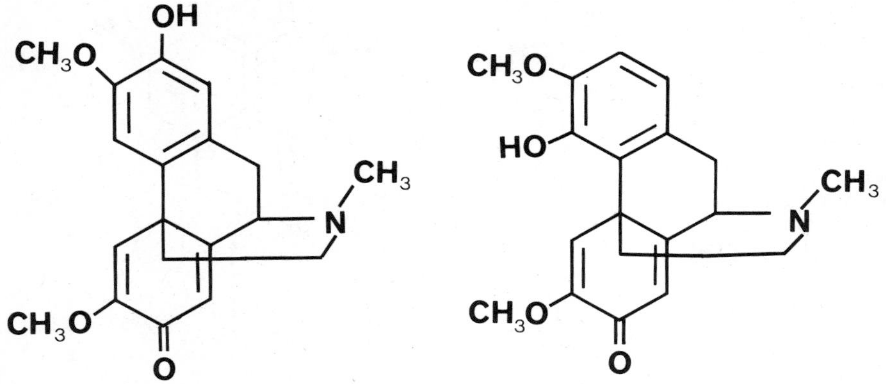

Scheme 1. Possible cyclization routes for a bis-dienone radical derived from reticuline (**20**).

the other substitution patterns noted previously. An alternative route that has attracted considerable attention proposes the formation of aporphines through proaporphine intermediates. Interestingly some of the earliest studies relevant to this route concerned the trisubstituted 1-btiq *N*-methylcoclaurine (**21**) and its conversion into the aporphines roemerine (**22**), in which ring D lacks oxygenation, and mecambroline (**23**), in which ring D retains C(10) substitution (Scheme 2). This conversion can be explained [19] by formation of the *ortho–para* diradical of **21** and its cyclization to the known proaporphine glaziovine (**24**). Protonation of the dienone carbonyl [C(10)] of **24** with consequent dienone–phenol rearrangement leads directly to the 1,2,10-substitution pattern of **23**, requiring only the subsequent formation of the methylenedioxy ring to give mecambroline. On the other hand, an initial reduction of the dienone carbonyl to the corresponding proaporphinol (**25**) can then lead, *via* protonation, to dehydration through a dienol–benzene rearrangement to leave a 1,2-substituted aporphine which again needs only formation of the methylenedioxy ring to give roemerine. As is clear from Scheme 2, the configuration at C(6a) [C(1) of the 1-btiq] is retained in biogenetic sequences of this type.

As well as the trisubstituted 1-btiq giving rise to a proaporphine it is equally possible to envisage a tetrasubstituted 1-btiq following the same route. This has been confirmed [19] by experiments that implicate the 1-btiq orientaline (**26**) as a precursor of the 1,2,11-substituted aporphine isothebaine (**28**), presumably through the intermediacy of the proaporphine orientalinone (**27**) and then through the proaporphinol (cf. **25**) and a dienol–benzene rearrangement (Scheme 3). Although Shamma and Guinaudeau [19] do not cite any example where this has yet been confirmed, direct protonation of **27** and a dienone–phenol rearrangement could offer an alternative route to the 1,2,10,11-substituted aporphines. Furthermore, cyclization of the *ortho–para* diradical of **26** could, in theory, yield **29** rather than **27** (Scheme 3).

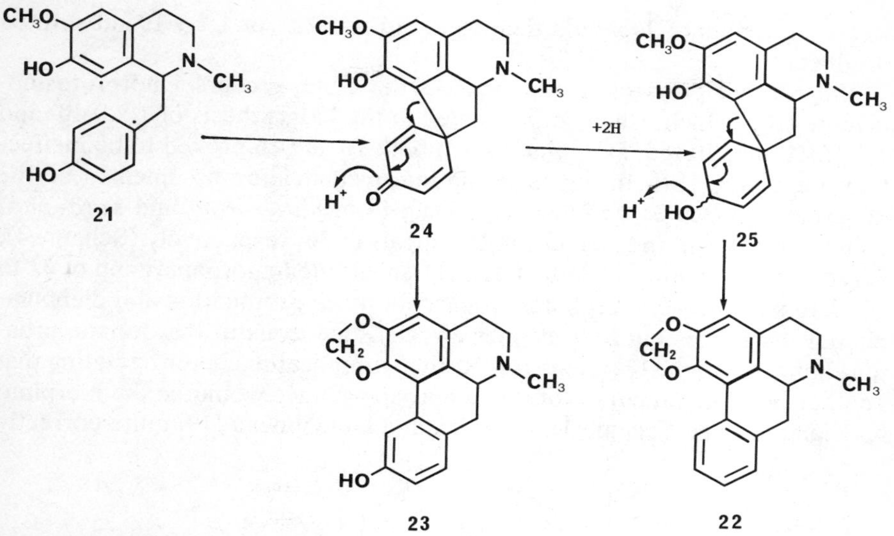

Scheme 2. Routes to 1,2- and 1,2,9-substituted aporphines *via* a proaporphine intermediate.

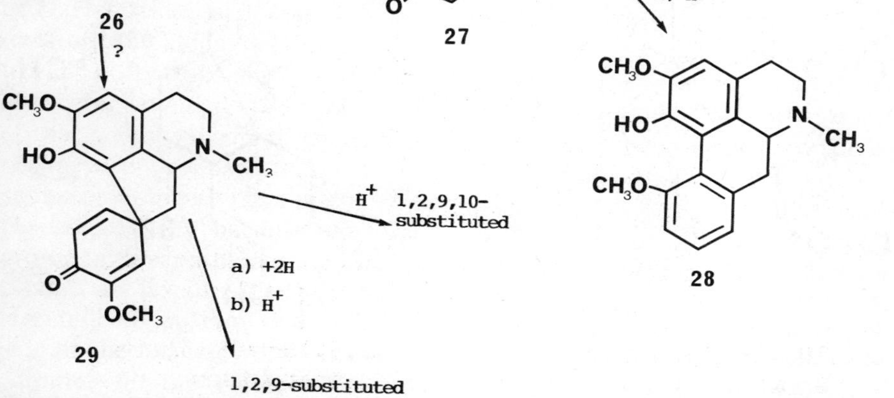

Scheme 3. Possible mechanisms for formation of 1,2,9,10-, 1,2,10,11-, 1,2,9- and 1,2,11-substituted aporphines from orientaline (**26**).

145

Rearrangements of **29** would then yield either 1,2,9- or 1,2,9,10-substituted products.

Yet another proposed route employs the 1-btiq precursor norprotosino-menine (**30**), which has been implicated in the biosynthesis of 1,2,9,10- and 1,2,10,11-substituted aporphines when both **20** and **26** proved to be ineffective precursors [19]. In this case the intermediates are assumed to be the neoproaporphines **31** and **32**, originating from *para–ortho* and *para–para* coupling between ring A and ring D radicals of **30**, respectively (Scheme 4). Direct transformation of **31** to 1,2,10,11-substituted aporphines and of **32** to 1,2,9,10-substituted aporphines would follow *via* protonation and dienone–phenol rearrangement in a manner directly analogous to that for the proaporphines. Cordell [17] points to a further complicating factor by noting that another possible precursor of the neoproaporphines would be the morphinanedienones (see Scheme 1). Shamma and Guinaudeau [19] quite correctly

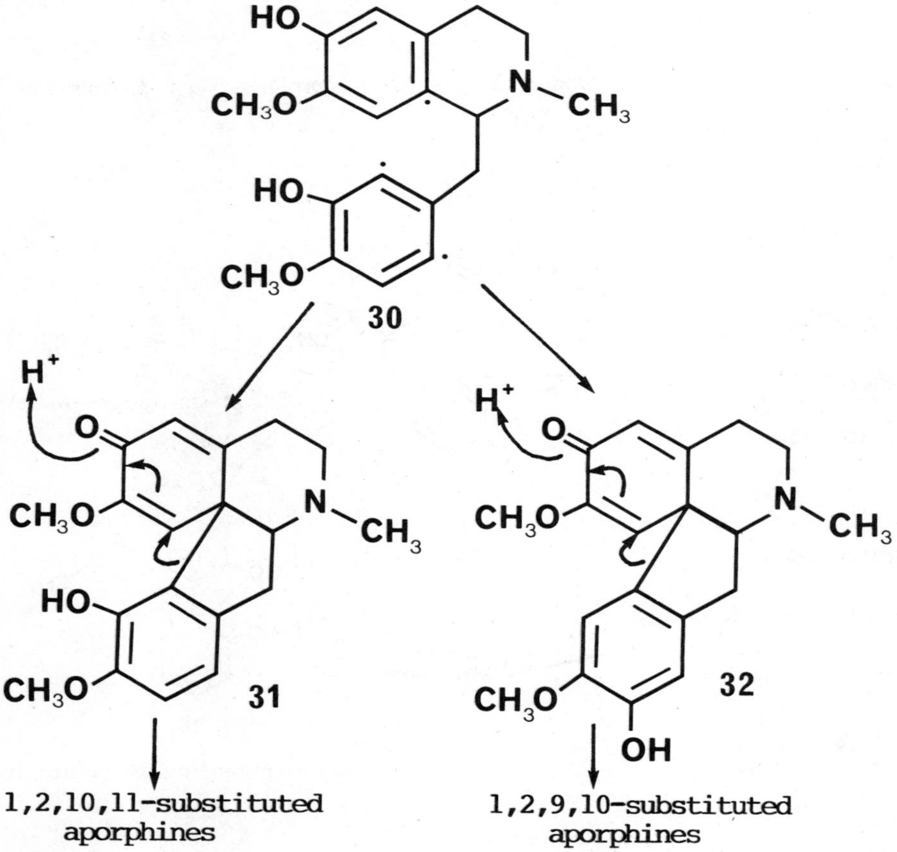

Scheme 4. Possible route to tetrasubstituted aporphines from norprotosinomenine (**30**) through neoproaporphine intermediates.

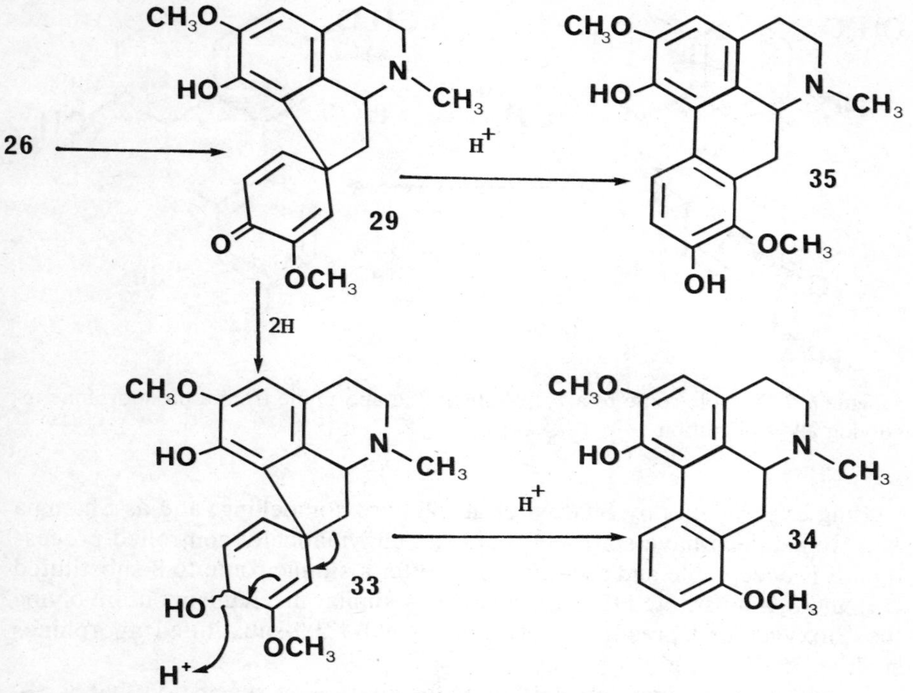

Scheme 5. Route to 1,2,8- and 1,2,8,9-substituted aporphines *via* proaporphines.

sound a cautionary note regarding the generality of the neoproaporphine pathway by pointing out that we could expect to see rearrangements involving a neoproaporphinol stage with subsequent loss of the C(2) substituent as well as the products of direct protonation. As yet, no aporphines lacking C(2) substituents have been reported to occur anywhere in the plant kingdom and neither for that matter have any alkaloids with the neoproaporphinoid skeleton.

Schemes (1–4) offer a choice of pathways to 1,2,9,10- and 1,2,10,11-substitution patterns and a route to 1,2-, 1,2,11-, 1,2,10-, and possibly 1,2,9-substituted aporphines. None of the schemes outlined offer routes to the unusual 1,2,8 or 1,2,8,9 patterns. This problem may be simply resolved. In a recent *in vivo* study Sharma et al. [22] reported that orientaline (26) was incorporated into the 8-substituted aporphine prestephanine (34), presumably through the proaporphine intermediate 29 (Scheme 5). Shamma and Guinaudeau [19] have indicated some reservations regarding the route proposed by Sharma et al. [22], being concerned both with the mechanism of the rearrangement in relation to C(6a) stereochemistry and particularly with the requirement of an alkyl migration rather than the normal aryl ring A migration; there is no precedence for this happening. The results of the

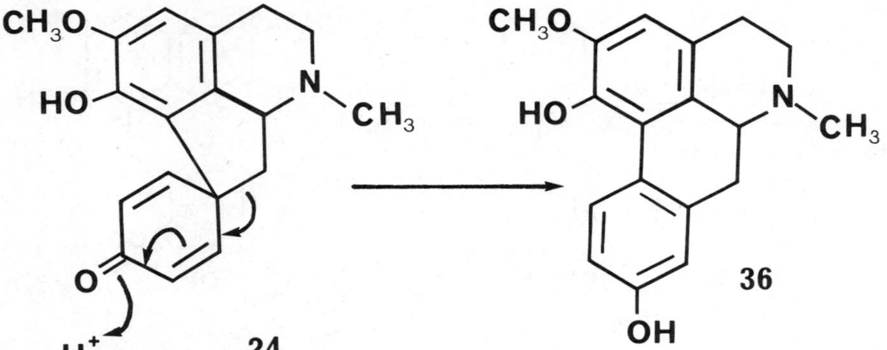

Scheme 6. Possible route to a 1,2,9-substituted aporphine from a proaporphine involving alkyl migration.

feeding experiments by Sharma et al. [22] are compelling, and as Shamma and Guinaudeau have stated, if this is an enzymatically controlled process then it is acceptable and as such would offer a simple route to 8-substituted (**34**) and 8,9-substituted (**35**) aporphines. A similar mechanism but involving the trioxygenated precursor (**24**) could yield 1,2,9-substituted aporphines such as **36** (Scheme 6).

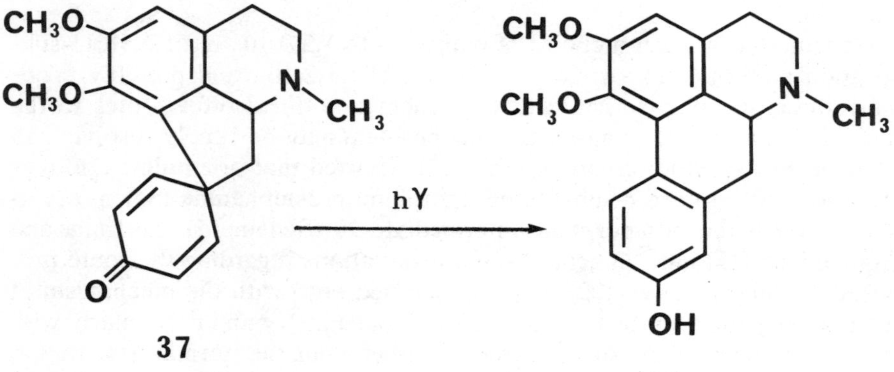

Shamma and Guinaudeau [19] offer the alternative suggestion that 8-, 9-, and 8,9-substitution patterns could possibly arise from photochemical modification of proaporphine intermediates such as **24** and **29**. There is some supporting evidence for this in the light-induced transformation of pronuciferine (**37**) into **38** in high yield, but as recognized by Shamma and Guinaudeau, the hypothesis that this is a pathway involved in normal biosynthesis must at present be considered speculative. One general argument against it is that the substitution patterns that are generated are rare but when present often assume major or at least highly significant percentages of the total extractable alkaloids. Such a pattern of distribution does not support a hypothesis that would imply random and widespread generation of these alkaloids.

There have been no *in vivo* studies on the biosynthesis of oxoaporphines but a rational sequence of events [19] would involve the stepwise oxidation of an identically substituted aporphine (**39**) through the dehydroaporphine (**40**) and then 4,5,6a,7-tetradehydroaporphine (**41**), which would be susceptible to oxidation at C(7) leading to the generally unstable 7-oxo-*N*-metho quaternary oxoaporphinium ion (**42**), which very readily loses the *N*-methyl group to give the oxoaporphine (**43**) (Scheme 7). There is evidence [23] that aporphines can undergo this process naturally on prolonged exposure to air and thus it must be considered possible that oxoaporphines can be formed without enzyme mediation. Observations made in the laboratory of two of the authors (AC and ML) suggest that the susceptibility of aporphines to oxidation depends to a considerable extent on the structure of the aporphine involved. For example, nonphenolic aporphines are much more readily oxidized than their phenolic counterparts whereas noraporphines appear to be particularly resistant to oxidation at C(7), probably because they are unable to form the quaternary *N*-methyltetradehydroaporphine intermediate (such as **42**).

The dehydroaporphine (**40**) is probably the key intermediate in the formation of 7-oxygenated and 7-alkylated compounds, the reactive electrophilic enamine group readily reacting with hydroxyl- or methyl-donating

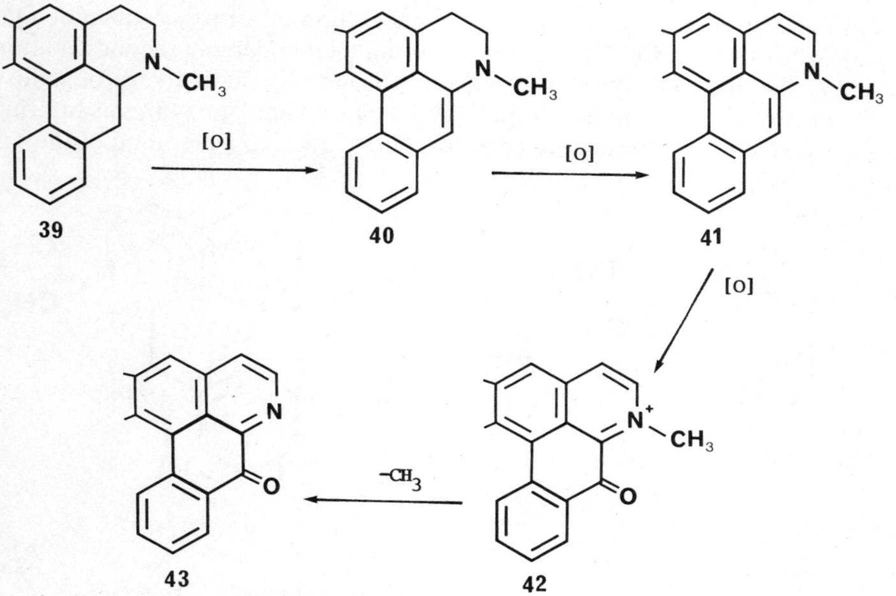

Scheme 7. Oxidation of aporphines to oxoaporphines.

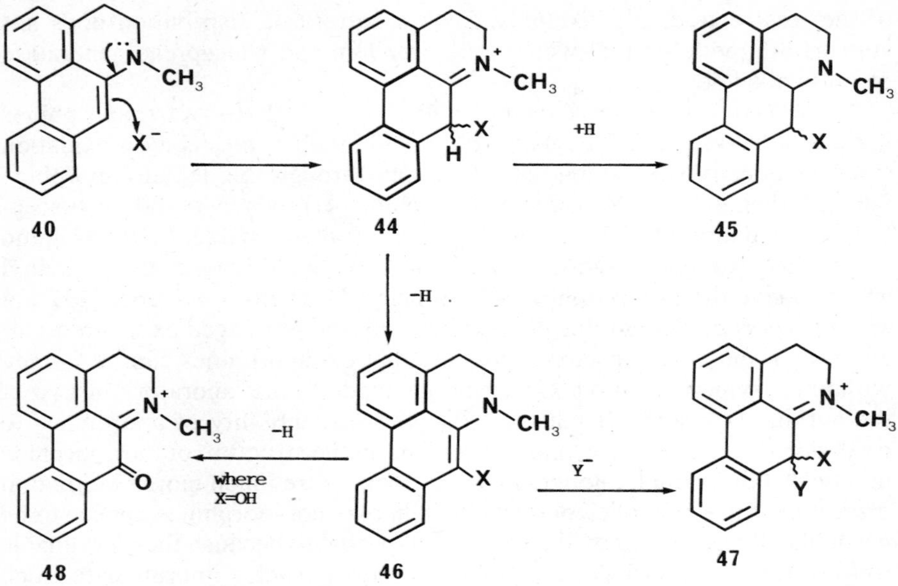

Scheme 8. General pathway to 7-substituted aporphine derivatives (X and Y = CH₃ and OH).

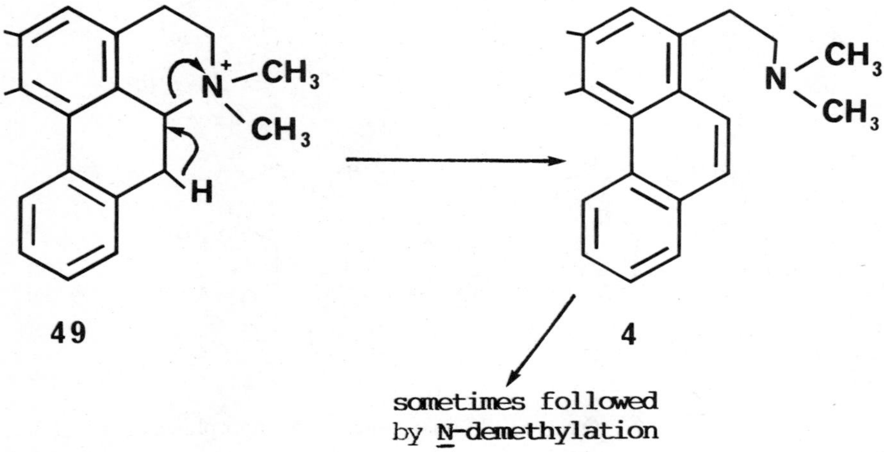

moieties leading to the corresponding 7-substituted imine (**44**). This imine can then either protonate at C(6a) to give the 7-substituted product (**45**) or can undergo another oxidation with regeneration of a further enamine (**46**) now substituted at C(7). The enamine **46** can then undergo a second addition to give the imine **47** now disubstituted at C(7) or, alternatively, through loss of a proton, the 4,5-dihydro-*N*-methyl oxoaporphinium ion (**48**), this offering a possible alternative route to oxoaporphines in which generation of the 4,5 double bond and N-demethylation are the final steps (Scheme 8). Among the 7-alkylated alkaloids of Section 5.2.3 there are many variations that can arise through the general Scheme 8 sequence. No attempt has been made here to indicate specific sequences leading to each variant.

The remaining substantial group of alkaloids to be considered, the phenanthrenes (4), offer no conceptual difficulties as they are obviously the product of a classic Hofmann elimination of the corresponding N,N-dimethyl quaternary aporphinium ion (49).

Other problems relating to biosynthetic pathways in the aporphinoids of the Annonaceae concern single or very small groups of closely related alkaloids and are discussed in the relevant parts of Section 5.

3. DISTRIBUTION OF APORPHINOID ALKALOIDS IN THE ANNONACEAE

Table 1 gives a list of all species of the Annonaceae that have been reported to contain aporphinoids to date. For each alkaloid, structural type is denoted by reference to its location in Section 5. The table also gives a list of original references for the report of each alkaloid from the species concerned, the geographical source of each species studied, and a code that is used to denote the genus in Table 10 in Section 6.

Table 1. Distribution of the Aporphinoid Alkaloids of the Annonaceae

Genus (origin)/Species[a]	Aporphinoid	Section	References
Alphonsea (Asia) ALP			
1. *A. ventricosa*	Glaucine	5.2.1	[24]
	Norglaucine	5.2.1	[24]
Anaxagorea (Amer.) ANA			
1. *A. dolichocarpa*	Anaxagoreine	5.2.4	[25]
	Asimilobine	5.2.1	[25]
2. *A. prinoides*	Anaxagoreine	5.2.4	[25]
	Asimilobine	5.2.1	[25]
Annona (Amer./Africa) ANN			
1. *A. acuminata* (Amer.)	Homomoschatoline	5.3	[26]
	(= methylmoschatoline)		
	Liriodenine	5.3	[26]
	Lysicamine	5.3	[26]
2. *A. cherimolia* (Amer.)	Anonaine	5.2.1	[27–30]
	Corytuberine	5.2.1	[29, 30]
	Isoboldine	5.2.1	[29, 30]
	Lanuginosine	5.3	[28–30]
	Liriodenine	5.3	[27–30]
	Nornantenine	5.2.1	[29, 30]
	Norushinsunine	5.2.4	[27]

Table 1. (*continued*)

Genus (origin)/Species[a]	Aporphinoid	Section	References
3. *A. crassiflora* (Amer.)	Anonaine	5.2.1	[31]
	Asimilobine	5.2.1	[31]
	Liriodenine	5.3	[31]
4. *A. cristalensis* (Amer.)	Liriodenine	5.3	[32]
5. *A. glabra* (Amer.)	Anolobine	5.2.1	[33, 34]
	Anonaine	5.2.1	[33, 34]
	Asimilobine	5.2.1	[33, 34]
	Isoboldine	5.2.1	[33, 34]
	Liriodenine	5.3	[33–36]
	N-Methylactinodaphnine	5.2.1	[37]
	Nornuciferine	5.2.1	[33, 34]
	Norushinsunine	5.2.4	[33, 34]
	Roemerine	5.2.1	[33, 34]
6. *A. montana* (Amer.)	Anonaine	5.2.1	[38, 39]
	Argentinine	5.6	[38, 39]
	Asimilobine	5.2.1	[39]
	Atherosperminine	5.6	[38, 39]
	Isoboldine	5.2.1	[39]
	Liriodenine	5.3	[38, 39]
	Xylopine	5.2.1	[38, 39]
7. *A. muricata* (Amer.)	Atherosperminine	5.6	[40, 41]
	Stepharine	5.1	[41]
8. *A. purpurea* (Amer.)	*O*-Demethylpurpureine	5.2.1	[42]
	Glaziovine	5.1	[42]
	Isocorydine	5.2.1	[42]
	Norpurpureine	5.2.1	[42]
	Oxoglaucine	5.3	[42]
	Oxopurpureine	5.3	[42]
	Purpureine	5.2.1	[42]
	Stepharine	5.1	[42]
9. *A. reticulata* (Amer.)	Anonaine	5.2.1	[43–46]
	Liriodenine	5.3	[45–47]
	Norushinsunine	5.2.4	[45, 46]
10. *A. squamosa* (Amer.)	Anolobine	5.2.1	[48]
	Anonaine	5.2.1	[48–51]
	Corydine	5.2.1	[50, 51]
	Glaucine	5.2.1	[50, 51]
	Isocorydine	5.2.1	[51]
	Lanuginosine	5.3	[52]
	Liriodenine	5.3	[48]
	Norcorydine	5.2.1	[50]
	Norisocorydine	5.2.1	[50]
	Norushinsunine	5.2.4	[48]
	Roemerine	5.2.1	[50]
	Xylopine	5.2.1	[52]
Artabotrys (Asia/Africa) ART			
1. *A. suaveolens* (Asia)	Artabotrinine	5.2.1	[53]
	(= isocorydine)	5.2.1	[53, 54]
	Suaveoline	5.2.1	[53, 54]

Table 1. (*continued*)

Genus (origin)/Species[a]	Aporphinoid	Section	References
Asimina (Amer.) ASI			
1. *A. triloba*	Anolobine	5.2.1	[55]
	Asimilobine	5.2.1	[55]
	Isocorydine	5.2.1	[55]
	Liriodenine	5.3	[55]
	Norushinsunine	5.2.4	[55]
Cananga (Asia) CAN			
1. *C. latifolia*	Liriodenine	5.3	[56]
2. *C. odorata*	Anonaine	5.2.1	[57]
	Eupolauridine	5.7	[57, 58]
	Liriodenine	5.3	[57]
	Roemerine	5.2.1	[57]
	Ushinsunine	5.2.4	[57]
Cleistopholis (Africa) CLE			
1. *C. patens*	Cleistopholine	5.7	[59]
	Eupolauridine	5.7	[59]
	Eupolauridine di-*N*-oxide	5.7	[59]
	Eupolauridine *N*-oxide	5.7	[59]
	Isomoschatoline	5.3	[60]
	Liriodenine	5.3	[59, 60]
	Onychine	5.7	[59]
Cymbopetalum (Amer.) CYM			
1. *C. brasiliense*	Asimilobine	5.2.1	[61]
	Magnoflorine	5.2.1	[61]
	Norushinsunine	5.2.4	[61]
Desmos (Asia) DES			
1. *D. tiebaghiensis*	Anonaine	5.2.1	[62]
	Asimilobine	5.2.1	[62]
	Boldine	5.2.1	[62]
	Glaziovine	5.1	[62]
	Isoboldine	5.2.1	[62]
	Laurotetanine	5.2.1	[62]
	N-Methyllaurotetanine	5.2.1	[62]
	Norushinsunine	5.2.4	[62]
Duguetia (Amer.) DUG			
1. *D. calycina*	Atherospuerminine	5.6	[63]
	Calycinine	5.2.1	[63]
	Duguecalyne (duguecaline)	5.2.3	[64]
	Duguenaine	5.2.3	[64]
	N-Formylputerine	5.2.1	[65]
	O-Methylpukateine	5.2.1	[63]
	Noratherospuerminine	5.6	[66]
	Obovanine	5.2.1	[63]
	Oxoputerine	5.3	[63]
	Puterine	5.2.1	[63]
	Xylopine	5.2.1	[63]
2. *D. eximia*	*O*-Methylmoschatoline	5.3	[67]
	Oxopukateine	5.3	[67]
	Oxoputerine	5.3	[67]

Table 1. (*continued*)

Genus (origin)/Species[a]	Aporphinoid	Section	References
3. *D. obovata*	Anolobine	5.2.1	[65]
	Buxifoline	5.2.1	[65]
	Calycinine	5.2.1	[65]
	Duguevanine	5.2.1	[65]
	N-Formylbuxifoline	5.2.1	[65]
	N-Formylduguevanine	5.2.1	[65]
	N-Formylxylopine	5.2.1	[65]
	Isolaureline	5.2.1	[65]
	N-Methylbuxifoline	5.2.1	[65]
	N-Methylcalycinine	5.2.1	[65]
	N-Methylduguevanine	5.2.1	[65]
	Oxobuxifoline	5.3	[65]
	Xylopine	5.2.1	[65]
4. *D. spixiana*	Duguespixine	5.2.3	[68, 69]
5. *D. species*	Dicentrine	5.2.1	[70]
	Duguetine	5.2.4	[70]
	Norglaucine	5.2.1	[70]
Enantia (Africa) ENA			
1. *E. chlorantha*	Argentinine	5.6	[71]
	Atherosperminine	5.6	[71]
	Lysicamine	5.3	[71]
	O-Methylmoschatoline	5.3	[71]
2. *E. pilosa*	Lanuginosine	5.3	[72]
	Liriodenine	5.3	[72]
	Oliveridine	5.2.4	[72]
	Oliveridine N-oxide	5.2.4	[72]
	Oliverine	5.2.4	[72]
	Oliverine N-oxide	5.2.4	[72]
3. *E. polycarpa*	Anonaine	5.2.1	[73]
	Atherospermidine	5.3	[73]
	Isoboldine	5.2.1	[73]
	Isocorydine	5.2.1	[73, 74]
	Liriodenine	5.3	[73]
	Lysicamine	5.3	[73]
	Magnoflorine	5.2.1	[73]
	Menisperine	5.2.1	[73]
	N-Methyllaurotetanine	5.2.1	[73]
	Nornuciferine	5.2.1	[73]
Fissistigma (Asia) FIS			
1. *F. oldhamii*	Anolobine	5.2.1	[75]
	Calycinine (fissistigine A) (fissoldine)	5.2.1	[75–77]
	Xylopine	5.2.1	[75–77]
Fusaea (*Fusea*) (Amer.) FUS			
1. *F. longifolia*	O-1-Acetylapoglaziovine	5.2.1	[78]
	Anonaine	5.2.1	[78]
	Asimilobine	5.2.1	[78]
	Fuseine	5.4	[79]
	Glaziovine	5.1	[78]
	3-Hydroxynuciferine	5.2.1	[78]

Table 1. (*continued*)

Genus (origin)/Species[a]	Aporphinoid	Section	References
	Isoboldine	5.2.1	[78]
	Lirinidine	5.2.1	[78]
	Liriodenine	5.3	[78, 79]
	Noroliveroline	5.2.4	[78]
	Oliveroline	5.2.4	[78]
Greenwayodendron (Africa) GRE			
1. *G. oliveri*			
G. suaveolens (see *Polyalthia*)			
Guatteria (Amer.) GUA			
1. *G. chrysopetala*	*O,N*-Dimethylliriodendronine	5.3	[80]
	Isoboldine	5.2.1	[80]
	Lanuginosine	5.3	[80]
	Liriodenine	5.3	[80]
	Lysicamine	5.3	[80]
	Nornuciferine	5.2.1	[80]
2. *G. cubensis*	Corydine	5.2.1	[81]
	Liriodenine	5.3	[81]
3. *G. discolor*	Argentinine	5.6	[82]
	Atherosperminine	5.6	[82–83]
	Guacolidine	5.2.3	[82]
	Guacoline	5.2.3	[82]
	Guadiscidine	5.2.3	[82]
	Guadiscine	5.2.3	[82, 84]
	Guadiscoline	5.2.3	[82, 84]
	Isocalycinine	5.2.1	[82]
	O-Methylcalycinine (discoguattine)	5.2.1	[82, 84]
	10-*O*-Methylhernovine	5.2.1	[83]
	O-Methylpukateine	5.2.1	[82, 83]
	Noratherosperminine	5.6	[83]
	Oxoisocalycinine	5.3	[82]
	Oxoputerine	5.3	[83]
	N-Oxyatherosperminine	5.6	[82, 83]
	Puterine	5.2.1	[82, 83]
	Xylopine	5.2.1	[83]
4. *G. elata*	Norlaureline	5.2.1	[85]
	Oxolaureline (lauterine)	5.3	[86]
	Oxoputerine	5.3	[86]
	Puterine	5.2.1	[85]
5. *G. goudotiana*	Dehydroneolitsine	5.2.2	[87]
	Goudotianine	5.2.3	[87]
	3-Hydroxynornuciferine	5.2.1	[87]
	Isoboldine	5.2.1	[87]
	Isodomesticine	5.2.1	[87]
	Lindcarpine	5.2.1	[87]
	Liriodenine	5.3	[87]
	N-Methyllaurotetanine	5.2.1	[87]
	Neolitsine	5.2.1	[87]
	Norisodomesticine	5.2.1	[87]
	Norpredicentrine	5.2.1	[87]

Table 1. (*continued*)

Genus (origin)/Species[a]	Aporphinoid	Section	References
6. *G. melosma*	3-Hydroxynornuciferine	5.2.1	[60]
	Isoboldine	5.2.1	[88, 89]
	Isomoschatoline	5.3	[60]
	Liriodenine	5.3	[60]
	Melosmidine	5.2.3	[88, 89]
	Melosmine	5.2.3	[88, 89]
	Oxoanolobine	5.3	[88–90]
7. *G. modesta*	Liriodenine	5.3	[91, 92]
	Roemerine	5.2.1	[91, 92]
8. *G. moralessi*	Corydine	5.2.1	[81]
9. *G. ouregou*	Dehydroformouregine	5.2.2	[93]
	Dehydro-*O*-methylisopiline	5.2.2	[93]
	Dehydronornuciferine	5.2.2	[94]
	Dihydromelosmine	5.2.3	[94, 95]
	Formouregine	5.2.1	[93]
	N-Formylnornuciferine	5.2.1	[93]
	Gouregine	5.7	[95, 96]
	Guattouregidine	5.2.3	[94, 95]
	Guattouregine	5.2.3	[94, 95]
	3-Hydroxynornuciferine	5.2.1	[93]
	3-Hydroxynuciferine	5.2.1	[93]
	Isopiline	5.2.1	[93, 95]
	Lirinidine	5.2.1	[93]
	Lysicamine	5.3	[93, 95]
	Melosmine	5.2.3	[95, 96]
	3-Methoxynuciferine	5.2.1	[93]
	N-Methylisopiline	5.2.1	[93]
	O-Methylisopiline	5.2.1	[93, 95]
	O-Methylmoschatoline	5.3	[93, 95]
	Norcepharadione B	5.4	[93]
	Nornuciferine	5.2.1	[93, 95]
	Nuciferine	5.2.1	[93]
	Ouregidione	5.4	[93]
	Oureguattidine	5.2.1	[95]
	Oureguattine	5.2.1	[93]
	Pentouregine	5.7	[93a]
	Subsessiline	5.3	[93, 95]
10. *G. psilopus*	Atherospermidine	5.3	[97]
	Guatterine	5.2.4	[97]
11. *G. saffordiana*	Lysicamine	5.3	[98]
	O-Methylmoschatoline	5.3	[98]
12. *G. sagotiana*	Anolobine	5.2.1	[104]
	Dehydroroemerine	5.2.2	[104]
	Dehydrostephalagine	5.2.2	[104]
	Duguespixine	5.2.3	[104]
	Elmerrillicine	5.2.1	[104]
	Guatterine	5.2.4	[104]
	Guatterine *N*-oxide	5.2.4	[104]
	3-Hydroxynornuciferine	5.2.1	[104]
	Lirinidine	5.2.1	[104]
	Liriodenine	5.3	[104]

Table 1. (*continued*)

Genus (origin)/Species[a]	Aporphinoid	Section	References
	N-Methylelmerrillicine	5.2.1	[104]
	O-Methylpachyconfine	5.2.4	[104]
	O-Methylpukateine	5.2.1	[104]
	Norlaureline	5.2.1	[104]
	Nornuciferine	5.2.1	[104]
	Noroliveroline	5.2.4	[104]
	Obovanine	5.2.1	[104]
	Oliveroline	5.2.4	[104]
	Oliveroline *N*-oxide	5.2.4	[104]
	Oxoanolobine	5.3	[104]
	Oxolaureline	5.3	[104]
	Oxoputerine	5.3	[104]
	Pachyconfine	5.2.4	[104]
	Pukateine	5.2.1	[104]
	Puterine	5.2.1	[104]
	Roemerine	5.2.1	[104]
	Trichoguattine	5.2.3	[104]
	Xylopine	5.2.1	[104]
13. *G. scandens*	Actinodaphnine	5.2.1	[99]
	Anolobine	5.2.1	[99]
	Asimilobine	5.2.1	[99]
	Atheroline	5.3	[99]
	Dicentrinone	5.3	[99]
	Guattescidine	5.2.3	[99–101]
	Guattescine	5.2.3	[99–101]
	Lanuginosine	5.3	[99]
	Laurotetanine	5.2.1	[99]
	Liriodenine	5.3	[99]
	O-Methylisopiline	5.2.1	[99]
	N-Methyllaurotetanine	5.2.1	[99]
	Nordicentrine	5.2.1	[99]
	Norpredicentrine	5.2.1	[99]
	Xylopine	5.2.1	[99]
14. *G. schomburgkiana*	Anolobine	5.2.1	[102, 103]
	Anonaine	5.2.1	[103]
	Belemine	5.2.3	[102, 103]
	Dehydroguattescine	5.2.3	[102, 103]
	N-Formylputerine	5.2.1	[103]
	Guadiscine	5.2.3	[102, 103]
	Guattescine	5.2.3	[102, 103]
	Lanuginosine	5.3	[103]
	Liriodenine	5.3	[103]
	N-Methylputerine	5.2.1	[103]
	(*O*-methylpukateine)		
	Oxoputerine	5.3	[103]
	Puterine	5.2.1	[103]
	Xylopine	5.2.1	[103]
G. subsessilis (reidentified as *Heteropetalum brasiliensis* [81])			
Heteropetalum (Amer.) HET			
1. *H. brasiliensis* (see also	*O*-Methylmoschatoline	5.3	[105]
Guatteria subsessilis [81])	Subsessiline	5.3	[105, 106]

Table 1. *(continued)*

Genus (origin)/Species[a]	Aporphinoid	Section	References
Hexalobus (Africa) HEX			
1. *H. crispiflorus*	Anonaine	5.2.1	[107]
	Asimilobine	5.2.1	[107]
	N-Carbamoylanonaine	5.2.1	[107]
	N-Carbamoylasimilobine	5.2.1	[107]
	N-Formylanonaine	5.2.1	[107]
	3-Hydroxydehydronuciferine	5.2.2	[107, 108]
	3-Hydroxynornuciferine	5.2.1	[107, 108]
	Liriodenine	5.3	[107]
	Nornuciferine	5.2.1	[107]
	Norstephalagine	5.2.1	[107, 108]
Isolona (Africa) ISO			
1. *I. campanulata*	Anonaine	5.2.1	[109]
	Liriodenine	5.3	[109]
	Nornuciferine	5.2.1	[109]
	Oliveridine	5.2.4	[109]
	Oliverine	5.2.4	[109]
	Oliverine N-oxide	5.2.4	[109]
2. *I. pilosa*	Anonaine	5.2.1	[110]
	Caaverine	5.2.1	[110]
	Isopiline	5.2.1	[110, 111]
	Nornuciferine	5.2.1	[110]
	Pronuciferine	5.1	[110]
	Roemerine	5.2.1	[110]
	Zenkerine	5.2.1	[110]
3. *I. zenkeri*	Caaverine	5.2.1	[110]
	Lirinidine	5.2.1	[110]
	N-Methylcrotsparine	5.1	[110]
	Zenkerine	5.2.1	[110]
Meiocarpidium (Africa) MEC			
1. *M. lepidotum*	Methoxyatherosperminine	5.6	[112]
	Methoxyatherosperminine N-oxide	5.6	[112]
Meiogyne (Asia) MEI			
1. *M. virgata*	Anonaine	5.2.1	[113]
	Asimilobine	5.2.1	[113]
	Cleistopholine	5.7	[113]
	Corytuberine	5.2.1	[113]
	Liriodenine	5.3	[113]
	Norushinsunine	5.2.4	[113]
Melodorum (Asia) MEL			
1. *M. punctulatum*	Asimilobine	5.2.1	[114]
	Liriodenine	5.3	[114]
	Norushinsunine	5.2.4	[114]
Mitrella (Asia) MIT			
1. *M. kentii*	Anonaine	5.2.1	[115]
	Asimilobine	5.2.1	[115]
	Liriodenine	5.3	[115]
Monanthotaxis (Africa) MOA			
1. *M. cauliflora*	Asimilobine	5.2.1	[116]

Table 1. (*continued*)

Genus (origin)/Species[a]	Aporphinoid	Section	References
	9-Hydroxy-1,2-dimethoxynor-aporphine	5.2.1	[116]
	Laurelliptine	5.2.1	[116]
	Nuciferine	5.2.1	[116]
Monodora (Africa) MON			
1. *M. angolensis*	Argentinine	5.6	[117]
	Crotsparine	5.1	[117]
	Sparsiflorine	5.2.1	[117]
	Wilsonirine	5.2.1	[117]
2. *M.* cf. *brevipes*	Isoboldine	5.2.1	[117, 118]
3. *M. tenuifolia*	Anolobine	5.2.1	[119, 120]
	Anonaine	5.2.1	[119, 120]
	Laurelliptine	5.2.1	[119–121]
	Liriodenine	5.3	[119, 120]
	Magnoflorine	5.2.1	[119, 120]
	Sparsiflorine	5.2.1	[119, 120]
	Stepharine	5.1	[119, 120]
Onychopetalum (Amer.) ONY			
1. *O. amazonicum*	Onychine	5.7	[122]
Orophea (Asia) ORO			
1. *O. enterocarpa*	Enterocarpam I	5.7	[241]
	Enterocarpam II	5.7	[241]
Pachypodanthium (Africa)—PAC			
1. *P. confine*	Guatterine	5.2.4	[123]
	Guatterine *N*-oxide	5.2.4	[123]
	Oliveroline	5.2.4	[123]
	Pachyconfine	5.2.4	[123]
2. *P. staudtii*	Liriodenine	5.3	[124, 125]
	N-Methylpachypodanthine	5.2.4	[124]
	Norpachystaudine	5.2.4	[124]
	Pachypodanthine	5.2.4	[124, 126]
	Pachystaudine	5.2.4	[124]
Polyalthia (Asia, Africa) POL			
1. *P. acuminata* (Asia)	Anolobine	5.2.1	[127]
	Anonaine	5.2.1	[127]
	Asimilobine	5.2.1	[127]
	Caaverine	5.2.1	[127]
	3-Hydroxynornuciferine	5.2.1	[127]
	Isoboldine	5.2.1	[127]
	Isopiline	5.2.1	[127]
	Liriodenine	5.3	[127]
	3-Methoxynuciferine	5.2.1	[127]
	O-Methylisopiline (*O*-methylnorlirinine)	5.2.1	[127]
	O-Methylmoschatoline	5.3	[127]
	Norannuradhapurine	5.2.1	[127]
	Norliridinine	5.2.1	[127]
	Nornuciferine	5.2.1	[127]
	Noroliveroline	5.2.4	[127]
	Norushinsunine	5.2.4	[127]

Table 1. (*continued*)

Genus (origin)/Species[a]	Aporphinoid	Section	References
	Stepharine	5.1	[127]
	Tuduranine	5.2.1	[127]
2. *P. cauliflora* var. *beccarii* (Asia)	Atherospermidine	5.3	[128]
	Beccapoline	5.5	[128, 129]
	Beccapolinium	5.5	[128, 129]
	Beccapolydione	5.5	[128]
	Boldine	5.2.1	[128]
	Dehydropredicentrine	5.2.2	[128]
	Liriodenine	5.3	[128]
	Lysicamine	5.3	[128]
	O-Methylmoschatoline	5.3	[128]
	Oxostephanine	5.3	[128, 129]
	Polybeccarine	5.5	[128]
	Predicentrine	5.2.1	[128]
	Thailandine	5.3	[128]
3. *P. emarginata* (Madagascar)	Anonaine	5.2.1	[130]
	Lanuginosine	5.3	[130]
	Liriodenine	5.3	[130]
4. *P. nitidissima* (Asia)	Liriodenine	5.3	[131, 132]
	Norushinsunine	5.2.4	[132]
	Ushinsunine	5.2.4	[132]
5. *P. oligosperma* (Madagascar)	Noroconovine	5.2.1	[130]
	Polygospermine	5.2.1	[130]
6. *P. oliveri* (Africa) (= *Greenwayodendron*)	Anonaine	5.2.1	[133]
	Lanuginosine	5.3	[133]
	Liriodenine	5.3	[133]
	N-Methylcorydine	5.2.1	[133]
	N-Methylpachypodanthine *N*-oxide	5.2.4	[133]
	Noroliveridine	5.2.4	[133]
	Oliveridine	5.2.4	[133, 134]
	Oliverine	5.2.4	[133, 134]
	Oliveroline	5.2.4	[133]
	Oliveroline *N*-oxide	5.2.4	[133]
	Pachypodanthine	5.2.4	[133]
7. *P. suaveolens* (Africa) (= *Greenwayodendron*)	Guatterine	5.2.4	[135]
	Lysicamine	5.3	[135]
	Noroliverine	5.2.4	[135]
	Oliveridine	5.2.4	[135]
	Oliveridine *N*-oxide	5.2.4	[136]
	Oliverine	5.2.4	[135, 136]
	Oliverine *N*-oxide	5.2.4	[136]
	Oliveroline	5.2.4	[135]
	Oxostephanine	5.3	[135, 136]
	Pachypodanthine	5.2.4	[135]
	Polyalthine	5.2.4	[135]
	Polysuavine	5.2.4	[135, 136]
Popowia (Asia) POP			
1. *P.* cf. *cyanocarpa*	Asimilobine	5.2.1	[137]
	Norcorydine	5.2.1	[137]
	Wilsonirine	5.2.1	[137]
2. *P. pisocarpa*	Asimilobine	5.2.1	[138]
	Bipowine	5.5	[138]
	Bipowinone	5.5	[138]

Table 1. (*continued*)

Genus (origin)/Species[a]	Aporphinoid	Section	References
	Norcorydine	5.2.1	[138]
	Wilsonirine	5.2.1	[138]
Pseudoxandra (Amer.) PSO			
1. *P. sclerocarpa*	Ushinsunine	5.2.4	[139]
Pseuduvaria (Asia) PSU			
1. *P.* cf. *dolichonema*	Glaucine	5.2.1	[131]
	Norglaucine	5.2.1	[131]
	Norpredicentrine	5.2.1	[131]
2. *P.* cf. *grandifolia*	Anonaine	5.2.1	[131]
	Liriodenine	5.3	[131]
	Nornuciferine	5.2.1	[131]
3. *P.* species	Anonaine	5.2.1	[131]
	Liriodenine	5.3	[131]
	Nornuciferine	5.2.1	[131]
Rollinia (Amer.) ROL			
1. *R. papilionella*	Lanuginosine	5.3	[140]
	Liriodenine	5.3	[140]
	Lysicamine	5.3	[140]
2. *R. sericea*	Atherospermidine	5.3	[141]
	Liriodenine	5.3	[141]
	O-Methylmoschatoline (homomoschatoline)	5.3	[141]
Schefferomitra (Asia) SCH			
1. *S. subaequalis*	Anolobine	5.2.1	[142]
	Anonaine	5.2.1	[131, 142]
	Aristolactam B2	5.7	[142, 143]
	Asimilobine	5.2.1	[142]
	Isoboldine	5.2.1	[142]
	Liriodenine	5.3	[131]
Unonopsis (Amer.) UNO			
1. *U. guatterioides*	Anonaine	5.2.1	[144]
	Asimilobine	5.2.1	[144]
	Liriodenine	5.3	[144]
	Lysicamine	5.3	[144]
	Norushinsunine	5.2.4	[144]
2. *U. stipitata*	Argentinine	5.6	[144]
	Stipitatine	5.6	[144]
	Thalicthuberine	5.6	[144]
Uvaria (Africa, Asia) UVA			
1. *U. chamae* (Africa)	Asimilobine	5.2.1	[145]
	Glaucine	5.2.1	[145]
	Glaziovine	5.1	[145]
	Isoboldine	5.2.1	[145]
	Pronuciferine	5.1	[145]
	Thaliporphine	5.2.1	[145]
Uvariopsis (Africa) UVI			
1. *U. congolana*	Uvariopsine	5.6	[146]
2. *U. guineensis*	Liriodenine	5.3	[11]
	8-Methoxyuvariopsine	5.6	[11]
	Noruvariopsamine	5.6	[11]
	Uvariopsamine	5.6	[11]
	Uvariopsamine *N*-oxide	5.6	[11]
	Uvariopsine	5.6	[11]

Table 1. (*continued*)

Genus (origin)/Species[a]	Aporphinoid	Section	References
3. *U. solheidii*	8-Methoxyuvariopsine	5.6	[11]
	Uvariopsine	5.6	[147]
Xylopia (Pantropical) XYL			
1. *X. brasiliensis* (Amer.)	Anonaine	5.2.1	[148]
	Lanuginosine	5.3	[148]
	Liriodenine	5.3	[148]
	Xylopine	5.2.1	[148]
2. *X. buxifolia* (Madagascar)	Anonaine	5.2.1	[149]
	Buxifoline	5.2.1	[149]
	Lanuginosine	5.3	[149]
	Liriodenine	5.3	[149]
	N-Methylasimilobine	5.2.1	[149]
	Nornuciferine	5.2.1	[149]
	Norstephalagine	5.2.1	[149]
	Pronuciferine	5.1	[149]
	Xylopine	5.2.1	[149]
3. *X. danguyella* (Madagascar)	Corydine	5.2.1	[149]
	Danguyelline	5.2.1	[149]
	Isoboldine	5.2.1	[149]
	Laurotetanine	5.2.1	[149]
	Norcorydine	5.2.1	[149]
	Norisocorydine	5.2.1	[149]
	Norisodomesticine	5.2.1	[149]
	Nornantenine	5.2.1	[149]
	Xyloguyelline	5.2.1	[149]
4. *X. discreta* (Amer.)	Xylopine	5.2.1	[150]
5. *X. frutescens* (Amer.)	Anolobine	5.2.1	[151]
	Anonaine	5.2.1	[152]
	Asimilobine	5.2.1	[152]
	Isoboldine	5.2.1	[151]
	Lanuginosine	5.3	[152]
	Laurotetanine	5.2.1	[152]
	Liriodenine	5.3	[152]
	N-Methyllaurotetanine	5.2.1	[152]
	Nantenine	5.2.1	[152]
	Nornantenine	5.2.1	[152]
	Nornuciferine	5.2.1	[152]
	Xylopine	5.2.1	[152]
6. *X. lemurica* (Madagascar)	Lanuginosine	5.3	[153]
7. *X. pancheri* (Asia)	Anonaine	5.2.1	[154]
	Liriodenine	5.3	[154]
	Norcorydine	5.2.1	[154]
	Roemerine	5.2.1	[154]
	Xylopine	5.2.1	[154]
8. *X. papuana* (Asia)	Anonaine	5.2.1	[155]
	Laurolitsine	5.2.1	[155]
	Roemerine	5.2.1	[155]
	Xylopine	5.2.1	[155]
9. *X. vielana* (Asia)	Liriodenine	5.3	[156]

[a] Asia refers to species from the Asian mainland, Malesian region, and Australasia.

4. STRUCTURE ELUCIDATION

Today the structure elucidation of aporphinoids is largely achieved by the use of spectral methods. In this section we briefly describe those features of the major spectral techniques that are of most general value in the identification of individual compounds.

4.1. Ultraviolet Spectroscopy

The UV spectra of aporphinoids are of considerable value in structure elucidation because differences within each group of aporphinoids depend primarily on the location rather than on the nature of substituents. Thus for the aporphines *s.s.* and for some other groups where the B and C rings contain no additional unsaturation, notably the 7-oxygenated aporphines, the chromophore can be considered as a twisted biphenyl system. All such aporphines are substituted at C(1) and C(2) but the differences in D ring substitution can have valuable diagnostic value, particularly because C(11) substitution increases the stress on the biphenyl. For example, aporphines without substitution at C(11) exhibit two unequal maxima at ~273–280 nm (larger log ϵ) and at 303–310 nm, whereas when C(11) is substituted the lower wavelength band is displaced to ~268–272 nm and the two maxima are of more equal intensity [157]. Some examples of maxima and log ϵ values for differently substituted aporphines are given in Table 2.

In the UV spectrum of an aporphine the presence of phenolic substituent(s) is, as expected, indicated by a bathochromic shift on the addition of base. More specifically the occurrence of a hydroxyl function at C(9) of an otherwise nonphenolic compound can cause both a bathochromic and a pronounced hyperchromic effect between 315 and 330 nm [158]. It has recently been observed [127] that a phenol at C(3), the comparable position to C(9) in ring A, produces the same effect.

The UV spectra of dehydroaporphines (Table 2) are characterized by the presence of a maximum or a pronounced shoulder between 250 and 265 nm which is attributable [15] to the additional 6a,7 double bond. Not surprisingly the same pattern is noted in the spectra of phenanthrenes but they also give rise to two further bands, of weak intensity, in the region 345–375 nm. Spectra of dehydroaporphines run in an acidic medium also exhibit these additional maxima.

In the oxoaporphines, where the aporphine nucleus is fully conjugated and planar, the resulting UV spectra are complex and immediately diagnostic, showing several maxima between 215 and 315 nm and two to three in the range 350–440 nm. One interesting feature is that under acidic conditions there is a marked bathochromic shift and increase in intensity in the long-wavelength maxima of all oxoaporphines. Chen et al. [159] have suggested that this is due to the formation of an imine/aromatic system or the

Table 2. UV Spectra of Some Representative Aporphinoids

Class/Compound	Substitution	λ_{max} (nm)[a]	log ε
Proaporphine			
Pronuciferine	1,2	230, 282	4.41, 3.49
Aporphines *s.s.*			
Nuciferine	1,2	230, 274, 312	4.15, 4.23, 3.57
Isopiline	1,2,3	220, 275, 292s, 310	4.39, 4.15, 3.91, 3.78
Anolobine	1,2,9	215, 238s, 280, 292, 320	4.21, 3.87, 4.05, 4.05, 3.33
Sparsiflorine	1,2,10	226, 266, 275, 310	4.54, 4.09, 4.23, 3.95
Pukateine	1,2,11	218, 272, 303	4.42, 4.10, 3.95
Oureguattidine	1,2,3,9	222, 282, 300s, 316	4.34, 4.25, 4.08, 4.01
Elmerrillicine	1,2,3,11	224, 240s, 268s, 276, 298	4.41, 4.23, 4.16, 4.20, 4.02
Norannuradhapurine	1,2,8,9	218s, 281, 298s, 317s	4.27, 3.87, 3.69, 3.46
Isoboldine	1,2,9,10	219, 268s, 280, 304, 313	4.58, 4.08, 4.16, 4.21, 4.18
Calycinine	1,2,9,11	222, 268, 278, 299	4.43, 4.12, 4.23, 4.08
Corydine	1,2,10,11	218, 262, 270, 302	4.19, 3.72, 3.70, 3.40
Purpureine	1,2,3,9,10	273s, 282, 303, 312s	4.26, 4.36, 4.33, 4.29
Duguevanine	1,2,3,9,11	222, 269s, 279, 297, 305s	4.48, 4.17, 4.28, 4.09, 4.05
Noroconovine	1,2,3,10,11	221, 274, 307s	4.49, 4.12, 3.68
Dehydroaporphines			
Dehydroroemerine	1,2	218, 256s, 262, 330	4.42, 4.74, 4.77, 4.21
Dehydropredicentrine	1,2,9,10	215, 243s, 262, 270s, 294s, 329, 380s	4.14, 4.28, 4.49, 4.45, 4.04, 3.89, 3.31
Oxoaporphines			
Liriodenine	1,2	247, 268, 309, 413	4.22, 4.13, 3.62, 3.82
Atheroline	1,2,9,10	244, 273, 292s, 355, 380s, 435	4.09, 4.17, 3.96, 3.90, 3.83, 3.62
Phenanthrenes			
Argentinine	1,2	232, 254, 278, 302, 314, 346, 368	4.26, 4.50, 4.01, 3.92, 3.92, 3.17, 3.17
Thalicthuberine	1,2,9,10	230, 263, 282, 322, 348, 366	4.08, 4.41, 3.92, 3.74, 3.22, 3.21

[a] s = shoulder.

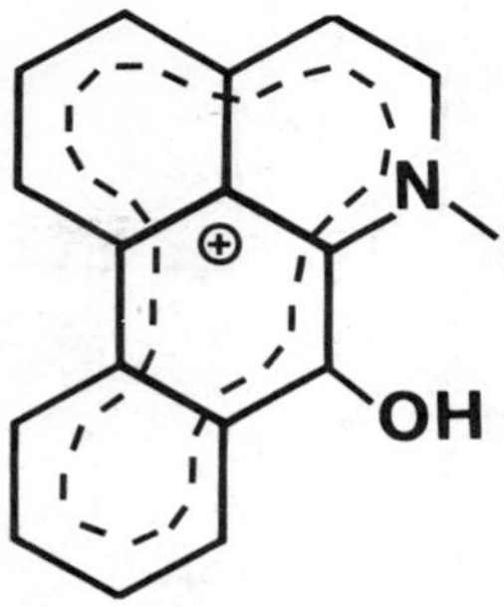

50

diprotonated moiety **50**, although the latter seems unlikely because the change occurs with acidic solutions that appear to be too dilute to cause diprotonation. The same effects can be observed in aporphines in which ring B is aromatized or where there is an imine function (see Section 5.2.3).

Finally, the proaporphines, with their dienone moiety, give their own characteristic spectra, showing maxima at ~230 and 290 nm, with occasional splitting of the 290-nm band. According to Sangster and Stuart [157] the 290-nm band shows a bathochromic shift in alkali.

4.2. Infrared Spectroscopy

The IR spectra of aporphinoids are generally of little diagnostic value. In the proaporphines the carbonyl band of the dienone system occurs between 1650 and 1675 cm^{-1}, whereas the carbonyl band in oxoaporphines is found in the 1650–1665 cm^{-1} region except in cases where the nitrogen is quaternized, when it is found at about 1625 cm^{-1}. The alkylaporphines (see Section 5.2.3), in which the nitrogen is involved in an imine function, exhibit a weak band between 1630 and 1645 cm^{-1}.

4.3. Mass Spectrometry

For the aporphinoids as a whole MS is often a relatively unrewarding technique. Determination of the molecular ion can usually be achieved without problem but it should be noted that chemical ionization is often more appropriate than electron impact for this purpose.

The MS of aporphines s.s. do not reveal any characteristic fragmentation pathways, losses primarily being of small fragments, for example, H˙, OH˙, CH$_3^˙$, or OCH$_3^˙$. Ring B will undergo a retro-Diels–Alder reaction (Scheme 9) to give loss of the nitrogen and C(5) as CH$_2$=N—R (if R = H, m/z 29; if R = CH$_3$, m/z 43) and this is of value in identifying the N-substituent.

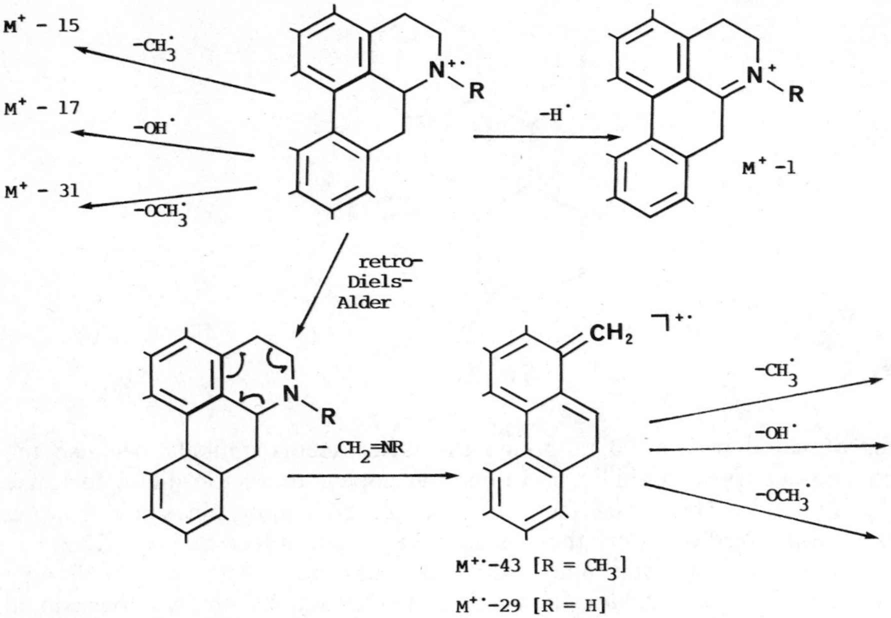

$M^+ - 15$ $-CH_3^{\cdot}$

$M^+ - 17$ $-OH^{\cdot}$

$M^+ - 31$ $-OCH_3^{\cdot}$

$-H^{\cdot}$

$M^+ -1$

retro-Diels-Alder

$- CH_2=NR$

$-CH_3^{\cdot}$

$-OH^{\cdot}$

$-OCH_3^{\cdot}$

$M^{+\cdot}-43$ [R = CH_3]

$M^{+\cdot}-29$ [R = H]

Scheme 9. Characteristic mass spectral fragmentation of a typical aporphine.

Aporphines substituted at 1,2,9,10 can be differentiated from those substituted at 1,2,10,11 by the relative intensities of $M^{+\cdot}$ and $M^{+\cdot}-1$ ions. For 1,2,9,10-substituted aporphines the latter fragment is generally the base peak whereas for 1,2,10,11-substituted aporphines the molecular ion is the base peak and $M^{+\cdot}-1$ rarely exceeds 50% relative abundance.

Greater interpretative power from MS can sometimes be achieved by analysis of trifluoroacetyl (TFA) or acetyl derivatives. Green et al. [160] have demonstrated that the isomeric 10,11-hydroxy/methoxy aporphines can be distinguished readily by MS of their corresponding TFA derivatives because of the preferential loss of the C(10) substitutent. Thus the 10-methoxy-11-TFA aporphine exhibits significant $M^{+\cdot}-CH_3$ and $M^{+\cdot}-CH_3-2H$ fragments whereas in the 10-TFA-11-methoxy aporphine these are replaced by ions for $M^{+\cdot}-CF_3CO$ and $M^{+\cdot}-CF_3CO-2H$. The N-TFA derivatives of noraporphines could be characterized by the occurrence of an ion $M^{+\cdot}-126$ due to the phenanthrene **51** remaining after a McLafferty rearrangement had led to fission of the N(6)/C(6a) bond followed by a second fission between C(4) and C(5) (Scheme 10).

The value of studying fragmentation of acetylated aporphines has recently been illustrated in the identification of 3-substituted noraporphines from *Hexalobus crispiflorus* [108]. MS of the N-acetylated derivative of the 3-methoxy noraporphine **52** (R = CH_3) led to a major ion at $M^{+\cdot}-72$ corresponding to **51**, but gave in addition the ion **53**, which can be derived through route a indicated in Scheme 11. Where the 3-substituent was an hydroxy group, MS of the *N,O*-diacetate (**52**, R = CH₃CO) revealed the additional ions **54** and **55** as well as one corresponding to **53**. The additional ions appear to arise through the facile loss of the *O*-acetyl group to give **54** (route b); the formation of such ions seems to be characteristic for 3-hydroxyaporphines. The mechanism by which these ions are generated is not immediately obvious.

The MS of oxoaporphines invariably show a strong molecular ion, other important features being ions for $M^{+\cdot} - CO$ and $M^{+\cdot} - CH_2O$. The relative abundance of these fragments has been used to differentiate between 9-methoxy- and 10-methoxyoxoaporphines [161] with the molecular ion dominating in the latter, whereas in the 9-methoxy compounds the $M^{+\cdot} - CH_2O$ fragment has a higher intensity than $M^{+\cdot} - CO$. For example, the MS of oxolaureline (1,2-methylenedioxy-10-methoxyoxoaporphine) shows the molecular ion, *m/z* 305, as the base peak, with *m/z* 277 ($-CO$), as the only other major fragment. By contrast, in the isomer lanuginosine (1,2-methylenedioxy-9-methoxyoxoaporphine) the loss of the CH_2O fragment is more likely than loss of CO and a major ion is observed at *m/z* 275.

Data available for the MS of dehydroaporphines suggest that the molecular ion is always the base peak. On the other hand, the characteristic feature of the MS of phenanthrenes is the cleavage of the aminoethyl side chain β to the nitrogen (Scheme 12), so generating a stable tropylium ion and, depending on the level of *N*-substitution, a base peak of *m/z* 58 (R = Me, Scheme 12) or *m/z* 44 (R = H, Scheme 12).

For the proaporphines also the molecular ion is normally the base peak. As an example the fragmentation pattern of pronuciferine is illustrated in Scheme 13. It should be noted [162] that the significant ions at $M^{+\cdot} - 29$ and $M^{+\cdot} - 43$ are in no way specific for proaporphines where low-resolution MS is employed, but that high-resolution MS permits differentiation of *m/z* 29 ($-CO^{\cdot}$, $-H^{\cdot}$) from the *m/z* 29 ($-CH_2{=}NH$) of noraporphines.

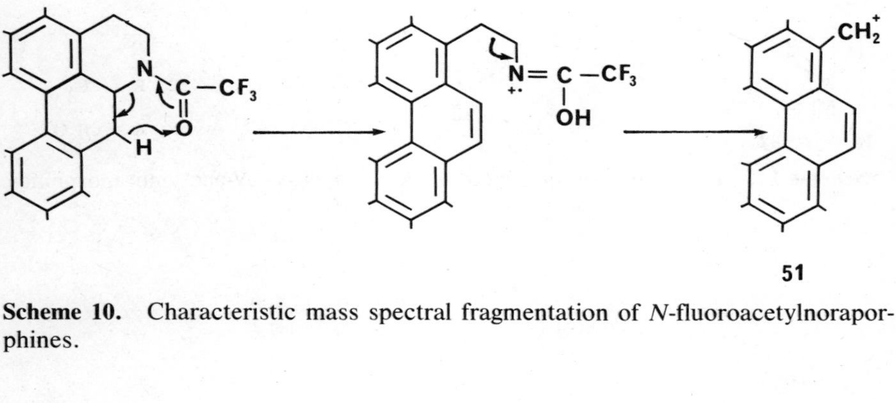

51

Scheme 10. Characteristic mass spectral fragmentation of *N*-fluoroacetylnoraporphines.

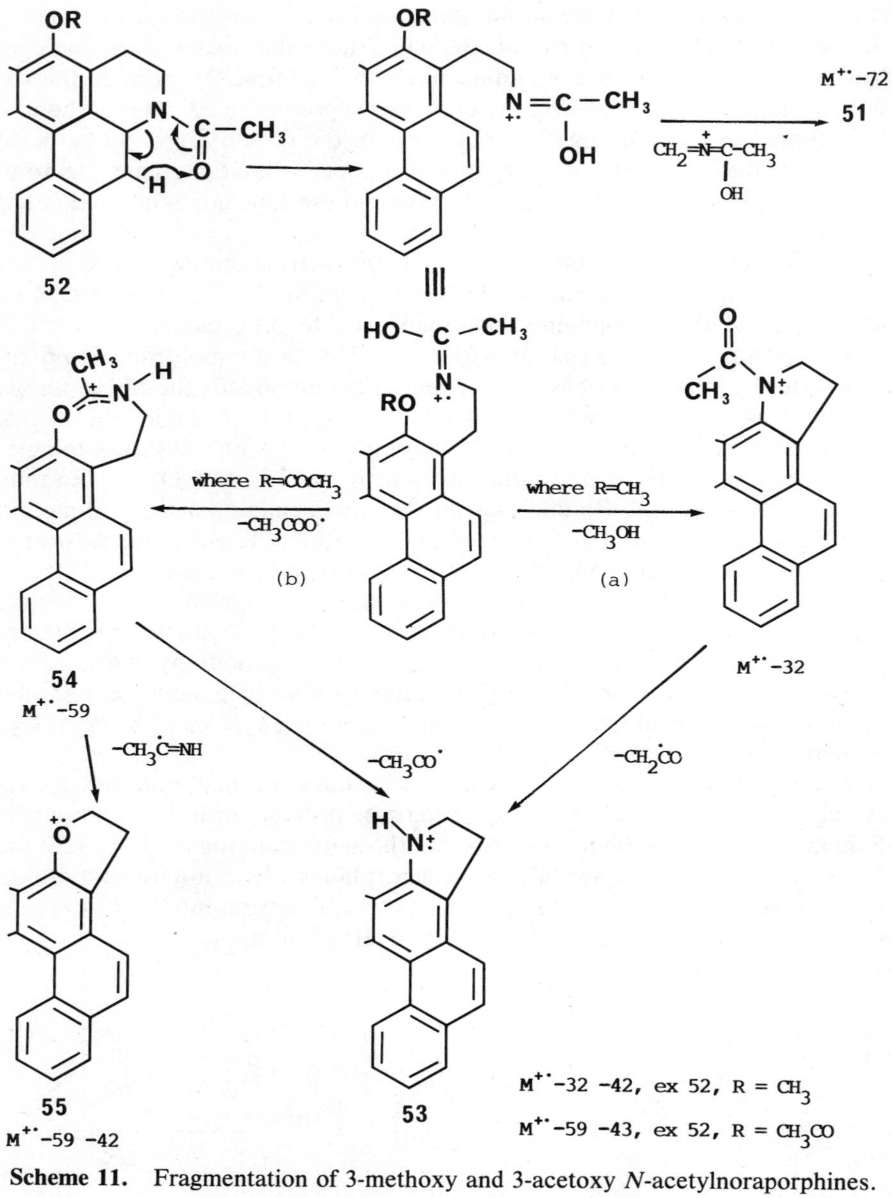

Scheme 11. Fragmentation of 3-methoxy and 3-acetoxy *N*-acetylnoraporphines.

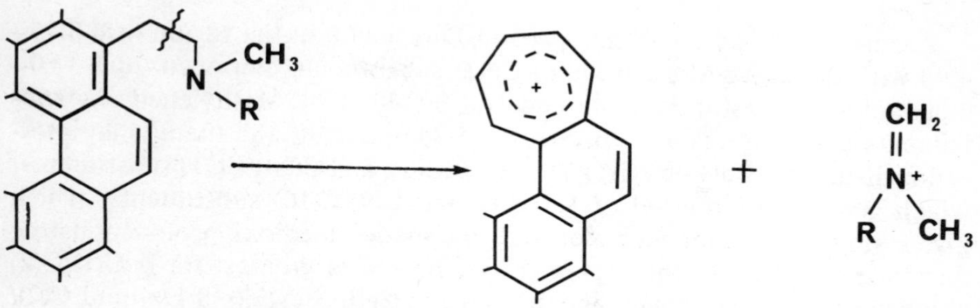

Scheme 12. Fragmentation of phenanthrenes.

4.4. ¹H Nuclear Magnetic Resonance

¹H NMR spectroscopy is extremely valuable in the structure elucidation of aporphinoids, two regions of the spectrum of particular note being those for methoxyl and aromatic protons. Unless otherwise noted, chemical shift values quoted below are for spectra taken in deuteriochloroform.

In the aporphines *s.s.* the noraporphines can readily be distinguished by the absence of a 3H singlet for the *N*-methyl group, found normally between δ 2.50 and 2.60 but at appreciably lower field in *N*-oxides. Signals for the aliphatic H(4), H(5), H(6a), and H(7) protons occur between δ 2.00 and 4.00 and have rarely been distinguished in published spectra. In aporphines substituted in rings B and/or C they become important in structure elucidation (see Sections 5.2.3 and 5.2.4).

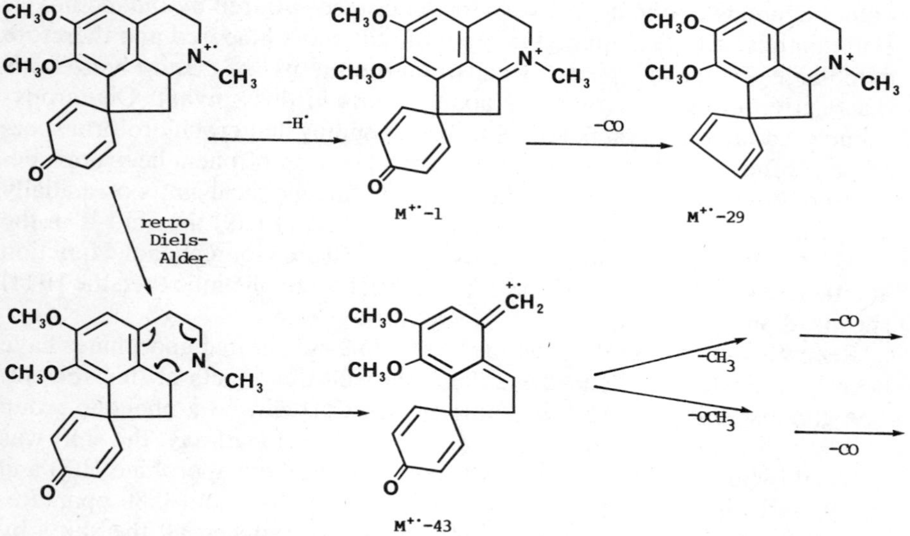

Scheme 13. Fragmentation of proaporphines.

Aromatic methoxyls on rings A and D resonate in the region of δ 3.75–4.00 with the notable exception of C(11) substituents that occur at δ 3.60–3.80 and C(1) substituents that occur at δ 3.40–3.70. Methylenedioxy substituents generally resonate between δ 5.90 and 6.10, and there is an interesting distinction between C(1)/C(2), C(2)/C(3), and C(10)/C(11) substituents, which give an AB quartet (J 1.5 Hz), and C(9)/C(10) substituents, which give a singlet. The nonequivalence of the methylenedioxy protons must be due to the torsion of the biphenyl system and is greatest for a C(1)/C(2) substituent (4–12 Hz apart) and smaller for C(10)/C(11) (~8 Hz) and C(2)/C(3) (2–4 Hz) [163].

Among the aromatic protons those for H(3) and H(11) are always readily assignable, the former being observed as a singlet between δ 6.50 and 6.70 and the latter, which is comparatively deshielded, between δ 7.50 and 8.20. The deshielding of H(11) is appreciably influenced by the presence of a methoxyl at C(10) and also by the C(1) substituent. For example, with a C(1)/C(2) methylenedioxy group H(11) will be in the range δ 7.45–7.85, but where C(1) is substituted with hydroxyl or methoxyl it will be in the range δ 7.80–8.20. The remaining D-ring protons occur within the limits set by H(3) and H(11). In cases where ring D is unsubstituted they are rarely differentiated, whereas in substituted D rings they often fail to show coupling patterns because of magnetic equivalence.

Comparison of spectra run in deuteriochloroform and other solvents, notably deuterobenzene and deuteropyridine, can often yield additional data valuable to the assignment of structure. Deuterobenzene is of particular use in studying methoxyl resonances because it causes a shielding that to a large extent depends on the degree of steric hindrance suffered by the methoxyl. For example, a C(1) methoxyl is generally the most hindered and therefore exhibits less shielding. A major constraint on the use of deuterobenzene is the relative insolubility of many aporphinoids in this solvent. Deuteropyridine spectra, when compared with corresponding deuterochloroform spectra, are often very valuable in locating the position of phenolic groups because of the strong deshielding they cause in the chemical shifts of spatially adjacent protons [164]. For protons H(3), H(8), and H(9) the shift is in the range of 0.25–0.40 ppm, for H(11) it is about 0.60 ppm for a phenolic function at either C(1) or C(10), and if both C(1) and C(10) are phenolic then the H(11) shift is about 0.90 ppm.

Recently deuteropyridine-induced shifts in 3-substituted aporphines have been investigated [95], and these illustrate well the effects of this solvent. The greatest shift on H(11) was noted when C(1) carried a phenolic group (~0.60 ppm) and where this was replaced by a C(1) methoxyl the shift was reduced by about 50%. In 1,9-dihydroxy-2,3-dimethoxyaporphine H(8) and H(10) were shifted by about 0.50 ppm and H(11) by about 0.80 ppm. Replacement of the C(1) hydroxyl with a methoxyl reduced all the shifts by about 50%.

With deuterated dimethyl sulfoxide and sodium deuterohydroxide phenolic aporphines form an anion [165], resulting in shifts in protons that depend upon their position relative to the anion; *ortho* protons are shielded 0.45–0.55 ppm, *meta* protons 0.25–0.40 ppm, and *para* protons 0.65–0.80 ppm. Protons of a nonphenolic aromatic ring undergo only mild shielding of ~0.10–0.25 ppm. A C(1) phenol causes a major deshielding in the order of 1 ppm at H(11) but in contrast a C(11) phenol affects only one of the protons of a 1,2-methylenedioxy group appreciably, and that by only about 0.20 ppm [65]. Replacement of deuterated dimethyl sulfoxide with deuteromethanol causes similar but weaker effects [65] but has the advantage that the alkaloid is more easily reclaimed.

Acetylation of phenolic aporphines and subsequent ^{1}H NMR in deuterochloroform is another technique that can be of value in identifying the position of phenolic groups. Acetoxy derivatives at C(1) in particular show a strong shielding in H(11) which can be as much as 0.50–0.60 ppm [166]. More generally protons adjacent to an acetoxy group can be expected to undergo a 0.20–0.35 ppm deshielding whereas a *para* proton can shift up to 0.15 ppm. There is often a small shielding observed for *meta* protons, on the order of 0.02–0.10 ppm, but this *meta* effect is not constant. An interesting example is presented by 9-hydroxyaporphines in which acetylation leads to a deshielding of 0.15–0.25 ppm for H(8) and H(10) and also causes a deshielding on the order of 0.10 ppm for H(11) [89,167].

Characteristic features of the spectra of dehydroaporphines include the deshielding of the N-methyl resonance to δ 2.95–3.10 and of H(11). For the latter the degree of deshielding depends on the C(1) substituent: where C(1)/C(2) is a methylenedioxy then H(11) generally occurs in the region δ 8.40–8.70 but where C(1) and C(2) carry hydroxy or methoxy substituents, H(11) has a chemical shift greater than δ 9.00, and generally in the range δ 9.30–9.50. Where ring D is unsubstituted chemical shifts of H(11) are at the extreme deshielded end of the range. As with the typical aporphines one aromatic proton is observed as a shielded singlet (~δ 6.50–6.60) but in dehydroaporphines this is attributable to H(7) rather than to H(3), which is generally inseparable from the D-ring protons. Other notable points are that methylenedioxy resonances occur as singlets and C(1) methoxyl groups are not as strongly shielded as in normal aporphines, features denoting the loss of the stressed biphenyl system.

Because of their poor solubility in most solvents ^{1}H NMR spectra of oxoaporphines have often been obtained in trifluoracetic acid (TFA). Because of the effect of TFA all signals are deshielded in comparison with aporphines *s.s.*, although relative shielding is still observed for H(3) (δ 7.50–7.65) and C(1) and C(11) methoxy substituents (δ 4.20–4.30). Methylenedioxy resonances are always observed as singlets, those for C(1)/C(2) being more deshielded (δ ~6.60) than those situated at C(9)/C(10) (δ ~6.20–6.30). The protons H(4) and H(5) resonate as a typical AB quartet (J ~6 Hz) at δ

7.90–8.40 and δ 8.50–8.80, respectively. The H(4) proton undergoes considerable deshielding to δ 8.80–8.90 when C(3) is substituted. Although H(11) is again generally the most deshielded proton, often resonating as low as δ 9.50, in 1,2-methylenedioxyoxoaporphines both H(5) and H(8) are more deshielded than H(11) [161].

The spectra of phenanthrenes share a number of features in common with those of dehydroaporphines, that is, H(11) is the most deshielded proton and its resonance position is influenced by the C(1) substituent, a C(1)/C(2) methylenedioxy occurs as a singlet at δ 6.20 and a C(9)/C(10) methylenedioxy at δ 6.05. A C(1) methoxy substituent is not comparatively shielded to the same extent as it is in aporphines *s.s.* The N-methyl protons resonate at δ 2.30–2.40 and although H(3) remains the most shielded proton it is relatively deshielded at δ 7.10–7.20. The H(6a) and H(7) protons form an AB quartet with *J* 9–10 Hz but often overlap with resonances for H(8), H(9), and H(10).

Proaporphines are characterized by the spirodienone system that makes up ring D, the protons of which form a classic AA'BB' system when unsubstituted, with H(9) and H(11) resonating at δ 6.10–6.60 and H(8) and H(12) at δ 6.70–7.30. *Ortho* coupling is of the order of 10 Hz but for *meta* coupling J_{9-11} (~1.5 Hz) is less than J_{8-12} (~2.5 Hz) [168]. High-field (360 MHz) studies of proaporphines have proved valuable in allowing differentiation of *syn* and *anti* D-ring protons [i.e., syn where H(6a) and the vinylic proton(s) are on the same side of the nucleus]. By a series of decoupling experiments it has been established that D-ring protons which lie *syn* exhibit a larger difference in chemical shifts (~0.7 ppm) than do those that lie *anti* (~0.5 ppm) [168]. As with other aporphinoids reviewed the N-methyl resonance is found between δ 2.40 and 2.50, H(3) is relatively shielded (δ 6.50–6.70), and a C(1) methoxyl substituent is more shielded than one in other positions.

4.5.   ¹³C Nuclear Magnetic Resonance

The ¹³C NMR spectra of a large number of aporphinoids have been published and in some cases correlations between observed chemical shifts and structure discussed [9,10,169–174].

In the aporphines *s.s.* the sp3 carbons of rings B and C are readily distinguishable. Although C(4) always resonates between 28 and 30 ppm, C(5), C(6a), and C(7) are influenced by the pattern of N-substitution. Thus in noraporphines C(5), C(6a), and C(7) resonate near 43, 53, and 37 ppm, respectively, whereas owing to the α-deshielding and β-shielding effects of N-methylation they are shifted to near 53, 62, and 34 ppm, respectively.

The aromatic carbons of rings A and D (not including ring junction carbons) can be divided into three groups with clearly differentiated chemical shifts. Shielded resonances at ~105–115 ppm arise owing to carbons adjacent to oxygen substituents. Deshielded resonances, generally in the range 155–

160 ppm, are attributable to a single oxygenated carbon on ring D or to other situations where both *ortho* positions are unsubstituted. Other O-substituted carbons are generally found in the range 145–150 ppm except for C(1) and, to a lesser extent, C(11), where they are less deshielded (~140–144 ppm). The central carbon of three adjacent oxygenated aromatic carbons [only C(2) in a 1,2,3-substituted alkaloid in the Annonaceae] is also less deshielded and can be difficult to distinguish from C(1) (see melosmine, Fig. 1). Remaining tertiary carbons are normally difficult to assign and occur in the region 125–

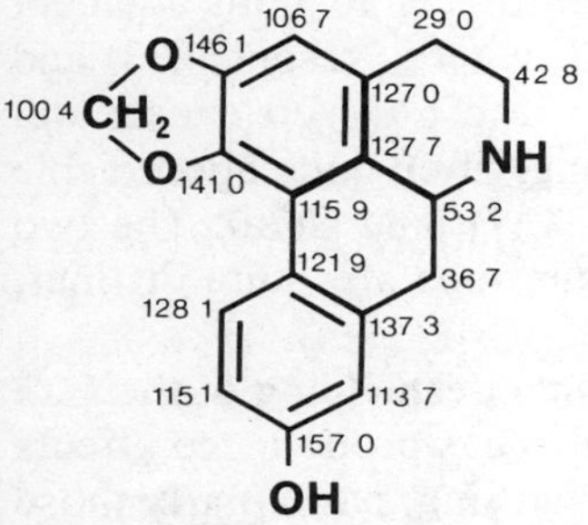

Anolobine
(DMSO)

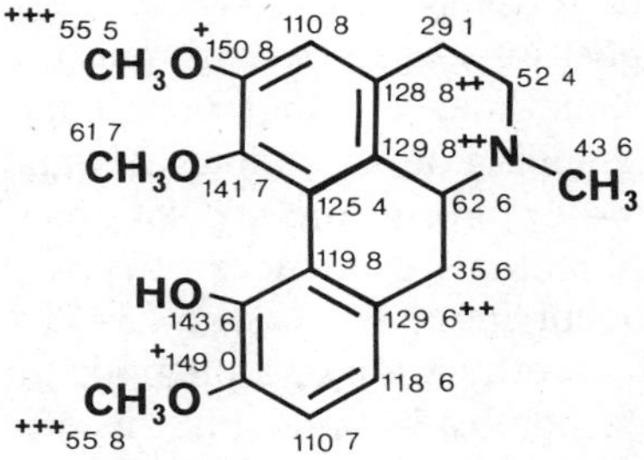

Isocorydine
(CDCl$_3$)

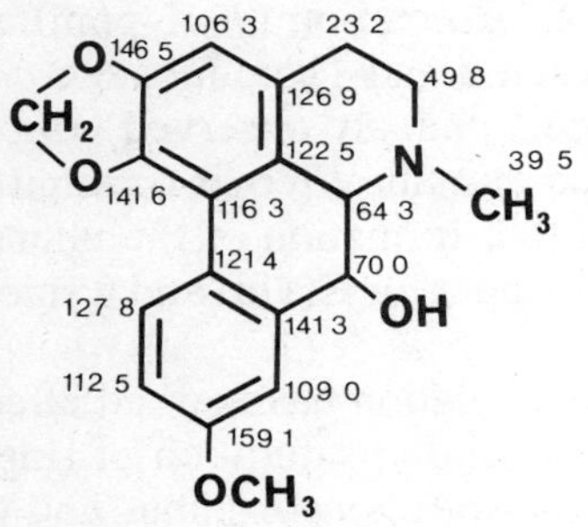

Oliveridine
(CDCl$_3$)

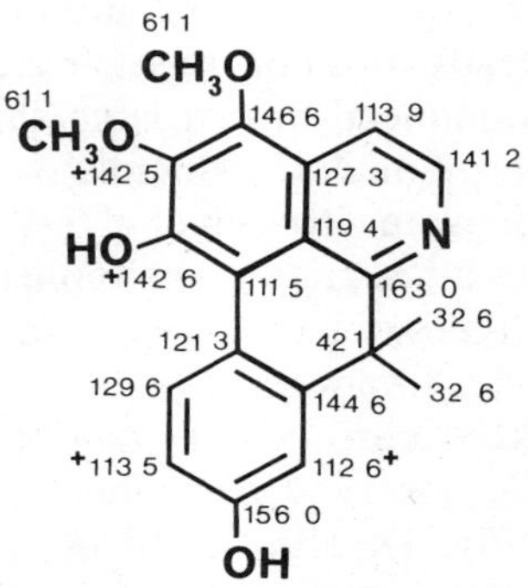

Melosmine
(CDCl$_3$)

Figure 1. ^{13}C NMR resonances for some typical aporphines of the Annonaceae (values on a structure with the same number of + signs are interchangeable).

130 ppm, but where *para* to an *O*-substituted position they are more shielded [i.e., C(8) in an 11-substituted aporphine].

The ring junction carbons in a simple 1,2-substituted aporphine are found between 119 and 133 ppm with C(1a) shielded by the adjacent C(1) oxygen and C(1b) to a lesser extent by being *para* to C(2). All are shielded to some extent owing to *ortho* or *para* substitution on rings A or D, and for C(11a) this can be by as much as 10 ppm in 9,11-oxygenated compounds.

The *N*-methyl substituent, where present, resonates between 43 and 44 ppm whereas methoxyl carbons are found either between 59 and 61 ppm or between 55 and 56 ppm. Deshielded methoxyl resonances are attributable to those positions where steric hindrance occurs owing to both adjacent positions being substituted [i.e., always C(1) and, where present, C(3) and sometimes at other positions]. Quaternization of the nitrogen causes deshielding of ~8 ppm at C(5) and C(6a) and shielding, often quite appreciable (4–9 ppm), at the γ positions C(1b), C(3a), C(4), C(7), and C(7a). The two *N*-methyl resonances of *N,N*-dimethylaporphinium ions are quite distinct, occurring at 43–44 and 53–55 ppm.

Methylation of a phenolic group often results in a deshielding of the *ipso* carbon by 3–4 ppm but this effect is variable. It also has pronounced effects on the resonances observed for other carbons in that ring, particularly those *ortho* and *para* to the position where methylation occurs. In polyoxygenated systems the shielding caused by methylation can be greater. Acetylation has an effect opposite to methylation. Data from Ronsch et al. [175] show that *O*-acetylation of 3-hydroxyglaucine (3-hydroxy-1,2,9,10-tetramethoxyaporphine) causes a shielding of 4.9 ppm of the *ipso* carbon, deshielding of 5.8 and 6.2 ppm for the two *ortho* carbons, and deshielding of 5.3 ppm for the *para* carbon.

For spectra of phenolic aporphines run in dimethyl sulfoxide useful information can often be obtained by comparison of the original spectrum with that produced on the addition of NaOD [174,175]. Formation of the phenate ion leads to a considerable deshielding of the *ipso* carbon (9–16 ppm) and a pronounced but variable shielding of between 2 and 12 ppm in the *para* carbon. Smaller deshieldings of between 2 and 5 ppm are observed *ortho* to the phenate. Ronsch et al. [175] have observed that this effect is transmitted to the other ring of the biphenyl system and that formation of the phenate of 3-hydroxyglaucine leads to a shielding of 3.2 ppm for C(11a) and a shielding of 1.8 ppm for C(9).

The occurrence of ring-C oxygenation and alkylation has marked effects on the observed resonances as does partial or total unsaturation of ring B (see Fig. 1). The ^{13}C NMR spectra of a number of oxoaporphines and dehydroaporphines have been analyzed by Marsaioli et al. [176] and by Castedo et al. [177], and a similar listing of proaporphines and their reduced D ring derivatives is also available [178]. Examples of the spectra obtained for these types of aporphinoids are given in Fig. 2.

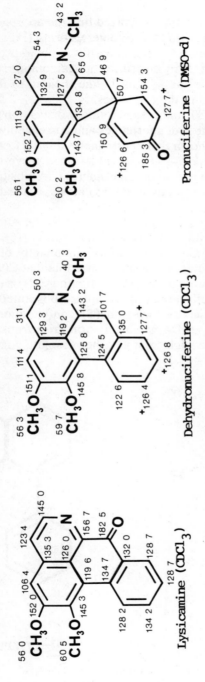

Figure 2. Examples of ^{13}C NMR resonances for oxoaporphines, dehydroaporphines, and proaporphines (values on a structure with the same number of + signs are interchangeable).

4.6. Optical Activity

It was Shamma [179] who first recognized that the aporphines *s.s.* were not planar but could exist in two stereoisomeric forms (Fig. 3). Because they contain only one asymmetric center [C(6a)] it is obvious that absolute configuration can be recognized by measuring optical rotation; all dextrorotatory compounds, measured at the sodium D line, belong to the (*S*) series and levorotatory compounds to the (*R*) series.

Circular dichroism [180] and optical rotatory dispersion [181] curves have been obtained for a number of aporphines and correlated with absolute configuration. It has been shown that aporphines exhibit a Cotton effect of high amplitude in the region 235–245 nm and that the curve is independent of substitution patterns on the aporphine nucleus. A positive Cotton effect requires the aporphine to belong to the (*S*) series, a negative Cotton effect denotes the (*R*) series.

Finally, it should be noted that there appears to be some correlation between absolute configuration and D-ring substitution [16]. Aporphines substituted at C(9)/C(10) or C(10)/C(11) are predominantly of the (*S*) configuration, whereas those with either a single substituent at C(9), C(10), or C(11) or completely unsubstituted in ring D are of either (*R*) or (*S*) configuration. This is thought [19] to reflect the origins of tetrasubstituted aporphines from precursors, notably reticuline, with an (*S*) configuration, whereas aporphines with a single or no D-ring substitution arise from precursors, such as *N*-methylcoclaurine, which exist in either absolute configuration.

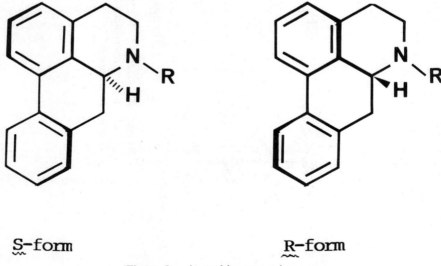

S-form   R-form

Figure 3. Aporphine stereoisomers.

5. CHEMISTRY OF THE APORPHINOIDS OF THE ANNONACEAE

In this section the aporphinoids that have been isolated from the Annonaceae are listed and important features of their chemistry are discussed, paying particular attention to recently isolated alkaloids and those with more un-unusual aporphinoid nuclei.

5.1. Proaporphines

The proaporphines isolated from the Annonaceae are listed in Table 3. The number of structures has not increased since the 1982 review [5] but both

Table 3. Occurrence and Distribution of Proaporphines

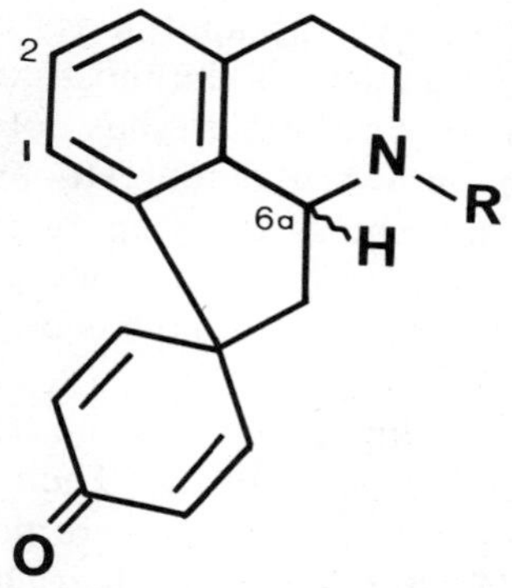

Name	C(1)	C(2)	NR	Sources
(+)-Crotsparine (**56**) (norglaziovine)	OH	OCH₃	H	*Monodora angolensis* [117]
Glaziovine[a] (**57**)	OH	OCH₃	CH₃	*Annona purpurea* [42] *Desmos tiebaghiensis* [62] *Fusaea longifolia* [78] *Uvaria chamae* [145]
N-Methylcrotsparine[a] (**58**)	OH	OCH₃	CH₃	*Isolona zenkeri* [110]
Pronuciferine (**59**)	OCH₃	OCH₃	CH₃	*Isolona pilosa* [110] *Uvaria chamae* [145] *Xylopia buxifolia* [149]
Stepharine (**60**)	OCH₃	OCH₃	H	*Annona muricata* [41] *Annona purpurea* [42] *Monodora tenuifolia* [119,120] *Polyalthia acuminata* [127]

[a] Enantiomers.

glaziovine (57) and pronuciferine (59) have been recorded from additional sources. Structure elucidation of proaporphines can normally be achieved by analysis of spectral data (see Section 4). In acidic conditions a proaporphine is readily converted to the corresponding aporphine [15] with retention of configuration at C(6a), but with inversion of optical activity during this process [182].

5.2. Aporphines

5.2.1. Aporphines *sensu stricto*. In all, a total of 90 different alkaloids with the typical aporphine structure have now been isolated from the Annonaceae; these are listed in Table 4. The most widespread basic substitution pattern, in terms of number of isolations, is that involving only positions C(1) and C(2), but the greatest diversity is seen in aporphines with the 1,2,9,10-substitution pattern. Among aporphines carrying one substituted position in ring D the 1,2,9 pattern is the most common. Both 1,2,10 and 1,2,11 patterns occur but the 1,2,8 pattern has not yet been found (but see Section 5.3). The 1,2,11 pattern is, to date, restricted to species of *Duguetia* and *Guatteria*. For the aporphines carrying two ring-D substituents the 1,2,9,10 pattern has been reported about twice as often as the 1,2,10,11 group. A single 1,2,8,9-substituted alkaloid has been noted (but see Section 5.6).

Perhaps the most interesting group are the 1,2,9,11-substituted aporphines; these are discussed in detail below. Their distribution is similar to the 1,2,11-substituted alkaloids in that they are primarily found in species of *Guatteria* and *Duguetia*, although one of their number, calycinine, is also reported from a species of *Fissistigma*. The only other position that is reported to be substituted is C(3). Approximately 25% of aporphines reported are substituted at C(3) (either methoxyl of hydroxyl), and these occur with all basic substitution patterns except the rare 1,2,10 and 1,2,8,9 types.

Substituents on rings A and D consist either of hydroxy, methoxy, or methylenedioxy groups, the only exception being a 1-acetoxy function in 1-acetoxyapoglaziovine (62). Methylenedioxy groups occur most commonly at C(1)/C(2) and are to be found in greater abundance among some of the basic patterns, notably the 1,2,9, 1,2,11, and 1,2,9,11 groups. In the typical 1,2,9,10- and 1,2,10,11-substituted aporphines of the Annonaceae, C(10) is always found as a methoxy, or occasionally in a methylenedioxy group, but never as a hydroxy function. This would be expected in the direct oxidative coupling of a 1-btiq precursor, where the radical-initiating phenolic group in ring C must finish at either C(9) or C(11) of the aporphine (see Scheme 1). The one exception to this rule is suaveoline (144), but the reported structure for this alkaloid is in doubt. One other alkaloid for which the proposed structure is open to question is 88, reported from *Monanthotaxis cauliflora* [116]. The C(9) hydroxy group in 88 was placed by reference to the antici-

Table 4.   Occurrence and Distribution of Aporphines _s.s._

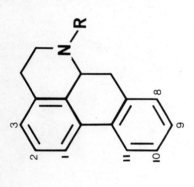

Name	C(1)	C(2)	C(3)	NR	C(8)	C(9)	C(10)	C(11)	Sources
Actinodaphnine (61)	O—CH₂—O		—	H	—	OH	—	OCH₃	—	_Guatteria scandens_ [99]
O-1-Acetylapoglaziovine (62)	OAc	OCH₃	—	CH₃	—	—	OH	—	—	_Fusaea longifolia_ [78]
Anolobine (63)	O—CH₂—O		—	H	—	OH	OH	—	—	_Annona glabra_ [33, 34]
										Annona squamosa [48]
										Asimina triloba [55]
										Duguetia obovata [65]
										Fissistigma oldhamii [75]
										Guatteria scandens [99]
										Guatteria schomburgkiana [102,103]
										Guatteria sagotiana [104]
										Monodora tenuifolia [119–120]
										Polyalthia acuminata [127]
										Schefferomitra subaequalis [142]
										Xylopia frutescens [151]

Table 4. (continued)

Name	C(1)	C(2)	C(3)	NR	C(8)	C(9)	C(10)	C(11)	Sources
Anonaine (**64**)	O—CH$_2$—O		H		—	—	—	—	*Annona cherimolia* [27–30]
									Annona crassiflora [31]
									Annona glabra [33,34]
									Annona montana [38,39]
									Annona reticulata [43–46]
									Annona squamosa [48–51]
									Cananga odorata [57]
									Desmos tiebaghiensis [62]
									Enantia polycarpa [73]
									Fusaea longifolia [78]
									Guatteria schomburgkiana [103]
									Hexalobus crispiflorus [107]
									Isolona campanulata [109]
									Isolona pilosa [110]
									Meiogyne virgata [113]
									Mitrella kentii [115]
									Monodora tenuifolia [119–120]
									Polyalthia acuminata [127]
									Polyalthia emarginata [130]
									Polyalthia oliveri [133]
									Pseuduvaria cf. *grandifolia* [131]
									Pseuduvaria sp. [131]
									Schefferomitra subaequalis [131, 142]
									Unonopsis guatterioides [144]
									Xylopia brasiliensis [148]
									Xylopia buxifolia [149]
									Xylopia frutescens [152]
									Xylopia pancheri [154]
									Xylopia papuana [155]

Asimilobine (65)   OCH₃ OH   —   H   —   —   —

Compound	Substituents
Asimilobine (65)	OCH₃, OH, —, H, —, —, —
Boldine (66)	OCH₃, OH, —, CH₃, —, OH, OCH₃
Buxifoline (67)	O—CH₂—O, OCH₃, H, OCH₃, —
Caaverine (68)	OH, OCH₃, —, H, —
Calycinine (69) (fissoldine) (fissistigine A)	O—CH₂—O, —, H, —, OCH₃, —, OH
N-Carbamoylanonaine (70)	O—CH₂—O, —, CONH₂, —
N-Carbamoylasimilobine (71)	OCH₃, OH, —, CONH₂, —

Anaxagorea dolichocarpa [25]
Anaxagorea prinoides [25]
Annona crassiflora [31]
Annona glabra [33,34]
Annona montana [39]
Asimina triloba [55]
Cymbopetalum brasiliense [61]
Desmos tiebaghiensis [62]
Fusaea longifolia [78]
Guatteria scandens [99]
Hexalobus crispiflorus [107]
Meiogyne virgata [113]
Melodorum punctulatum [114]
Mitrella kentii [115]
Monanthotaxis cauliflora [116]
Polyalthia acuminata [127]
Popowia cf. cyanocarpa [137]
Popowia pisocarpa [138]
Schefferomitra subaequalis [142]
Unonopsis guatterioides [144]
Uvaria chamae [145]
Xylopia frutescens [152]
Desmos tiebaghiensis [62]
Polyalthia cauliflora var. beccarii [128]
Duguetia obovata [65]
Xylopia buxifolia [149]
Isolona pilosa [110]
Isolona zenkeri [110]
Polyalthia acuminata [127]
Duguetia calycina [63]
Duguetia obovata [65]
Fissistigma oldhamii [75–77]
Hexalobus crispiflorus [107]
Hexalobus crispiflorus [107]

181

Table 4. (*continued*)

Name	C(1)	C(2)	C(3)	NR	C(8)	C(9)	C(10)	C(11)	Sources
Corydine (**72**)	OH	OCH$_3$	—	CH$_3$	—	—	OCH$_3$	OCH$_3$	*Annona squamosa* [50,51] *Guatteria cubensis* [81] *Guatteria moralessi* [81] *Xylopia danguyella* [149]
Corytuberine (**73**)	OH	OCH$_3$	—	CH$_3$	—	—	OCH$_3$	OH	*Annona cherimolia* [29,30] *Meiogyne virgata* [113]
Danguyelline (**74**)	OCH$_3$	OH[a]	OCH$_3$[a]	H	—	—	OCH$_3$	OH	*Xylopia danguyella* [149]
O-Demethylpurpureine (**75**)	OCH$_3$	OH[a]	OCH$_3$[a]	CH$_3$	—	OCH$_3$	OCH$_3$	—	*Annona purpurea* [42]
Dicentrine (**76**)	O—CH$_2$	—O	—	CH$_3$	—	OCH$_3$	OCH$_3$	—	*Duguetia* sp. [70]
Discoguattine (**77**) (*O*-methylcalycinine)	O—CH$_2$	—O	—	H	—	OCH$_3$	—	OCH$_3$	*Guatteria discolor* [82,84]
Duguevanine (**78**)	O—CH$_2$	—O	OCH$_3$	H	—	OCH$_3$	—	OH	*Duguetia obovata* [65]
Elmerrillicine (**79**)	O—CH$_2$	—O	OCH$_3$	H	—	—	—	OH	*Guatteria sagotiana* [104]
Formouregine (**80**)	OCH$_3$	OCH$_3$	OCH$_3$	CHO	—	—	—	—	*Guatteria ouregou* [93]
N-Formylanonaine (**81**)	O—CH$_2$	—O	—	CHO	—	—	—	—	*Hexalobus crispiflorus* [107]
N-Formylbuxifoline (**82**)	O—CH$_2$	—O	OCH$_3$	CHO	—	OCH$_3$	—	—	*Duguetia obovata* [65]
N-Formylduguevanine (**83**)	O—CH$_2$	—O	OCH$_3$	CHO	—	OCH$_3$	—	OH	*Duguetia obovata* [65]
N-Formylnornuciferine (**84**)	OCH$_3$	OCH$_3$	—	CHO	—	—	—	—	*Guatteria ouregou* [93]
N-Formylputerine (**85**)	O—CH$_2$	—O	—	CHO	—	—	—	OCH$_3$	*Duguetia calycina* [65] *Guatteria schomburgkiana* [103]
N-Formylxylopine (**86**)	O—CH$_2$	—O	—	CHO	—	OCH$_3$	—	—	*Duguetia obovata* [65]
Glaucine (**87**)	OCH$_3$	OCH$_3$	—	CH$_3$	—	OCH$_3$	OCH$_3$	—	*Alphonsea ventricosa* [24] *Annona squamosa* [50,51] *Pseuduvaria* cf. *dolichonema* [131] *Uvaria chamae* [145]
1,2-Dimethoxy-9-hydroxynoraporphine (**88**)	OCH$_3$	OCH$_3$	—	H	—	OH	—	—	*Monanthotaxis cauliflora* [116]
3-Hydroxynornuciferine (**89**)	OCH$_3$	OCH$_3$	OH	H	—	—	—	—	*Guatteria goudotiana* [87] *Guatteria melosma* [60] *Guatteria ouregou* [93]

Compound							Species [ref.]
3-Hydroxynuciferine (90) (lirinine)	OCH₃	OCH₃	OH	CH₃	—	—	*Guatteria sagotiana* [104]
							Hexalobus crispiflorus [107,108]
							Polyalthia acuminata [127]
Isoboldine (91)	OH	OCH₃	—	CH₃	OH	OCH₃	*Fusaea longifolia* [78]
							Guatteria ouregou [93]
							Annona cherimolia [29,30]
							Annona glabra [33,34]
							Annona montana [39]
							Desmos tiebaghiensis [62]
							Enantia polycarpa [73]
							Fusaea longifolia [78]
							Guatteria chrysopetala [80]
							Guatteria goudotiana [87]
							Guatteria melosma [88,89]
							Monodora cf. *brevipes* [117,118]
							Polyalthia acuminata [127]
							Schefferomitra subaequalis [142]
							Uvaria chamae [145]
							Xylopia danguyella [149]
							Xylopia frutescens [151]
Isocalycinine (92)	O—CH₂—O		—	H	OH	OCH₃	*Guatteria discolor* [82]
Isocorydine (93)	OCH₃	OCH₃	—	CH₃	OCH₃	OH	*Annona purpurea* [42]
							Annona squamosa [51]
							Artabotrys suaveolens [53,54]
							Asimina triloba [55]
Isodomesticine (94)	OCH₃	OH	—	H	O—CH₂—O		*Enantia polycarpa* [73,74]
Isolaureline (95) (*N*-methylxylopine)	O—CH₂—O		—	CH₃	OCH₃	—	*Guatteria goudotiana* [87]
Isopiline (96)	OH	OCH₃	OCH₃	H	—	—	*Duguetia obovata* [65]
Laurelliptine (97) (norisoboldine)	OH	OCH₃	—	H	OH	OCH₃	*Guatteria ouregou* [93,95]
							Isolona pilosa [110,111]
							Polyalthia acuminata [127]
							Monanthotaxis cauliflora [116]
							Monodora tenuifolia [119–121]

183

Table 4. (*continued*)

Name	C(1)	C(2)	C(3)	NR	C(8)	C(9)	C(10)	C(11)	Sources
Laurolitsine (**98**) (norboldine)	OCH₃	OH	—	H	—	OH	OCH₃	—	*Xylopia papuana* [155]
Laurotetanine (**99**)	OCH₃	OCH₃	—	H	—	OH	OCH₃	—	*Desmos tiebaghiensis* [62] *Guatteria scandens* [99] *Xylopia danguyella* [149] *Xylopia frutescens* [152]
Lindcarpine (**100**)	OCH₃	OH	—	H	—	—	OCH₃	OH	*Guatteria goudotiana* [87] *Fusaea longifolia* [78] *Guatteria ouregou* [93] *Guatteria sagotiana* [104]
Lirinidine (**101**)	OH	OCH₃	—	CH₃	—	—	—	—	*Isolona zenkeri* [110]
Magnoflorine (**102**)	OH	OCH₃	—	(CH₃)₂	—	—	OCH₃	OH	*Cymbopetalum brasiliense* [61] *Enantia polycarpa* [73] *Monodora tenuifolia* [119–120]
Menisperine (**103**)	OCH₃	OCH₃	—	(CH₃)₂	—	—	OCH₃	OH	*Enantia polycarpa* [73] *Guatteria ouregou* [93] *Polyalthia acuminata* [127]
3-Methoxynuciferine (**104**)	OCH₃	OCH₃	OCH₃	CH₃	—	—	—	—	
N-Methylactinodaphnine (**105**)	O—CH₂—O		—	CH₃	—	OH	OCH₃	—	*Annona glabra* [37]
N-Methylasimilobine (**106**)	OCH₃	OH	—	CH₃	—	—	—	—	*Xylopia buxifolia* [149]
N-Methylbuxifoline (**107**)	O—CH₂—O		OCH₃	CH₃	—	OCH₃	—	—	*Duguetia obovata* [65]
N-Methylcalycinine (**108**)	O—CH₂—O		—	CH₃	—	OCH₃	—	OH	*Duguetia obovata* [65]
N-Methylcorydine (**109**)	OH	OCH₃	—	(CH₃)₂	—	—	OCH₃	OCH₃	*Polyalthia oliveri* [133]
N-Methylduguevanine (**110**)	O—CH₂—O		OCH₃	CH₃	—	OCH₃	—	OH	*Duguetia obovata* [65]
N-Methylelmerrillicine (**111**)	O—CH₂—O		OCH₃	CH₃	—	—	OCH₃	OCH₃	*Guatteria sagotiana* [104]
10-*O*-Methylhernovine (**112**)	OCH₃	OH	—	H	—	—	OCH₃	OCH₃	*Guatteria discolor* [83]
N-Methylisopiline (**113**)	OH	OCH₃	OCH₃	CH₃	—	—	—	—	*Guatteria ouregou* [93]
O-Methylisopiline (**114**) (*O*-methylnorlirinine)	OCH₃	OCH₃	OCH₃	H	—	—	—	—	*Guatteria ouregou* [93,95] *Guatteria scandens* [99] *Polyalthia acuminata* [127]
N-Methyllaurotetanine (**115**)	OCH₃	OCH₃	—	CH₃	—	OH	OCH₃	—	*Desmos tiebaghiensis* [62] *Enantia polycarpa* [73] *Guatteria goudotiana* [87]

Table (rotated 90°). Aporphine alkaloids — substituent data and plant sources.

Compound	Substituents (as printed, left → right)	Source species [Ref.]
O-Methylpukateine (**116**) (*N*-methylputerine)	O—CH₂—O · — · CH₃ · — · — · OCH₃	*Guatteria scandens* [99]; *Xylopia frutescens* [152]; *Duguetia calycina* [63]
Nantenine (**117**)	OCH₃ · OCH₃ · O—CH₂—O · CH₃	*Guatteria discolor* [82,83]; *Guatteria schomburgkiana* [103]; *Guatteria sagotiana* [104]
Neolitsine (**118**)	O—CH₂—O · O—CH₂—O · CH₃	*Xylopia frutescens* [152]; *Guatteria goudotiana* [87]
Norannuradhapurine (**119**)	O—CH₂—O · H · OH · OCH₃	*Polyalthia acuminata* [127]
Norcorydine (**120**)	OH · OCH₃ · H · — · OCH₃ · OCH₃	*Annona squamosa* [50]; *Popowia* cf. *cyanocarpa* [137]; *Popowia pisocarpa* [138]; *Xylopia danguyella* [149]; *Xylopia pancheri* [154]
Nordicentrine (**121**)	O—CH₂—O · H · OCH₃ · OCH₃	*Guatteria scandens* [99]; *Alphonsea ventricosa* [24]
Norglaucine (**122**)	OCH₃ · OCH₃ · H · OCH₃ · OCH₃	*Duguetia* sp. [70]
Norisocorydine (**123**)	OCH₃ · OCH₃ · H · — · OCH₃ · OH	*Pseuduvaria* cf. *dolichonema* [131]; *Annona squamosa* [50]; *Xylopia danguyella* [149]
Norisodomesticine (**124**)	OCH₃ · OH · H · O—CH₂—O	*Guatteria goudotiana* [87]; *Xylopia danguyella* [149]
Norlaureline (**125**)	O—CH₂—O · H · OCH₃	*Guatteria elata* [85]; *Guatteria sagotiana* [104]
Norliridinine (**126**)	OCH₃ · OH · OCH₃ · H	*Polyalthia acuminata* [127]
Normantenine (**127**)	OCH₃ · OCH₃ · H · O—CH₂—O	*Annona cherimolia* [29,30]; *Xylopia danguyella* [149]; *Xylopia frutescens* [152]
Nornuciferine (**128**)	OCH₃ · OCH₃ · H	*Annona glabra* [33,34]; *Enantia polycarpa* [73]; *Guatteria chrysopetala* [80]; *Guatteria ouregou* [93]; *Guatteria sagotiana* [104]; *Hexalobus crispiflorus* [107]; *Isolona campanulata* [109]; *Isolona pilosa* [110]

185

Table 4. (*continued*)

Name	C(1)	C(2)	C(3)	NR	C(8)	C(9)	C(10)	C(11)	Sources
Noroconovine (**129**)	OCH₃	OCH₃	OCH₃	H	—	—	OCH₃	OH	*Polyalthia acuminata* [127] *Pseuduvaria* cf. *grandifolia* [131] *Pseuduvaria* sp. [131] *Xylopia buxifolia* [149] *Xylopia frutescens* [152]
Norpredicentrine (**130**)	OCH₃	OH	—	H	—	OCH₃	OCH₃	—	*Polyalthia oligosperma* [130] *Guatteria goudotiana* [87] *Guatteria scandens* [99] *Pseuduvaria* cf. *dolichonema* [131]
Norpurpureine (**131**)	OCH₃	OCH₃	OCH₃	H	—	OCH₃	OCH₃	—	*Annona purpurea* [42] *Hexalobus crispiflorus* [107,108] *Xylopia buxifolia* [149]
Norstephalagine (**132**)	O—CH₂—O		OCH₃	H	—	—	—	—	*Guatteria ouregou* [93]
Nuciferine (**133**)	OCH₃	OCH₃	—	CH₃	—	—	—	—	*Monanthotaxis cauliflora* [116]
Obovanine (**134**)	O—CH₂—O		—	H	—	—	—	OH	*Duguetia calycina* [63] *Guatteria sagotiana* [104]
Oureguattidine (**135**)	OH	OCH₃	OCH₃	H	—	OH	—	—	*Guatteria ouregou* [95]
Oureguattine (**136**)	OCH₃	OCH₃	OCH₃	H	—	OH	—	—	*Guatteria ouregou* [93]
Polygospermine (**137**)	OCH₃	OCH₃	OCH₃	H	—	—	O—CH₂—O		*Polyalthia oligosperma* [130]
Predicentrine (**138**)	OCH₃	OH	—	CH₃	—	OCH₃	OCH₃	—	*Polyalthia cauliflora* var. *beccarii* [128]
Pukateine (**139**)	O—CH₂—O		—	CH₃	—	—	—	OH	*Guatteria sagotiana* [104]
Purpureine (**140**) (thalicsimidine)	OCH₃	OCH₃	OCH₃	CH₃	—	OCH₃	OCH₃	—	*Annona purpurea* [42]
Puterine (**141**)	O—CH₂—O		—	H	—	—	—	OCH₃	*Duguetia calycina* [63] *Guatteria discolor* [82,83] *Guatteria elata* [85] *Guatteria schomburgkiana* [103] *Guatteria sagotiana* [104]
Roemerine (**142**)	O—CH₂—O		—	CH₃	—	—	—	—	*Annona glabra* [33,34] *Annona squamosa* [50]

Compound							
Sparsiflorine (143)[b]	OH	OCH$_3$	—	H	—	OH	—
Suaveoline (144)[b]	OCH$_3$	OCH$_3$	—	CH$_3$	—	OH	OH
Thaliporphine (145)	OH	OCH$_3$	—	CH$_3$	—	OCH$_3$	OCH$_3$
Tuduranine (146)	OCH$_3$	OCH$_3$	—	H	—	OH	—
Wilsonirine (147)	OH	OCH$_3$	—	H	—	OCH$_3$	OCH$_3$
Xyloguyelline (148)	OCH$_3$	OH[a]	OCH$_3$[a]	H	—	O—CH$_2$—O	—
Xylopine (149)	O—CH$_2$—O	—	H	—	OCH$_3$	—	
Zenkerine (150)	OH	OCH$_3$	—	H	—	OCH$_3$	—

Sources:

Cananga odorata [57]
Guatteria modesta [91,92]
Guatteria sagotiana [104]
Isolona pilosa [110]
Xylopia pancheri [154]
Xylopia papuana [155]
Monodora angolensis [117]
Monodora tenuifolia [119,120]
Artabotrys suaveolens [53,54]
Uvaria chamae [145]
Polyalthia acuminata [127]
Monodora angolensis [117]
Popowia cf. *cyanocarpa* [137]
Popowia pisocarpa [138]
Xylopia danguyella [149]
Annona montana [38,39]
Annona squamosa [52]
Duguetia calycina [63]
Duguetia obovata [65]
Fissistigma oldhamii [75–77]
Guatteria discolor [83]
Guatteria scandens [99]
Guatteria schomburgkiana [103]
Guatteria sagotiana [104]
Xylopia brasiliensis [148]
Xylopia buxifolia [149]
Xylopia discreta [150]
Xylopia frutescens [152]
Xylopia pancheri [154]
Xylopia papuana [155]
Isolona pilosa [110]
Isolona zenkeri [110]

[a] Positions may be reversed.
[b] This structure is doubtful.

pated change in the UV spectrum on addition of alkali. However, methylation produced an alkaloid apparently identical (by comparison of published data) with *O*-methyllirinine, which was then thought to be 1,2-methoxy-9-hydroxynoraporphine but which has now been reassigned a 1,2-methoxy-3-hydroxy substitution pattern [10]. Unfortunately **88** was only isolated in trace amounts and no NMR data are available, but its UV characteristics strongly suggest that structure **88** is correct and that comparability of UV, IR, and m.p. data to those for *O*-methyllirinine was purely fortuitous.

Substitution of N(6) is normally either *N*-methyl (aporphines) or N-H (noraporphines). Quaternary *N*,*N*-dimethylaporphinium ions are rather rare in the Annonaceae (only three recorded: **102**, **103**, **109**) but ability to methylate to the quaternary level may be more widespread than this suggests in view of the greater difficulty in isolating quaternary alkaloids and their probable role as intermediates in the formation of phenanthrenes. So far, no *N*-oxides have been recorded among the aporphines *s.s.* of the Annonaceae, but they are known both in the 7-hydroxyaporphines and among other aporphinoids. Examples of unusual oxidized derivatives of *N*-methyl compounds are seen in the *N*-carbamoyl (**70, 71**) and *N*-formyl (**80–86**) noraporphines, isolated notably from *Hexalobus crispiflorus*, but also from species of *Duguetia* and *Guatteria*. These also occur among the 7-alkylaporphines and can be implicated in the formation of pentacyclic aporphines such as duguenaine (see Section 5.2.3).

One particularly interesting feature of the *N*-formyl noraporphines is their occurrence with both an (*E*) and a (*Z*) relationship between the formyl proton and H(6a). This phenomenon has been observed in the ^{1}H NMR spectrum of a number of alkaloids [65,107] with two distinct signals for the aldehyde proton and H(6a) protons clearly visible, and occasionally also for H(5$_{eq}$). The (*E*) form, which always represents the minor isomer of the mixture, exhibits the more deshielded aldehyde signal (δ 8.30–8.40) and the more shielded H(6a) signal (δ ~4.50). By contrast, in the more abundant (*Z*) form the aldehyde proton is more shielded by up to 0.12 ppm and H(6a) is more deshielded by about 0.5 ppm. The mass spectra of *N*-formylnoraporphines exhibit the same type of ring-B fragmentation as previously noted for *N*-acetylnoraporphines, resulting in the formation of the major ion $M^{+\cdot}-58$, through loss of $CH_2{=}N^+H(CHO)$. The structure of *N*-formylnoraporphines can also be readily studied by their preparation from the corresponding noraporphines, using formic acid/acetic anhydride, or by reduction to the corresponding aporphine using lithium aluminum hydride [65]. Likewise, *N*-carbamoylnoraporphines have been prepared from the corresponding noraporphines by means of treatment with chlorosulfonyl isocyanate in ether at room temperature [107]. Interestingly, *N*-formylnoraporphines have at present been isolated only from the genera *Duguetia*, *Guatteria*, and *Hexalobus*.

The commonest individual alkaloids are among the simplest, notably the C(1), C(2) substituted anonaine (**64**), asimilobine (**65**), and nornuciferine

(128), of which all three are noraporphines. Anolobine (63) and xylopine (149), 1,2,9-substituted noraporphines, have also been recorded quite widely. The most widespread tetrasubstituted aporphine is isoboldine (91).

Although many of the aporphines noted in Table 4 can be regarded as typical and their identification and biogenesis have been covered adequately in Sections 4 and 2, respectively, two groups do require further discussion, the 1,2,11- and 1,2,9,11-substituted aporphines. Although the 1,2,11-substituted compounds (79, 85, 116, 134, 139, 141) are reasonably well known, having been recorded from several families [8–10], the 1,2,9,11 pattern (69, 77, 78, 83, 92, 108, 110) appears, to date, to be unique to the Annonaceae. Although no synthesis of the 1,2,9,11-substituted aporphines has yet been reported, the spectroscopic evidence on which their structures are based [63,65,82], notably the use of dimethyl sulfoxide and dimethyl sulfoxide/ NaOD NMR shifts, leaves no doubt with regard to their authenticity. Within the Annonaceae the distribution of the two types appears to be interlinked, inasmuch as the 1,2,9,11 pattern has so far been recorded from *Duguetia calycina*, *D. obovata*, *Guatteria discolor*, and *Fissistigma oldhamii*. Of the five species that have yielded the 1,2,11 pattern, two are *D. calycina* and *G. discolor*, and the others are the allied species *G. elata*, *G. schomburgkiana*, and *G. trichostemon*. The concentration of these substitution patterns in alkaloids from *Duguetia* and *Guatteria* is confirmed if the comparison is extended to include 7-alkylaporphines and oxoaporphines.

The 1,2,9,11-substituted aporphines, with their very unusual *meta*-substituted D ring, offer a challenge to the general biosynthetic pathways proposed for the aporphines (see Section 2). Possible routes for their formation are numerous and several have been outlined by Debitus [183]. We suggest that the following are worthy of consideration.

1.   Routes involving 1-btiq precursors already modified so as to give the 1,2,9,11 pattern on cyclization.

(a)   The simplest solution would be a pathway involving cyclization of a 1-btiq alkaloid with *meta* substitution in ring C (Scheme 14). If this is the pathway involved then it could explain the interesting dichotomy between *Guatteria* on the one hand and *Duguetia* and *Fissistigma* on the other. In *Guatteria* C(11) carries a methoxy substituent and C(9) a hydroxy substituent, indicative of *ortho–para* coupling, whereas in the other genera the hydroxy unit in ring D is at C(11), indicative of *ortho–ortho* coupling. Unfortunately this pathway does not appear to offer an alternative route to the 1,2,11-substituted aporphines. This situation makes it difficult to accept this pathway in view of the obvious link in distribution between the two alkaloid types.

(b)   Aporphines trisubstituted in ring D have been isolated from the Lauraceae and are presumed to be derived *via* cyclization of a 3′,4′,5′-substituted 1-btiq precursor, probably through a proaporphine intermediate. If the aporphine were to arise through the proaporphine (151) *via* its proa-

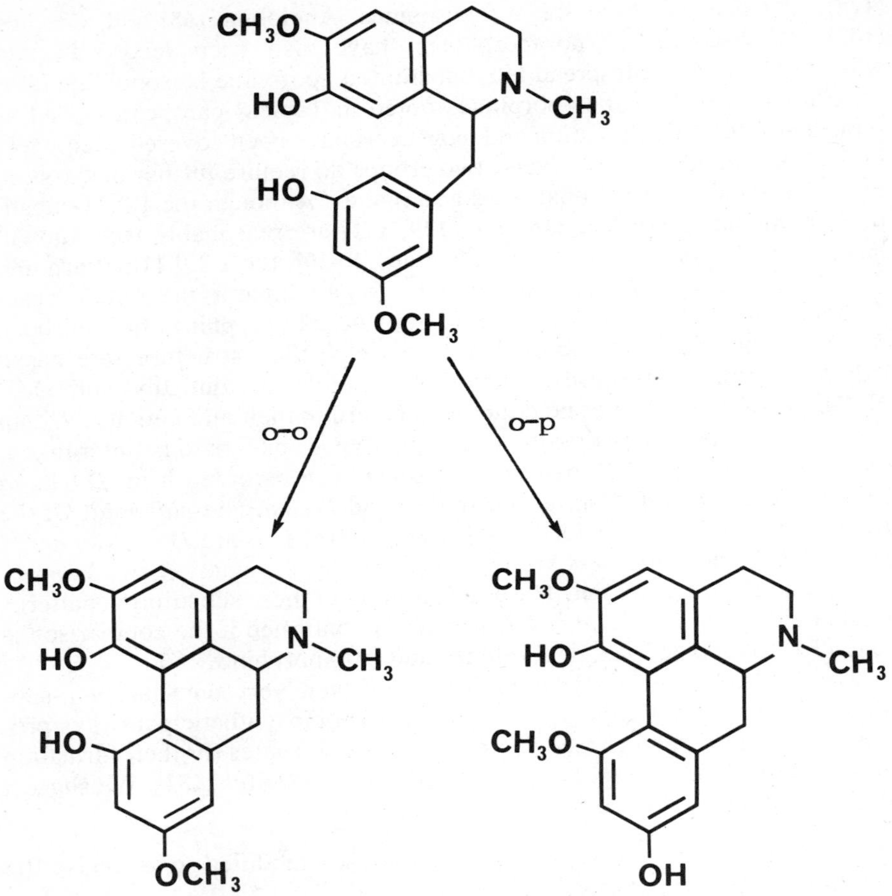

Scheme 14. Possible direct route to the 1,2,9,11-substituted aporphines from a 1-btiq precursor *meta* substituted in ring D.

porphinol with a dienol–benzene rearrangement (as in Scheme 2), then it would result in an alkaloid with the 1,2,9,11 pattern.

(c) A further possibility involves the light-catalyzed rearrangement of **151** through a cyclopropane intermediate, one of the routes proposed by Shamma and Guinaudeau [19] to explain the formation of a number of the rarer D-ring substitution patterns among the aporphines. The mechanism proposed by Debitus [183] is depicted in Scheme 15, but appears more suitable for the explanation of an 8,10 pattern in ring D than the required 9,11 pattern because the latter required an alkyl migration over the normally favored aryl migration.

2. Ring D oxygenation after cyclization of the 1-btiq precursor or after formation of the aporphine.

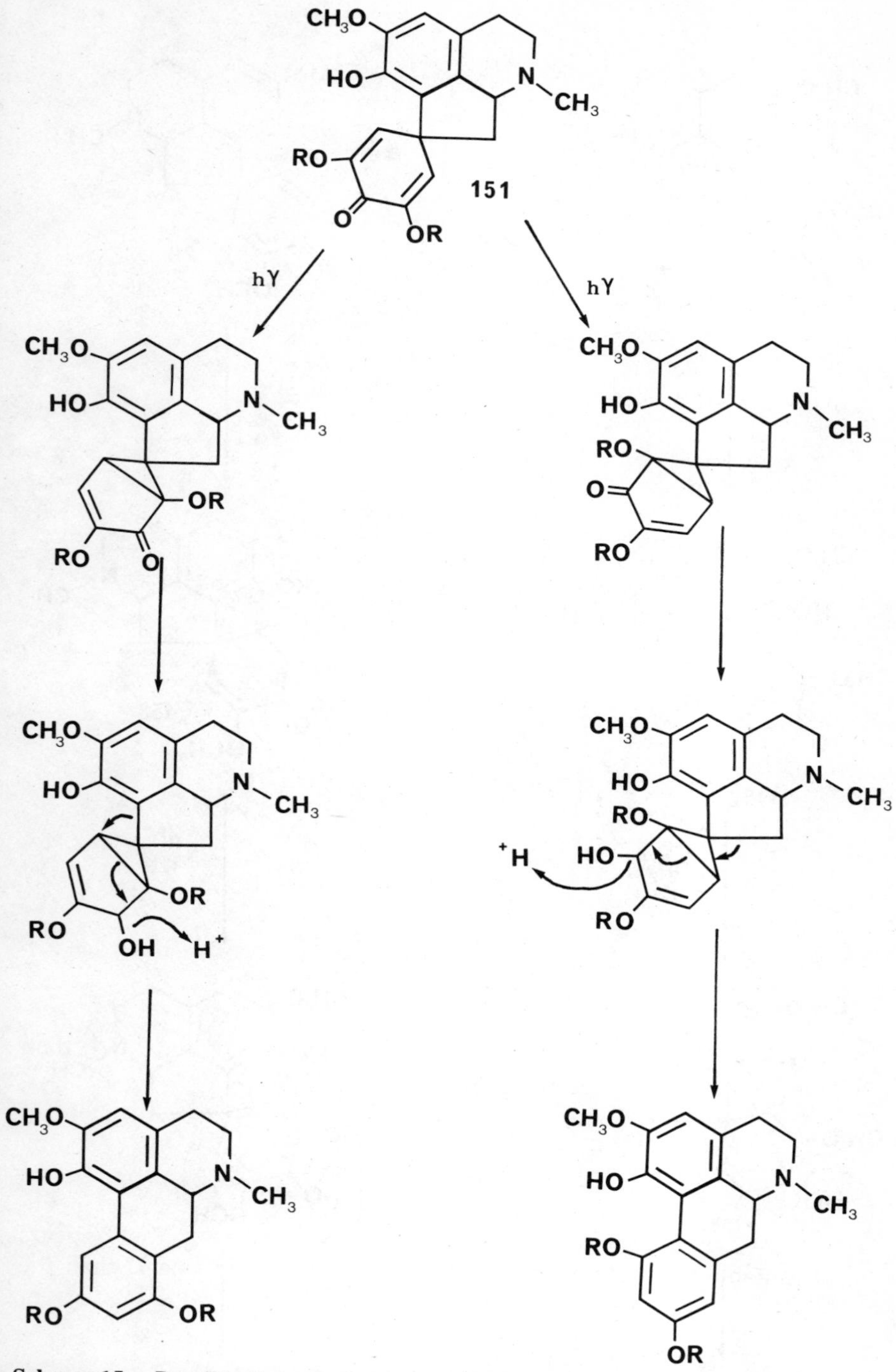

Scheme 15. Possible light-catalyzed routes to aporphines *meta* substituted in ring D from a 3′,4′,5′-trisubstituted 1-btiq precursor.

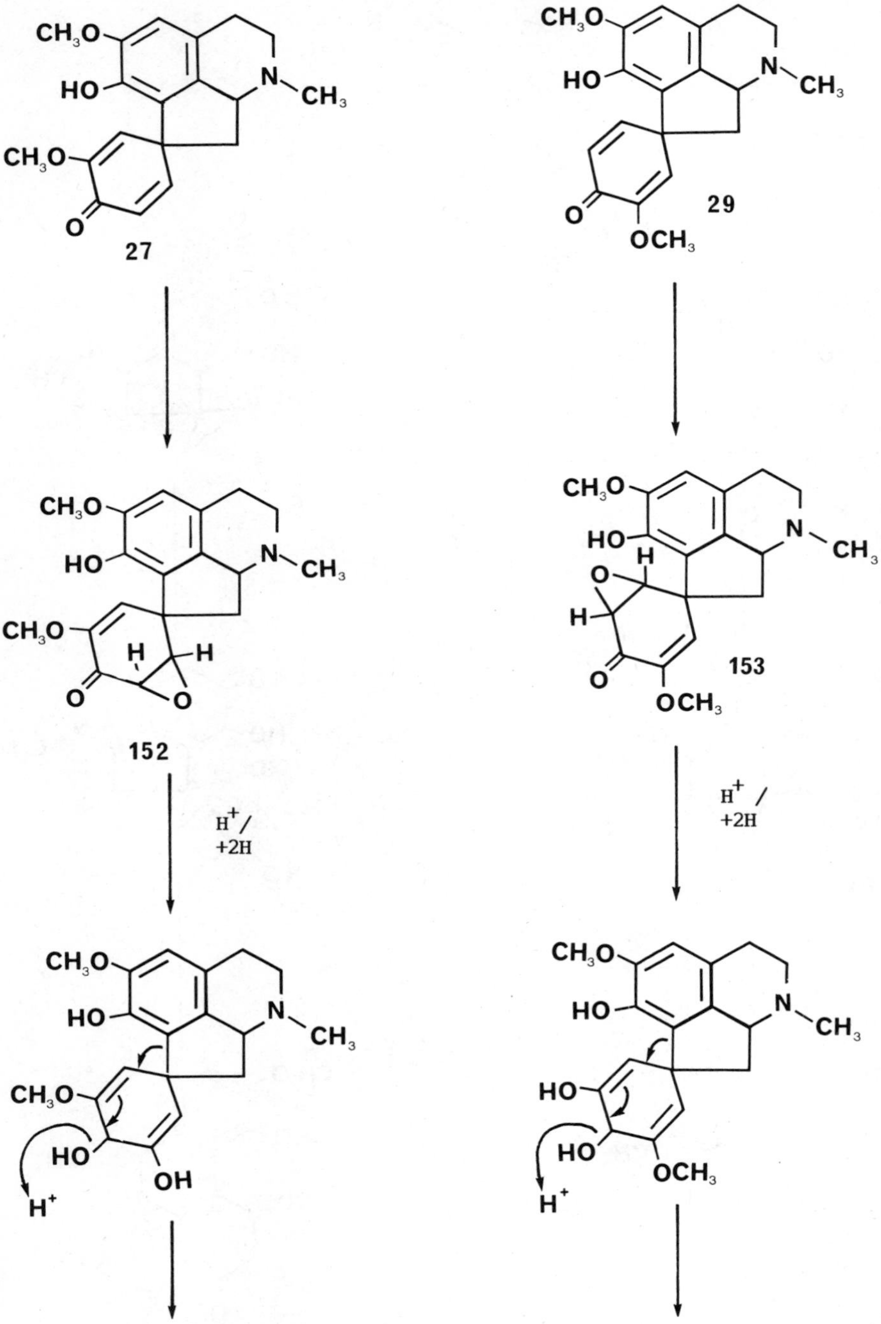

Scheme 16. Possible routes to 1,2,9,11-substituted aporphines through the epoxidation of proaporphine intermediates.

192

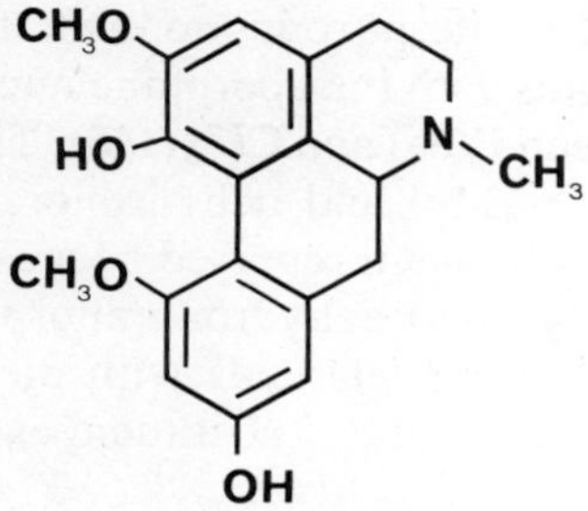

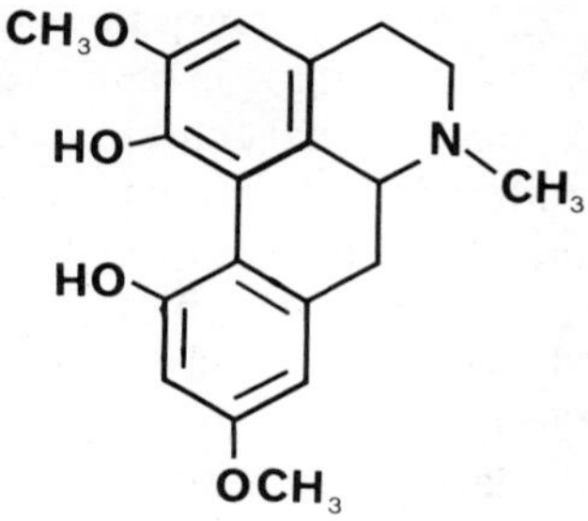

Scheme 16. (*continued*)

(a) Aporphinoids with the 1,2,9,11 pattern are invariably accompanied by 1,2,9-substituted aporphinoids and generally by 1,2,11-substituted aporphinoids. It is possible that C(11) oxidation occurs on a preformed 1,2,9-aporphinoid skeleton or perhaps C(9) oxidation on a 1,2,11-substituted skeleton.

(b) Oxidation of an intermediate proaporphine following one of the normal routes would also be plausible. Such a pathway could be analogous to Scheme 3 but with oxidation of ring D through addition of an epoxide across the free double bond at either the aporphine or proaporphinol stage, for example, 27 → 152, 29 → 153. An enol–benzene rearrangement could then lead to conversion of 152 and 153 into 1,2,9,11-substituted aporphines, with 152 being the more likely intermediate in *Duguetia* and 153 the more likely in *Guatteria*, for these lead to hydroxyl groups in the correct positions (Scheme 16). This hypothesis has the virtue of being able to explain not only the 1,2,9,11 pattern but also the 1,2,9 and 1,2,11 patterns.

(c) Following the theme of a light-catalyzed reaction (see Scheme 15), then epoxidation of a proaporphine-derived intermediate 154 with only two original D-ring substituents could lead to the 1,2,9,11 pattern.

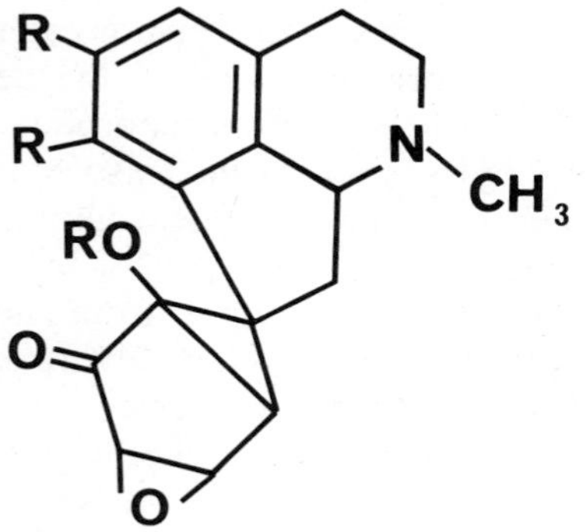

154

5.2.2. Dehydroaporphines. As pointed out in the introduction, the term dehydroaporphine is restricted to those alkaloids with the aporphine nucleus and an additional center of unsaturation between C(6a) and C(7) (**155**). These include the dehydroaporphines *s.s.* (**155**, R = Me) and dehydronoraporphines (**155**, R = H), the latter having only rarely been reported because of their relative instability. Recently a single *N*-formyl dehydronoraporphine (**155**, R = CHO) has also been isolated. Other aporphinoids with the unsaturation pattern of dehydroaporphines but with an alkyl substituent at C(7) are treated separately in Section 5.2.3.

As recently as 1982, no alkaloids fitting the above requirements for inclusion as dehydroaporphines had been reported from the Annonaceae [5]. However, in the past 2 years eight have been recorded (Table 5) out of a total of 25 known dehydroaporphines [8–10]. An examination of Table 5 reveals that to date the distribution of this group is restricted to just three genera, *Guatteria*, *Hexalobus*, and *Polyalthia*, with D-ring substitution restricted to C(9) and C(10). Several of the dehydroaporphines are C(3) substituted.

Of the eight alkaloids, **158** and **160–163** do not require particular comment. On the other hand, dehydro-*O*-methylisopiline (**157**) and dehydronornuciferine (**159**), both from *Guatteria ouregou* [93], deserve attention as the only two naturally occurring dehydronoraporphines, other than the tentative report of dehydroanonaine from *Nelumbo nucifera* (Nympheaceae), based on GLC/MS evidence only [184].

Dehydroformouregine (**156**), also from *Guatteria ouregou* [93], is the first natural *N*-formyl dehydroaporphine, its identity being deduced from the study of spectral data. The UV spectrum was typical of dehydroaporphines (Table 2) and the IR (ν_{max} 1675 cm^{-1}) indicated the amide carbonyl. The ^{1}H NMR spectrum revealed several important features, notably the resonance of H(4) and H(5) protons as two (2H) triplets ($J = 6$ Hz) at δ 3.23 and 4.13 and H(7) as a deshielded singlet at δ 7.38. The formyl proton also occurred as a singlet, at δ 8.90, thereby indicating that **156** exists in a single conformation, unlike the *N*-formylnoraporphines (Section 5.2.1). The presence of the formyl group appears to confer a greater stability on **156** than is seen in the dehydronoraporphines or even the dehydroaporphines. Although **156** is the only *N*-formyl dehydronoraporphine noted here, two other alkaloids, duguespixine and trichoguattine, also have the *N*-formyl dehydronoraporphine nucleus but with additional alkylation at C(7); they are included in Section 5.2.3.

Dehydroaporphines have been obtained from the corresponding aporphines by a number of oxidative methods, involving permanganate, mercury salts, iodine, and DDQ [185], and most recently with good yields using 1,4-benzoquinone in diethylene glycol dimethyl ether [186]. Catalytic dehydrogenation of the aporphine, using mild conditions such as palladium–charcoal under reflux in acetonitrile, also gives good yields [187]. Although these methods are generally satisfactory for nonphenolic aporphines, their phe-

Table 5. Occurrence and Distribution of Dehydroaporphines

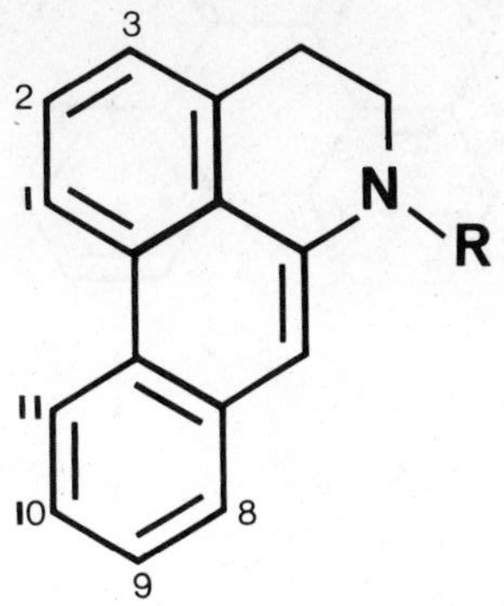

Name	C(1)	C(2)	C(3)	NR	C(8)	C(9)	C(10)	C(11)	Sources
Dehydroformouregine (**156**)	OCH$_3$	OCH$_3$	OCH$_3$	CHO	—	—	—	—	*Guatteria ouregou* [93]
Dehydro-*O*-methylisopiline (**157**)	OCH$_3$	OCH$_3$	OCH$_3$	H	—	—	—	—	*Guatteria ouregou* [93]
Dehydroneolitsine (**158**)	O—CH$_2$—O		—	CH$_3$	—	O—CH$_2$—O		—	*Guatteria goudotiana* [87]
Dehydronornuciferine (**159**)	OCH$_3$	OCH$_3$	—	H	—	—	—	—	*Guatteria ouregou* [93]
Dehydropredicentrine (**160**)	OCH$_3$	OH	—	CH$_3$	—	OCH$_3$	OCH$_3$	—	*Polyalthia cauliflora* var. *beccarii* [128]
Dehydroroemerine (**161**)	O—CH$_2$—O		—	CH$_3$	—	—	—	—	*Guatteria sagotiana* [104]
Dehydrostephalagine (**162**)	O—CH$_2$—O		OCH$_3$	CH$_3$	—	—	—	—	*Guatteria sagotiana* [104]
3-Hydroxydehydronuciferine (**163**)	OCH$_3$	OCH$_3$	OH	CH$_3$	—	—	—	—	*Hexalobus crispiflorus* [107,108]

nolic counterparts and noraporphines give a range of secondary products, even under the mildest conditions. Dehydrogenation of the chloramine derivative of a noraporphine using sodium ethylate has recently been used successfully to make the corresponding dehydronoraporphine [69]. Various routes to the dehydroaporphines and dehydronoraporphines commencing from 1-btiq precursors have also been reported [188–192].

Dehydroaporphines can be reduced to the corresponding aporphines either by catalytic hydrogenation or by zinc–hydrochloric acid reduction. Nonphenolic dehydroaporphines (155, R = CH$_3$) are oxidized under mild acid conditions (pH 6) to the corresponding oxoaporphine (164) [185], whereas under alkaline conditions air oxidation gives not only 164 but also, in low yields, the corresponding 4,5-dioxo derivative (165) and aristolactam

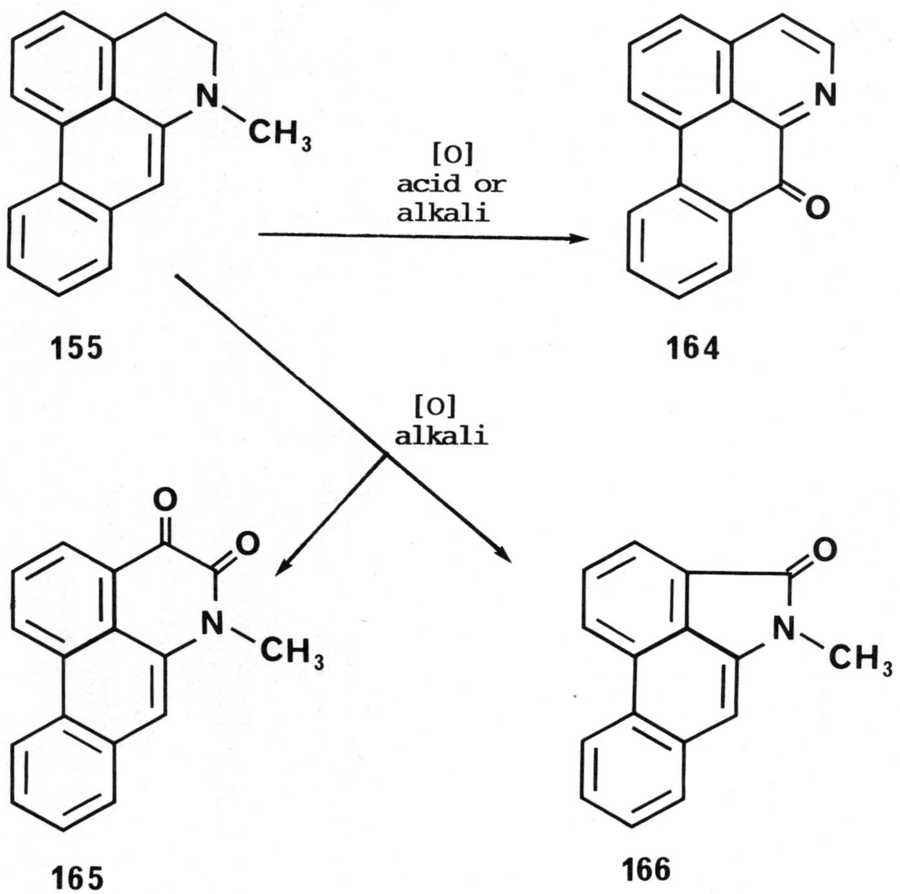

Scheme 17.   Products of the oxidation of dehydroaporphines under acid and alkaline conditions.

(166) [193] (Scheme 17). Finally, Cava and others have demonstrated the enamine-type character of dehydroaporphines, as is evidenced by their behavior on protonation [194], Reimer–Tiemann formylation [195], acylation [196], and Michael addition [197,198]. This reactivity is of particular value with respect to alkylation at C(7) [195,199] and its value in the synthesis of 7-alkyl aporphinoids and 7,7'-dimeric aporphinoids is discussed in Sections 5.2.3 and 5.5, respectively.

5.2.3. 7-Alkylaporphinoids.

In the last few years several species of *Guatteria* have yielded a novel type of aporphinoid characterized by alkylation at C(7) and with a variable degree of oxidation in ring B or ring C; either 4,5,6,6a-tetradehydro (167), or 6,6a-dehydro (168) or 6a,7-dehydro with a methyl substituent at C(7) (169). Related to these are two aporphinoids from *Duguetia* that are characterized by the presence of an additional oxazine ring linking N(6) and C(7) (170). The general 7-alkylaporphinoid structures (167–170), the alkaloids isolated to date, and their sources are given in Table 6. In the four *Guatteria* species for which both stem barks and leaves have been examined, 7-alkylaporphinoids have been isolated only from stem barks.

Melosmine and Guadiscine Types. These alkaloids are characterized by a 7-*gem*-dimethyl group, and by the presence of a pyridine (melosmine type) or dihydropyridine (guadiscine type) B ring. Melosmine (167c), from *Guatteria melosma* [88,89], was the first of the 7-alkyl aporphinoids to be identified. The UV spectrum indicated it was phenolic but its complexity suggested that it was not a typical aporphine or oxoaporphine. The ^{1}H NMR spectrum had some characteristics of a 1,2,3,9-substituted oxoaporphine with an AB quartet ($J = 6$ Hz) at δ 8.50 and 7.71 for H(5) and H(4) of an aromatic B ring. The most striking feature was a 6H singlet at δ 1.73 for a *gem*-dimethyl group which, given an aporphine nucleus and aromatic B ring, could only be placed at C(7). The proposed structure 167c was supported by ^{1}H NMR studies on the aporphine generated by zinc–hydrochloric acid reduction and on various O- and N-methylated derivatives and in particular the O,O-diacetyl derivative. The latter allowed the assignment of the methoxy groups to C(2) and C(3) and the hydroxy groups to C(1) and C(9). The structure (167c) was later confirmed by X-ray analysis and melosmine has more recently been isolated from a second species, *Guatteria ouregou* [95,96].

N,O,O-Trimethyltetrahydromelosmine (171) has been synthesized [200] by the route depicted in Scheme 18. The amine (172) and acid (173) were condensed by means of a Schotten–Baumann reaction and the resulting ketoamine 174 was cyclized using the conditions of a Bischler–Napieralski reaction. The resulting imine (175) was reduced by NaBH$_4$ to the amine (176), which was N-methylated to the 1-btiq (177) and this then cyclized to 171 using thallium trifluoroacetate. The final compound was identical with

Table 6. Occurrence and Distribution of 7-Alkyl Aporphinoids

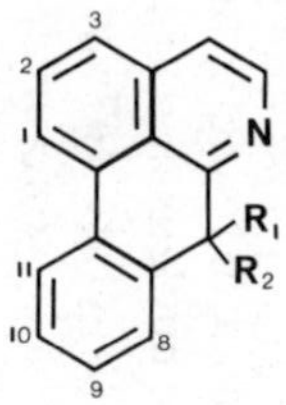

Name	C(1)	C(2)	C(3)	NR	C(7)	C(9)	C(10)	C(11)	Sources
Belemine (169a)	O—CH₂—O		—	CH₃	CH₃	OH	—	—	*Guatteria schomburgkiana* [102,103]
Dehydroguattescine (167a)	O—CH₂—O		—	—	CH₃,OH	OCH₃	—	—	*Guatteria schomburgkiana* [102,103]
Dihydromelosmine (168a)	OH	OCH₃	OCH₃	—	(CH₃)₂	OH	—	—	*Guatteria ouregou* [94,95]
Duguecalyne (170a)	O—CH₂—O		—	—	—	—	—	OCH₃	*Duguetia calycina* [64]
Duguenaine (170b)	O—CH₂—O		—	—	—	—	—	—	*Duguetia calycina* [64]
Duguespixine (169b)	OCH₃	OH	—	CHO	CH₃	—	—	—	*Duguetia spixiana* [68,69] *Guatteria sagotiana* [104]
Goudotianine (169c)	OCH₃	OH	OCH₃	CH₃	CH₃	OH	—	—	*Guatteria goudotiana* [87]

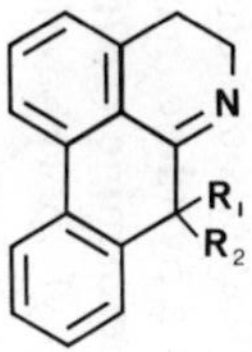

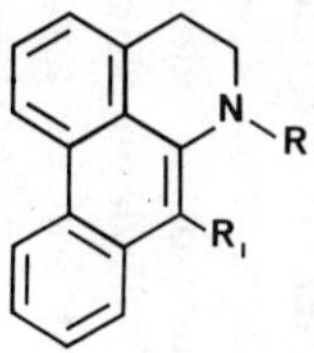

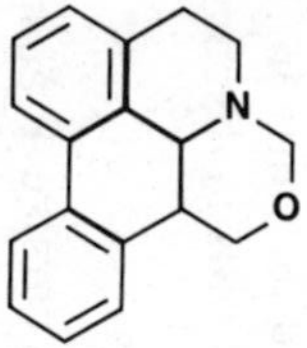

198

Compound								Source
Guacolidine (168b)	O—CH₂—O		—	CH₃,OH	OH	—	OCH₃	*Guatteria discolor* [82]
Guacoline (168c)	O—CH₂—O		—	CH₃,OH	OCH₃	—	OCH₃	*Guatteria discolor* [82]
Guadiscidine (168d)	O—CH₂—O		—	(CH₃)₂	OH	—	—	*Guatteria discolor* [82]
Gaudiscine (168e)	O—CH₂—O		—	(CH₃)₂	OCH₃	—	—	*Guatteria discolor* [82,84]; *Guatteria schomburgkiana* [102,103]
Guadiscoline (168f)	O—CH₂—O		—	(CH₃)₂	OCH₃	—	OCH₃	*Guatteria discolor* [82,84]
Guattescidine (168g)	O—CH₂—O		—	CH₃,OH	OH	—	—	*Guatteria scandens* [99–101]
Guattescine (168h)	O—CH₂—O		—	CH₃,OH	OCH₃	—	—	*Guatteria scandens* [99–101]; *Guatteria schomburgkiana* [102,103]
Guattouregidine (168i)	OH	OCH₃ OCH₃	—	CH₃,OH	OH	—	—	*Guatteria ouregou* [94,95]
Guattouregine (168j)	OH	OCH₃ OCH₃	—	CH₃,OH	OCH₃	—	—	*Guatteria ouregou* [94,95]
Melosmidine (167b)	OCH₃	OCH₃ OCH₃	—	(CH₃)₂	OH	—	—	*Guatteria melosma* [88,89]
Melosmine (167c)	OH	OCH₃ OCH₃	—	(CH₃)₂	OH	—	—	*Guatteria melosma* [88,89]; *Guatteria ouregou* [95,96]
Trichoguattine (169d)	O—CH₂—O		CHO	CH₃	—	—	—	*Guatteria sagotiana* [104]

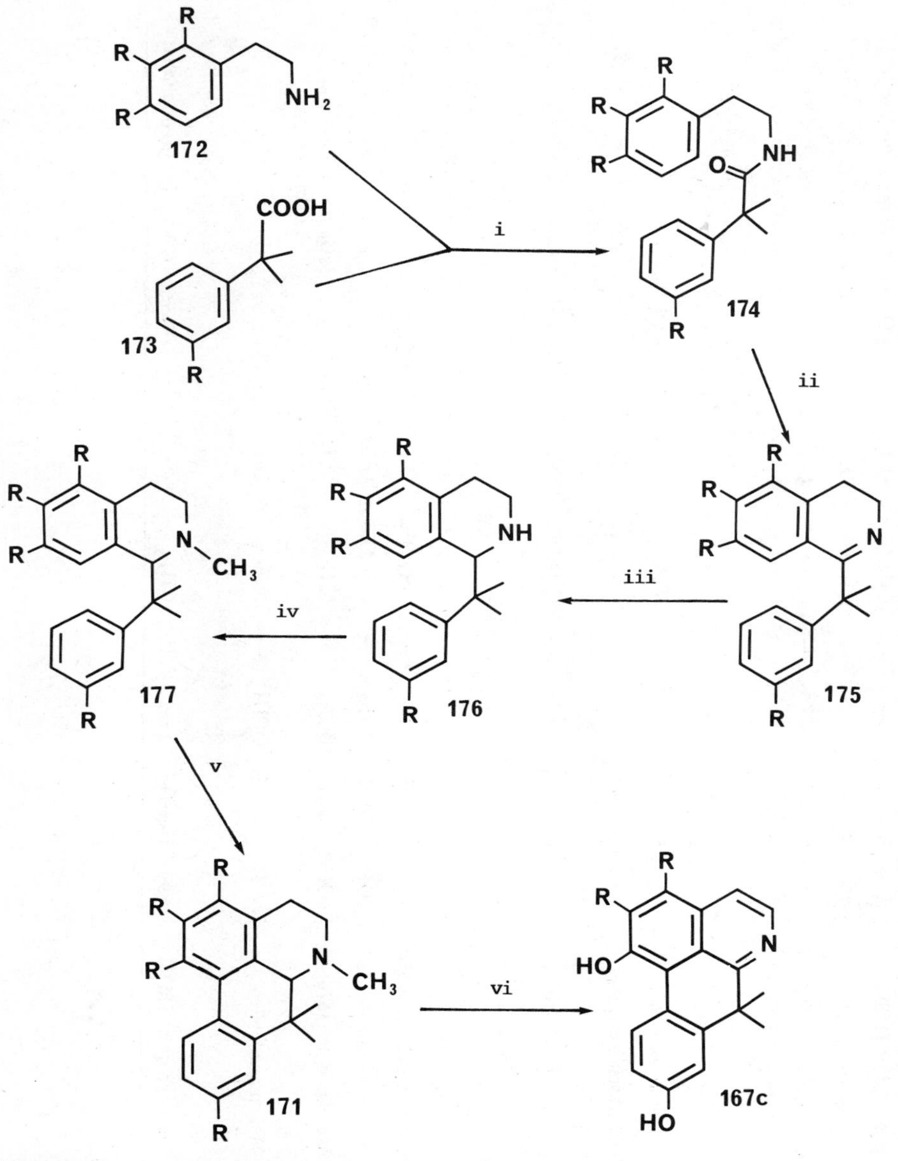

Scheme 18. Synthesis of *N,O,O*-trimethyltetrahydromelosmine (**171**), R = OCH₃. (i) SOCl₂/dry benzene, 50°, 3 hr; (ii) POCl₃/dry benzene, reflux, 1.5 hr; (iii) NaBH₄/ EtOH, room temp., 2 hr; (iv) HCHO/EtOH, NaBH₄; (v) thallium trifluoroacetate/ TFA/CH₂CL₂, BF₃/EtOH, 40°, 3 hr; (vi) CH₂N₂/Et₂O, Zn/AcOH, HCHO, NaBH₄.

that obtained when natural melosmine was O-methylated, reduced, and then N-methylated.

Melosmidine (167b), also isolated from *Guatteria melosma* [88,89], differed from melosmine only in the presence of an additional methyl substituent on one of the hydroxy groups. The ^{1}H NMR spectrum of O-acetylmelosmidine revealed the loss of the C(1) phenolic group, allowing the assignment of structure (167b).

Guadiscine (168e), guadiscidine (168d), dihydromelosmine (168a), and guadiscoline (168f), all isolated only from *Guatteria* species (Table 6), are a closely allied group of alkaloids distinguished by the presence of a 6,6a double bond and a *gem*-dimethyl substituent at C(7). Guadiscine (168e) was the first representative to be isolated, initially from *G. discolor* [82,84] and later from *G. schomburgkiana* [102,103]. As for the melosmine group the ^{1}H NMR spectrum revealed a singlet (6H) at δ 1.50 for the equivalent C(7) methyl groups and a pattern of aromatic protons indicative of 1,2,9 substitution, with the C(1)/C(2) substituent a methylenedioxy group resonating at δ 6.07 as a singlet, thus indicating the planarity conferred upon the aporphinoid ring system by the introduction of the additional unsaturation. The imine nature of the double bond in 168e was indicated by the absence of an NH resonance in the ^{1}H NMR spectrum, a bathochromic effect in the UV spectrum under acid conditions (see Section 4.1) and a band at 1660 cm^{-1} in the IR spectrum. Reduction with sodium borohydride yielded dihydroguadiscine, for which the ^{1}H NMR spectrum showed the characteristics of a noraporphine, with a notable feature being the nonequivalence of the C(7) methyls, which were now observed at δ 0.91 and 1.52. The ^{13}C NMR spectrum of 168e was assigned by reference to spectra for xylopine (149) and melosmine (167c) [84].

Guadiscidine (168d), isolated from *G. discolor* [82], was identified as the 9-O-demethyl derivative of 168e by chemical interconversion. Dihydromelosmine (168a) from *G. ouregou* [94,95] was recognized as a dihydroxy 1,2,3,9-substituted member of this group and the hydroxy functions assigned to C(1) and C(9) on the basis of reduction with sodium borohydride to tetrahydromelosmine, identical with the product obtained by reduction of melosmine (167c) with zinc–hydrochloric acid, and by comparison of ^{1}H NMR spectra run in deuterochloroform and deuteropyridine.

The final alkaloid in this group, guadiscoline (168f) isolated from *G. discolor* [82,84], differed from guadiscine (168e) by the presence of additional substitution at C(11), giving a 1,2,9,11-substitution pattern. A major distinguishing feature was the appearance of an AB system in the ^{1}H NMR spectrum for H(8) and H(10) (J = 2.8 Hz). The ^{13}C NMR spectrum of 168f correlated well with data for 168e and for the 1,2,9,11-substituted aporphine discoguattine (*O*-methylcalycinine) (77). This additional member of the 1,2,9,11-substituted group is also from a *Guatteria* species, in agreement with observations made in the aporphines section (5.2.1) concerning their limited distribution in the genera *Duguetia*, *Guatteria*, and *Fissistigma*.

Guattescine and Dehydroguattescine Types. These alkaloids are characterized by the presence at C(7) of one methyl and one tertiary hydroxyl group and by either a single double bond (6,6a) (guattescine type) or a fully aromatized B ring (dehydroguattescine type).

For the guattescine type six alkaloids are known, constituting three pairs that differ in having either methoxy or hydroxy functions at C(9). These are guattescine (**168h**) from *G. scandens* [99–101], *G. schomburgkiana* [102,103] and guattescidine (**168g**) from *G. scandens* [99–101], guattouregine (**168j**) and guattouregidine (**168i**) from *G. ouregou* [94,95], and guacoline (**168c**) and guacolidine (**168b**) from *G. discolor* [82].

Guattescine (**168h**) gave the bathochromic shift in acid in the UV spectrum and a weak band at 1648 cm^{-1} in the IR spectrum indicative of its imine structure. The ^{1}H NMR spectrum was generally comparable with that of the 1,2,9-substituted noraporphine xylopine (**149**), notable differences being a 3H singlet at δ 1.47 for the C(7) methyl group and a relative deshielding of H(8) to δ 7.43. Surprisingly the signal for the methylenedioxy substituent at C(1)/C(2) appeared as an AB quartet despite the supposed planarity conferred on the aporphine nucleus by the 6,6a bond (cf. **168e**). Likewise, although the H(4) and H(5) protons in **168e** were equivalent, giving rise to a pair of triplets, in **168h** they gave rise to a complex multiplet. The absence of planarity was confirmed by an X-ray analysis [101].

Treatment of **168h** with acetic anhydride in a methanol–dichloromethane solution yielded the *O*-acetyl derivative (**178**). It is suggested (Scheme 19) that this selective acetylation of the tertiary alcohol may be through the formation of an acylimine which interacts with the C(7) hydroxyl to give a pentacyclic oxazolidine intermediate (**179**) from which **178** is generated. By contrast, conversion of **168h** to dihydroguattescine (**180**) followed by treatment with acetic anhydride in pyridine leads to the specific acetylation of the secondary amine to give **181** (Scheme 19), for the formation of **179** is not feasible from **180**.

The structure of **168h** was confirmed by X-ray studies [101]. These established that the 7-methyl substituent is axial and the 7-hydroxy substituent equatorial. In the crystal form, molecules of **168h** pack in pairs linked by hydrogen bonds between N(6)–HO(7') and N(6')–HO(7), the two constituents of the pair being of different chirality and therefore giving a racemic product. In reality, **168h** isolated from *G. scandens* is a partially racemic product from the dextrorotatory form, and by successive recrystallizations the pure racemate is obtained [101].

Guattescidine (**168g**) was identified as the 9-demethyl analog of guattescine (**168h**) by methylation with diazomethane. The optical activity of the resulting semisynthetic **168h** was found to have the opposite sign to that of the natural alkaloid. A comparison [99] of the CD curves of the dihydro derivatives of **168h** and **168g** showed that the former exhibited a positive Cotton effect at 235 nm [6a(R)] whereas the latter showed a negative Cotton effect [6a(S)].

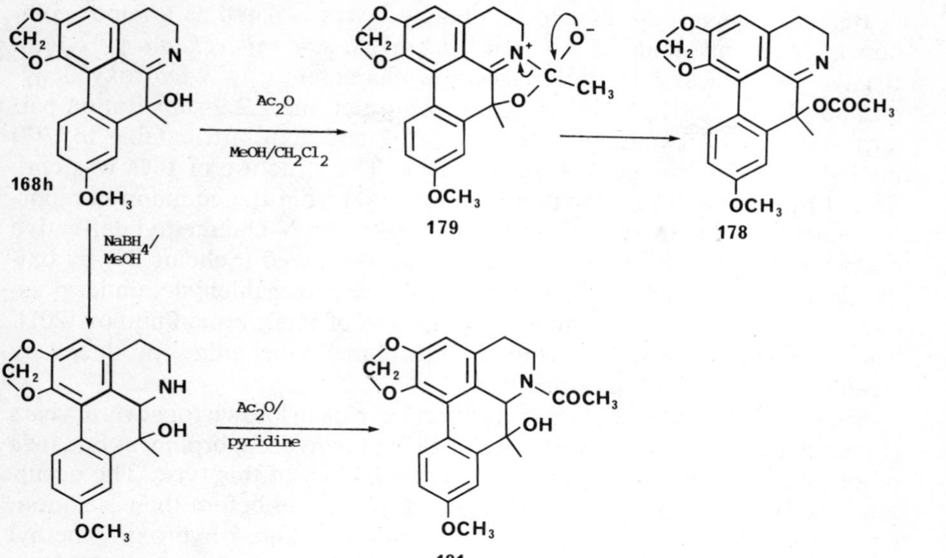

Scheme 19. Acetylation of guattescine (**168h**) and dihydroguattescine (**180**).

Guattouregine (**168j**) and guattouregidine (**168i**) differed from **168h** and **168g**, respectively, only by the presence of an additional methoxy substituent. Based on a full analysis of spectral data and the preparation of a number of derivatives the additional substituent was placed at C(3) [95]. Acetylation of these alkaloids and their dihydro derivatives showed the same distinctive pathways noted in Scheme 19. Guacoline (**168c**) and guacolidine (**168b**) are isomeric with **168j** and **168i** but with the additional methoxy function at C(11); they form part of a large spectrum of 1,2,9,11-substituted aporphinoids from *G. discolor* [82], others being the noraporphines discoguattine (**77**) and isocalycinine (**92**), the oxoaporphine oxoisocalycinine (**262**) and guadiscoline (**168f**).

Dehydroguattescine (**167a**) is unique, being the only 7-hydroxy-7-methyl aporphinoid in which the B ring has been oxidized to the pyridine level. It has been isolated only from the bark of *G. schomburgkiana* [102,103] and its structure established from the close similarity of spectral data with those for guattescine (**168h**), which has been isolated from the same source. A synthesis of **167a** has been achieved from the common noraporphine anolobine (**63**) and is outlined below.

Belemine type. Recently four alkaloids with a dehydroaporphine skeleton modified by the addition of a 7-methyl group have been isolated from species of *Duguetia* and *Guatteria*; belemine (**169a**), goudotianine (**169c**), duguespixine (**169b**), and trichoguattine (**169d**).

Belemine (**169a**), the first to be obtained, was isolated as a minor component of the alkaloids of the stem bark of *G. schomburgkiana* [102,103]. It gave the typical UV and IR spectral characters of a 9-hydroxy dehydroaporphine; the ^{1}H NMR spectrum confirmed the 1,2,9-substitution pattern and showed two 3H singlets at δ 2.57 and 2.78, attributable to C(7) methyl and *N*-methyl groups, respectively. The structure of **169a** was confirmed by synthesis of *O*-methylbelemine (**182**) from the common noraporphine anolobine (**63**) [102]. Conversion of **63** to its *N*-*O*-dimethyl derivative (isolaureline, **95**) using normal methods was followed (Scheme 20) by oxidation and methylation of C(7) of **95** to **182** using formaldehyde, under pressure, at 105° for 72 hr according to the method of Mollov and Philipov [201]. The product proved to be identical with *O*-methylbelemine, produced by treatment of **169a** with diazomethane.

Synthetic 7-methyl dehydroaporphines have been known for several years [195,199] by virtue of the reactivity of C(7) in dehydroaporphines; but **169a** represented the first example of a natural alkaloid of this type. The occurrence of belemine-type alkaloids had been presumed before their isolation, ever since the earlier discovery of 7,7-dimethyl and 7-hydroxy-7-methyl aporphinoids in *Guatteria*. The *in vitro* route from the 7-methyldehydroaporphinoids to 7-hydroxy-7-methylaporphinoids has recently been demons-

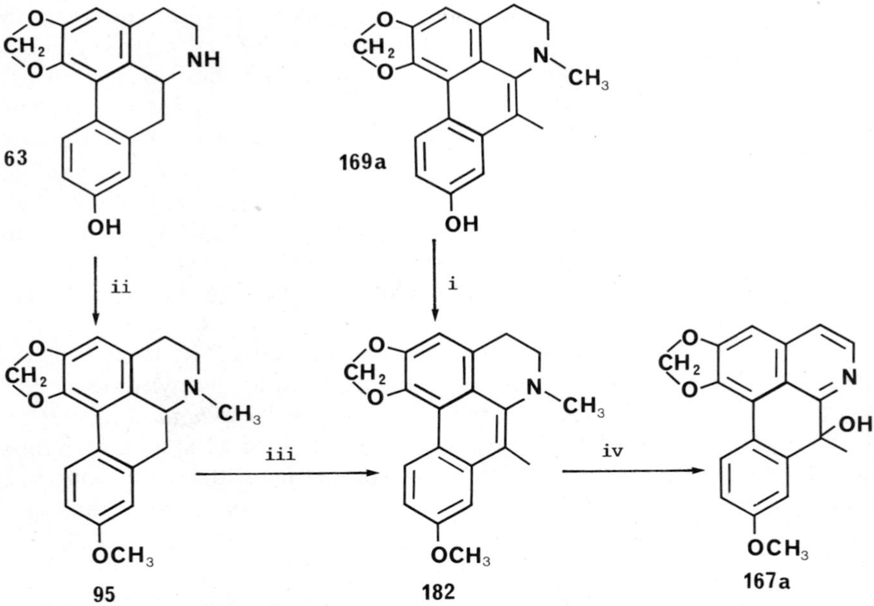

Scheme 20. Synthesis of dehydroguattescine (**167a**) from anolobine (**63**) *via* *O*-methylbelemine (**182**). (i) CH_2N_2/Et_2O, room temp., 24 hr; (ii) HCHO, $NaBH_4$, CH_2N_2/Et_2O; (iii) HCHO/MeOH, 105°, 72 hr; (iv) *m*-$ClC_6H_4CO_3H/CH_2Cl_2$, 5°, 1 hr.

trated [102,103] by the conversion of **182** to dehydroguattescine (**167a**) in 15% yield by treatment with *m*-chloroperbenzoic acid (Scheme 20).

Goudotianine was obtained from the Colombian species *G. goudotiana* and the structure **169c** has been proposed. This structure rests primarily on the ¹H NMR spectrum and associated *nOe* results [87]. The placement of the A-ring hydroxy group at C(2) rather than C(3) is based on the anticipated instability of an aporphinoid with both C(3) and C(9) phenolic groups [87].

Duguespixine (**169b**) was first isolated from the Colombian species *Duguetia spixiana* [68,69]. Although the spectral data clearly established it as a 2-hydroxy-1-methoxy dehydroaporphine with methyl and formyl substituents at N(6) and C(7), they failed to distinguish between **169b** and its isomer **183** because the methyl resonance at δ 3.28 was deshielded beyond the level expected for either N(6) or C(7), presumably because of the presence of the *peri* formyl function. To resolve this problem 7-formyldehydronuciferine (**184**) was synthesized from the aporphine nuciferine (**133**). Comparison of (**184**) with the *O*-methyl derivative of duguespixine showed them to be different and therefore structure **169b** for duguespixine was confirmed. Reduction of **184** by sodium cyanoborohydride in acidic medium [195] yielded 7-methyldehydronuciferine (**185**), an analog of belemine not yet recorded from nature [68]. Duguespixine has recently been isolated again, from the leaves of the Guyanese species *G. sagotiana* [104], together with another alkaloid of this group, trichoguattine (**169d**), which is the 1,2-methylenedioxy analog of **169b**.

Duguenaine type.   Duguenaine (**170b**) and duguecalyne (**170a**) represent a novel class of pentacyclic aporphinoids isolated recently from the Guy-

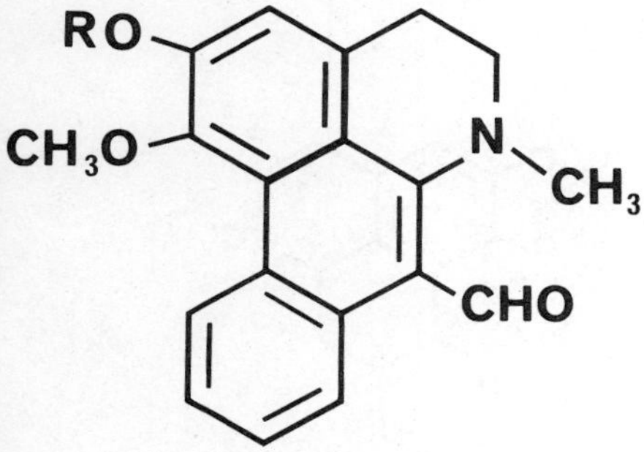

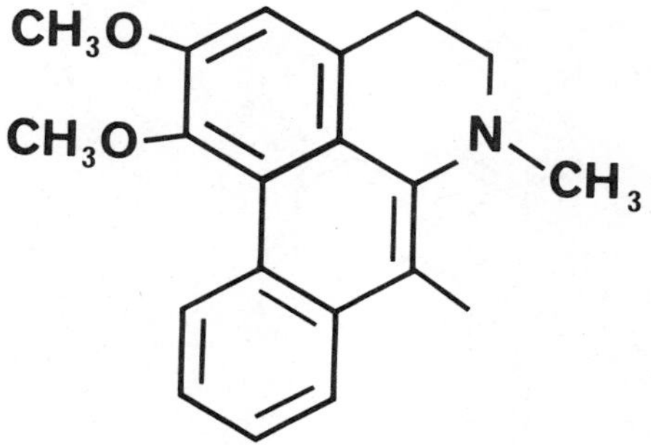

183   R = H

184   R = CH₃

185

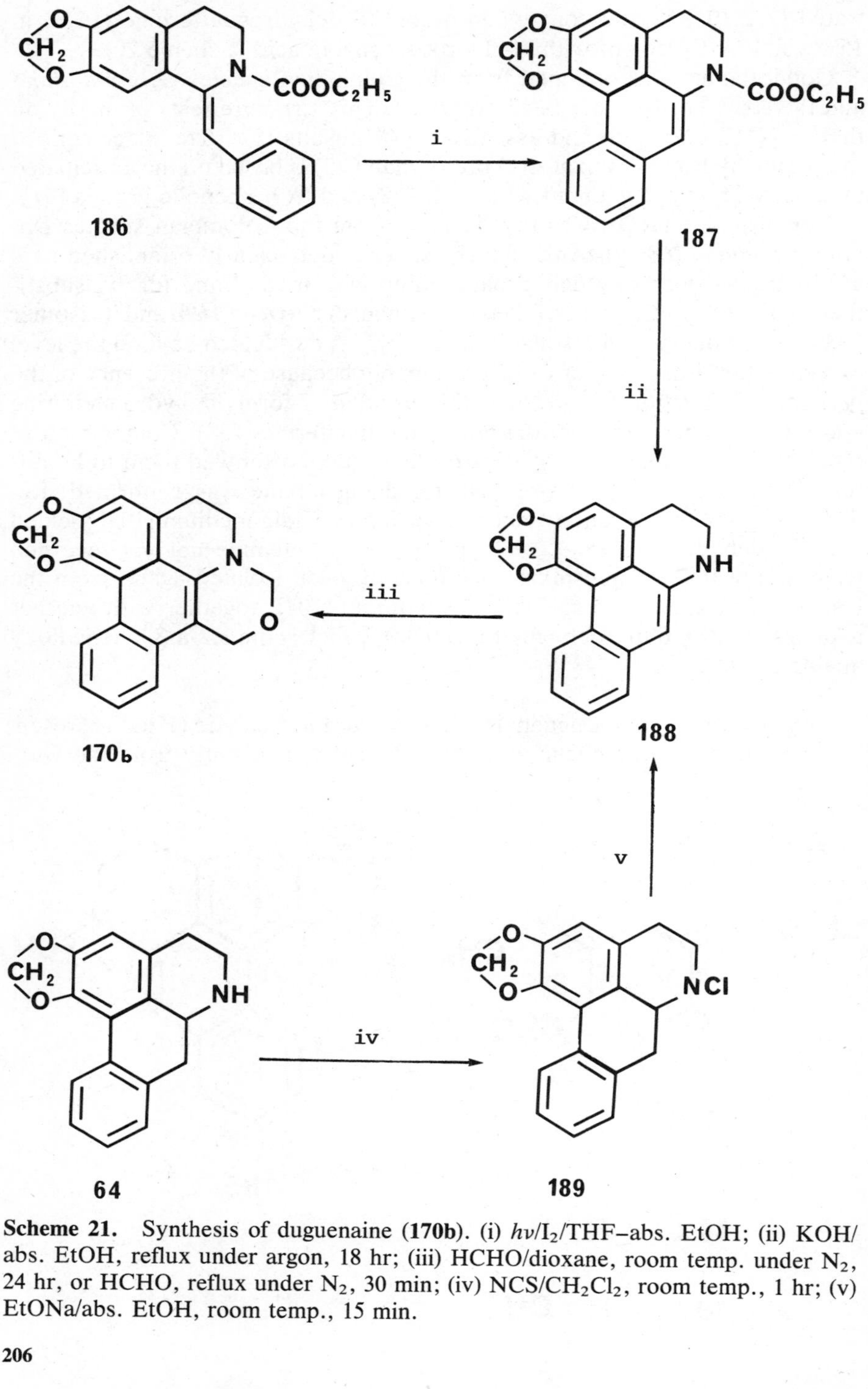

Scheme 21. Synthesis of duguenaine (**170b**). (i) $h\nu/I_2/THF$–abs. EtOH; (ii) KOH/ abs. EtOH, reflux under argon, 18 hr; (iii) HCHO/dioxane, room temp. under N_2, 24 hr, or HCHO, reflux under N_2, 30 min; (iv) NCS/CH_2Cl_2, room temp., 1 hr; (v) EtONa/abs. EtOH, room temp., 15 min.

anese tree *Duguetia calycina* [64], the additional ring arising as an oxazine from N(6),C(7)-substituted dehydroaporphines. The dehydroaporphine nucleus of **170b** was realized from the UV spectrum and the presence of a methylenedioxy substituent in the A ring (singlet, indicating planarity) and an unsubstituted D ring established from ^{1}H NMR, thus requiring the placement of the third oxygen in the additional ring system. The oxazine nature was also indicated by the ^{1}H NMR spectrum in which two pairs of methylene protons resonated as singlets at δ 4.70 and 5.22. The ^{13}C NMR spectrum of **170b** was in accord with the proposed structure. Duguecalyne (**170a**) differed from **170b** only by the presence of an additional methoxy group, which was readily assigned to C(11) [64].

The synthesis of **170b** has been reported by two groups, both using routes which can be considered to be biomimetic (see below) in that a dehydroaporphine was prepared and then exposed to electrophilic attack at N(6) and C(7) with formaldehyde. The route (Scheme 21) followed by Lenz and Koszyk [192] commenced with the methylenedioxy enamide (**186**), which was photocyclized to the corresponding 6-ethoxycarbonyl dehydroaporphine (**187**). Saponification of **187** yielded the known dehydroanonaine (**188**) and treatment of this with formaldehyde yielded **170b**. The route followed by Debourges et al. [12,69] differed only in that **188** was synthesized directly from the corresponding noraporphine anonaine (**64**) *via* its chloramine (**189**) (Scheme 21). In addition to **170b**, a range of other duguenaine derivatives (**190a–c**) have been synthesized [192].

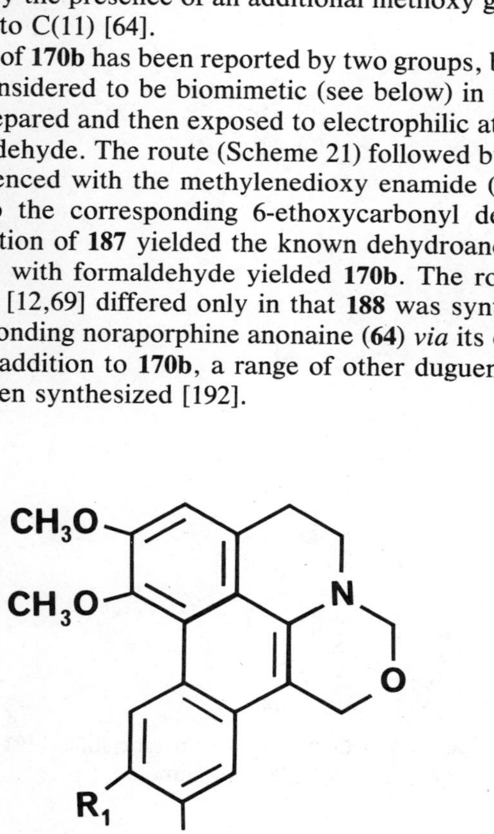

	R_1	R_2
190a	H	H
190b	OCH_3	OCH_3
190c	$O-CH_2-O$	

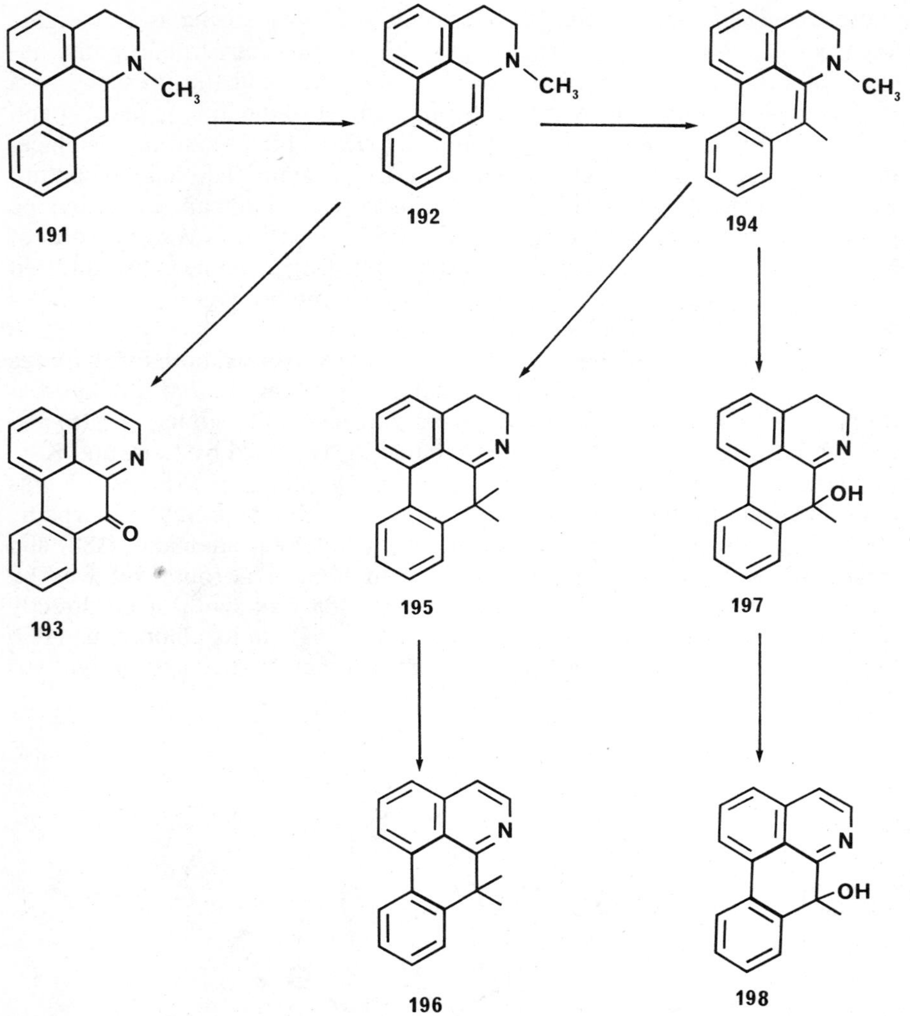

Scheme 22. A possible biogenetic pathway from aporphines (**191**) to melosmine type (**196**) and dehydroguattescine type (**198**) aporphinoids.

Possible biogenetic pathways to 7-alkyl aporphines. The range of different 7-alkyl aporphinoids isolated from *Guatteria* and *Duguetia* and discussed in this section are now sufficient to suggest plausible routes for their *in vivo* formation from either aporphines (Scheme 22) or noraporphines (Scheme 23).

Aporphines (e.g., **191**, Scheme 22) are readily oxidized, initially to the corresponding dehydroaporphine (**192**), this generally being the first stage of the complete oxidation to oxoaporphines (**193**) in the Annonaceae. How-

ever, it is well established that dehydroaporphines are sufficiently nucleophilic at C(7) to allow the introduction of suitable substituents [195]. It is,
therefore, not unreasonable to postulate the presence in the plant of enzyme
systems able to catalyze the addition of a 1C unit (from *S*-methyl adenosyl
methionine) to yield the belemine-type 7-methyl dehydroaporphines (**194**).
These can in turn act as intermediates for the formation of 7,7-dimethyl 6,6a-
dehydroaporphines of the guadiscine type (**195**) and via further oxidation,
the melosmine type (**196**). Alternatively, addition of a hydroxy at C(7) of

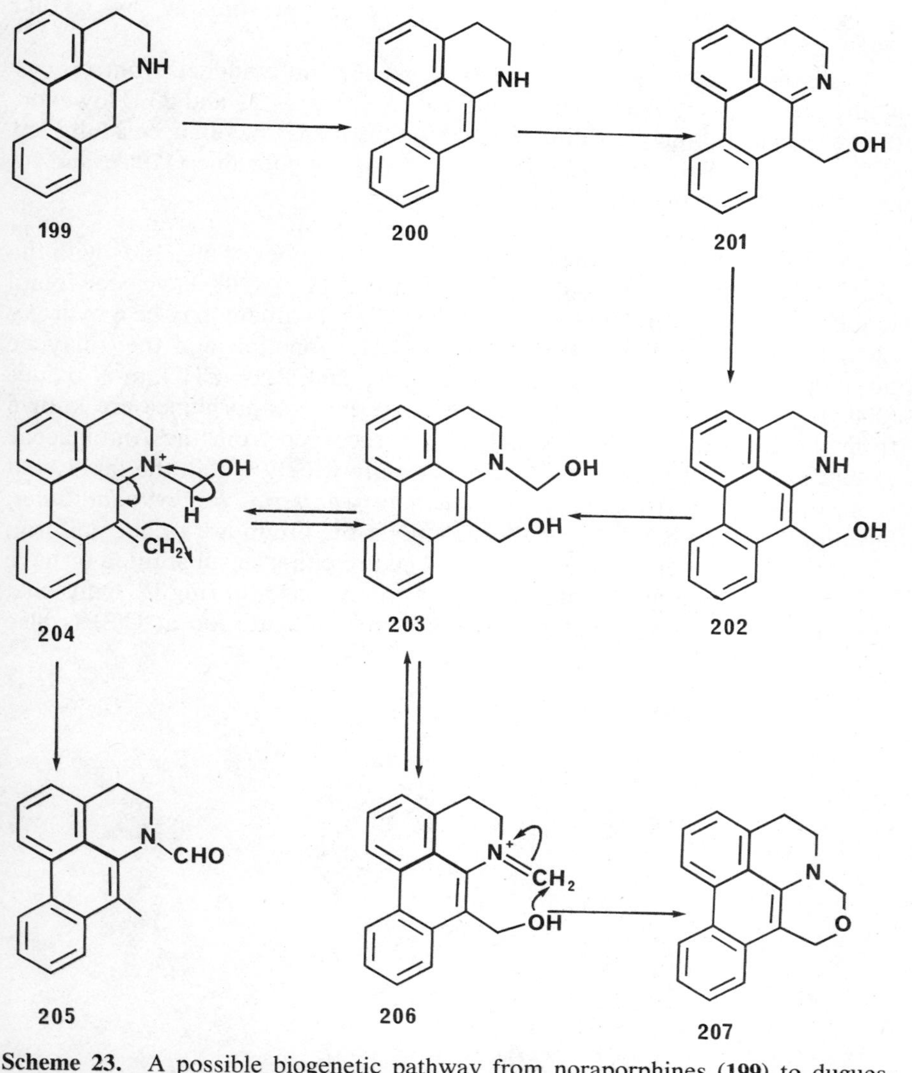

Scheme 23. A possible biogenetic pathway from noraporphines (**199**) to duguespixine type (**205**) and duguenaine type (**207**) aporphinoids.

194 rather than a second methyl could lead to the guattescine type (**197**) and then the dehydroguattescine type (**198**) [102].

Noraporphines (e.g., **199**, Scheme 23) are likewise readily oxidized to the highly reactive dehydronoraporphines (**200**). The addition, in two stages, of formyl or carbamoyl units to give the *bis*-methylol intermediate (**203**), *via* **201** and **202**, can be followed by development along two different paths. Firstly, reduction of the C(7) substituent can lead to *N*-formyl 7-methyl dehydroaporphines of the duguespixine type (**205**) through the intermediate (**204**) or, by means of dehydration of N(6),C(7) methylol substituents to the pentacyclic duguenaine-type alkaloids (**207**), presumably *via* (**206**) [69,192,202].

It should be stressed that there is at present no evidence from *in vivo* studies to support the pathways outlined in Schemes 22 and 23. However, they are based largely on biomimetic syntheses that have successfully led to belemine (**169a**), dehydroguattescine (**167a**), duguenaine (**170b**), and its homologs **190a–c** [69,102,192].

5.2.4. 7- and 4,7-Oxygenated Aporphines. A number of alkaloids with the normal aporphine nucleus but with oxygenation at C(7) (**208**) have been found in the Annonaceae. In these alkaloids the C(7) substituent may be a hydroxy (**208**, R_2 = H) or methoxy (**208**, R_2 = CH_3) function and they may be noraporphines (**208**, R_1 = H), tertiary amines (**208**, R_1 = CH_3), or *N*-oxides (**208**, R_1 = CH_3—N^+O^-). In all, 30 7-oxygenated aporphines are known from the plant kingdom, of which 24 are recorded from the Annonaceae (Table 7). They appear to be distributed quite widely within the family but seem to occur most regularly in the genera *Guatteria*, *Pachypodanthium*, and *Polyalthia*. Substitution patterns among the group are rather uniform; most have a 1,2-methylenedioxy group and are either unsubstituted or have only a C(9) substituent (methoxy except in one case) in ring D. Only one, duguetine (**210**), has two ring-D substituents. Substitution at C(3) is also

208

Table 7. Occurrence and Distribution of 7-Oxygenated Aporphines

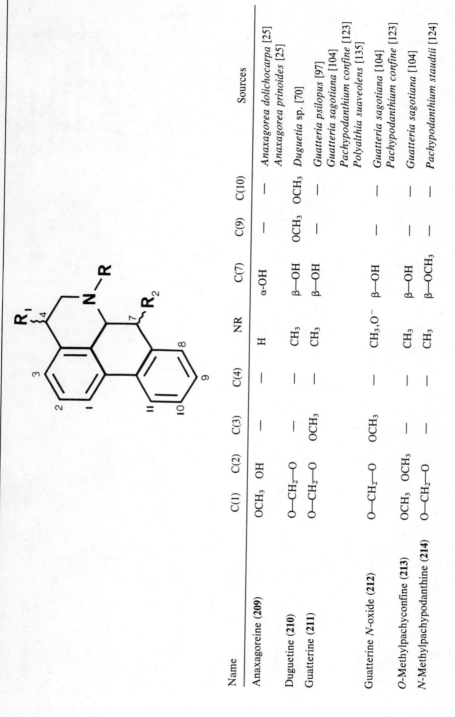

Name	C(1)	C(2)	C(3)	C(4)	NR	C(7)	C(9)	C(10)	Sources
Anaxagoreine (209)	OCH_3	OH	—	—	H	α—OH	—	—	*Anaxagorea dolichocarpa* [25] *Anaxagorea prinoides* [25]
Duguetine (210)	O—CH_2—O		—	—	CH_3	β—OH	OCH_3	OCH_3	*Duguetia* sp. [70]
Guatterine (211)	O—CH_2—O		OCH_3	—	CH_3	β—OH	—	—	*Guatteria psilopus* [97] *Guatteria sagotiana* [104] *Pachypodanthium confine* [123] *Polyalthia suaveolens* [135]
Guatterine N-oxide (212)	O—CH_2—O		OCH_3	—	CH_3,O^-	β—OH	—	—	*Guatteria sagotiana* [104] *Pachypodanthium confine* [123]
O-Methylpachyconfine (213)	OCH_3	OCH_3	—	—	CH_3	β—OH	—	—	*Guatteria sagotiana* [104]
N-Methylpachypodanthine (214)	O—CH_2—O		—	—	CH_3	β—OCH_3	—	—	*Pachypodanthium staudtii* [124]

Table 7. (*continued*)

Name	C(1)	C(2)	C(3)	C(4)	NR	C(7)	C(9)	C(10)	Sources
N-Methylpachypodanthine N-oxide (**215**)	O—CH$_2$—O		—	—	CH$_3$,O$^-$	β—OCH$_3$	—	—	*Polyalthia oliveri* [133]
Noroliveridine (**216**)	O—CH$_2$—O		—	—	H	β—OH	OCH$_3$	—	*Polyalthia oliveri* [133]
Noroliverine (**217**)	O—CH$_2$—O		—	—	H	β—OCH$_3$	OCH$_3$	—	*Polyalthia suaveolens* [135]
Noroliveroline (**218**)	O—CH$_2$—O		—	—	H	β—OH	—	—	*Fusaea longifolia* [78] *Guatteria sagotiana* [104] *Polyalthia acuminata* [127]
Norpachystaudine (**219**)	O—CH$_2$—O		—	β—OH	H	β—OCH$_3$	—	—	*Pachypodanthium staudtii* [124]
Norushinsunine (**220**) (michelalbine)	O—CH$_2$—O		—	—	H	α—OH	—	—	*Annona cherimola* [27] *Annona glabra* [33,34] *Annona reticulata* [45,46] *Annona squamosa* [53] *Asimina triloba* [55] *Cymbopetalum brasiliense* [61] *Desmos tiebaghiensis* [62] *Meiogyne virgata* [113] *Melodorum punctulatum* [114] *Polyalthia acuminata* [127] *Polyalthia nitidissima* [132] *Unonopsis guatterioides* [144]
Oliveridine (**221**)	O—CH$_2$—O		—	—	CH$_3$	β—OH	OCH$_3$	—	*Enantia pilosa* [72] *Isolona campanulata* [109] *Polyalthia oliveri* [133,134] *Polyalthia suaveolens* [135]

Compound							Source [Ref.]
Oliveridine *N*-oxide (**222**)	O—CH₂—O	—	CH₃,O⁻	β—OH	OCH₃	—	*Enantia pilosa* [72] *Polyalthia suaveolens* [136]
Oliverine (**223**)	O—CH₂—O	—	CH₃	β—OCH₃	OCH₃	—	*Enantia pilosa* [72] *Isolona campanulata* [109] *Polyalthia oliveri* [133,134] *Polyalthia suaveolens* [135,136]
Oliverine *N*-oxide (**224**)	O—CH₂—O	—	CH₃,O⁻	β—OCH₃	OCH₃	—	*Enantia pilosa* [72] *Isolona campanulata* [109] *Polyalthia suaveolens* [136]
Oliveroline (**225**)	O—CH₂—O	—	CH₃	β—OH		—	*Fusaea longifolia* [78] *Guatteria sagotiana* [104] *Pachypodanthium confine* [123] *Polyalthia oliveri* [133] *Polyalthia suaveolens* [135]
Oliveroline *N*-oxide (**226**)	O—CH₂—O	—	CH₃,O⁻	β—OH	—	—	*Guatteria sagotiana* [104] *Polyalthia oliveri* [133]
Pachyconfine (**227**)	OCH₃ OH	—	CH₃	β—OH	—	—	*Guatteria sagotiana* [104] *Pachypodanthium confine* [123]
Pachypodanthine (**228**)	O—CH₂—O	—	H	β—OCH₃	—	—	*Pachypodanthium staudtii* [124,126] *Polyalthia oliveri* [133]
Pachystaudine (**229**)	O—CH₂—O	β-OH	CH₃	β—OCH₃	—	—	*Polyalthia suaveolens* [135] *Pachypodanthium staudtii* [124]
Polyalthine (**230**)	O—CH₂—O	OCH₃	CH₃	β—OH	OCH₃	—	*Polyalthia suaveolens* [135]
Polysuavine (**231**)	O—CH₂—O	—	CH₃	β—OCH₃	OH	—	*Polyalthia suaveolens* [135,136]
Ushinsunine (**232**)	O—CH₂—O	—	CH₃	α—OH	—	—	*Cananga odorata* [57] *Polyalthia nitidissima* [132] *Pseudoxandra sclerocarpa* [139]

rather rare but the number of *N*-oxides is larger than in most groups of alkaloids dealt with. Two of the alkaloids, norpachystaudine (**219**) and pachystaudine (**229**), are unusual in that they also carry a hydroxy substituent at C(4).

A striking feature of all natural 7-oxygenated aporphines is that they possess the same absolute configuration (*S*) at C(6a). Note that this notation is reversed in comparison to aporphines *s.s.* as the presence of the C(7) oxygen changes the sequence of priorities [203]. The presence of another chiral center at C(7) prohibits the direct use of specific rotation as an indicator of C(6a) configuration. The C(6a) configuration can, however, still be readily obtained from circular dichroism by making use of the Cotton effect in the region 235–245 nm, a negative effect indicating that C(6a) is *S* [H(6aβ)] independent of other substituents. In this context all 7-hydroxyaporphines so far isolated are levorotatory whereas all 7-methoxyaporphines are dextrorotatory [8–10], even though they all have the same C(6a) configuration.

By contrast the configuration at C(7) is variable, with both 6a(*S*),7(*R*) (*erythro*) and 6a(*S*),7(*S*) (*threo*) types known. In the Annonaceae only three of the isolated alkaloids, anaxagoreine (**209**), norushinsunine (**220**), and ushinsunine (**232**), have the 6a(*S*),7(*R*) configuration, the remainder being 6a(*S*),7(*S*), i.e., with the 7-substituent β and a *trans* relationship between H(6a) and H(7). Other than in the Annonaceae 7-hydroxyaporphines have been found in the Ranalean families Eupomatiaceae, Lauraceae, Magnoliaceae, Menispermaceae, and Monimiaceae [8–10], supporting the contention [204] that they represent an archaic group of alkaloids. In all cases other than in the Annonaceae they have the 6a(*S*),7(*R*) configuration [205].

7-Hydroxyaporphines. This group presents a number of diagnostic spectral characteristics. The mass spectrum exhibits the normal fragmentation of an aporphine and, in addition, an ion at $M^+ - 18$ owing to the loss of the elements of water from the molecular ion [189]. In the ^1H NMR spectrum a most important feature is the presence of an AB quartet for the H(6a) and H(7) protons, the former resonating (in deuteriochloroform) in the region of δ 3.30–3.60 and the latter between 4.50 and 4.80, the exact positions being influenced by C(7) stereochemistry and the *N*-substituent. The coupling constant allows unequivocal assignment of H(6a),H(7) stereochemistry, giving a value on the order of 12 Hz for *trans* and 2.5–3.0 Hz for *cis*. The presence of a C(7) substituent leads to a deshielding of H(8) which is particularly noticeable when the spectrum is run in deuteriopyridine [70]. The ^{13}C NMR spectra of these alkaloids are characterized by the resonance at ~70 ppm for C(7) and the occurrence of the 7-substituent always causes the anticipated β shift of +2 and +5 ppm for C(6a) and C(7a), respectively, and, where it is present, causes a shift of −4 ppm in the *N*-methyl resonance [171].

Oliveridine (**221**) can be taken as a typical example of this group both in its spectral characteristics and its reactivity [134]. The ^1H NMR spectrum exhibited the typical AB quartet centered at δ 3.40 and 4.55 ($J = 12.5$ Hz) with H(8) deshielded to δ 7.30 as a doublet (J = 2.8 Hz) in comparison with a resonance position of δ 6.80 in the aporphine analog isolaureline (95).

Oxidation of 221 with chromium trioxide in pyridine yielded the corresponding oxoaporphine, lanuginosine (255). When acetylated (Scheme 24) 221 gave its 7-acetyl derivative (221a), in which the acetyl methyl resonated at δ 2.30 and H(6a) and H(7) had been deshielded to δ 3.77 and 6.06, respectively. The acetyl group proved to be very labile because of its benzylic nature, passage through a column of alumina or boiling in methanol for a short time being sufficient to regenerate 221. Reaction of 221 with phosphorus oxychloride in pyridine caused dehydration to the corresponding dehydroaporphine (233) which on hydrogenation in acetic acid in the presence of Adams catalyst gave the aporphine isolaureline (95) (Scheme 24).

Attempts to reduce 221 directly to 95 using Adams catalyst in ethanol and palladium–carbon in methanol and acetic acid failed [206]. However, an analogous Clemmensen reduction using ushinsunine (232), which has the opposite, cis, stereochemistry, did lead to a racemic mixture of the corresponding aporphine [207,208], whereas reduction of 232 dissolved in hydrogen bromide, in the presence of a platinum catalyst, led to the optically active aporphine with unchanged C(6a) configuration [209].

Finally, synthesis of the oliveridine methiodide (234) followed by a Hofmann reaction gave rise to a vinylphenanthrene derivative (235), resulting

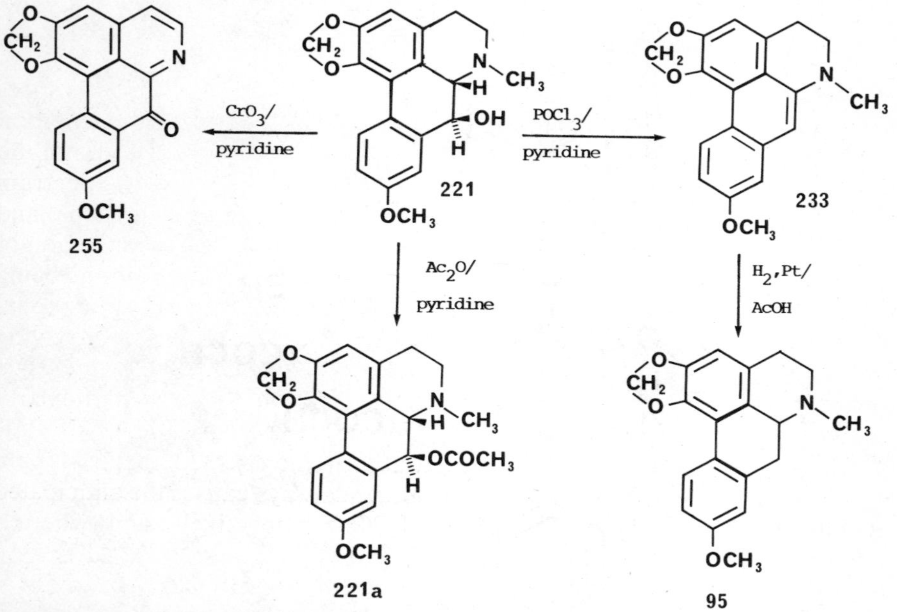

Scheme 24. Aporphinoids derived from oliveridine (221) by simple synthetic reactions.

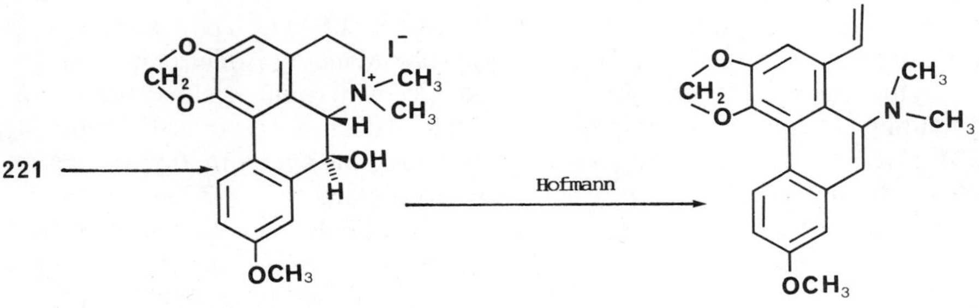

221 ⟶

234

Hofmann ⟶

235

from dehydration and opening of ring B between C(5) and N(6) [206]. The products of this procedure are different from those found with aporphines *s.s.* (Section 5.6), which lead to dimethylaminoethylphenanthrenes through the opening of ring B between N(6) and C(6a). This reaction appears to be independent of the relative C(6a), C(7) configuration [207,210].

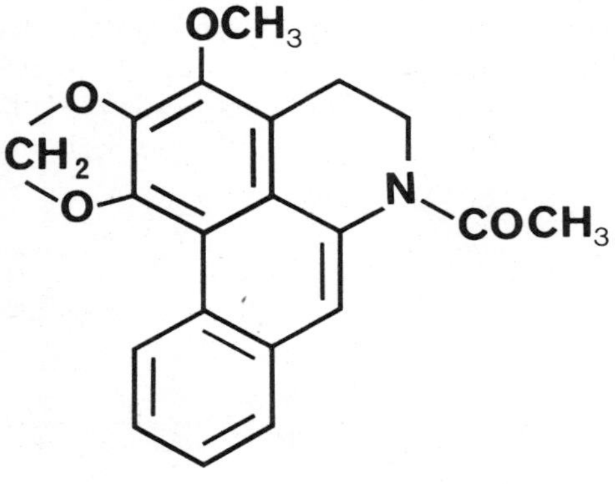

236

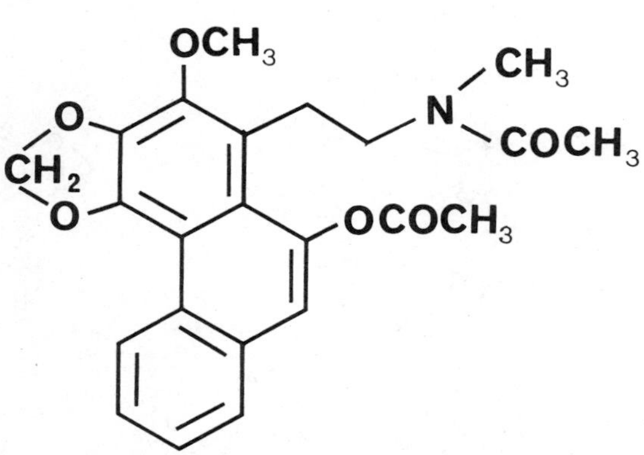

237

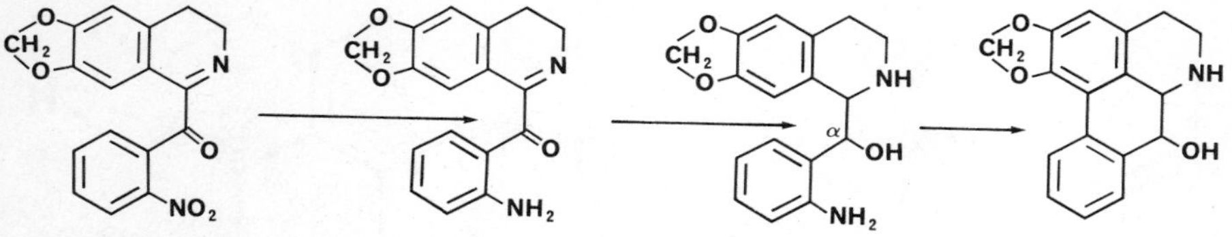

Scheme 25. Synthesis of (±)-norushinsunine (**220**).

Whereas acetylation of 7-hydroxyaporphines normally proceeds without problems, in the case of guatterine (**211**) attempts to acetylate using gentle conditions failed. The adoption of a more vigorous procedure (reflux with acetic anhydride) led to the formation of the N-acetyldehydronoraporphine (**236**) and the N-acetylphenanthrene (**237**) [97].

Syntheses of some 7-hydroxyaporphines have been reported, this work having been given impetus by suggestions that they have some pharmacological activity, notably in relation to Parkinson's disease [191,211,212].

Synthesis of 7-hydroxyaporphines such as ushinsunine (**232**) in which there is a *cis* configuration between H(6a) and H(7) does not present any difficulty, and methods involving Pschorr cyclizations have been reported by several authors [189,191,209,213]. For example (Scheme 25), catalytic hydrogenation of the nitrobenzoyl-3,4-dihydroisoquinoline (**238**) with Raney nickel yields the aminobenzoyldihydroisoquinoline (**239**), reduction of which with sodium borohydride affords the α-hydroxyaminobenzyltetrahydroisoquinoline (**240**). Pschorr cyclization of **240** gives (+)-norushinsunine (**220**) without any trace of the H(6a),H(7)-*trans* diastereoisomer because of the highly selective nature of the sodium borohydride reduction [209]. A variation of this method [189,191] uses the intermediate (**241**) through a Pschorr cyclization to give the oxazoloaporphine (**242**), reduction of which

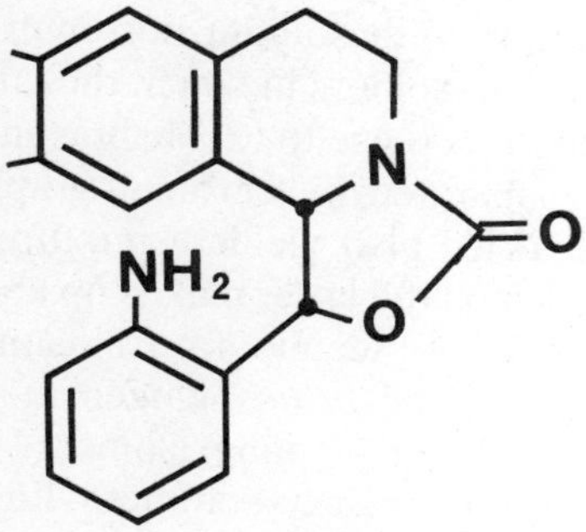

241

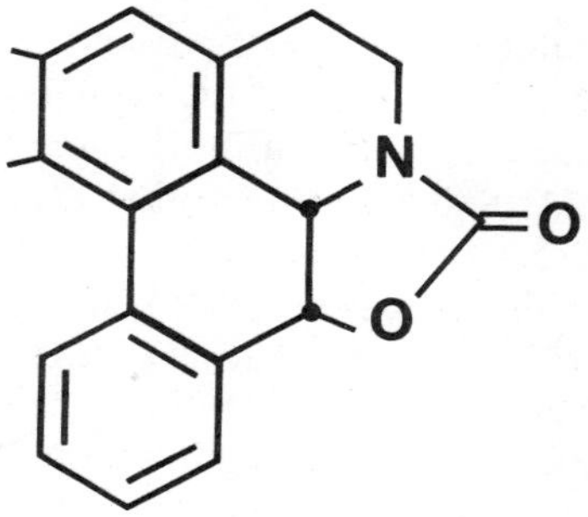

242

243

244

with lithium aluminum hydride yields the 7-hydroxynoraporphine (**243**, R = H) with *cis* stereochemistry. Treatment of the noraporphine with methyl-lithium yields the corresponding aporphine (**243**, R = Me).

The first method leading to the *trans* series of 7-hydroxyaporphines was described by Chackalamannil and Dalton [213]. Reduction of the salt of the *N*-methyloxoaporphinium ion (**244**) with a large excess of potassium boro-hydride in methanol gave a mixture of products containing 60% of the *cis*-7-hydroxyaporphine. By contrast, hydrogenation of **244** in methanol using a platinum oxide catalyst, followed immediately by potassium borohydride reduction in methanol, furnished in addition to a majority of the *cis*-7-hy-droxyaporphine a small quantity of the *trans* isomer, in a mixture of (*R*)–(*R*) and (*S*)–(*S*) configurations.

Kessar et al. [214,215] have shown that it is possible to obtain a mixture in which the *trans* isomer is enhanced by using a methylimine intermediate (Scheme 26). Treatment of the ketoimine (**245**) with iodomethane gave the quaternary derivative (**246**) which on reduction with sodium borohydride yields a mixture of the *cis* (**247a**) and *trans* (**247b**) isomers in which the latter makes up about 40%. Pschorr cyclization converts these to (±)-ushinsunine (**232**) and (±)-oliveroline (**225**), which are separated by chromatography. Direct reduction of **245** with sodium borohydride also yields a mixture of alcohols, **248a** (*erythro*) and **248b** (*threo*), of which the latter makes up about 20%. Photocyclization of **248** leads to a mixture of racemic norushinsunine (**220**) and noroliveroline (**218**). The presence of the *ortho* halogen in **245** appears to influence the proportion of the *threo* (*trans*-forming) isomer [212]. An analogous reaction sequence employing the necessary ketomethylimine or ketoimine gives racemic oliveridine (**221**) and its C(7) epimer or racemic noroliveridine (**216**), respectively [212]. 7-Epioliveridine has yet to be iso-lated from a natural source.

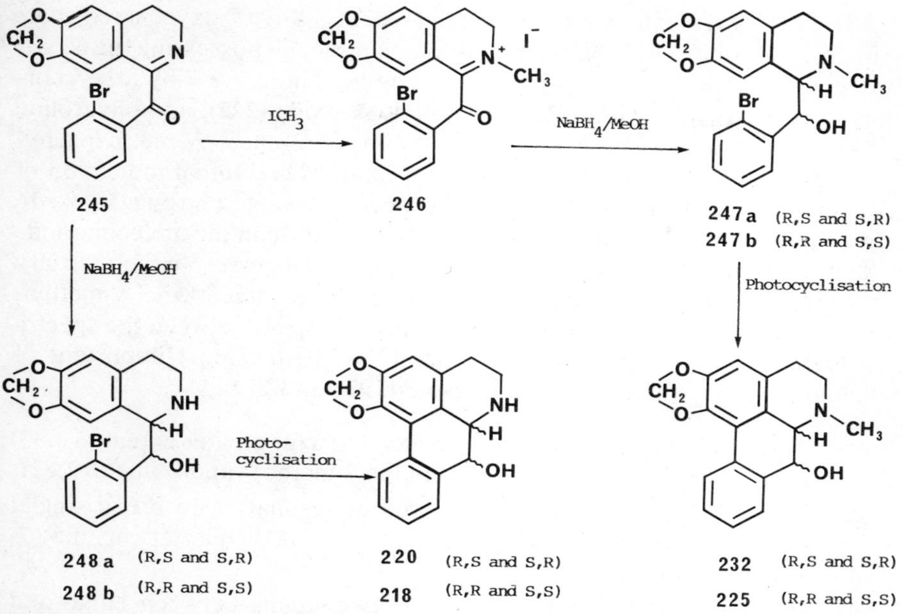

Scheme 26. Synthesis of racemic oliveroline (±**225**) and ushinsunine (±**232**), and of their nor-derivatives ±**218** and ±**220**.

7-Methoxyaporphines. Naturally occurring 7-methoxyaporphines have been reported only from the Annonaceae (Table 7) and possess an absolute configuration 6a(S), 7(S) that incorporates a *trans* relationship [205]. The ^{1}H NMR spectrum of 7-methoxyaporphines shows the additional methoxy function in the region δ 3.60–3.70 (δ 3.75 in noraporphines). The characteristic AB quartet for H(6a) and H(7) (J ~12 Hz) is found at about δ 3.65 and 4.30, a difference of ~0.65 ppm. This contrasts with the situation in 7-hydroxyaporphines with the same configurations, where the difference in resonance positions of H(6a) and H(7) is approximately 1.15 ppm. In 7-methoxynoraporphines, this difference is reduced to about 0.45 ppm, and in the corresponding 7-hydroxynoraporphines to about 0.85 ppm [133]. The deshielding of H(8) is much less pronounced in 7-methoxy than in 7-hydroxy alkaloids, that is, δ 7.10 in **223**, δ 7.30 in **221** [134].

The ^{13}C NMR spectrum of oliverine (**223**) is virtually identical with that of oliveridine (**221**) other than for the effects anticipated on methylation. The C(7) carbon is deshielded by about 11 ppm, resonating at 81.5 ppm in **223**. The methoxyl carbon resonates at 55 ppm [171].

Little has been described concerning the chemistry of the 7-methoxy-aporphines. A Hofmann reaction on the methiodide of oliverine (**223**) leads to the same vinylphenanthrene (**235**) produced from the methiodide of its 7-hydroxy analog oliveridine methiodide (**234**) [206].

7-Hydroxy- and 7-methoxyaporphine N-oxides. Five 7-oxygenated aporphine N-oxides have been found in the Annonaceae, all possessing the 6a(S),
7(S) configuration that is unique to the family. Three are 7-hydroxyapor-
phines, guatterine N-oxide (212), oliveridine N-oxide (222), and oliveroline
N-oxide (226), and the other two 7-methoxyaporphines, N-methylpachy-
podanthine N-oxide (215) and oliverine N-oxide (224). Initial indication of
N-oxidation usually originates from the occurrence of a strong $M^{+\cdot} - 16$
fragment in the mass spectrum that is often stronger than the molecular ion.

The ^1H NMR spectrum is generally similar to the corresponding tertiary
7-oxygenated aporphines, but with significant deshielding of the N-methyl,
H(6a) and H(7) resonances. Clear differences do appear between the spectra
of 7-methoxy and 7-hydroxy aporphines [72,133], the most important of
which are demonstrated by a comparison of 224 and 222:

(a)   The N-methyl resonance is deshielded to a far greater extent (δ 3.33)
      in the 7-hydroxyaporphine than in the 7-methoxyaporphine (δ 2.82).

(b)   The C(1)/C(2) methylenedioxy protons resonate as a broad singlet
      in the 7-methoxyaporphine but as an AB quartet in the 7-
      hydroxyaporphine.

(c)   In the 7-methoxyaporphine the *trans* coupling between H(6a) and
      H(7) is reduced to about 5.5 Hz.

(d)   A singlet appears at about δ 2.7 in the 7-methoxyaporphine but is
      lost on addition of deuterium oxide.

The above modifications in the spectra of 7-methoxyaporphine N-oxides
indicate that the nucleus tends toward planarity but this is not the case in
the 7-hydroxyaporphine N-oxides. The replaceable signal must reflect sol-
vation with water, presumably chelated between the oxygens on N(6) and
C(7). Such chelation could bring about the approach to planarity in the aporphine nucleus [72,133].

Identification of individual N-oxides is generally confirmed by zinc–hydrochloric acid reduction to the corresponding 7-oxygenated aporphine or
by N-oxidation of the latter with hydrogen peroxide to give the synthetic
N-oxide. Finally, it should be noted that N-oxides often occur in appreciable
amounts and it seems unlikely that they are artifacts.

4,7-Dioxygenated aporphines. At present only two alkaloids of this type
have been found in the Annonaceae, pachystaudine (229) and norpachystaudine (219), both isolated only from *Pachypodanthium staudtii* [124]. A
number of other 4-oxygenated aporphines have been recorded, but not from
the Annonaceae [8–10].

The structure of pachystaudine (229) was deduced primarily from the ^1H
NMR spectrum, which clearly established a 1,2-methylenedioxy substituent
and an unsubstituted ring D. A C(7) methoxy was indicated by a singlet at

δ 3.66 and *trans* H(6a) and H(7) protons by an AB quartet (J = 12 Hz) centered at δ 3.53 and 4.26. The empirical formula needed the presence of an additional hydroxy substituent, which was required to be alcoholic and not a phenolic hydroxy. This was confirmed by the ^{1}H NMR spectrum which revealed a double doublet (J = 10, 5.5 Hz) for an axial oxymethine proton. The placement of the alcohol at C(4) was suggested by the deshielding of H(3) to δ 6.93, and this was confirmed by preparation of the acetyl derivative (249), in which H(4) resonated at δ 6.00 and H(3) was shielded to δ 6.66 [124].

Circular dichroism measurements confirmed that 229 retained the 6a(S) configuration typical of all 7-oxygenated aporphines and assignment of 7(S) followed from this. Assignment of (S) configuration to C(4) was made by comparison of the ^{1}H NMR spectrum with those of other C(4) oxygenated aporphines in which the configuration of C(4) was known [216,217]. It has been shown that the relative stereochemistry of C(4) and C(6a) can be established from the resonance position of H(4); in the region δ 4.5–4.6 in the form of a triplet (J ~2.5 Hz) for the *syn* series [H(4) and H(6a) on the same side of the nucleus] and in the region of δ 4.8–4.9 as a double doublet (J ~10, 5 Hz) where H(4) and H(6a) are *anti* [218,219]. From this the (S) assignment for C(4) follows readily.

The identity of the corresponding noraporphine, norpachystaudine (219), was established by N-methylation to give 229 [124].

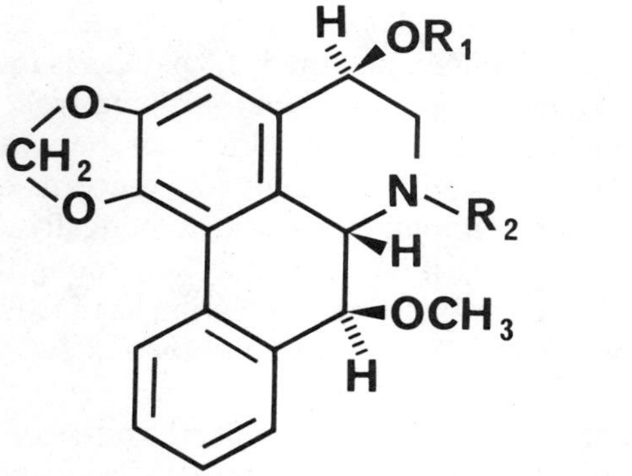

	R_1	R_2
219	H	H
229	H	CH_3
249	$COCH_3$	CH_3

Biogenesis of 7-oxygenated Aporphines. No studies directly concerned with the formation of these alkaloids have been reported. Their derivation may be envisaged through general pathways discussed in Section 2 (Scheme 8) and in Section 5.2.3 for the 7-alkyl aporphinoids. On the other hand, it is also feasible to consider that oxygenation of C(4) and C(7) occurs early in the biogenetic pathway, even before the cyclization of the aporphine nucleus [19]. In support of this idea, note that oxygenation at the carbon corresponding to C(4) is not uncommon in some other types of isoquinoline alkaloids and that 1-btiq alkaloids in which the α-carbon [C(7) of aporphines] is oxygenated are known. The possibility that the *cis* and *trans* forms of the 7-oxygenated aporphines do not have the same biogenetic origin should not be overlooked.

5.3. Oxoaporphines

The oxoaporphines isolated from the Annonaceae now number 20 (Table 8). To some extent variation in structure follows that of the aporphines *s.s.* Thus oxoaporphines substituted only in ring A [including C(3)] are the most common numerically and most widely distributed, followed by 1,2,9-substituted compounds. The most widespread individual oxoaporphine is liriodenine (**256**) which is analogous to anonaine, which is in turn the most widely distributed aporphine *s.s.* (see Table 3). Among oxoaporphines with a single ring-D substituent the 1,2,11 pattern is again at present restricted to alkaloids from species of *Duguetia* and *Guatteria*.

By contrast to the aporphines *s.s.* the 1,2,8 pattern is found in the alkaloids oxostephanine (**267**) and thailandine (**269**), isolated from species of *Polyalthia*. However, the most noteworthy difference between the substitution patterns of oxoaporphines and aporphines *s.s.* concerns compounds bearing two ring-D substituents. Although these are numerically common among the aporphines *s.s.*, they are strikingly uncommon among the oxoaporphines, with 1,2,9,10-substituted oxoaporphines having been isolated only from *Annona purpurea* and *Guatteria scandens* and the 1,2,10,11 pattern being unrecorded to date.

Worthy of specific mention are the recent reports of two *N*-methylated quaternary oxoaporphinium ions, *N,O*-dimethylliriodendronine (**253**) from *Guatteria chrysopetala* [80] and thailandine (**269**) from *Polyalthia cauliflora* [128]. Thailandine is a typical quaternary carrying an external counterion (depicted here as OH⁻) but *N,O*-dimethylliriodendronine appears to exist in the form of a zwitterion.

Methods for the synthesis of oxoaporphines, which are well documented and are mentioned incidentally in Sections 5.2.4 and 5.4, are not treated in detail here. Briefly, the normal method for complete synthesis is *via* Pschorr cyclization of the fully aromatized B-ring analog of the α-keto benzylisoquinoline **239** (Scheme 25). An example of this is the synthesis of subsessiline

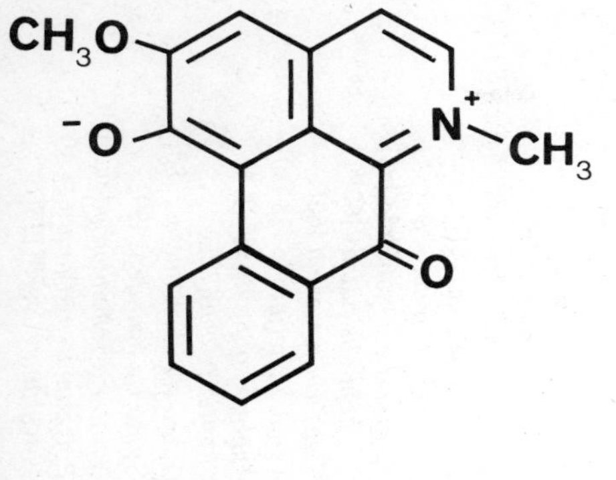

253

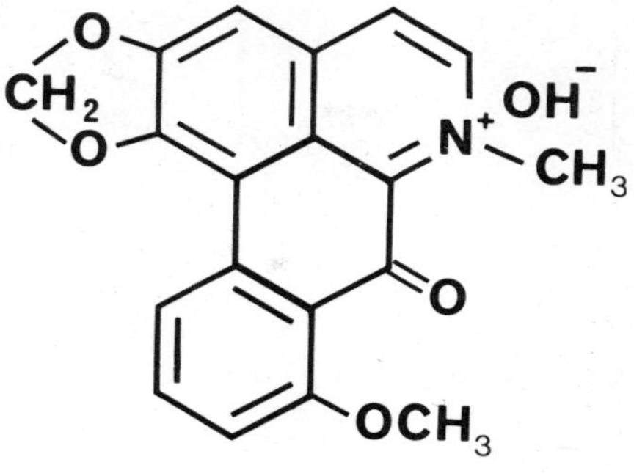

269

(268) [105], which confirmed placement of the phenolic function at C(9) rather than C(3) (Scheme 27). Nonphenolic noraporphines can be converted to the corresponding oxoaporphines by means of oxidation with iodine, which gives yields superior to those obtained with chromium trioxide in pyridine [185]. Aporphines are oxidized to oxoaporphines by a range of

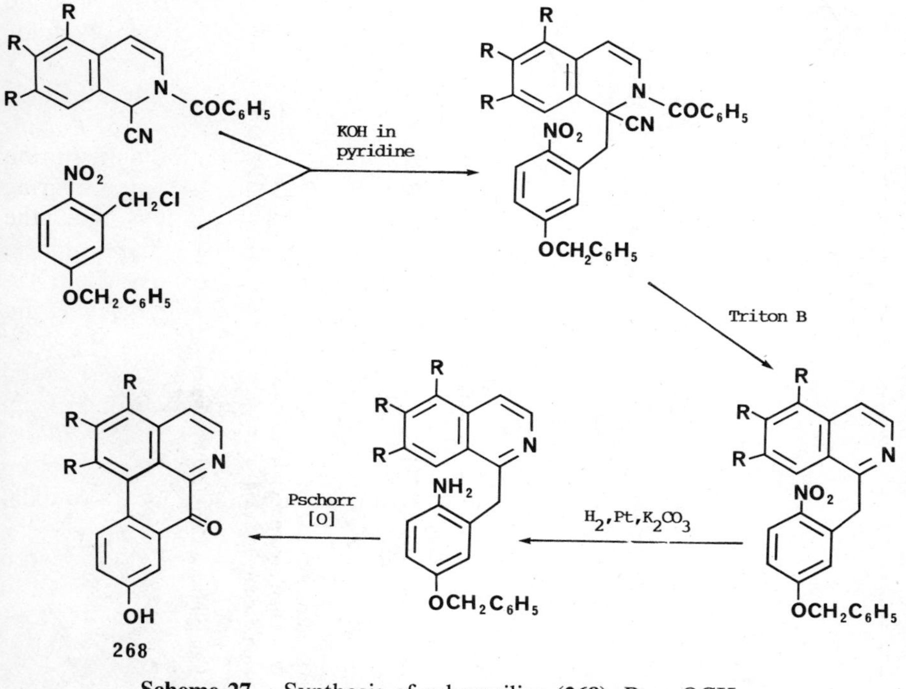

Scheme 27. Synthesis of subsessiline (268), R = OCH₃.

Table 8. Occurrence and Distribution of Oxoaporphines

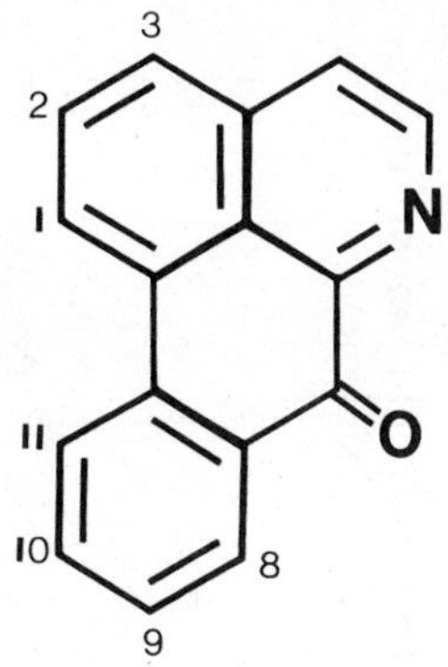

Name	C(1)	C(2)	C(3)	NR	C(8)	C(9)	C(10)	C(11)	Sources
Atheroline (250)	OCH₃	OCH₃	—	—	—	OH	OCH₃	—	*Guatteria scandens* [99] *Enantia polycarpa* [73] *Guatteria psilopus* [97] *Polyalthia cauliflora* var. *beccarii* [128] *Rollinia sericea* [141]
Atherospermidine (251)	O—CH₂—O		OCH₃	—	—	—	—	—	
Dicentrinone (252)	O—CH₂—O		—	—	—	OCH₃	OCH₃	—	*Guatteria scandens* [99]
N,O-Dimethylliriodendronine (253)	O⁻	OCH₃	—	CH₃	—	OCH₃	OCH₃	—	*Guatteria chrysopetala* [80]
Isomoschatoline (254)	OCH₃	OCH₃	OH	—	—	—	—	—	*Cleistopholis patens* [60] *Guatteria melosma* [60]
Lanuginosine (255) (oxoxylopine)	O—CH₂—O		—	—	—	OCH₃	—	—	*Annona cherimolia* [28–30] *Annona squamosa* [52] *Enantia pilosa* [72] *Guatteria chrysopetala* [80] *Guatteria scandens* [99] *Guatteria schomburgkiana* [103] *Polyalthia emarginata* [130] *Polyalthia oliveri* [133]

Liriodenine (**256**) (oxoushinsunine)

O—CH₂—O

—	
—	
—	
—	

Rollinia papilionella [140]
Xylopia brasiliensis [148]
Xylopia buxifolia [149]
Xylopia frutescens [152]
Xylopia lemurica [153]
Annona acuminata [26]
Annona cherimolia [27–30]
Annona crassiflora [31]
Annona cristalensis [32]
Annona glabra [33–36]
Annona montana [38,39]
Annona reticulata [45–47]
Annona squamosa [48]
Asimina triloba [55]
Cananga latifolia [56]
Cananga odorata [57]
Cleistopholis patens [59,60]
Enantia pilosa [72]
Enantia polycarpa [73]
Fusaea longifolia [78,79]
Guatteria chrysopetala [80]
Guatteria cubensis [81]
Guatteria goudotiana [87]
Guatteria melosma [60]
Guatteria modesta [91,92]
Guatteria scandens [99]
Guatteria schomburgkiana [103]
Guatteria sagotiana [104]
Hexalobus crispiflorus [107]
Isolona campanulata [109]
Meiogyne virgata [113]
Melodorum punctulatum [114]
Mitrella kentii [115]
Monodora tenuifolia [119,120]

Table 8. (*continued*)

Name	C(1)	C(2)	C(3)	NR	C(8)	C(9)	C(10)	C(11)	Sources
									Pachypodanthium staudtii [124,125]
									Polyalthia acuminata [127]
									Polyalthia cauliflora var. *beccarii* [128]
									Polyalthia emarginata [130]
									Polyalthia nitidissima [131,132]
									Polyalthia oliveri [133]
									Pseuduvaria cf. *grandifolia* [131]
									Pseuduvaria sp. [131]
									Rollinia papilionella [140]
									Rollinia sericea [141]
									Schefferomitra subaequalis [131]
									Unonopsis guatterioides [144]
									Uvariopsis guineensis [11]
									Xylopia brasiliensis [148]
									Xylopia buxifolia [149]
									Xylopia frutescens [152]
									Xylopia pancheri [154]
									Xylopia vielana [156]
Lysicamine (**257**) (oxonuciferine)	OCH$_3$	OCH$_3$	—	—	—	—	—	—	*Annona acuminata* [26]
									Enantia chlorantha [71]
									Enantia polycarpa [73]
									Guatteria chrysopetala [80]
									Guatteria ouregou [93,95]
									Guatteria saffordiana [98]
									Polyalthia cauliflora var. *beccarii* [128]
									Polyalthia suaveolens [135]
									Rollinia papilionella [140]
									Unonopsis guatterioides [144]

Compound	1	2	3	9	10	11	Sources
O-Methylmoschatoline (**258**) (homomoschatoline) (liridine)	OCH₃	OCH₃	OCH₃	—	—	—	*Annona acuminata* [26]; *Duguetia eximia* [67]; *Enantia chloranha* [71]; *Guatteria ouregou* [93,95]; *Guatteria saffordiana* [98]; *Heteropetalum brasiliensis* [81,105]; *Polyalthia acuminata* [127]; *Polyalthia cauliflora* var. *beccarii* [128]; *Polyalthia sericea* [141]
Oxoanolobine (**259**)	O—CH₂—O		—	OH	—		*Guatteria melosma* [88–90]
Oxobuxifoline (**260**)	O—CH₂—O		OCH₃	—	OCH₃	—	*Guatteria sagotiana* [104]
Oxoglaucine (**261**) (*O*-methylatheroline)	OCH₃	OCH₃	—	OCH₃	OCH₃	—	*Duguetia obovata* [65]; *Annona purpurea* [42]
Oxoisocalycinine (**262**)	O—CH₂—O		—	OH	—	OCH₃	*Guatteria discolor* [82]
Oxolaureline (**263**) (lauterine)	O—CH₂—O		—	—	OCH₃	—	*Guatteria elata* [86]; *Guatteria sagotiana* [104]
Oxopukateine (**264**)	O—CH₂—O		—	—	OH		*Duguetia eximia* [67]
Oxopurpurine (**265**)	OCH₃	OCH₃	—	OCH₃	—		*Annona purpurea* [42]
Oxoputerine (**266**)	O—CH₂—O		—	—	OCH₃	OCH₃	*Duguetia calycina* [63]
Oxostephanine (**267**)	O—CH₂—O		OCH₃	—	—		*Duguetia eximia* [67]; *Guatteria discolor* [83]; *Guatteria elata* [86]; *Guatteria schomburgkiana* [103]; *Guatteria sagotiana* [104]; *Polyalthia cauliflora* var. *beccarii* [128,129]
Subsessiline (**268**)	OCH₃	OCH₃	OCH₃	—	OH	—	*Polyalthia suaveolens* [135,136]; *Guatteria ouregou* [93,95]
Thailandine (**269**)	O—CH₂—O		CH₃	OCH₃	—	OCH₃	*Heteropetalum brasiliensis* [81,105,106]; *Polyalthia cauliflora* var. *beccarii* [128]

reagents (lead tetraacetate, manganese dioxide, chromium trioxide, iodine, ceric ammonium nitrate), the resulting reaction generally yielding a mixture of oxoaporphines and dehydroaporphines, the relative amounts of which vary with the reaction conditions [15,16,185,220]. Semisynthesis can also be achieved by chromium trioxide in pyridine oxidation of the corresponding 7-hydroxyaporphine (Scheme 24) or by photooxidation of the corresponding dehydroaporphine in methanol (Scheme 29). Reduction with zinc amalgam converts an oxoaporphine to the corresponding noraporphine; using the oxoaporphine methiodide generates the corresponding aporphine.

5.4. 5- and 4,5-Oxoaporphines

This is a small group of aporphinoids characterized by oxidation in ring B and, in the 4,5-dioxoaporphines, by a 6a,7 double bond. At present only a single 5-oxoaporphine is known, fuseine (270), and it has been reported only from the Annonaceae. The 4,5-dioxoaporphines include 10 alkaloids [8–10], of which only two have been reported from the Annonaceae, norcepharadione-B (271) and its 3-methoxy derivative ouregidione (272). A third dioxoaporphine is known from the Annonaceae, as part of the dimeric aporphinoid beccapolydione (295), and is discussed in Section 5.5.

Fuseine (270).   This alkaloid has been isolated only once, in trace amounts, from the wood of the Brazilian species *Fusaea longifolia* [79]. Its structure was established by comparison of spectral data with that for anonaine (64), the analogous noraporphine. Notable features allowing placement of the carbonyl at C(5) were a band at ν_{max} 1680 cm^{-1} in the IR spec-

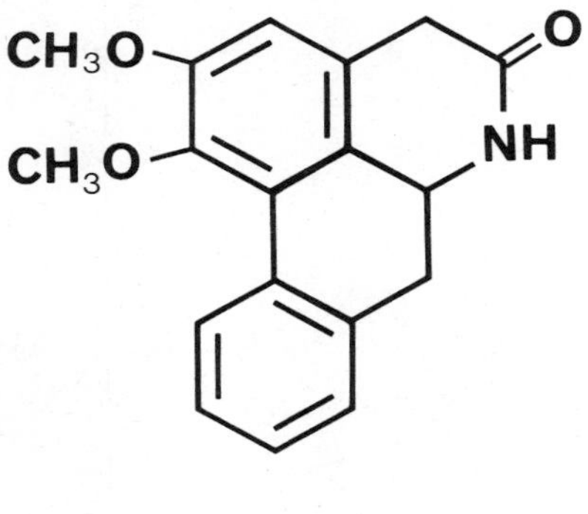

270

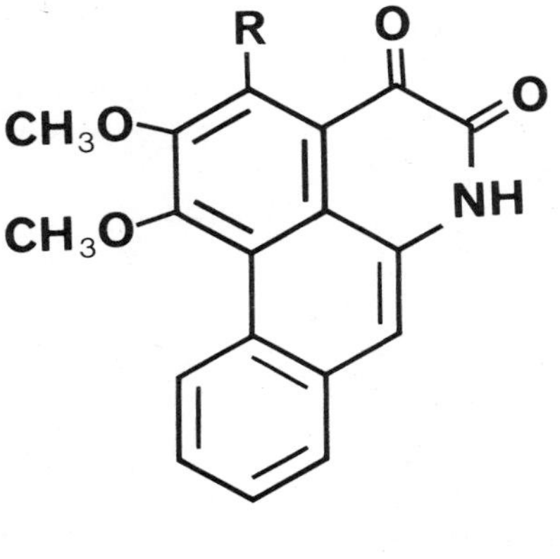

271   R = H

272   R = OCH$_3$

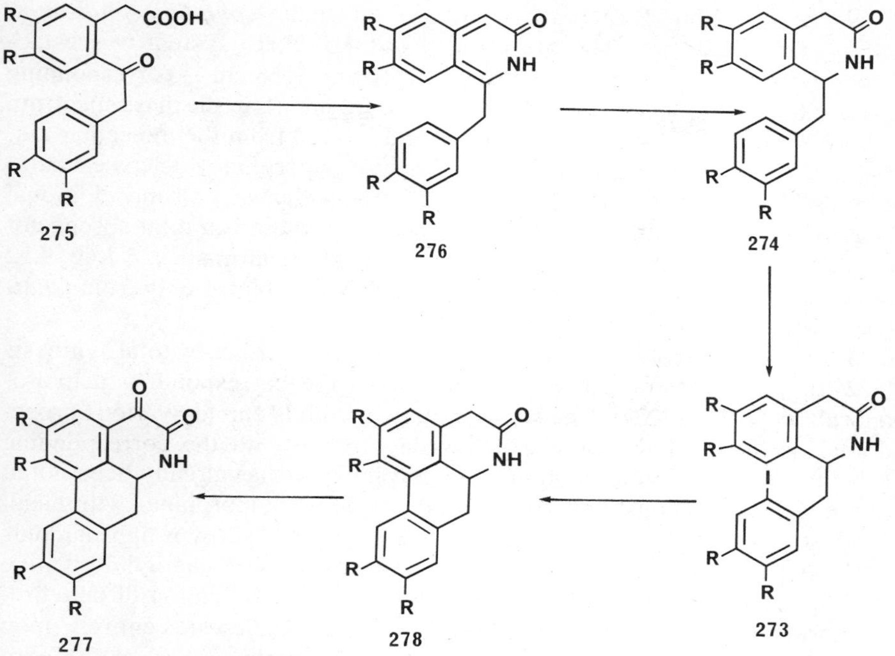

Scheme 28. Synthesis of norpontevedrine (**277**), R = OCH$_3$.

trum for a lactam and a fragment in the mass spectrum owing to the loss of HNCO through a retro-Diels–Alder opening of ring B. The presence of an amide group was also in agreement with the insolubility of **270** in hydrochloric acid. The configuration at C(6a) is unknown.

Synthesis of **270** has not yet been achieved but there has been a promising attempt by Castedo et al. [221]. Their approach (Scheme 28) is based on a photochemical cyclization of halogenated benzylisoquinolin-3-ones such as **273**, which is readily available through halogenation of **274**, which can in turn be obtained from the keto acid (**275**) *via* the lactam (**276**). Irradiation of **273** produced a complex reaction mixture from which the 4,5-dioxonoraporphine norpontevedrine (**277**) was obtained in a yield of about 10%. The alkaloid **277** must be formed by oxidation of C(4) of the corresponding 5-oxoaporphine (**278**) and although Castedo et al. [221] failed to isolate the latter, it seems likely that with modification of the reaction conditions 5-oxoaporphines will be obtained from this pathway.

4,5-Dioxoaporphines. Both norcepharadione-B (**271**) and ouregidione (**272**) have been found only in the leaves of *Guatteria ouregou* and in small amounts [93]. Like all 4,5-dioxoaporphines they are orange-red in color and strongly fluorescent.

Norcepharadione-B (271) was first reported from *Stephania cepharantha* (Menispermaceae) [9]. The presence of the 4,5-diketo system was established from bands in the IR spectrum at 1650 and 1668 cm^{-1} corresponding to an amide and adjacent carbonyl and major fragments in the mass spectrum at *m/z* 279 and 264 owing to loss of CO and HNCO from the molecular ion. The UV spectrum of the 4,5-dioxo-6a,7-dehydroaporphine nucleus is characterized by the typical bands of a phenanthrene with a very strong additional band at about 440 nm. In the ^{1}H NMR spectrum there is a total absence of aliphatic protons and considerable deshielding of the aromatics, δ 7.60–9.52 in deuteriochloroform. Ouregidione (272) has been isolated only from *Guatteria ouregou* [93].

Several 4,5-dioxoaporphines have been prepared, either by total synthesis [222–225] or by semisynthesis, usually from the corresponding dehydroaporphine [193,226,227]. The semisynthetic methods employ various procedures for the oxidation of the dehydroaporphine with the corresponding substitution to the required 4,5-dioxoaporphine. As has already been noted (Section 5.2.2, Scheme 17), air oxidation of dehydroaporphines with alkali catalyst gives 4,5-dioxo as well as 7-oxo and aristolactam-type aporphinoids [193]. A simple synthesis of cepharadione-B (279), the *N*-methyl derivative of 271, has been described [226] involving the photooxidation of dehydronuciferine (280). The course of photooxidation of 280 depends entirely upon solvent polarity (Scheme 29). In methanol 280 yhields ~35% of the oxoaporphine lysicamine (257), but no detectable 279. By contrast, in a nonpolar solvent (hexane), oxidation of the enamine system of 280 is slowed appreciably, allowing oxygen attack on ring B to become competitive and giving a small amount of 279. A comparable enzyme oxidation of 280 to 279 seems a likely biogenetic route [226].

Norcepharadione B (271) has been synthesized by two routes (Scheme 30). The first is based on an intermolecular Diels–Alder cyclization between benzyne and the 1-methylene-3,4-dioxoisoquinoline derivative (281), which had in turn been prepared from the isoquinolinedione 282. The reaction afforded 271 in 40% yield [223]. The second route involved photocyclization of the bromobenzylidene isoquinolinedione 283, which had been obtained

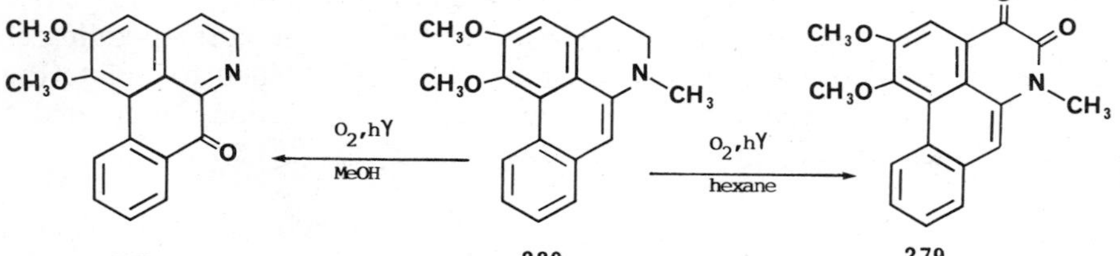

Scheme 29. Influence of solvent polarity on the course of photo-oxidation of dehydronuciferine (280); synthesis of cepharadione B (279).

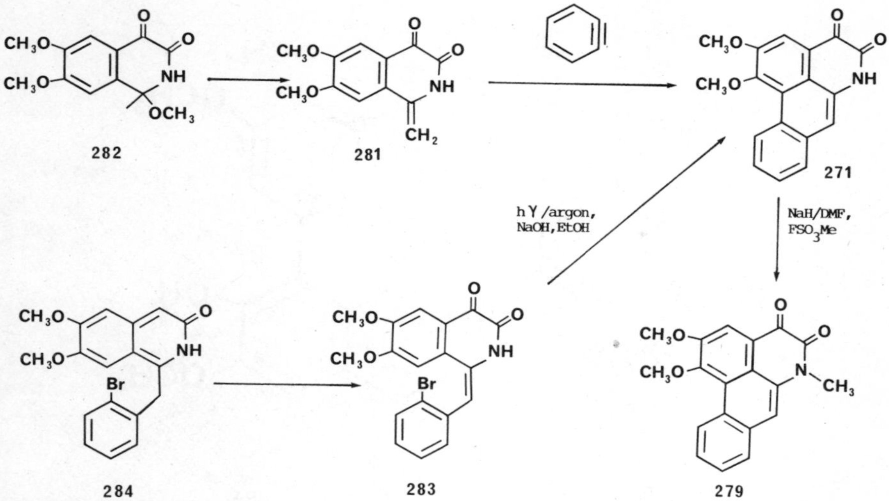

Scheme 30. Syntheses of the 4,5-dioxoaporphines norcepharadione B (**271**) and cepharadione B (**279**).

by two alternative methods of controlled oxidation of the bromobenzylisoquinolin-3-one **284**. Irradiation of an ethanolic solution of **283** under argon gave **271** in a 45% yield [224]. N-methylation of **271** using sodium hydride in DMF with the addition of methyl fluorosulfonate gave the corresponding N-methyl derivative (**279**) in 65% yield [224].

Although not established by *in vivo* experiments, the 4,5-dioxoaporphines probably represent an intermediate stage in the oxidation of aporphines *s.s.*, through dehydroaporphines and, possibly, 4-hydroxyaporphines, to the aristolactams (Section 5.7).

5.5. Dimeric Aporphines

In the past 5 years two reviews of the dimeric aporphinoids have been published [228,229]. The dimers may be divided into a number of different groups depending upon the form of the monomeric units, for example, aporphine–benzylisoquinoline dimers and their oxidation products, proaporphine–benzylisoquinoline dimers, aporphine–pavine dimers, and bisaporphines, which are dimeric C–C linked aporphinoids. Only dimers of the last group are known from the Annonaceae and, at present, they are unique to the family. They may be divided into two subgroups, depending on whether the monomers are linked C(7) to C(7′) or C(4) to C(7′).

7,7′-Linked Dimers.   These consist of the two alkaloids bipowine (**285**) and bipowinone (**286**), at present known only from the stem bark of the

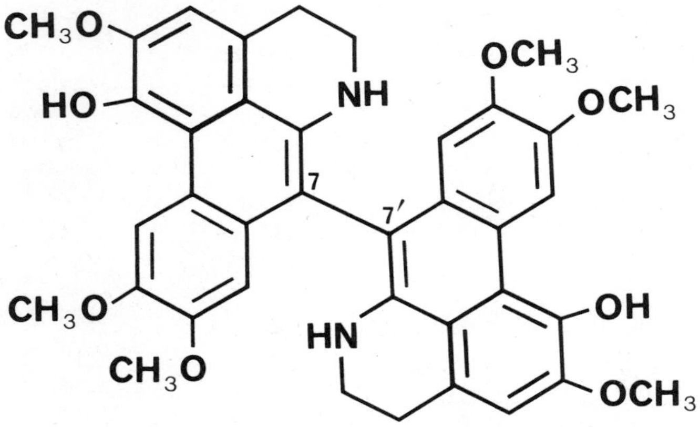

285

Indonesian species *Popowia pisocarpa* [138]. The symmetrical nature of **285** was indicated by the mass spectrum [M$^+$ 648 (100%), *m/z* 324 (M/2, 50%) and by the ^{1}H NMR spectrum, which resembled that of a dehydronorapor- phine but lacked any signal corresponding to H(7). The UV spectrum was also typical of a dehydroaporphine. The ^{1}H NMR spectrum of **285** revealed resonances at δ 3.46 and 6.62, for 9/9′-methoxy and H(8/8′) aromatic protons, shielded by the 7–7′ bond.

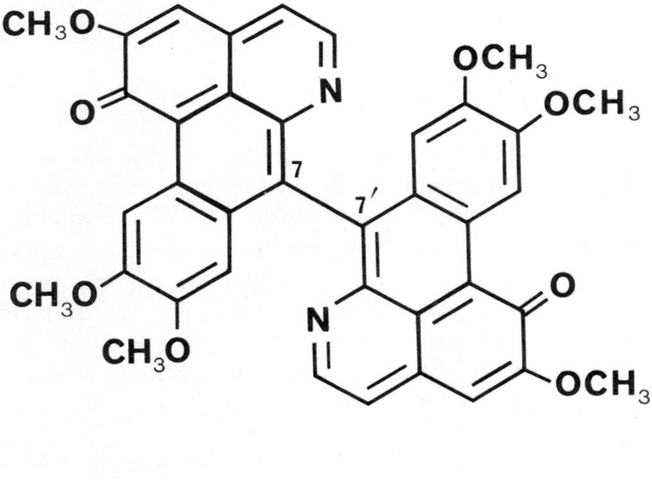

286

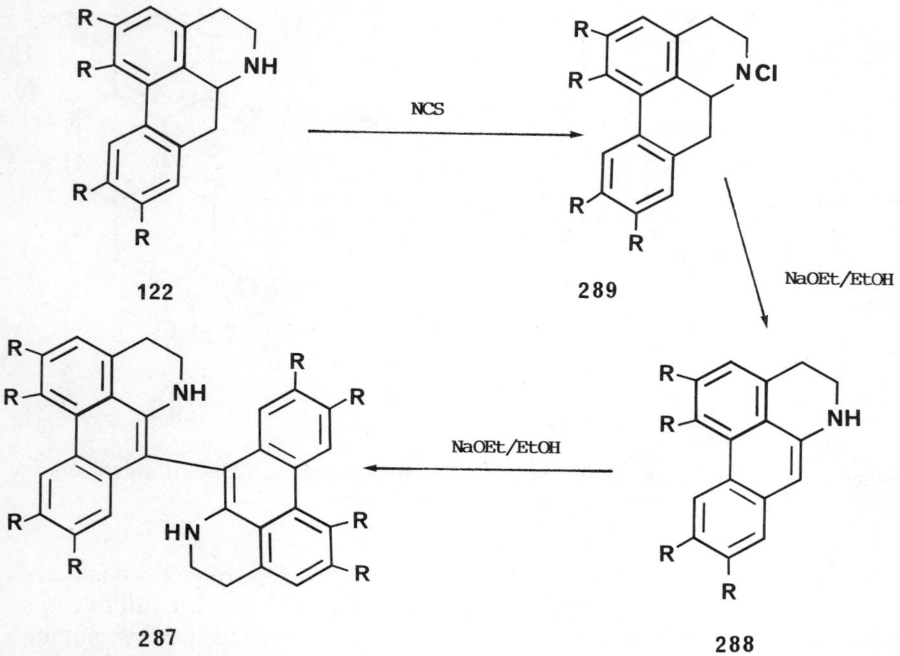

Scheme 31. Synthesis of bisdehydronorglaucine (**287**), R = OCH₃.

Dimers with this linkage have been synthesized by oxidation, under various conditions, of dehydroaporphines [230,231]. In all reported examples they are *N*-methyl dimeric derivatives. To examine further the structure of **285**, bisdehydronorglaucine (**287**) was prepared [138] by dimerization of dehydronorglaucine (**288**) prepared from norglaucine (**122**) *via* the chloramine (**289**) (Scheme 31). The highly reactive **288** dimerized readily, either through treatment with excess sodium ethanolate or simply by spontaneous oxidation at room temperature of a solution in a mixture of methanol and dichloromethane. The synthetic **287** showed comparable spectral characteristics to **285**, with the exception that in the ¹H NMR spectrum there was an additional singlet (6H) at δ 3.99 assignable to 1 and 1′ methoxyls.

An attempt to synthesize **285** direct from wilsonirine (**147**) was unsuccessful (Scheme 32). Preparation of the chloramine of **147** using *N*-chlorosuccinimide, followed by treatment with sodium ethanolate in ethanol, did not yield dehydrowilsonirine but gave a 1-oxoaporphine quinone (**290**), a type of compound synthesized previously [232,233].

The identical 1-oxoaporphine quinone structure is found in the second isolated dimer, bipowinone (**286**). Oxidation of **285** in air or in solution in dichloromethane yields **286**, both confirming that the phenolic groups in **285** must be at C(1) and C(1′) and also suggesting the possibility that **286** is actually an artifact formed from **285**.

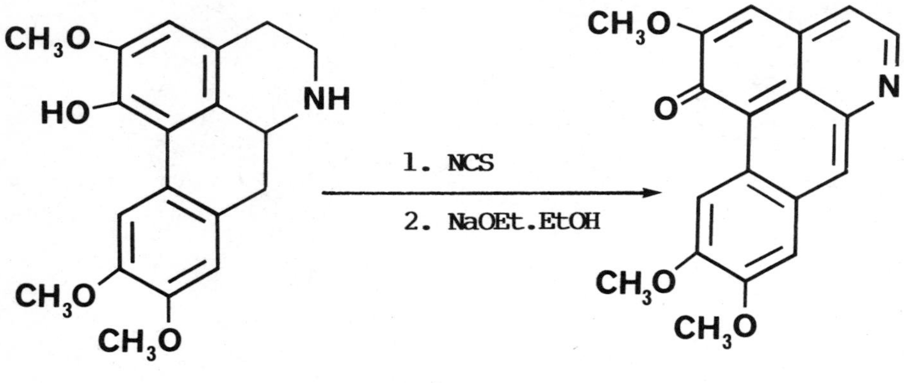

Scheme 32. Formation of the quinonoid 1-oxoaporphine (**290**) from wilsonirine (**147**).

4,7′-Dimeric Aporphinoids. Four dimers of this type have been isolated, all from the stem bark of the Indonesian species *Polyalthia cauliflora* var. *beccarii* [128,129]. The first to be isolated and identified was beccapoline (**291**) [128,129]. Its dimeric nature was apparent from the mass spectrum, M^+ 608 with no fragments between m/z 562 and 305. The UV spectrum indicated a highly conjugated system in which the inclusion of a conjugated carbonyl was revealed by a bathochromic shift in acidic media and also by the presence of a band at 1650 cm^{-1} in the IR spectrum.

The ^{1}H NMR spectrum revealed a series of signals comparable with those for the oxoaporphine oxostephanine (**267**), except that the resonance for H(4) was absent and H(5) was observed as a singlet at δ 8.75. This monomer (designated unit A) was, therefore, attached to the second monomer (unit B) through C(4). The second unit was identified as 4,5,6a,7-tetradehydro-stephanine, with H(4′) and H(5′) resonating as an AB quartet ($J = 7.5$ Hz) at δ 5.90 and 6.12. The absence of a signal attributable to H(7′) in unit B indicated the second linkage position, and this was supported by shielding of the N′-methyl resonance (δ 2.34) and 8′-methoxy resonance (δ 2.95). An examination of a Dreiding model of **291** shows that with a 7′ link there is steric hindrance to rotation of the monomers with both the N′-methyl and 8′-methoxy lying in the shielding cone from the nucleus of unit A.

The structure **291** was confirmed by reduction with zinc–hydrochloric acid (Scheme 33), which gave desoxydecahydrobeccapoline **292**, the mass spectrum of which gave fragments at m/z 294 and 308 for norstephanine (unit A) and stephanine (unit B), respectively.

Beccapolinium (**293**) had an analysis [128,129] agreeing with one methyl more than **291** and gave a similar ^{1}H NMR spectrum except for the presence of an additional singlet (3H) at δ 4.92, attributable to the N-methyl of an

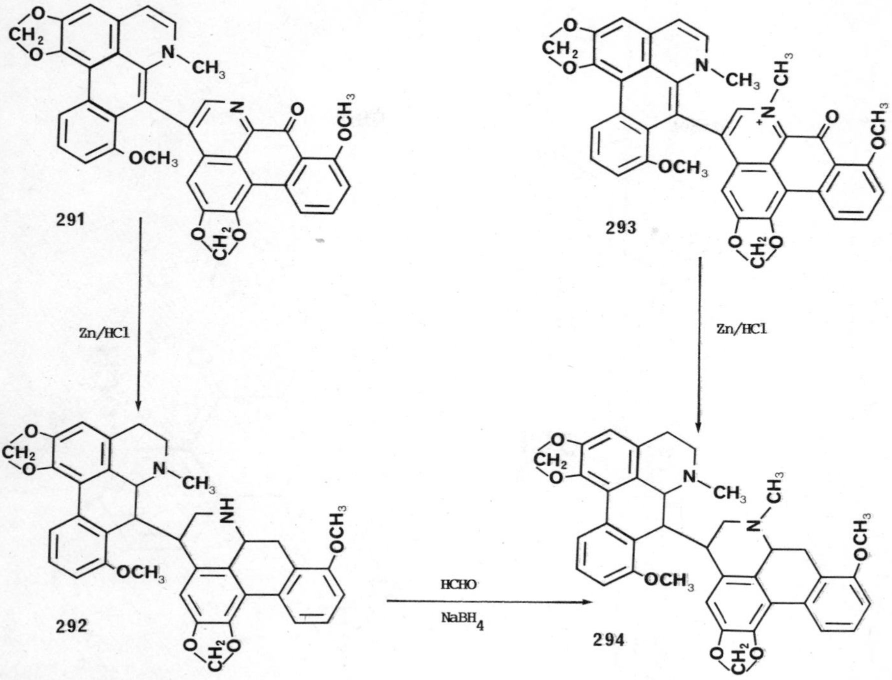

Scheme 33. Interrelationship between beccapoline (**291**) and beccapolinium (**293**).

oxoaporphinium ion such as thailandine (**269**). The UV spectra of **291** and **293**, run in acidic conditions, were identical. In **293**, therefore, the oxo-aporphine unit (unit A) differs in being an N-methyl quarternary compound whereas unit B remains the same as in **291**. This structure was confirmed by reduction of **293** with zinc–hydrochloric acid (Scheme 33) to yield a bi-stephanine dimer **294**, which was identical with the product of N-methylation of the dimer **292** formed on the identical reduction of **291**.

Beccapolydione (**295**) [128] contained two more oxygens but two fewer hydrogens than **291**. Major diagnostic features were carbonyl bands at 1665 and 1650 cm^{-1} in the IR spectrum and in the ^{1}H NMR spectrum signals indicative of an oxostephanine unit linked through C(4) (unit A). In comparison with **291**, the AB system for H(4′) and H(5′) was absent and H(3′) showed strong deshielding to δ 8.09, suggesting the presence of a carbonyl at C(4). That unit B was a 4,5-dioxoaporphine was further supported by the occurrence of a fragment at m/z 307 corresponding to loss of CO, a common feature of 4,5-dioxoaporphines (Section 5.4). The relative deshielding of the N′-methyl resonance to δ 2.66 and the 8′-methoxy resonance to δ 3.09, presumably due to anisotropic effects, is also in agreement with the proposed structure.

The final alkaloid of the group, polybeccarine (**296**), differs from **291** in

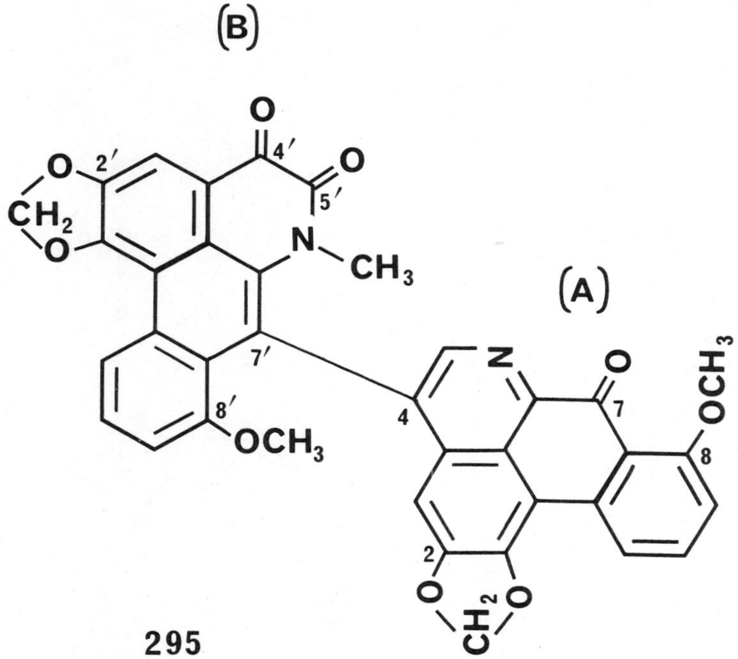

(B)

(A)

295

the absence of one of the methoxy substituents [128]. The highly shielded position of the remaining methoxyl (δ 2.99) in **296** requires its placement at C(8′) and thus the oxoaporphine moiety in **296** must be liriodenine (**256**) rather than oxostephanine (**267**).

No *in vivo* studies on the biogenesis of these dimers have yet been carried

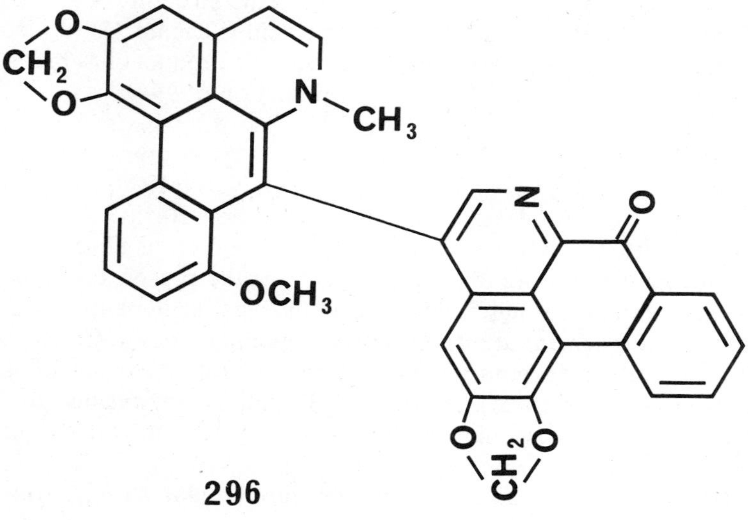

296

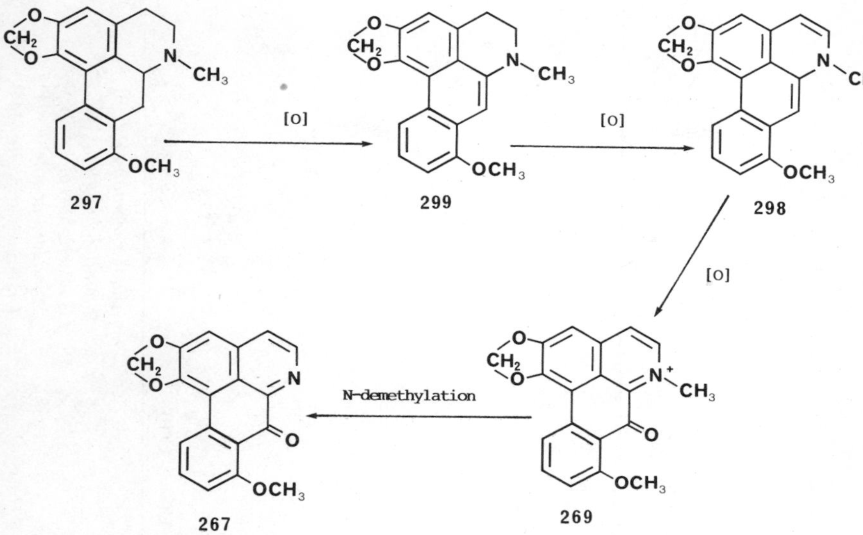

Scheme 34. Biogenetic pathway from stephanine (**297**) to oxostephanine (**267**).

out. However, it has been proposed [12,19] that stephanine (**297**) and its oxidized derivatives are likely precursors of these dimers (Scheme 34). Guinaudeau et al. [234] suggest that **297** can be converted to thailandine (**269**) as an intermediate in its oxidation to oxostephanine (**267**), both of which have been isolated from *Polyalthia cauliflora* var. *beccarii*. A probable intermediate in the route to **269** and **267** from **297** would be tetradehydrostephanine (**298**), formed *via* dehydrostephanine (**299**).

Immediate precursors of dimerization are likely to be thailandine (**269**) or a similar *N*-methyl quaternary oxoaporphinium ion, in which C(4) is electron deficient, and the tetradehydroaporphine (**298**) or dehydroaporphine (**299**), in which C(7) is suitably electron rich. If the dehydroaporphine is involved, then oxidation at 4,5 probably proceeds simultaneously with linkage. Such a pathway (Scheme 35) would envisage initial formation of beccapolinium (**293**), with beccapoline (**291**) resulting from facile *N*-demethylation of **293** and beccapolydione (**295**) in turn formed by oxidation of the B ring of **291**.

5.6. Phenanthrenes

The phenanthrenes remain a relatively small group of alkaloids in the Annonaceae, now numbering 13 structures, none of which are very common (Table 9). The numbering system used in this review follows that of the aporphine precursors (see **4**) so that comparison with the other groups of alkaloids discussed is simplified. The majority are substituted only in ring A [including C(3)] and vary from secondary amines to quaternary *N,N*-

Table 9. Occurrence and Distribution of Phenanthrenes

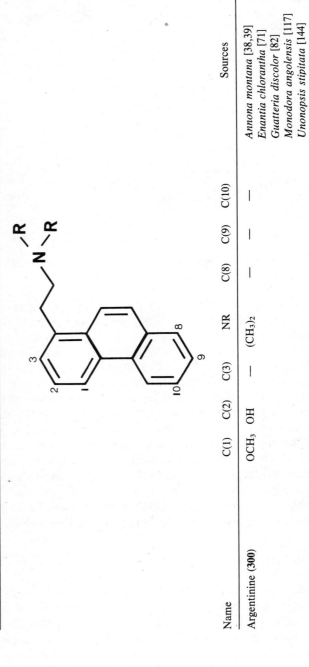

Name	C(1)	C(2)	C(3)	NR	C(8)	C(9)	C(10)	Sources
Argentinine (**300**)	OCH₃	OH	—	(CH₃)₂	—	—	—	*Annona montana* [38,39] *Enantia chlorantha* [71] *Guatteria discolor* [82] *Monodora angolensis* [117] *Unonopsis stipitata* [144]

Compound								Source
Atherospermine (301)	OCH₃	OCH₃	—	(CH₃)₂	—	—	—	Annona montana [38,39]; Annona muricata [40,41]; Duguetia calycina [63]; Enantia chlorantha [71]; Guatteria discolor [82,83]
Atherospermine N-oxide (302)	OCH₃	OCH₃	—	(CH₃)₂,O⁻	—	—	—	Guatteria discolor [82,83]
Methoxyatherospermine (303)	OCH₃	OCH₃	OCH₃	(CH₃)₂	—	—	—	Meiocarpidium lepidotum [112]
Methoxyatherospermine N-oxide (304)	OCH₃	OCH₃	OCH₃	(CH₃)₂,O⁻	—	—	—	Meiocarpidium lepidotum [112]
8-Methoxyuvariopsine (305)	O—CH₂—O		—	(CH₃)₂	OCH₃	OCH₃	—	Uvariopsis guineensis [11]; Uvariopsis solheidii [11]
Noratherospermine (306)	OCH₃	OCH₃	—	CH₃,H	—	—	—	Duguetia calycina [66]; Guatteria discolor [83]
Noruvariopsamine (307)	OCH₃	OCH₃	—	CH₃,H	OCH₃	OCH₃	—	Uvariopsis guineensis [11]
Stipitatine (308)	OH	OCH₃	OH	(CH₃)₂	—	—	—	Unonopsis stipitata [144]
Thalicthuberine (309)	OCH₃	OCH₃	—	(CH₃)₂	—	O—CH₂—O	—	Unonopsis stipitata [144]
Uvariopsamine (310)	OCH₃	OCH₃	—	(CH₃)₂	OCH₃	OCH₃	—	Uvariopsis guineensis [11]
Uvariopsamine N-oxide (311)	OCH₃	OCH₃	—	(CH₃)₂,O⁻	OCH₃	OCH₃	—	Uvariopsis guineensis [11]; Uvariopsis congolana [146]
Uvariopsine (312)	O—CH₂—O		—	(CH₃)₂	—	OCH₃	—	Uvariopsis guineensis [11]; Uvariopsis solheidii [147]

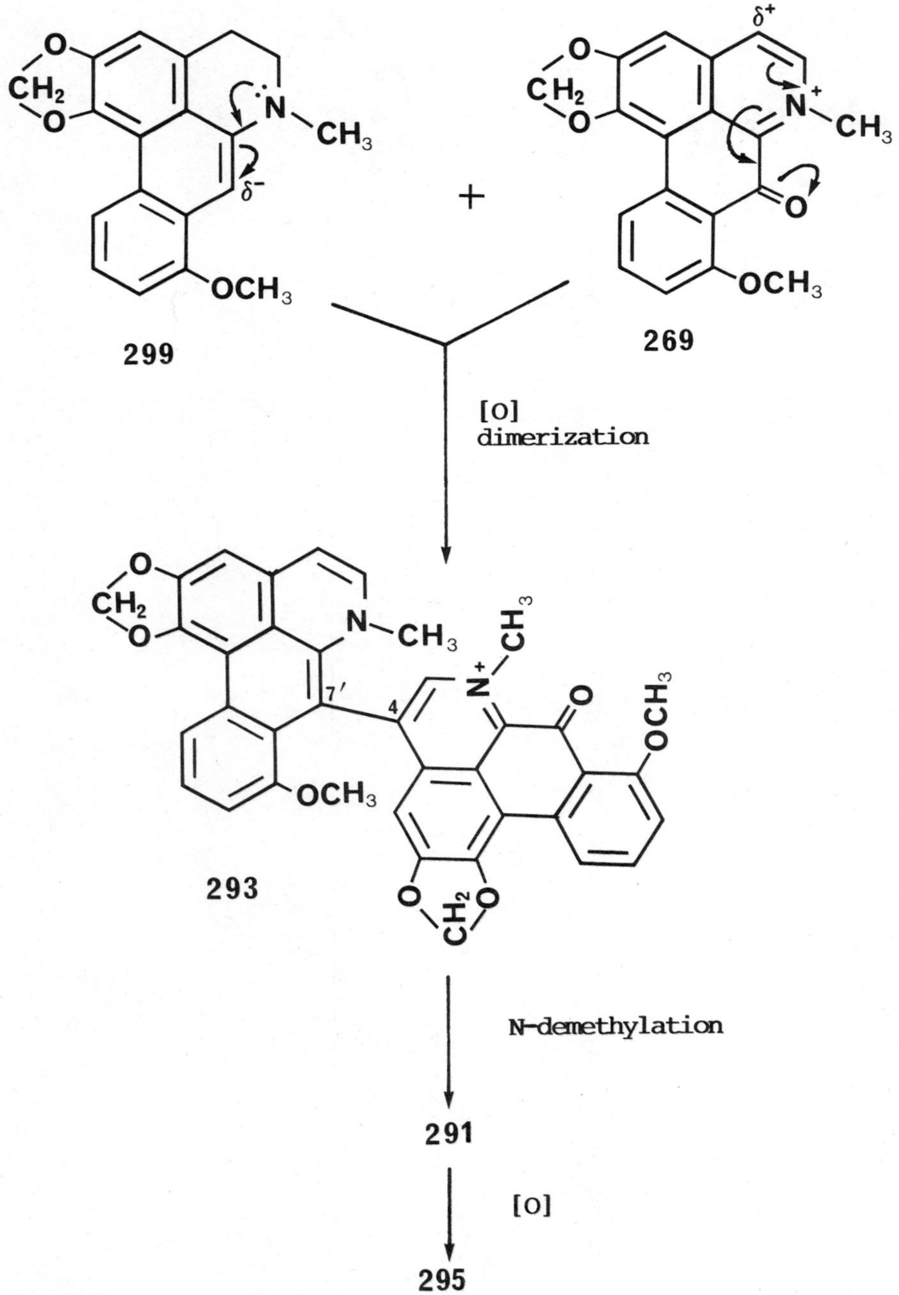

Scheme 35. Possible biogenetic pathway for the formation of 4,7'-dimeric aporphinoids.

dimethyl-*N*-oxides. Until recently phenanthrenes with ring-D substitution (aporphine ring numbering) were restricted to three species of the genus *Uvariopsis*, but thalicthuberine (**309**), from *Unonopsis stipitata*, has recently been added to this group [144].

By contrast to some other groups of alkaloids reviewed, methylenedioxy substitution at C(1)/C(2) appears to be uncommon among the phenanthrenes. Major spectral features relevant to their identification have already been discussed (Section 4), but attention should be drawn to the highly deshielded nature of H(11) (below δ 9.00), except where C(10) is substituted.

Recent investigation of the American species *Unonopsis stipitata* [144] has yielded three phenanthrenes including the novel stipitatine and the 1,2,9,10-substituted thalicthuberine (**309**). The identity of **309** was confirmed by its synthesis from nantenine (**117**) by the route outlined in Scheme 36

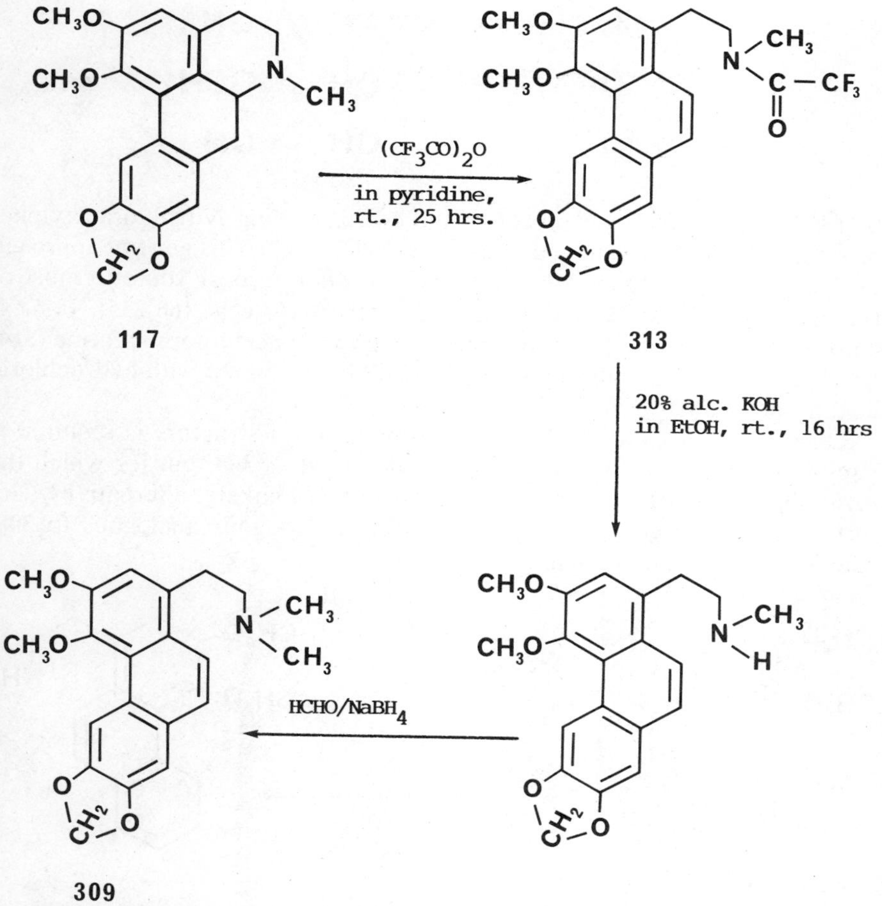

Scheme 36. Synthesis of thalicthuberine (**309**) from nantenine (**117**).

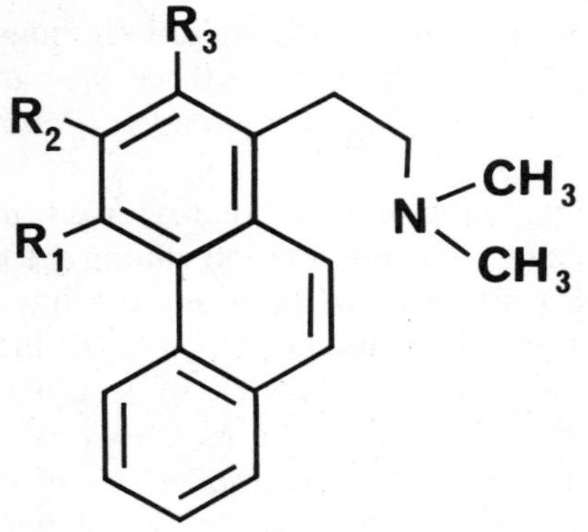

	R_1	R_2	R_3
308	OH	OCH₃	OH
316	OH	OH	OCH₃
317	OCH₃	OH	OH

[83] in which conversion of **117** to the corresponding *N*-trifluoracetylphenanthrene (**313**) was achieved in close to 100% yield. This general approach to the synthesis of phenanthrenes seems to offer an easy route to most of the alkaloids of this class isolated from the Annonaceae [66,235]. For example, Hufford et al. [236] report conversion of *N*-acetylnornuciferine (**314**) to the corresponding norphenanthrene (**315**) by treatment with hydrochloric acid in methanol.

One phenanthrene for which a rigorous proof of structure is required is stipitatine, which has tentatively been identified as **308**, but for which the *ortho* diphenols **316** and **317** remain plausible, if unlikely, alternatives. Unfortunately the required aporphine starting compounds analogous to **308**, **316**, and **317** are not available at present.

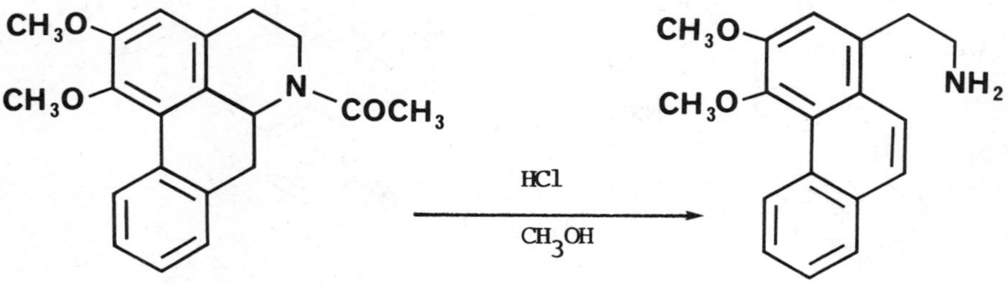

314   315

5.7. Miscellaneous Aporphinoid-derived Alkaloids

In this section we group those remaining alkaloids that appear to be derived through the aporphine biogenetic pathway but that fail to be classified satisfactorily in any of the preceding sections. Some, such as pentouregine (318), still obviously possess the aporphine nucleus, whereas others, such as eupolauridine (330), have clearly undergone drastic modification through processes of oxidative cleavage or rearrangement and have been considered [19] to represent products of the catabolism of aporphines in the plant.

Pentouregine (318). Pentouregine was one of the minor phenolic alkaloids isolated from the leaves of *Guatteria ouregou* [93a]. The presence of a monophenolic noraporphine nucleus was established from spectral data and the presence of the methyleneoxy bridge between C(1) and C(11) was apparent from the ^{1}H NMR spectrum, which revealed an AB quartet centered at δ 4.94 and 5.15 (*J* 13.5 Hz) for the methylene and failed to show any signal for H(11). Placement of the phenolic group at C(2) was based on the comparison of the mass spectra of the *O,N*-diacetyl derivative [93a]. It should be noted that 318 is the first example of a 1,11-oxymethylene aporphine in the Annonaceae, previous reports being restricted to species of *Thalictrum* (Ranunculaceae) and a species of *Phellodendron* (Rutaceae) [8–10].

Two possible pathways leading to the methyleneoxy bridge have been proposed [237]; the most likely involves oxidation of the 1-methoxy group of an aporphine (Scheme 37) to the alcohol (319) and then to the oxonium ion (320) which cyclizes to form the additional C(1)/C(11) ring (321) [19].

Aristolactams (aristololactams). The aristololactams (usually incorrectly called aristolactams [238]) are a small group of nonbasic aporphinoids

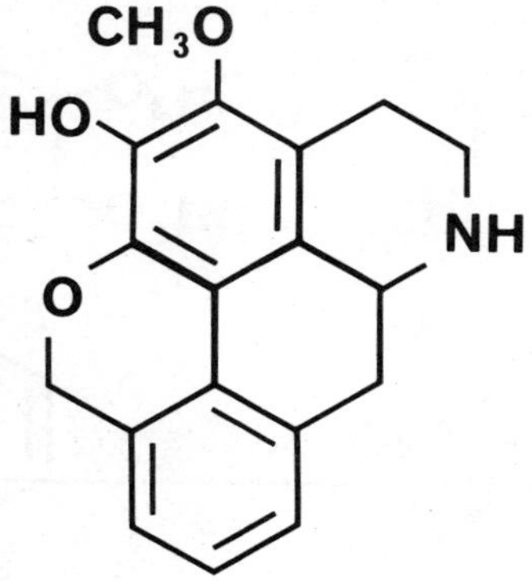

318

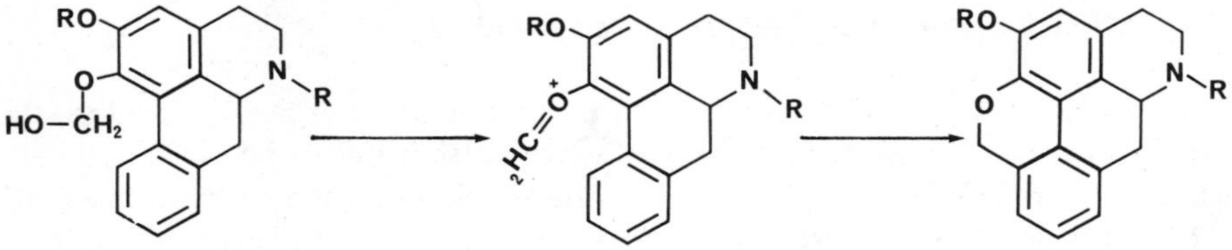

Scheme 37. A possible biogenetic pathway leading to methyleneoxy-bridged aporphines.

usually found in company with the aristolochic acids in the Aristolochiaceae but also reported from the Annonaceae, Menispermaceae, and Monimiaceae [239]. The aristolactam skeleton is represented by **322** with, to date, all known compounds being secondary amines, no *N*-methyl compounds having been recorded [19]. Two numbering systems, one conventional and the other biogenetic, are in current use. In this review the alternative biogenetic numbering is adopted to permit direct comparison with the aporphine precursors.

Aristolactam B2 (cepharanone-B) (**323**), previously known from *Aristolochia argentina* and *Stephania cepharantha* [239], has also been isolated from *Schefferomitra subaequalis* [142,143]. It shows the UV spectrum of a phenanthrene chromophore and a band at 1720 cm^{-1} in the IR spectrum for the lactam. The mass spectrum is characterized by loss of CO and HCN and the ^{1}H NMR spectrum, run in deuteriodimethyl sulfoxide, gives distinct signals for H(3) (δ 7.85), H(7) (δ 7.13), and H(11) (δ 9.12). A synthesis of **323** has been reported [240].

Very recently two more aristolactam-type alkaloids, enterocarpam-I (**324**) and enterocarpam-II (**325**), have been reported from the ether-soluble frac-

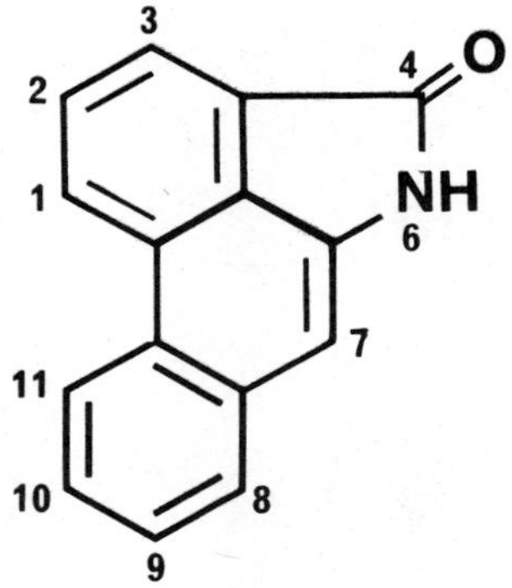

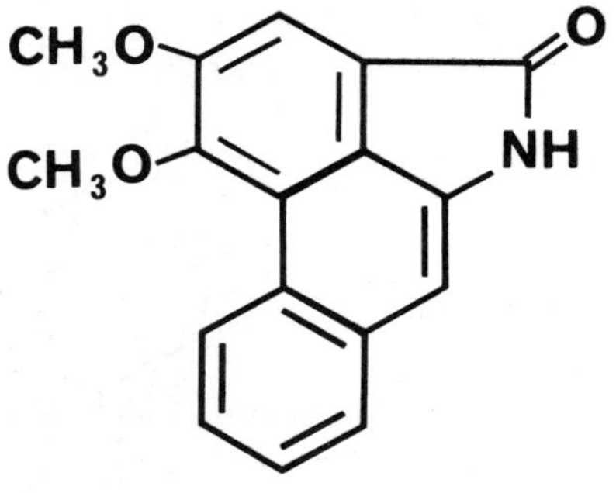

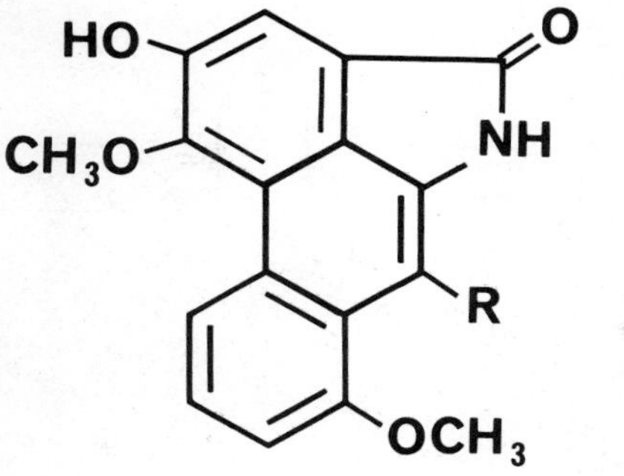

324 R = OCH₃

325 R = H

tion of *Orophea enterocarpa* [241]. Identification is based on both chemical and spectral analysis but no details are yet at hand.*

Biogenetically the aristolactams probably are derived *via* oxidative cleavage of a 4,5-dioxoaporphine (**326**). Two potential routes can be considered [19] and are outlined in Scheme 38. First, **326** could suffer decarbonylation to generate the lactam (**327**), which could then go on to form the corresponding aristolochic acid (**328**). Second, **326** could be oxidized directly to **328**, with reduction and dehydration of **328** then giving rise to **327**. The second route appears the most likely in view of the fact that no natural aristolactams with an *N*-methyl substituent are known. The presence of a 7-methoxy substituent in enterocarpam-I (**324**) raises the possibility that it is formed by a somewhat different route, perhaps through the comparable oxidative cleavage of ring B of an oxoaporphine, such as has been proposed for eupolauramine (**329**), an alkaloid from *Eupomatia laurina* of the allied family Eupomatiaceae [238].

Oxidation products of oxoaporphines. Several alkaloids with structures far removed from the normal aporphine nucleus, and at first sight from each other, have been isolated from the Annonaceae. Enough information is available now, however, to indicate that they share a common origin in an oxoaporphine nucleus that has undergone oxidative cleavage of ring A. The alkaloids concerned are:

(a) The diazafluoranthene or indenonaphthyridine derivative eupolauridine (**330**), an alkaloid originally isolated from *Eupomatia laurina*

---

* After this review was completed, the necessary data were published in *Phytochemistry* **25**, 965 (1986).

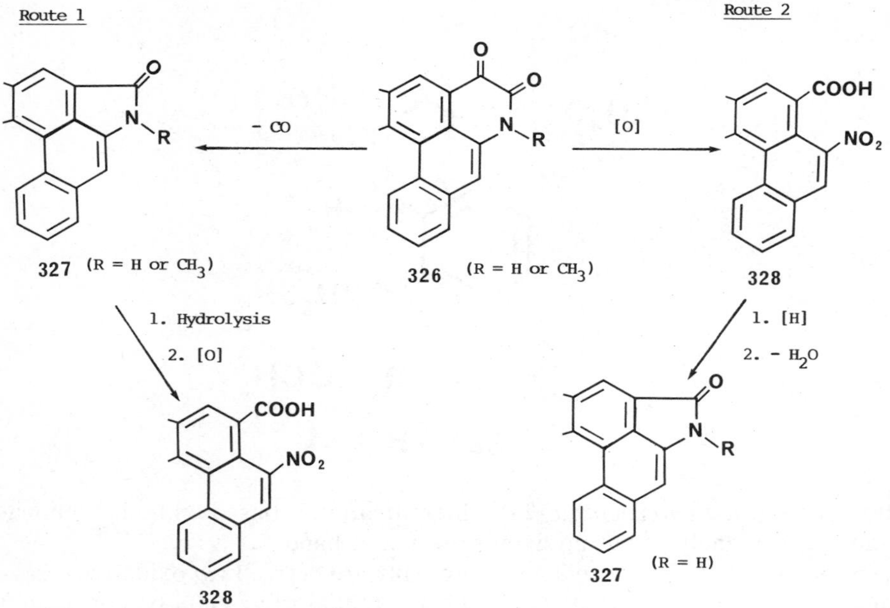
Route 1
Route 2
- CO
[O]
327 (R = H or CH3)
326 (R = H or CH3)
328
1. Hydrolysis
2. [O]
1. [H]
2. - H2O
328
327 (R = H)
COOH
NO2
N
R
O

Scheme 38. Two putative biogenetic routes to the aristolactams.

(Eupomatiaceae) [242,243], and later from *Cananga odorata* [57,58] and *Cleistopholis patens* [59], both from the Annonaceae. From *Cleistopholis patens* **330** was accompanied by its *N*-oxide (**331**) and its di-*N*-oxide (**332**) [59].

(b) An azafluorenone, onychine (**333a**) was first reported from the trunk wood of the Brazilian annonaceous species *Onychopetalum amazonicum* [122] and has recently been reported again in the root bark of *Cleistopholis patens* [59].

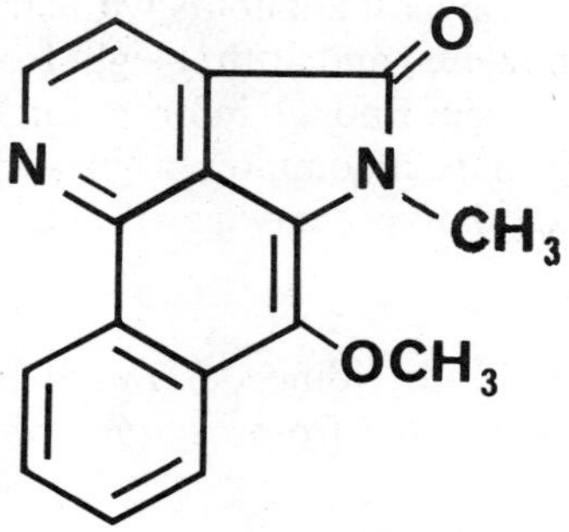
N
O
N
CH3
OCH3

329

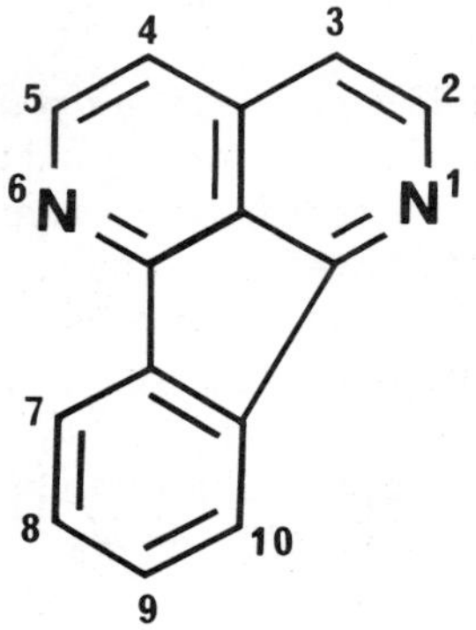
4
3
5
2
6N
N1
7
8
9
10

330

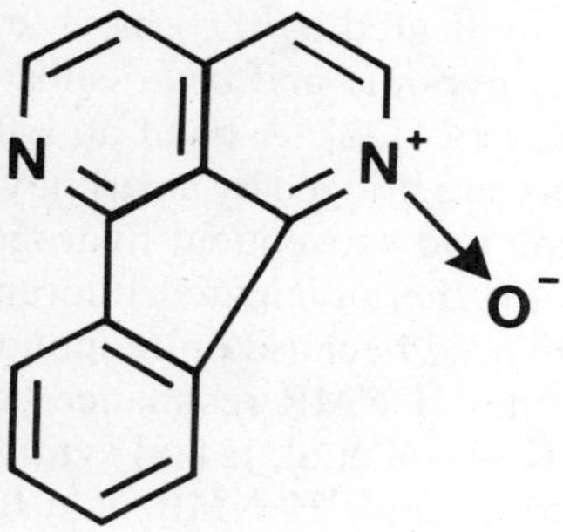

331 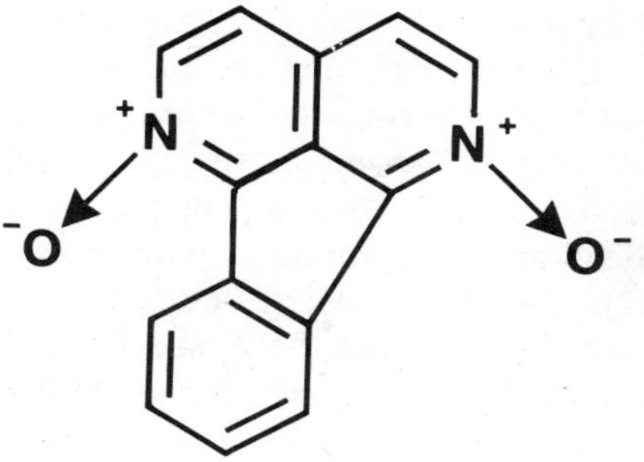   **332**

(c)   The azaanthraquinone alkaloid cleistopholine (**334**), isolated at about
      the same time from the root bark of *Cleistopholis patens* [59] and
      the stem bark of *Meiogyne virgata* [113].

Eupolauridine (**330**) has a unique skeleton lacking any oxygen function.
Its molecular formula $C_{14}H_8N_2$ and high-intensity UV absorption suggested
an extended aromatic chromophore, and the absence of a significant MS
fragmentation indicated its stability. The NMR spectra reveal the symmetry
of the molecule by showing equivalent signals for four pairs of protons in
the ^1H spectrum and for the carbon nuclei except C(3a) and C(3b) in the ^{13}C
spectrum [59]. The proposed structure for **330** has been confirmed by synthesis [243]. The identities of the *N*-oxide (**331**) and di-*N*-oxide (**332**) were
established by analysis of spectral data and chemical interconversion. The
NMR spectra show an interesting variation from the symmetrical **330** through
the asymmetric **331** (14 signals in ^{13}C NMR and 8 signals in ^1H NMR) to
the symmetrical **332** (8 signals in ^{13}C NMR and 4 signals in ^1H NMR) [59,244].

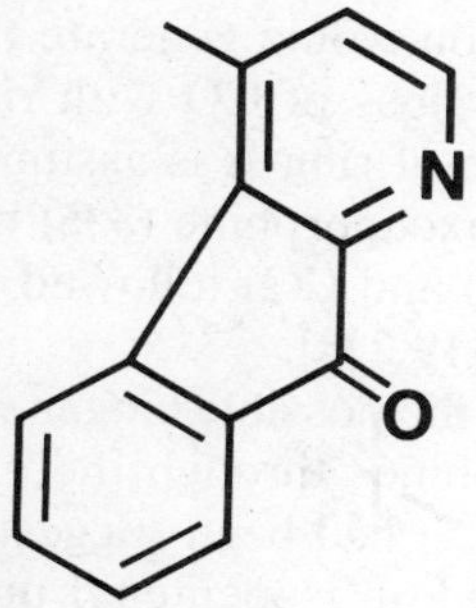

333b 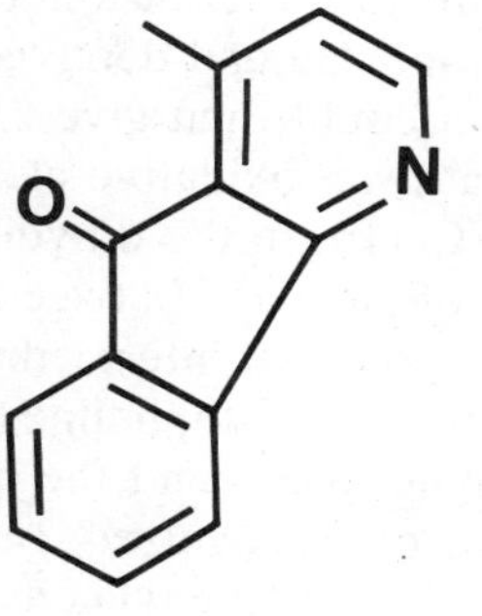   **333a** 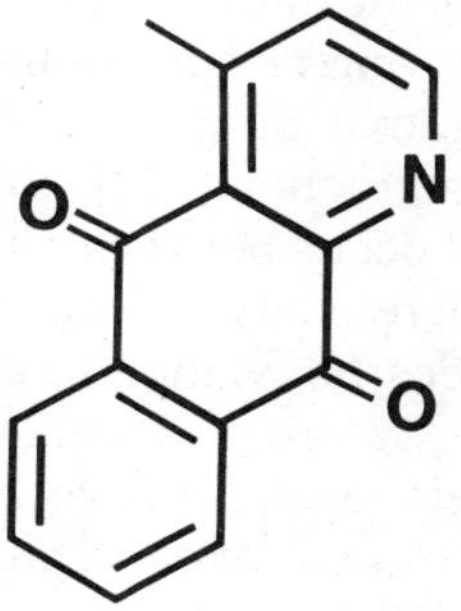   **334**

The azafluorenone skeleton of onychine was indicated by the characteristic UV spectrum and the presence of γ-methylpyridine and *ortho*-substituted aromatic rings from the IR and NMR spectra [122]. A band at 1703 cm^{-1} in the IR spectrum suggested that the rings were bridged by a carbonyl. The structure 333b was proposed after reduction and subsequent hydrogenolysis gave a compound with characteristics for the anticipated fluorene [122]. The alternative structure 333a was not favored because on reduction to dihydroonychine there was little change in the ^{1}H NMR resonance position of the methyl substituent. Subsequently Koyama et al. [244a] synthesized both 333a and 333b and found that the m.p. and ^{1}H NMR data for natural onychine agreed with that for 333a and differed markedly from that for 333b and that, accordingly, onychine was actually 333a. As pointed out by the Japanese group, 333a had actually been synthesized by Bowden et al. [243] prior to its isolation, as part of their route to eupolauridine. The recent report of onychine from *Cleistopholis patens* [59] also indicates structure 333b but was based on its co-identity with published data [122], and this isolate must also be assigned structure 333a. The ^{13}C NMR spectrum of 333a has recently been published [59].

Cleistopholine (334) differs from 333a only by the presence of an additional carbonyl (ν_{max} 1695, 1685 cm^{-1}). The ^{1}H NMR spectrum confirmed the presence of the γ-methylpyridine system and an *ortho*-substituted aromatic ring. Two structures were possible, the γ-diketone (334) or the corresponding α-diketone, the former being supported by the relative symmetry of the chemical shifts for the aromatic nucleus in both ^{1}H and ^{13}C NMR spectra, which were comparable to data published for anthraquinones [59].

Given the three alkaloids 330, 333a, and 334, a hypothesis for their formation can be suggested starting from proposals made [19,238] regarding the biogenesis of eupolauramine (329) and eupolauridine (330). It has been proposed that the oxoaporphine liriodenine (256), which occurs in all plants producing 330, or its possible 1,2-hydroxy precursor (335) is converted to the azaanthraquinone acid (336) by oxidative cleavage of ring A, probably *via* the intermediate 337 (Scheme 39). Transamination of 336 to 338, followed by ring closure to the pyridone 339 and deoxygenation would generate the azaoxoaporphine 340, which could then give 330 by loss of CO with ring contraction [238]. In this pathway oxidative cleavage of ring A is assumed to occur between C(1) and C(11b) in the dihydroxyoxoaporphine (335) but there is also the possibility of cleavage between C(1) and C(2) followed by decarboxylation to generate potential intermediates [19,238].

Before the recent discovery of cleistopholine (334) the possible biogenetic relationship between onychine (333a) and the aporphines through the oxidation of oxoaporphines was not recognized. Indeed, it had been suggested [122] that it could be formed from phenylalanine and an isopentenyl unit. However, we can now recognize a likely route to 333a from liriodenine (256) through the intermediacy of 334 (Scheme 39). Thus the initial product of ring A fission (337) could be converted to the azaanthraquinone aldehyde 341, a

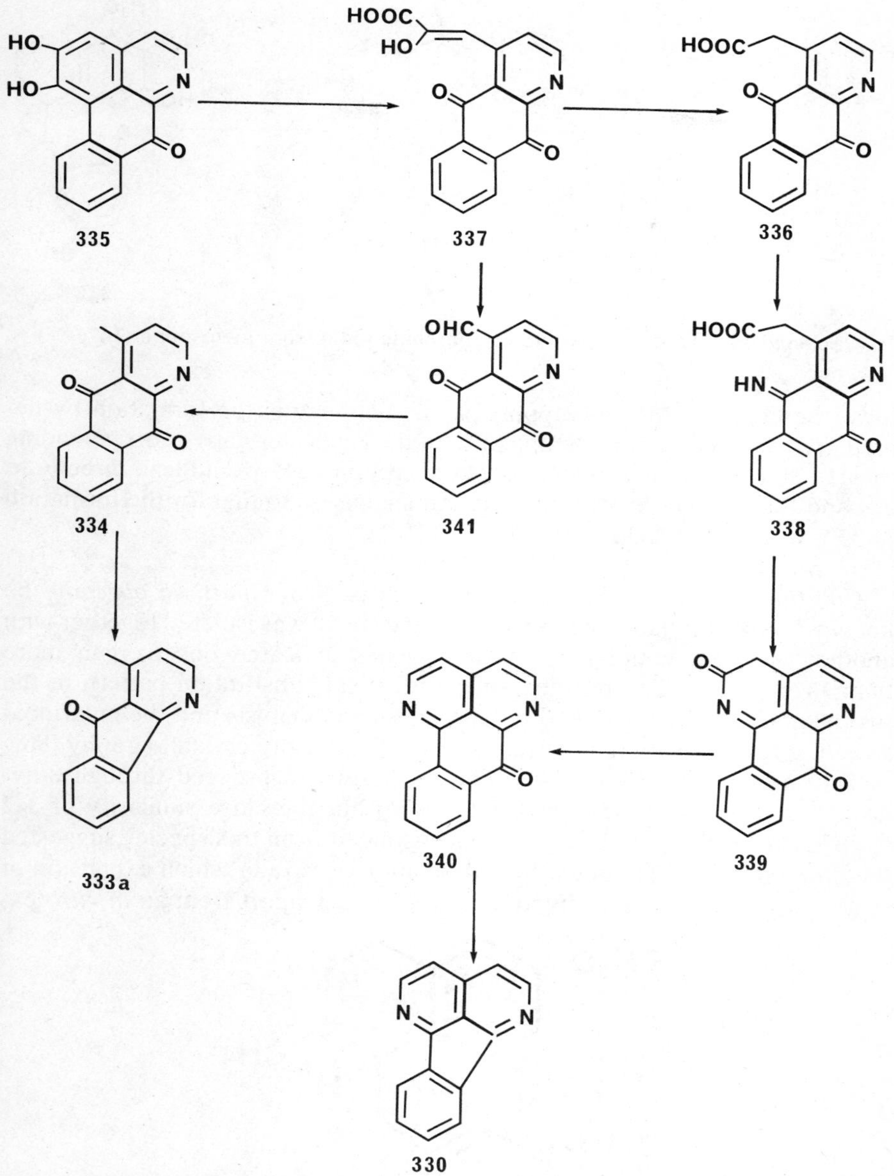

Scheme 39. A possible biogenetic pathway to eupolauridine (**330**), onychine (**333**), and cleistopholine (**334**).

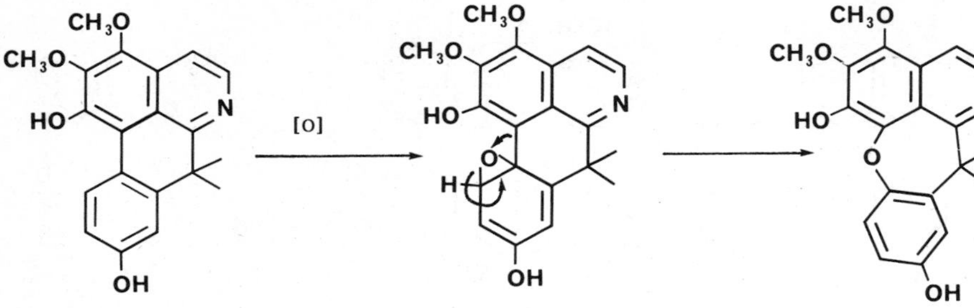

Scheme 40. Direct formation of gouregine (**342**) from melosmine (**167c**).

lower homolog of **336**. In support of this pathway is the formation by degradative oxidation of liriodenine (**256**) of the acidic derivative corresponding to **341** [245]. Reduction of the aldehydic group of **341** would lead directly to **334** and this in turn, by decarbonylation analogous to that for the formation of **330**, would give **333a**.

Cularine-like alkaloids. From the stem bark of *Guatteria ouregou* the known 7,7-dimethylaporphinoid melosmine (**167c**) was isolated together with another alkaloid, gouregine, whose analysis indicated one oxygen more than **167c** [95,96]. The presence of an identical substitution pattern in the two alkaloids was established by spectroscopic analysis and the cularinoid structure (**342**) for gouregine was confirmed by X-ray crystallography [96].

The *in vivo* formation of cularines is known to proceed through 8-hydroxybenzylisoquinoline intermediates [16] but the close similarity of **342** to **167c** and some of the other alkaloids isolated from this species suggested the possibility that **342** might be a derivative of **167c** in which expansion of ring C had occurred. This hypothesis received support from an *in vitro* ex-

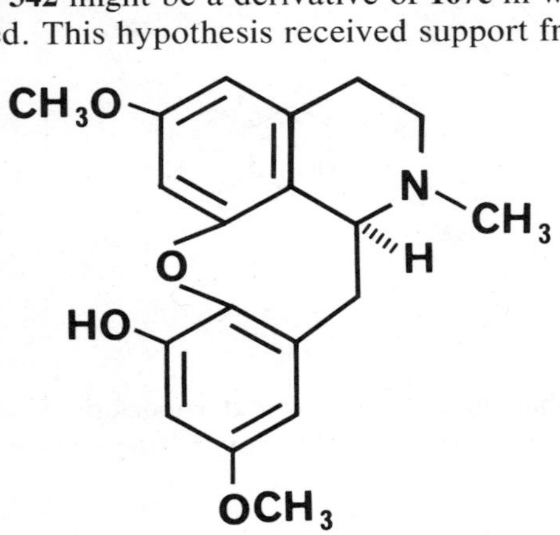

344

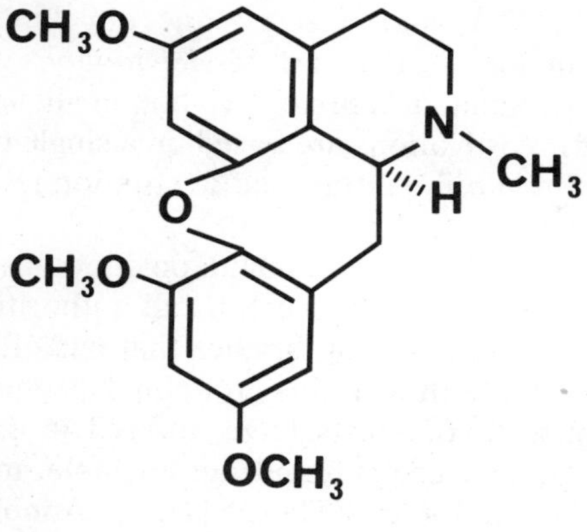

345

periment [96] in which treatment of **167c** with Fenton reagent (FeSO₄/H₂O₂/ H₂SO₄) gave **342** in 90% yield, the epoxide **343** being a likely intermediate (Scheme 40).

Very recently two further cularine-like alkaloids, fissistigine-B (**344**) and fissistigine-C (**345**), have been reported from *Fissistigma oldhamii* together with the known 1,2,9,11-substituted aporphine calycinine (**67**) [76]. Unfortunately, no further information on the alkaloids is available, but if these structures are correct they represent interesting examples of aporphinoids that could be formed from 1-btiq precursors meta-substituted in both rings A and D. Other biogenetic routes to these alkaloids, involving epoxidation of ring D in a proaporphine intermediate or proaporphines with tetrasubstituted D rings, can be envisaged but it seems fruitless to speculate on this problem until the structures of **344** and **345** are firmly established.*

6.   CHEMOTAXONOMIC CONSIDERATIONS

From the approximately 120 genera and more than 2000 species that are generally considered to make up the Annonaceae, less than 50 genera and 200 species appear in the chemical literature at all. Even this fact overstates our level of knowledge of the phytochemistry of the family as many of the reported studies are at best fragmentary. For example, most studies report work on only one part of the plant, most commonly the stem bark, and on one type of secondary compound. However, some comparative studies have illustrated how different the chemical profiles of different plant parts can be, for example, 7-alkylaporphines from stem bark but not leaves of *Guat-*

---

* After this review was completed, it was shown that fissistigines-B and -C are morphinandienones. See *Chem. Abstr.* **104,** 168645 (1986).

teria species (Section 5.2.3), and cleistopholine, onychine, and eupolauridine from the root bark but not the stem bark of *Cleistopholis patens* [59]. Nothing is yet known about seasonal or diurnal variation in alkaloid content because the vast majority of investigations are based on a single collection. Similarly there is little known regarding intraspecific variation in alkaloid production in any annonaceous species.

At the suprafamilial level the taxonomic position of the Annonaceae has not been seriously questioned inasmuch as all authorities regard it as part of the supposedly archaic group of families that has often been termed the Ranalean complex [246], although this grouping has encompassed a prolific nomenclature because taxonomists have differed in exact delimitation of groupings, so that Annonaceae is found, for example, in the Magnoliidae of Takhtajan [2], the Annoniflorae of Thorne [1], the Annonales of Hutchinson [247], and the Magnoliiflorae of Dahlgren [248]. 1-Btiq alkaloids are widely associated with this group of taxa [6,7,249] although they are not universal within it and are known for some non-Ranalean taxa [248]. Comparison of aporphinoids of the Annonaceae with those from other families of the Annoniflorae suggest particularly close links with Eupomatiaceae, Aristolochiaceae, Monimiaceae, and at a less specific level Magnoliaceae, Lauraceae, and Ranunculaceae. The Myristicaceae, which is thought to be particularly close to the Annonaceae, appears to lack the range of alkaloids produced by the latter, and as yet too little is known regarding the nonalkaloidal constituents of the Annonaceae to make clear statements about other areas where the chemistry of the two families overlap.

By contrast, the internal systematics of the Annonaceae is in a rather unsatisfactory state and a working group has recently been set up [250] with a view to improving the taxonomy of the family. As a contribution to that multidisciplinary study in this final section we intend to examine our present knowledge of the distribution of aporphinoids within the Annonaceae in order to ascertain whether they show correlation with any of the extant systems of classification for this large family. It must be remembered that this exercise is concerned solely with aporphinoids and does not take into account, except by passing reference, other potentially interesting groups of alkaloids (bis-1-btiq's, berberines) or nonalkaloidal (diterpene, flavonoid) compounds.

At present there are three extant classification systems that can be correlated with the known distribution of aporpohinoids. These are the traditional systems of Fries [4] and Hutchinson [3] and the more recent classification proposed by Walker that is based primarily on palynological evidence [251–253] with some supporting karyological data [254]. The three systems are outlined in Table 10 together with the placements of the genera for which data on aporphinoids are available. The system proposed by Sinclair [255] is not included because it is based primarily on Asian taxa and therefore can not be considered to be a comprehensive treatment, although in reality it closely resembles that of Hutchinson. Likewise, in her treatment

Table 10. Classification Systems for the Annonaceae

System	Subfamily Tribe Subtribe/group	Genera Included
Fries [4]	1. Annonoideae	
	1. Uvarieae	
	1. Uvaria	UVA
	2. Duguetia	DUG, FUS, PAC
	3. Asimina	ASI, CLE, PSO
	4. Hexalobus	HEX
	5. Guatteria	GUA, HET
	2. Unoneae	
	1. Desmos	ALP, DES, MEC
	2. Polyalthia	CAN, MEI, POL
	3. Unonopsis	ONY, UNO
	4. Xylopia	ANA, FIS, MEL, MIT, XYL
	5. Artabotrys	ART, ENA
	6. Orophea	ORO, POP, PSU, SCH
	7. Annona	ANN, ROL
	8. Trigynaea	CYM
	9. Monanthotaxis	MOA, UVI
	2. Monodoroideae	ISO, MON
Hutchinson [3]	1. Annonoideae	
	1. Uvarieae	CLE, DUG, FUS, GUA, PAC, PSO, UVA
	2. Miliuseae	CYM, HET, ORO
	3. Unoneae	
	1. Xylopiineae A	ALP, ANA, ART, ASI, CAN, DES, FIS, HEX, MEC, MEI, MEL, MIT, MOA, ONY, POL, POP, PSU, SCH, UNO, XYL
	Xylopiineae B	UVI
	Xylopiineae C	ENA
	2. Annonineae	ANN, ROL
	2. Monodoroideae	ISO, MON
Walker [251]	1. Malmea	
	1. Malmea	ENA, ONY, PSO, UNO
	2. Uvaria	ALP, ART, CLE, DES, DUG, FIS, MEI, MEL, MIT, MOA, ORO, POL, POP, PSU, UVA
	3. Guatteria	GUA, HET
	2. Fusaea	ANA, CAN, FUS, GRE,[a] MEC, XYL
	3. Annona	
	1. Hexalobus	HEX, ISO, MON, UVI
	2. Asimina	ASI
	3. Annona	ANN, ROL
	4. Cymbopetalum	CYM

[a] GRE = *Greenwayodendron* (separated from *Polyalthia*).

of the Annonaceae for the *Flore du Gabon,* Le Thomas [256] generally follows the system of Hutchinson. The systems of Fries and Hutchinson also
share many points in common, notably the separation of *Isolona* and *Monodora* into a separate subfamily because of universally united carpels. By
contrast, in Walker's system these genera are included in the Annona
subfamily, in the tribe Hexalobus.

In looking for patterns of distribution among the aporphinoids several
criteria can be examined:

(a)   The distribution of different skeletal types of aporphinoids, using the
      divisions in Section 5.
(b)   The distribution of different ring-D substitution patterns.
(c)   The occurrence of C(3) oxidation.
(d)   The level of *N*-substitution that is found (NH, NMe, $N^+(Me)_2$,
      $N^+(Me)O^-$).

The results of such an analysis show that there is no strong correlation with
any of the three classifications. If we look first at different structural classes
among the aporphinoids, we see that proaporphines have been recorded from
12% of species reported to produce aporphinoids, and are notable in *Isolona/
Monodora,* which are segregated in the systems of both Fries and Hutchinson (Table 10). Aporphines *s.s.* are reported from 71% of species and
oxoaporphines from 69%, but in neither case does there appear to be any
particular pattern. The 7-alkylaporphinoids are restricted to Walker's Mal-
mea subfamily and Hutchinsons' and Fries' Uvarieae tribes, but this only
reflects placement of *Guatteria* and *Duguetia.* The 7-oxygenated aporphines
occur in 27% of species and are widely distributed. If the *cis* and *trans*
isomers are separated then the latter appear to dominate in the Uvarieae
tribes of Fries and, in particular, of Hutchinson.

Considering substitution patterns, the 1,2 pattern is very widespread (81%
of species), followed by the 1, 2,9 (40%), 1,2,9,10 (33%), and 1,2,10,11 (19%).
In none of these are any patterns discernible. Among the rarer types the
1,2,9,11 and 1,2,11 patterns are generally restricted to the taxonomic groups
including *Guatteria* and *Duguetia.* The 1,2,10 type is restricted to the Hexalobus group of the Annona subfamily in Walker's system. Substitution at
C(3) has been reported in 31% of species and appears to be twice as common
in the Uvarieae tribes of Fries and Hutchinson as it is in the rest of the
family. No patterns are recognized in the distribution of different types of
N-substitution. The *N*-formyl substituent is at present too rare to be considered of general taxonomic value but its distribution in *Guatteria* and *Duguetia* is a further indication of the affinities between these genera. The only
other known source of *N*-formyl aporphines in the family is *Hexalobus,*
which is included in the same tribe, Uvarieae, as *Guatteria* and *Duguetia*

by Fries but which is widely separated from those two genera in the other systems.

A comparison based on division along geographic lines (simply American, African, or Asian origin—excluding Madagascan species) yields more interesting differences (Table 11), the major of which may be summarized as follows. Aporphines *s.s.* are reported from only about half of the African species but are in a greater proportion (~75%) from the other two areas. African species are the major source of phenanthrenes (29% of species) and these alkaloids are, to date, absent from Asian species. Likewise *N,N*-di-

Table 11. Geographical Distribution of Possible Chemotaxonomic Features of the Aporphinoids (Values Expressed as % of Total Species Examined from that Area

Character Considered	Geographical Area		
	Africa	America	Asia[a]
Structural types			
Proaporphines	24	7	9
Aporphines *s.s.*	52	73	82
7-Alkylaporphinoids	0	22	0
7-Oxygenated aporphines	29	29	39
trans only	29	7	5
cis only	0	21	29
Oxoaporphines	52	80	64
Phenanthrenes	29	12	0
Substitution Patterns[b]			
1,2	86	83	82
1,2,9	33	51	23
1,2,10	19	7	5
1,2,11	0	15	0
1,2,9,10	29	29	45
1,2,10,11	14	17	23
1,2,8,9	14	0	5
1,2,9,11	0	7	5
3	38	37	9
NH	57	63	77
NMe	71	63	59
N(Me)$_2$	33	17	5
N$^+$Me,O$^-$	38	5	0
Number of species included in survey	21	42	21

[a] Asia includes both Asian and Australasian taxa. Species from Madagascar are omitted from the survey.

[b] These patterns also include aporphinoids with additional C(3), C(4), C(5), and C(7) substitution; C(3)-substituted aporphinoids are also classified separately.

methyl and N-oxides have been found far more widely among African species than those from the other areas. American species are the only sources of 7-alkyl aporphinoids and alkaloids with the 1,2,11 substitution pattern, although not the 1,2,9,11 pattern. The normal tetrasubstituted aporphines (1,2,9,10 and 1,2,10,11), on the other hand, tend to occur somewhat more commonly among Asian species. At first sight 7-oxygenated aporphines appear to be generally distributed. However, separation into *cis* and *trans* types reveals an interesting geographical split. The *trans* type, at present unique to the Annonaceae, occurs widely in African species, rarely in American species, and has as yet been reported from only one Asian species. By contrast, the *cis* type, which occurs in a number of other plant families, is found commonly in Asian species, to a lesser extent in American species, and never, to date, in African species. This distinction is particularly interesting with regard to the *Greenwayodendron/Polyalthia* group; the two African species (sometimes segregated as *Greenwayodendron*) both produce *trans* 7-oxygenated aporphines, whereas one Asian species, *P. nitidissima*, has yielded the *cis* type. Unfortunately the situation is confused by another Asian species, *P. acuminata,* which is at present unique in being the only species to yield both *cis* and *trans* types and is also the only Asian species to produce the *trans* type. This dichotomy appears to have possible evolutionary implications for the Annonaceae (see below). Finally, 3 substitution appears to be much more common among African and American taxa than in those of Asian origin.

At the generic level the correlation found between the alkaloid content of congeneric species is not noticeably better than at the suprageneric level. Where genera occur on more than one continent there can be large differences, for example, in *Xylopia,* in which three African taxa have been found to produce a large range of diterpenes [257–260] but no alkaloids, whereas species from America, Madagascar, and Asia all produce alkaloids but, so far, no diterpenes have been reported. The same dichotomy may exist in *Annona,* where alkaloids and diterpenes occur in American species, but from the one African species so far examined, *A. senegalensis,* only diterpenes have been reported [5]. The situation in *Greenwayodendron/Polyalthia* is comparable, only the two African species having yielded indolosesquiterpene alkaloids, whereas African, Madagascan, and Asian species yield aporphinoids [13].

There are some equally large differences between species that appear to be more closely linked. For example, in the genus *Cleistopholis* in West Africa one species, *C. patens,* produces a range of aporphinoids (see Table 3) but another, *C. staudtii,* produces only bis-1-btiq dimers [261]. Similar bis-1-btiq-producing species occur in other genera where aporphinoids are found widely, for example, *Isolona hexaloba* [110] and *Guatteria megalophylla* [262]. In the same vein, there are within the aporphinoid-producing species a number that appear to be specialists at producing one type of product whereas other species, often in the same genus, produce a range of different types. Some examples of the specialist type are phenanthrene specialists such as *Uvariopsis* spp., *Unonopsis stipitata,* and *Meiocarpidium lepidotum,* and 7-oxygenated aporphine specialists, *Pachypodanthium* spp. and *Polyalthia suaveolens.* The same may be true of some genera; for example, present knowledge would suggest that *Xylopia* is an aporphine *s.s.* specialist.

From data presently available, therefore, it seems that the distribution of aporphinoids offers little of value in support of any of the extant classifications. Why should this be so when the range of aporphinoids produced are apparently so diverse? It is possible that this situation is simply because the various subclasses of aporphinoids merely reflect different stages on a biogenetic pathway open, to some extent, to all species, and that the levels attained in a species in a given instance reflect extrinsic factors, not some specific pattern due to the biogenetic limitations of that species. This suggestion will be confirmed or denied only after more extensive sampling of individual species, which should be done as a matter of urgency.

What does seem to be the case, however, is that distribution of aporphinoids shows a greater degree of correlation with geographical distribution than with available taxonomic schemes. Why this pattern should be is unclear but Thorne [1] does make the point that "of all living angiosperm families, at least, these (Annonaceae, Myristicaceae, Canellaceae) seem to be the most likely survivors of such possible continental rifting and rafting." If this idea is true then African and American taxa would have been linked more recently with each other than with those from Asia and Australasia, and such a linkage could be reflected in the apparent bias of some alkaloid types to a distribution in Africa and America and not in Asia. The alkaloid patterns do not necessarily support the contention of Walker [254] that the Annonaceae are of American or possibly African origin, or the more general suggestion of an origin in west Gondwanaland [263]. Alkaloid distribution appears to be largely neutral on this point, with one possible exception being the distribution of the *cis* and *trans* types of 7-oxygenated aporphines. This observation could be interpreted in terms of a common origin for all 7-oxygenated aporphines in American species with later dispersal of the *trans* type to Africa and of the *cis* type to Asia. Such a hypothesis could then explain the apparent geographic bias in alkaloid distribution, but it does not explain the heterogeneity observed between closely allied species. Such heterogeneity may well need no further explanation other than the fact that the Annonaceae is a very primitive family in which individual species have almost certainly developed in isolation over considerable time. Given this state of affairs, then, it is to be expected that a reticulate pattern of extant secondary metabolic profiles will have developed, just as has been observed for morphological and karyological features of potential taxonomic value [264].

7. ALPHABETICAL LIST OF APORPHINOIDS OF THE ANNONACEAE (INCLUDING SYNONYMS)

Name	Structure	Section
Actinodaphnine	61	5.2.1
O-1-Acetylapoglaziovine	62	5.2.1
Anaxagoreine	209	5.2.4
Anolobine	63	5.2.1
Anonaine	64	5.2.1
Argentinine	300	5.6
Aristolactam B2	323	5.7
Asimilobine	65	5.2.1
Atheroline	250	5.3
Atherospermidine	251	5.3
Atherosperminine	301	5.6
Atherosperminine N-oxide	302	5.6
Beccapoline	291	5.5
Beccapolinium	293	5.5
Beccapolydione	295	5.5
Belemine	169a	5.2.3
Bipowine	285	5.5
Bipowinone	286	5.5
Boldine	66	5.2.1
Buxifoline	67	5.2.1
Caaverine	68	5.2.1
Calycinine	69	5.2.1
Canangine	330	5.7
N-Carbamoylanonaine	70	5.2.1
N-Carbamoylasimilobine	71	5.2.1
Cepharanone B	323	5.7
Cleistopholine	334	5.7
Corydine	72	5.2.1
Corytuberine	73	5.2.1
(+)-Crotsparine	56	5.1
Danguyelline	74	5.2.1
Dehydroformouregine	156	5.2.2
Dehydroguattescine	167a	5.2.3
Dehydro-O-methylisopiline	157	5.2.2
Dehydroneolitsine	158	5.2.2
Dehydronornuciferine	159	5.2.2
Dehydropredicentrine	160	5.2.2
Dehydroroemerine	161	5.2.2
Dehydrostephalagine	162	5.2.2
O-Demethylpurpureine	75	5.2.1
Dicentrine	76	5.2.1
Dicentrinone	252	5.3
Dihydromelosmine	168a	5.2.3
N,O-Dimethylliriodendronine	253	5.3

Name	Structure	Section
Discoguattine	77	5.2.1
Duguecalyne	170a	5.2.3
Duguenaine	170b	5.2.3
Duguespixine	169b	5.2.3
Duguetine	210	5.2.4
Duguevanine	78	5.2.1
Elmerrillicine	79	5.2.1
Enterocarpam-I	324	5.7
Enterocarpam-II	325	5.7
Eupolauridine	330	5.7
Eupolauridine di-N-oxide	332	5.7
Eupolauridine N-oxide	331	5.7
Fissistigine A	69	5.2.1
Fissistigine B	344	5.7
Fissistigine C	345	5.7
Fissoldine	69	5.2.1
Formouregine	80	5.2.1
N-Formylanonaine	81	5.2.1
N-Formylbuxifoline	82	5.2.1
N-Formylduguevanine	83	5.2.1
N-Formylnornuciferine	84	5.2.1
N-Formylputerine	85	5.2.1
N-Formylxylopine	86	5.2.1
Fuseine	270	5.4
Glaucine	87	5.2.1
Glaziovine	57	5.1
Goudotianine	169c	5.2.3
Gouregine	342	5.7
Guacolidine	168b	5.2.3
Guacoline	168c	5.2.3
Guadiscidine	168d	5.2.3
Guadiscine	168e	5.2.3
Guadiscoline	168f	5.2.3
Guatterine	211	5.2.4
Guatterine N-oxide	212	5.2.4
Guattescidine	168g	5.2.3
Guattescine	168h	5.2.3
Guattouregidine	168i	5.2.3
Guattouregine	168j	5.2.3
Homomoschatoline	258	5.3
3-Hydroxydehydronuciferine	163	5.2.2
9-Hydroxy-1,2-dimethoxynoraporphine	88	5.2.1
3-Hydroxynornuciferine	89	5.2.1
3-Hydroxynuciferine	90	5.2.1
Isoboldine	91	5.2.1
Isocalycinine	92	5.2.1
Isocorydine	93	5.2.1

Name	Structure	Section
Isodomesticine	**94**	5.2.1
Isolaureline	**95**	5.2.1
Isomoschatoline	**254**	5.3
Isopiline	**96**	5.2.1
Lanuginosine	**255**	5.3
Laurelliptine	**97**	5.2.1
Laurolitsine	**98**	5.2.1
Laurotetanine	**99**	5.2.1
Lauterine	**263**	5.3
Lindcarpine	**100**	5.2.1
Liridine	**258**	5.3
Lirinidine	**101**	5.2.1
Lirinine	**90**	5.2.1
Liriodenine	**256**	5.3
Lysicamine	**257**	5.3
Magnoflorine	**102**	5.2.1
Melosmidine	**167b**	5.2.3
Melosmine	**167c**	5.2.3
Menisperine	**103**	5.2.1
Methoxyatherosperminine	**303**	5.6
Methoxyatherosperminine *N*-oxide	**304**	5.6
3-Methoxynuciferine	**104**	5.2.1
Methoxyuvariopsine	**305**	5.6
N-Methylactinodaphnine	**105**	5.2.1
N-Methylasimilobine	**106**	5.2.1
O-Methylatheroline	**261**	5.3
N-Methylbuxifoline	**107**	5.2.1
N-Methylcalycinine	**108**	5.2.1
O-Methylcalycinine	**77**	5.2.1
N-Methylcorydine	**109**	5.2.1
N-Methylcrotsparine	**58**	5.1
N-Methylduguevanine	**110**	5.2.1
N-Methylelmerrillicine	**111**	5.2.1
10-*O*-Methylhernovine	**112**	5.2.1
N-Methylisopiline	**113**	5.2.1
O-Methylisopiline	**114**	5.2.1
N-Methyllaurotetanine	**115**	5.2.1
O-Methylmoschatoline	**258**	5.3
O-Methylnorlirinine	**114**	5.2.1
O-Methylpachyconfine	**213**	5.2.4
N-Methylpachypodanthine	**214**	5.2.4
N-Methylpachypodanthine *N*-oxide	**215**	5.2.4
O-Methylpukateine	**116**	5.2.1
N-Methylputerine	**116**	5.2.1
N-Methylxylopine	**95**	5.2.1
Michelalbine	**220**	5.2.4

Name	Structure	Section
Nantenine	117	5.2.1
Neolitsine	118	5.2.1
Norannuradhapurine	119	5.2.1
Noratherosperminine	306	5.6
Norboldine	98	5.2.1
Norcepharadione B	271	5.4
Norcorydine	120	5.2.1
Nordicentrine	121	5.2.1
Norglaucine	122	5.2.1
Norglaziovine	56	5.1
Norisoboldine	97	5.2.1
Norisocorydine	123	5.2.1
Norisodomesticine	124	5.2.1
Norlaureline	125	5.2.1
Norliridinine	126	5.2.1
Norlirinine	89	5.2.1
Nornantenine	127	5.2.1
Nornuciferine	128	5.2.1
Noroconovine	129	5.2.1
Noroliveridine	216	5.2.4
Noroliverine	217	5.2.4
Noroliveroline	218	5.2.4
Norpachystaudine	219	5.2.4
Norpredicentrine	130	5.2.1
Norpurpureine	131	5.2.1
Norstephalagine	132	5.2.1
Norushinsunine	220	5.2.4
Noruvariopsamine	307	5.6
Nuciferine	133	5.2.1
Obovanine	134	5.2.1
Oliveridine	221	5.2.4
Oliveridine N-oxide	222	5.2.4
Oliverine	223	5.2.4
Oliverine N-oxide	224	5.2.4
Oliveroline	225	5.2.4
Oliveroline N-oxide	226	5.2.4
Onychine	333	5.7
Ouregidione	272	5.4
Oureguattidine	135	5.2.1
Oureguattine	136	5.2.1
Oxoanolobine	259	5.3
Oxobuxifoline	260	5.3
Oxoglaucine	261	5.3
Oxoisocalycinine	262	5.3
Oxolaureline	263	5.3
Oxonuciferine	257	5.3

Name	Structure	Section
Oxopukateine	264	5.3
Oxopurpureine	265	5.3
Oxoputerine	266	5.3
Oxostephanine	267	5.3
Oxoushinsunine	256	5.3
Oxoxylopine	255	5.3
Pachyconfine	227	5.2.4
Pachypodanthine	228	5.2.4
Pachystaudine	229	5.2.4
Pentouregine	318	5.7
Polyalthine	230	5.2.4
Polybeccarine	296	5.5
Polygospermine	137	5.2.1
Polysuavine	231	5.2.4
Predicentrine	138	5.2.1
Pronuciferine	59	5.1
Pukateine	139	5.2.1
Purpureine	140	5.2.1
Puterine	141	5.2.1
Roemerine	142	5.2.1
Sparsiflorine	143	5.2.1
Stepharine	60	5.1
Stipitatine	308	5.6
Suaveoline	144	5.2.1
Subsessiline	268	5.3
Thailandine	269	5.3
Thalicsimidine	140	5.2.1
Thalicthuberine	309	5.6
Thaliporphine	145	5.2.1
Trichoguattine	169d	5.2.3
Tuduranine	146	5.2.1
Ushinsunine	232	5.2.4
Uvariopsamine	310	5.6
Uvariopsamine N-oxide	311	5.6
Uvariopsine	312	5.6
Wilsonirine	147	5.2.1
Xyloguyelline	148	5.2.1
Xylopine	149	5.2.1
Zenkerine	150	5.2.1

REFERENCES

1. R. F. Thorne, *Aliso* **8**, 147 (1974).
2. A. Takhtajan, *Flowering Plants, Origin and Dispersal,* Oliver and Boyd, Edinburgh, 1969.
3. J. Hutchinson, *The Genera of Flowering Plants,* Vol. 1, University Press, Oxford, 1964.

4. R. E. Fries, "Annonaceae," in *Die Naturlichen Pflanzenfamilien*, 2nd ed., Vol. 17aII, A. Engler and K. Prantl, Eds., Dunker and Humblot, Berlin, 1959.

5. M. Leboeuf, A. Cavé, P. K. Bhaumik, B. Mukherjee, and R. Mukherjee, *Phytochemistry* **21**, 2783 (1982).

6. K. Kubitzki, *Taxon* **18**, 360 (1969).

7. R. Hegnauer, *Chemotaxonomie der Pflanzen*, Vol. 3, Birkhaüser, Basel, 1964.

8. H. Guinaudeau, M. Leboeuf, and A. Cavé, *Lloydia* **38**, 275 (1975).

9. H. Guinaudeau, M. Leboeuf, and A. Cavé, *J. Nat. Prod.* **42**, 325 (1979).

10. H. Guinaudeau, M. Leboeuf, and A. Cavé, *J. Nat. Prod.* **46**, 761 (1983).

11. M. Leboeuf and A. Cavé, *Phytochemistry* **11**, 2833 (1972).

12. A. Cavé, "Annonaceae Alkaloids," in *The Chemistry and Biology of Isoquinoline Alkaloids*, J. D. Phillipson, M. F. Roberts, and M. H. Zenk, Eds., Springer, Berlin, 1985.

13. P. G. Waterman, "The Indolosesquiterpene Alkaloids of the Annonaceae," in *Alkaloids: Chemistry and Biological Perspectives*, Vol. 3, S. W. Pelletier, Ed., Wiley-Interscience, Wiley, New York, 1985, pp. 91–111.

14. P. G. Waterman and I. Muhammad, *J. Chem. Soc., Chem. Commun.* **1984**, 1280.

15. M. Shamma, *The Isoquinoline Alkaloids, Chemistry and Pharmacology*, Academic, London, 1972.

16. M. Shamma and J. L. Moniot, *Isoquinoline Alkaloids Research 1972–1977*, Plenum, New York, 1978.

17. G. A. Cordell, *Introduction to Alkaloids, a Biogenetic Approach*, Interscience, New York, 1981.

18. *The Alkaloids, Specialist Periodical Reports*, Vols. 1–13, The Chemical Society, London, 1971–1983.

19. M. Shamma and H. Guinaudeau, *Tetrahedron* **40**, 4795 (1984).

20. H. M. Schumacher, M. Rüffer, N. Nagakura, and M. H. Zenk, *Planta Med.* **48**, 212 (1983).

21. O. Prakash, D. S. Bhakuni, and R. S. Kapil, *J. Chem. Soc., Perkin Trans. 1*, **1979**, 1515.

22. V. Sharma, S. Jain, D. S. Bhakuni, and R. S. Kapil, *J. Chem. Soc., Perkin Trans. 1*, **1982**, 1153.

23. M. Tomita, S. T. Lu, S. J. Wang, C. H. Lee, and T. H. Shih, *J. Pharm. Soc. Japan* **88**, 1143 (1968).

24. P. K. Mahanta, R. K. Mathur, and K. W. Gopinath, *Indian J. Chem.* **13**, 306 (1975).

25. R. Hocquemiller, S. Rasamizafy, C. Moretti, H. Jacquemin, and A. Cavé, *Planta Med.* **41**, 48 (1981).

26. I. Borup-Grochtmann and D. G. I. Kingston, *J. Nat. Prod.* **45**, 102 (1982).

27. A. Urzua and B. K. Cassels, *Rev. Latinoam. Quim.* **8**, 133 (1977).

28. A. Villar del Fresno and J. L. Rios Canavate, *J. Nat. Prod.* **46**, 438 (1983).

29. A. Villar, J. L. Rios Canavate, and M. Mares, *Farm. Tijdschr. Belg.* **61**, 299 (1984).

30. A. Villar, M. Mares, J. L. Rios Canavate, and D. Cortes, *J. Nat. Prod.* **48**, 151 (1985).

31. R. Hocquemiller, A. Cavé, H. Jacquemin, A. Touché, and P. Forgacs, *Plant. Med. Phytother.* **16**, 4 (1982).

32. J. Faust, A. Preiss, H. Ripperger, D. Sandoval, and K. Schreiber, *Pharmazie* **36**, 713 (1981).

33. T. H. Yang and C. M. Chen, *Taiwan Yao Hsueh Tsa Chih* **25**, 1 (1973); through *Chem. Abstr.* **84**, 102339 (1976).

34. T. H. Yang and C. M. Chen, *Proc. Nat. Sci. Counc. Republ. China* **7**, 177 (1974); through *Chem. Abstr.* **86**, 127149 (1977).

35. D. Warthen, E. L. Gooden, and M. Jacobson, *J. Pharm. Sci.* **58**, 637 (1969).

36. M. Y. Haggag and K. F. Taha, *Egypt. J. Pharm. Sci.* **22**, 153 (1981); through *Chem. Abstr.* **100**, 135828 (1984).

37. T. H. Yang, C. M. Chen, and S. S. Kuan, *J. Chin. Chem. Soc.* (*Taipei*) **18**, 133 (1971).

38. T. H. Yang and C. M. Chen, *Proc. Nat. Sci. Counc. Republ. China* **3**, 63 (1979); through *Chem. Abstr.* **90**, 200312 (1979).

39. M. Leboeuf, A. Cavé, P. Forgacs, R. Tiberghien, J. Provost, A. Touché, and H. Jacquemin, *Plant. Med. Phytother.* **16**, 169 (1982).

40. G. Aguilar-Santos, J. R. LIbrea, and A. C. Santos, *Philipp. J. Sci.* **96**, 399 (1967).

41. M. Leboeuf, C. Legueut, A. Cavé, J. F. Desconclois, P. Forgacs, and H. Jacquemin, *Planta Med.* **42**, 37 (1981).

42. P. E. Sonnet and M. Jacobson, *J. Pharm. Sci.* **60**, 1254 (1971).

43. A. C. Santos, *Philipp. J. Sci.* **43**, 561 (1930).

44. K. W. Gopinath, T. R. Govindachari, B. R. Pai, and N. Viswanathan, *Chem. Ber.* **92**, 776 (1959).

45. T. H. Yang, C. M. Chen, and H. H. Kong, *Taiwan K'o Hsueh* **24**, 99 (1970); through *Chem. Abstr.* **75**, 45637 (1971).

46. T. H. Yang, C. M. Chen, and H. H. Kong, *Pei I Hsueh Pao,* **1973**, 130; through *Chem. Abstr.* **81**, 60846 (1974).

47. B. Anjaneyulu, V. Babu Rao, A. K. Ganguly, T. R. Govindachari, B. S. Joshi, V. N. Kamat, A. H. Manmade, P. A. Mohamed, A. D. Rahimtula, A. K. Saksena, D. S. Varde, and N. Viswanathan, *Indian J. Chem.* **3**, 237 (1965).

48. T. H. Yang and C. M. Chen, *J. Chin. Chem. Soc.* (*Taipei*) **17**, 243 (1970).

49. F. R. Reyes and A. C. Santos, *Philipp. J. Sci.* **44**, 409 (1931).

50. D. S. Bhakuni, S. Tewari, and M. M. Dahr, *Phytochemistry* **11**, 1819 (1972).

51. R. V. K. Rao, N. Murty, and J. V. L. N. Rao, *Indian J. Pharm. Sci.* **40**, 170 (1978).

52. P. K. Bhaumik, B. Mukherjee, J. P. Juneau, N. S. Bhacca, and R. Mukherjee, *Phytochemistry* **18**, 1584 (1979).

53. G. Barger and L. J. Sargent, *J. Chem. Soc.* 991 (1939).

54. E. Schlittler and H. U. Huber, *Helv. Chim. Acta* **35**, 111 (1952).

55. M. Tomita and M. Kozuka, *J. Pharm. Soc. Japan* **85**, 77 (1965).

56. Y. Y. Siv, *Trav. Lab. Matière Méd. Pharm. Gal. Fac. Pharm. Paris* **56**, 87 (1971); *Chem. Abstr.* **77**, 143762 (1972).

57. M. Leboeuf, J. Streith, and A. Cavé, *Ann. Pharm. Fr.* **33**, 43 (1975).

58. M. Leboeuf and A. Cavé, *Lloydia* **39**, 459 (1976).

59. P. G. Waterman and I. Muhammad, *Phytochemistry* **24**, 523 (1985).

60. S. Abd-El Atti, H. A. Ammar, C. H. Phoebe, Jr., P. L. Schiff, Jr., and D. J. Slaktin, *J. Nat. Prod.* **45**, 476 (1982).

61. A. Cavé, D. Debourges, G. Lewin, C. Moretti, and C. Dupont, *Planta Med.* **50**, 517 (1984).

62. M. Leboeuf, A. Cavé, M. El Tohami, J. Pusset, P. Forgacs, and J. Provost, *J. Nat. Prod.* **45**, 617 (1982).

63. F. Roblot, R. Hocquemiller, H. Jacquemin, and A. Cavé, *Plant. Med. Phytother.* **12**, 259 (1978).

64. F. Roblot, R. Hocquemiller, and A. Cavé, *C. R. Acad. Sci.* Ser II **293**, 373 (1981).

65. F. Roblot, R. Hocquemiller, A. Cavé, and C. Moretti, *J. Nat. Prod.* **46**, 862 (1983).

66. M. Leboeuf, F. Bévalot, and A. Cavé, *Planta Med.* **38**, 33 (1980).

67. O. R. Gottlieb, A. F. Magalhaes, E. G. Magalhaes, J. G. S. Maia, and A. J. Marsaioli, *Phytochemistry* **17,** 837 (1978).

68. D. Debourges, R. Hocquemiller, and A. Cavé, "Communication to the International Symposium on the Chemistry and Biology of Isoquinoline Alkaloids," London, April 1984.

69. D. Debourges, R. Hocquemiller, A. Cavé, and J. Lévy, *J. Nat. Prod.* **48,** 310 (1985).

70. C. Casagrande and G. Ferrari, *Farmaco. Ed. Sci.* **25,** 442 (1970).

71. M. Hamonnière, M. Leboeuf, A. Cavé, and R. R. Paris, *Plant. Med. Phytother.* **9,** 296 (1975).

72. M. Nieto, A. Cavé, and M. Leboeuf, *Lloydia* **39,** 350 (1976).

73. A. Jossang, M. Leboeuf, and A. Cavé, *Planta Med.* **32,** 249 (1977).

74. M. Leboeuf and A. Cavé, *Plant. Med. Phytother.* **6,** 87 (1972).

75. S. T. Lu and Y. C. Wu, *Heterocycles* **20,** 813 (1983).

76. C. Xu, P. Xie, Y. Zhu, N. Sun, and X. Liang, *Zhongyao Tongbao* **7,** 30 (1982); *Chem. Abstr.* **97,** 212639 (1982).

77. C. Xu, P. Xie, Y. Zhu, N. Sun, and X. Liang, *Zhongcaoyao* **14,** 148 (1983); *Chem. Abstr.* **99,** 19659 (1983).

78. M. Leboeuf, A. Cavé, C. Martin, J. Provost, and P. Forgacs, unpublished work.

79. R. Braz, S. J. Gabriel, C. M. R. Gomes, O. R. Gottlieb, M. D. G. A. Bichara, and J. G. S. Maia, *Phytochemistry* **15,** 1187 (1976).

80. R. Hocquemiller, S. Rasamizafy, C. Moretti, and A. Cavé, *Plant. Med. Phytother.* **18,** 165 (1984).

81. M. Diaz, C. Schreiber, and H. Ripperger, *Rev. Cubana Farm.* **15,** 93 (1981).

82. R. Hocquemiller, C. Debitus, F. Roblot, A. Cavé, and H. Jacquemin, *J. Nat. Prod.* **47,** 353 (1984).

83. M. El Tohami, *Thèse de Doctorat D'Etat,* Université de Paris-Sud, Chatenay-Malabry, 1984.

84. R. Hocquemiller, C. Debitus, F. Roblot, and A. Cavé, *Tetrahedron Lett.* **23,** 4247 (1982).

85. C. C. Hsu, R. H. Dobberstein, G. A. Cordell, and N. R. Farnsworth, *Lloydia* **40,** 505 (1977).

86. C. C. Hsu, R. H. Dobberstein, G. A. Cordell, and N. R. Farnsworth, *Lloydia* **40,** 152 (1977).

87. L. Castedo, J. A. Granja, A. Rodriguez de Lera, and M. C. Villaverde, "Communication to the International Symposium on the Chemistry and Biology of Isoquinoline Alkaloids," London, April 1984.

88. C. H. Phoebe, Jr., Ph.D. Thesis, University of Pittsburgh, 1980.

89. V. Zabel, W. H. Watson, C. H. Phoebe, Jr., J. E. Knapp, P. L. Schiff, Jr., and D. J. Slatkin, *J. Nat. Prod.* **45,** 94 (1982).

90. C. H. Phoebe, Jr., P. L. Schiff, Jr., J. E. Knapp, and D. J. Slatkin, *Heterocycles* **14,** 1977 (1980).

91. H. A. Ammar, J. E. Knapp, P. L. Schiff, Jr., and D. J. Slatkin, *J. Nat. Prod.* **42,** 696 (1979).

92. H. A. Ammar, P. L. Schiff, Jr., and D. J. Slatkin, *J. Nat. Prod.* **47,** 392 (1984).

93. D. Cortes, R. Hocquemiller, M. Leboeuf, A. Cavé, and C. Moretti, *J. Nat. Prod.* **49,** (1986), in press.

93a. D. Cortes, R. Hocquemiller, M. Leboeuf, and A. Cavé, *Phytochemistry* **24,** 2776 (1985).

94. M. Leboeuf, D. Cortes, R. Hocquemiller, and A. Cavé, *C. R. Acad. Sci.* **295,** ser. II, 191 (1982).

95. M. Leboeuf, D. Cortes, R. Hocquemiller, and A. Cavé, *Planta Med.* **48**, 234 (1983).

96. M. Leboeuf, D. Cortes, R. Hocquemiller, A. Cavé, A. Chiaroni, and C. Riche, *Tetrahedron* **38**, 2889 (1982).

97. W. M. Harris and T. A. Geissman, *J. Org. Chem.* **30**, 432 (1965).

98. J. A. Garbarino, W. Petzall, and J. Salazar, *Rev. Latinoam. Quim.* **15**, 67 (1984).

99. R. Hocquemiller, S. Rasamizafy, A. Cavé, and C. Moretti, *J. Nat. Prod.* **46**, 335 (1983).

100. R. Hocquemiller, S. Rasamizafy, and A. Cavé, *Tetrahedron* **38**, 911 (1982).

101. A. Chiaroni, C. Riche, R. Hocquemiller, S. Rasamizafy, and A. Cavé, *Tetrahedron* **39**, 2163 (1983).

102. D. Cortes, A. Ramahatra, H. Dadoun, and A. Cavé, *C. R. Acad. Sci.* Ser. II **299**, 311 (1984).

103. D. Cortes, A. Ramahatra, A. Cavé, J. de Carvalho Bayma, and H. Dadoun, *J. Nat. Prod.* **48**, 254 (1985).

104. S. Rasamizafy, R. Hocquemiller, A. Cavé, and H. Jacquemin, *J. Nat. Prod.* **49**, (1986), in press.

105. M. Hasegawa, M. Sojo, A. Lira, and C. Marquez, *Acta Cient. Venez.* **23**, 165 (1972); *Chem. Abstr.* **79**, 42716 (1973).

106. J. W. Skiles and P. Cava, *J. Org. Chem.* **44**, 409 (1979).

107. H. Achenbach, C. Renner, and I. Addae-Mensah, *Liebigs Ann. Chem.* 1623 (1982).

108. H. Achenbach, C. Renner, J. Worth, and I. Addae-Mensah, *Liebigs Ann. Chem.* 1132 (1982).

109. R. Hocquemiller, P. Cabalion, J. Bruneton, and A. Cavé, *Plant. Med. Phytother.* **12**, 230 (1978).

110. R. Hocquemiller, P. Cabalion, A. Fournet, and A. Cavé, *Planta Med.* **50**, 23 (1984).

111. R. Hocquemiller, P. Cabalion, A. Bouquet, and A. Cavé, *C. R. Acad. Sci.* **285**, ser. II, 447 (1977).

112. M. Leboeuf, A. Fournet, A. Bouquet, and A. Cavé, *Plant. Med. Phytother.* **11**, 284 (1977).

113. D. Tadić, B. K. Cassels, M. Leboeuf, and A. Cavé, *Phytochemistry*, in press.

114. I. R. C. Bick and N. W. Preston, *Aust. J. Chem.* **24**, 2187 (1971).

115. J. Ellis, E. Gellert, and R. E. Summons, *Aust. J. Chem.* **25**, 2735 (1972).

116. P. G. Waterman and K. Pootakahm, *Planta Med.* **37**, 247 (1979).

117. M. Leboeuf and A. Cavé, *Plant. Med. Phytother.* **8**, 147 (1974).

118. M. Leboeuf, J. Parello, and A. Cavé, *Plant. Med. Phytother.* **6**, 112 (1972).

119. A. I. Spiff, F. K. Duah, D. J. Slatkin, and P. L. Schiff, Jr., "Communication to the Annual Meeting of the American Society of Pharmacognosy," Pittsburgh, 1982.

120. A. I. Spiff, F. K. Duah, D. J. Slatkin, and P. L. Schiff, Jr., *Planta Med.* **50**, 455 (1984).

121. L. A. Djakoure, D. Kone, and L. L. Douzoua, *Ann. Univ. Abidjan, Ser C* **14**, 49 (1978); *Chem. Abstr.* **93**, 41533 (1980).

122. M. E. L. de Almeida, R. Fo. Braz, M. V. von Bulow, O. R. Gottlieb, and J. G. S. Maia, *Phytochemistry* **15**, 1186 (1976).

123. F. Bévalot, M. Leboeuf, A. Bouquet, and A. Cavé, *Ann. Pharm. Fr.* **35**, 65 (1977).

124. F. Bévalot, M. Leboeuf, and A. Cavé, *Plant. Med. Phytother.* **11**, 315 (1977).

125. K. Sarpong, D. K. Santra, G. J. Kapadia, and J. W. Wheeler, *Lloydia* **40**, 616 (1977).

126. F. Bévalot, M. Leboeuf, and A. Cavé, *C. R. Acad. Sci.* Ser. C **282**, 865 (1976).

127. M. H. Abu Zarga and M. Shamma, *J. Nat. Prod.* **45**, 471 (1982).

128. A. Jossang, M. Leboeuf, A. Cavé, T. Sévenet, and K. Padmawinata, *J. Nat. Prod.* **47**, 504 (1984).

129. A. Jossang, M. Leboeuf, and A. Cavé, *Tetrahedron Lett.* **23,** 5147 (1982).

130. H. Guinaudeau, A. Ramahatra, M. Leboeuf, and A. Cavé, *Plant. Med. Phytother.* **12,** 166 (1978).

131. S. R. Johns, J. A. Lamberton, C. S. Li, and A. A. Sioumis, *Aust. J. Chem.* **23,** 423 (1970).

132. A. Jossang, M. Leboeuf, P. Cabalion, and A. Cavé, *Planta Med.* **49,** 20 (1983).

133. M. Hamonnière, M. Leboeuf, and A. Cavé, *Phytochemistry* **16,** 1029 (1977).

134. M. Hamonnière, M. Leboeuf, and A. Cavé, *C. R. Acad. Sci.* Ser. C **278,** 921 (1974).

135. A. Cavé, H. Guinaudeau, M. Leboeuf, A. Ramahatra, and J. Razafindrazaka, *Planta Med.* **33,** 243 (1978).

136. C. M. Hasan, T. M. Healey, P. G. Waterman, and C. H. Schwalbe, *J. Chem. Soc., Perkin Trans. I,* **1982,** 2807.

137. S. R. Johns, J. A. Lamberton, C. S. Li, and A. A. Sioumis, *Aust. J. Chem.* **23,** 363 (1970).

138. A. Jossang, M. Leboeuf, A. Cavé, and T. Sévenet, *J. Nat. Prod.* **49** (1986), in press.

139. D. Cortes, R. Hocquemiller, A. Cavé, and J. Saez, *J. Nat. Prod.* **49** (1986), in press.

140. T. T. Dabrah and A. T. Sneden, *J. Nat. Prod.* **46,** 436 (1983).

141. R. M. Brash and A. T. Sneden, *J. Nat. Prod.* **46,** 437 (1983).

142. E. Gellert and R. Rudtzats, *Aust. J. Chem.* **25,** 2477 (1972).

143. S. F. Dyke and E. Gellert, *Phytochemistry* **17,** 599 (1978).

144. M. El Tohami, M. Leboeuf, and A. Cavé, "Communication to the International Symposium on the Chemistry and Biology of Isoquinoline Alkaloids," London, April 1984.

145. M. Leboeuf and A. Cavé, *Plant. Med. Phytother.* **14,** 143 (1980).

146. A. Bouquet and A. Fournet, *Plant. Med. Phytother.* **6,** 149 (1972).

147. A. Bouquet, Ad. Cavé, A. Cavé, and R. R. Paris, *C. R. Acad. Sci.* **271,** ser. C, 1100 (1970).

148. C. Casagrande and G. Merotti, *Farmaco. Ed. Sci.* **25,** 799 (1970).

149. R. Hocquemiller, A. Cavé, and A. Raharisololalao, *J. Nat. Prod.* **44,** 551 (1981).

150. J. Schmutz, *Helv. Chim. Acta* **42,** 335 (1959).

151. F. Allera, M. Leboeuf, and A. Cavé, unpublished results.

152. M. Leboeuf, A. Cavé, J. Provost, P. Forgacs, and H. Jacquemin, *Plant. Med. Phytother.* **16,** 253 (1982).

153. M. Nieto, M. Leboeuf, and A. Cavé, *Phytochemisry* **14,** 2508 (1975).

154. M. Nieto, T. Sévenet, M. Leboeuf, and A. Cavé, *Planta Med.* **30,** 48 (1976).

155. S. R. Johns, J. A. Lamberton, and A. A. Sioumis, *Aust. J. Chem.* **21,** 1883 (1968).

156. T. Ngo Van, T. Dong Viet, and M. Nguyen Tuyet, *Tap San, Hoa-Hoc* **12,** 46 (1974): through *Chem. Abstr.* **83,** 55666 (1975).

157. A. W. Sangster and K. L. Stuart, *Chem. Rev.* **65,** 69 (1965).

158. M. Shamma, S. Y. Yao, B. R. Pai, and R. Charubala, *J. Org. Chem.* **36,** 3253 (1971).

159. C. H. Chen, H. M. Chang, and E. B. Cowling, *Phytochemistry* **15,** 547 (1976).

160. J. F. Green, G. N. Jham, J. L. Neumeyer, and P. Vouros, *J. Pharm. Sci.* **69,** 936 (1980).

161. A. Urzua and B. K. Cassels, *Heterocycles* **4,** 1881 (1976).

162. L. Dolejš, *Collect. Czech. Chem. Commun.* **39,** 571 (1974).

163. M. Shamma and J. L. Moniot, *Experientia* **32,** 282 (1976).

164. G. Severini Ricca and C. Casagrande, *Gazz. Chim. Ital.* **109,** 1 (1979).

165. K. G. R. Pachler, R. R. Arndt, and W. H. Baarschers, *Tetrahedron* **21,** 2159 (1965).

166. M. Shamma and J. L. Moniot, *Tetrahedron Lett.* 2291 (1974).

167. S. F. Hussain, M. T. Siddiqui, G. Manikumar, and M. Shamma, *Tetrahedron Lett.* **21**, 723 (1980).

168. V. Fajardo, H. Guinaudeau, V. Elango, and M. Shamma, *J. Chem. Soc., Chem. Commun.* 1350 (1982).

169. M. Shamma and D. M. Hindenlang, *Carbon-13 NMR Shift Assignments of Amines and Alkaloids,* Plenum, New York, 1979.

170. A. J. Marsaioli, F. de A. M. Reis, A. F. Magalhaes, E. A. Rúveda, and A. M. Kuck, *Phytochemistry* **18**, 165 (1979).

171. L. M. Jackman, J. C. Trewella, J. L. Moniot, M. Shamma, R. L. Stephens, E. Wenkert, M. Leboeuf, and A. Cavé, *J. Nat. Prod.* **42**, 437 (1979).

172. D. W. Hughes and D. B. MacLean, "The ^{13}C-NMR Spectra of Isoquinoline Alkaloids," in *The Alkaloids,* Vol. 18, R. G. A. Rodrigo, Ed., Academic, New York, 1981.

173. T. A. Broadbent and E. G. Paul, *Heterocycles* **20**, 863 (1983).

174. L. Castedo, R. Riguera, and F. J. Sardina, *An. Quim., Ser. C* **77**, 138 (1981).

175. H. Ronsch, A. Preiss, K. Schreiber, and H. Fernandez de Cordoba, *Liebigs Ann. Chem.* 744 (1983).

176. A. J. Marsaioli, A. F. Magalhaes, E. A. Rúveda, and F. de A. M. Reis, *Phytochemistry* **19**, 995 (1980).

177. L. Castedo, F. Riguera, and J. Sardina, *An. Quim., Ser. C* **78**, 103 (1982).

178. G. Severini Ricca and C. Casagrande, *Org. Magn. Reson.* **9**, 8 (1977).

179. M. Shamma, *Experientia* **16**, 484 (1960).

180. B. Ringdahl, R. P. K. Chan, J. C. Craig, M. P. Cava, and M. Shamma, *J. Nat. Prod.* **44**, 80 (1981).

181. J. C. Craig and S. K. Roy, *Tetrahedron* **21**, 395 (1965).

182. D. S. Bhakuni and M. M. Dhar, *Experientia* **24**, 10 (1968).

183. C. Debitus, Thèse de Doctorat, Université de Paris-Sud, Orsay, 1983.

184. J. I. Kunitomo, Y. Yoshikawa, S. Tanaka, Y. Imori, K. Isoi, Y. Masada, K. Hashimoto, and T. Inoue, *Phytochemistry* **12**, 699 (1973).

185. M. P. Cava, A. Venkateswarlu, M. Srinivasan, and D. L. Edie, *Tetrahedron* **28**, 4299 (1972).

186. S. Philipov, O. Petrov, and N. Mollov, *Arch. Pharm (Weinheim)* **315**, 498 (1982).

187. M. P. Cava, D. L. Edie, and J. M. Saa, *J. Org. Chem.* **40**, 3601 (1975).

188. M. P. Cava, M. J. Mitchell, S. C. Havlicek, A. Lindert, and R. J. Spangler, *J. Org. Chem.* **35**, 175 (1970).

189. J. L. Neumeyer and F. E. Granchelli, *Tetrahedron Lett.* 5261 (1970).

190. T. Kametani, S. Shibuya, and S. Kano, *J. Chem. Soc. Perkins Trans. 1,* **1973**, 1212.

191. F. E. Granchelli and J. L. Neumeyer, *Tetrahedron* **30**, 3701 (1974).

192. G. R. Lenz and F. J. Koszyk, *J. Chem. Soc. Perkin Trans. I* **1984**, 1273.

193. J. I. Kunitomo, Y. Murakami, and M. Akasu, *J. Pharm. Soc. Japan* **100**, 337 (1980).

194. A. Venkateswarlu and M. P. Cava, *Tetrahedron* **32**, 2079 (1976).

195. J. M. Saa and M. P. Cava, *J. Org. Chem.* **42**, 347 (1977).

196. J. M. Saa and M. P. Cava, *J. Org. Chem.* **43**, 1096 (1978).

197. M. D. Menachery, J. M. Saa, and M. P. Cava, *J. Org. Chem.* **46**, 2584 (1981).

198. M. D. Menachery, P. Carroll, and M. P. Cava, *Tetrahedron Lett.* **24**, 167 (1983).

199. S. Philipov, O. Petrov, and N. Mollov, *Arch. Pharm. (Weinheim)* **314**, 1034 (1981).

200. H. A. Ammar, P. L. Schiff Jr., and D. J. Slatkin, *Heterocycles* **20**, 451 (1983).

201. N. Mollov and S. Philipov, *Chem. Ber.* **112**, 3737 (1979).

202. D. Debourges, D. Cortes, R. Hocquemiller, J. Lévy, and A. Cavé, "Communication to the CNRS International Symposium—Chimie des Substances Naturelles: Etat et Perspectives," Gif-sur-Yvette, 1984.

203. R. S. Cahn, C. K. Ingold, and V. Prelog, *Experientia* **12**, 81 (1956).

204. A. Urzua and B. K. Cassels, *Lloydia* **41**, 98 (1978).

205. H. Guinaudeau, M. Shamma, B. Tantisewie, and K. Pharadai, *J. Nat. Prod.* **45**, 355 (1982).

206. M. Hamonnière, Thèse de Doctorat D'Etat, Université de Paris-Sud, Chatenay-Malabry, 1976.

207. T. H. Yang, *J. Pharm. Soc. Japan* **82**, 798 (1962).

208. S. K. Talapatra, A. Patra, and B. Talapatra, *Tetrahedron* **31**, 1105 (1975).

209. J. I. Kunitomo, M. Miyoshi, E. Yuge, T. H. Yang, and C. M. Chen, *Chem. Pharm. Bull.* **19**, 1502 (1971).

210. K. L. Wert, S. Chackalamannil, E. Miller, D. R. Dalton, D. E. Zacharias, and J. P. Glusker, *J. Org. Chem.* **47**, 5141 (1982).

211. A. Quevauviller and M. Hamonnière, *C. R. Acad. Sci.* **284**, ser. D, 93 (1977).

212. S. V. Kessar, T. Mohammad, and Y. P. Gupta, *Indian J. Chem.* **22B**, 321 (1983).

213. S. Chackalamannil and D. R. Dalton, *Tetrahedron Lett.* **21**, 2029 (1980).

214. S. V. Kessar, Y. P. Gupta, V. S. Yadav, M. Narula, and T. Mohammad, *Tetrahedron Lett.* **21**, 3307 (1980).

215. S. V. Kessar, Y. P. Gupta, and T. Mohammad, *Indian J. Chem.* **20B**, 984 (1981).

216. H. Guinaudeau, M. Leboeuf, M. Debray, A. Cavé, and R. R. Paris, *Planta Med.* **27**, 304 (1975).

217. J. Hartenstein and G. Satzinger, *Angew. Chem. Int. Ed. Engl.* **16**, 730 (1977).

218. W. D. Smolnycki, J. L. Moniot, D. L. Hindenlang, G. A. Miana, and M. Shamma, *Tetrahedron Lett.* 4617 (1978).

219. D. P. Allais and H. Guinaudeau, *Heterocycles* **20**, 2055 (1983).

220. L. Castedo, E. Quiñoa, and R. Riguera, *An. Quim.* **78**, 171 (1982).

221. L. Castedo, R. J. Estevez, F. Rodriguez, and C. Villaverde, "Communication to the International Symposium on the Chemistry and Biology of Isoquinoline Alkaloids," London, April 1984.

222. L. Castedo, R. Suau, and A. Mouriño, *Tetrahedron Lett.* 501 (1976).

223. L. Castedo, E. Guitian, J. M. Saa, and R. Suau, *Tetrahedron Lett.* **23**, 457 (1982).

224. L. Castedo, R. J. Estevez, J. M. Saa, and R. Suau, *J. Heterocycl. Chem.* **19**, 1319 (1982).

225. L. Castedo, J. M. Saa, R. Suau, and R. J. Estevez, *An. Quim.* **79**, 329 (1983).

226. J. M. Saa, M. J. Mitchell, and M. P. Cava, *Tetrahedron Lett.* 601 (1976).

227. V. B. Chervenkova, N. M. Mollov, and S. Paszyc, *Phytochemistry* **20**, 2285 (1981).

228. H. Guinaudeau, M. Leboeuf, and A. Cavé, *J. Nat. Prod.* **42**, 133 (1979).

229. H. Guinaudeau, M. Leboeuf, and A. Cavé, *J. Nat. Prod.* **47**, 565 (1984).

230. M. Gerecke, R. Borer, and A. Brossi, *Helv. Chim. Acta* **58**, 185 (1975).

231. L. Castedo, R. Riguera, J. M. Saa, and R. Suau, *Heterocycles* **6**, 677 (1977).

232. M. P. Cava, I. Noguchi, and K. T. Buck, *J. Org. Chem.* **38**, 2394 (1973).

233. S. M. Kupchan and A. J. Liepa, *J. Am. Chem. Soc.* **95**, 4062 (1973).

234. H. Guinaudeau, M. Shamma, B. Tantisewie, and K. Pharadai, *J. Chem. Soc. Chem. Commun.*, 1118 (1981).

235. R. A. Olofson, J. T. Martz, J-P. Senet, M. Piteau, and T. Malfroot, *J. Org. Chem.* **49**, 2081 (1984).

236. C. D. Hufford, A. S. Sharma, and B. O. Oguntimein, *J. Pharm. Sci.* **69**, 1180 (1980).

237. M. Shamma and J. L. Moniot, *Heterocycles* **2**, 427 (1974).

238. W. C. Taylor, *Aust. J. Chem.* **37**, 1095 (1984).

239. D. B. Mix, H. Guinaudeau, and M. Shamma, *J. Nat. Prod.* **45**, 657 (1982).

240. R. Crohare, H. A. Priestap, M. Fariña, M. Sedola, and E. A. Rúveda, *Phytochemistry* **13**, 1957 (1974).

241. T. Mahmood, K. C. Chan, M. H. Park, Y. N. Han, and B. H. Han, "Communication to the Fifth Asian Symposium on Medicinal Plants and Spices," Seoul, August 1984.

242. B. F. Bowden, E. Ritchie, and W. C. Taylor, *Aust. J. Chem.* **25**, 2659 (1972).

243. B. F. Bowden, K. Picker, E. Ritchie, and W. C. Taylor, *Aust. J. Chem.* **28**, 2681 (1975).

244. I. Muhammad, Ph.D. Thesis, University of Strathclyde, Glasgow, 1985.

244a. J. Koyama, T. Sugita, Y. Suzuta, and H. Irie, *Heterocycles* **12**, 1017 (1979).

245. W. I. Taylor, *Tetrahedron* **14**, 42 (1961).

246. A. Cronquist, *The Evolution and Classification of Flowering Plants,* Nelson, London, 1968.

247. J. Hutchinson, *Evolution and Phylogeny of Flowering Plants,* Academic, London, 1969.

248. R. M. T. Dahlgren, *Bot. J. Linn. Soc.* **80**, 91 (1980).

249. C. M. Andrade da M. Rezende, O. R. Gottlieb, and M. C. Marx, *Biochem. Syst. Ecol.* **3**, 63 (1975).

250. Anon., *Taxon* **32**, 528 (1983); *ibid* **33**, 801 (1984).

251. J. W. Walker, *Contrib. Gray Herb.* **202**, 3 (1971).

252. J. W. Walker, *Grana* **11**, 45 (1971).

253. J. W. Walker, *Bot. J. Linn. Soc.* **65**, 173 (1972).

254. J. W. Walker, *Taxon* **21**, 57 (1972).

255. J. Sinclair, *Gard. Bull. Singapore* **14**, 149 (1955).

256. A. Le Thomas, "Annonaceae", in *Flore du Gabon,* A. Aubreville, Ed., Museum National d'Histoire Naturelle, Paris, 1969.

257. C. M. Hasan, T. M. Healey, and P. G. Waterman, *Phytochemistry* **21**, 177 (1982).

258. C. M. Hasan, T. M. Healey, and P. G. Waterman, *Phytochemistry* **21**, 1365 (1982).

259. C. M. Hasan, T. M. Healey, and P. G. Waterman, *Phytochemistry* **21**, 2134 (1982).

260. C. M. Hasan, T. M. Healey, and P. G. Waterman, *Phytochemistry* **24**, 192 (1985).

261. P. G. Waterman and I. Muhammad, *Planta Med.* **50**, 282 (1984).

262. C. Galeffi, G. B. Marini-Bettolo, and D. Vecchi, *Gazz. Chim. Ital.* **105**, 1207 (1975).

263. A. Le Thomas, *Revue de Paléobiologie* 119 (1984).

264. W. Sauer and F. Ehrendorfer, *Plant. Syst. Evol.* **146**, 47 (1984).

Chapter Four

The *Thalictrum* Alkaloids: Chemistry and Pharmacology

Paul L. Schiff, Jr.
Department of Pharmaceutical Sciences
School of Pharmacy
University of Pittsburgh
Pittsburgh, Pennsylvania 15261

*Many times a day I realize how much my own outer
and inner life is built upon the labors of my
fellow men, both living and dead, and how earnestly
I must exert myself in order to give in return as
much as I have received. My peace of mind is often
troubled by the depressing sense that I have borrowed
too heavily from the work of other men.*

ALBERT EINSTEIN

CONTENTS

---

1.   INTRODUCTION*

The *Thalictrum* alkaloids are phenylalanine- or tyrosine-derived bases that occur as a variety of diverse but related monomeric or dimeric structural variants. In 1948, there were fewer than 10 alkaloids that had been isolated from the genus whereas today there are approximately 200. Although the interest has been worldwide, laboratories in Bulgaria (Dutschewska and Mollov), Japan (Furukawa, Tomimatsu, Tomita), the United States (Beal, Doskotch, Kupchan, Shamma), and the USSR (Yunusov) have been most active. The advent of improved isolation techniques and the utilization of high-resolution proton magnetic resonance spectrometry and mass spectrometry have spurred the isolation of an ever-increasing number of alkaloids from this genus. Today, with the addition of carbon magnetic resonance spectrometry and selected nuclear Overhauser enhancement ^{1}H NMR spectrometry, the elucidation of structure of very small quantities of alkaloids is possible, thus allowing a greater understanding of the overall anabolism/catabolism of these bases.

Thalictrum is an ancient name of doubtful origin which may derive from the Greek *thallo* (to grow green). *Thalictrum* (meadow rue), a genus of plants of the Ranunculaceae (Crowfoot family), includes a wide range of tall, erect, perennial herbs that are mostly dispersed in temperate and cold countries at altitudes ranging from 200 to 2000 m above sea level [1]. Ranunculus, the Latin word for "little frog," refers to the habitat of many of the species in ponds and wet areas [2]. There are at least 90 species within the genus and an early monograph was published by Lecoyer in 1885 in which 69 species were described [3]. A number of treatises were subsequently published concerning botanical description of the genus [4–9].

Crude extracts of *Thalictrum* species have been used as folk medicinals throughout the world for centuries. The literature abounds with references to the use of these extracts in the treatment of numerous conditions, some of which include snakebite, jaundice, leprosy, rheumatism, external and

---

* This chapter is dedicated to two very special people. First, to my dear wife Lois, the joy of my life and the partner of my soul; and second, to Professor Jack L. Beal, teacher, scholar and dear friend, who first aroused my interest in these alkaloids some 25 years ago.

internal infections, and pregnancy-induced vomiting in countries as culturally diverse as China, Russia, Japan, India, the United States, and Canada [10].

The purpose of this chapter is to review the *Thalictrum* alkaloids with particular reference to the techniques of isolation and structure elucidation and to the individual pharmacology of these bases. To this end, the alkaloids are classified as monomeric or dimeric and discussed within a particular chemical group based on the nucleus of the alkaloid. Finally, three other reviews concerning the *Thalictrum* alkaloids have been published and include works by Tomimatsu [11–13], Mollov et al. [14], and Schiff and Doskotch [10].

2. MONOMERIC ALKALOIDS

2.1. Aporphines

The following reviews and references are particularly relevant to the isolation and identification of aporphine alkaloids:

1. Tabular reviews [10,15–19]
2. Annual reviews [20–31]
3. Other reviews [10–13,32,33]
4. Book chapters [14,34–37,241,242]

The numbering of the aporphine ring (**1**) is according to accepted practice [15].

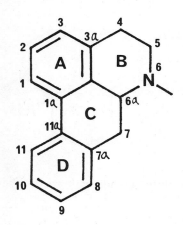

1

2.1.1.   1,2,10-Trioxygenated Aporphines

2.1.1.1.   N,O,O,-Trimethylsparsiflorine.   Tabular review [17, alkaloid No. 261]; UV [17,38]; IR [17,38]; ^{1}H NMR [17,38,39].

N,O,O-Trimethylsparsiflorine (**2**), $C_{20}H_{23}NO_3$ (325.1677), m.p. 124–126°, was first isolated from *Thalictrum foliolosum* DC in 1980–1981 [38]. The UV spectrum, which possessed maxima at 216 nm (log ε 4.56), 275 nm (log ε 4.01), and 298 nm (log ε 4.13) [38] lacked a bathochromic shift in alkali, and was suggestive of a nonphenolic aporphine [15–17]. The ^{1}H NMR spectrum indicated the presence of one N-methyl group as a singlet at δ 2.57; three methoxy groups as singlets at δ 3.69 [C(1)], 3.84 [C(2) or C(10)], and 3.89 [C(10) or C(2)]; and four aromatic protons as a singlet at δ 6.64 [H(3)], a double doublet at δ 6.80 (J = 3, 8 Hz) [H(9)], a doublet at δ 7.16 (J = 8 Hz) [H(8)], and a doublet at δ 8.03 (J = 3 Hz) [H(11)] [17]. The most salient feature of the ^{1}H NMR spectrum was the aromatic proton region which showed *ortho* and *meta* coupling of the C(9) proton and was suggestive of monosubstitution in ring D [15–17,36]. In addition, the presence of one N-methyl group and three aromatic methoxy groups (one of which was at a higher field (δ 3.69), thus indicating C(1) substitution [15–17,36]) further supported the structural assignment as a 1,2,10-trimethoxylated aporphine. The mass spectrum showed a molecular ion at m/z 325 and other fragment ions at m/z 324(M—H), 310(M—CH$_3$), 294(M—OCH$_3$), and 282(M—[CH$_2$=NCH$_3$]) characteristic for an aporphine [15–17,36]. The alkaloid failed to undergo acetylation (acetic anhydride and pyridine) or methylation (diazomethane) but did form a crystalline methiodide (methyl iodide/methanol) that was identical with authentic N,O,O-trimethylsparsiflorine methiodide. The lack of a report of either a specific rotation or a CD spectrum combined with the paucity of literature CD spectra of 1,2,10-trioxygenated aporphines precludes the assignment of absolute configuration of this al-

2

kaloid. *N,O,O*-Trimethylsparsiflorine (**2**) failed to exhibit antimicrobial activity when tested against Gram-positive and Gram-negative bacteria as well as numerous fungi [38], *N*-methylsparsiflorine iodide (1,10-*O,O*-demethyl-**2**) elicited a hypotensive response in animals [40].

2.1.2. 1,2,9,10-Tetraoxygenated Aporphines

2.1.2.1. Glaucine (O-Methylthalicmidine). Tabular review [15, alkaloid No. 59]; UV [15], IR [15], ^{1}H NMR [15], ^{13}C NMR [17,41,42], MS [43,44].

Glaucine (**3**), $C_{21}H_{25}NO_4$ (355.1783), m.p. 120–121°, $[\alpha]_D$ +116° (c 0.75, MeOH) and $[\alpha]_D$ −114.5° (CHCl$_3$) [15], was the first aporphine to have been isolated from nature and the first to have been synthesized [34]. (+)-Glaucine was first isolated from nature in 1893 and in a pure state from *Glaucium flavum* Crantz. (Papaveraceae) in 1901 [34]. This aporphine was first reported in a *Thalictrum* species, *T. minus* L., in 1952 [45].

Glaucine has been demonstrated to produce narcosis and convulsions in animals [36]. Both glucine methiodide and glaucine ethiodide caused a decrease in motility, in addition to inducing dyspnea, respiratory depression, and death in rabbits [46]. The LD$_{50}$ of glaucine methiodide was 59 mg/kg after subcutaneous adminstration and 4.8 mg/kg after intravenous administration. Similarly, the LD$_{50}$ of glaucine ethiodide was 400 mg/kg after subcutaneous administration and 10.3 mg/kg after intravenous administration [46]. In dogs, both glaucine methiodide and ethiodide in dosages of 0.01–2.0 mg/kg decreased the blood pressure by 15–70% over varying periods of time [46]. Glaucine also produced a hypotension and respiratory depression when administered intravenously to cats [47,48].

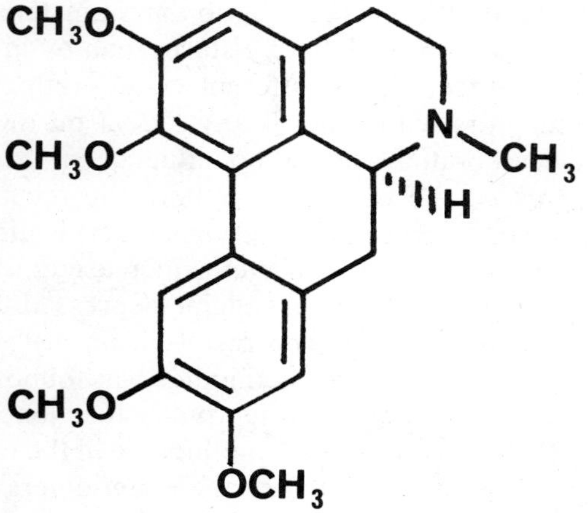

$\underset{\sim}{\textbf{3}}$

Oral administration of a Bulgarian glaucine preparation (Glaovent) to rats and cats in a daily dosage of 12 mg/kg for 6 months produced a decrease in blood glucose of 70–90 mg%. A dose-dependent increase in glycogen storage by hepatic cells occurred with no apparent toxicity [49]. Adult male rats were treated with glaucine by gavage in a dose of 75 or 150 mg/kg for 6 weeks. Animals receiving the higher dose exhibited a transient decrease in body weight, decreased motor activity after 3 weeks of dosing, and decreased grip strength after 3 and 6 weeks of dosing. The lower dose decreased forelimb grip strength after 6 weeks of dosing. The alkaloid had no effects on other measures of motor function or reactivity to noxious or nonnoxious stimula. The neurobehavioral functions of the glaucine-treated rats were no different from those of controls after 5 weeks postdosing. Histopathological examination revealed a brown discoloration of the thyroid, brain, and muscles with the appearance of brown pigments in the cell bodies of brainstem and spinal cord neurons [50]. Glaucine has also exhibited antitussive properties greater than or equal to that of codeine [51–54], and (+)-glaucine has been used extensively in clinical practices in Bulgaria and Poland [54]. A comparison of the antitussive and toxic effects of (±)-glaucine phosphate with codeine showed that regardless of route of administration, the safety margins were approximately the same for the antitussive effect but codeine appeared more toxic than glaucine. The respiratory and cardiovascular effects of the alkaloids were similar but somewhat less pronounced with (+)-glaucine. (+)-Glaucine prolonged hexobarbital sleeping time but lacked analgesic and anticonvulsant effects. Levallorphan had no effect on the antitussive effect of (±)-glaucine but blocked that of codeine. Finally, glaucine possessed a mild muscle-relaxant activity [54].

A study of the sites of antitussive action of (±)-glaucine phosphate showed that smaller doses could be administered to suppress coughing in cats when administered by routes leading to the brainstem, such as the vertebral artery and the cerebromedullary cistern, than by intravenous routes. The alkaloid had no effect on the afferent cough–reflex pathway or pulmonary stretch receptors with little, if any, effect on the efferent cough–reflex pathway. Decerebration exerted no influence on the antitussive effect but definitely depressed the potentials of both the recurrent and internal intercostal nerves evoked by superior laryngeal nerve stimulation. In deafferentated and decerebrate cats, (±)-glaucine increased the spontaneous discharges of the phrenic nerve whereas codeine depressed them. Apparently, the antitussive action of (±)-glaucine results from a direct action on the cough center [55]. (+)-(S)-Glaucine methiodide was found to exhibit a subtle but statistically significant increased neuromuscular junction blocking potency as compared to (−)-(R)-glaucine methiodide in the in vivo cat tongue–hypoglossal nerve preparation. Each of the enantiomers was only a weak inhibitor of acetylcholinesterase [56].

Although glaucine was found to be an inhibitor of bovine hepatic glutamate dehydrogenase [57], it was devoid of butrylcholinesterase inhibition

[58]. Recently, the LD_{50} of glaucine in mice was established to be 150–180 mg/kg after intraperitoneal administration and 510–620 mg/kg after intragastric administration. Chronic intragastric administration of doses of 5–75 mg/kg for 3 months produced no obvious toxic effects and the drug also appeared to lack allergizing, mutagenic, embryotoxic, or teratogenic properties [59].

The microbial biotransformation of glaucine by *Streptomyces griseus* afforded norglaucine and predicentrine (1,9,10-trimethoxy-2-hydroxyaporphine) [60]. However, *Fusarium solani* induced the production of didehydroglaucine (6a,7-dehydroglaucine) from (+)-(*S*)-glaucine but not from the enantiomeric (−)-(*R*)-glaucine, thus suggesting an apparent substrate enantiospecificity [60].

Glaucine has also been isolated from *Thalictrum baicalense* Turcz. [61], *T. fendleri* Engelm. ex Gray [62], *T. filamentosum* Maxim. [63], *T. foetidum* L. [64,192], *T. minus* L. [65,66], and *T. sachalinense* Lecoyer. [67], and has been quantified in *T. minus* L. via HPLC [66].

2.1.2.2. Isoboldine (N-Methyllaurelliptine). Tabular review [15, alkaloid No. 40], UV [15,70], IR [15], ^{1}H NMR [15,70], ^{13}C NMR [41,42], MS [15,43], CD [70].
Isoboldine (**4**), $C_{19}H_{21}NO_4$ (327.1471), m.p. 122–123°, $[\alpha]_D^{13}$ +54° (c 0.2, EtOH) [15], was first isolated from *Nandina domestica* Thunb. (Berberidaceae) in 1962 [35] and was later shown to be one of the insect feeding inhibitory substances in the leaves of *Cocculus trilobus* [69]. The first reported isolation of (+)-(*S*)-isoboldine in the genus *Thalictrum* was from *T. alpinum* L. in 1980 [70]. Shortly thereafter, the alkaloid was isolated from *T. foetidum* L. [64,192].

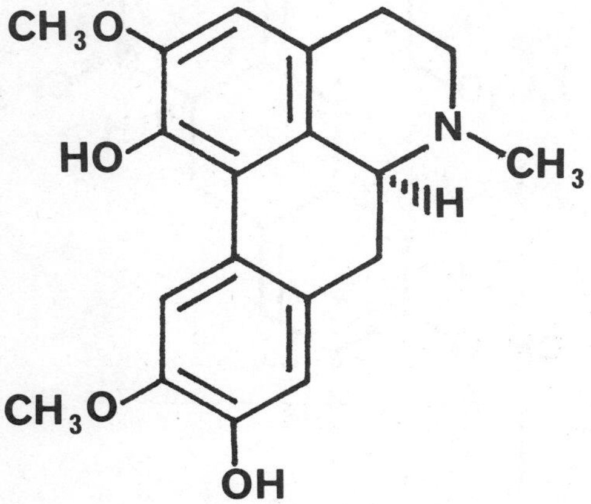

4

  2.1.2.3. N-Methyllaurotetanine. Tabular review [15, alkaloid No. 55],
UV [15,71,72], ^{1}H NMR [15,72], ^{13}C NMR [42], MS [15,72], CD [73].
N-Methyllaurotetanine (**5**), $C_{20}H_{23}NO_4$ (341.1627), m.p. 109° [72], 237–238°
(HBr salt) [15]; $[\alpha]_D$ +88° (c 0.64, CHCl$_3$) [15]; $[\alpha]_D^{25}$ +103° (c 0.3, MeOH)
[72], was first isolated from *Litsea citrata* (Lauraceae) in 1933 [34]. This
aporphine was first reported as a constituent of *Thalictrum* spp. when it was
isolated from the tops of *T. revolutum* DC. in 1977 [68] and subsequently
from *T. dioicum* L. [72] and the roots of *T. revolutum* DC. [74].
  N-Methyllaurotetanine (**5**) depressed the mean arterial pressure and force
of myocardial contraction in the anesthetized rhesus monkey at an intravenous dose of 10 mg/kg [75]. The alkaloid also increased carotid artery
blood flow and heart rate in the monkey while producing an immediate negative inotropic and chronotropic response in isolated, hemoperfused dog
hearts [75]. In fact, many of the cardiovascular actions of the aporphine–
benzylisoquinoline dimeric alkaloid thalicarpine (**192**) can be ascribed to *N*-
methyllaurotetanine (**5**), which is the aporphine-half of thalicarpine [75]. *N*-
Propyllaurotetanine has been shown to exhibit cardiac antifibrillar activity
[76]. Finally, *N*-Methyllaurotetanine (**5**) was an inhibitor of the growth of
the microorganism *Scenedesmus obliquus* [77].

  2.1.2.4. N-Methylnantenine. Tabular review [16, alkaloid No. 202],
UV [16,78], ^{1}H NMR [16,67,78], MS [67].
N-Methylnantenine (**6**), $C_{21}H_{24}N^+O_4Cl^-$ (354.1705), m.p. 213–214°
(CHCl$_3$); $[\alpha]_D$ +39° (EtOH) was first isolated from *Thalictrum polygamum*
Muhl. in 1975 [78]. The UV spectrum was characteristic of a 1,2,9,10-

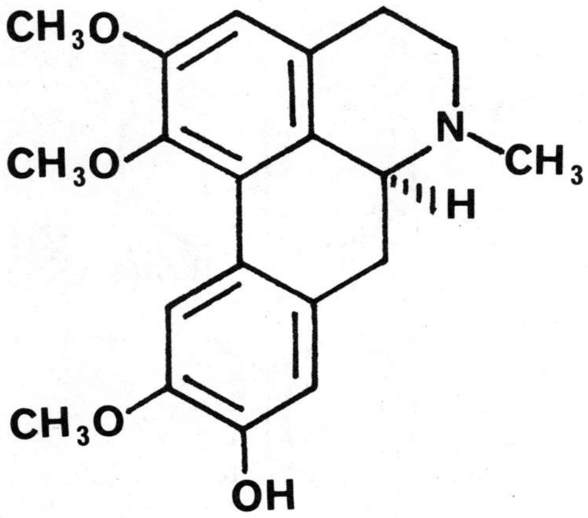

5

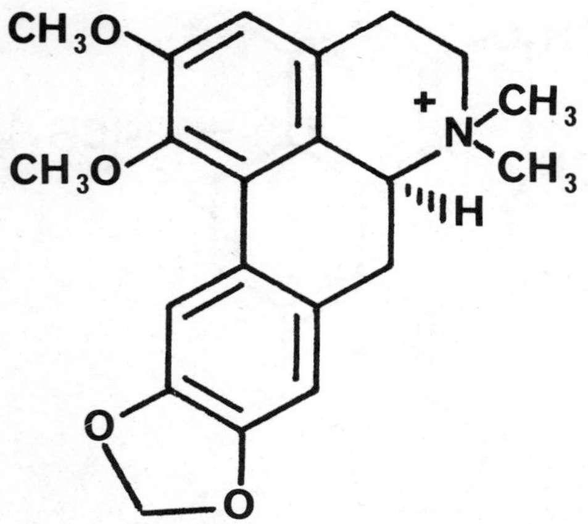

6

tetraoxygenated aporphine and showed maxima at 225 nm (log ϵ 4.24), 278 nm (sh) (log ϵ 3.65), 285 nm (log ϵ 3.79), 310 nm (log ϵ 3.94), and 320 nm (sh) (log ϵ 3.84) [78]. The ^{1}H NMR spectrum demonstrated the presence of a quaternary N-methyl function as two singlets at δ 3.15 and 3.72, two methoxy groups as singlets at δ 3.65 [C(1)] and 3.85 [C(2)], one methylenedioxy group as a broad singlet at δ 5.83, and three aromatic protons as singlets at δ 6.70, 6.95, and 7.70 [H(11)] [78]. The mass spectrum of the iodide salt showed the apparent molecular ion at m/z 353 (M^+-HI), 295, 251, 209, and 58 (100%) [67]. A direct comparison with ($\pm$)-nantenine methochloride prepared from a synthetic sample of nantenine confirmed the identification [78].

The presence of N-methylnantenine in *T. polygamum* suggests an important role in the possible biogenesis of thalphenine (**15**). It is possible that the methyleneoxy bridge of thalphenine could be formed via an intermediate oxonium ion at C(1) of nantenine or N-methylnantenine via the net loss of a hydride ion [78].

In 1978, N-methylnantenine iodide was isolated from *Thalictrum sachalinense* Lecoyer. [67]. Although the UV spectrum was similar to that reported by Shamma and Moniot [78], the ^{1}H NMR spectrum (CDCl$_3$) contained some significant differences. In particular, the chemical shifts for the quaternary methylamino groups were observed as two singlets at δ 3.02 (3H) and 3.05 (3H) whereas the mid- to higher-field aromatic protons were found as two singlets at δ 6.64 and 6.79. The latter signals were reported at δ 6.70 and 6.95 by Shamma and Moniot [78] but the remaining signals were consistent with those reported earlier by these authors [78]. The identity was confirmed via Hofmann degradation to thalicthuberine (2,3-methylenedioxy-5,6-dimethoxy-8-(2-dimethylaminoethyl)phenanthrene (**7**) [67].

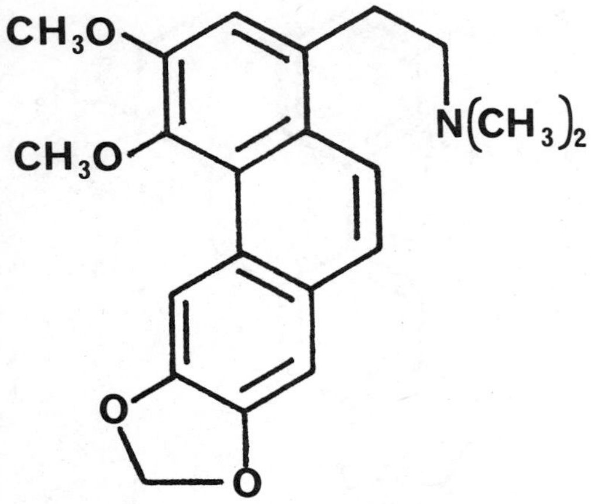

$$\underset{\sim}{\textbf{7}}$$

2.1.2.5. Thaliporphine (Thalicmidine, O-Methylisoboldine). Tabular review [15, alkaloid No. 44], UV [15,68,79], ^{1}H NMR [15,68,79,80], ^{13}C NMR [16,41,42], MS [15,81], CD [68].

Thalicmidine (**8**), $C_{20}H_{23}NO_4$ (341.1627), m.p. 170–172° (MeOH) [15]; $[\alpha]_D^{17}$ +44° (c 0.97, EtOH) [82]; $[\alpha]_D^{17}$ +41° (c 1.0, CHCl$_3$) [82]; $[\alpha]_D$ +110° (EtOH) [83], was first isolated from *Thalictrum minus* L. in 1950 [84] and shortly thereafter postulated to be 1,9,10-trimethoxy-2-hydroxyaporphine (**9**) on the basis of oxidation with KMnO$_4$ to *m*-hemipinic acid (**10**) and Hofmann degradation of *O*-methylthalicmidine methiodide to 2,3,5,6-tetramethoxy-8-vinylphenanthrene (**11**) [45].

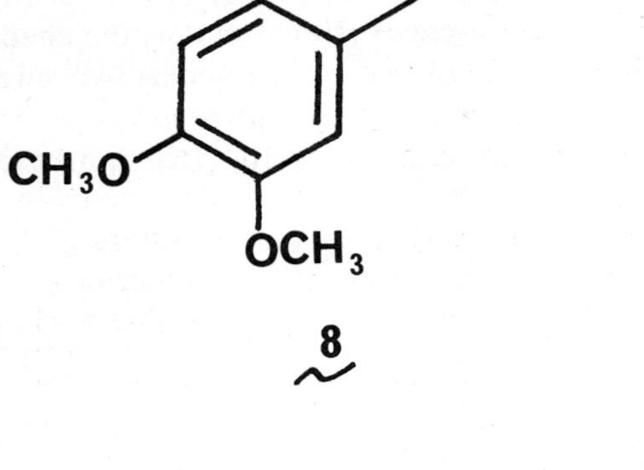

$$\underset{\sim}{\textbf{8}}$$

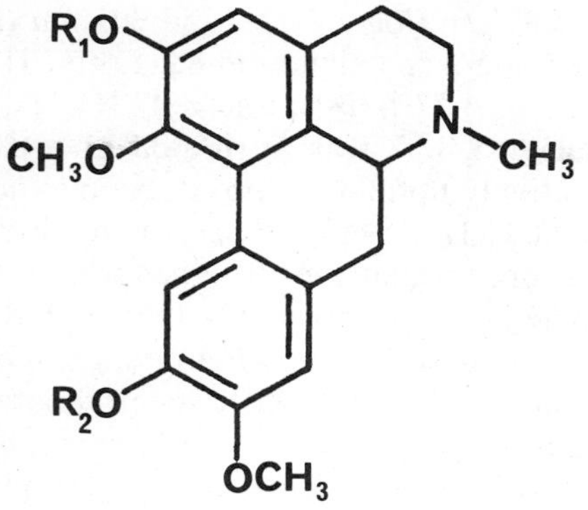

9 R₁=H, R₂=CH₃
12 R₁=CH₃, R₂=H

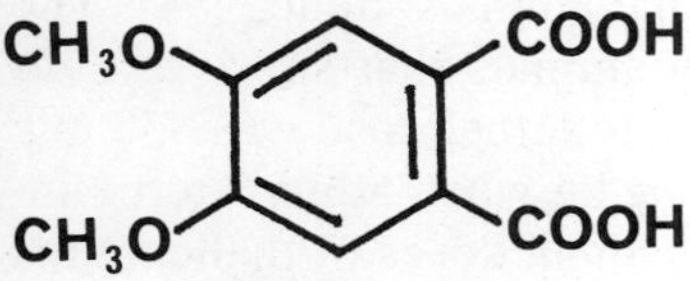

10

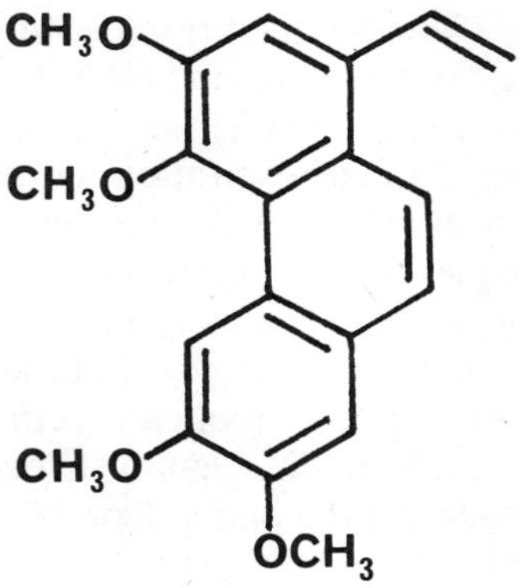

11

Some 10 years later a series of conflicting papers began to appear which postulated several different structures for this monophenolic, trimethoxylated 1,2,9,10-oxygenated aporphine. In 1962, 1,2,9-trimethoxy-10-hydroxyaporphine (12) [85] was postulated as the most probable structure for thalicmidine. Just 1 year later, another report appeared in agreement with the assignment of the phenolic group to C(10) [86]. Both these papers utilized indirect and comparative data which were, however, consistent for the period in arriving at a structural postulation. In 1967, further indirect evidence was cited for the assignment of the phenolic group to C(2) [87].

Thaliporphine (8) was isolated from *Thalictrum fendleri* in 1967 [62,79]. This dextrorotatory aporphine exhibited a UV spectrum with maxima at 220 nm (log ϵ 4.52), 280 nm (log ϵ 4.12), and 305 nm (log ϵ 4.12); the ^{1}H NMR spectrum showed important signals at δ 2.53 (NCH$_3$), 3.83 (OCH$_3$), 3.89 (2x OCH$_3$), 6.51 [H(3)], 6.77 [H(8)], and 8.07 [H(11)] [79]. The presence of a downfield signal at δ 8.07 was attributed to a C(11) proton whereas the absence of a relatively upfield methoxy group indicated the presence of the phenolic group at C(1) [15–17]. Because treatment of thaliporphine with diazomethane afforded glaucine (3), thaliporphine was assigned as 1-hydroxy-2,9,10-trimethoxyaporphine (8) [79] and this conclusion was confirmed by direct comparison with an authentic synthetic sample [88,89]. In 1968, a consideration of the ^{1}H NMR spectral data [80] and mass spectrum [81] of thalicmidine from *T. minus* L. led to an identical structural assignment with the phenolic group being placed at C(1) [80]. The mass spectrum showed the molecular ion and base peak at M$^+$ m/z 341 (100%), with other important fragment ions at m/z 340(92) (M-1), 326(36) (M-15), 310(28) (M-31), 298(50) (M-43), 283(11) (M-43-15), 267(57) (M-43-31), and 236(21) (M-43-31-31) [81]. Finally, the identity of thalicmidine with thaliporphine was unequivocally established in 1969 by direct comparison [UV, IR, ^{1}H NMR, TLC (six different solvent systems)] [83].

Thaliporphine was later synthesized via several different routes [90–93].

The administration of thalicmidine in subcutaneous doses of 100–350 mg/kg to white mice and rabbits produced motility disturbances, occasional head trembling, inapparent weakening of muscle strength in the lower extremeties, forced respiration, convulsions, and in some animals, apnea [94]. The duration of effects was about 45–70 min, with animals that survived subsequently reverting to normal status. The absolute lethal dose was 450 mg/kg when administered subcutaneously and 10 mg/kg when administered intravenously, with death following in 10–40 sec. Small doses of thalicmidine methiodide provoked a definite hypotensive response [94].

Thalicmidine has also been reisolated several times from *T. minus* L. [95,96] and isolated from *T. alpinum* L. [70] and *T. foetidum* L. [64,192].

2.1.2.6.   Thalicmidine N-Oxide (Thaliporphine N-Oxide).   Tabular review [15, alkaloid No. 45], UV [15,96], IR [15,96], ^{1}H NMR [15,96], MS [15,96].

Thalicmidine N-Oxide (13), C$_{20}$H$_{23}$NO$_5$ (357.1576), m.p. 192–193° dec., was first isolated from *Thalictrum minus* L. in 1972 [96]. The UV spectrum was similar to that of thalicmidine (8) and showed maxima at 227 nm (log ϵ 4.42), 282 nm (log ϵ 4.10), and 308 nm (log ϵ 4.07) [96]. The mass spectrum showed the parent ion at m/z 357 (6%), some 16 mass units higher than that of thalicmidine, with other fragment ions at m/z 341(100%) (M-16), 340(97) (M-17), 326(11) (M-31), and 298(69) (M-59) [96]. The ^{1}H NMR spectrum (TFA) indicated the presence of an oxygen-bearing methylamino function as a singlet at δ 3.08, three methoxy groups as a singlet at δ 3.55 (9H), and three aromatic protons as singlets at δ 6.42, 6.54, and 7.77 [96]. A consideration of these data and of the solubility properties and TLC behavior of the alkaloid

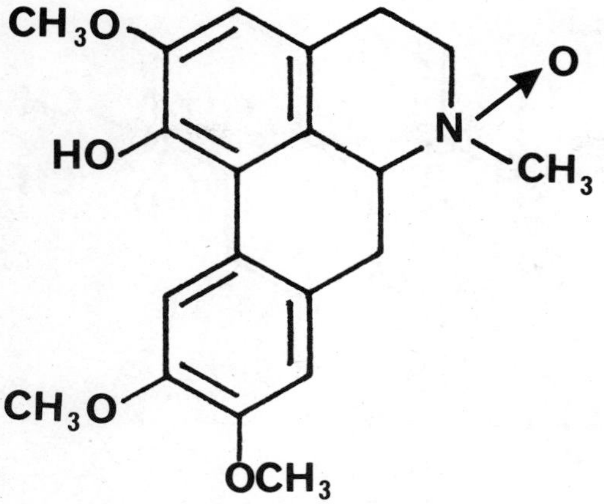

13

strongly suggested its identity as thalicmidine N-oxide (**13**). This was confirmed by reduction of the alkaloid with zinc–hydrochloric acid to thalicmidine (**8**) and by oxidation of thalicmidine with hydrogen peroxide to thalicmidine N-oxide (**13**).

Although it is customary to ascertain if newly isolated alkaloid N-oxides may be artifacts by subjecting solutions of the parent alkaloid to similar isolation techniques as were utilized (standing in solvents or solvent mixtures containing various column and/or TLC adsorbents), this was not done. Furthermore, neither the specific rotation of the isolated N-oxide nor that of its reduction product was reported, thus precluding the potential assignment of stereochemistry.

2.1.2.7. Xanthoplanine. Tabular review [15, alkaloid No. 56], UV [15,223], ^{1}H NMR [223], ^{13}C NMR [16,164], MS[223], ORD[209].
Xanthoplanine (**14**), $C_{21}H_{26}N^+O_4$ (356.1861), m.p. 218–220° dec. (chloride) [15], 190° (iodide) [223]; $[\alpha]_D^{26}$ +53° [c 0.28, MeOH (iodide)] [223], was first isolated from *Xanthoxylum planispium* (Rutaceae) in 1961 [15] and first found in a *Thalictrum* species, *T. foliolosum* DC. in 1983 [223].

2.1.3. 1,2,9,10-Tetraoxygenated-11-methylene Bridged Aporphines

2.1.3.1. Thalphenine. Tabular review [15, alkaloid No. 114], UV [15,97,98], ^{1}H NMR [15,97,98], MS [15,97,98].
Thalphenine (**15**), $C_{21}H_{22}N^+O_4$ (352.1549), was first isolated as its chloride salt, m.p. 185–186° (MeOH–acetone), 180–182° dec. (picrate); $[\alpha]_D$ +69° (c 1.3, EtOH), from *Thalictrum polygamum* Muhl. in 1972 [97]. The UV spectrum showed maxima at 221 nm (log ε 4.32), 230 nm (sh) (log ε 4.21), 280 nm (sh) (log ε 3.69), 288 nm (log ε 3.83), 317 nm (log ε 3.97), and 328

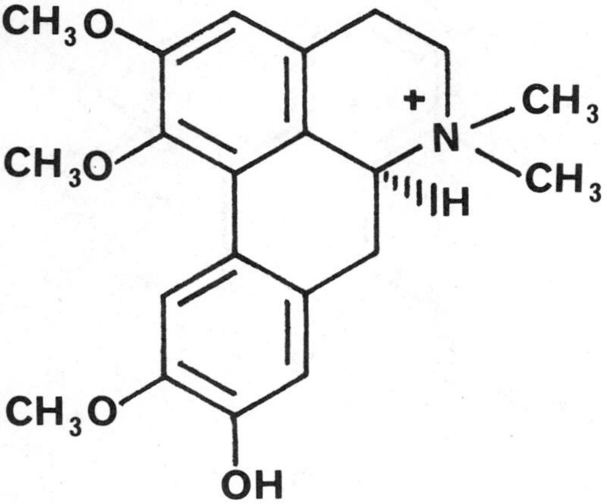

14

nm (sh) (log ϵ 3.87) and somewhat resembled other aporphines [15]. The ^{1}H NMR spectrum (DMSO) indicated the presence of a quaternary dimethyl-amino function as two singlets at δ 3.05 and 3.45, one methoxy group as a singlet at δ 3.76, one methylenedioxy group as a doublet at δ 6.02 (2H, J_{gem} = 2.5 Hz), two aromatic protons as singlets at δ 6.79 [H(10)] and 6.82 [H(3)] and an AB quartet at δ 5.00 (2H) (J_{gem} = 14 i.c.s. 28 Hz) indicative of an Ar–CH$_2$OAr function [97]. The absence of a signal at around δ 8.0 suggested the absence of a C(11) proton [15–17]. The mass spectrum showed intense fragment ions at m/z 351 (M-1), 293 [M-1-CH$_2$N$^+$(CH$_3$)$_2$], and 250 (293-CH$_3$—CO). Hofmann degradation of thalphenine (**15**) with hot methanolic

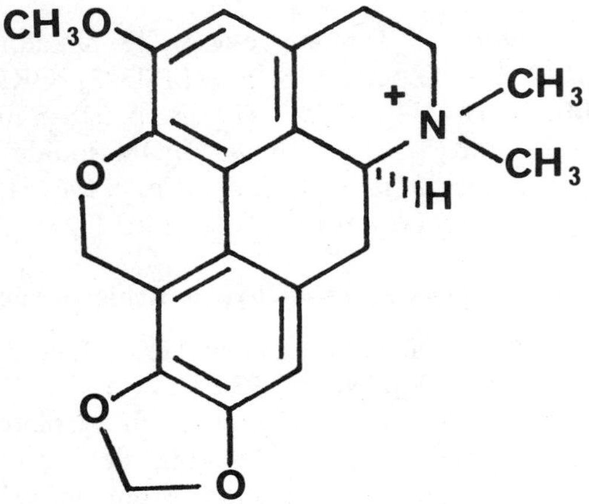

15

potassium hydroxide afforded the optically inactive thalphenine methine (16), which was subsequently isolated from the nonquaternary alkaloid fraction of the same plant [97]. The mass spectrum of thalphenine was, in fact, that of the Hofmann elimination product produced by thermal decomposition in the heated inlet of the spectrometer. A single-crystal X-ray diffraction analysis of thalphenine iodide, m.p. 198–199° (H₂O–acetone) confirmed its structure [97]. The absolute configuration of thalphenine was derived from its positive rotation and confirmed crystallographically by the anomalous dispersion method [97]. Thalphenine was subsequently synthesized via a procedure utilizing, as a key step, the photolysis of the appropriately sub-stituted bromophenolic benzyltetrahydroisoquinoline [99,100]. Thalphenine

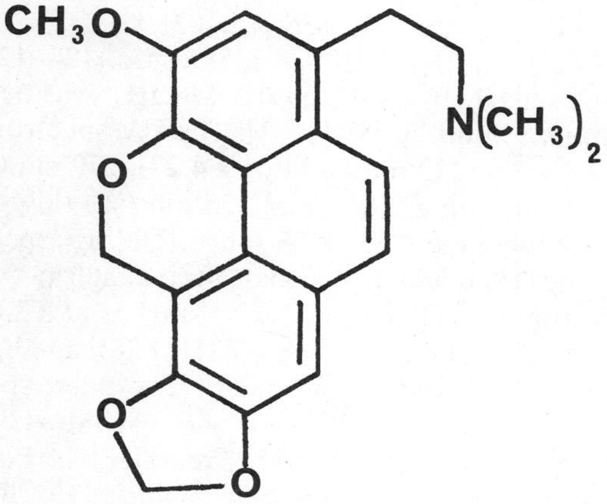

16

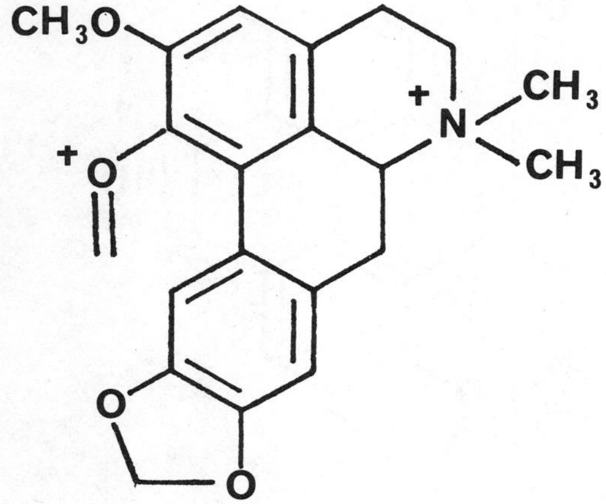

17

was the first example of an aporphine alkaloid with a methyleneoxy bridge and it was postulated that this bridge may be biosynthesized via the intermediacy of an oxonium ion (17) derived from nantenine or *N*-methylnantenine (6) via the net loss of a hydride ion [78].

Thalphenine was found to exhibit hypotensive properties in rabbits [101] and to possess weak in vitro antimicrobial activity against *Staphylococcus aureus* [98,101], *Mycobacterium smegmatis* [98,101], and *Candida albicans* [101].

Thalphenine has also been isolated from *Thalictrum revolutum* DC. [101], *T. rugosum* Ait. (*T. glaucum* Desf.) [98], and *T. minus* L. race B [102].

2.1.3.2. Bisnorthalphenine.   Tabular review [15, alkaloid No. 113], UV [15,103,104], ^{1}H NMR [15,103,104], MS [15,103].
Bisnorthalphenine (18), $C_{19}H_{17}NO_4$ (323.1158), m.p. 124–125° (Et_2O), 238° ($Et_2O–MeOH$)(HCl salt); $[\alpha]_D^{25}$ +81° (c 1.12, MeOH), was first isolated from *Thalictrum polygamum* Muhl. in 1974 [103]. The UV spectrum showed maxima at 220 nm (log ϵ 4.30), 232 nm (sh) (log ϵ 4.27), 279 nm (log ϵ 3.88), 289 nm (log ϵ 3.90), 316 nm (log ϵ 3.80), and 325 nm (sh) (log ϵ 3.74) and was similar to that of thalphenine (15) [103]. The ^{1}H NMR spectrum was also similar to thalphenine (15) but lacked signals for a quaternary dimethylamino function. The spectrum was characterized by singlets at δ 3.88 (3H, OCH_3), 6.00 (2H, CH_2O_2), 6.55 [1H, H(3)], and 6.63[1H, H(8)] and by an AB quartet at δ 5.14(2H, J = 14 Hz, $ArCH_2OAr$) [103]. The mass spectrum showed intense ions at M$^+$ *m/z* 323, 322 (M—H), and 294 (M—CH_2=NH), characteristic of a noraporphine [15–17,103]. Treatment of bisnorthalphenine with methyl iodide afforded (+)-thalphenine iodide (15). Thus bisnorthal-

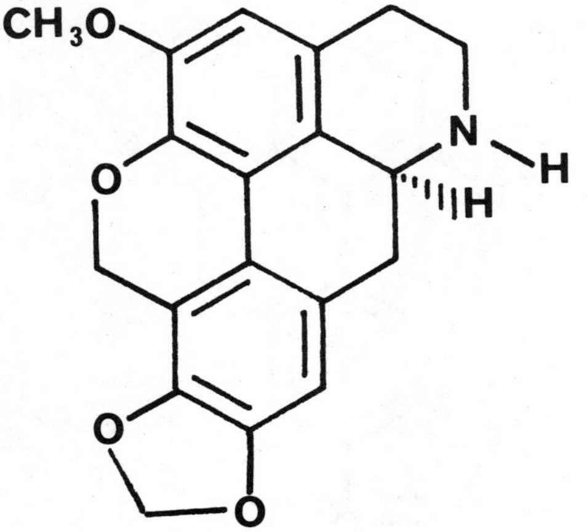

18

phenine (18) was the second aporphine characterized as containing a meth-yleneoxy bridge.

Bisnorthalphenine (18) was screened for antimicrobial activity using a standardized in vitro screen and found to possess some activity against *Staphylococcus aureus, Mycobacterium smegmatis,* and *Candida albicans* [68].

Bisnorthalphenine has been isolated from only one other *Thalictrum* species, *T. revolutum* DC. [104].

2.1.3.3. N-Demethylthalphenine. Tabular review [16, alkaloid No. 213], UV [16,100,104], ^{1}H NMR [16,100,104], MS [100].

N-Demethylthalphenine (19), $C_{20}H_{19}NO_4$ (337.1314), m.p. 179.5–180.5° (MeOH), $[\alpha]_D^{25}$ +104° (c 0.18, MeOH), was first isolated from *Thalictrum revolutum* DC. tops in 1977 [104]. The UV spectrum resembled that of thalphenine (15) and displayed maxima at 233 nm (sh) (log ϵ 4.13), 277 nm (sh) (log ϵ 3.61), 288 nm (log ϵ 3.76), 314 nm (log ϵ 3.87) and 325 nm (sh) (log ϵ 3.85) [104]. The ^{1}H NMR spectrum indicated the presence of one N-methyl group at δ 2.53 (s), one methoxy group at 3.89 (s), one ArOCH_2Ar function at δ 4.90 and 5.45 (q, J = 13.5 Hz), one methylenedioxy group at δ 5.94 and 5.98 (q, J = 1 Hz), and two aromatic protons at δ 6.53 (s) and 6.64 (s), and was also similar to that of thalphenine (15) [104]. The alkaloid was identified as N-dimethylthalphenine (18) [104] by comparison with a sample prepared previously as an intermediate in the total synthesis of thalphenine (15) [100].

N-Demethylthalphenine (19) was found to exhibit in vitro antimicrobial activity against *Staphylococcus aureus, Mycobacterium smegmatis,* and *Candida albicans* [68].

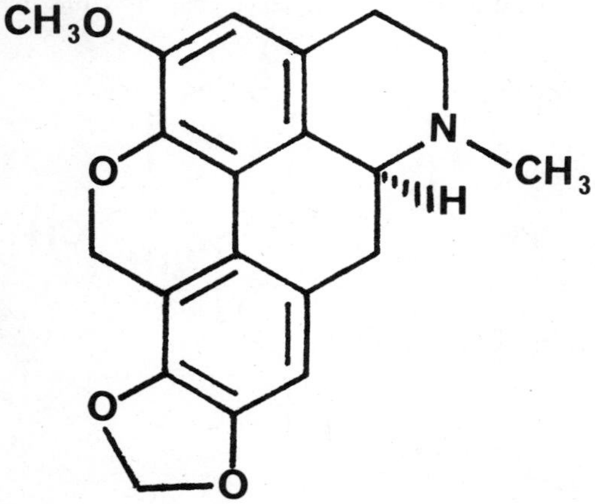

19

N-Dimethylthalphenine (**19**) has also been found in the fruits of *Thalictrum revolutum* DC. [105].

2.1.4. 1,2,10,11-Tetraoxygenated Aporphines

2.1.4.1.   Corydine.   Tabular review [15, alkaloid No. 74], UV [15,71,72], ^{1}H NMR [15,72], ^{13}C NMR [42], MS [15,43,44,72].
Corydine (**20**), $C_{20}H_{23}NO_4$ (341.1627), m.p. 148°, $[\alpha]_D^{25}$ +204° (c 0.6, EtOH) [15], was first isolated from *Corydalis tuberosa* (Papaveraceae) in 1902 [34]. The aporphine was first isolated from a *Thalictrum* species, *T. dioicum* L. (whole plant) in 1973 [106] and later reisolated from the same species (whole plant) in 1978 [72]. It has not been reported as a constituent of any other *Thalictrum* species to date.

Corydine exhibits irritant and central nervous system depressant properties, and its hydrochloride salt possesses some antitumor effect against sarcoma in hybrid mice [107]. Corydine hydrochloride produced respiratory stimulation and a brief hypotensive response in dogs in a dose of 0.25–3.0 mg/kg. In addition, the alkaloid also increased the convulsant effects of corazole when administered subcutaneously in a dose of 30–60 mg/kg in mice [108]. Corydine has an adrenolytic effect [109,110] and acts cholinergically, although weakly [46,111]. Corydine was evaluated as a potential antitussive but had little or no activity [112]. The methiodide and ethiodide salts caused a decrease in motility, dyspnea, and repiratory arrest in mice [46]. The LD$_{50}$ values for the methiodide and ethiodide were 2.81 and 3.7 mg/kg, respectively, after subcutaneous administration and 2.24 and 3.25 mg/kg after intravenous administration [46]. Muscle relaxation was observed in rabbits at dosages of 0.1–2.0 mg/kg for the methiodide and 0.25–7.0 mg/ kg for the ethiodide salt. Finally, both the methiodide and the ethiodide in doses of 1–4 mg/kg produced a hypotensive response in dogs [46]. Additional

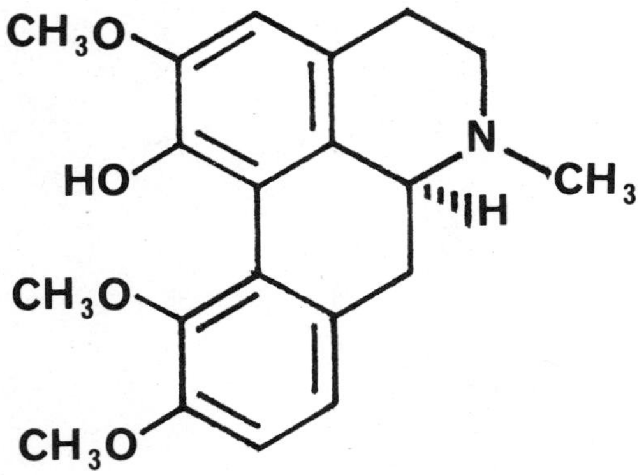

20

studies confirmed the hypotensive effect of corydine methiodide in dogs (0.5–1.0 mg/kg IV) and also demonstrated a blockage of nerve impulse transmission through the superior cervical ganglia of cats. Furthermore, large doses acted to block neuromuscular transmission in frogs and rabbits [113]. (+)-(S)-Corydine methiodide exhibited a slight increase in neuromuscular junction blocking potency in the cat tongue–hypoglossal nerve preparation as compared to its enantiomer [56].

2.1.4.2. Isocorydine. Tabular review [15, alkaloid No. 85], UV [15,71], ^{1}H NMR [15], ^{13}C NMR [16,41,42], MS [15,43], CD [17,73]. Isocorydine (**21**), $C_{20}H_{23}NO_4$ (341.1627), m.p. 185°, $[\alpha]_D^{33}$ +215° (c 1.0, CHCl$_3$) [15], was first isolated from *Corydalis lutea* (L.) DC. (Papaveraceae) in 1942 [34] and first reported in a *Thalictrum* species, *T. aquilegifolium* L., in 1970 [114]. Isocorydine was found to possess an adrenolytic effect [109,116] and to act cholinergically [46,111]. Isocorydine hydrochloride increased the convulsant effects of corazole when the alkaloid was administered subcutaneously in a dosage of 30–60 mg/kg in mice [108]. (+)-(S)-Isocorydine methiodide was found to be a slightly more potent neuromuscular blocker than its (R) enantiomer in the cat tongue–hypoglossal nerve preparation [56]. Isocorydine methochloride produced apnea and cardiac failure in various animals [115–117]. Isocorydine was evaluated for antitussive activity but found to be virtually inactive [112]. Likewise, isocorydine was established to produce only weak butyrylcholinesterase inhibition in vitro. This activity is postulated to be the result of hydrogen bonding between the hydroxyl group of isocorydine and some catalytically significant groups (such as the imidazole moiety of histidine) in the esteratic site of the enzyme [58].

(+)-(S)-Isocorydine has been isolated from only one other *Thalictrum* species, *T. urbaini* Hayata in 1977 [118].

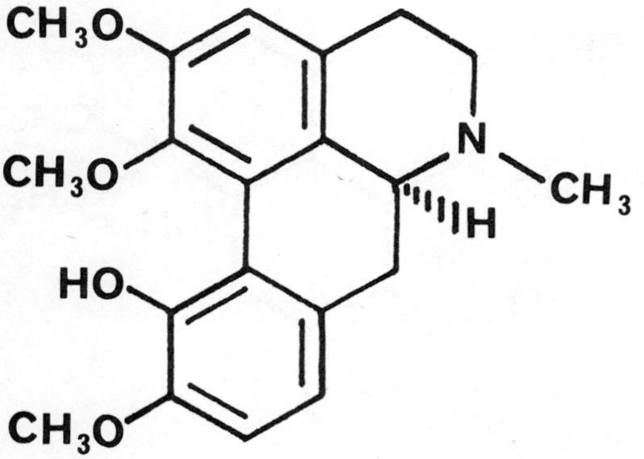

21

2.1.4.3. Magnoflorine (Thalictrine). Tabular review [15, alkaloid No. 72], UV [15,119,120], ¹H NMR [15,119,120], ¹³C NMR [16,164], MS [15,120], ORD [209], CD [17,73,120].

Magnoflorine (**22**), $C_{20}H_{24}N^+O_4$ (342.1705), m.p. 248–249° dec. (iodide) [121], $[\alpha]_D^{28.5}$ +204° (c 0.20, MeOH)(iodide) [121]; m.p. 257–258° dec. (perchlorate) [122], $[\alpha]_D^{33}$ +218° (c 0.78, MeOH) (perchlorate) [122]; m.p. 236–237° dec. (chloride) [121], $[\alpha]_D^{30.5}$ +215° (c 0.2, MeOH) (chloride) [121], was first isolated from *Thalictrum* species, *T. thunbergii* DC. in 1956 [123]. Although magnoflorine has been isolated numerous times from various *Thalictrum* species as well as many other genera from at least eight different plant families [15–17], there existed the possibility for confusion of magnoflorine with its positional isomer *N,N*-dimethyllindcarpine (**23**). This was resolved in 1980 when Stermitz et al. demonstrated that magnoflorine (**22**), prepared by selective *O*-demethylation of *N*-methylisocorydine (**24**), and *N,N*-dimethyllindcarpine (**23**), prepared by quaternization of *N*-methyllindcarpine (**25**), could be readily distinguished via their chromatographic and ¹H NMR spectrometric properties [119]. In addition, Hemingway et al. in 1981 showed the significance of pH in determining the ¹H NMR spectrum of magnoflorine by adding DCl and NaOD to DMSO or CD₃OD solutions of magnoflorine and in each case observing two different spectra [120]. Interestingly, the "acidic" spectrum in CD₃OD appears identical with that reported for *N,N*-dimethyllindcarpine [124].

Magnoflorine was characterized pharmacologically as producing less curare-like activity than roemerine (1,2-methylenedioxyaporphine) hydroxide and less hypotensive effects than *O*-methylisocorydine (1,2,10,11-tetramethoxyaporphine) methiodide [125,126].

Magnoflorine has been isolated from numerous *Thalictrum* species including *T. alpinum* L. [70], *T. baicalense* Turcz. [127], *T. calabricum* Spreng. (colorimetric determination—[128]), *T. dasycarpum* Fisch. and Lall [129],

22

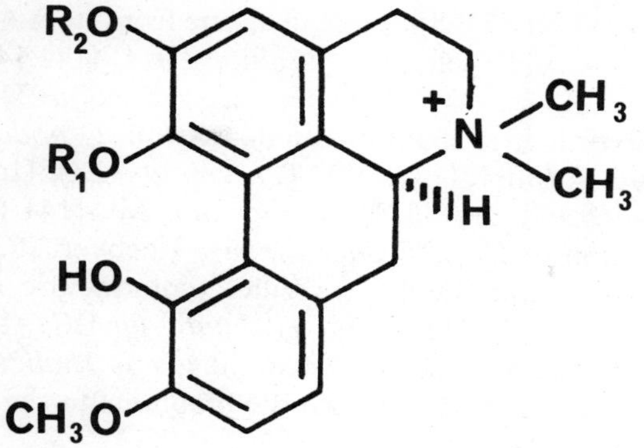

$$\underset{\sim}{22} \quad R_1 = H, \ R_2 = CH_3$$

$$\underset{\sim}{23} \quad R_1 = CH_3, \ R_2 = H$$

$$\underset{\sim}{24} \quad R_1 = R_2 = CH_3$$

T. dasycarpum Fisch. and Lall var. *hypoglaucum* (Rydb.) Boivin [122], *T. dioicum* L. [130], *T. fauriei* Hayata [131], *T. fendleri* Engelm ex Gray [62,132], *T. flavum* L. [133], *T. foetidum* L. [64,134,192], *T. foliolosum* DC. [38,135,223,347], *T. isopyroides* C.A.M. [134], *T. javanicum* B1. [136], *T. longipedunculatum* E. Nik. [134], *T. longistylum* DC. [137], *T. lucidum* TLC/ paper chromatographic detection—[138,139], *T. minus* L. [140,141,181] (col-

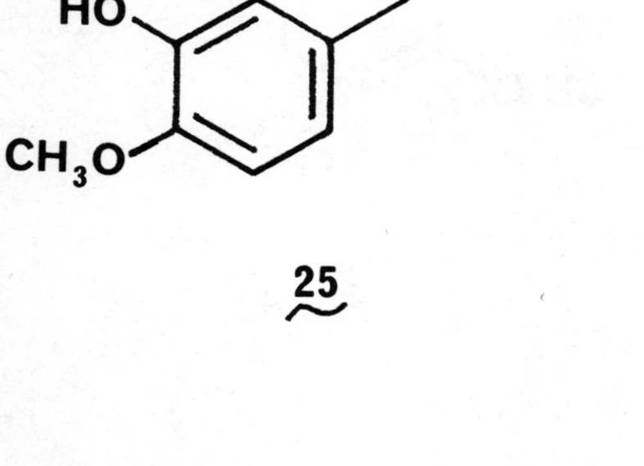

$$\underset{\sim}{25}$$

orimetric determination—[128]), (callous tissue from stem—[142]), *T. minus* race B [143], *T. minus* L. var. *adiantifolium* Hort. [144,145], *T. minus* L. var. *microphyllum* Boiss. [146], *T. minus* L. var. *minus* [147, 148], *T. petaloideum* L. (colorimetric determination—[128]), *T. podocarpum* Humb. [301], *T. polygamum* Muhl. [149,212], *T. revolutum* DC. [101,368], *T. rochebrunianum* Franc. and Sav. [121], *T. rugosum* Ait. [144,150–152] (colorimetric determination—[128]), *T. sachalinense* Lecoyer. [67], *T. simplex* L. [153,154], *T. squarrosum* Steph. ex Willd. (colorimetric determination— [128]), *T. strictum* Ledeb. [155], and *T. thunbergii* DC. [123,142,156,157]. No other alkaloid occurs so commonly in the genus *Thalictrum* as this quaternary aporphine. Indeed, it is probably ubiquitous to the genus.

2.1.5. 1,2,3,9,10-Pentaoxygenated Aporphines

2.1.5.1. Baicalidine. UV[61], IR[61], ^{1}H NMR[61], MS[61]. Baicalidine (**26**), $C_{21}H_{23}NO_5$ (369.1576), m.p. 146–147° (acetone), $[\alpha]_D$ +55° (MeOH), was isolated from *Thalictrum baicalense* Turcz. in 1982 [61]. The UV spectrum was characteristic of a 1,2,9,10- or a 1,2,3,9,10-oxygenated aporphine and showed maxima at 217, 242, 291, 306, and 318 nm; the mass spectrum supported this supposition with fragment ions at M$^+$ *m/z* 369, 368 (M—H), 354 (M—CH$_3$), 326 (M—[CH$_2$=NCH$_3$]), and 311 (M—[CH$_2$=NCH$_3$]—CH$_3$) [61]. The ^{1}H NMR spectrum indicated the presence of one *N*-methyl group at δ 2.44 (3H, s), three methoxy groups at δ 3.76 (3H, s) and 3.81 (6H, s), one methylenedioxy group at δ 5.89 (1H, d) and 5.9 (1H, d) (this latter signal misprinted as 5.4 in ref. 61), and two

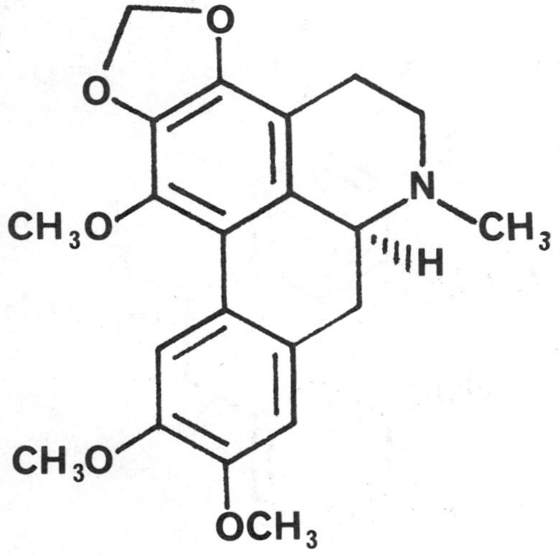

26

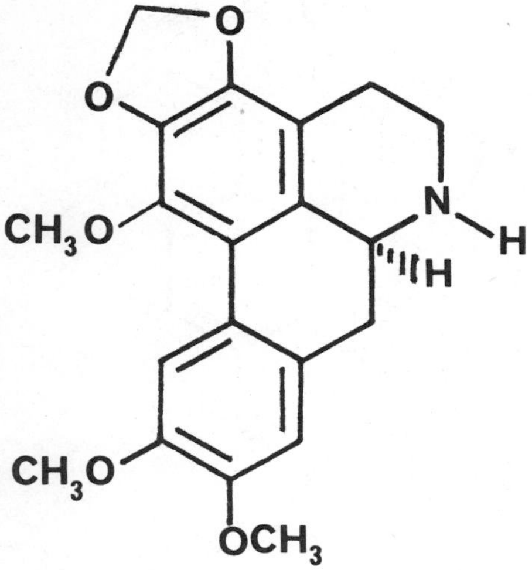

$$\underset{\displaystyle \sim}{27}$$

aromatic protons at δ 6.64 (1H, s) [H(8)] and 7.71 (1H, s) [H(11)] [61]. A direct comparison (m.p., TLC, IR) with an authentic sample of N-methyl-baicaline (26) prepared via Hess methylation of baicaline (27) [127] confirmed the identification [61]. Interestingly, the ^{1}H NMR spectrum of N-methyl-baicaline (26) [prepared from baicaline (27)] differed somewhat from that of the natural baicalidine (N-methylbaicaline). In particular, the chemical shifts for the methoxy groups of the former were observed as a singlet (9H) at δ 3.84 [127], not as two distinct singlets at δ 3.76 (3H) and 3.81 (6H), as reported for the latter [61].

 2.1.5.2. *Baicaline*. Tabular review [17, alkaloid No.296], UV [17,127], ^{1}H NMR [127], MS [127].
Baicaline (27), $C_{20}H_{21}NO_5$ (355.1420), m.p. 169–170° (Et$_2$O), [α]$_D$ +48° (MeOH), was isolated from *Thalictrum baicalense* Turcz. in 1981 [127]. The UV spectrum was like those of other 1,2,9,10- or 1,2,3,9,10-oxygenated apor-phines and showed maxima of 220, 246, 287, 303, and 315 nm [127]. The ^{1}H NMR spectrum suggested the presence of three methoxy groups as a singlet at δ 3.91 (9H), one methylenedioxy group as a broad singlet at δ 6.03 (2H), and two aromatic protons as singlets at δ 6.82 [H(8)] and 7.94 [H(11)] [127]. The mass spectrum displayed the parent ion at M$^+$ m/z 355 with other diagnostic fragment ions at m/z 354 (M—H), 340 (M—CH$_3$), and 326 (M—[CH$_2$=NH]) [127]. These spectral data were characteristic of a 1,2,3,9,10-pentaoxygenated noraporphine [15–17]. Treatment of baicaline with formaldehyde–formic acid afforded N-methylbaicaline (26) and ace-

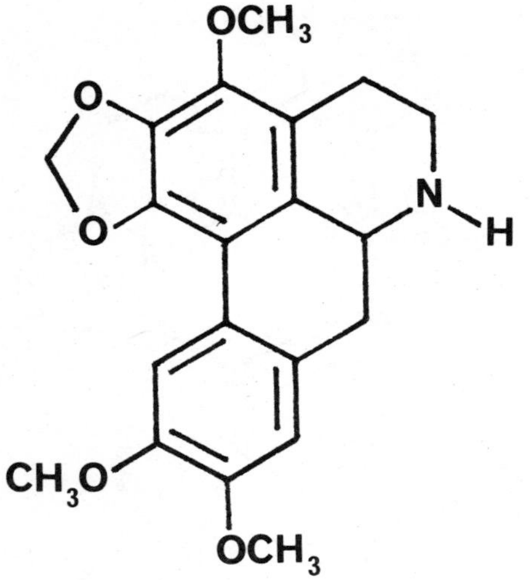

$$\underset{\sim}{\textbf{28}}$$

tylation with acetic anhydride–pyridine produced a *N*-acetyl derivative, with the amide function being observed as a singlet at δ 2.12 in the ¹H NMR spectrum. The presence of the methylenedioxy function as two closely separated one-proton doublets at about δ 5.9 in *N*-methyl- and *N*-acetylbaicaline and the two characteristic singlet aromatic protons, one of which was down-

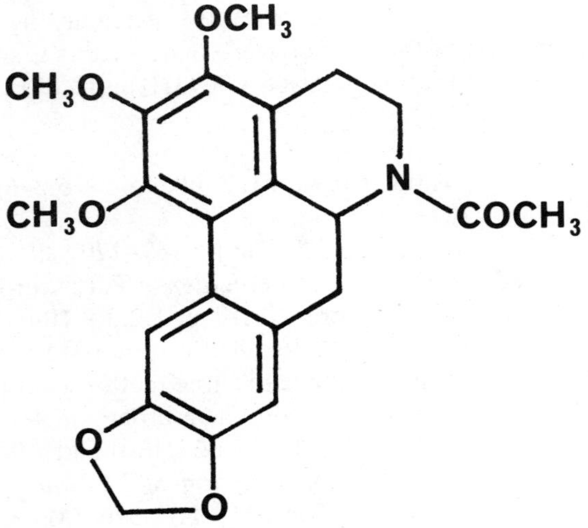

$$\underset{\sim}{\textbf{29}}$$

field (δ 7.7–7.9), suggested that baicaline was 1,9,10-trimethoxy-2,3-methylenedioxynoraporhine (**27**). This was confirmed by a direct comparison of northalicmine (1,2-methylenedioxy-3,9,10-trimethoxynoraporphine) (**28**) with baicaline, which showed them to be different, as did the same comparison between *N*-acetyl-1,2,3-trimethoxy-9,10-methylenedioxynoraporphine (**29**) and *N*-acetylbaicaline. Therefore, baicaline was identified as 1,9,10-trimethoxy-2,3-methylenedioxynoraporphine (**27**) and assigned the (*S*) configuration owing to a strong positive Cotton effect at 240 nm in its CD spectrum [127].

2.1.5.3.   O-Methylcassyfiline (Northalicmine, O-Methylcassythine, Hexahydrothalicminine).   Tabular review [15, alkaloid No. 108], UV [15], ^{1}H NMR [15].

O-Methylcassyfiline (**30**), $C_{20}H_{21}NO_5$ (355.1420), m.p. 150–152°, $[\alpha]_D^{15}$ +16.4° (c 1.0, CHCl$_3$) [15], was first isolated from *Cassytha americana* (*C. filiformis* L.) (Lauraceae) in 1968 [158], although traces of the alkaloid were detected in *C. filiformis* in 1966 [159].

This noraporphine was isolated from *Thalictrum strictum* Ledeb. in 1975 and identified by conversion to thalicmine (**31**) [160]. To date, *O*-methylcassyfiline has not been isolated from any other *Thalictrum* species except *T. strictum* [160,161].

2.1.5.4.   Octeine (Thalicmine, N,O-Dimethylcassyfiline).   Tabular Review [15, alkaloid, No. 109], UV [15,62,162], IR [162,163], ^{1}H NMR [15,62,80], ^{13}C NMR [164], MS [62,81], ORD [165], CD [73].

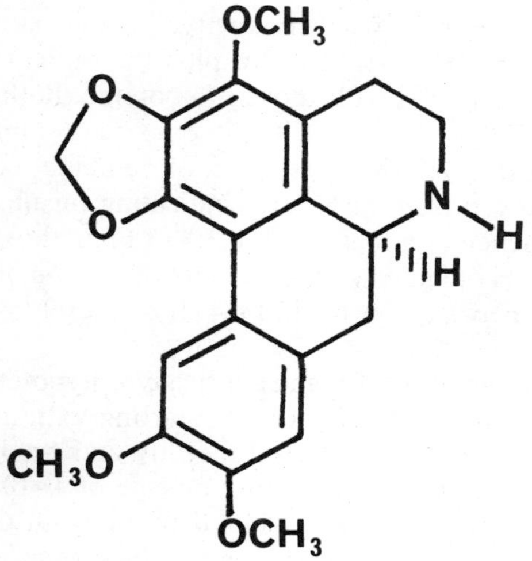

30

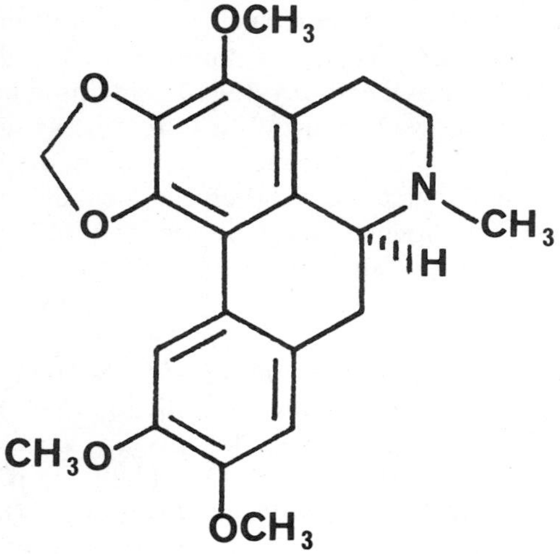

31

Ocoteine (**31**), $C_{21}H_{23}NO_5$ (369.1576), m.p. 140–142° [15], $[\alpha]_D^{25}$ +36.2° (c 0.92, EtOH) was first isolated from *Ocotea puberula* Nees (Lauraceae) in 1951 [166]. The alkaloid was also isolated from *Thalictrum minus* L. about the same time and was named thalicmine [84]. Thalicmine was first assigned as 1,2-methylenedioxy-3,10,11-trimethoxyaporphine (**32**) on the basis of degradative reactions, including the Hofmann degradation, which appeared to afford 3,4,7-trimethoxy-5,6-methylenedioxy-8-vinylphenanthrene (**33**) [45]. However, synthesis of racemic **32** afforded a product with distinct differences in the UV spectrum when compared to the natural product [167]. Because the UV and ^{1}H NMR spectra plus the optical rotation and ORD values of thalicmine agreed with those of ocoteine, both alkaloids were assigned as 1,2-methylenedioxy-3,9,10-trimethoxyaporphine (**31**) [168]. The salient features of the ^{1}H NMR spectrum of ocoteine were the presence of a low-field aromatic proton at δ 7.57, indicating unsubstitution at C(1) or C(11), and the absence of a high-field (δ 3.4–3.6) methoxy group, indicating the absence of a C(1) methoxy function [169–171]. Finally, an unequivocal synthesis of ocoteine was reported in 1964 [168], thereby settling its structure as **31**.

Thalicmine was noted to possess antitussive, hypotensive, spasmolytic, and adrenolytic effects in dogs and cats starting with a dose of 0.5 mg/kg [172]. Thalicmine ethiodide decreased motility and produced muscle relaxation when administered to rabbits in a dosage of 1–10 mg/kg [46]. In addition, thalicmine also reduced the blood pressure in dogs when given in doses of 0.1–2.0 mg/kg without substantial changes in respiration [46]. Intravenous administration of thalicmine to dogs at a dosage of 0.5 mg/kg produced antitussive effects for 30–45 min; at 1 mg/kg, the alkaloid inhibited

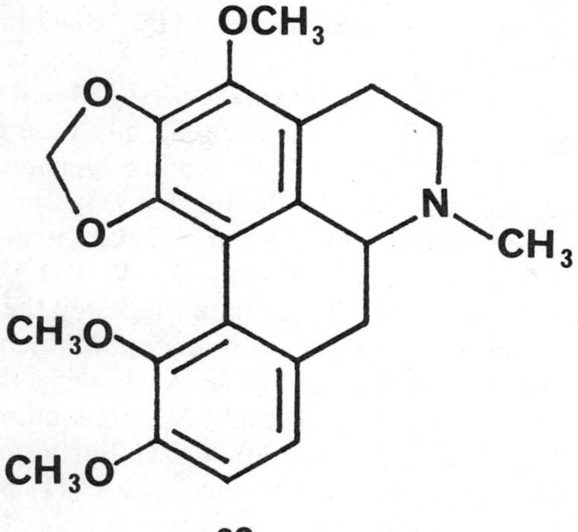

32

coughing for 2 hr and lowered arterial pressure by 20–40 mm Hg [112]. The antitussive effect of thalicmine was estimated to exceed that of codeine by 1.5–2 times [112]. Thalicmine has also been noted to produce a hypergly-cemia in rats by increasing the blood sugar level about 30% [173]. Finally, ocoteine was established to be an inhibitor of the growth of the microor-ganism *Scenedesmus obliquus* [77].

Ocoteine has also been isolated from *Thalictrum fendleri* Engelm. ex Gray [62], *T. isopyroides* C.A.M. [174], *T. simplex* L. [153], and *T. strictum* Ledeb. [155,160,161], as well as being reisolated from *T. minus* L. [95,96,181]. Finally, a quantitative HPLC determination of ocoteine in *T. minus* L. has also been reported [68].

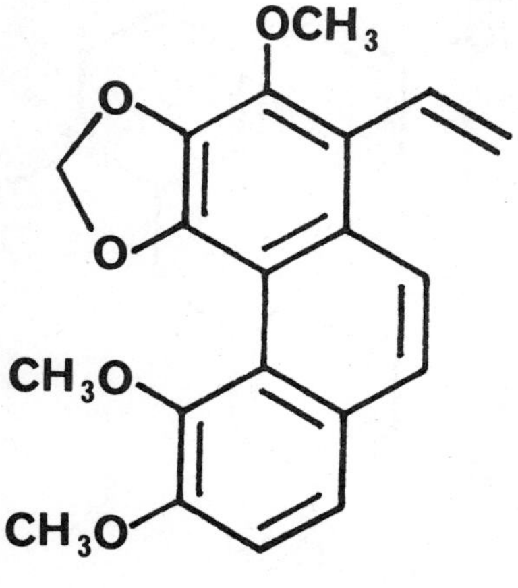

33

2.1.5.5. *Preocoteine*. Tabular review [15, alkaloid No. 96], UV [15,62,79,155], ¹H NMR [15,62,79,155], MS [15,79,155].

Preocoteine (34), $C_{21}H_{25}NO_5$ (371.1733), was first isolated from *Thalictrum fendleri* Engelm. ex Gray in 1967 as a phenolic oily base [62,79]. The UV spectrum was characteristic of a 1,2,3,9,10-pentaoxygenated aporphine [168] with maxima at 278 nm (log ϵ 4.16), 302 nm (log ϵ 4.22) and 312 nm (log ϵ 4.16) [79]. The mass spectrum showed the molecular ion at m/z 371 [79,155] with other fragment ions at m/z 370 (M–H), 356 (M–CH₃), 340 (M–OCH₃), and 328 (M–[CH₂=NCH₃]) [155]. These data suggested the presence of four methoxy groups and one hydroxy group on the aporphine ring [79]. The ¹H NMR spectrum supported this assumption by showing the four methoxy groups as two singlets at δ 3.88 (6H) and 3.92, thus eliminating positions C(1) and C(11) as sites of methoxylation [169–171]. Furthermore, a downfield signal at δ 7.99 (1H) suggested the presence of a C(11) proton [169–171]; the only other aromatic proton in the spectrum [δ 6.74 (1H, s)] was characteristic of a C(8) proton in this system [168]. Because naturally occurring aporphines are substituted at C(1), the phenolic group was placed at that position and preocoteine assigned as 1-hydroxy-2,3,9,10-tetramethoxyaporphine (34) [79]. Methylation of preocoteine produced *O*-methylpreocoteine (35), whose UV spectrum was clearly different from an authentic sample of 1,2,8,9,10-tetramethoxyaporphine (36), thus precluding the possibility that ocoteine was 1-hydroxy-2,8,9,10-tetramethoxyaporphine (37) [79].

Preocoteine has also been isolated from *Thalictrum strictum* Ledeb. [155] and identified by conversion to *O*-methylpreocoteine (thalicsimidine) (35). However, note that the ¹H NMR methoxy group resonances of the preocoteine from *T. strictum* were reported as three singlets at δ 3.80, 3.85, and

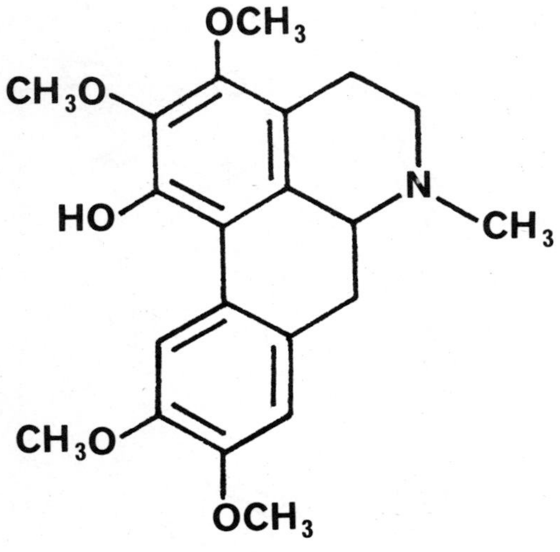

34

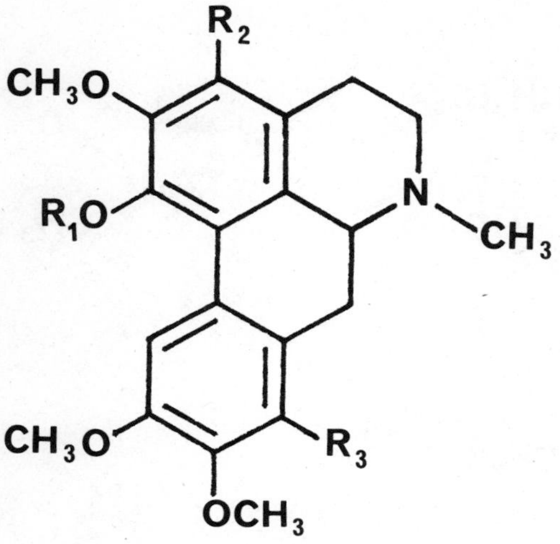

$$\underset{\sim}{\textbf{35}} \quad R_1 = CH_3, R_2 = OCH_3, R_3 = H$$

$$\underset{\sim}{\textbf{36}} \quad R_1 = CH_3, R_2 = H, R_3 = OCH_3$$

$$\underset{\sim}{\textbf{37}} \quad R_1 = R_2 = H, R_3 = OCH_3$$

3.89 (integration not detailed) [155], which is clearly different from the values reported for the preocoteine of *T. fendleri* [79]. The aromatic protons did, however, possess very close chemical shifts [δ 6.76 (1H, s) and 7.91 (1H, s)] to those of the *T. fendleri* preocoteine [79].

2.1.5.6. *Preocoteine N-Oxide.* Tabular review [15, alkaloid No. 97], UV [15,96], IR [15,96], ^{1}H NMR [15,96], MS [15,96].
Preocoteine *N*-oxide (**38**), $C_{21}H_{25}NO_6$ (387.1682), m.p. 199–200° dec., was first isolated from *Thalictrum minus* L. in 1972 [96]. The UV spectrum was characteristic of a 1,2,3,9,10-pentaoxygenated aporphine [62,79,162] with maxima at 226 nm (log ε 4.52), 282 nm (log ε 4.01), and 306 nm (log ε 4.11) [96]. The mass spectrum showed a weak molecular ion at *m/z* 387 with the loss of 16 a.m.u. to the base peak at *m/z* 371 (100%) and other fragment ions at *m/z* 370 (59%) (M—O—H), 356 (17) (M—O—CH$_3$), and 328 (43) (M—O—[CH$_2$=NCH$_3$]) characteristic of an aporphine [43]. The ^{1}H NMR spectrum (TFA) indicated the presence of one *N*-oxide methyl group at δ 3.10, four methoxy groups at δ 3.55 (3H, s) and 3.73 (9H, s), and two aromatic protons at δ 6.56 [1H, s, H(8)], and 7.74 [1H, s, H(11)] and was similar to that obtained for thalicmidine *N*-oxide (**13**) [96], except for the presence of

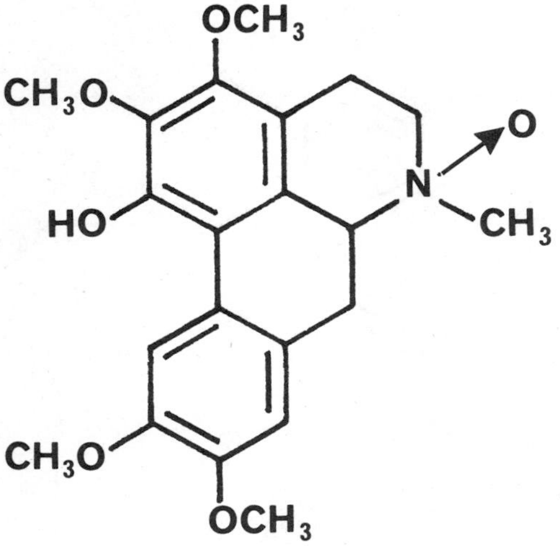

38

an additional methoxy group. Finally, structural confirmation was accomplished via reduction (Zn/HCl) of preocoteine *N*-oxide (**38**) to preocoteine (**34**) [96].

2.1.5.7. *Thalicsimidine (O-Methylpreocoteine, Purpureine).*

Tabular review [15, alkaloid No. 100], UV [15,153,154,176], ^{1}H NMR [15,80,176,177], MS [15,81,177].

Thalicsimidine (**35**), $C_{22}H_{27}NO_5$ (385.1889), m.p. 131–132° (EtOH) [154]; $[\alpha]_D$ +20.26° (c 1.38, CHCl$_3$) [154]; $[\alpha]_D$ +57.85° (c 0.96, EtOH) [154], was first isolated from *Thalictrum simplex* L. in 1967 [154]. The hydrobromide (m.p. 250–252° dec.), sulfate (m.p. 198–202° dec.), and picrate (m.p. 141–150°) salts were prepared and the UV spectrum with maxima at 220, 280, and 300 nm suggested the aporphine nature of the alkaloid [154]. One year later, the ^{1}H NMR and mass spectra of the alkaloid were reported [177]. The ^{1}H NMR spectrum demonstrated the presence of one *N*-methyl group at δ 2.47 (3H, s); five methoxy groups as four singlets at δ 3.64 (3H) [C(1)], 3.82 (3H) [C(2)], 3.85 (6H) [C(9) and C(10)], and 3.88 (3H) [[C(3)]; and two aromatic protons as singlets at δ 6.70 [H(8)] and 7.89 [H(11)] [177]. The mass spectrum showed the molecular ion at M$^+$ *m/z* 385 and other important fragment ions at *m/z* 384 (M—H), 370 (M—CH$_3$), 354 (M—OCH$_3$), 342 (M—[CH$_2$=NCH$_3$]), 327 (M—[CH$_2$=NCH$_3$]–CH$_3$), 311 (M—[CH$_2$=NCH$_3$]–OCH$_3$), and 280 (M—[CH$_2$=NCH$_3$]–2OCH$_3$) [177]. These data and the low specific rotation indicated that thalicsimidine was probably 1,2,3,9,10-pentamethoxyaporphine (**35**). Of particular importance in the ^{1}H NMR spectrum was the high-field methoxy singlet at δ 3.64 [C(1)]

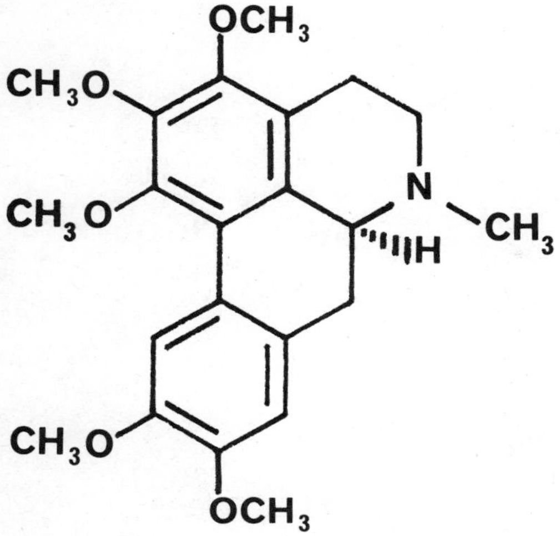

OCH₃ ... (chemical structure **35**)

and the low-field aromatic proton singlet at δ 7.89 [C(11)] [177]. A total synthesis of racemic thalicsimidine was reported in 1969 [178] and of (+)-(S)- thalicsimidine (35) in 1971 [176], confirming the earlier structural assignment [177]. In addition, two other reports concerning the synthesis of thalicsimidine were published in 1971 [179,180].

Thalicsimidine has also been isolated from *Thalictrum filamentosum* Maxim. [63], *T. minus* L. [181], and *T. strictum* Ledeb. [155], as well as being reisolated from *T. simplex* L. [153].

2.1.5.8. Thalisopynine (Thalisopinine). Tabular review [17, alkaloid No. 295], UV [175,176], ¹H NMR [175,176], MS [175].

Thalisopynine (**39**), $C_{21}H_{25}NO_5$ (371.1733) oil [175], m.p. 186–187° (MeOH) [176]; $[\alpha]_D$ +45° (c 0.13, MeOH) [175]; $[\alpha]_D$ +108° (c 0.17, MeOH) [176], was first isolated from *Thalictrum isopyroides* C.A.M. in 1978 [175]. The UV spectrum was characteristic of a 1,2,3,9,10-pentaoxygenated phenolic aporphine [168] and showed maxima at 282, 305, and 316 nm (sh) with a bathochromic and hyperchromic shift on addition of methanolic KOH [176], characteristic of phenolic function at C(9) [71]. The mass spectrum showed a molecular ion at m/z 371 and other characteristic fragmentation ions at m/z 370 (M—H), 356 (M—CH₃), 340 (M—OCH₃), and 328 (M—[CH₂=NCH₃]), and was typical of a monophenolic, tetramethoxylated aporphine [15–17,175]. The ¹H NMR spectrum indicated the presence of one N-methyl group at 2.49 (3H, s); four methoxy groups at δ 3.65 [3H, s, C(1)], 3.82 (3H, s), 3.85 (3H, s), and 3.90 (3H, s); and two aromatic protons at δ 6.73 [1H, s, H(8)] and 7.85 [1H, s, H(11)] [175]. The high-field methoxy group (δ 3.65) was characteristic of a C(1) methoxy whereas the low-field

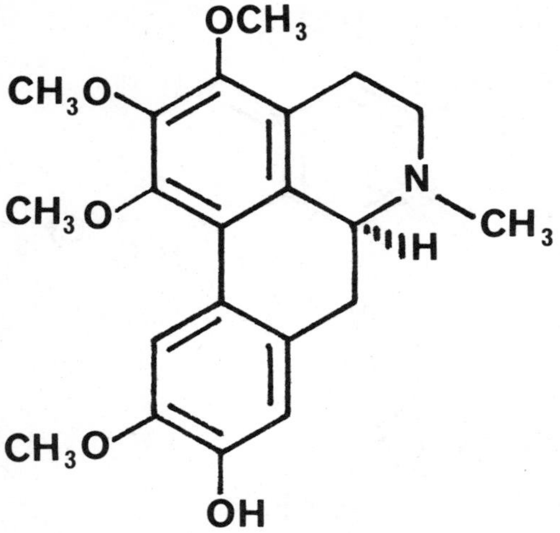

$$\underset{\sim}{\mathbf{39}}$$

aromatic proton (δ 7.85) was diagnostic of a H(11) proton [15–17,175]. Methylation of thalisopynine with diazomethane afforded thalicsimidine (**35**), confirming the monophenolic nature and oxygenation pattern of thalisopynine [175]. Acetylation of thalisopynine in the usual fashion afforded an *O*-acetyl ester. The ^{1}H NMR spectrum of this ester showed a paramagnetic shift of the H(8) and H(11) aromatic proton signals by 0.13 and 0.17 ppm, respectively, characteristic of a C(9) or C(10) phenolic group in disubstituted C(9,10) aporphine ring D [159]. The placement of the phenolic group at C(9) was established via a consideration of the NOE effect between the C(10) aromatic methoxy group and the H(11) proton [175].

2.1.6. 1,2,3,10,11-Pentaoxygenated Aporphines

   2.1.6.1. Oconovine.   Tabular   review   [15,   alkaloid   No.   102],   UV [15,118,182], ^{1}H NMR [15,118,182], MS [118].
Oconovine (**40**), $C_{21}H_{25}NO_5$ (371.1733), amorphous, $[\alpha]_D^{27}$ +156° (c 0.2, CHCl$_3$), was first isolated from *Ocotea* species in 1968 [182]. In 1976, oconovine was isolated from *Thalictrum urbaini* Hayata [118] and has not been reported in any other *Thalictrum* species to date.

2.2. Dehydroaporphines

The following reviews and references are particularly relevant to the isolation and identification of dehydroaporphine alkaloids:

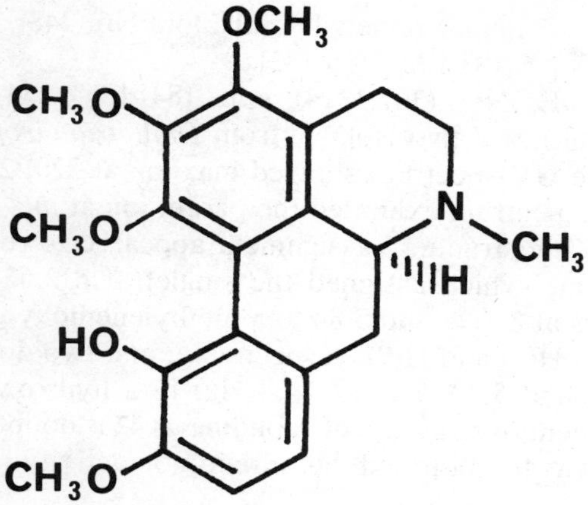

40

1. Tabular reviews [15–17]
2. Annual reviews [22–31]
3. Other reviews [13]
4. Book chapters [36,37,241,242]

The numbering of the dehydroaporphine (41) ring is according to accepted practice [15].

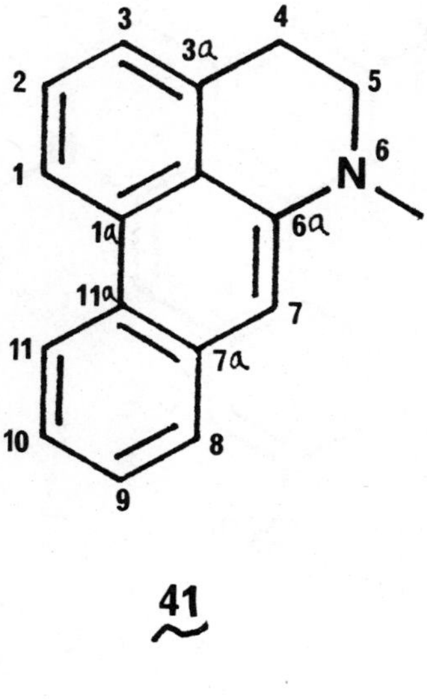

41

2.1.1.   Cabudine.   Tabular review [16, alkaloid No. 248], UV [16,183], IR [16,183], ^{1}H NMR [16,183], MS [16,183].

Cabudine (**42**), $C_{20}H_{19}NO_4$ (337.1314), m.p. 184–185°, 254–255° (HCl salt), 194–196° (CH_3I salt), was first isolated from *Thalictrum isopyroides* C.A.M. in 1975 [183]. The UV spectrum showed maxima at 220, 280, 291, and 320 nm and the mass spectrum exhibited the parent ion at m/z 337 and the base peak at m/z 336. The structural assignment appeared to rely greatly on the ^{1}H NMR spectrum, which assigned the singlet at δ 2.53 to an *N*-methyl function; doublets at δ 5.78 and 5.84 to a methylenedioxy group; singlets at δ 6.42 and 6.52 to H(7) and H(8), respectively; and two doublets at δ 4.65–4.88 (*J* = 8 Hz) and 5.20–5.42 (*J* = 8 Hz) to a hydroxymethyl group at C(3). The assignment of structure of cabudine as **42** is doubtful at the present time. Cabudine was found to exhibit adrenolytic activity [184].

2.2.2.   Dehydrothalicmine (Dehydroocoteine).   Tabular review [15, alkaloid No. 159], UV [15,185], IR [15,185], ^{1}H NMR [15,185], MS [15,185].

Dehydrothalicmine (**43**), $C_{21}H_{21}NO_5$ (367.1420), m.p. 190–191° (CHCl$_3$–EtOH), was first isolated from *Thalictrum isopyroides* C.A.M. in 1971 and was the first dehydroaprophine found in *Thalictrum* spp. [185]. The UV spectrum showed maxima at 267 nm (log ε 4.60) and 337 nm (log ε 3.93); the IR spectrum showed strong absorption in the 1590–1640 cm^{-1} region [185]. These data were consistent for a dehydroaporphine [15–17]. The ^{1}H NMR spectrum exhibited signals for one downfield *N*-methyl group [δ 2.96 (3H, s)], three methoxy groups [δ 3.95 (6H, s) and 4.05 (3H, s)], one methyl-

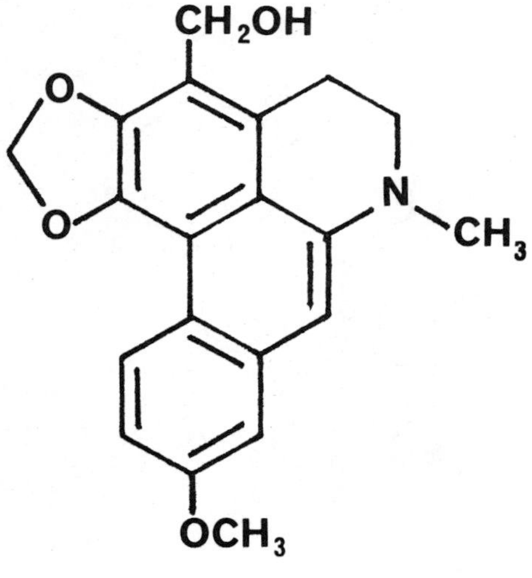

42

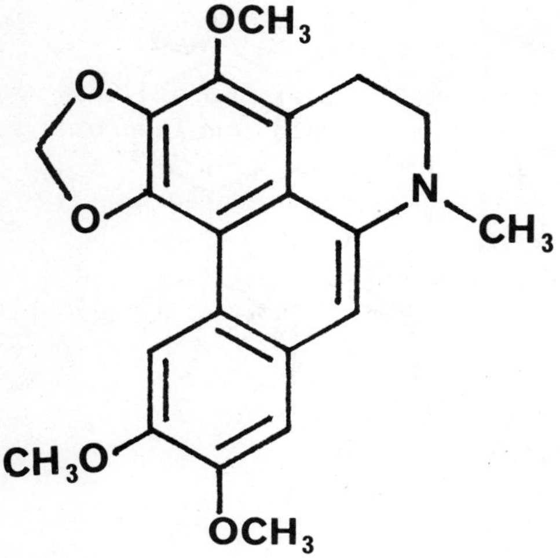

43

enedioxy group [δ 6.10 (2H, s)], and three singlet aromatic protons [δ 6.52 (1H), 6.98 (1H), 8.18 (1H)], one of which was downfield at δ 8.18 [C(11)], thus suggesting 1,2,3,9,10-pentaoxygenation. The mass spectrum supported this suggestion and showed the parent ion and base peak at m/z 367 and lacked the M-1 and M-43 fragment ions characteristic of aporphines [15–17]. The structure was confirmed by oxidation of thalicmine (ocoteine) (**31**) with potassium permanganate in acetone to afford authentic dehydrothalicmine, identical with the natural product.

2.3. Oxoaporphines

The following reviews and references are particularly relevant to the isolation and identification of oxoaporphine alkaloids:

1. Tabular reviews [10,15–19]
2. Annual reviews [20–31]
3. Other reviews [10–13]
4. Book chapters [14,186–188,241,242]

The numbering of the oxoaporphine (**44**) ring is according to the accepted practice [15].

2.3.1. Corunnine. Tabular review [15, alkaloid No. 134], UV [15,189], IR [15,189], ^{1}H NMR [15,189].

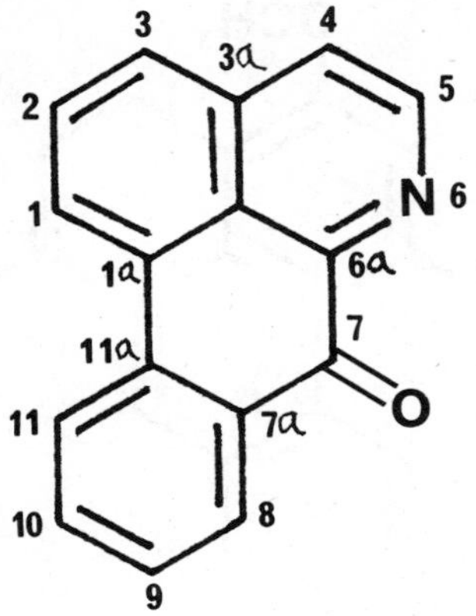

44

Corunnine (**45**), $C_{20}H_{17}NO_5$ (351.1107), m.p. 255–257°, was first isolated from *Glaucium flavum* Cr. var. *vestitum* (Papaveraceae) in 1971 [189]. The alkaloid crystallized as violet needles and was green in neutral or alkaline solution and red in acidic solution [189].

Corunnine has been isolated from *Thalictrum foetidum* L. [64,192] and *T. minus* L. [181] to date.

2.3.2. Oxoglaucine (*O*-Methylatheroline). Tabular review [15, alkaloid No. 124], UV [15], IR [15], ^{1}H NMR [15], ^{13}C NMR [17,190], MS [15].

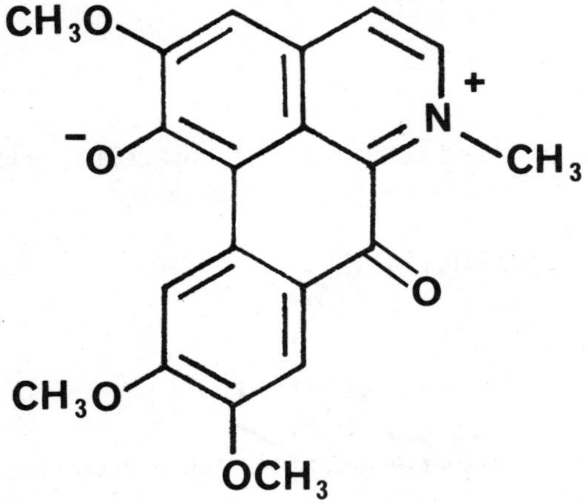

45

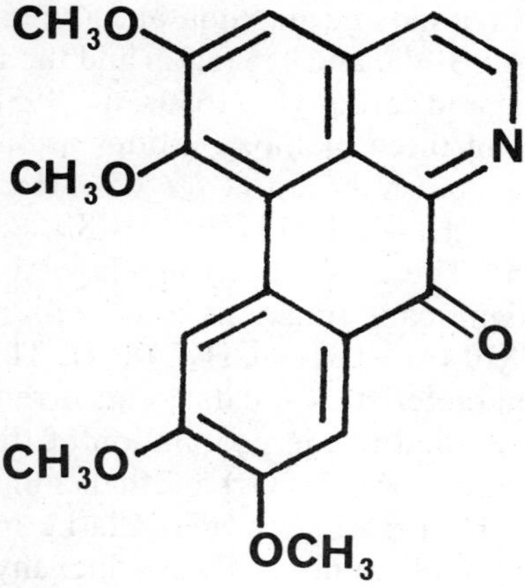

$$\underset{\sim}{46}$$

Oxoglaucine (**46**), $C_{20}H_{17}NO_5$ (351.1107), m.p. 225–226° dec. [15], was first isolated from *Liriodendron tulipifera* L. (Magnoliaceae) in 1960 [191] and has only recently (1983) been found in *Thalictrum foetidum* L. [192].

2.3.3. Thalicminine. Tabular review [15, alkaloid No. 130], UV [15,95], IR [15,95], ^{1}H NMR [15,95], MS [15,193].
Thalicminine (**47**), $C_{20}H_{15}NO_6$ (365.0899), m.p. 263–265° (CHCl$_3$) [95], was first isolated from the roots of *Thalictrum minus* L. in 1966 [95]. The UV

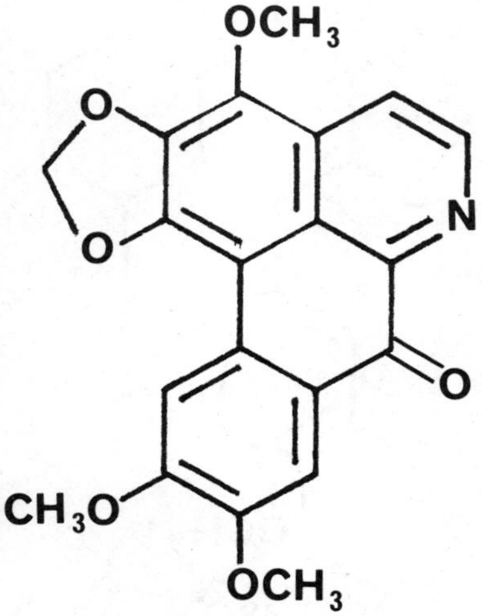

$$\underset{\sim}{47}$$

spectrum (EtOH–CHCl$_3$) showed maxima at 252 nm (log ϵ 4.29), 282 nm (log ϵ 4.43), 364 nm (log ϵ 3.91), and 456 (3.72) and the IR spectrum indicated the presence of a conjugated carbonyl at 1650 cm^{-1} [95]. The ^{1}H NMR (TFA) indicated the presence of three methoxy groups as singlets at δ 4.20, 4.25, and 4.55; one methylenedioxy group at δ 6.65 (2H, s); and four aromatic protons at δ 8.10 [1H, s, H(4)], 8.35 [1H, s, H(8)], 8.88 [1H, s, H(5)], and 8.94 [1H, s, H(11)] [15]. The mass spectrum showed the parent ion at M$^+$ m/z 365 with other significant fragment ions at m/z 350 (M—CH$_3$), 322 (M—CH$_3$—CO), and 320 (M—CH$_3$—CH$_2$O) [15]. The orange color of the alkaloid, its spectral characteristics and its co-occurrence with the aporphine ocoteine (thalicmine) (31) led to the postulation of thalicminine as an oxo-aporphine [95]. Reduction (Zn/H$_2$SO$_4$) of thalicminine afforded hexahy-drothalicminine (48), which on Eschweiler–Clarke methylation (HCOOH/HCHO) gave ocoteine (thalicmine) (N-methylhexahydrothalicminine) (31) [95]. Hofmann degradation of N-methylhexahydrothalicminine methiodide in the usual manner afforded des-N-methylthalicmine (des-N-methyloco-teine). Finally, oxidation of a previously isolated authentic sample of thal-icmine (ocoteine) (31) with either potassium permanganate–acetone or chromic anhydride–pyridine afforded thalicminine (47), thus confirming the structural assignment [95].

Administration of thalicminine (47) to dogs in a dosage of 0.25–3.0 mg/kg produced respiratory stimulation and a brief hypotensive effect. At 100 mg/kg, the alkaloid prolonged the hypnotic effect of chloral hydrate in mice [108].

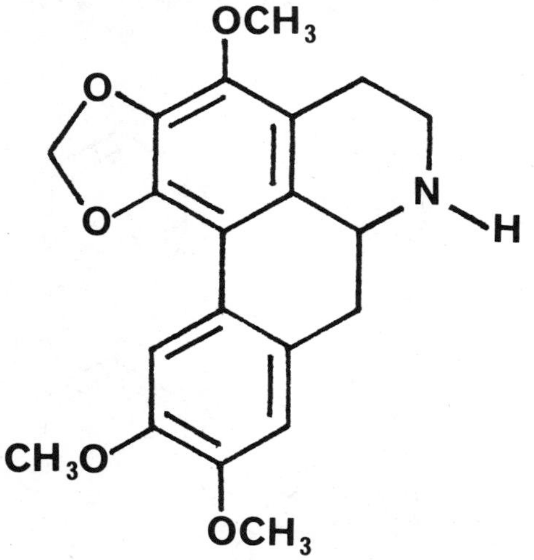

48

Thalicminine has also been isolated from *Thalictrum dioicum* L. [193], *T. isopyroides* C.A.M. [174], *T. simplex* L. [153], and *T. strictum* Ledeb. [160,161], and reisolated from *T. minus* [96,181].

2.4. Benzylisoquinolines

The following reviews and references are particularly relevant to the isolation and identification of benzylisoquinoline alkaloids:

1. Tabular reviews [10,18,19]
2. Annual reviews [20–22,24,194–201]
3. Other reviews [10–13]
4. Book chapters [14,202–205]

The numbering of the benzylisoquinoline (**49**) ring is according to accepted practice [204].

2.4.1. Laudanidine. Tabular review [18], UV [68], ^{1}H NMR [68], ORD [206].

Laudanidine (**50**), $C_{20}H_{25}NO_4$ (343.1783), m.p. 181–182° (MeOH–Me$_2$CO) [68], $[\alpha]_D^{28}$ + 134° (c 0.51, MeOH) [68], was first isolated from opium in the late nineteenth century as its levorotatory isomer ($[\alpha]_D^{18}$ − 100.6°, CHCl$_3$) [202]. In 1969, (+)-laudanidine was isolated from *Thalictrum dasycarpum* Fisch. and Lall [207] and was later found in *T. revolutum* DC. [68] some 8 years later. The alkaloid failed to exhibit antimicrobial activity when tested against a standard set of organisms [68]. In a study now almost 100 years

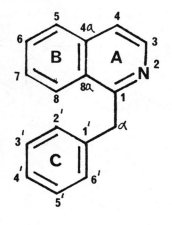

49
~

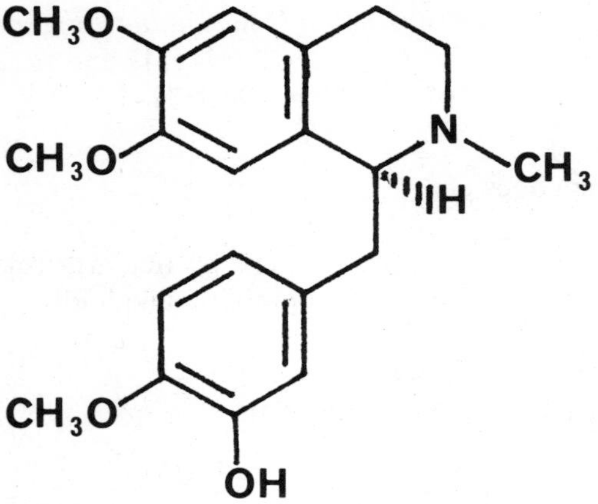

50

old laudanidine was observed to induce convulsions and caused paralysis in frogs and pigeons [208]. The alkaloid also produced an increase in respiration in rabbits, dogs, and cats, and tetany was observed on increasing dosage [208].

2.4.2. N-Methylarmepavine.   Tabular review [18], UV [105], IR [105], ^{1}H NMR [105], ORD [209].

N-Methylarmepavine (**51**), $C_{20}H_{26}O_3N^+$ [328.1913], m.p. 248–249° dec. (MeOH) (chloride salt) [105], was first isolated from a natural source (*Xanthoxylum inerme* [*Fagara boninensis*]) in 1972 [210]. The alkaloid was isolated from *Thalictrum revolutum* DC. in 1980 as the optically inactive (both

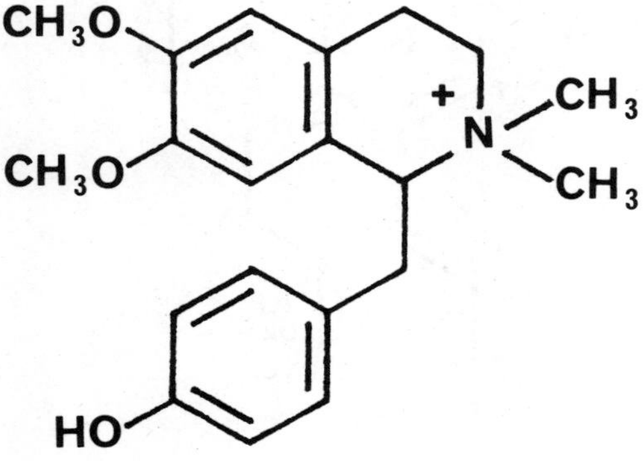

51

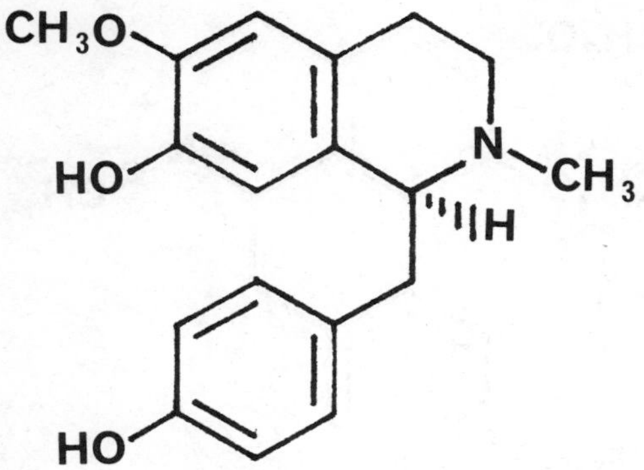

$(+)$-52

by $[\alpha]_D$ and CD) chloride salt [105]. The identity was confirmed via the quaternization (CH$_3$I) of authentic armepavine followed by conversion (anion-exchange resin) to the chloride salt and direct comparison with the isolate [105].

2.4.3. N-Methylcoclaurine. Tabular review [18], UV [68,72], IR [68], ^{1}H NMR [68,72], MS [68,72], CD [68].

(+)-N-Methylcoclaurine (52), C$_{18}$H$_{21}$NO$_3$ (299.1521), m.p. 91–94° (CHCl$_3$) [68], $[\alpha]_D^{25}$ + 101° (c 0.25, MeOH) [68], was isolated from *Thalictrum revolutum* DC. tops in 1977 [68] and fruits in 1980 [105]. In the meantime, its enantiomer ($[\alpha]_D^{25}$ −89° (c 0.6, MeOH)) was isolated from *T. dioicum* L. in 1978 [72]. (−)-N-Methylcoclaurine had a weak inhibitory effect on indirectly stimulated contractions of a frog sciatic nerve–sartorius muscle preparation [complete blockage at a dosage of 257 μM as compared to 1.47 μM for (+)-tubocurarine] and very little effect on directly stimulated muscular contractions (50% inhibitory dose of 211 μM compared to 3.09 μM for (+)-tubocurarine) [211].

2.4.4. N-Methylpalaudinium. UV [212], ^{1}H NMR [212], MS [212].

N-Methylpalaudinium chloride (53), C$_{20}$H$_{22}$O$_4$N$^+$ (340.1547), m.p. 232–233° (MeOH–Me$_2$CO), was isolated from *Thalictrum polygamum* Muhl. in 1972 [212]. The UV spectrum showed maxima at 228 nm (log ε 4.40), 256 nm (log ε 4.75), 286 nm (log ε 3.86), 315 nm (log ε 4.02), 330 nm (sh) (log ε 3.89), and 350 nm (sh) (log ε 3.75) and was very similar to N-methylpapaverine (54). The ^{1}H NMR spectrum (DMSO) was very informative and showed the presence of three methoxy groups as singlets at δ 3.75 [3H, C(4)], 4.03 [3H, C(7)], and 4.11 [3H, C(6)]; one quaternary N-methyl group (δ 4.38, 3H, s), seven aromatic protons, and one phenolic hydroxy group (δ 9.25, s, D$_2$O exchangeable). The chemical shifts and coupling patterns of the aromatic

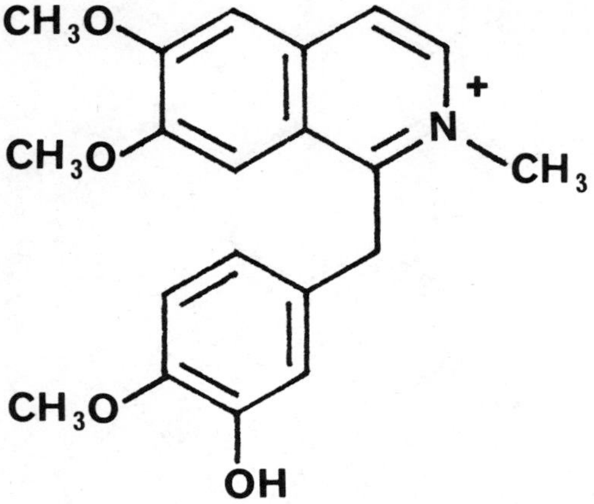

53

protons were quite characteristic for the C(5) and C(8) protons appearing as singlets at δ 7.90 and 7.80, respectively; the C(3) and C(4) protons were observed as an AB quartet at δ 8.32 and 8.65 ($J = 7$ Hz). The benzyl ring protons were found as a broad singlet [δ 6.60, C(2′)] and a pair of doublets [δ 6.55, $J = 8$ Hz, C(6′) and δ 6.88, $J = 8$ Hz, C(5′)]. The ^{1}H NMR spectrum was also very similar to that of *N*-methylpapaverine (**54**). *N*-methylpalaudinium formed a monoacetate under the usual conditons, confirming the existence of a single phenolic group. Reduction of *N*-methylpalaudinium

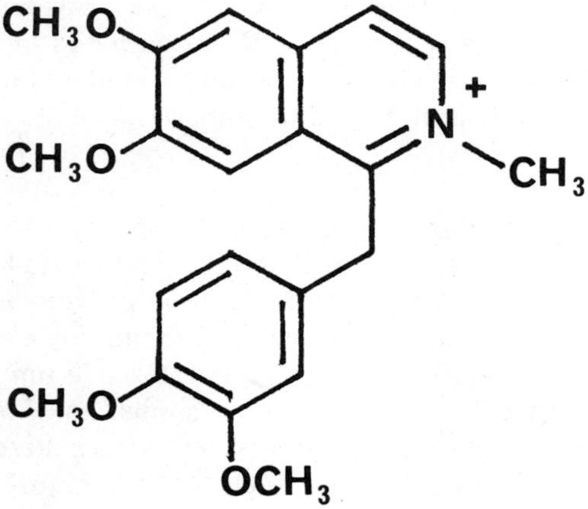

54

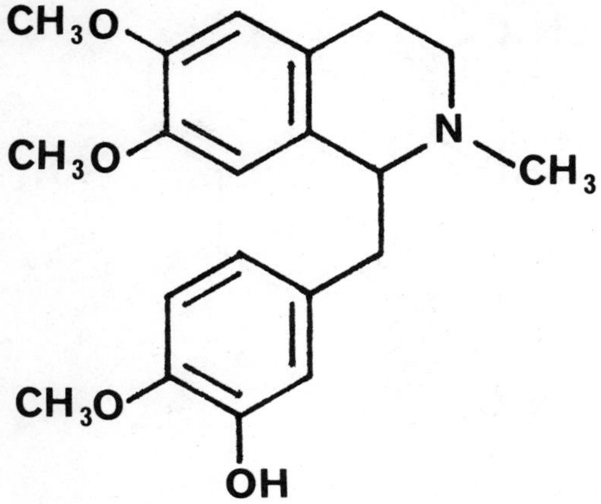

55

with sodium borohydride gave (±)-laudanine (**55**), thus placing the phenolic group at C(3') and establishing the structure of *N*-methylpalaudinium (**53**). Additional confirmation was obtained by the methylation (CH₃I) of palaudine (**56**), followed by conversion (ion exchange) to the chloride salt, which was identical with natural *N*-methylpalaudinium (**53**).

2.4.5.   Reticuline.   Tabular review [18], UV [68], ¹H NMR [68], ¹³C NMR [214], MS [68, 292], CD [68].

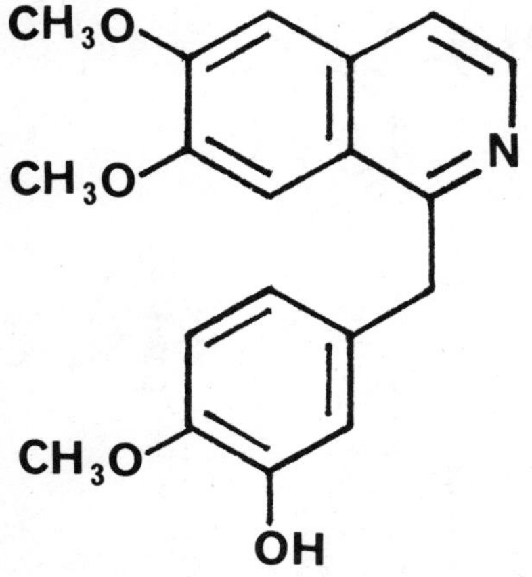

56

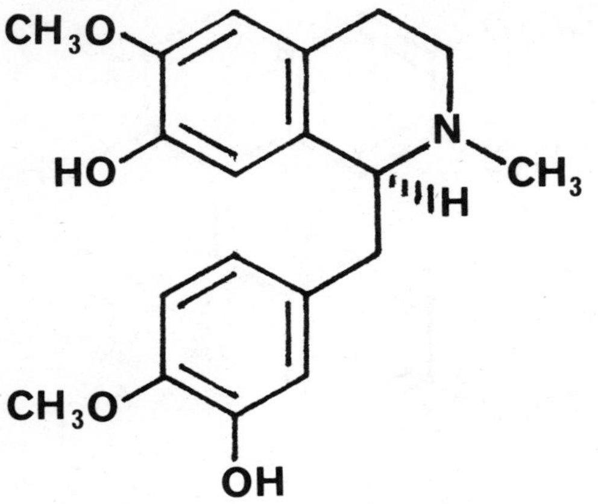

57

Reticuline (**57**), $C_{19}H_{23}NO_4$ (329.1627) was first isolated from *Annona reticulata* (Annonaceae) in 1959 [213] and first found in a *Thalictrum* species, *T. revolutum* DC., as its (+) isomer ([α]$_D^{25}$ +100° [c 0.78 (MeOH)] in 1977 [68]. (+)-(S)-reticuline was subsequently isolated from *T. minus* L. race B in 1978 (oxalate salt, m.p. 156.5–158°) (free base [α]$_D^{25}$ +139° (c 0.19, MeOH) [215] and *T. foliolosum* DC. in 1982 (hydrochloride salt, [α]$_D^{28}$ +72° (c 0.5, H_2O) [38]. The alkaloid was found to be devoid of microbial inhibiting properties when tested against a standard set of organisms [68].

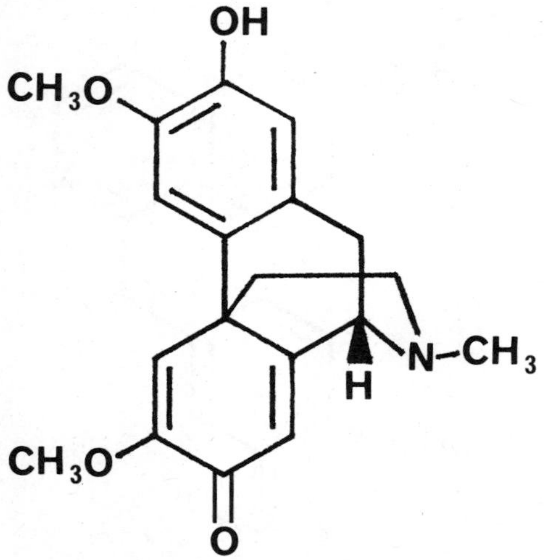

58

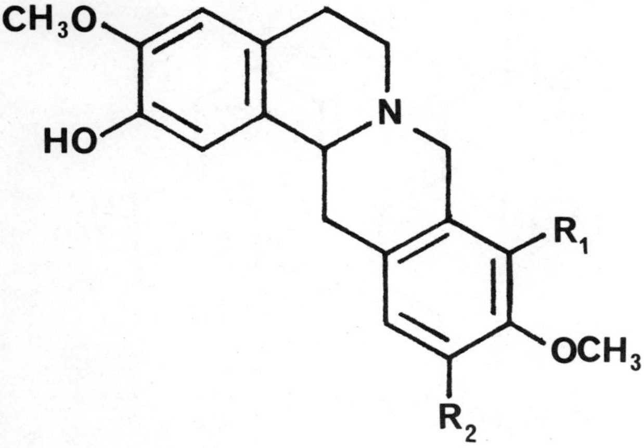

$$\underset{\sim}{59} \quad R_1 = H, \ R_2 = OH$$
$$\underset{\sim}{60} \quad R_1 = OH, \ R_2 = H$$

Incubation of ($\pm$)-reticuline with rat liver homogenate resulted in the biotransformation of the alkaloid into the morphinandienone pallidine (58), the aporphine isoboldine (4), and the tetrahydroprotoberberines coreximine (59) and scoulerine (60) [216].

2.4.6. Takatonine. Tabular review [18], UV [217], IR [217], ^{1}H NMR [217,219,221], MS [217].

Takatonine (61), $C_{21}H_{24}N^+O_4$ (354.1705), was first isolated from *Thalictrum thunbergii* DC. ("Takato-gusa"—a Japanese commercial crude drug) as its

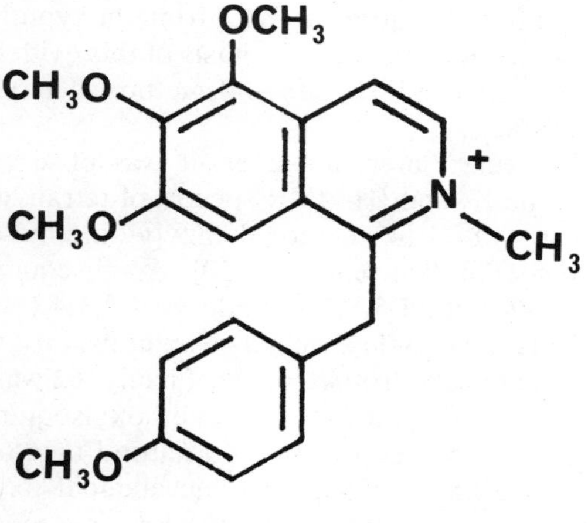

$$\underset{\sim}{61}$$

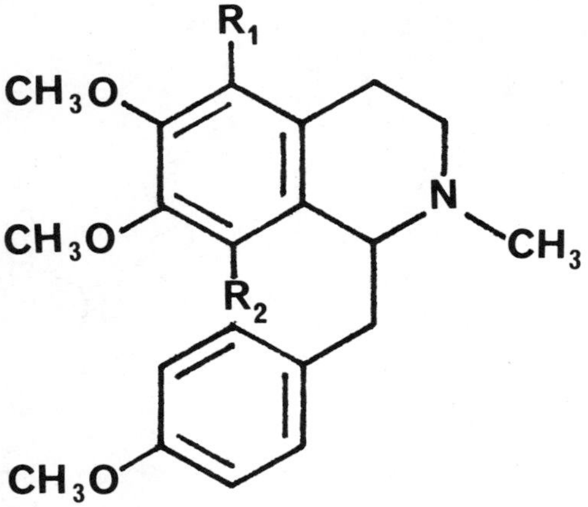

$$\underset{\sim}{62} \quad R_1 = OCH_3, R_2 = H$$
$$\underset{\sim}{63} \quad R_1 = H, R_2 = OCH_3$$

iodide salt, m.p. 192–193° (MeOH), in 1959 [218]. Reduction of takatonine with zinc–acetic acid gave the tetrahydro derivative (62) (hydrochloride, m.p. 186–187° (EtOH–petrol); hydrobromide, m.p. 184–185°; picrate, m.p. 141.5–143°) which was subsequently oxidized with potassium permanganate to anisic acid (*p*-methoxybenzoic acid). Hofmann degradation of tetrahydrotakatonine gave the oily methyl methine which was also oxidized with potassium permanganate (acetone) to anisic acid and an amine. The amine afforded a Hofmann product characterized as 2-vinyl-4,5,6-trimethoxybenzoic acid after oxidation to 3,4,5-trimethoxyphthalic acid via potassium permanganate (acetone). On the basis of this evidence, the structure of takatonine was proposed to be 1-(4′-methoxybenzyl)-2-methyl-6,7,8-trimethoxyisoquinoline.

Approximately 5 years later, however, it was observed that the TLC behavior and both the IR and ^{1}H NMR spectra of tetrahydrotakatonine differed from those of 1-(4′-methoxybenzyl)-2-methyl-6,7,8-trimethoxy-1,2,3,4-tetrahydroisoquinoline (63) [219,220]. A subsequent synthesis of 1-(4′-methoxybenzyl)-2-methyl-5,6,7-trimethoxy-1,2,3,4-tetrahydroisoquinoline (62) prepared via the classic Bischler–Napieralski route showed it to be identical with natural tetrahydrotakatonine. Finally, 62 was dehydrogenated to 1-(4′-methoxybenzyl)-2-methyl-5,6,7-trimethoxyisoquinoline (61) (m.p. 181–182°), which was likewise identical to natural takatonine. Takatonine was the first example of benzylisoquinoline alkaloid oxygenated at C(5). Takatonine was synthesized some 10 years later via a modified Pomeranz–Fritsch reaction [m.p. 178–180° (MeOH–C_6H_6)] [221].

Salient features of the spectral data for takatonine include the UV maxima at 226, 265, 315, and 365 nm [212] plus the ^{1}H NMR (CDCl$_3$) signals, which show the methoxy groups as four singlets at δ 3.76, 4.01, 4.08, and 4.14; the quaternary N-methyl at δ 4.68 (s); the C-ring protons in a typical A$_2$B$_2$ pattern [δ 6.82 (d, J = 9 Hz) and δ 7.20 (d, J = 9 Hz)]; the C(8) proton as a singlet at δ 7.44 and the B-ring protons as a low-field pair of doublets at δ 8.32 (J = 9 Hz) [H(4)] and δ 9.20 (J = 9 Hz) [H(3)] [217]. The mass spectrum showed the M$^+$ at m/z 354 (21%) and other intense fragment ions at m/z 353 (78%), 338 (48), and 121 (100) [217].

Takatonine iodide exhibited weak atropine-like and papaverine-like spasmolytic activity on isolated smooth muscle preparation of guinea pig ileum and mouse small intestine [222]. Takatonine has also been reisolated from *Thalictrum thunbergii* DC. [157] and isolated from *T. minus* L. var. *microphyllum* Boiss. [m.p. 181–182° (MeOH)] [217].

2.4.7. Tembetarine (N-Methylreticuline). Tabular review [18], UV [223], IR [223], ^{1}H NMR [223], ORD [209].
Tembetarine (**64**), C$_{20}$H$_{26}$N$^+$O$_4$ (344.1862), m.p. 148–152° (chloride) [223]; [α]$_D^{30}$ +135° (c 1.45, MeOH) [223], was first isolated from *Fagara* species (Rutaceae) in 1964 [224] and first reported in a *Thalictrum* species, *T. foliolosum* DC., in 1983 [223].

Tembetarine chloride was found to have an LD$_{50}$ of 63 mg/kg in rats, with clonic convulsions preceding death. In addition, the alkaloid produced a transient hypotension and bradycardia in anesthetized cats and dogs while potentiating the blood pressure responses to (−)-norepinephrine in these same animals when given in doses of 1–5 mg/kg. However, the alkaloid did

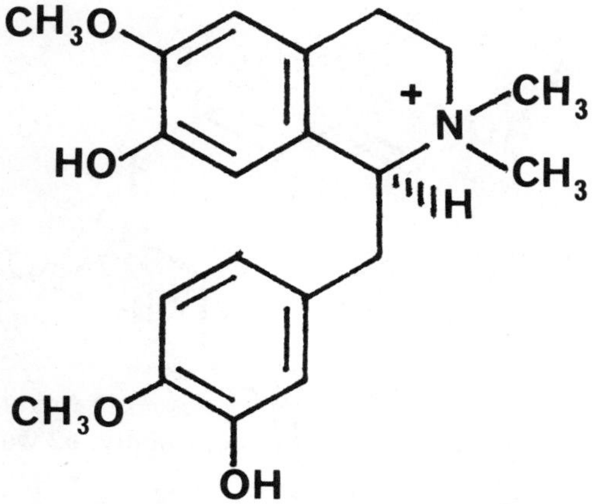

64

produce a diminished response to carotid occlusion in dogs, suggesting sympathetic nervous system mediation of the activity of tembetarine. Finally, tembetarine chloride potentiated the cumulative dose–response curves for (−)-norepinephrine and angiotensin in the vas deferens from both untreated and reserpine-treated rats on exposure to concentrations of 10^{-5} M of the alkaloid [225].

2.4.8.   Thalifendlerine.   Tabular review [18], UV [132], ^{1}H NMR [132], MS [132], ORD [226].

Thalifendlerine (**65**), $C_{20}H_{25}NO_4$ (343.1783), m.p. 177–178° [132]; $[\alpha]_D^{25}$ −108° (MeOH) [132], was first isolated from *Thalictrum fendleri* Engelm. ex Gray in 1965 [62,132]. The UV spectrum was typical of a simple benzyltetrahydroisoquinoline with a single maximum at 282 nm (log ϵ 3.49) [132]. The ^{1}H NMR spectrum was very informative and showed the presence of one N-methyl group at δ 2.49 (3H, s); three methoxy groups at δ 3.51 (3H, s) and 3.85 (6H, s); and five aromatic protons, four of which were present in a typical A_2B_2 system (ring C) centered at δ 6.6 and 6.85 ($J = 8.4$ Hz) and the fifth as an upfield singlet at δ 5.83 [H(8)] [132]. The mass spectrum exhibited the molecular ion at m/z 341 with other important fragment ions at m/z 236 (**66**) and 107 (**67**), the former representing the isoquinoline portion and the latter the benzyl portion of the alkaloid, and resulting from mass fragmentation via cleavage at the bond alpha to both the nitrogen atom and the benzyl carbon [132]. Additional proof of the 5,6,7-substitution resulted when the corresponding 6,7,8-trimethoxylated isomer was synthesized and found to be different from thalifendlerine [132]. Ultimately, methylation of thalifenderline with diazomethane afforded a monomethyl ether (m.p. 195–197°), which was identical to tetrahydrotakatonine (**62**), thus establishing the

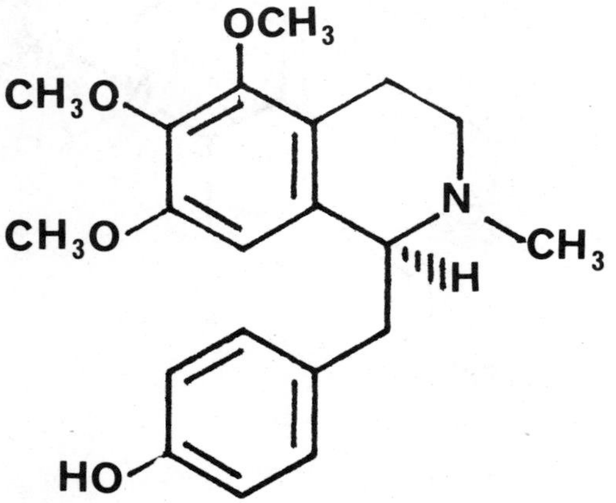

65

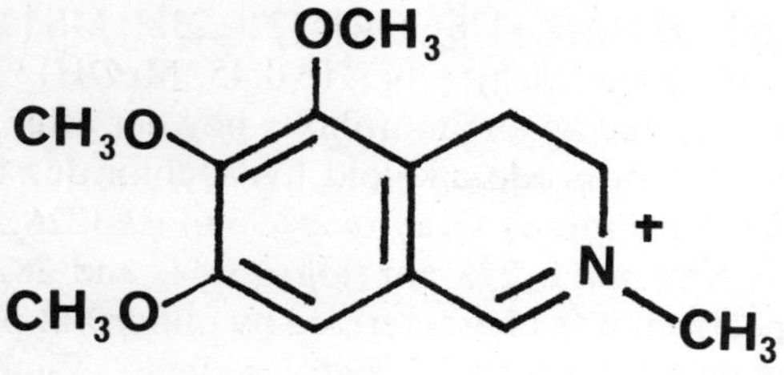

66

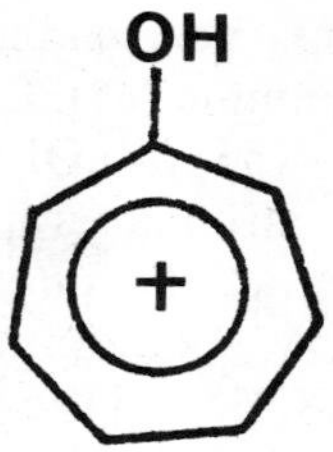

67

structure of the thalifendlerine as **65**. Thalifendlerine shows a negative Cotton effect in the ORD indicating the (*R*) configuration [226–228]. Finally, (±)-thalifendlerine was also synthesized in 1968 [229].

2.4.9. Thalmeline. IR [230], ^{1}H NMR [230].
Thalmeline (**68**), $C_{19}H_{23}NO_4$ (329.1627), m.p. 87° (amorphous); m.p. 210–212° (oxalate); $[\alpha]_D^{18} -31°$ (c 0.6, CHCl$_3$), was first isolated from *Thalictrum minus* L. ssp. *elatum* in 1970 [230]. The pale-yellow, amorphous phenolic alkaloid was characterized by a ^{1}H NMR spectrum that showed the presence of one *N*-methyl group at δ 2.55 (3H, s), two methoxy groups at δ 3.50 (3H, s) and 3.83 (3H, s), one high-field aromatic proton at δ 5.57 (1H, s), and an A_2B_2 system containing four aromatic protons as two doublets at δ 6.65 (2H) and 6.90 (2H) (*J* = 8.5 Hz). The alkaloid was identified by comparison with its enantiomer, which was obtained by sodium in liquid ammonia cleavage of thalfoetidine (**370**) [230,231].

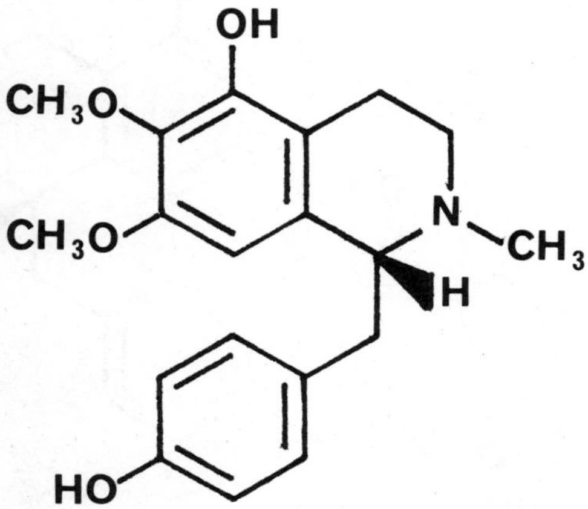

68

2.4.10.   Veronamine.   UV [226], ¹H NMR [226], ORD [226,232], MS [232]. Veronamine (**69**), $C_{26}H_{35}NO_8$ (489.2363), $[\alpha]_D$ $-145°$ (c 0.45, MeOH) [226]; $[\alpha]_D$ $-155°$ (MeOH) [232]; was isolated as an amorphous powder after partition chromatography of a mixture of crude alkaloid hydrochlorides from an extract of *Thalictrum fendleri* Engelm. ex Gray over cellulose [226,232]. The UV spectrum showed two maxima at 275 nm (log ϵ 3.44) and 282 nm (log ϵ 3.43) and the ¹H NMR spectrum was characterized by important peaks at δ 1.21 (J = 6.5 Hz) for a *C*-methyl doublet, δ 2.49 for a singlet *N*-methyl group, δ 3.85 (9H) for three methoxy groups, δ 5.85 for a characteristic high-field aromatic proton [C(8)], and δ 5.42 as a broad doublet for the anomeric proton [226]. Hydrolysis of veronamine with hot dilute sulfuric acid afforded ($-$)-thalifendlerine (ORD) (**65**) and L-($-$)-rhamnose, the latter characterized as its diethyl dithioacetal [226,232]. Synthesis of ($-$)-veronamine was achieved via a modified Koenigs–Knorr reaction by condensation of 2,3,4-tri-*O*-acetyl-α-L-rhamnopyranosyl bromide with the dry potassium salt of natural ($-$)-thalifendlerine (**65**) in dry acetone. Hydrolysis of the acetate esters of the crude reaction product with sodium methoxide in methanol, followed by purification, afforded ($-$)-veronamine (85% yield) [226]. Because it is known that the Koenigs–Knorr reaction occurs with Walden inversion at the anomeric center when alkali is used, and because the triacetyl bromide derivative of rhamnose was of the alpha configuration, it follows that the resulting ($-$)-veronamine (**69**) must have beta configuration at the anomeric carbon. ($-$)-Veronamine is the first glycosidic alkaloid to have been isolated from a *Thalictrum* species and only the second glycosidic benzylisoquinoline isolated from a higher plant, the first being latericine from various *Papaver* species [18].

69

Intravenous (0.1–1.0 mg/kg) and intra-arterial (0.25–1.0 mg/kg) administration of graded doses of (−)veronamine to anesthetized dogs produced a dose-related decrease in mean arterial blood pressure and a reduction in the pressure of the perfused hind limb. The duration of the systemic hypotension varied from 1 to 3 min with lower doses and from 5 to 20 min with higher doses. Bradycardia outlasted the systemic hypotension and was observed along with reduction in left ventricular systolic pressure and left ventricular *dP/dt*. Some premature ventricular contractions occurred. The systemic hypotension and negative chronotropic activities of veronamine involve cholinergic mechanisms, for these effects were blocked by bilateral cervical vagotomy or pretreatment with atropine. In addition, the experiments using the perfused hind limb indicate that veronamine may also possess noncholinergic, smooth muscle-relaxing properties [233].

These experiments do not indicate the site of action of veronamine. Whether the systemic hypotension and bradycardia are the result of stimulation of afferent nerve endings or baroreceptors, in a manner similar to that produced by veratrum alkaloids, or are due to a direct action in the central nervous system remains to be determined [233].

2.5. Oxobenzylisoquinolines

The following reviews and references are relevant to the isolation and identification of oxobenzylisoquinoline alkaloids:

1. Tabular reviews [18,19]
2. Annual reviews [20–22,24,194–201]
3. Book chapters [202–204]

The numbering of the oxobenzylisoquinoline ring (70) is according to accepted practice and is similar to that of the benzylisoquinoline ring.

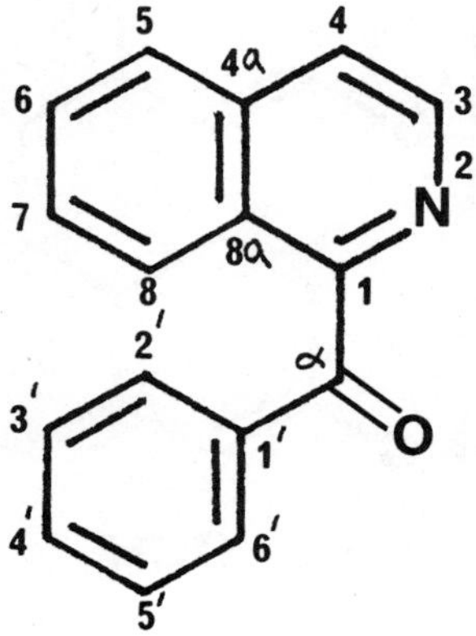

70

2.5.1. Rugosinone.   UV [234], IR [234], ^{1}H NMR [234], MS [234].
Rugosinone (**71**), $C_{19}H_{15}NO_6$ (353.0899), m.p. 223–224° (EtOAc), was isolated from extracts of *Thalictrum rugosum* Ait. (*T. glaucum* Desf.) in 1980
[234]. The UV spectrum showed maxima at 237 nm (log ϵ 4.57), 298 nm (log
ϵ 4.13), and 329 nm (log ϵ 3.95) with a bathochromic shift in sodium methoxide whereas the IR spectrum showed conjugated carbonyl absorption at
1627 cm^{-1} [234]. The ^{1}H NMR spectrum was very informative and showed
a similarity to that of papaveraldine (**72**) [234]. In particular, the ^{1}H NMR
spectrum of rugosinone lacked both methylene and methine signals and
showed the presence of six aromatic protons, two of which were singlets (δ
7.15 [H(8)] and 7.38 [H(5)] and the remainder were doublets: δ 6.41 (1H, J
= 9 Hz) [H(5′)], 7.20 (1H, J = 9 Hz) [H(6′)], 7.61 (1H, J = 5 Hz) [H(4)],
and 8.43 (1H, J = 5 Hz) [H(3)]. These relationships were confirmed via
decoupling experiments and the lower-field signals with their particular
coupling constants were like those of the H(3)/H(4) protons in papaveraldine
(**72**) [234]. In addition, there was a very-low-field, D_2O-exchangeable proton
at δ 12.40, characteristic of a hydrogen-bonded phenol. The methoxy groups
were observed as singlets at δ 3.91 (3H) and 3.95 (3H) and the methylenedioxy group was a singlet at δ 6.11 (2H). The mass spectrum showed the
molecular ion at m/z 353 and this, coupled with IR carbonyl bond at 1627
cm^{-1} (some 30 cm^{-1} lower than that of papaveraldine); the very-low-field,
D_2O-exchangeable proton; the presence of important mass-spectral fragment
ions at m/z 172 (isoquinoline portion) and 181 (benzyl portion); and a consideration of the other spectral data and biosynthetic implications suggested
that rugosinone was best represented as **71** [234]. This was confirmed by
total synthesis using two similar pathways, each of which utilized the same
Reissert compound [235]. Rugosinone was subsequently isolated from *T.
foliolosum* DC. [223].

71

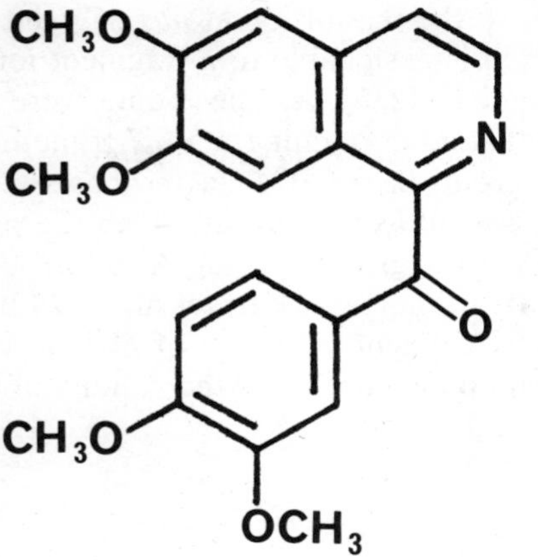

72

2.5.2. Thalmicrinone. Thalmicrinone (**73**), $C_{20}H_{19}NO_5$ (353.1263), m.p. 130–132° (abs. EtOH), was isolated from *T. minus* L. var. *microphyllum* Boiss. in 1982 [217]. The UV spectrum was characterized by maxima at 235, 250 (sh), 295, and 350 nm, and the IR spectrum showed bands for a conjugated carbonyl at 1660 (C=O) and 1620 cm^{-1} (C=C) [217]. The ^{1}H NMR spectrum indicated the presence of four aromatic methoxy groups as singlets (δ 3.89, 3.93, 4.03, 4.08); a singlet aromatic proton at δ 7.38; an A_2B_2 system at δ 6.96 (d, 2H, J = 9 Hz) and δ 7.96 (d, 2H, J = 9 Hz); and a low-field AB system at δ 8.01 (d, 1H, J = 6 Hz) [H(4)] and 8.48 (d, 1H, J = 6 Hz)

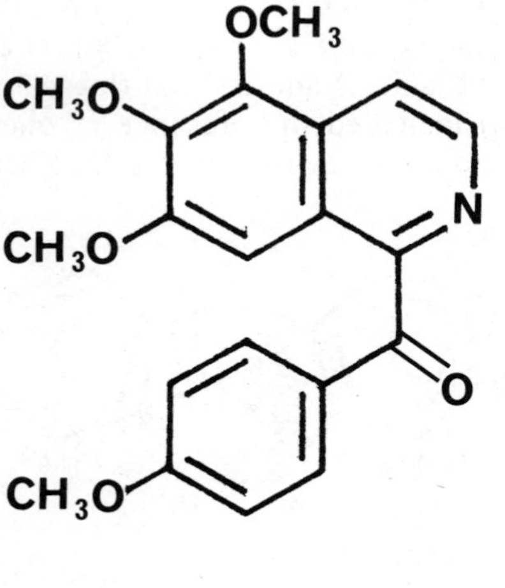

73

[H(3)]. The MS showed the parent ion at m/z 353 and the base peak at m/z 135, the latter of which is assignable to a fragment ion representing the 4'-methoxyoxobenzyl moiety [236]. All these data were indicative of an oxobenzylisoquinoline alkaloid containing a 5,6,7-trimethoxy sequence in ring A and a 4-methoxy group in ring C. On the basis of these spectral data, thalmicrinone was assigned as **73** and, although the possibility of other arrangements of the methoxy groups in ring A cannot be excluded, the presence of takatonine (**61**) in the same extract may add further support to this structural assignment. A recent synthesis of thalmicrinone utilizing the appropriate Reissert compound confirmed the validity of the above assignment [236a].

2.6. Isopavines

The following reviews and references are particularly relevant to the isolation and identification of isopavine alkaloids:

1. Tabular reviews [10,18,19,237]
2. Annual reviews [20–22,24,194–201]
3. Other reviews [10–13]
4. Book chapters [14,240–242,280,281]

The numbering of the isopavine (**74**) ring is according to accepted practice [237].

2.6.1. Thalidicine.

Tabular review [237, alkaloid No. 24], UV [130,237], IR [130,237], ¹H NMR [130,237], MS [130,237].
(−)Thalidicine (**75**), $C_{19}H_{21}NO_4$ (327.1471), m.p. 200° (EtOH–Et₂O), was isolated from *Thalictrum dioicum* L. in 1976 [130]. The UV spectrum showed maxima at 221 nm (log ε 4.38) and 290 nm (log ε 3.96), and the ¹H NMR spectrum (DMSO-d₆) indicated the presence of one *N*-methyl group at δ 2.26, six methylene and methine protons plus one phenolic proton at δ 3.06–

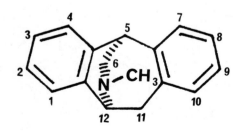

74

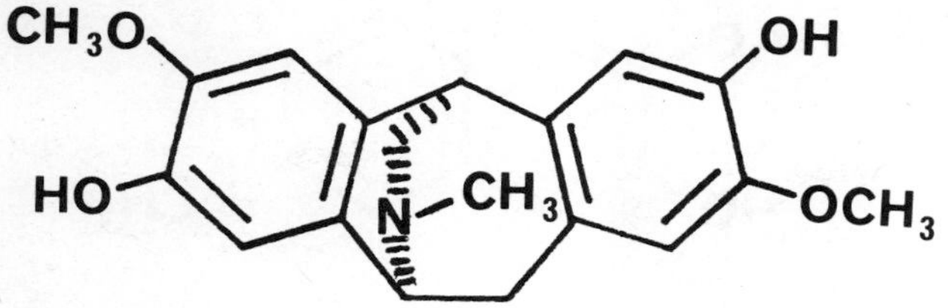

75

3.6 (7H, m), two methoxy groups at δ 3.70 (s), and four aromatic protons as four singlets at δ 6.36, 6.60, 6.66, and 6.73. A low-field singlet at δ 8.58 (1H) was assigned to a phenolic proton. The mass spectrum showed the molecular ion at m/z 327(12%), the base peak at m/z 190 (6-methoxy-7-hydroxy-2-methylisoquinoline fragment ion), and a significant fragment ion at m/z 284 (M-43). Thalidicine was assigned as isopavine **75** on the basis of available spectral data but the structural determination is incomplete and thalidicine, may in reality, be identical with (−)-thalidine (**76**).

2.6.2. Thalidine. Tabular review [237, alkaloid No. 25], UV [237,243], ^{1}H NMR [237,243], MS [237,243], CD [72,237].

Thalidine (**76**), $C_{19}H_{21}NO_4$ (327.1471), m.p. 205–207° (MeOH), $[\alpha]_D^{25}$ −172° (c 0.7, MeOH), was isolated from *Thalictrum dioicum* L. in 1976 [243]. The UV spectrum showed maxima at 250 nm (sh) (log ε 4.08) and 291 nm (log ε 4.30) with a bathochromic shift in alkali and was consistent with that of pavines and isopavines [238]. The ^{1}H NMR spectrum (TFA) was characterized by six well-defined singlets at δ 3.00 (3H, NCH$_3$), 3.99 (6H, 2OCH$_3$), and 6.70, 6.83, 7.00 and 7.12 (4 × 1H) (ArH). The mass spectrum exhibited the parent ion at m/z 327 with significant fragment ions being observed at m/z 284 (M-43) and 190. The facile loss of 43 a.m.u. (—[CH$_2$=NCH$_3$]) from the parent ion is characteristic of an isopavine versus a pavine alkaloid [238] and the fragment ion at m/z 190 is suggestive of a monomethoxy, mono-

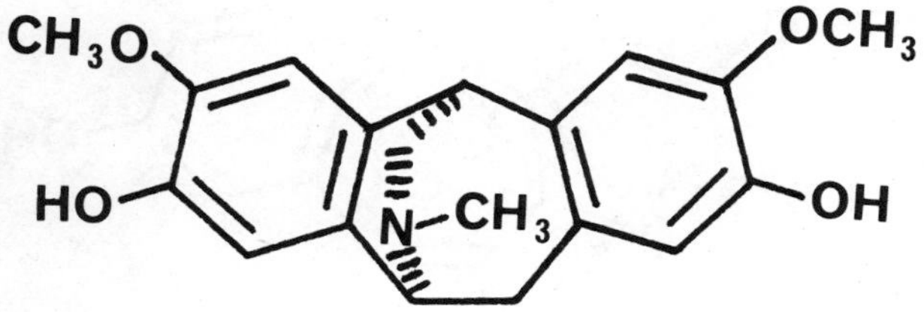

76

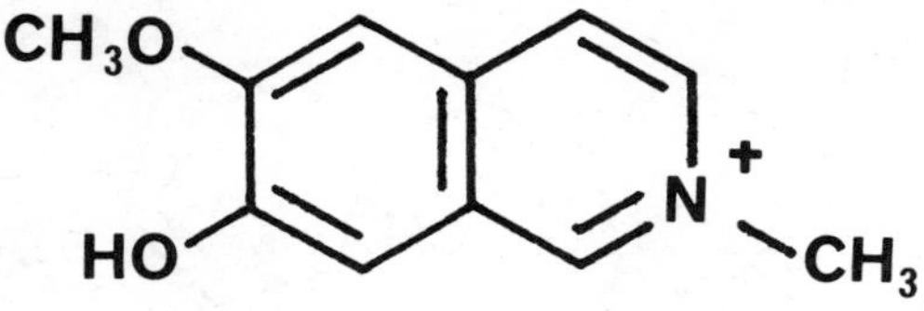

77

hydroxy-*N*-methylisoquinoline ion of type **77**. Methylation of thalidine with CH_2N_2 gave a mixture of thalisopavine (**78**) and *O*-methylthalisopavine (**79**), which were separated by preparative TLC. These data indicated that thalidine could be represented by either **76** or **80**. On the basis of biogenetic considerations involving the likely derivation of isopavine alkaloids from reticuline (**57**) or a reticuline analog, (±)-isopavine **76** was synthesized and found to be identical with natural (−)-thalidine (**76**) [234,284]. Because naturally occurring thalidine is levorotatory, as are all other known naturally occurring isopavines, and because the absolute configuration of levorotatory isopavines has been determined via the aromatic chirality method [244], the configuration of thalidine was assigned as in **76**. (−)-Thalidine was subsequently reisolated from *T. dioicum* L. in 1978 [72].

2.6.3. Thalisopavine.   Tabular review [237, alkaloid No. 23], UV [207,237], IR [207,237], ^{1}H NMR [207,237], MS [207,237].
Thalisopavine (**78**), $C_{20}H_{23}NO_4$ (341.1627), m.p. 211–212° (EtOH), $[\alpha]_D^{25}$ −210° (c 0.21, $CHCl_3$) was isolated from *Thalictrum dasycarpum* Fisch. and Lall. in 1969 [207] and was the first isopavine alkaloid to be found in the genus *Thalictrum*. The UV spectrum showed a single maximum at 289 nm (log ε 4.06) with a bathochronic shift in alkali. The ^{1}H NMR spectrum indicated the presence of one *N*-methyl group (δ 2.48), three methoxy groups (δ 3.86, 9H), and four aromatic protons [δ 6.54, 6.61, and 6.75 (2H)], with signals at δ 4.90 (br, 1H) and δ 2.60–3.80 (m, 6H) being attributable to a

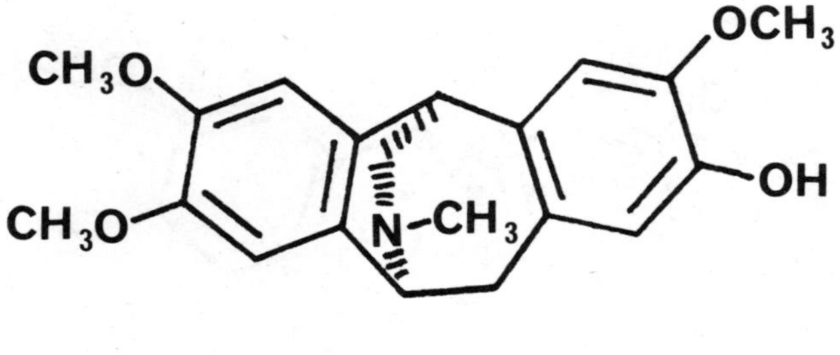

78

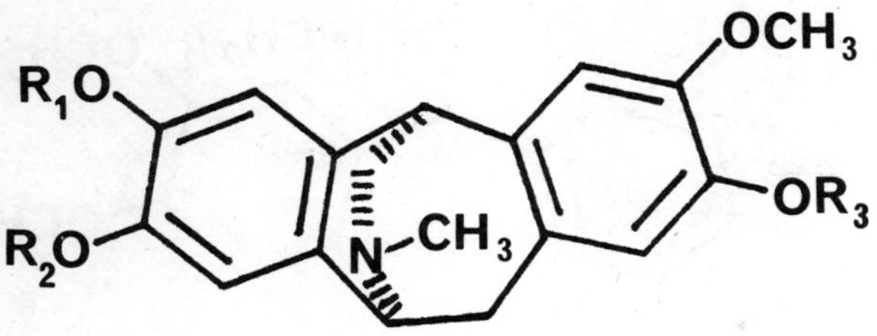

$$78 \quad R_1 = R_2 = CH_3, R_3 = H$$

$$79 \quad R_1 = R_2 = R_3 = CH_3$$

$$80 \quad R_1 = R_3 = H, R_2 = CH_3$$

phenolic hydroxy group and methylene/methine protons, respectively. The mass spectrum was characterized by a parent ion at m/z 341 and a base peak at m/z 204, the latter attributable to the dimethoxy-N-methylisoquinolinium ion **81**. The great similarity in the alphatic and aromatic proton region of the ^{1}H NMR spectrum of thalisopavine to that of the isopavine alkaloid amurensine (**82**) plus a consideration of the other spectral data indicated that thalisopavine is an isopavine alkaloid. Confirmation of this occurred when

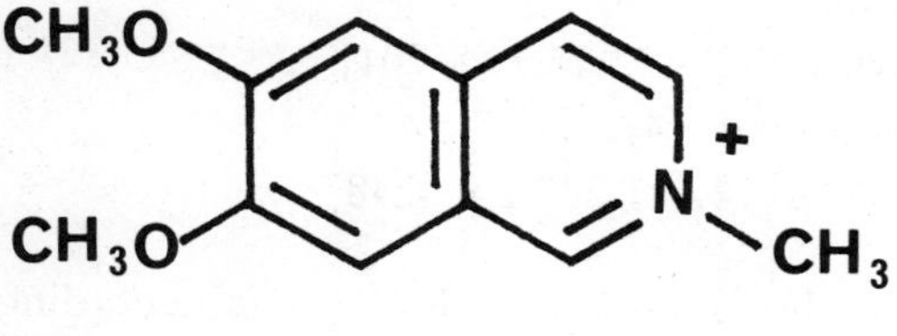

81

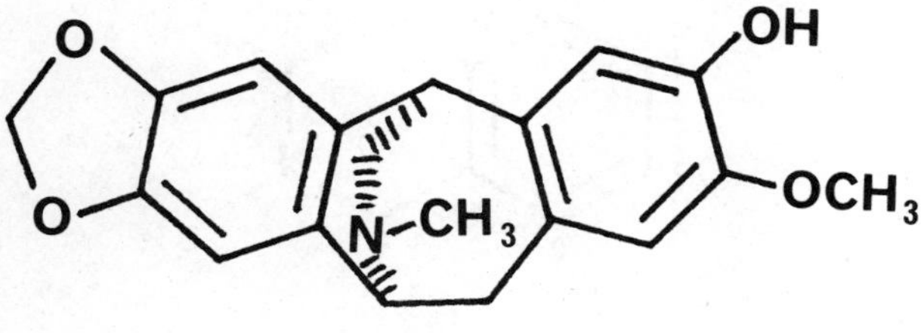

82

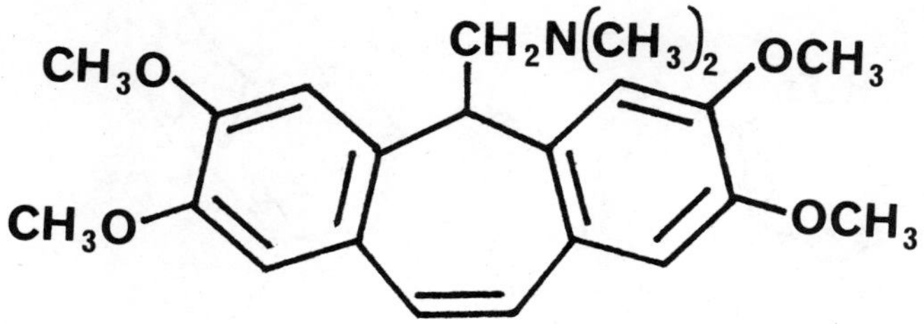

83

methylation of thalisopavine with diazomethane afforded an *O*-methyl ether which was converted to its methiodide salt and degraded via the Hofmann method to *N*-methylisopavine methine (**83**). The position of the phenolic hydroxy group in thalisopavine was thought to be at C(9) after a consideration of the ^{1}H NMR spectral data for thalisopavine (**78**) and its *O*-methyl ether as well as amurensine (**82**) and its *O*-methyl ether, and this was confirmed by synthesis of (±)-thalisopavine along classical lines [207].

2.7. Isoquinolines

The following reviews and references are particularly relevant to the isolation and identification of isoquinoline alkaloids:

1. Tabular reviews [10,18,19]
2. Annual reviews [20–22,24,194–201]
3. Other reviews [10–14]
4. Book chapters [14,241,242,245–248]

The numbering of the isoquinoline (**84**) ring is according to the accepted practice [247].

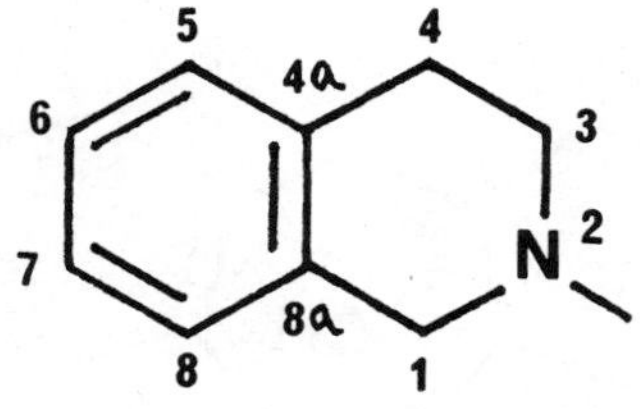

84

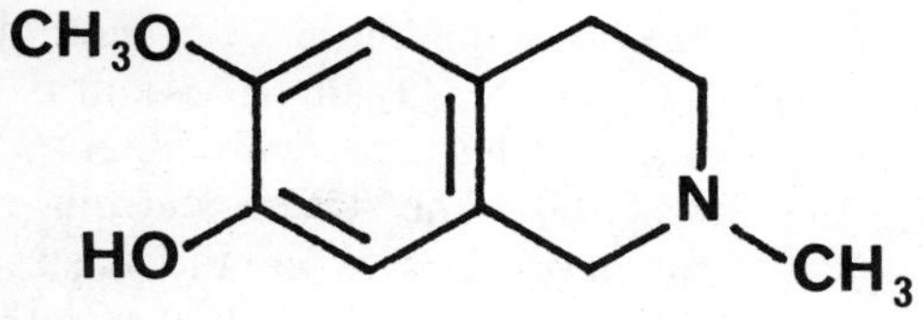

CH₃O, HO — structure **85**

$$\underset{\sim}{85}$$

2.7.1. Corypalline. Tabular review [18], UV [249], IR [249], ¹H NMR [249], MS [249].

Corypalline (**85**), $C_{11}H_{15}NO_2$ (193.1103), m.p. 168° (MeOH–Et₂O), was first isolated from *Corydalis aurea* (Papaveraceae) and *C. pallida* in 1937 [250]. It was first found in a *Thalictrum* species, *T. dasycarpum* Fisch. and Lall. in 1969 [207] and later in *T. rugosum* Ait. [234] and *T. uchiyamai* Nakai [249].

2.7.2. O-Methylcorypalline. UV [72], ¹H NMR [72], ¹³C NMR [252,253], MS [72].

O-Methylcorypalline (**86**), $C_{12}H_{17}NO_2$ (207.1259), m.p. 69–70° (MeOH) [72], was first isolated from *Nelumbo nucifera* (Nymphaceae) in 1970 [251]. The alkaloid has been found in only one *Thalictrum* species, *T. dioicum* L. [72]. In general, simple 6,7-dioxygenated isoquinolines such as corypalline (**85**) and O-methylcorypalline are characterized by UV spectra that show maxima around 230 nm (log ε 3.7) and 283 nm (log ε 3.5) [247]. The ¹H NMR spectra are relatively uncomplicated and show methoxy signals around δ 3.85, singlet aromatic protons around δ 6.5–6.6, the C(1) methylene protons as a singlet around δ 3.5, and the C(3)/C(4) methylene protons as a broad multiplet in the region of δ 2.5–3.0 [247,254]. The mass spectra are marked by intense fragment ions (frequently the base peak) due to retro-Diels–Alder type cleavages (M⁺—[CH₂=NH] or M⁺—[CH₂=NCH₃]) [72].

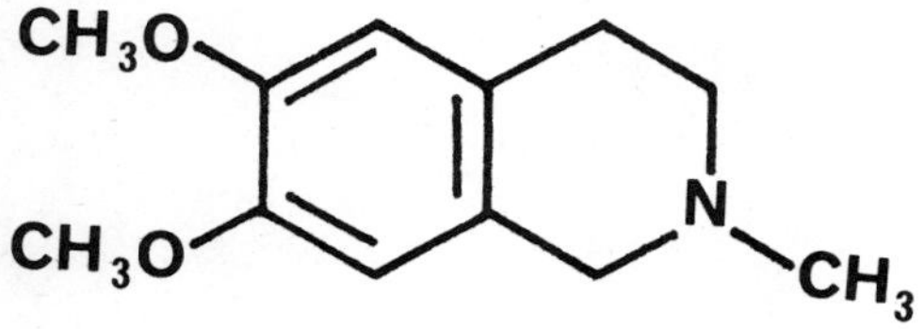

CH₃O, CH₃O — structure **86**

$$\underset{\sim}{86}$$

2.7.3.   N-Methyl-6,7-dimethoxyisoquinolinium.   *N*-Methyl-6,7-dimethoxy-isoquinolinium chloride (**81**), $C_{12}H_{14}N^+O_2$ (204.1024), m.p. 185.5–186.5 (petrol–EtOH), was isolated as a colorless, optically inactive compound from *T. revolutum* DC. in 1980 [105]. The UV spectrum showed maxima at 253 nm (log ϵ 4.91) and 310 nm (log ϵ 3.95) and the mass spectrum showed the molecular ion at *m/z* 204 (10%) and base peak at *m/z* 189. The ^{1}H NMR spectrum (MeOH-d$_4$) was indicative of quaternary isoquinoline and showed peaks at δ 4.07 and 4.12 for two methoxy groups and δ 4.43 (quaternary *N*-methyl). The aromatic protons were observed as singlets at δ 7.64 [H(5)] and 7.71 [H(8)] plus a low-field AB quartet at δ 8.20 and 8.35 [*J* = 7 Hz, H(3) and H(4)] and a lower-field singlet at 9.38 [H(1)]. The structure was assigned with certainty when reduction of the quaternary base with sodium borohydride afforded *O*-methylcorypalline (**86**) [105].

2.8.   Isoquinolones

The following reviews and references are particularly relevant to the isolation and identification of isoquinolone alkaloids:

1.   Tabular reviews [10,18,19,255]
2.   Annual reviews [20–22,24,194–201]
3.   Other reviews [10–14]
4.   Book chapters [14,241,242,245,246,256,257]

The numbering of the isoquinolone (**87**) ring is according to the accepted practice [255].

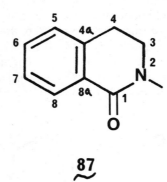

87

2.8.1.   1-Oxo-1,2,3,4-tetrahydroisoquinolines

2.8.1.1.   N-Methylcorydaldine.   Tabular review [255, alkaloid No. 5], IR [232, 255], ^{1}H NMR [232, 255], ^{13}C NMR [258], MS [232, 255].

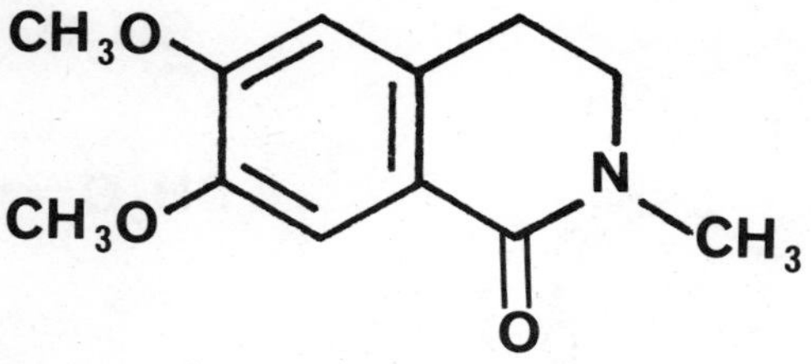

88

N-Methylcorydaldine (**88**), $C_{12}H_{15}NO_3$ (221.1052), was isolated from *Thalictrum fendleri* Engelm. ex Gray as a colorless oil in 1971 [232]. Although it was initially believed that the alkaloid was not new, because a base of the same structure (but with no physical constants) had been isolated from *Hernandia ovigera* L. (Hernandiaceae) some 5 years earlier [259], the structure of that base was subsequently determined to be 1-oxo-2-methyl-6,7-dimethoxy-1,2-dihydroisoquinoline (**89**) [260] and thus N-methylcorydaldine (**88**) was indeed a new alkaloid, with its source being *T. fenderi.*

The IR spectrum of N-methylcorydaldine showed a strong lactam carbonyl absorption at 1634 cm^{-1} and the ¹H NMR spectrum indicated the presence of a lactam N-methyl group (δ 3.10), two methoxy groups (δ 3.85, 3.90), two aromatic protons (δ 6.58 [H(5)] and 7.58 [H(8)] all as singlets, and two adjacent methylene groups as symmetrical triplets at δ 2.89 (*J* = 7 Hz) [H(4)] and 3.90 (*J* = 7 Hz) [H(3)] [232]. The mass spectrum was quite revealing with the ions at *m/z* 178 (**90**) (100%) (M—[CH₂=N—CH₃]), 150 (**91**) (**90**—CO), 135 (**90**—CH₃), and 107 (**90**—CH₃—CO), suggesting a retro-Diels–Alder type of fragmentation [232]. The structure was assigned with certainty by a direct comparison with a laboratory oxidative degradation product of the bisbenzylisoquinoline alkaloid cissampareine (**92**) [232,261].

N-Methylcorydaldine (**88**) was found to be devoid of cytotoxicity when measured in monolayer KB cell cultures via the degree of inhibition of protein synthesis [262].

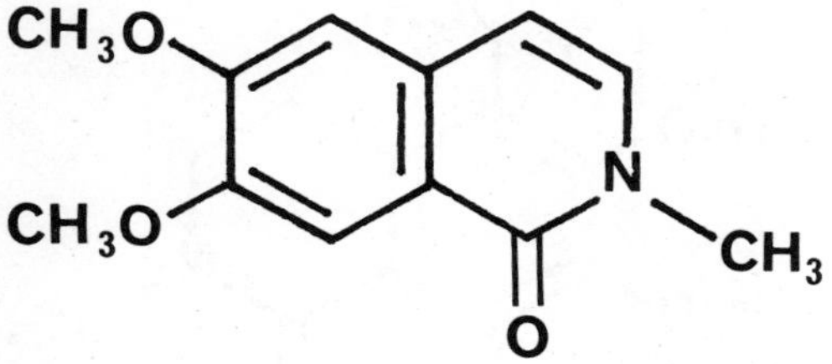

89

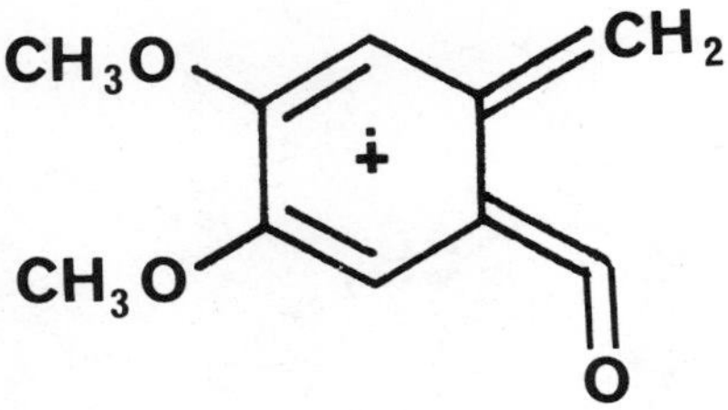

90

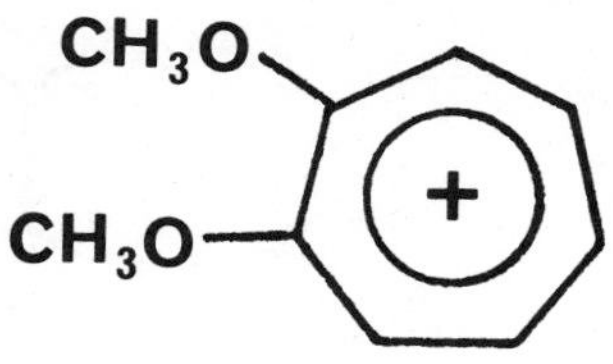

91

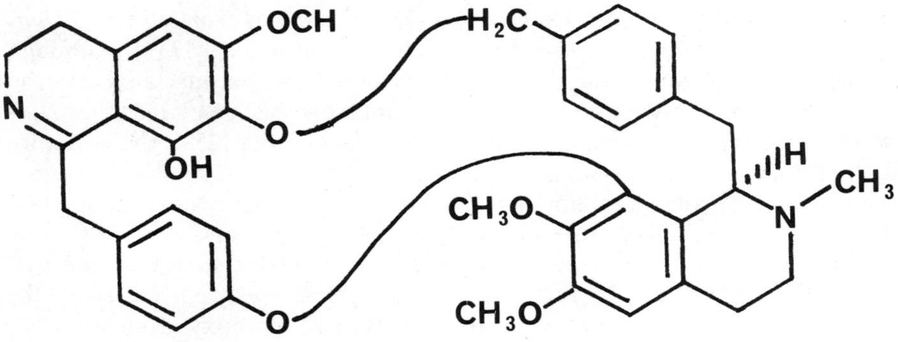

92

2.8.1.2.   *N-Methylthalidaldine*.   Tabular review [255, alkaloid No. 7],
IR [232,255], ¹H NMR [232,255], MS [232,255].
N-Methylthalidaldine (**93**), $C_{13}H_{17}NO_4$ (251.1158), m.p. 104–106° [263], was
isolated as a colorless, optically inactive oil from *Thalictrum fendleri* Engelm. ex Gray in 1971 [232]. The IR spectrum was characterized by a strong
tertiary lactam carbonyl band at 1634 cm⁻¹ and the ¹H NMR spectrum was

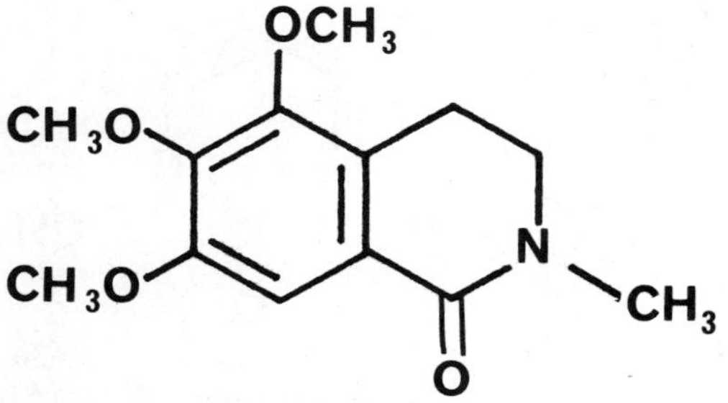

93

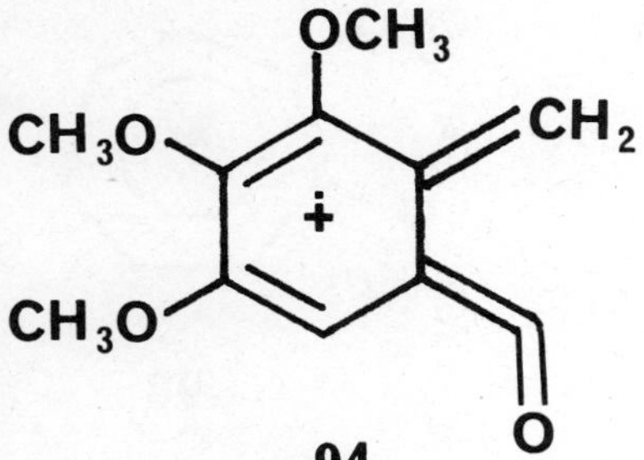

$$94 \qquad 95$$

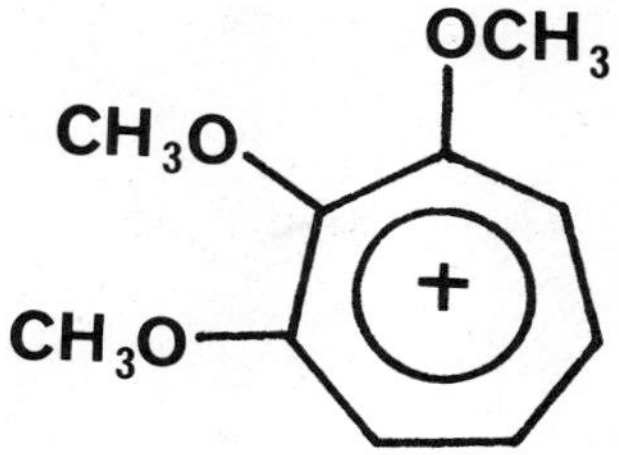

very similar to that of N-methylcorydaldine (**88**) and showed singlets for one lactam N-methyl group (δ 3.11), three methoxy groups (δ 3.83, 3.86, 3.88), and one aromatic proton (δ 7.43). In addition, two symmetrical triplets at δ 2.90 (J = 7 Hz) [H(4)] and 3.50 (J = 7 Hz) [H(3)] were assigned to adjacent methylene groups in the lactam. The mass spectrum also showed great similarity to that of N-methylcorydaldine (**88**) and was characterized by important fragment ions at m/z 221 (100%) (M-30), 208 (M—[CH$_2$=N—CH$_3$]) (**94**), 180 (**94**—CO) (**95**), 165 (**94**—CO—CH$_3$), 150 (**94**—CO—2CH$_3$), 135 (**94**—CO—3CH$_3$), and 107 (**94**—2CO—3CH$_3$). Although these data suggest either a 5,6,7-trimethoxy or a 6,7,8-trimethoxy arrangement in the aromatic ring of 1-oxo-2-methyltetrahydroisoquinoline system, the former was considered more likely because of the proclivity of production of alkaloids with this oxygenation pattern by various *Thalictrum* species. Hence oxidation of naturally occurring ($-$)-thalifendlerine (**65**) with potassium permanganate in acetone gave a tertiary lactam that was identical with naturally occurring N-methylthalidaldine (**93**), thus unequivocally setting the structure of the latter [232].

2.8.1.3. Noroxyhydrastinine. Tabular review [255, alkaloid No. 8], UV [255,264], IR [255,264], ^{1}H NMR [255,264], ^{13}C NMR [255,266], MS [255,264].
Noroxyhydrastinine (**96**), C$_{10}$H$_9$NO$_3$ (191.0582), m.p. 182–183° (MeOH), was isolated from *Thalictrum minus* L. var. *adiantifolium* Hort. in 1969

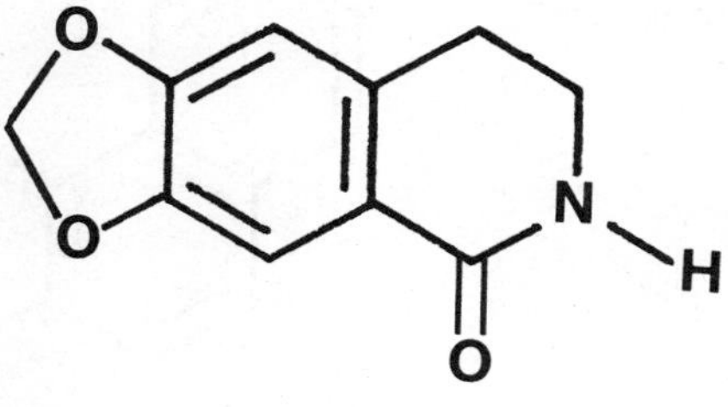

$$96$$

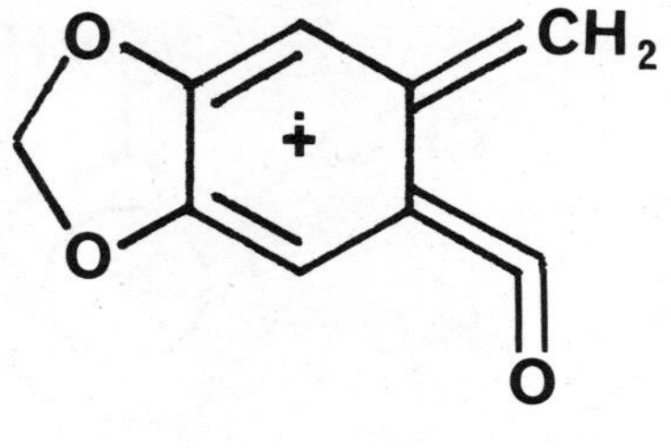

97

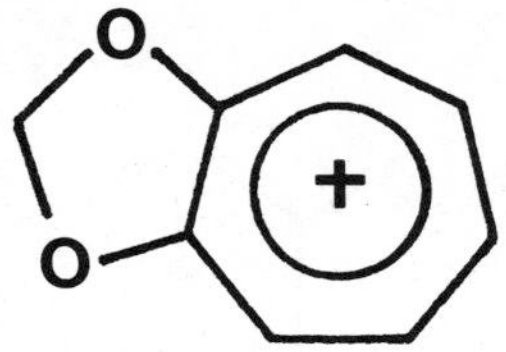

98

[145,264]. The UV spectrum was characterized by three maxima at 222.5 nm (log ϵ 4.31), 261 nm (log ϵ 3.58), and 304 nm (log ϵ 3.67); the IR spectrum showed a prominent band at 1670 cm^{-1} characteristic of a secondary δ-lactam. The ^{1}H NMR spectrum was relatively simple and was characterized by three singlets representing one methylenedioxy group (δ 6.02) and two aromatic protons (δ 6.69 [H(5)] and δ 7.58 [H(8)]) and by two symmetrical triplets at δ 2.88 ($J = 7$ Hz) [H(4)] and δ 3.53 ($J = 7$ Hz) [H(3)] for adjacent methylene groups in a lactam ring. The mass spectrum was most informative and was characterized by an intense parent ion at m/z 191 (80%) and fragment ions at m/z 162 (**97**) (M—[CH$_2$=NH]), 134 (100%) (**98**) (**97**—CO), 104 (**98**—CH$_2$O), and 76 (**98**—CH$_2$O—CO), reflecting the retro-Diels–Alder mode of fragmentation. Oxidation of berberine chloride (**99**) with alkaline potassium permanganate produced noroxyhydrastinine (**96**), thus establishing the structure of the isolated lactam. This was the first reported isolation of noroxyhydrastinine from a natural source even though the compound had long been known in the literature as an oxidation product of berberine (**99**). Noroxyhydrastinine (**96**) was found to be devoid of antimicrobial activity when tested against a routine bank of microorganisms [234]. Noroxyhydrastinine has also been isolated from *Thalictrum alpinum* L. [70], *T.*

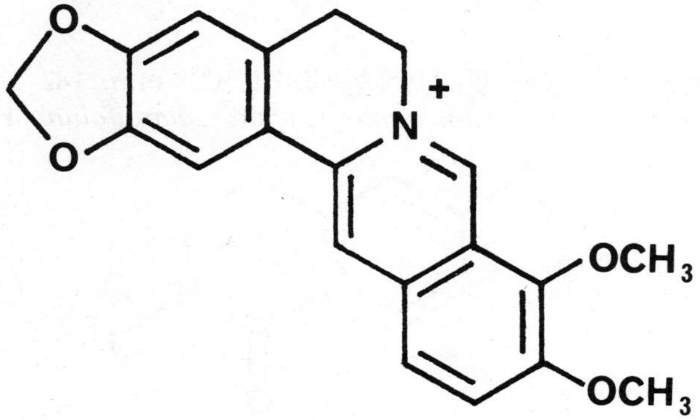

99

foliolosum DC. [268], *T. minus* L. ssp. *elatum* [230], and *T. rugosum* Ait [234].

   2.8.1.4.   Thalflavine.   Tabular review [255, alkaloid No. 13], UV [137,255], IR [137,255,269], ^{1}H NMR [137], MS [137,255,269].

Thalflavine (**100**), $C_{12}H_{13}NO_4$ (235.0845), m.p. 132–133° (Me$_2$CO), 137–139° (CHCl$_3$) [270], was isolated from *Thalictrum flavum* in 1970 [269]. The UV spectrum was characterized by two maxima at 216 nm (log ϵ 4.34) and 280 nm (log ϵ 4.13) [137]. The IR spectrum had a major absorption band at 1625 cm^{-1} [269] (1640 cm^{-1} [137]) corresponding to a tertiary lactam. The ^{1}H NMR spectrum was like that of other isoquinolones and was characterized by a series of uncoupled singlets at δ 3.05 (lactam *N*-methyl) (δ 3.13 [137], 3.84 (one OCH$_3$) (δ 3.92 [137]), 5.95 (CH$_2$O$_2$) (δ 6.03 [137]), and 7.32 (1 ArH) (δ 7.40 [137]). In addition, symmetrical triplets representative of coupled adjacent methylene groups in a δ-lactam were found at δ 2.77 [*J* = 7 Hz, H(4)], δ 2.86 [*J* = 6.5 Hz] [137]), and δ 3.95 [*J* = 8 Hz, H(3)] (δ 3.54 [*J* =

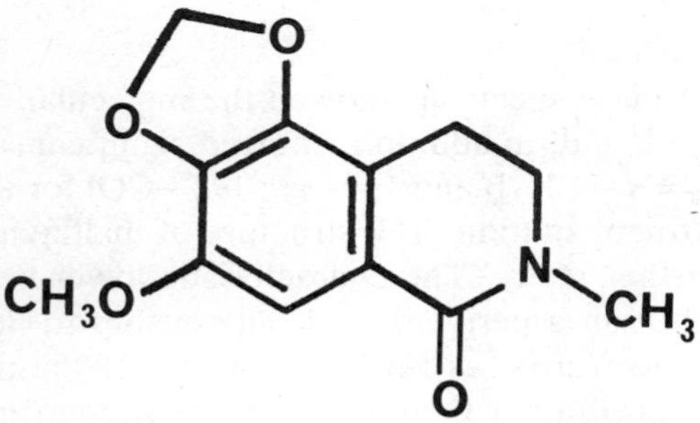

100

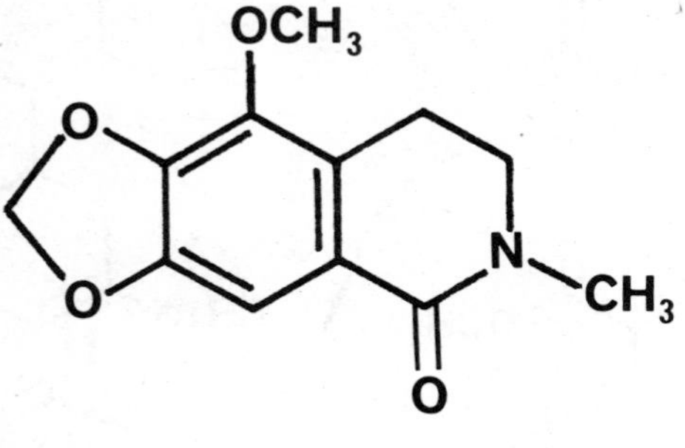

100 a

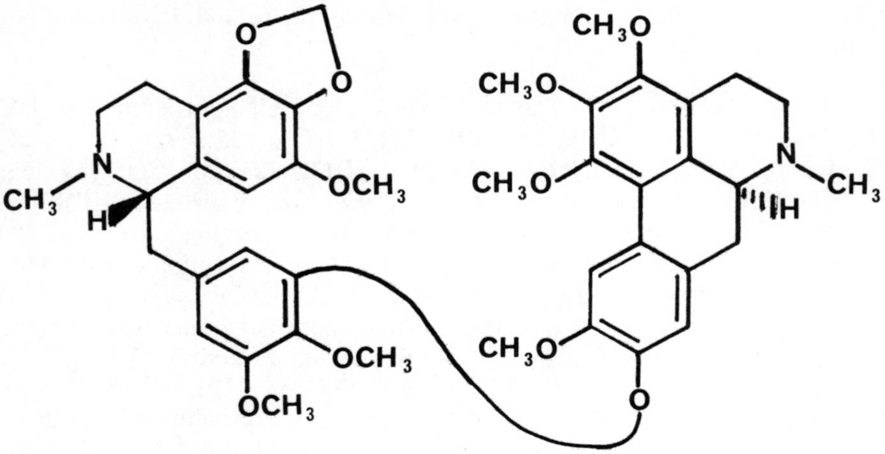

101

6.5 Hz] [137]). The mass spectrum showed the molecular ion at m/z 235 to be also the base peak and, in addition, showed significant fragment ions at m/z 192 (M—[CH$_2$═N—CH$_3$]) and 164 (m/z 192—CO) for the characteristic retro-Diels–Alder fragmentation. The structure of thalflavine was proposed as **100a** without further study. The characteristic lower-field singlet of the H(8) proton; which is in a peri-carbonyl relationship to the C(1) carbonyl group, could have been used as further evidence of the placement of the aromatic proton at C(8) rather than C(5). Apparent confirmation of the validity of thalflavine as **100a** occurred when oxidation (KMnO$_4$ in acetone)

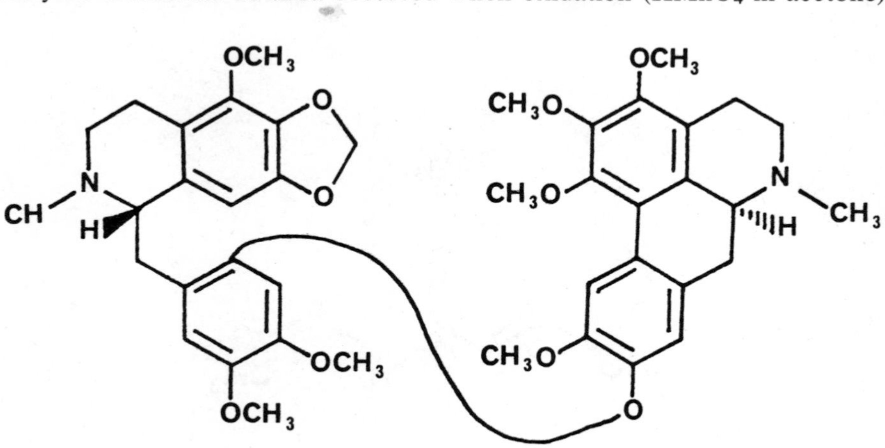

101a

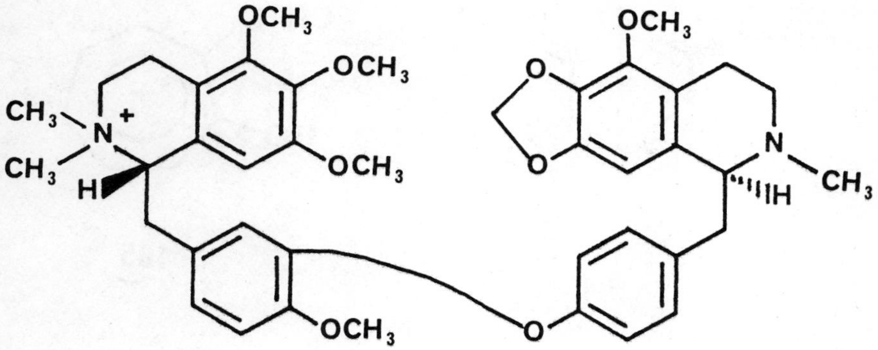

102

of what was the assigned as thalmelatidine (**101a**) produced thalflavine, the structure of the latter of which was apparently confirmed by synthesis [270]. However, a recent communication revises the structure of thalmelatidine to **101** [270a]. This was based on extensive ^{1}H NMR studies, particularly double irradiation techniques and NOE experiments. Hence the structure of thalflavine must be represented as **100**, since thalflavine is a direct oxidation product of thalmelatidine (**101**). In addition, a similar oxidation of thalistyline (**102**), some 7 years later, also produced thalflavine (**100**) [137].

2.8.1.5. Thalifoline. Tabular review [255, alkaloid No. 14], UV [255,264], IR [255,264], ^{1}H NMR [255,264], MS [255,264], X ray [265]. Thalifoline (**103**), $C_{11}H_{13}NO_3$ (207.0895), m.p. 210–211° (MeOH), was isolated from *Thalictrum minus* L. var. *adiantifolium* Hort. in 1969 [145,264], along with the previously described noroxyhydrastinine (**96**). The UV spectrum was very similar to that of noroxyhydrastinine and showed maxima at 223.5 nm (log ϵ 4.41), 261 nm (log ϵ 3.87), and 302 nm (log ϵ 3.77) but with a marked bathochromic shift in the presence of alkali [264]. The IR spectrum

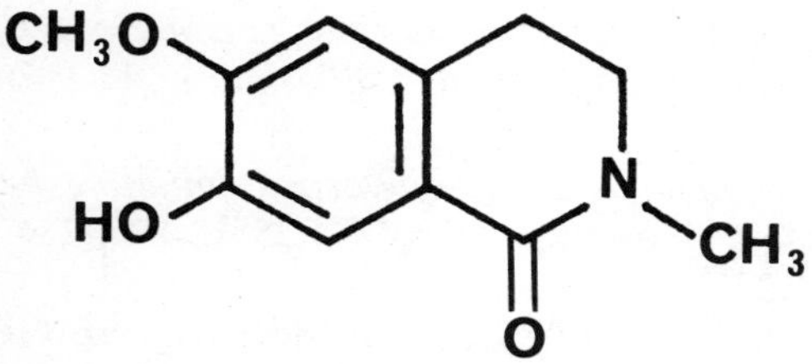

103

104

105

was characterized by a strong carbonyl absorption at 1640 cm^{-1}, typical of a tertiary δ-lactam [264]. The ^{1}H NMR spectrum was notable in its simplicity and showed the presence of four singlets at δ 3.77 (lactam *N*-methyl), 4.03 (3H, methoxy group), δ 6.63 [1H, ArH, H(5)], and 7.68 [1H, ArH, H(8)] [264]. The mass spectrum showed a prominent molecular ion at *m/z* 207 with fragment ions at *m/z* 164 (**104**) (M—[CH$_2$=N—CH$_3$]) 136 (**105**) (**104**—CO), 121 (**105**—CH$_3$), and 93 (**105**—CH$_3$—CO) and was again reminiscent of the familiar retro-Diels–Alder type of fragmentation [264]. Although the data indicated that thalifoline was a 6,7-dioxygenated-2-methyl-1,2,3,4-tetrahydro-1-oxoisoquinoline containing one methoxy and one hydroxy group at the C(6)/C(7) or C(7)/C(6) positions, synthesis was necessary to confirm the placement of the methoxy group. A seven-step synthesis beginning with vanillin (3-methoxy-4-hydroxybenzaldehyde) produced thalifoline (**103**) and thus established the position of the methoxy group at C(6) [264].

2.8.2. 1-Oxo-1,2-dihydroisoquinolines

2.8.2.1. N-Methyl-6,7-Dimethoxy-1,2-Dihydroisoquinolin-1-one. Tabular review [255, alkaloid No. 6], UV [70,175,255], IR [70,175,255], ^{1}H NMR [70,175,255], MS [70,175,255].

N-Methyl-6,7-dimethoxy-1,2-dihydroisoquinolin-1-one (**89**), C$_{12}$H$_{13}$NO$_3$ (219.0895), m.p. 104–105° (Et$_2$O) [175], 112–113° (CHCl$_3$–MeOH) [70], was first isolated from *Hernandia ovigera* L. (Hernandiaceae) in 1966 [259] and was synthesized via an isocoumarin intermediate in 1978 [267]. The alkaloid was first isolated from a *Thalictrum* species, *T. isopyroides* C.A.M., in 1978 [175] and subsequently isolated from *T. alpinum* L. in 1980 [70].

2.8.2.2. 6,7-Methylenedioxy-1,2-dihydroisoquinoline-1-one. Tabular review [255, alkaloid No. 10], UV [234,255], IR [234,255], ^{1}H NMR [234,255], MS [234,255].

6,7-Methylenedioxy-1,2-dihydroisoquinolin-1-one (**106**), C$_{10}$H$_7$NO$_3$ (189.0426), m.p. 268–270° dec. (CHCl$_3$), was first isolated as a natural product from *Thalictrum rugosum* Ait. in 1980 [234], although it had been previously synthesized in 1969 [271]. The UV spectrum was more complex than

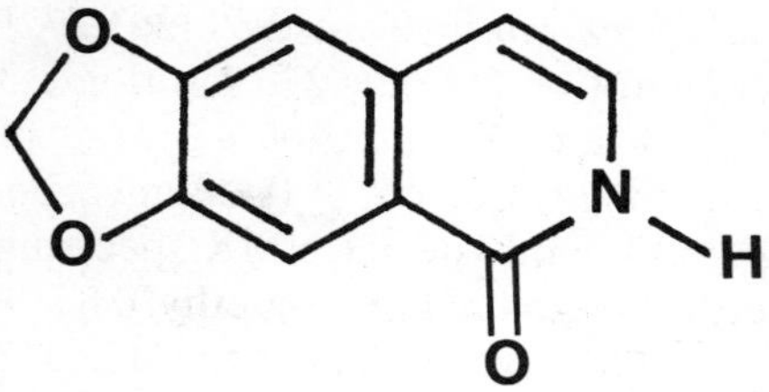

106

that of its 3,4-dihydro analog noroxyhydrastinine (**96**) and showed a series of maxima at 268 nm (sh) (log ϵ 3.60), 282 nm (log ϵ 3.58), 293 nm (log ϵ 3.59), 312 nm (sh) (log ϵ 3.37), 326 nm (log ϵ 3.47), and 340 nm (log ϵ 3.36), thus suggesting enhanced conjugation. The IR spectrum showed character-istic bands for a conjugated, secondary δ-lactam at 3400 cm^{-1}(NH) and 1660 cm^{-1} (carbonyl). The ^{1}H NMR spectrum indicated the loss of adjacent meth-yene groups as in noroxyhydrastinine (**96**) and their replacement with olefinic protons at δ 6.42 [d, J = 7 Hz, H(4)] and 7.01 [brd, J = 7 Hz, H(3)]. In addition, the presence of singlets at δ 6.08 (2H), 6.89 [H(5)], and 7.76 [H(8)] suggested the presence of a 6,7-methylenedioxy function and *para* protons in an aromatic ring. The mass spectrum (EI) reflected the added conjugation in showing the molecular ion and base peak at *m/z* 189, with few fragment ions of significant intensity and was supported by the appropriate M + 1 peak at *m/z* 190 in the CI mass spectrum (isobutane). The assignment of structure was entirely via interpretation of the spectroscopic data and is consistent with that data. No trivial name was assigned to the new alkaloid [234].

   2.8.2.3.   Thalactamine.   Tabular review [255, alkaloid No. 12], UV [217,255,263], IR [217,255,263], ^{1}H NMR [217,255,263], MS [217,255,263]. Thalactamine (**107**), $C_{13}H_{15}NO_4$ (249.1001), m.p. 112–114° [217,263], was

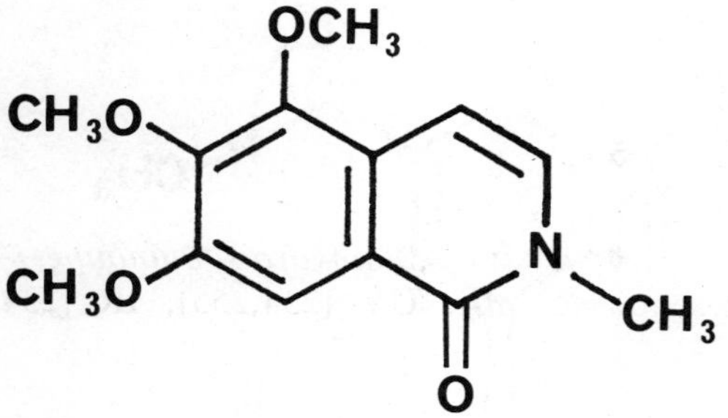

107

first isolated from *Thalictrum minus* L. in 1969 [263,272]. The UV spectrum showed maxima at 247 nm (log ε 4.62), 270 nm (log ε 3.60), 281 nm (log ε 3.68), 293 nm (log ε 3.74), and 332 nm (log ε 3.48) while the IR spectrum was characterized by bands at 1660 cm^{-1} (lactam carbonyl) and 1620 cm^{-1} (conjugated double band) [263]. The ^{1}H NMR spectrum was characterized by four distinct singlets at δ 3.45 (lactam *N*-methyl), δ 3.90 (6H, two methoxy groups), 3.93 (3H, one methoxy group) and 7.56 (1H, aromatic proton) [H(8)] and an AB-type quartet at δ 6.58 [H(4)] and 6.91 [H(3)] with *J* = 12 Hz [263]. Later ^{1}H NMR spectral data generally agreed with these assignments except that the H(3) and H(4) protons were assigned as a pair of doublets at δ 7.01 (*J* = 7.5 Hz) and 6.72 (*J* = 7.5 Hz), respectively [217]. The mass spectrum supported the hypothesis of a trimethoxylated *N*-methyl-1,2-dihydroisoquinolin-1-one structure with the parent ion at *m/z* 249. Finally, high-pressure (80 atm) hydrogenation at 80°C of thalactamine afforded the dihydro derivative **93** [263], whose structure was confirmed by synthesis [263,267] and which was later isolated as a new alkaloid, named *N*-methylthalidaldine (**93**), from *T. fendleri* Engelm. ex Gray in 1971 [232]. Thalactamine has also been isolated from *T. minus* L. var. *microphyllum* Boiss. [217].

2.9. Morphinandienones

The following reviews and references are particularly relevant to the isolation and identification of morphinandienone alkaloids.

1. Tabular reviews [18,19,273]
2. Annual reviews [20–22,24,194–201]

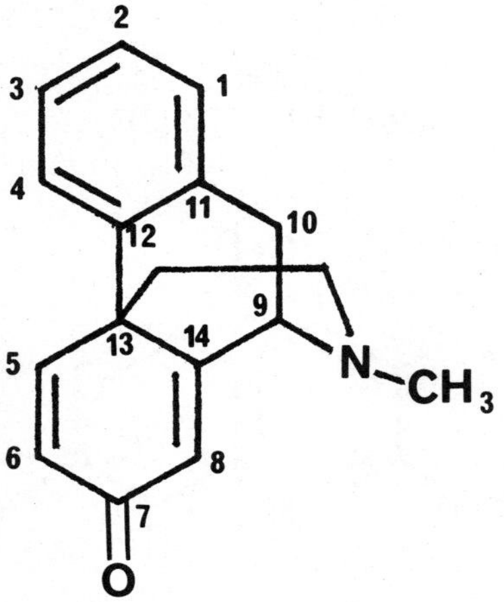

108

3. Other reviews [273]
4. Book chapters [240–242,274–276]

The numbering of the morphinan (108) ring is according to accepted practice [273].

2.9.1. Pallidine. Tabular review [273], UV [72,273], IR [72,273], ^{1}H NMR [72,273], MS [72].

Pallidine (58) [(S)-2-hydroxy-3,6-dimethoxymorphinandienone], $C_{19}H_{21}NO_4$ (327.1471), amorphous, $[\alpha]_D^{25}$ − 19° (c 0.3, MeOH) [72], was first isolated from *Corydalis pallida* var. *tenuis* Yatabe (Papaveraceae) in 1969 [277]. The spectral data of this alkaloid are typical of the class and include UV maxima at 210 nm (log ϵ 4.23), 239 nm (log ϵ 3.79), and 282 nm (log ϵ 3.52) and strong IR bands at 1668, 1640, and 1620 cm^{-1} for the cross-conjugated cyclohexadienone system [72]. The ^{1}H NMR spectrum shows the presence of a N-methyl group at δ 2.43 (s); two methoxy groups at δ 3.78 [s, C(6)] and δ 3.87 [s, C(3)]; four aromatic protons at δ 6.28 [s, H(8)], 6.40 [s, H(1) or H(4)], 6.65 [s, H(4) or H(1)], and 6.80 [s, H(5)], with a cluster of aliphatic protons from δ 1.66 to 3.80 (m, 8H) and the phenol as a broad singlet at δ 5.83 [72]. The mass spectrum shows a prominent parent ion and other significant fragment ions at m/z 312 (M—CH$_3$), 299 (M—CO), and 284 (100%) (M—[CH$_2$=NCH$_3$]) [72].

Reports of the pharmacological activity of morphinandienone alkaloids are not abundant in the literature. However, studies on O-methylflavinantine, (R)-2,3,6-trimethoxy-morphinandienone (109), the enantiomeric O-methyl ether of pallidine (58), have shown that this alkaloid inhibits the

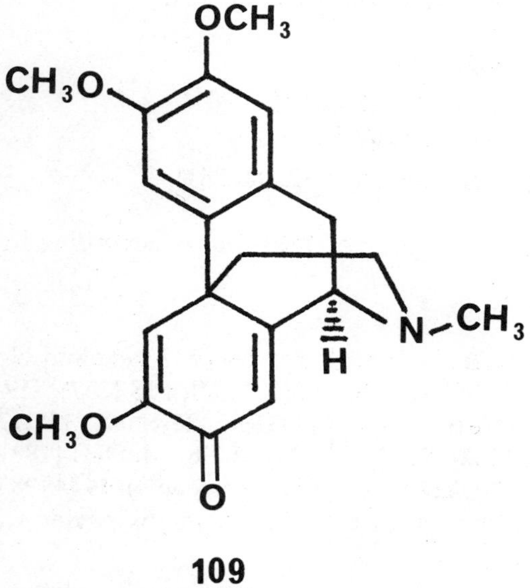

109

peristaltic reflex in guinea pig ileum, an action that was not antagonized by nalorphine [278]. Furthermore, it was found that the alkaloid possessed centrally mediated anti-nociceptive activity similar to that of morphine [278]. This was demonstrated in mouse hot plate and abdominal constriction tests using various nociceptive agents which included 5-hydroxytryptamine, acetylcholine, bradykinin, prostaglandin E_1, formic acid, and phenylquinone. The potency of the alkaloid in the hot plate test was about 10 times less than that of morphine and the effect was naloxone reversible. Morphine was considerably more potent (78–650×) than *O*-methylflavinantine in the abdominal constriction test, depending on the nociceptive agent used, but with a higher dose of nalozone necessary to reverse the response to formic acid. Pretreatment of mice with reserpine reduced and α-methyl-*p*-tyrosine potentiated the antinociceptive effects of both morphine and *O*-methylflavinantine in the hot plate test [278].

Pallidine (**58**) was first isolated from *Thalictrum dioicum* L. in 1973 when it was identified via preparation of its *O*-acetyl ester [106] and later reisolated from the same plant in 1976 [130] and 1978 [72]. Finally, it was most recently isolated from *T. faberi* Ulbr. in 1984 [279]. This alkaloid is the only morphinandienone to have been isolated from a *Thalictrum* species to date and was found in the same species (*T. dioicum* L.) growing in central Pennsylvania [72,106] in the United States and around Montreal [130].

2.10. Pavines

The following reviews and references are particularly relevant to the isolation and identification of pavine alkaloids:

1. Tabular reviews [10,18,19,237]
2. Annual reviews [20–22,24,194–201]
3. Other reviews [10–13]
4. Book chapters [14,240–242,280,281]

The numbering of the pavine (**110**) ring is according to accepted practice [237].

2.10.1. Argemonine.   Tabular review [237, alkaloid No. 1], UV [237], IR [237], ^{1}H NMR [104,237], ^{13}C NMR [237], MS [237], ORD [237], CD [237]. Argemonine (*N*-Methylpavine) (**111**), $C_{21}H_{25}NO_4$ (355.1783), m.p. 156–157° (Et$_2$O–EtOAc) [282], $[\alpha]_D^{26}$ −200° (c 0.98, MeOH) [282], was first isolated from *Argemone hispida* Gray (Papaveraceae) in 1944 [283]. However, it was some 20 years before its structure as *N*-methylpavine was settled with certainty [280].

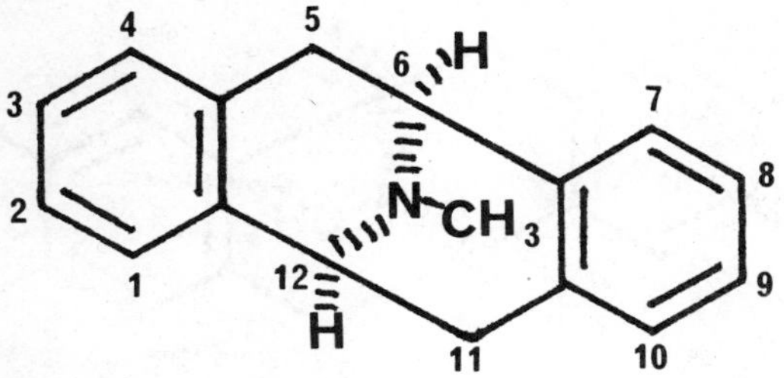

110

Argemonine was found to exhibit weak in vitro antimicrobial activity against the acid-fast microorganism *Mycobacterium smegmatis* [68]. The stereochemical isomers of *N,N'*-decamethylenebis(*N*-methylpavinium iodide) were used as probes in the evaluation of curarimimetic neuromuscular junction blockade in the mouse inclined screen assay and the cat tongue–hypoglossal nerve technique. The (*R,R*) isomer showed a modest, statistically significant increase in potency over the (*S,S*) isomer in the cat assay, with no difference in the mouse assay [285]. (−)-Argemonine (**111**) was first isolated from a *Thalictrum* species, *T. dasycarpum* Fisch. and Lall., in 1968 [282] and subsequently from *T. minus* L. [63,65,181], *T. revolutum* DC. [104,105], and *T. strictum* Ledeb. [155,160].

2.10.2. Bisnorargemonine. Tabular review [237, alkaloid No. 7], UV [237,286], IR [237,286], ^{1}H NMR [237], MS [237], ORD [237], CD [237]. Bisnorargemonine (**112**), $C_{19}H_{21}NO_4$ (327.1471), m.p. 244–246° (EtOH) [207], $[\alpha]_D^{25}$ −265.8° (c 0.158, MeOH) [286], was first isolated from *Argemone munita* subsp. *rotundata* (Papaveraceae) in 1960 and given the name rotundine [286] (a name previously designated for another alkaloid isolated from *Stephania rotunda* Loureiro (Menispermaceae) in 1944 [286a]). The alkaloid has been isolated from only one *Thalictrum* species, *T. dasycarpum* Fisch. and Lall., in 1969 [207].

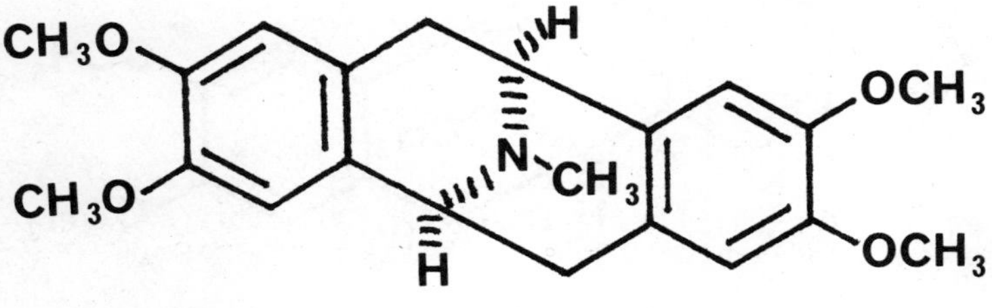

111

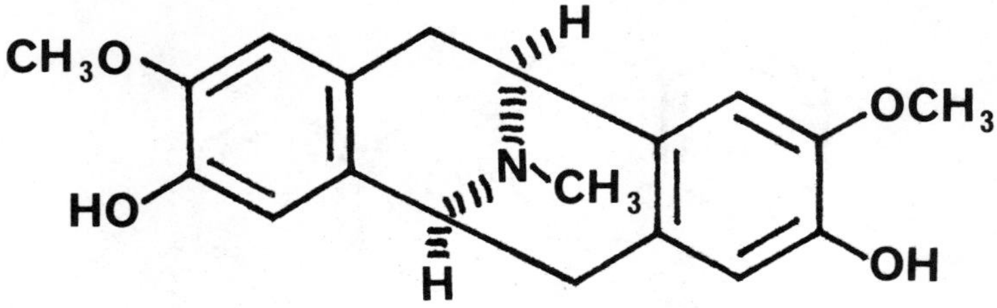

112

2.10.3. Eschscholtzidine. Tabular review [237, alkaloid No. 8], UV [237], ¹H NMR [104,237], MS [237], ORD [237], CD [237].
Eschscholtzidine (**113**), $C_{20}H_{21}NO_4$ (339.1471), amorphous [237], oil [104], $[\alpha]_D^{24}$ − 194.2° (c 1.56, MeOH) [286], was first isolated from *Eschscholtzia californica* Cham. (Papaveraceae) in 1966 [287]. Eschscholtzidine was observed to be only weakly active in in vitro testing against *Mycobacterium smegmatis* [68]. This pavine was isolated from *Thalictrum revolutum* DC. in 1977 [104] and 1980 [105] and from *T. minus* L. in 1980 [65].

2.10.4. Isonorargemonine. Tabular review [237, alkaloid No. 6], UV [68,237], ¹H NMR [68,237], CD [68,237].
Isonorargemonine (**114**), $C_{20}H_{23}NO_4$ (341.1627), m.p. 219–221° (MeOH–H₂O) [237,288], $[\alpha]_D^{25}$ − 202° (c 3.31, CHCl₃) [237], was first isolated from *Argemone gracilenta* Greene (Papaveraceae) in 1969 [288]. The alkaloid was isolated from *Thalictrum revolutum* DC. in 1977 and found to be devoid of in vitro antimicrobial activity when tested against a standard set of microorganisms [68].

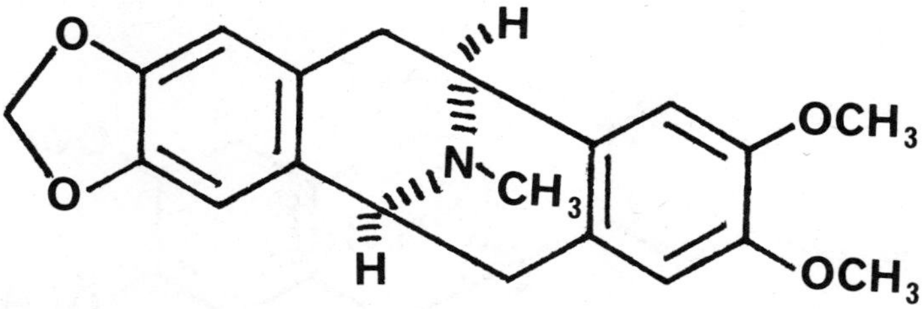

113

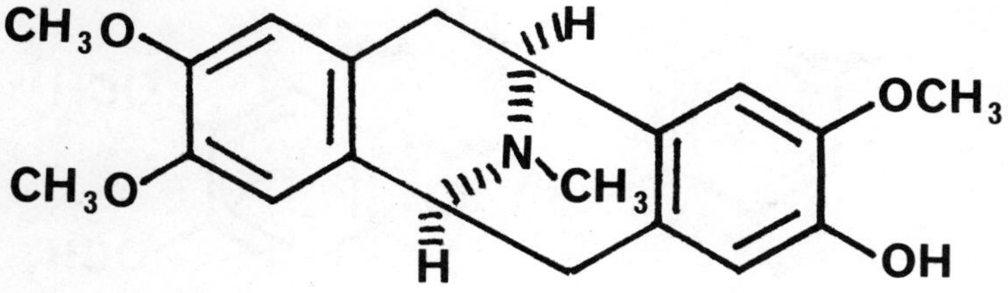

114

2.10.5. N-Methylargemonine.

Tabular review [232, alkaloid No. 3], UV [105,237], IR [237], ^{1}H NMR [105,237], MS [105,237], CD [105,237].

N-Methylargemonine (**115**), $C_{22}H_{28}N^+O_4$ (370.2018), m.p. 170–172° (abs. EtOH) (chloride) [105], 149–159° (MeOH) (chloride) [105], 272–273° dec. (MeOH) (iodide) [105], 274–275° (MeOH) (perchlorate) [237]; $[\alpha]_D^{25}$ −170° (c 2.81, CHCl$_3$) (hydroxide) [237], $[\alpha]_D^{25}$ −200 ± 6° (c 0.14, MeOH) (iodide) [237], was first isolated from *Argemone gracilenta* Greene (Papaveraceae) as its hydroxide salt in 1969 [288].

(S,S)- and (R,R)-N-Methylargemonine iodide (N-methylpavine methiodide) were used as probes for stereochemical preferences for curarimimetic neuromuscular junction blockade in various assays, with the result that the (S,S) isomer had only a slightly enhanced activity over the (R,R) isomer [56].

N-Methylargemonine chloride was isolated from *Thalictrum revolutum* DC. in 1980 [105] and from *T. minus* L. in 1983 [181].

2.10.6. N-Methyleschscholtzidine.

Tabular review [237, alkaloid No. 10], UV [105,237], ^{1}H NMR [105,237], CD [105,237].

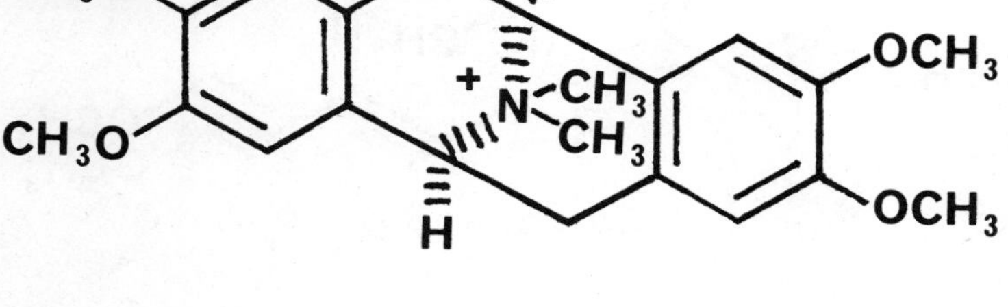

115

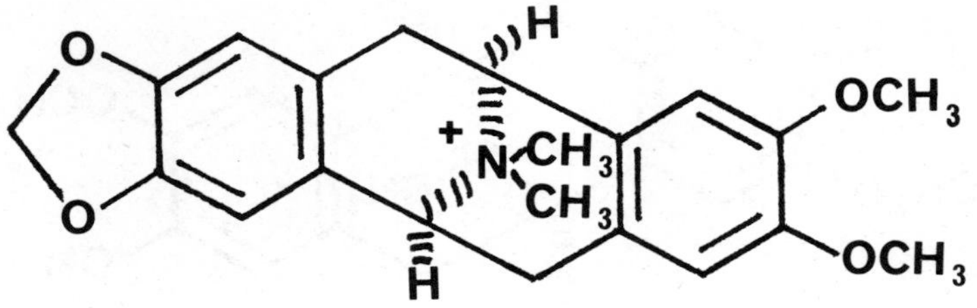

116

N-Methyleschscholtzidine (**116**), $C_{21}H_{24}N^+O_4$ (354.1705), was first isolated as its amorphous chloride salt, $[\alpha]_D^{20} - 170°$ (c 0.26, MeOH), from *Thalictrum revolutum* DC. in 1980 [105]. The UV spectrum was characterized by two maxima at 257 nm (log ϵ 3.46) and 289 nm (log ϵ 3.74) with a shoulder at 230 nm (log ϵ 3.91). The ^{1}H NMR spectrum showed major signals at δ 3.68 (brs, $^+N(CH_3)_2$), 3.79 (s, OCH_3), 3.86 (s, OCH_3), 5.89 (brs, CH_2O_2), 6.48 (s, 1ArH), 6.80 (s, 1ArH), and 6.85 (s, 1ArH) with a broad doublet centered at δ 5.57 (2H). The structural assignment was made after a comparison of the ^{1}H NMR spectrum of *N*-methyleschscholtizidine iodide (prepared from eschscholtizidine (MeI–MeOH) with that of the isolate [105].

2.10.7. Norargemonine. Tabular review [237, alkaloid No. 5], UV [237], IR [237], ^{1}H NMR [237], MS [237], ORD [237], CD [237].
Norargemonine (**117**), $C_{20}H_{23}NO_4$ (341.1627), m.p. 239–242° (Me$_2$CO) [207], $[\alpha]_D^{25} - 150°$ (c 0.34, CHCl$_3$) [237] was first isolated from *Argemone hispida* Gray (Papaveraceae) in 1944 [283]. The alkaloid was isolated from *Thalictrum dasycarpum* Fisch. and Lall. in 1969 [207].

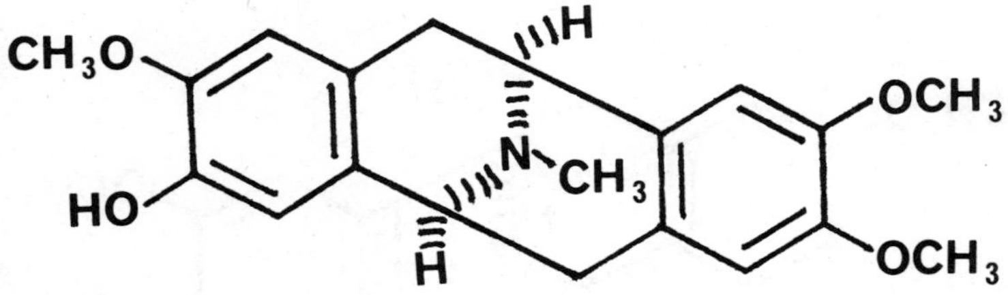

117

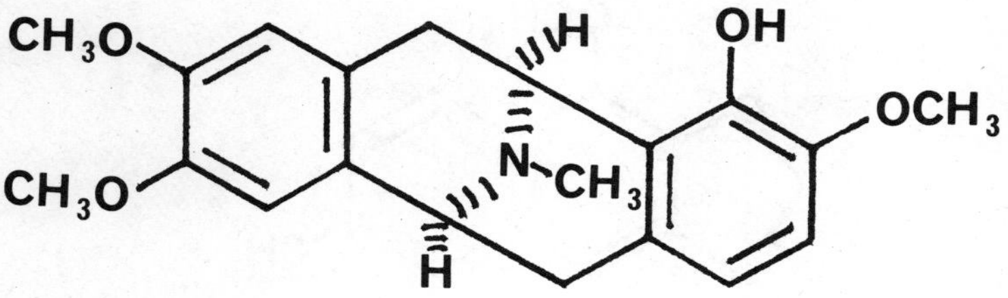

118

2.10.8. Platycerine.

Tabular review [237, alkaloid, No. 17], UV [68,237], IR [237], ^{1}H NMR [68,237], MS [237], CD [68,237].

Platycerine (**118**), $C_{20}H_{23}NO_4$ (341.1627), m.p. 130–132° (Et$_2$O) [237], amorphous [68]; $[\alpha]_D^{28}$ − 305° (c 0.2, MeOH) [68], was first isolated from *Argemone platyceras* Link and Otto (Papaveraceae) in 1963 [289]. The alkaloid was isolated from *Thalictrum revolutum* DC. in 1977 [68] and 1980 [105] and was found to be devoid of in vitro antimicrobial activity against standard microorganisms [68].

2.10.9. 2,3-Methylenedioxy-4,8,9-trimethoxypavinane.

Tabular review [237, alkaloid No. 21], UV [161,237], ^{1}H NMR [161,237], MS [161,237].

2,3-Methylenedioxy-4,8,9-trimethoxypavinane (**119**), $C_{21}H_{23}NO_5$ (369.1576), m.p. 144–145° (Et$_2$O) [161], $[\alpha]_D^{22}$ − 174° (c 0.977, MeOH) [161], was isolated from *Thalictrum strictum* Ledeb. in 1976 [155,161]. The UV spectrum was characterized by a single absorption at 287 nm (log ϵ 3.84) and the ^{1}H NMR spectrum showed singlets at δ 2.46 (*N*-methyl), 3.72 (one methoxy group), 3.80 (two methoxy groups), 6.23, 6.36, and 6.54 (each one aromatic proton) [161]. In addition, the presence of a methylenedioxy group

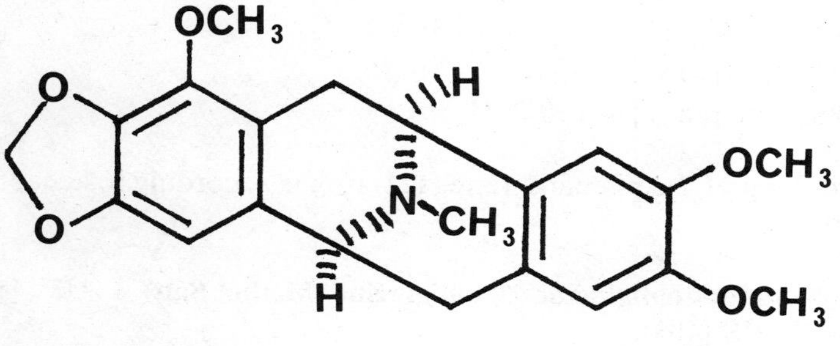

119

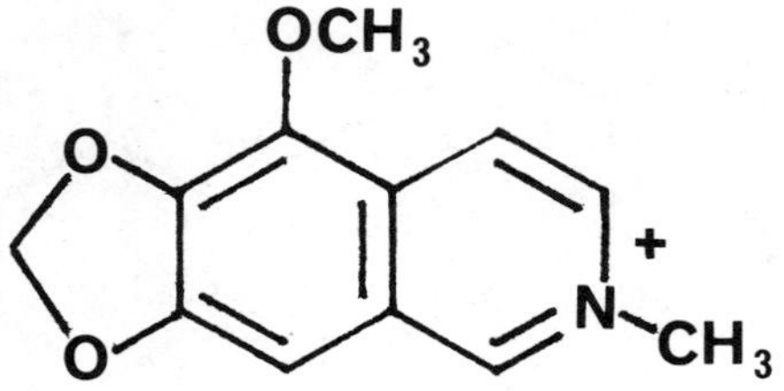

OCH$_3$

120

was noted as a doublet of doublets at δ 5.75 (1H, J = 1.5 Hz) and 5.80 (1H, J = 1.5 Hz), with a cluster of signals from δ 2.40 to 4.05 representing six methylene/methine protons [161]. The mass spectrum showed peaks at m/z 369 (M$^+$), 368 (M-1), and 354 (M-15), with strong diagnostic fragment ions at m/z 218 (70%) and 204 (100%) as represented by **120** and **81**, respectively [161]. On the basis of these spectral data, biogenetic considerations, and the previous isolation of 1,2-methylenedioxy-3-methoxy type alkaloids, including ocoteine (thalicmine) (**31**), *O*-methylcassyfiline (northalicmine) (**30**), and thalicminine (**47**), from *T. strictum* Ledeb. [60], the newly isolated pavine alkaloid was assigned structure **119** as the most probable choice, with no trivial name being cited [161]. Confirmation of the validity of this assigment awaits degradative and/or synthetic studies.

2.11. Phenanthrenes

The following reviews and references are particularly relevant to the isolation and identification of phenanthrene alkaloids:

1. Tabular reviews [10,15–19]
2. Annual reviews [20–22,24,294–301]
3. Other reviews [10–13]
4. Book chapters [14,290,291]

The numbering of the phenanthrene (**121**) ring is according to accepted practice [290].

2.11.1. *N*-Methylthaliglucine (Thaliglucine Metho Salt).   UV [103], ^{1}H NMR [103], MS [103].

N-Methylthaliglucine (**122**), $C_{22}H_{24}N^+O_4$ (366.1705), m.p. 249–250° (Et$_2$O–MeOH), was first isolated as its chloride salt from *Thalictrum polygamum* Muhl. in 1974 [103]. The UV spectrum was very similar to that of thaliglucine (**123**) [91] and showed a series of maxima at 233 nm (sh) (log ε 4.22), 260

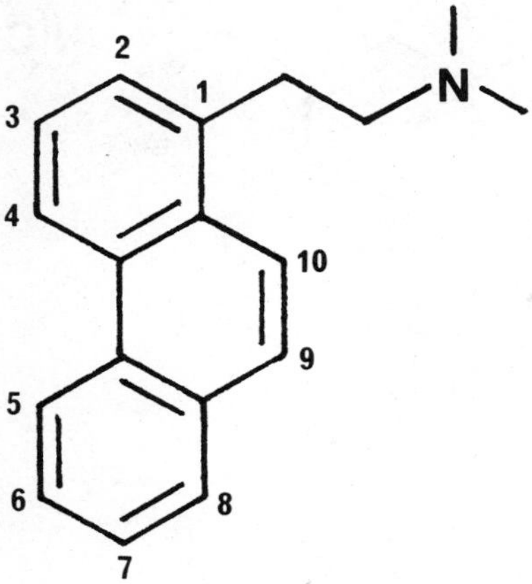

121

nm (log ε 4.42), 272 nm (log ε 4.46), 282 nm (log ε 4.45), 295 nm (sh) (log ε 4.23), 326 nm (log ε 3.82), 340 nm (log ε 3.71), 359 nm (log ε 3.32), and 370 nm (log ε 3.32). In addition, the ^{1}H NMR spectrum also resembled that of thaliglucine (**123**) [97], except for the addition of a singlet (9H) at δ 3.27 for the quaternary *N*-methyl group in *N*-methylthaliglucine. Other signals in the spectrum of *N*-methylthaliglucine were observed at δ 3.99 (3H, s, OCH$_3$), 5.53 (2H, s, ArCH$_2$O), 6.08 (2H, s, CH$_2$O$_2$), 7.08 [1H, s, H(2)], 7.15 [1H,

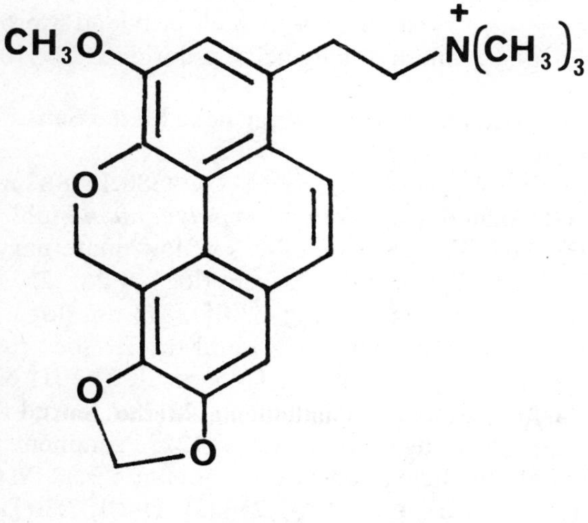

122

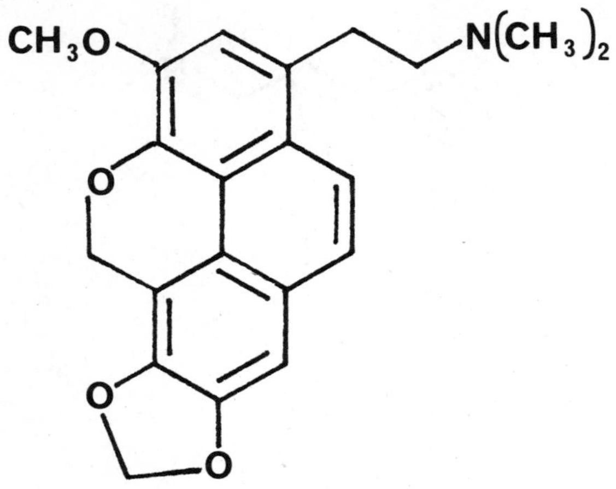

123

s, H(8)], and 7.47 [ABq, J = 9 Hz, 2H, H(9) and H(10)]. Finally, the mass spectrum of *N*-methylthaliglucine was also similar to that of thaliglucine (**123**) [97] and was characterized by facile β cleavage, with the spectrum of latter exhibiting a base peak at m/z 58 [CH$_2$=$\overset{+}{\text{N}}$(CH$_3$)$_2$], whereas the former was characterized by a base peak at m/z 73 [CH$_2$—$\overset{+}{\text{N}}$(CH$_3$)$_3$]. Confirmation of structure of *N*-methylthaliglucine chloride (**122**) was affected via quaternization of thaliglucine (**123**) with methyl iodide followed by conversion (anion-exchange resin) to the chloride, which provided a product identical to naturally occurring *N*-methylthaliglucine chloride (**122**) [103].

2.11.2. *N*-Methylthaliglucinone (Thaliglucinone Metho Salt).   UV [103], IR [103], ^{1}H NMR [103].

N-Methylthaliglucinone (**124**), $C_{22}H_{22}N^+O_5$ (380.1498), m.p. 274–275° (MeOH), was first isolated from *Thalictrum polygamum* Muhl. as its chloride salt in 1974 [103]. The UV spectrum of this yellow quaternary base showed maxima at 225 nm (sh) (log ε 4.15), 237 nm (log ε 4.26), 257 nm (sh) (log ε 4.36), 267 nm (log ε 4.50), 288 nm (log ε 3.85), 313 nm (log ε 3.99), 333 nm (sh) (log ε 3.71), and 400 nm (log ε 3.61) and the IR spectrum displayed a prominent carbonyl absorption band at 1735 cm^{-1}. The ^{1}H NMR spectrum was similar to that of *N*-methylthaliglucine (**122**) but lacked the singlet at δ 5.53 for the methylene bridge (ArCH_2O) of **122**. Prominent singlets in the spectrum of *N*-methylthaliglucinone were found at δ 3.32 [9H, N$^+$(CH$_3$)$_3$], 4.10 [3H, OCH$_3$], 6.34 (2H, CH$_2$O$_2$), 7.25 [1H, H(2)], 7.36 [1H, H(8)], and an AB quartet was observed at δ 7.65 [2H, J = 9 Hz, H(9) and H(10)]. Unequivocal structural confirmation was provided by conversion of thal-

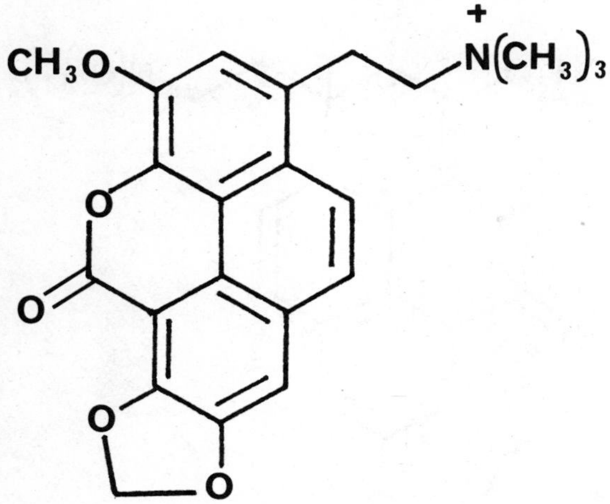

124

iglucinone (**125**) to *N*-methylthaliglucinone chloride via quaternization with methyl iodide followed by anion exchange via resin [103].

2.11.3. Thalflavidine. Tabular review [15, alkaloid No. 174], UV [293], IR [293], ^{1}H NMR [293], MS [293].

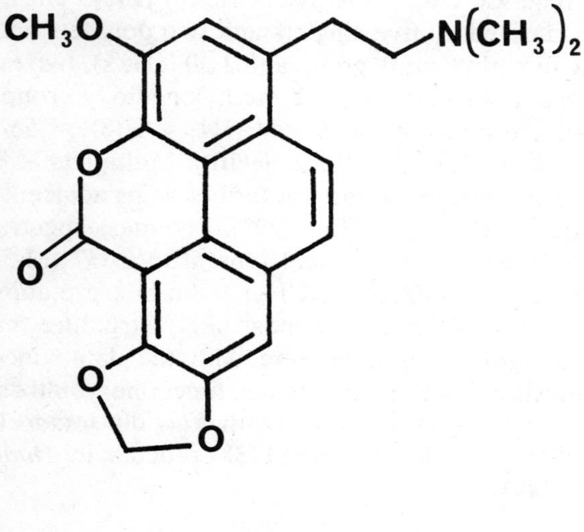

125

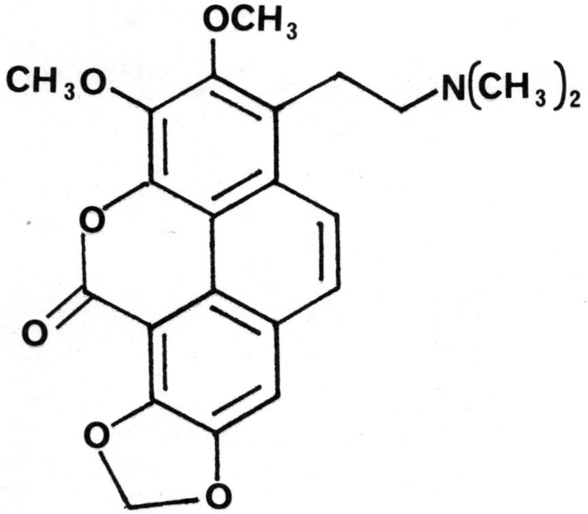

126

Thalflavidine (**126**), $C_{22}H_{21}NO_6$ (395.1369), m.p. 219–220° (MeOH), was iso-lated as a yellow, optically inactive base from *Thalictrum flavum* L. in 1973 [293]. The UV spectrum was characterized by four maxima at 254 nm (log ϵ 4.38), 263 nm (log ϵ 4.49), 296 nm (log ϵ 3.94), and 400 nm (log ϵ 3.60); the IR spectrum displayed prominent absorption bands at 1730 cm^{-1} (α-pyrone carbonyl) and 1060/950 cm^{-1} (methylenedioxy) [293]. The ^{1}H NMR spec-trum was characterized by five singlets and two doublets and indicated the presence of one dimethylamino group at δ 2.30 (6H, s), two methoxy groups at δ 3.82 (3H, s) and 4.04 (3H, s), one methylenedioxy group at δ 6.29 (2H, s), and three aromatic protons at δ 7.44 [1H, s, H(8)], 7.56 [1H, d, J = 9 Hz, H(9)], and 7.81 [1H, d, J = 9 Hz, H(10)]. Multiplets at δ 2.50 and 3.26 were assigned to methylene groups, the former being adjacent to an aromatic ring and the latter to a nitrogen atom [293]. The mass spectrum showed the parent ion at m/z 395 with prominent ions at m/z 337 (M-58), 322 (M-58-CH$_3$), 279, and 58 (100%) [CH$_2$=$\overset{+}{\text{N}}$(CH$_3$)$_2$]. An interpretation of these data led to the postulation of **126** as the most likely structure for thalflavidine. Although this assignment is consistent with the data, unequivocal proof awaits the synthesis of thalflavidine. In due time, one would expect the report of the presence of the deoxy analog of thalflavidine, namely, **127**, just as thaliglucine (**123**) and thaliglucinone (**125**) co-occur in *Thalictrum polyga-mum* Muhl. [97,149].

2.11.4. Thalicsine (Thalixine). Tabular review [15, alkaloid No. 173], UV [15,297], IR [15,297], ^{1}H NMR [15,297], MS [15,297].

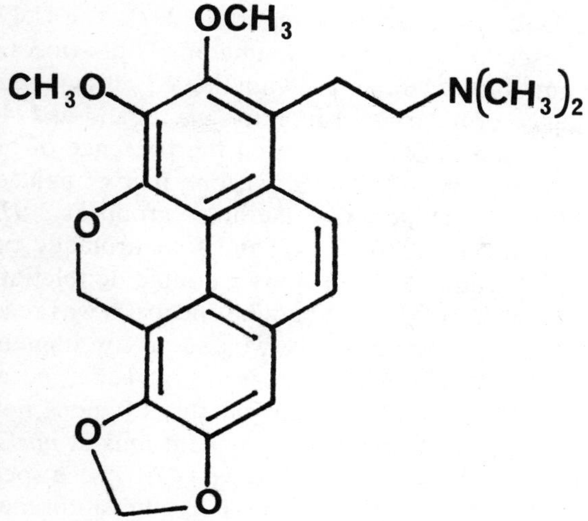

127

Thalicsine (**128**), $C_{21}H_{19}NO_5$ (365.1263), m.p. 193–194° [297], was first isolated from *Thalictrum simplex* L. in 1959 [294,295]. The alkaloid was incompletely characterized at that time and was later reisolated from *T. simplex* L. in 1970 [153], after having been found in *T. flavum* L. in 1965 [296]. It was not until its isolation from *T. longipedunculatum* E. Nik. in 1975 that

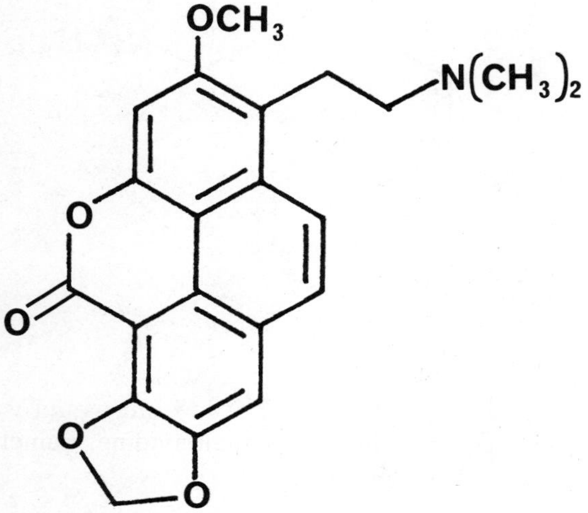

128

serious characterization studies were reported [297]. The UV spectrum was notable by the presence of multiple maxima at 237 nm (log ϵ 4.22), 265 nm (log ϵ 4.48), 313 nm (log ϵ 3.96), and 390 nm (log ϵ 3.60), and the IR spectrum showed the presence of a carbonyl function via a band at 1740 cm^{-1} [297]. The ^{1}H NMR spectrum (CDCl$_3$) indicated the presence of one quaternary dimethylamino function at δ 2.31 (6H, s), one pair of adjacent methylene groups at δ 2.40–3.30 (4H, m), one methoxy group at 3.97 (3H, s), one methylenedioxy group at 6.22 (2H, s), and four aromatic protons as two singlets at δ 7.11 (1H) and 7.25 (1H) plus a double doublet at δ 7.13 (1H, J = 9 Hz) and 7.48 (1H, J = 9 Hz). When the spectrum was recorded in TFA, the aromatic protons again appeared as two one-proton singlets and two one-proton doublets (J = 9 Hz) but were more easily visible, because there was a greater spread in the individual chemical shifts (values not given) [297]. The mass spectrum showed important fragment ions at m/z 365 (M$^+$), 307 (M—[CH$_2$=N(CH$_3$)$_2^+$)] and 58(CH$_2$=N(CH$_3$)$_2^+$)]. These spectral data suggested that thalicsine was a dimethylaminoethylphenanthrene alkaloid containing an α-pyrone ring. The placement of the methylenedioxy group at C(6)/C(7) was made because of the presence of this group at a downfield position, probably due to the effect of the neighboring carbonyl group. The similarity in spectral data to that of thaliglucinone (**125**), plus the nonidentity of the two alkaloids, prompted the authors to assign thalicsine as **128** [297]. However, because it is known that phenanthrene alkaloids originate from aporphines and that no naturally occurring aporphines are known to be unsubstituted at C(2) [15–17] [the position corresponding to C(3) in a phenanthrene], this structural assignment tends to be both highly unlikely and

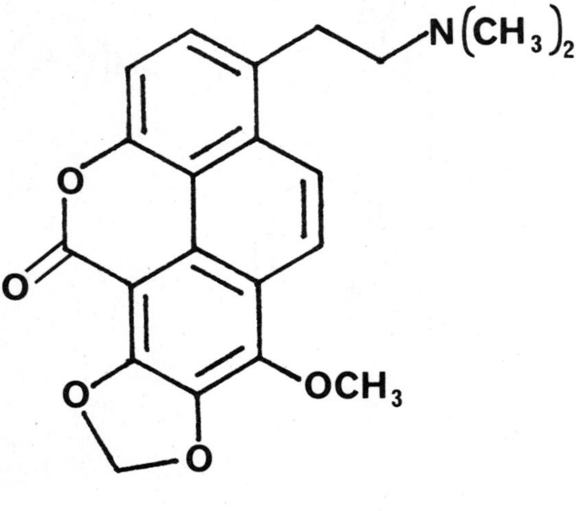

129

biogenetically untenable [25]. The same would be true if thalicsine were assigned structure **129** (also, one would expect ortho coupling of the C(2)/ C(3) protons). Hence one is left with the impression that the most likely possibility is the presence of the methylenedioxy group at one of C(6), C(7), or C(8) positions. Arguments can be made against each of these cases, however, and the final structural assignment of thalicsine remains unsettled at this time.

2.11.5. Thalicthuberine.   Tabular review [15, alkaloid No. 169], UV [15,155], IR [15], ^{1}H NMR [15,155,234], MS [155].

Thalicthuberine (**130**), $C_{21}H_{23}NO_4$ (353.1627), m.p. 126–127° [298], 117–118° (MeOH) [155], 126° (Et$_2$O) [234], 202–203° (HCl salt) [155], 202–210° (HCl salt) [298], 206–207° (oxalate salt) [298], was first isolated from *Thalictrum thunbergii* DC. in 1959 [298] and was the first phenanthrene to be isolated from a *Thalictrum*. The UV spectrum of thalicthuberine was characterized by numerous maxima at 261 nm (log ϵ 4.84), 285 nm (log ϵ 4.50), 310 nm (log ϵ 4.32) and 345 nm (log ϵ 3.50) [15,298] (also 264, 285, 313, 327, 348, and 366 nm [155]) and was typical of a substituted phenanthrene. Assignment of structure was via Hofmann degradation of *O*-methyldomesticine (**131**) to afford a product identical to thalicthuberine [298]. Some years later the ^{1}H NMR spectrum of thalicthuberine was determined and showed the dimethylamino group at δ 2.63 (6H, s); the methoxy groups at δ 3.88 (3H, s) and 3.99 (3H, s); the methylenedioxy group at 6.03 (2H, s); the aromatic protons at δ 7.13 [1H, s, H(2)], 7.45, and 7.71 [2H, ABq, J = 9 Hz, H(9) and H(10)], and 9.10 [1H, s, H(5)]; plus the methylene groups between δ 2.4 and 3.4 (4H, ABCDm) [234]. The mass spectrum of thalicthuberine is typical of the

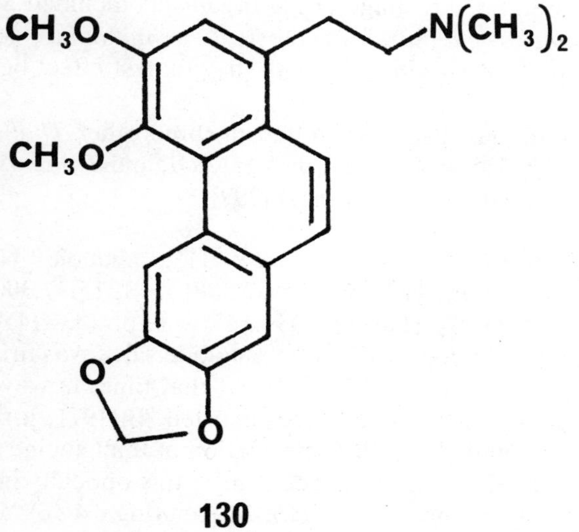

130

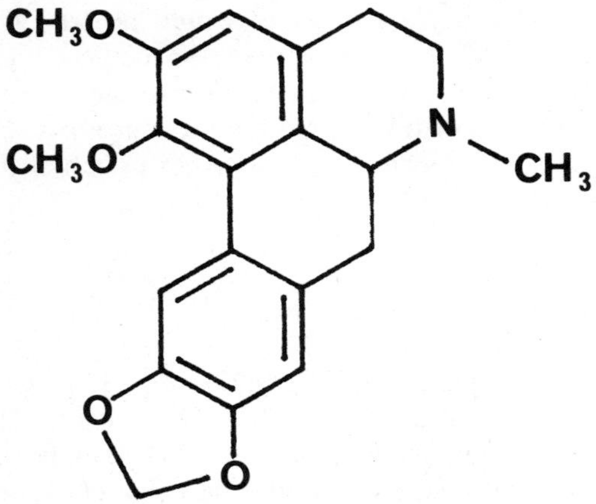

131

phenanthrenes of this group and is characterized by the facile loss of the dimethylaminomethylene ion followed by progressive losses of methyl radical and carbon monoxide. The resulting spectrum shows fragment ions at m/z 353 (M^+), 295 ($M—[CH_2=N(CH_3)_2]$), 280 (m/z 295—CH_3), 252 (m/z 295—CH_3—CO), 237 (m/z 295—2CH_3—CO), 209 (m/z 295—2CH_2—2CO), 179 (m/z 209—CH_2O), 151 (m/z 209—CH_2O—CO), and 58 (100%) ($CH_2=N^+(CH_3)_2$) [155,290].

Thalicthuberine was found to inhibit the growth of a wide variety of microorganisms in in vitro testing. These organisms included *Staphylococcus aureus, Salmonella gallinarum, Klebsiella penumoniae, Mycobacterium smegmatis,* and *Candida albicans,* with the greatest effect being against the last two [234].

Thalicthuberine has also been found in three other *Thalictrum* species, *T. strictum* Ledeb. [155], *T. rugosum* Ait. [234], and *T. minus* L. ssp. *majus* (also known as *T. minus* ssp. *elatum*) [299].

2.11.6. Thaliglucine. Tabular review [15, alkaloid No. 171], UV [15,97,300], IR [15,300], ^{1}H NMR [15,97,300], MS [15,97,300].

Thaliglucine (**123**), $C_{21}H_{21}NO_4$ (351.1471), m.p. 143–145° (Et_2O) [114], 122° (EtOH) [97], 300° dec. (EtOH) (methiodide salt), was first isolated from *Thalictrum rugosum* Ait. in 1970 [114]. At that time, however, no spectral data nor structural postulations were reported. In 1971, a report from the same laboratory detailed the characterization of thaliglucine as a new phenanthrene alkaloid [300]. The UV spectrum of this optically inactive alkaloid showed maxima at 263 nm (log ϵ 5.00), 289 nm (log ϵ 4.96), and 321 nm (log ϵ 4.8) [221 nm (log ϵ 4.26), 250 nm (log ϵ 4.41), 260 nm (log ϵ 4.36), 272 nm (sh) (log ϵ 4.13), 287 nm (sh) (log ϵ 3.81), 317 nm (log ϵ 3.36), 350 nm (log ϵ 3.36) and 370 nm (log ϵ 3.36) [97]] and IR bands at 1610 cm^{-1} (aromatic) and 915 cm^{-1} (CH_2O_2) [300]. The ^{1}H NMR spectrum was most informative and contained six singlets and two doublets. The former signals accounted for the presence of one dimethylamino group (δ 4.00, 6H), one methyleneoxy group (δ 5.52, 2H), one methylenedioxy group (δ 6.03, 2H), and two aromatic protons [δ 7.04 and 7.10 (1H each)]. The double doublet was observed at δ 7.40 (1H, J = 9 Hz) and 7.73 (1H, J = 9 Hz) [300]. The mass spectrum showed the molecular ion at m/z 351 and this, considered with the other spectral data, supported the assumption that thaliglucine was a dimethylaminoethylphenanthrene derivative. Hofmann degradation of N-methylthaliglucine (**122**) afforded the des-N-methine **132**, m.p. 136–138°, whose ^{1}H NMR and mass spectrum were consistent with **132**. Oxidation of thaliglucine with potassium dichromate–acetic acid afforded thaliglucinone (**125**), whose structure was assigned principally via a consideration of the IR and ^{1}H NMR spectral data [300]. In particular, the appearance of an absorption band in the IR spectrum at 1740 cm^{-1} plus the disappearance of the two-proton singlet at δ 5.53 in the ^{1}H NMR spectrum suggested the generation of an ester (such as an α-pyrone) at the methyleneoxy bridge. Furthermore, the chemical shift of the methylenedioxy group in thaliglucinone was at δ 6.30, whereas the same chemical shift in thaliglucine is at δ 6.05, thereby indicating the proximity of the generated carbonyl group to the methylenedioxy function. In addition, the chemical shift of the methoxy function of thaliglucinone (δ 4.00) was not significantly different from that of the same group in thaliglucine (δ 3.95), further supporting the proximity of the carbonyl function in thaliglucinone to the methylenedioxy group. The mass spectrum of thal-

132

iglucinone showed a parent ion at m/z 365, reflecting the replacement of two protons with an oxygen atom. Thaliglucinone (**125**) was also isolated, albeit in a very small quantity, from the same extract [300].

Confirmation of the structure of thaliglucine did not occur until 1972, when the phenanthrene was isolated along with thalphenine (**15**) from *Thalictrum plygamum* Muhl. [97]. At that time, an X-ray crystallographic study of thalphenine iodide unequivocally established its structure as **15**. Because thalphenine methine was generated as a Hofmann degradation product from thalphenine and isolated from the same extract, the structure of thaliglucine could be firmly assigned as **123**. Salient features of the ^{1}H NMR spectrum of thaliglucine (thalphenine methine) were the singlet at δ 5.56 (2H) for the methyleneoxy bridge and the double irradiation of the methoxy resonance at δ 4.00, which resulted in a 18% NOE of the δ 7.11 aromatic proton, allowing unambiguous assignment of the latter to the aromatic proton at H(2) [97]. Finally, synthesis of thalphenine (**15**) and thaliglucine (**123**) via a key photolytic intermediate confirmed the structure of these alkaloids [100]. Furthermore, because thaliglucinone (**125**) is an oxidation product of thaliglucine (**123**), the structure of the former is likewise delineated.

2.11.7. Thaliglucinone.   Tabular review [15, alkaloid No. 172], UV [98,146], IR [98,300], ^{1}H NMR [98,146,300], MS [146,300].

Thaliglucinone (**125**), $C_{21}H_{19}NO_5$ (365.1263), m.p. 126–128° (Et$_2$O) [300], 180–182° (Et$_2$O) [149], 197° (MeOH) [98], 115–116° then solidifies and remelts at 188–190° (MeOH) [139], 254–256° dec. (MeOH) (HCl salt) [149], was first isolated as a yellow crytalline substance in a very small quantity from *Thalictrum rugosum* Ait. in 1971 [300]. The establishment of structure followed that of thaliglucine (**123**) and is discussed in the preceding section. However, a summary of the spectral data for thaliglucinone is worthy of citation, for these data are not completely described above. The UV spectrum of thaliglucinone showed maxima at 238 nm (log ϵ 4.46), 256 nm (log ϵ 4.56), 264 nm (log ϵ 4.70), 287 nm (log ϵ 3.67), 312 nm (log ϵ 4.16), and 390 nm (log ϵ 3.80) [98]. The ^{1}H NMR spectrum was characterized by signals at δ 2.40 [6H, s, N(CH$_3$)$_2$], 4.08 (3H, s, OCH$_3$), 6.32 (2H, s, CH$_2$O$_2$), 7.27 (1H, s, ArH), 7.43 (1H, s, ArH), 7.47 (1H, d, J = 9 Hz) and 7.81 (1H, d, J = 9 Hz) [98]. The mass spectrum showed the base peak at m/z 58, for the facile loss of CH$_2$=N(CH$_3$)$_2^+$ [146].

Thaliglucinone (**125**) was found to possess antimicrobial properties, as demonstrated by the inhibition of growth of six standard test organisms of varying morphology [98,139,149]. In addition, the alkaloid lowered the blood pressure of normotensive dogs and rabbits [101,137,139,215] when given in a dosage of 0.5–1.75 mg/kg via intrajugular injection [215].

Thaliglucinone (**125**) has also been isolated from *Thalictrum longistylum* DC. [137], *T. lucidum* L. [139], *T. minus* L. var. *adiantifolium* [145] (see notation in [149]), *T. minus* L. var. *microphyllum* Boiss. [146], *T. minus* L. race B [215], *T. podocarpum* Humb. [301], *T. polygamum* Muhl. [149], *T.*

revolutum DC. [101], *T. rochebrunianum* Franc. and Sav. [302] (see notation in [149]), and *T. rugosum* Ait. [98,152] (see notation in [149]).

2.12. Protoberberines

The following reviews and references are particularly relevant to the isolation and identification of protoberberine and tetrahydroprotoberberine alkaloids:

1. Tabular reviews [10,18,19]
2. Annual reviews [20–22,24,194–201]
3. Other reviews [10–13]
4. Book chapters [14,208,241,303,304,407,408]

The numbering of the protoberberine (**133**) and tetrahydroprotoberberine (**134**) rings is according to accepted practice [303].

2.12.1. Berberine. Tabular review [10,18,19], UV [305,306], IR [306], ^{1}H NMR [307], ^{13}C NMR [308], MS [309].
Berberine (**135**), $C_{20}H_{18}N^+O_4$ (336.1236), m.p. 260–262° dec. (MeOH) (iodide) [145], 205° (MeOH) (chloride) [62], 239–240° (picrate) [310], was first isolated from *Zanthoxylum caribaeum* (Rutaceae) in 1826 and named "xanthopicrit" [303]. Nine years later the isolation of this quaternary alkaloid, then designated berberine, from *Berberis vulgaris* (Berberidaceae) was reported [303].

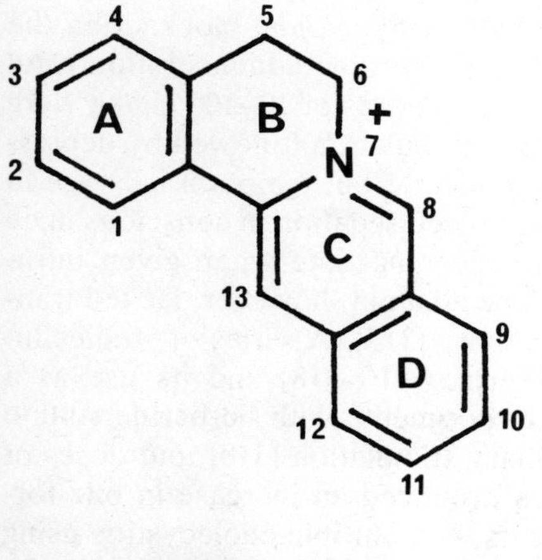

133

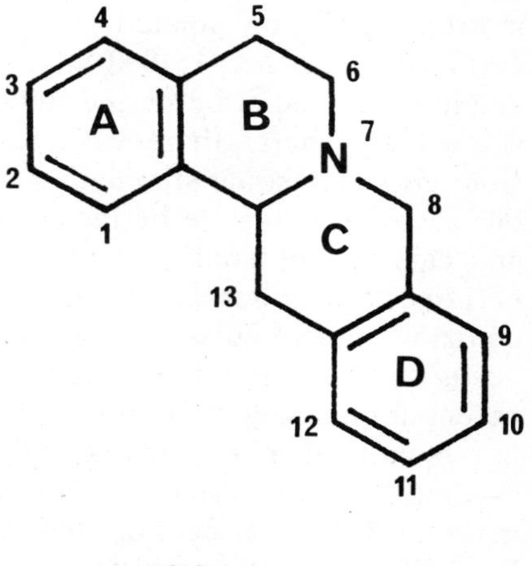

134

135

Berberine is well-known to possess a wide range of interesting biological activities, and many of the studies describing these are found in several excellent reviews [208,311]. The LD_{50} of berberine bisulfate in mice was found to be 27.5 mg/kg on intraabdominal administration [312]. Intravenous administration to dogs (above 0.2 mg/kg) and cats (above 0.05 mg/kg) produced temporary hypotension and respiratory stimulation [312]. The alkaloid was shown to possess anti-curariform and cholinesterase-inhibiting activities in various animals [312]. Berberine sulfate tended to be spasmogenic at lower concentrations (20–100 µg/mL) on the isolated guinea pig ileum [313]. The alkaloid also potentiated pentobarbital hypnosis in mice at a dose of 10 mg/kg, produced a mild hypotension in dogs (1–5 mg/kg), and showed a negative inotropic effect on isolated rabbit heart with a myocaridal blockade in the perfused frog heart (1:1000) [313]. Acute and chronic administration (oral and intravenous) of berberine sulfate to cats in doses of 10–100 mg/kg were studied. In general, there was a respiratory stimulation followed by depression, gastrointestinal stimulation, and a moderate but temporal increase in bile secretion [314]. Berberine chloride produced sedation in conscious mice and cats and potentiated pentobarbitone sleeping time when given intraperitoneally in a dosage of 5–40 mg/kg. The alkaloid, however, lacked tranquilizing, anticonvulsant, or analgesic activity [315]. A series of studies investigated the choleretic effect of berberine [316–318] and its use as a potential cholagogue in humans [316]. Experiments with berberine sulfate in rats and guinea pigs revealed little biliary stimulation [318], but doses of 1–5 mg/kg of berberine chloride to dogs produced an increase in bile formation [317]. Treatment of human patients with chronic cholecystitis using berberine chloride or sulfate in an oral dosage of 5–20 mg three times daily before meals for 24–48 hr improved clinical symmptomatology while decreasing bile bilirubin levels and increasing gallbladder bile volume [316]. In a series of experiments spanning 30 years, the effect of berberine on the myocardium was studied [319–324]. Berberine depressed the vagus and stimulated the myocardium in the rabbit heart while stimulating the bronchii and compressing pulmonary vessels [319]. In studies using canine isolated heart–lung preparation, isolated cat heart, and isolated cat and rabbit atria, berberine sulfate was found to be a myocardial stimulant in small doses and a depressant in larger doses. The cardioinhibitory action of acetylcholine and pilocarpine was potentiated by small doses of berberine but antagonized by moderate and large doses [320]. Pretreatment of rabbits with berberine chloride inhibited the myocardial arhythmic as well as hypertensive effects of epinephrine. Furthermore, acetylcholine-induced bradycardia was increased but histamine action was not altered [322]. Methoxyphenamine and methamphetamine-induced tachycardia was inhibited by berberine, and diisopropyl fluorophosphate and neostigmine bradycardia were reversed [323]. Other studies have focused on the deposition of berberine in various organs and tissues after administration [325–329]. Administration (subcutaneous, intraperitoneal, intravenous, or oral) of varying doses of berberine sulfate to rats resulted in the concentration of berberine principally in the heart, pancreas, liver, and omentum. However, 24 hr later, only the pancreas and lipid still retain appreciable amounts of the alkaloid [325]. A similar study in healthy and cholera-infected infant rabbits (dosed orally with berberine sulfate) demonstrated gastrointestinal absorption of berberine, with infected animals having lower blood levels but higher tissue levels of the drug than noninfected animals [326]. Berberine is known to inhibit the oxidative decarboxylation of yeast pyruvate [330] and the oxidation of alanine in rat kidney homogenates [331]. Furthermore, the alkaloid inhibits the actions of cholinesterase in rabbit spleen and pseudocholinesterase in horse serum [332] plus cellular respiration in ascitic tumors and tissue cultures [333,334]. The adrenal concentration of ascorbic acid (used as a criterion for ACTH secretion) is decreased by subcutaneous administration of berberine (5–10 mg/kg) but this effect is abolished by hypophysectomy. However, pretreatment with sodium pentobarbitone totally inhibited the adrenal depleting action of the alkaloid [335].

Berberine was found to inhibit immune hemolysis using a hemolytic system composed of rabbit serum containing a high titer of sheep cell-immune hemolysin and sheep erythrocytes, plus fresh guinea pig serum as complement [336]. Berberine had no significant activity on the production of plasmaoblasts and plasma cells in rabbits [337]. There was no effect on the number of erythrocytes and leukocytes nor on hemoglobin level upon administration of berberine to rabbits. However, in phenylhydrazine or toluenediamine-induced anemic rabbits, berberine exerted an antianemic effect [338]. Recently, the effects of berberine on in vitro acetylcholinesterase activity and on isolated rat duodenum and guinea pig ileum have been studied. Berberine was found to be an effective inhibitor of this enzyme (50%

inhibition caused by a concentration of 9.8×10^{-7} moles/L of berberine). Furthermore, isolated rat duodenum and guinea pig ileum contractions were strengthened [339]. These results tend to agree with the observation that compounds with a quaternary nitrogen atom act mostly as inhibitors of the anionic site of acetylcholinesterase [339]. In a similar study, berberine was found to exhibit only weak butrylcholinesterase inhibition [58]. Berberine was found to inhibit the locally induced inflammation by cholera toxin but was not effective in reducing the inflammation induced by carrageenin or formalin [340]. In a rather recent study, berberine sulfate produced a reversible hypotension in anesthetized rats that was not inhibited by aprotinin but did not decrease histamine aerosol-induced bronchospasm in guinea pigs or influence gastric acidity and secretion in rats and dogs. The alkaloid also potentiated apomorphine-induced emesis in dogs while lowering the rectal temperature of normal rats [341]. Berberine has long been known to possess a rather wide range of antimicrobial activity and numerous papers have been published on this subject [208]. In general, the alkaloid has antibacterial, trypanocidal, antiamebic, antifungal, anthelmintic, leishmanicidal, and tuberculostatic properties [208,313,342–344,357]. Berberine sulfate has been employed in fluorescent cytochemistry on blood cells, cells from lymph nodes, and chromosome preparations. Chromatin appeared with a brilliant yellow color and regions in lymph nodes which were unstained were assumed to contain undifferentiated or neoplastic cells containing a large amount of euchromatin. Regions with an intense yellow coloration probably corresponded to the localization of satellite or repetitive DNA [345]. Berberine sulfate has also been used in the exfoliative cytodiagnosis of carcinoma on the cervical uterus via fluorochroming the cells [346]. Some studies have demonstrated the antiblastic properties of berberine [208].

Finally, the biological activity of quaternary isoquinoline alkaloids may be explained by the addition of the biological nucleophile to the iminium bond or by binding of the quaternary alkaloid to DNA by intercalation of the quaternary cation. Another possible mechanism may be an interaction of the quaternary cation with anionic binding site of the biopolymers [339]. Berberine may be an effective antidiarrheal agent in *E. coli* heat-stable enterotoxin-mediated secretory diarrhea as shown from studies with *E. coli* heat-stable enterotoxin-perfused porcine jujunal segments [356]. An intraperitoneal dose of 25.6 mg/kg of berberine chloride protected mice from hypoxyia and antagonized chloroform- and aconitine-induced arrhythmias but not ouabain-induced arrhythmias [358]. Berberine failed to exert any direct vasodilator effect on isolated rabbit pulmonary and cat coronary arteries but reversed α-adrenergic-mediated vasoconstriction in both preparations. The alkaloid antagonized both norepinephrine- and ergonovine-induced constriction in rabbit pulmonary arteries and ergonovine-induced constriction in feline pulmonary arteries. Berberine also failed to block vasoconstriction of pulmonary smooth muscle by α-adrenergic independent agents. Hence berberine induces vasodilation in two different types of smooth muscle, apparently by α-adrenergic blocking mechanisms [359]. Berberine showed inhibition of reverse transcriptase activity of DNA tumor viruses in the presence of polyriboadenylic acid–oligodeoxythymidylic acid and other related template primers. This indicated an inhibition of enzyme activity by interaction with the template primer and competition with the template primer-binding site of the enzyme. Time course inhibition studies indicated the instant termination of DNA synthesis by berberine when the alkaloid was added after the polymerization initiation [360]. It is known that berberine does possess antitumor activity against experimental tumors [361], but lacked activity against the P-388 murine lymphocytic leukemia [362].

The first apparent isolation of berberine from a *Thalictrum* species was as the iodide salt from *T. foliolosum* DC. in 1941 [347]. The alkaloid was subsequently reisolated from *T. foliolosum* DC. [38,223,348] and has been isolated from *T. alpinum* L. [70], *T. baicalense* Turcz. [61,127], *T. dasy-carpum* Fisch. and Lall. var. *hypoglaucum* (Rydb.) Boivin [122], *T. dioicum* L. [72,193], *T. faberi* Ulbr. [349], *T. fendleri* Engelm. ex Gray [62,132] *T. flavum* L. [133,296 (thalsine = berberine)], *T. foetidum* L. [134,192], *T. javanicum* Bl. [136], *T. longistylum* DC. [137], *T. lucidum* L. [138,139], *T. minus* L. [65,66,134,140,141,181,272,310], *T. minus* L. race B [143], *T. minus* L. var. *adiantifolium* Hort. [144,145], *T. minus* L. var. *elatum* [262,350,351] (now probably better designated *T. minus* L. ssp. *majus* [229]), *T. minus* L. var. *hypoleucum* [callus tissue-[142] (also called *T. thunbergii* DC. [142]), *T. minus* var. *microphyllum* Boiss. [146], *T. minus* L. var. *minus* [147,148], *T. pedunculatum* Edgew. [352], *T. podocarpum* Humb. [301], *T. polygamum* Muhl. [149,212], *T. revolutum* DC. [101,105,368], *T. rochebrunianum* Franc. and Sav. [121], *T. rugosum* Ait. [150–152,353] (sometimes called *T. glaucum* Desf. [354]), *T. sachalinense* Lecoyer. [67], *T. strictum* Ledeb. [155], *T. thunbergii* DC. [157] (also called *T. minus* L. var. *hypoleucum* [142]), and *T. tuberiferum* Maxim. [355].

2.12.2. Berberrubine. Tabular review [18,19], UV [72], ^{1}H NMR [72,307]. Berberrubine (**136**), $C_{19}H_{16}N^+O_4$ (322.1079), dec. 280° (chloride) [363], was

136

apparently first isolated from *Berberis vulgaris* L. (Berberidaceae) in 1929 [see ref. 16 in 149]. The alkaloid was first isolated from a *Thalictrum* species, *T. polygamum* Muhl. in 1973, via treatment of an acetone solution of the alkaloid with excess methyl iodide to produce berberine (135) [149]. Furthermore, the isolate was identical to a sample of berberrubine prepared from berberine chloride [149]. Berberrubine was subsequently isolated from *T. dioicum* L. in 1978 [72].

2.12.3. Columbamine. Tabular review [10,18,19], UV [98,305,306], IR [98,306], ¹H NMR [307].
Columbamine (137), $C_{20}H_{20}N^+O_4$ (338.1392), m.p. 239.5–240.5° (H₂O) (chloride) [306], 240–241° dec. (MeOH) (chloride) [98], 206–208° (MeOH) (iodide) [102], 222–225° (MeOH) (iodide) [223], mp 228–228.5° (H₂O) (iodide) [306], was first isolated from *Jateorhiza columba* Miers (Menispermaceae) in 1906 [303].

The inhibition of butrylcholinesterase activity in human serum by columbamine chloride was very weak and probably affected by the presence of the phenolic hydroxy group. This group may account for potential hydrogen bonding between the hydroxy of the alkaloid and some catalytically significant groups (most probably the imidazole moiety of histidine) in the esteratic site of the enzyme [58]. Intravenous injection of columbamine into rats resulted in deposition of the alkaloid in the pancreas, as determined via in vivo pancreatic fluorescence [327]. Columbamine exhibited some in vitro inhibition of *Mycobacterium smegmatis* and *Candida albicans* [139].

The first isolation of columbamine from the genus *Thalictrum* was from *T. rugosum* Ait. in 1965 as the iodide salt [151]. The alkaloid was later reisolated from *T. rugosum* Ait. in 1976 [98] as well as having been found in *T. alpinum* L. [70], *T. foliolosum* DC. [223], *T. javanicum* Bl. [136], *T.*

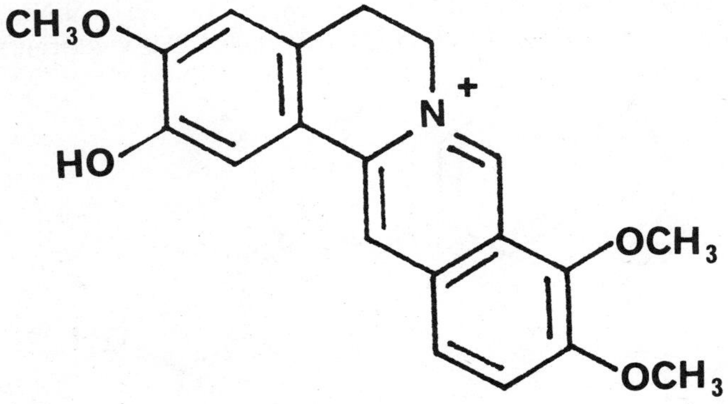

137

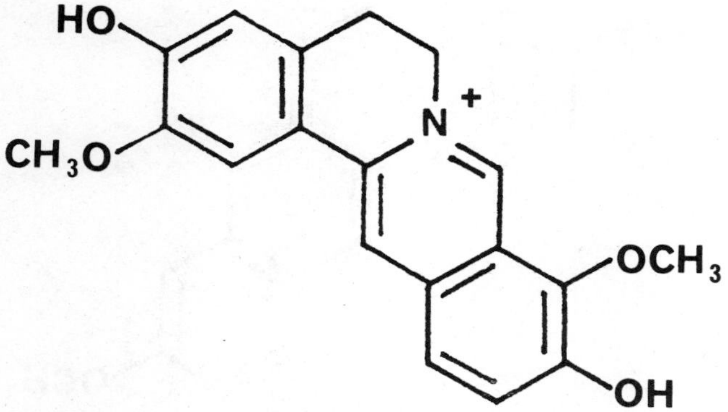

138

longistylum DC. [137], *T. lucidum* L. [139], *T. minus* L. var. *hypoleucum* (callous tissue—[142]) (also called *T. thumbergii* DC. [142]) (TLC identification only), *T. minus* L. race B [102], *T. podocarpum* Humb. [301], and *T. revolutum* DC. [101].

2.12.4. Dehydrodiscretamine.   UV [223,364], IR [223].

Dehydrodiscretamine (**138**), $C_{19}H_{18}N^+O_4$ (324.1236), was first isolated as pale orange needles of the chloride salt, m.p. 217–220° dec. (MeOH—Et$_2$O) from *Corydalis tashiori* Makino (Papaveraceae) in 1981 [364]. This quaternary protoberberine was isolated from *Thalictrum foliolosum* DC. in 1983 [223] and this remains the only report of this alkaloid in the genus to date and only the second reported isolation from a plant.

2.12.5. Dehydrodiscretine.   UV [131,365], IR [131,365], ^{1}H NMR [131,365], CI–MS [131].

Dehydrodiscretine (**139**), $C_{20}H_{20}N^+O_4$ (338.1392), was first isolated as fine, reddish-brown needles of the chloride salt, m.p. 230–234° (MeOH) from *Thalictrum fauriei* Hayata in 1980 [131]. The alkaloid possessed a UV spectrum typical of a protoberberine with maxima at 243 nm (log ε 4.24), 264.5 nm (log ε 4.21), 289 nm (log ε 4.48), 310 nm (sh) (log ε 4.36), 341 nm (log ε 4.19) and 379 nm (sh) (log ε 3.82), with a bathochromic shift in alkali [131]. The ^{1}H NMR spectrum (CD$_3$OD) indicated the presence of the three methoxy groups as three singlets at δ 3.89, 4.01, and 4.07; two adjacent methylene groups as two triplets, the first at δ 3.05 and the second obscured by the solvent peak at around δ 3.4; and six aromatic protons as five singlets at δ 6.53 [H(4)], 7.33 [H(12)], 7.43 (2H) [H(1) and H(9)], 8.27 [H(13)], and 9.07 [H(8)] [131]. The CI mass spectrum (isobutane) showed the molecular ion

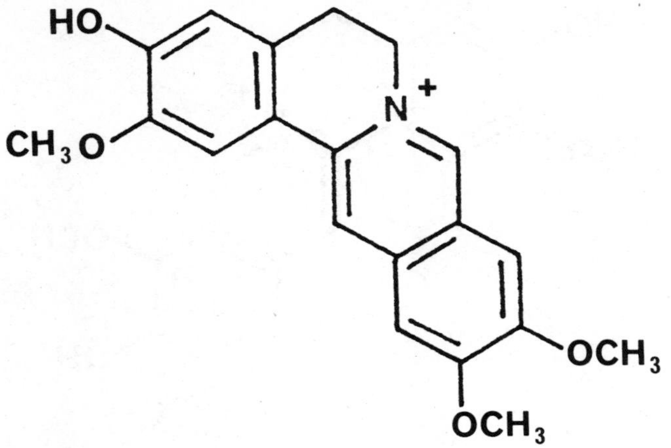

139

at m/z 338 with ions at m/z 339 and 353 representing thermal disproportionation [131]. Reduction of dehydrodiscretine with sodium borohydride gave discretine (**140**), identical to an authentic synthetic sample by direct comparison. The major mode of fragmentation of discretine in the EI mass spectrum is via a retro-Diels–Alder process to yield characteristic fragment ions at m/z 164 (100%) (**141**), 178 (**142**), and 176 (**143**). The reductive conversion of dehydrodiscretine to discretine plus the oxidation [Hg(OAc)$_2$] of discretine to a product identical to the isolate confirmed the structure of dehy-

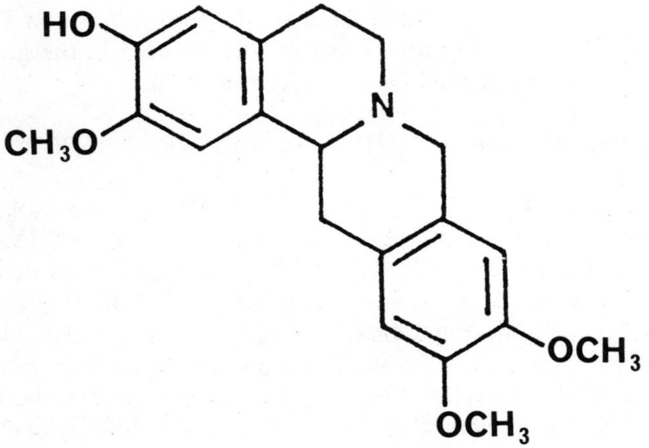

140

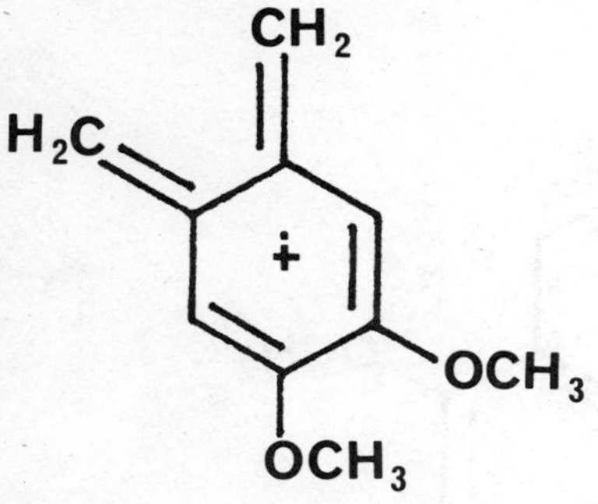

141

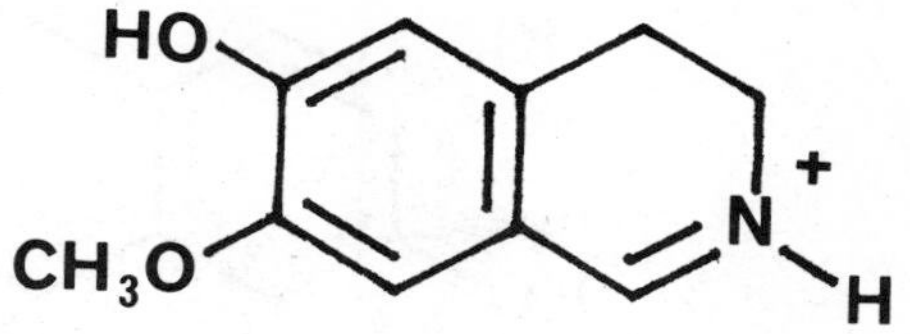

142

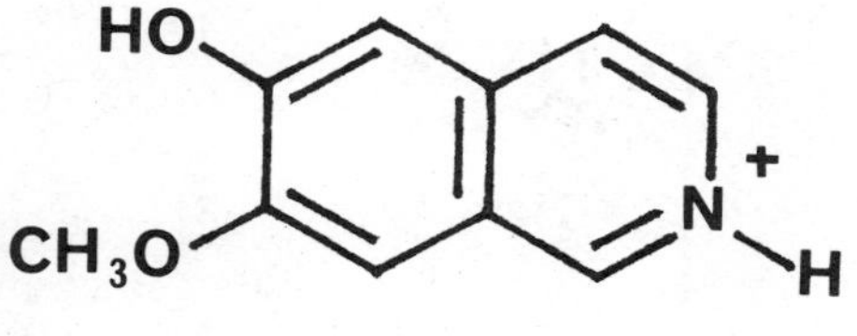

143

drodiscretine as **139** [131]. Finally, the synthesis of dehydrodiscretine utilizing a standard synthetic protocol for this series was recently reported [365].

2.12.6. Demethyleneberberine. UV [136], IR [136], ^{1}H NMR [136,307]. Demethyleneberberine (**144**), $C_{19}H_{18}N^+O_4$ (324.1236), was first isolated from *Thalictrum javanicum* Bl. as the chloride salt, m.p. 225°, in 1983 [136]. The UV spectrum was typical of a protoberberine and showed maxima at 235 nm (log ε 4.63), 270 nm (log ε 4.52), and 325 nm (log ε 4.49) with a bathochromic shift in both alkaline and acidic media [136]. The ^{1}H NMR spectrum (solvent not specified) indicated the presence of two methoxy groups at δ 3.87 (s) and 4.01 (s); plus six aromatic protons at δ 6.41 [s, H(4)], 7.61 [s, H(1)] and 7.91 (d, J = 8.5 Hz, H(12)], 8.12 (d, J = 8.5 Hz, H(11)], 8.70 [s, H(13)], and 9.66 [s, H(8)] [136]. The mass spectrum of the tetrahydro derivative (NaBH$_4$) showed the molecular ion at m/z 327 (72%) with fragment ions at m/z 164 (92%) as represented by both **145** and **146**, m/z 162 (**147**) and m/z **149** (100%) (**146**-CH$_3$), representing the retro-Diels–Alder type electron impact fragmentation [136]. Treatment of demethyleneberberine with CH$_2$N$_2$ afforded palmatine (**148**), thus confirming the skeletal structure and

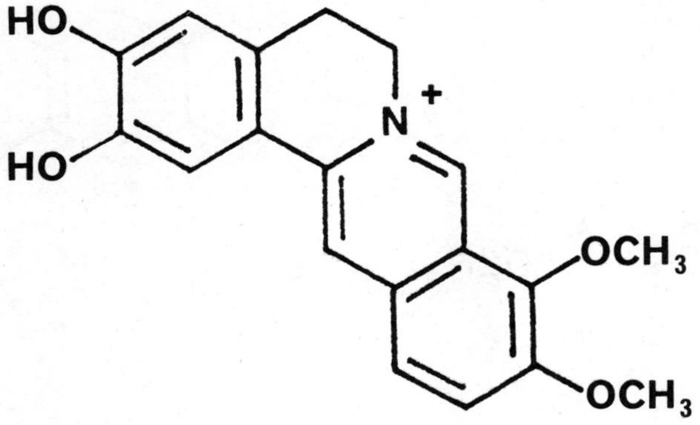

144

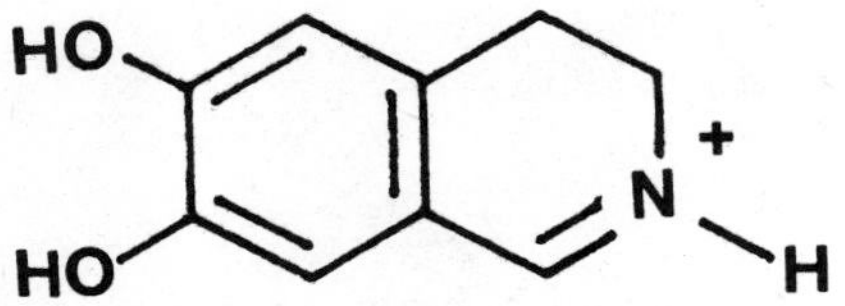

145

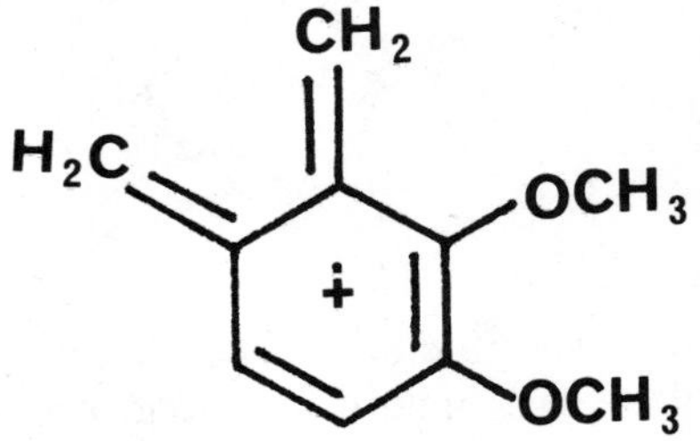

146

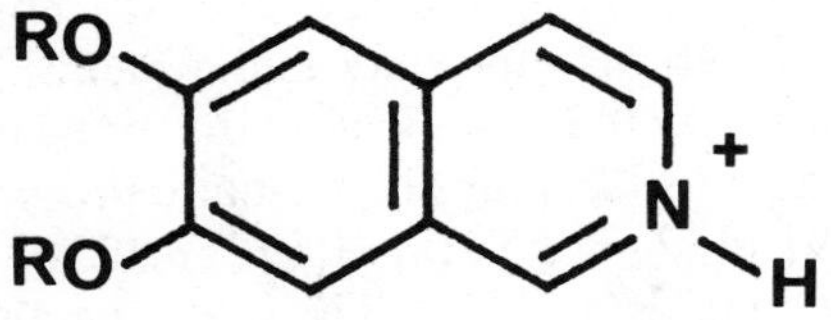

147 R = H

149 R = C₂H₅

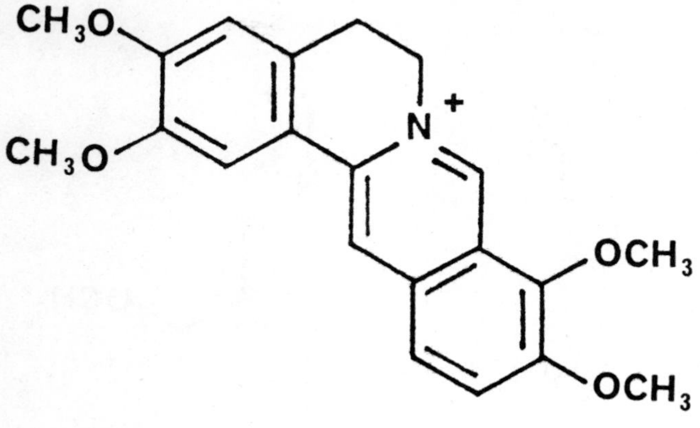

148

the 2,3,9,10-tetraoxygenation of demethyleneberberine. Acetylation of demethyleneberberine (Ac_2O/pyr) furnished an *O,O*-diacetyl ester as an oil ([1H] NMR-(CCl_4)—δ 2.07 [6H, s, $2OCOCH_3$]). Finally, ethylation of demethyleneberberine with diazoethane and reduction ($NaBH_4$) of the product afforded *O,O*-diethyltetrahydrodemethyleneberberine. A consideration of the mass spectral fragment ions of this derivative, particularly the isoquiniolinium ion **149**, established the position of the phenolic hydroxyl groups of demethyleneberberine in ring A and settled its structure as **144** [136].

2.12.7. Deoxythalidastine. UV [98,366,367], IR [98], [1H] NMR [98,366]. Deoxythalidastine (**150**), $C_{19}H_{14}N^+O_4$ (320.0923), was first isolated as fine red needles of its chloride salt, darkening about 210° (dec.) (MeOH), from

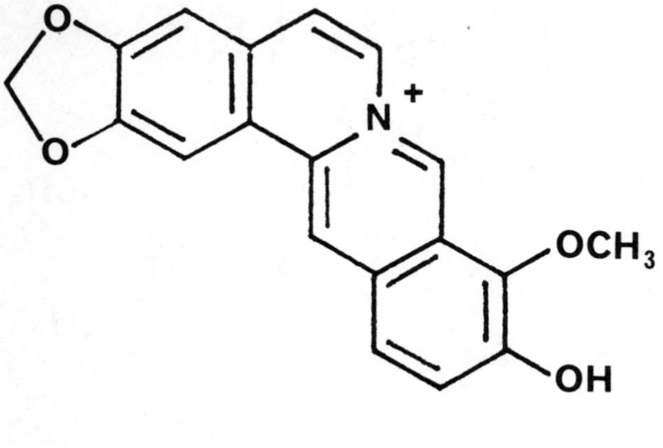

150

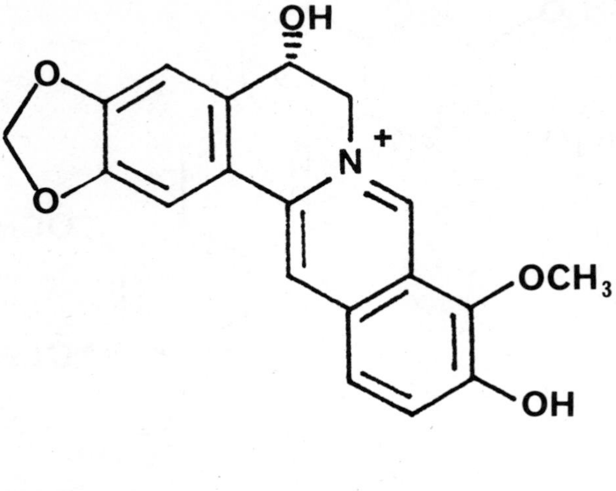

151

Thalictrum polygamum Muhl. in 1973 [149], although the alkaloid had been prepared via dehydration of thalidastine (**151**) in 1965 during a study of the structure determination of **151** [366]. In that study, thalidastine chloride (**151**) was converted to deoxythalidastine chloride (**150**) by heating in vacuo at 85°, or on standing at room temperature for several weeks, or by heating at 100° in 2N HCl for 15 min [366]. The deoxythalidastine thus obtained displayed UV maxima at 247 nm (log ϵ 4.24), 270 nm (log ϵ 4.18), 278 nm (log ϵ 4.18), 308 nm (log ϵ 4.08), 348 nm (log ϵ 3.71), and 463 nm (log ϵ 3.68) and

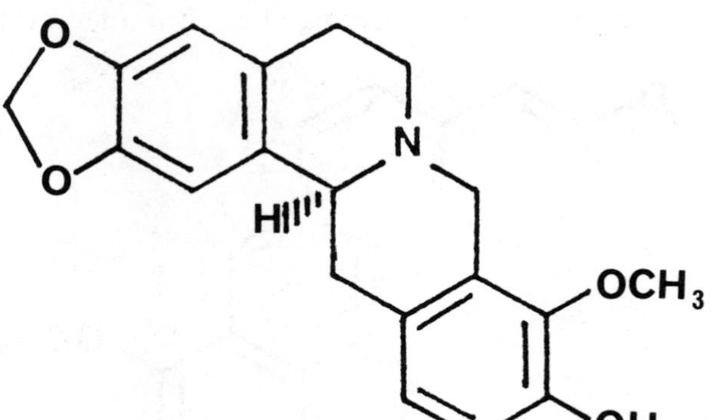

152

possessed a relatively simple ^{1}H NMR spectrum (TFA) characterized by singlets at δ 4.32 (3H, OCH$_3$), 6.34 (2H, CH$_2$O$_2$), 7.43 (1H, ArH), 8.03 (1H, ArH), 8.09 (1H, ArH), 8.22 (1H, ArH), 9.37 (1H, ArH) and 9.88 (1H, ArH), with two poorly defined doublets at δ 7.92 (1H, J = 7.2 Hz) and 8.69 (1H, J = 7.2 Hz) [366]. Because catalytic reduction (Adams catalyst) of deoxythalidastine chloride furnished tetrahydrothalifendine (**152**), the structure of deoxythalidastine was then unequivocally assigned as **150** [366]. Deoxythalidastine bromide was also synthesized by the quaternization of the oxime of 6,7-methylenedioxyisoquinoline-1-carboxaldehyde with 3-benzyloxy-2 methoxybenzyl bromide, followed by cyclization with hydrobromic acid [367]. Deoxythalidastine failed to inhibit the growth of a series of microorganisms in a standardized in vitro test [149]. Deoxythalidastine has since been isolated from *Thalictrum minus* L. var. *hypoleucum* (*T. thumbergii* DC.) (callus tissue [142]), *T. revolutum* DC. [101], and *T. rugosum* Ait. [98].

In retrospect, naming deoxythalidastine (**150**) as dehydrothalidastine would probably have been more consistent because, in reality, deoxythalidastine is a dehydration product of thalidastine (**151**).

2.12.8. Jatrorrhizine. Tabular review [10,18,19], UV [305], ^{1}H NMR [307].

Jatrorrhizine (**153**), C$_{20}$H$_{20}$N$^+$O$_4$ (338.1392), m.p. 204–206° dec. (MeOH) (chloride) [101], 208–210° dec. (MeOH) (iodide) [70], was apparently first isolated from *Jateorhiza columba* (Menispermaceae) in 1906 [303].

Jatrorrhizine was found to inhibit the growth of *Mycobacterium smegmatis* in in vitro antimicrobial testing [139] and to possess some inhibitory action against the growth of *Saccharomyces carlsbergensis* [343]. Intraperitoneal administration of jatrorrhizine chloride in a dosage of 25.6 mg/kg

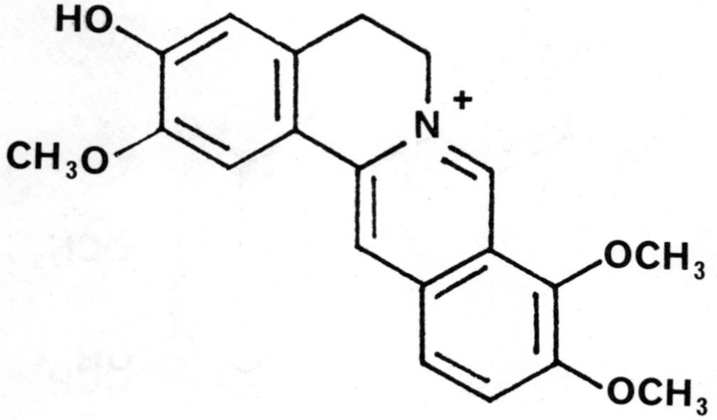

153

failed to protect mice from drug-induced hypoxia. However, chloroform-induced myocardial arrhythmias were inhibited by an intraperitoneally administered dose of 15.8 mg/kg of jatrorrhizine chloride [358]. The human serum butrylcholinesterase-inhibiting activity of jatrorrhizine was very weak and likely affected by the presence of the phenolic hydroxy in the molecule. The phenol may be responsible for potential hydrogen bonding between the hydroxy group in the alkaloid and some catalytically significant groups in the esteratic enzyme site. The imidazole group of histidine is considered to be one such catalytically significant group [58]. The first isolation of jatrorrhizine from a *Thalictrum* species was from *T. foliolosum* DC. as its tetrahydro derivative in 1952 [348]. The alkaloid was subsequently reisolated from the same species [223] and has also been isolated from *T. alpinum* L. [70], *T. fendleri* Engelm. ex Gray [62,132], *T. javanicum* Bl. [136], *T. longistylum* DC. [137], *T. lucidum* L. [138,139], *T. minus* L. [140], *T. minus* L. race B [102], *T. minus* L. var. *hypoleucum* (callus tissue [142]) (also called *T. thunbergii* DC. [142]), *T. minus* L. var. *microphyllum* Boiss. [146], *T. podocarpum* Humb. [301], *T. revolutum* DC. [101,368], and *T. rochebrunianum* Franc. and Sav. [121].

2.12.9. Oxyberberine (Berlambine). Tabular review [18,19], UV [143], IR [268], ^{1}H NMR [268], ^{13}C NMR [369], MS [268].

Oxyberberine (**154**), $C_{20}H_{17}NO_5$ (351.1107), m.p. 199–201° (MeOH) [70], was apparently first isolated from *Berberis thunbergii* DC. var. *Maximowiczu* Franch. (Berberidaceae) in 1930 [370], although it had been known as an oxidative product (KMnO$_4$) of berberine (**135**) since 1890 [303]. The alkaloid berlambine isolated from *Berberis lambertii* in 1953 [371] was later shown to be identical with oxyberberine [372].

Oxyberberine (**154**) does not share the same UV chromophore as the more

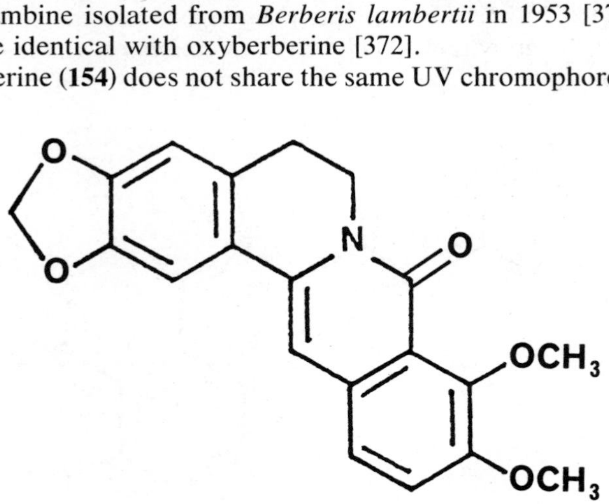

154

frequently occurring quaternary protoberberine alkaloids, for the former exhibits UV maxima at 225 nm (log ∈ 4.73), 285 nm (log ∈ 3.89), and 342 nm (log ∈ 4.44) [143], with the carbonyl band being observed at 1645 cm^{-1} (KBr) [143] to 1650 cm^{-1} [268] in the IR spectrum.

The first isolation of oxyberberine (154) from a *Thalictrum* species was from *T. minus* L. race B in 1972 [143]. The alkaloid was subsequently isolated from *T. alpinum* L. [70], *T. foliolosum* DC. [268], *T. longistylum* DC. [137], *T. lucidum* L. [139], *T. podocarpum* Humb. [301], and *T. rugosum* Ait. [234].

2.12.10. Palmatine. Tabular review [10,18,19], UV [305], IR [306], ^{1}H NMR [307].

Palmatine (148), $C_{21}H_{22}N^+O_4$ (352.1549), m.p. 250° dec. (MeOH) (chloride) [301], 245° dec. (MeOH) (iodide) [301], was first isolated from *Jateorhiza columba* Miers (*J. palmata* Miers) (Menispermaceae) in 1902 [303].

Palamatine was found to inhibit the actions of cholinesterase in rabbit spleen and pseudocholinesterase in horse serum [332]. The adrenal concentration of ascorbic acid in rats (used as a criterion for ACTH secretion) was decreased by subcutaneous administration (25–50 mg/kg) of palmatine but this effect was abolished by hypophysectomy. Pretreatment of the rats with sodium pentobarbital (intraperitoneal—40 mg/kg) did not inhibit the adrenal ascorbic acid-depleting effects of palmatine. However, this same depletion was completely inhibited by prior intraperitoneal injection of morphine (20 mg/kg) to rats under sodium pentobarbital anesthesia. Finally, the rats did not develop a tolerance to the pituitary gland stimulation induced by palmatine after a dosage of 70 mg/kg day for 6 days [335]. Palmatine inhibited the effect of epinephrine, as evaluated by the perfusion preparations of Laewen–Trendelenburg, on toad hind leg and isolated rat seminal vesicle, as well as on the blood pressure of rabbits. The alkaloid has an ACTH-like and anticholinergic effect and is bactericidal [373]. Palmatine was found to inhibit the growth of *Staphylococcus aureus, Klebsiella pnemoniae, Mycobacterium smegmatis,* and *Candida albicans* [139], as well as *Saccharomyces carlsbergensis* [343].

Palmatine was first isolated from a *Thalictrum* species, *T. foliolosum* DC., in 1952 as its tetrahydro derivative [348] and later reisolated as the quaternary salt [38,223]. Palmatine has also been isolated from *T. alpinum* L. [70], *T. javanicum* Bl. [136], *T. longistylum* DC. [137], *T. lucidum* L. [138,139], *T. minus* L. race B [143], *T. minus* L. var. *hypoleucum* (also called *T. thunbergii* DC.) (callus tissue) (TLC identification only) [142], *T. minus* L. var. *microphyllum* Boiss. [146], *T. podocarpum* Humb. [301], and *T. revolutum* DC. [101].

2.12.11. Thalidastine. Tabular review [10,18,19], UV [223,366], ^{1}H NMR [223].

Thalidastine (151), $C_{19}H_{16}N^+O_5$ (338.1028), >230° dec. (chloride) [366], 232° dec. (MeOH) (chloride) [223]; $[\alpha]_D^{25}$ +138° (MeOH) [366]; $[\alpha]_D^{23}$

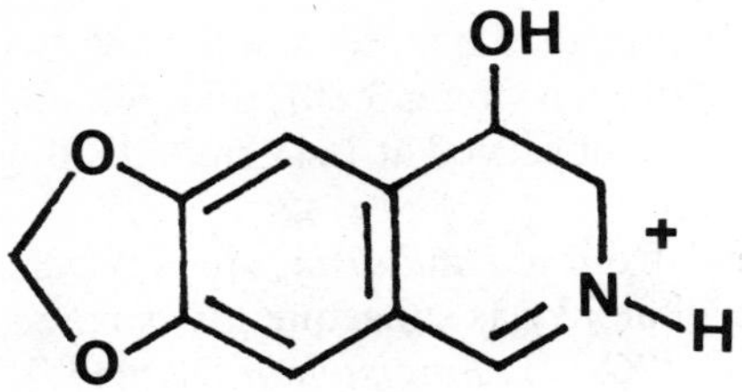

155

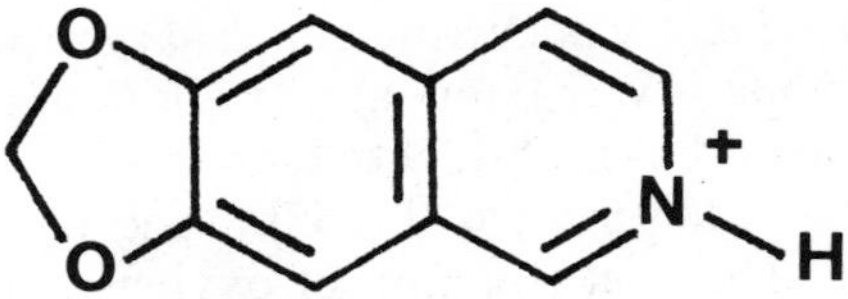

156

+105° (c 0.6, MeOH) [223], was first isolated as its yellow-brown chloride salt from *Thalictrum fendleri* Engelm. ex Gray in 1965 [62,366]. The UV spectrum resembled that of other quaternary protoberberines and possessed three maxima at 233 nm (log ϵ 4.17), 269 nm (log ϵ 4.16) and 348 nm (log ϵ 4.10) [366] (also at 425 nm (log ϵ 3.36) [223]) and showed a bathochromic shift with a concurrent change to a deep red color on the addition of alkali. Thalidastine was readily dehydrated to deoxythalidastine (**150**) on heating in vacuo at 85°, or standing at room temperature for several weeks or heating in 2*N* HCl at 100° for 15 min (preferred) [366]. Reduction of thalidastine chloride with Adams catalyst in ethanol afforded the tetrahydro derivative (M$^+$ *m/z* 341) whose mass spectral fragmentation was characteristic of this group with important retro-Diels–Alder fragment ions at *m/z* 192 (**155**), 174 (**156**), 150 (**157**), and 135 (**157**—CH₃) [366]. Because the UV spectrum of deoxythalidastine chloride was typical of a 2,3,9,10-substituted dehydroberberinium salt rather than a 2,3,10,11-substituted dehydroberberinium salt, the former substitution was assigned to thalidastine. Finally, deoxythalidastine bromide was different from dehydroberberrubinium bromide (**158**), and the former was converted to the known tetrahydrothalifendine (**152**) on reduction with Adams catalyst, thus establishing the structure of thalidastine (**151**). Thalidastine, which occurs to the extent of 0.7 g/g of berberine (**135**),

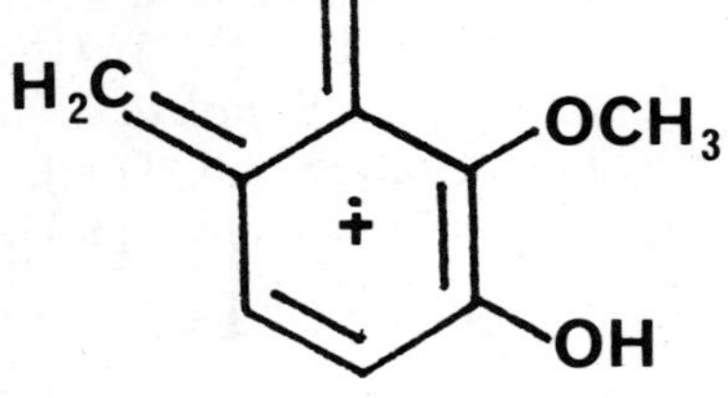

157

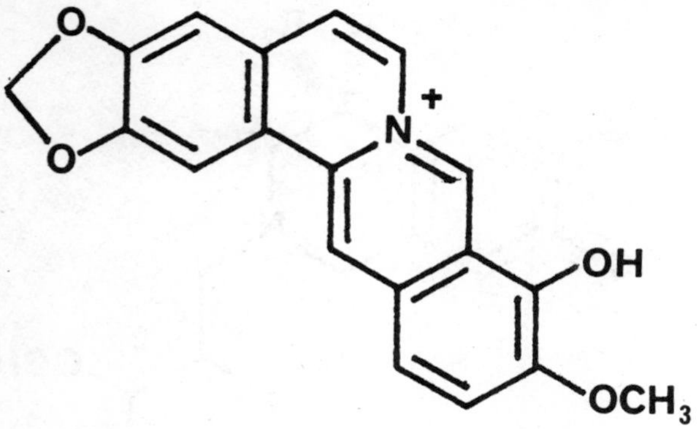

158

was only the second 5-hydroxylated protoberberine to have been found in nature, the first being berberastine (**159**) from *Hydrastis canadensis* (Menispermaceae) [374].

It was some 15 years later before the absolute configuration of thalidastine was established in an elegant paper by Abu Zarga and Shamma [375]. Briefly, jatrorrhizine chloride (**153**) was reduced (NaBH$_4$) to (±)-tetrahydrojatrorrhizine (**160**), which was resolved into the enantiomers with (−)-*O*-*O*-di-*p*-toluoyl-*d*-tartaric acid. Oxidation of the dextrorotatory isomer with lead tetracetate in glacial acetic acid afforded a mixture of diastereomeric C(5) monoacetates. This mixture was hydrolyzed and the two alcohols **161** (major) and **162** (minor) characterized via ^{1}H NMR spectroscopic differences in the chemical shifts and multiplicity of the H(4) and H(5) protons.

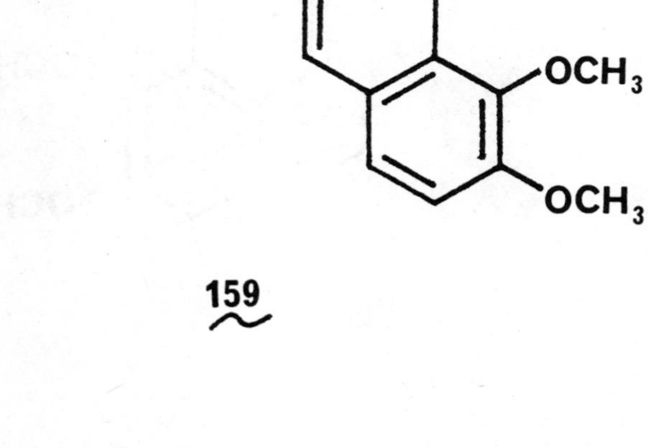

159

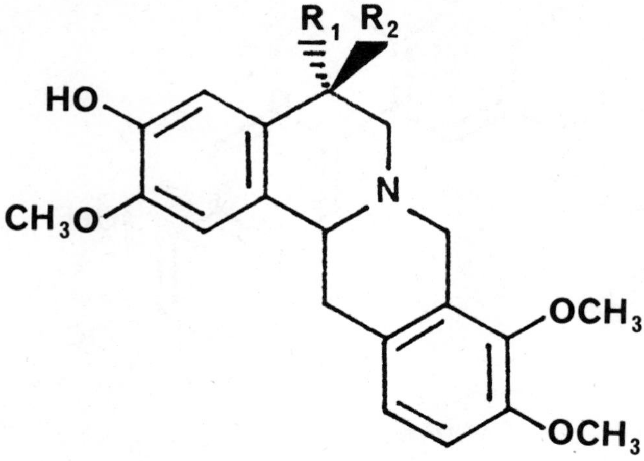

160 $R_1 = R_2 = H$
161 $R_1 = OH, R_2 = H$
162 $R_1 = H, R_2 = OH$

Finally, oxidation of **161** with iodine in ethanol afforded the alcoholic quaternary protoberberinum salt **163**, $[\alpha]_D^{25}$ +152° (c 0.25, MeOH), which must possess the same absolute configuration as both thalidastine (**151**) and berberastine (**159**).

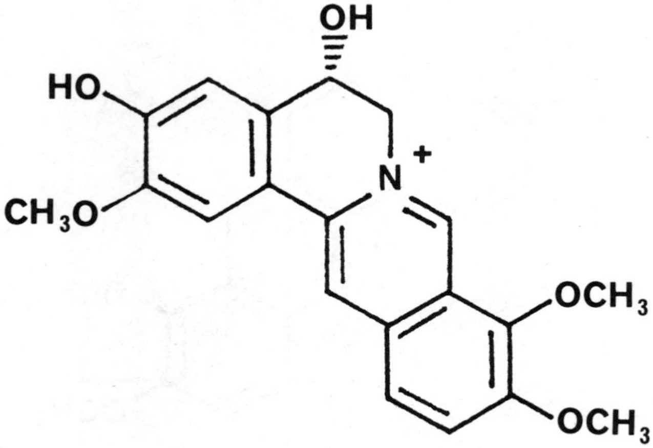

163

Thalidastine has also been isolated from *Thalictrum foliolosum* DC. [223] and *T. minus* L. var. *hypoleucum* (also known as *T. thunbergii* DC.) (callus tissue) [142].

2.12.12. Thalifaurine. UV [131,365], IR [131,365], ¹H NMR [131,365], MS [131].

Thalifaurine (**164**), $C_{19}H_{16}N^+O_4$ (322.1079), m.p. 258–260° [131], 259–261° dec. (conc. HCl) [305], was isolated from *Thalictrum fauriei* Hayata in 1979 [131]. The UV spectrum was characteristic of a quaternary 2,3,10,11-substituted protoberberine with maxima at 242 nm (log ε 4.34), 263 nm (log ε 4.29), 291 nm (log ε 4.56), 310 nm (sh) (log ε 4.46), 341 nm (log ε 4.25), and 380 nm (sh) (log ε 3.85) with a bathochromic shift (with hyperchromic effect) in both weak and strong alkali [131,365]. The ¹H NMR spectrum (CD₃OD) possessed eight singlets at δ 4.02 (3H, OCH₃), 6.33 (2H, CH₂O₂), 6.87 [1H, ArH, H(4)], 7.51 [2H, ArH, H(1) and H(12)], 7.64 [1H, ArH, H(9)], 8.59 [1H, ArH, H(13)], and 9.25 [1H, ArH, H(8)] with two triplets at δ 3.17 [2H, J = 6 Hz, H(5)] and 4.74 [2H, J = 6 Hz, H(6)] [131,365]. The CI–MS (isobutane) showed the molecular ion at m/z 322 and the base peak at m/z 320; ions at m/z 323 (**165**) and 337 (**166**) were the result of thermal disproportionation of the molecular ion [376]. Furthermore, ions at m/z 178 (**142**), 177 (**167**), and 176 (**143**) were consistent for the presence of monomethoxy, monophenolic ring A, and an ion at m/z 148 (**168**) indicated that ring D was the site of the methylenedioxy group [131]. These data suggested that thalifaurine was either 2-methoxy-3-hydroxy-10,11-methylenedioxyprotoberberine (**164**) or 2-hydroxy-3-methoxy-10,11-methylenedioxyprotoberberine

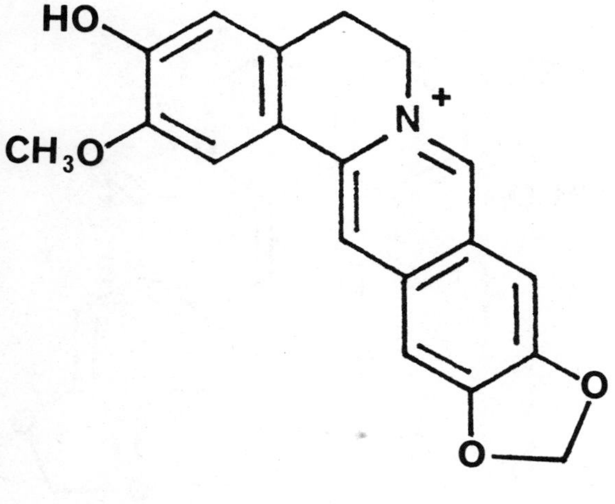

164

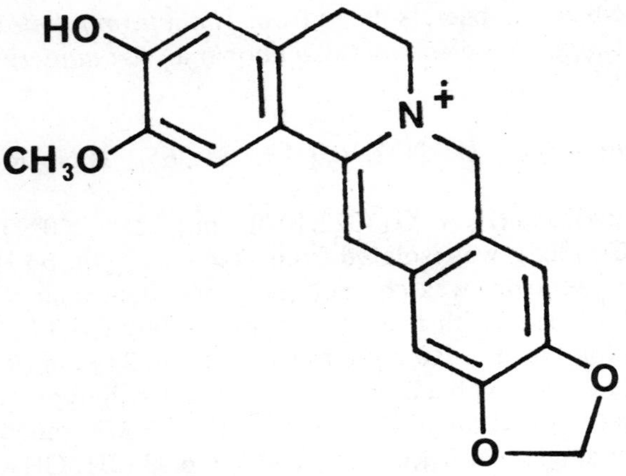

165

(169). Both UV spectral considerations and the fact that the latter is the known dehydropseudocheilanthfoline (169) and is different from the isolate preclude the former possibility. Reduction (NaBH$_4$) of thalifaurine afforded tetrahydrothalifaurine (170), m.p. 142–144° (CHCl$_3$–MeOH) [131], 154–156° (CHCl$_3$–MeOH) [365]; UV λ_{max}^{MeOH} 230 nm (sh) (log ϵ 3.72) and 290 nm (log ϵ 3.52) [365]; NMR (CDCl$_3$), δ 2.45–4.10 (9H, m, AlH), 3.85 (3H, s, OCH$_3$), 5.86 (2H, s, CH$_2$O$_2$), 6.56 (2H, s, ArH), 6.64 (1H, s, ArH), and 6.81 (1H,

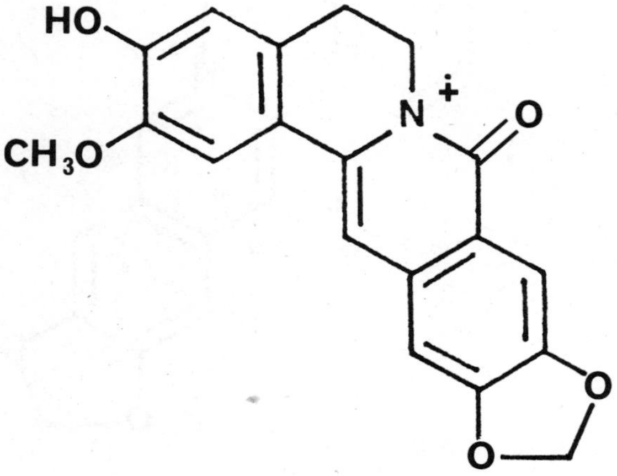

166

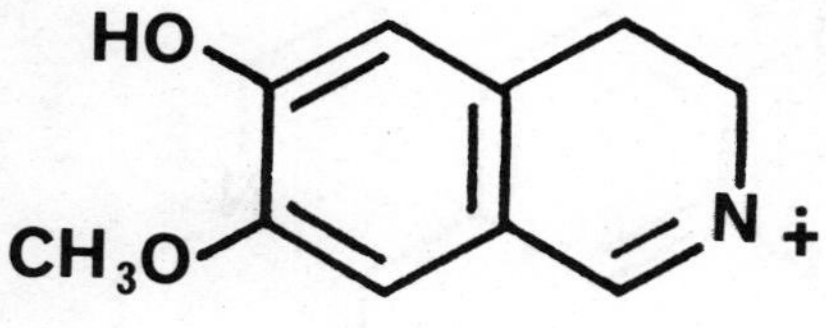

167

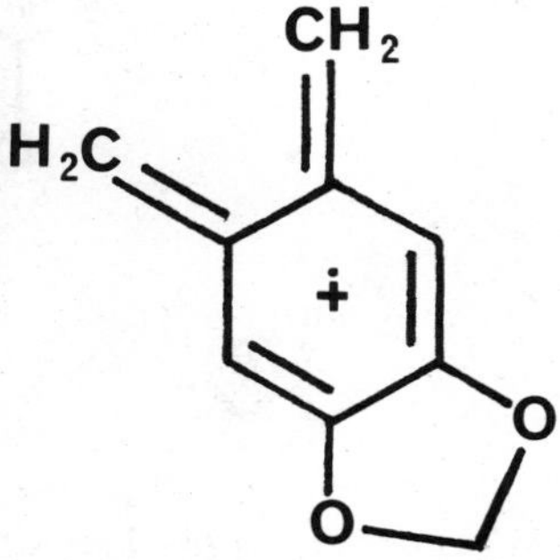

168

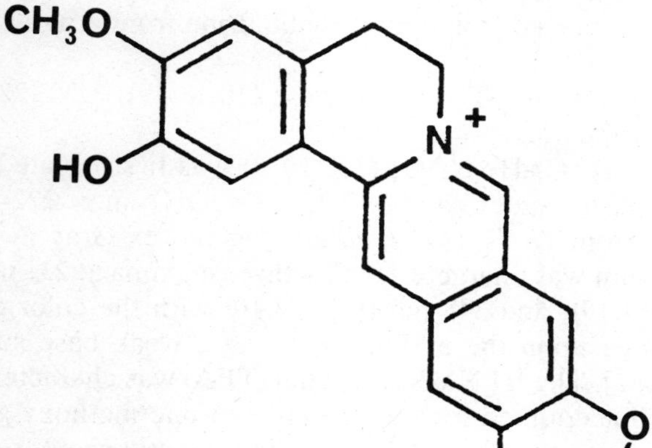

169

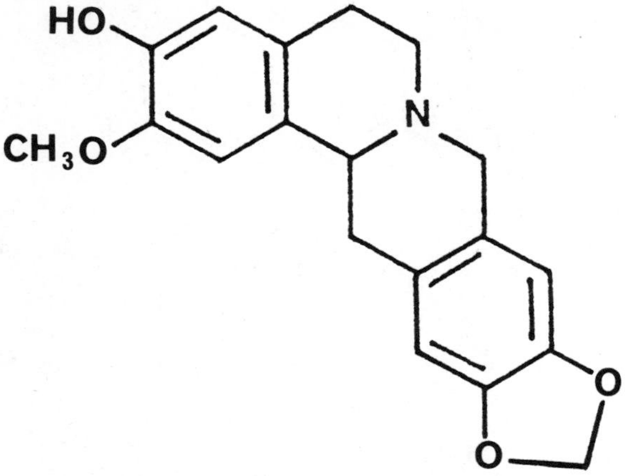

170

s, ArH) [131,365]; MS, M$^+$ m/z 325, 178, 176, and 148 (100), which was identical with authentic 2-methoxy-3-hydroxy-10,11-methylenedioxytetrahydroprotoberberine (**170**) [377], thus settling the structure of thalifaurine (**164**). Further proof was obtained by an unequivocal synthesis of tetrahydrothalifaurine (**170**) using classical methods [365] and the subsequent oxidation [Hg(OAc)$_2$/HOAc] of this product to thalifaurine (**164**) [365]. To date, this is the only reported isolation of thalifaurine from a natural source.

2.12.13. Thalifendine. Tabular review [10,18,19], UV [72,132,305], ^{1}H NMR [72,132,307].
Thalifendine (**171**), C$_{19}$H$_{16}$N$^+$O$_4$ (322.1079) was first isolated as its yellow crystalline chloride, sintering above 230° (MeOH); m.p. 222–225° (MeOH) (iodide) [223], from *Thalictrum fendleri* Engelm. ex Gray in 1965 [62,132]. The UV spectrum was characterized by three maxima at 231 nm (log ϵ 4.17), 269 nm (log ϵ 4.15), and 348 nm (log ϵ 4.10) with the color changing from yellow to orange upon the addition of even a weak base such as sodium bicarbonate [132]. The ^{1}H NMR spectrum (TFA) was characterized by seven singlets, which accounted for the presence of one methoxy group (δ 4.28), one methylenedioxy group (δ 6.15) and six aromatic protons (δ 6.95 [H(4)], 7.52 [H(1)], 7.95 (2H) [H(11) and H(12)], 8.52 [H(13)], and 9.55 [H(8)]) [132,307]. Reduction of thalifendine with Adams catalyst afforded tetrahydrothalifendine (**152**), m.p. 209–211°, [MS, M$^+$ m/z 325, 176 (**172**), and 150 (**157**)], and methylation (CH$_2$N$_2$) of the latter gave tetrahydroberberine (**173**), thus establishing the skeletal structure and positions of oxygenation of thalifendine, as well as fixing the position of the methylenedioxy group in ring

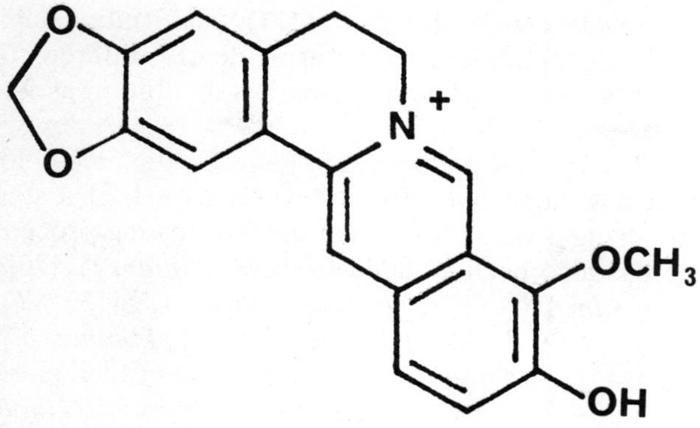

171

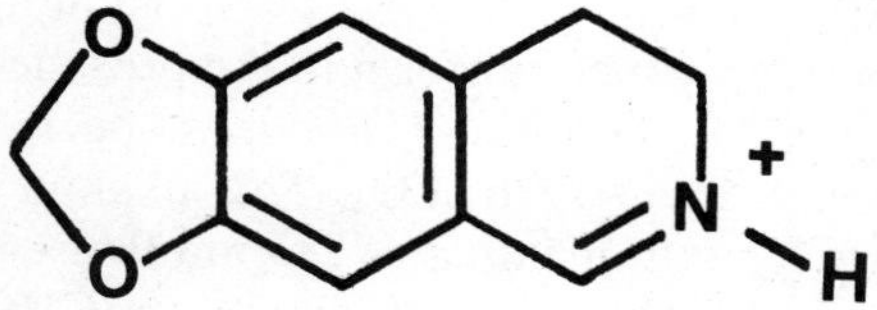

172

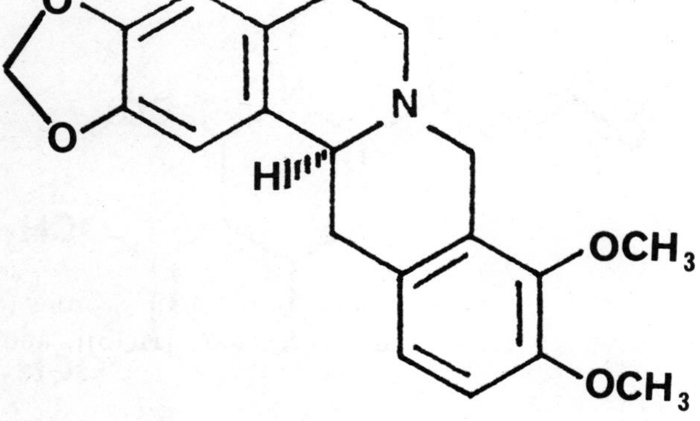

173

A. Finally, the structure of thalifendine (**171**) was firmly established when it was found that thalifendine chloride was clearly different from berberrubine chloride (**136**), thus settling the ring D substitution as 9-methoxy-10-hydroxy in thalifendine [132].

Thalifendine (1.0 mg/kg) produced no change in the mean blood pressure of normotensive dogs and failed to inhibit the growth of a standard set of microorganisms using a well established in vitro testing procedure [139].

Thalifendine has also been isolated from *T. alpinum* L. [70], *T. dioicum* L. [72], *T. foliolosum* DC. [223], *T. longistylum* DC. [137], *T. lucidum* L. [139], *T. minus* L. [140], *T. minus* L. race B [102], *T. minus* L. var. *adiantiofolium* Hort. [145], *T. minus* L. var. *hypoleucum* (callus tissue—[142]) (also called *T. thunbergii* DC. [142]), *T. podocarpum Humb.* [301], *T. polygamum* Muhl. [149,212], *T. revolutum* DC. [101], and *T. rugosum* Ait. [98].

2.12.14. 8-Trichloromethyldihydroberberine (Berberine-chloroform). UV [139,146], IR [139], ^{1}H NMR [139,146], ^{13}C NMR [369], MS [146].
8-Trichloromethyldihydroberberine (**174**), $C_{21}H_{18}NO_4Cl_3$ (454.4826), m.p. 182–184° dec. (CHCl$_3$) [139], 178° (C$_6$H$_6$) [378], is an artifact produced by the reaction of berberine with chloroform in the presence of alkali [378] and has long been known [303]. The first *Thalictrum* species from which it was isolated was *T. lucidum* L. in 1976 [139]. The alkaloid is characterized by a UV spectrum (CHCl$_3$) with maxima at 283 nm (log ϵ 3.97), 362 nm (log ϵ 4.01), and 380 nm (log ϵ 3.98) and an IR spectrum (CHCl$_3$) with significant absorptions at 3000, 1622, 1605, 1505, 1050, and 950 cm^{-1} [139]. The ^{1}H NMR spectrum shows peaks at δ 3.85 (3H, s, OCH$_3$), 3.95 (3H, s, OCH$_3$), 5.65 [1H, s, AlH, H(8)], 5.92 (2H, s, CH$_2$O$_2$), 6.08 [1H, s, ArH, H(13)], 6.60 (1H, s, ArH), 6.80 (1H, d, J = 8 Hz, ArH), 7.00 (1H, d, J = 8 Hz, ArH),

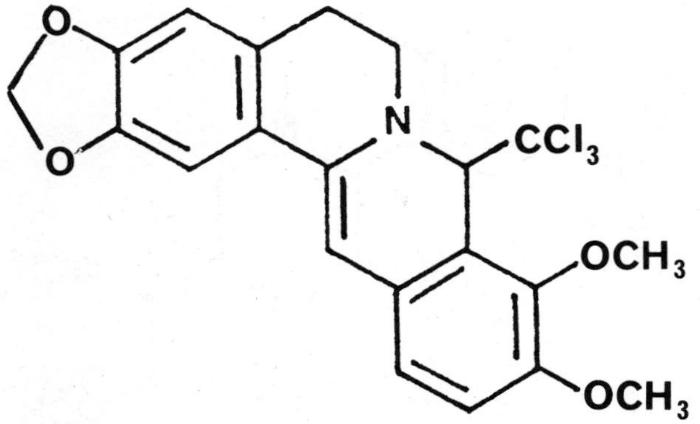

174

and 7.16 (1H, s, ArH) [139]. The alkaloid may be converted to berberine (135) iodide by treatment of an acetone solution with methyl iodide for three days [139]. Furthermore, this artifact may be prepared by stirring a suspension of berberine chloride in chloroform and ammonium hydroxide for 1 day and crystallizing the artifact from the chloroform layer [139].

8-Trichloromethyldihydroberberine had no effect on the blood pressure of normotensive dogs when administered in doses of 1 and 2 mg/kg [139]. The artifactual alkaloid did, however, exhibit in vitro antimicrobial activity against *Mycobacterium smegmatis* and *Candida albicans* [139].

Other *Thalictrum* species from which this artifact have been isolated include *T. longistylum* DC. [137], *T. minus* L. var. *microphyllum* [146], and *T. podocarpum* Humb. [301].

2.13. Tetrahydroprotoberberines

2.13.1. *N*-Methyltetrahydroberberine (*N*-Methylcanadine).

Tabular review [10,18,19], UV [305,380], IR [380,382], ^{13}C NMR [381].
N-Methyltetrahydroberberine (*N*-Methylcanadine) (175), $C_{21}H_{24}N^+O_4$ (354.1705), was apparently first isolated from *Zanthoxylum brachyacanthum* (Rutaceae) in 1913 as the α-(−)-chloride salt [379], m.p. 235–237° [382], $[\alpha]_D$ −134.2° [382]. Intravenous administration of (−)-canadine methochloride (*N*-methylcanadine chloride) at a dosage of 2 mg/kg produced a transient decrease (30–40 mm Hg) in systolic blood pressure of pentobarbital-anesthetized cats. There was also a concurrent bradycardia (20–30 beats/min) but no significant action on the nictitating membrane. These inhibitory effects on blood pressure and heart rate persisted for approximately 20 min. The alkaloid also inhibited the responses to stimulation of the superior cer-

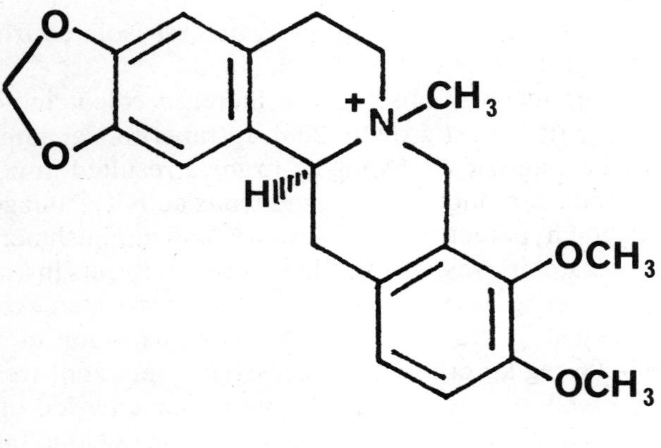

175

vical sympathetic nerve and to the response to dimethylphenylpiperazinium. Stimulation of the nictitating membrane during the period of ganglionic blockage (lasting 10 min) failed to elicit a response. The alkaloid potentiated the responses of (−)-norepinephrine but failed to effect angiotensin-induced responses. Higher doses (5 mg/kg) of (−)-canadine methochloride also produced a hypotensive response (40 mm Hg decrease) that lasted for 30 min. In addition, this dosage of alkaloid potentiated the blood pressure responses to (−)-norepinephrine, histamine, and angiotensin, while diminishing responses to carotid occlusion. It did not, however, alter the response to methacholine chloride and vagal stimulation on blood pressure. In in vitro experiments, the alkaloid reversibly potentiated the cumulative dose response to (−)-norepinephrine on isolated rat vas deferens and isolated reserpinized rat vas deferens. The alkaloid also reversibly potentiated the angiotensin-induced (5 mg/kg) contractile response in the isolated rodent vas deferens. Finally, the LD_{50} of (−)-N-methylcanadine chloride in female mice was established as 84 (± 2.8) mg/kg after intraperitoneal administration. Death was characterized by clonic convulsions [225].

N-Methylcanadine was first isolated from a *Thalictrum* species, *T. minus* L., in 1967 as L-canadine β-methochloride (m.p. 191–193° dec. and $[\alpha]_D$ − 158° [380]) and subsequently from the same species as L-canadine β-methohydroxide (m.p. 190–192° (Et$_2$O–MeOH) and $[\alpha]_D^{21}$ − 128.3° (c 0.4 EtOH) [65].

2.13.2. Tetrahydroberberine (Canadine).   Tabular review [10,18,19], UV [305], IR [383], ^{1}H NMR [98,384,385], ^{13}C NMR [386], MS [385,387], CD [388].

Tetrahydroberberine (canadine) (**173**), $C_{20}H_{21}NO_4$ (339.1471), occurs most commonly in nature as its levorotatory (*S*) isomer (m.p. 134–135° (MeOH), $[\alpha]_D$ − 298° (CHCl$_3$) [389]) and was first isolated from *Hydrastis canadensis* (Berberidaceae) in 1894 [390].

Tetrahydroberberine was observed to increase colchicine-mediated antimitosis in in vivo fibroblast growth [208]. Intraperitoneal administration of the alkaloid in a dosage of 40–65 mg/kg to mice resulted in potentiation of hexobarbital narcosis, reduction in spontaneous activity, antagonism of amphetamine-induced hyperactivity and toxicity, and diminishment of the mescaline-induced scratching response. Administration to cats in a subcutaneous dosage of 10–20 mg/kg partially abolished the motor-defense conditioned reflex and prolonged its latent period. Oral administration to monkeys in a dosage of 180–206 mg/kg produced successively sedation, tremor, and catatonia within 30–60 min, which disappeared over a period of 5 hr. Tetrahydroberberine aggravated strychnine-induced convulsions but slightly inhibited metrazole-induced convulsions in mice and apomorphine-induced emesis in dogs. Synchronization of electroencephalogram waves in the rabbit cerebral and subcerebral cortices was observed after an intravenous dosage of 30 mg/kg. The alkaloid produced transient hypotension and tachypnea in anesthetized rats and also possessed sedative-tranquilizing and analgesic effects [391]. The methyl orange dye method was used to measure the absorption (49–85%) of tetrahydroberberine from the gastrointestinal tract of mice after oral administration of an oral dose of 60 mg/kg. Intravenous administration to rabbits in a dosage of 60 mg/kg resulted in blood protein binding (70%) with high concentrations in the brain, lungs, kidneys, and liver. Administration of an intraperitoneal dose of 60 mg/kg to rabbits resulted in recovery of only 15% of the drug in the urine after 48 hr. Incubation with liver slices for 1–8 hr in the presence of oxygen results in the in vitro metabolism of the alkaloid.

The lipid–water partition coefficient of the alkaloid is more than 750 times greater than that of berberine (135). Passive transfer is likely an important mechanism for metabolism of the alkaloid [392]. In another study, repeated subcutaneous administration of tetrahydroberberine in a dosage of 50–100 mg/kg decreased the conditioned reflexes and the orientation reaction in mice and induced analgesia, both effects being inhibited by phenamine administration [393]. The alkaloid also possessed ataractic effects in rats at a dosage of 30 mg/kg [394] and eliminated the atrophine-induced psychosis in dogs when administered intravenously (10 mg/kg) [395]. (−)-Tetrahydroberberine-(+)-camphorsulfonate possessed sedative activity in mice at an oral dosage of 25 mg/kg with a LD_{50} of >1500 mg/kg and few side effects [396]. The alkaloid possessed few or no antitussive effects in dogs [112]. Tetrahydroberberine hydrochloride was found to have a low toxicity with no embryotoxic or carcinogenic properties. The alkaloid was principally characterized by sedative-tranquilizing activity but was also observed to produce muscle relaxant, hypothermic, aggressive-behavior- and conditioned-behavior-inhibiting as well as cerebral self-discharge-inhibiting activities. Sedation appeared to be mediated via central adrenolytic activity [397,398]. Intraperitoneal administration of tetrahydroprotoberberine to rats in a dosage of 100 mg/kg produced a 28% decrease in acetylcholine levels in the rodent striatum as compared with controls. However, no change in the acetylcholine levels in the hypocampus was observed. The alkaloid was considered to be a dopamine receptor antagonist [399]. Administration of tetrahydroberberine to mice produced transitory relaxation of uterine contractions at a minimum effective concentration of 5.0×10^{-6} g/ml [311].

Another study showed that tetrahydroberberine antagonized apomorphine- and L-dopa-induced rodent stereotyped behavior, suggesting a blockade of postsynaptic dopamine receptors. The alkaloid produced increased levels of corporal striatal homovanillic acid and L-dopa but this effect was antagonized by apomorphine. Enhancement of tyrosine hydroxylase activity of the alkaloid via a feedback mechanism that blocked the presynaptic dopamine receptor was a likely possibility [400]. Tetrahydroberberine (optical activity unknown) has been isolated from only one Thalictrum species, T. acteaefolium [401].

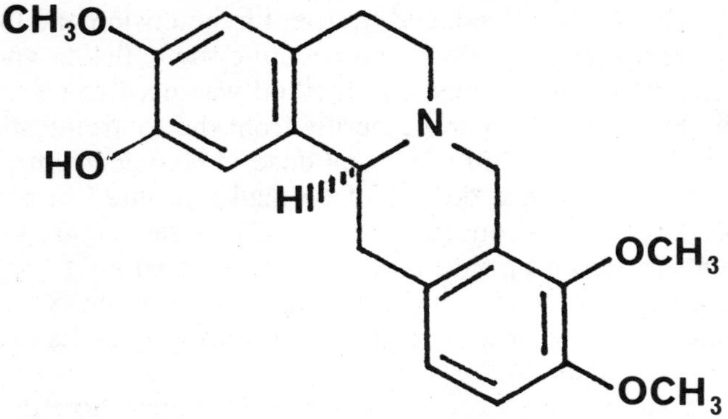

$$176$$

2.13.3. Tetrahydrocolumbamine (Isocorypalmine).

Tabular review [18,19], UV [72,305], ^{1}H NMR [72,98,384], MS [72,387], CD [72].

Tetrahydrocolumbamine (isocorypalmine) (176), $C_{20}H_{23}NO_4$ (341.1627), m.p. 243° (MeOH) [72], $[\alpha]_D^{25}$ −291° (c 0.3, MeOH) [72], was first isolated from nature as its dextrorotatory isomer in 1926 but subsequently as its more commonly occurring levorotatory isomer in 1938 [402].

The alkaloid possessed anticholinergic activity in the small intestine [403]. (−)-Tetrahydrocolumbamine has been isolated from only one *Thalictrum* species, *T. dioicum* L. [72].

2.13.4. Tetrahydrothalifendine.

Tabular review [10,18,19], UV [232,305], IR [232], ^{1}H NMR [232,384], ORD [232], MS [132,232].

Tetrahydrothalifendine (152), $C_{19}H_{19}NO_4$ (325.1314), m.p. 210–211°, $[\alpha]_D$ −175° (MeOH), was first isolated from *Thalictrum fendleri* Engelm. ex Gray in 1971 [232]. The UV spectrum was typical of tetrahydroprotoberberine and showed maxima at 209 nm (log ε 4.8) and 282 nm (log ε 3.77) with a bathochromic shift in alkali and positive ferric chloride test [232]. The mass spectrum was characterized by peaks at *m/z* 176 (172), 150 (157), 149 (157—H), and 135 (157—CH$_3$) and was typical of the retro-Diels–Alder type fragmentation of tetrahydroprotoberberine [132,232]. The ^{1}H NMR spectrum (CDCl$_3$) revealed the presence of one methoxy group (δ 3.82) and one methylenedioxy group (δ 5.92) as singlets and four aromatic protons as a complex multiplet from δ 6.50–7.08. Reduction of naturally occurring thalifendine (171) with Adams catalyst afforded racemic tetrahydrothalifendine, identical to the natural product, except for optical activity. The (*S*) configuration was assigned to the naturally occurring alkaloid based on its negative ORD Cotton effect with a trough at 306 nm (− 1650) by analogy with other tetrahydroprotoberberine alkaloids of established configuration [232]. Later research has shown that the sign of the intense 206–210 nm Cotton effect in the CD of tetrahydroprotoberberine correlates directly with absolute configuration and is independent of both substitution pattern and pH [404]. To date, the alkaloid has not been isolated from any other *Thalictrum* species.

2.14. Protopines

The following reviews and references are particularly relevant to the isolation and identification of protopine alkaloids:

1. Tabular reviews [10,18,19,405]
2. Annual reviews [20–22,24,194–201]
3. Other reviews [10–14]
4. Book chapters [14,208,240–242,409,410]

The numbering of the protopine ring (**177**) is according to accepted practice [405].

2.14.1.   Allocryptopine (Thalictrimine, β-Allocryptopine).   Tabular review [405, alkaloid No. 2], UV [104,405], IR [405], ^{1}H NMR [405,411], ^{13}C NMR [405,411,412], MS [405,413].

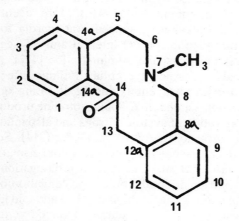

177

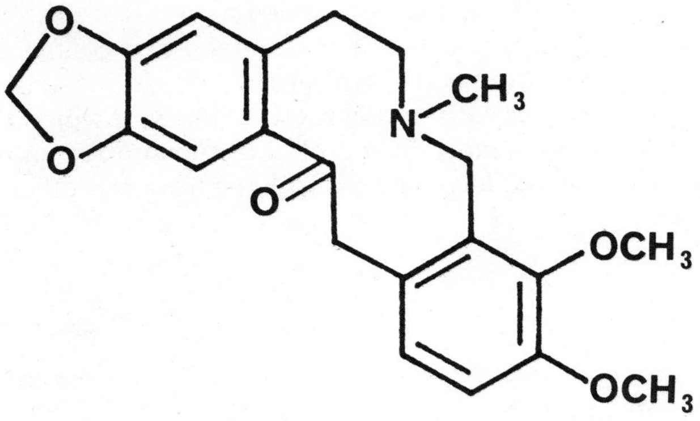

178

Allocryptopine (**178**), $C_{21}H_{23}NO_5$ (369.1576), α-allotropic form m.p. 160–161° (Me_2CO) [104], β-allotropic form m.p. 169–170° (Me_2CO) [153], is a commonly occurring alkaloid in *Corydalis* species (Fumariaceae) and numerous genera of the Family Pappaveraceae [405]. It was first isolated from *Chelidonium majus* L. (Papaveraceae) in 1890 [208].

Almost 100 years ago it was observed that atrial fibrillation frequently resulted in conversion to normal sinus rhythm. Ventricular arrhythmias were sometimes perpetuated by administration of the alkaloid, with serious and sometimes fatal consequences [415–416]. More recent studies in rabbits demonstrated that the alkaloid produced bradycardia followed by tachycardia, with a disturbance in the PQ interval and ST segment of the electrocardiogram, after intramuscular administration [417]. Additional studies with various animals confirmed the antiarrhythmic effects of the alkaloid when administered in an intravenous dosage of 10 mg/kg and also showed there was no loss of potassium ion. α-Allocryptopine produced a bradycardia and prolonged the duration of systole in frogs and also induced a bradycardia in cats and rabbits at a dosage of 10–20 mg/kg [414]. Subsequent studies over the next five decades showed that the alkaloid also affected many other properties of myocardial tissue, including a prolongation of the refractory period, a slowed ventricular conduction, a prolongation of the atrioventricular conduction time, and a vagal inhibition. The alkaloid inhibited aconitine-induced atrial fibrillation and mechanically induced ventricular fibrillation in dogs, when administered in a dosage of 4–10 mg/kg [414]. These antiarrhythmic effects were also observed in human subjects on intramuscular administration of the alkaloid in doses of 60–120 mg to patients with both atrial flutter and on the myocardium of aconitine-induced arrhythmic animals [418]. Other studies confirmed these antiarrhythmic effects [419,420].

α-Allocryptopine was further observed to increase the amplitude of uterine contractions in pregnant rabbits and guinea pigs [421–425]. Clinical trials in patients in labor whose uterine contractions were weak or who had subinvolution of the uterus confirmed the utility of the alkaloid as a clinically useful oxytocic. The alkaloid was administered via three intramuscular injections of 2% solution (2 ml) given at 1 hr intervals. The regulation of labor was established within 20–30 min after the first injection and complete cervical dilation occurred in 5–7 hr [426]. Finally, the alkaloid fails to inhibit bovine glutamate dehydrogenase [57] but does possess local anesthetic properties [208].

β-Allocryptopine (**178**) was first isolated from a *Thalictrum* species, *T. minus* L., in 1959, when the alkaloid was called thalictrimine [427]. It was subsequently reisolated from the same species in 1962 [428] and in 1965 [429], when it was ultimately noted to be identical with β-allocryptopine [429]. The alkaloid has also been isolated from *Thalictrum amurense* Maxim. [63], *T. contortum* L. [63], *T. revolutum* DC. [104,105], and *T. simplex* L. [153,154].

2.14.2. Cryptopine (Thalisopyrine). Tabular review [405, alkaloid No. 3], UV [405], IR [409], ^{1}H NMR [409], ^{13}C NMR [405,430], MS [405,413]. Cryptopine (**179**), $C_{21}H_{23}NO_5$ (369.1576), m.p. 222–223° (CHCl$_3$–EtOH) [405], was first isolated from the thebaine fraction of the opium alkaloids in 1867 [406].

Early pharmacological studies on the action of cryptopine in rabbits and frogs showed that it possessed a direct myocardial activity, reflected by a nonvagally mediated bradycardia with ultimate arrest in diastole. The alkaloid consistently decreased the tone, rate, and amplitude of contractions in perfused frog, toad, rabbit, and dog (10 mg/kg) hearts while prolonging the effective refractory period of the isolated guinea pig atrium. A dosage

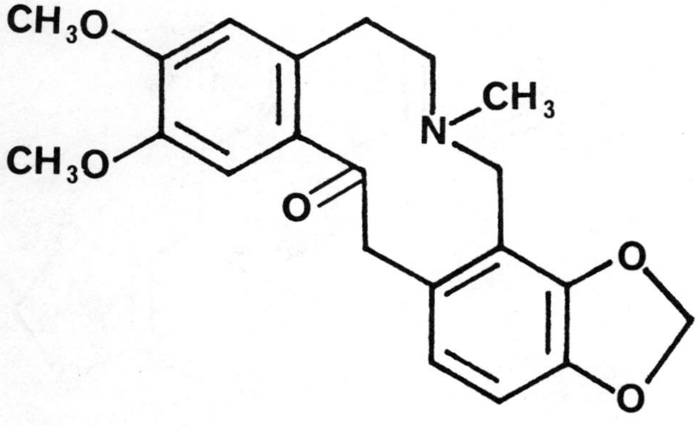

179

of 5×10^{-5} moles/kg to an intact animal produced a prolongation of the QRS, PR, and QT intervals of the electrocardiogram [414]. Furthermore, the alkaloid was observed to depress higher nervous centers, induce spinal paralysis in frogs and convulsions in mammals [406], and stimulate in vivo and in vitro uterine contractions [431]. Other studies in dogs demonstrated the production of hypotension, myocardial inhibition, respiratory stimulation, vagal inhibition, and decrease in the oculocardiac reflux; a skeletal muscle relaxation was observed in rabbits. The LD_{100} in guinea pigs was found to be 160 mg/kg after subcutaneous administration [208].

Cryptopine was first isolated from a *Thalictrum* species, *T. isopyroides* C.A.M., in 1959, when it was named thalisopyrine [432]. However, its identity with cryptopine was confirmed in 1968, when the alkaloid was isolated from *T. flavum* L. [133]. Finally, the alkaloid was reisolated from *T. isopyroides* C.A.M. in 1968 [174].

2.14.3.   Protopine.   Tabular review [405, alkaloid No. 1], UV [234,405], IR [234,405], ^{1}H NMR [234,405], ^{13}C NMR [412,430], MS [405,413].
Protopine (**180**), $C_{20}H_{19}NO_5$ (353.1263), m.p. 203–204° (MeOH) [234], was first isolated from opium in 1871 [406]. Early pharmacological research demonstrated that protopine produced bradycardia, ventricular fibrillation, increased coronary artery flow, and a prolongation of effective refractory period in whole animal and isolated frog/guinea pig myocardial preparations [414]. Later work showed that the alkaloid also possessed smooth muscle relaxant, hypotensive, and antimicrobial (Gram-positive bacilli) activities. Protopine was found to be cytotoxic but produced little inhibition of in vivo tumor growth [208]. The effects of protopine on the myocardium and on the induction of cardiac arrhythmias have been studied [433,434]. Considerable

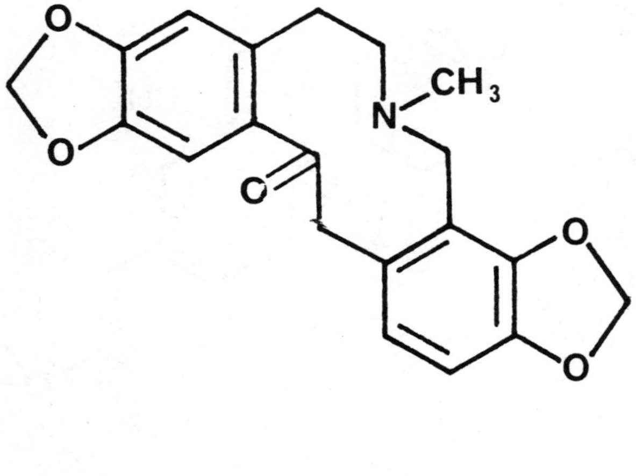

180

interest focused on the oxytocic effect of protopine in which the alkaloid produced prompt and powerful uterine contractions in both excised mammalian uterine strips and excised strips of human pregnant uterus. The alkaloid had to be administered intravenously (0.5 mg/kg) to be active, as oral administration produced a very irregular effect. A concentration of 1:1,000,000 induces a sharp increase in the tonus of excised human uterus, with powerful rhythmic contractions. This activity is not influenced by either atropine or ergotoxine [431]. Finally, protopine was found to be an effective inhibitor of blood platelet aggregation in rat blood platelets [435a] but failed to inhibit bovine glutamate dehydrogenase in vitro [57].

Protopine has only been isolated from one species of *Thalictrum*, *T. rugosum* Ait. [234].

2.14.4. Protothalipine. Tabular review [405, alkaloid No. 6], UV [98,405], IR [98,405], ^{1}H NMR [98,405], MS [98,405].

Protothalipine (**181**), $C_{21}H_{25}NO_5$ (371.1733), m.p. 195–196° dec. (MeOH), was isolated from *Thalictrum rugosum* Ait. in 1976 [98]. The UV spectrum of this optically inactive alkaloid showed maxima at 232 nm (log ϵ 4.60) and 282 nm (log ϵ 3.91) with a bathocromic shift in alkali; the IR spectrum (CHCl$_3$) was characterized by a benzenoid conjugated carbonyl band at 1655 cm^{-1} and a phenolic hydroxy group at 3540 cm^{-1} [98]. The ^{1}H NMR spectrum suggested the presence of one high-field *N*-methyl group at δ 1.87 (3H, s); two adjacent methylene groups as a symmetrical multiplet at δ 2.67 (2H, m) [C(6)] and 2.90 (2H, m) [C(5)]; two other methylene groups at δ 3.71 (2H, s) [C(8) or C(13)] and 3.75 (2H, s) [C(13) or C(8)]; three methoxy groups at δ 3.90 (9H, s); one hydroxy group at δ 4.07 (brs, D$_2$O exchangeable); and four aromatic protons at δ 6.68–7.05 (4H, m). The mass spectrum exhibited a typical retro-Diels–Alder type fragmentation and showed the molecular

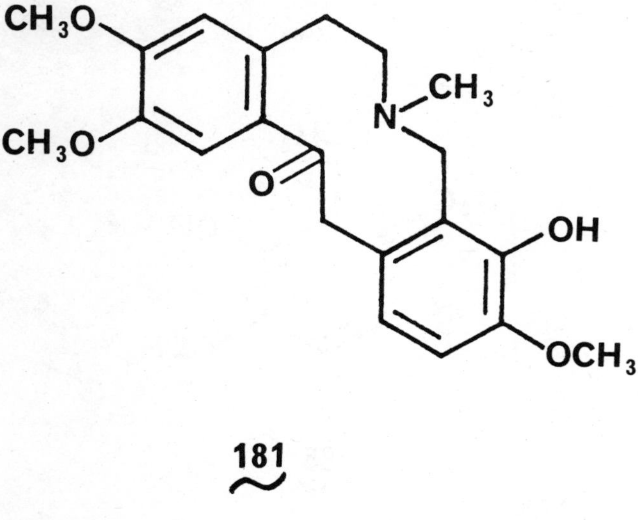

181

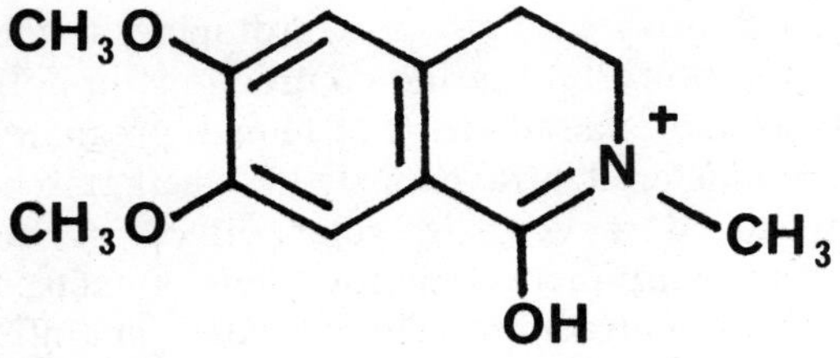

182

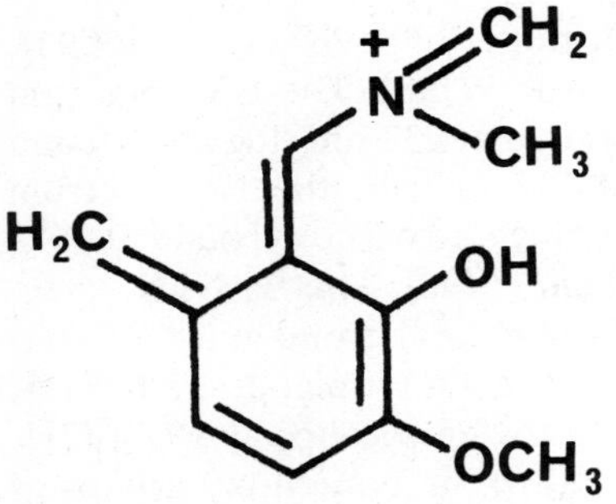

183

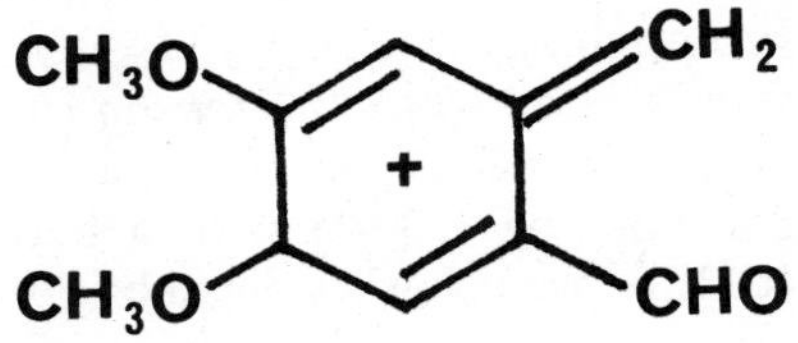

184

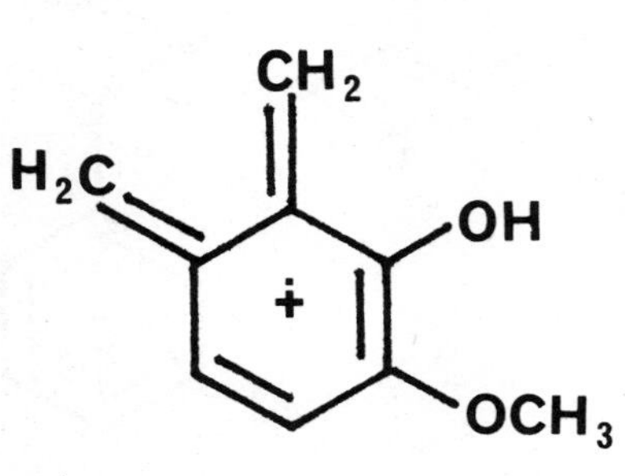

185

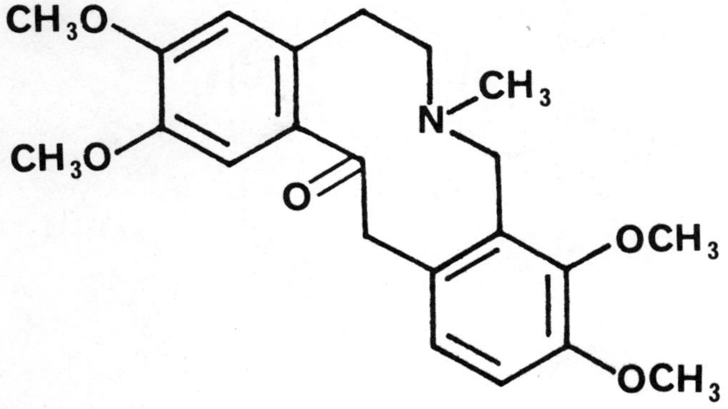

$\underset{\thicksim}{\underline{186}}$

ion at m/z 371 (6%) with other significant fragment ions at m/z 222 (100%) (**182**), 192 (33%) (**183**), 179 (72%) (**184**), 150 (86%) (**185**), 135 (26%) (**185**—CH$_3$), and 107 (27%) (**185**—CH$_3$—CO) [98,413].

These spectral data [particularly the mass spectral fragment ion at m/z 150 (**185**)] were characteristic of a protopine-alkaloid containing two methoxy groups in the aromatic ring adjacent to the carbonyl group plus one methoxy group in one hydroxy group in the other aromatic ring. Treatment of protopine with ethereal CH$_2$N$_2$ afforded muramine (**186**), thereby confirming the oxygenation pattern. Aromatic solvent-induced chemical shift (ASIS) differences were then measured for protothalipine, muramine (*O*-methylprotothalipine), and a series of tetrahydroprotoberberine alkaloids in CDCl$_3$, CDCl$_3$-C$_6$D$_6$ (1:1), and C$_5$D$_5$N, with the results suggesting the placement of the single phenolic hydroxy group of protothalipine (**181**) at C(9) [98]. Attempts to synthesize protothalipine were unsuccessful [98] and the alkaloid has not been isolated from another natural source to date. Finally, protothalipine was found to exhibit weak in vitro antimicrobial activity against *Staphylococcus aureus* and *Mycobacterium smegmatis* [98].

2.14.5.   Thalictricine (Thalictrisine).   Tabular review [405, alkaloid No. 7], UV [405,435], IR [405,435], ^{13}C NMR [412], MS [405,435].

Thalictricine (**187**), C$_{20}$H$_{21}$NO$_5$ (355.1420), m.p. 261–263° (MeOH), was isolated from *Thalictrum simplex* L. in 1968 [435]. The UV spectrum of the optically inactive alkaloid was characterized by a maximum at 288 nm (log ϵ 3.95), and the IR spectrum showed a prominent absorption band at 3640 cm^{-1} (hydroxy) and 1640 cm^{-1} (carbonyl). The MS showed the molecular ion at m/z 335 with other characteristic fragment ions at m/z 269, 207, 206 (100%), 192, and 150. These data were characteristic of a protopine-like

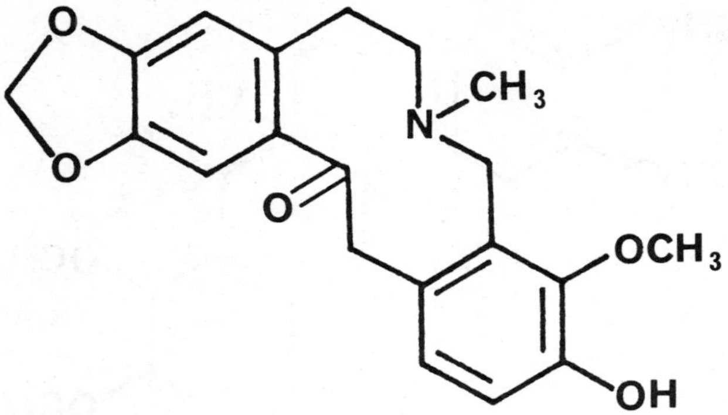

187

alkaloid containing a methylenedioxy group in the aromatic ring adjacent to the carbonyl group plus one methoxy group and one hydroxy group in the other aromatic ring [405,413]. Methylation of thalictricine with ethereal CH_2N_2 afforded β-allocryptopine (**178**), thereby fixing the positions of oxygenation and other functionality. Because the physicochemical properties of thalictricine differed from those of hunnemanine (**188**) [405] and both thalictricine and hunnemanine were converted to β-allocryptopine when methylated, thalictricine was assigned as **187**.

Thalictricine has been isolated from only one other species of *Thalictrum*, *T. amurense* Maxim. [63].

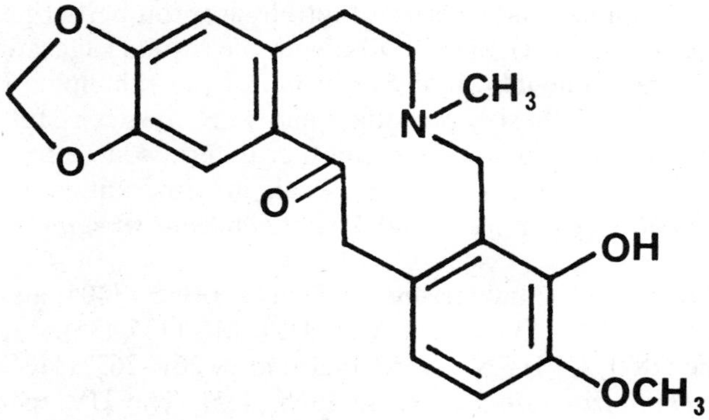

188

3. DIMERIC ALKALOIDS

3.1. Aporphine–Benzylisoquinolines

The following reviews and references are particularly relevant to the isolation and identification of aporphine-benzylisoquinoline alkaloids:

1. Tabular reviews [10,18,19,436,437]
2. Annual reviews [22,24–30,194]
3. Other reviews [10–14]
4. Book chapters [14,438–442]

The numbering of the aporphine-benzylisoquinoline ring (189) is according to accepted practice [436].

3.1.1. Adiantifoline. Tabular review [436, alkaloid No. 16], UV [143,436,443], ¹H NMR [143,436,443], MS [436,443], CD [436,443].
Adiantifoline (190), $C_{42}H_{50}N_2O_9$ (726.3516), m.p. 143.5–144° (EtOH), $[\alpha]_D^{28}$ +90° (c 0.11, MeOH), was first isolated from *Thalictrum minus* L. var. *adiantifolium* Hort. in 1968 [145,443]. The UV spectrum showed maxima at 283 nm (log ε 4.51), 302 nm (log ε 4.39), and 312 nm (log ε 4.34) and was characteristic of an aporphinoid system [443]. The ¹H NMR spectrum revealed the dimeric nature of the alkaloid and showed the presence of two N-methyl groups as singlets at δ 2.44 and 2.47; eight methoxy groups as six singlets at δ 3.59, 3.78 (9H), 3.82, 3.89, 3.94, and 3.96; and six aromatic protons as four singlets at 6.24, 6.55 (2H), 6.60 (2H), and 8.08. The mass

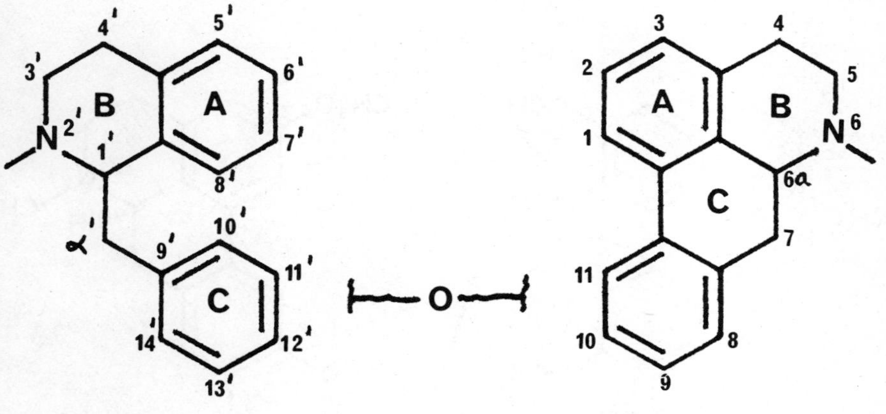

189

OCH3
CH3O
OCH3
OCH3
N
CH3
H
CH3O
N
CH3
H
OCH3
OCH3
CH3O
O

190

CH3O
CH3O
N+
CH3

191

OCH3
OCH3
N
CH3
H
CH3O
CH3O
N
CH3
H
OCH3
OCH3
CH3O
O

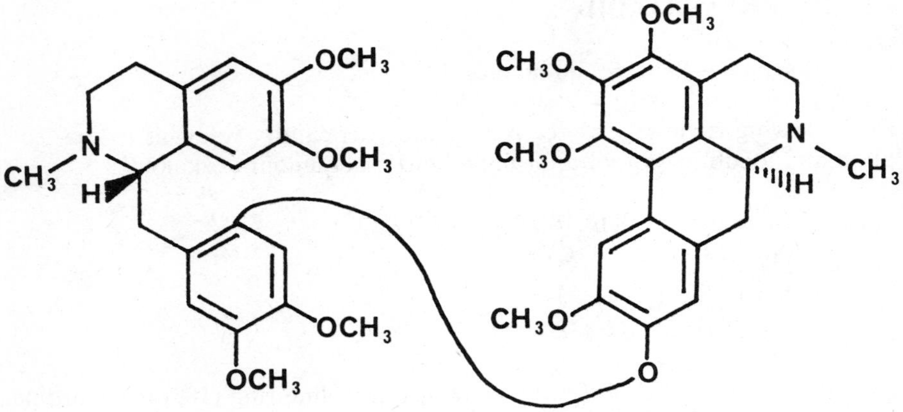

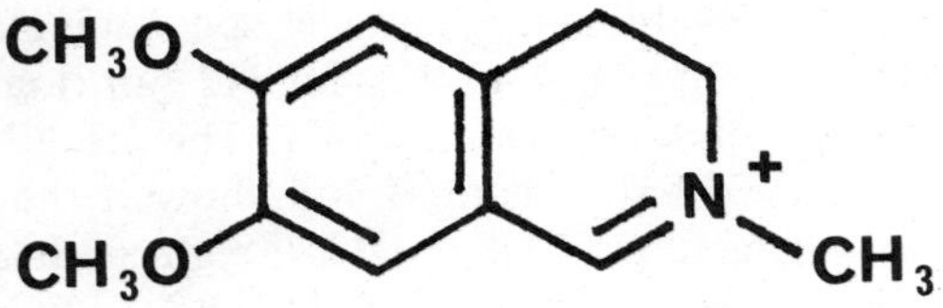

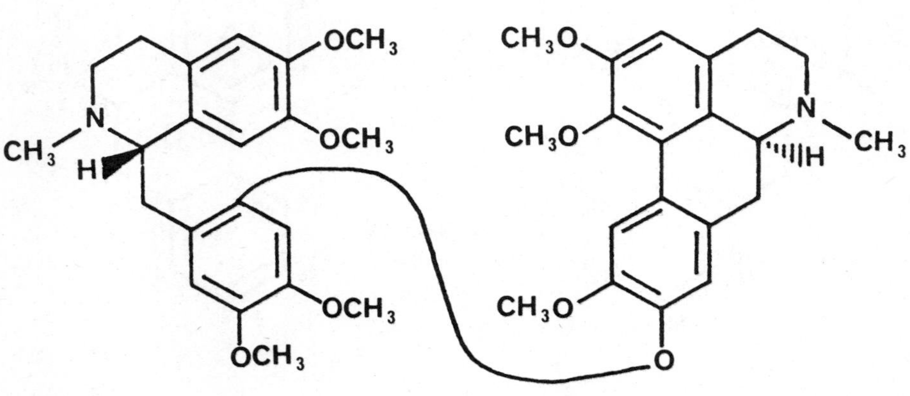

192

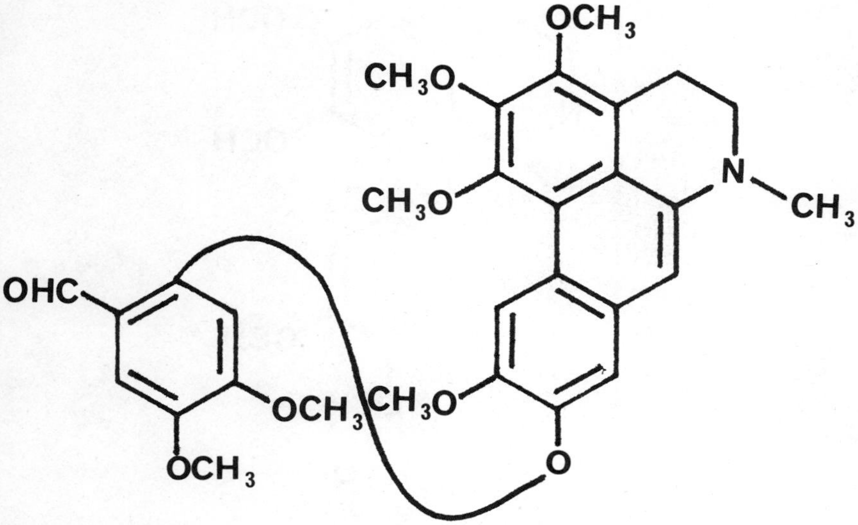

193

spectrum showed no parent ion, with the first discernible peaks at m/z 521, 520, 519 (all about 2%) and an intense fragment ion at m/z 206 (**191**). Because the first discernible cluster of ions in the mass spectrum of thalicarpine (**192**) was found at m/z 492, 491, and 490 with a sizable ion at m/z 206, the spectral data for adiantifoline were suggestive of a thalicarpine-like alkaloid bearing one extra methoxy group on either the aporphine ring or the benzyl portion of the benzylisoquinoline system. Additional support for this assignment was gained by oxidation of adiantifoline with potassium permanganate in acetone to afford the aldehyde **193** and by reductive cleavage with the liquid NH_3 to yield 6'-hydroxylaudanosine (**194**) and ketone **195**, the latter two products also being obtained via a similar cleavage of thalicarpine (**192**). The position of the terminus of the diphenyl ether bridge in the aporphine ring of adian-tifoline was proposed to be at C(9), as in thalicarpine (**192**), with the as-sumption of the loss of a methoxy group at C(1) in the aporphine ring during cleavage. Finally, because a low-field proton (δ 8.08) characteristic of a H(11) proton in the aporphine ring was present in the ^{1}H NMR spectrum of adian-tifoline and because biosynthetic and steric considerations suggest that C(8) is an unlikely position for the terminus of the biphenyl ether in the aporphine side of the alkaloid, the terminus was proposed to be at C(9), as in thali-carpine, with the final methoxy group of adiantifoline being located at C(3). A comparison of the CD curves of thalicarpine (**192**) and adiantifoline (**190**) ($[\theta]_{241}$ +234,000, $[\theta]_{275}$ -31,200, and $[\theta]_{305}$ -33,800) suggested identical

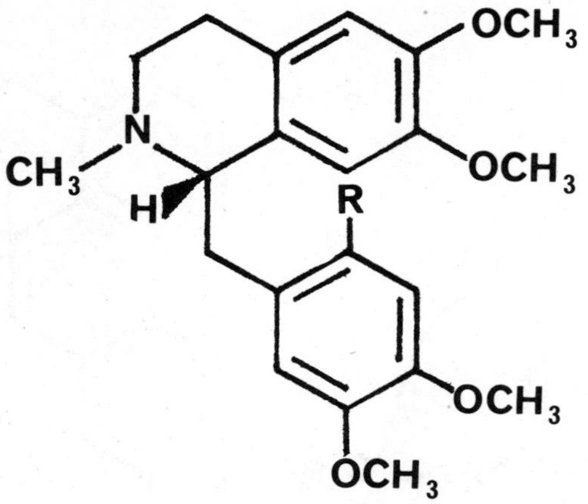

194 R = OH

196 R = Br

stereochemistry, with adiantifoline being assigned as **190** [443]. A total synthesis of adiantifoline (**190**) involving the coupling of (+)-(*S*)-6′-bromolaudanosine (**195**) and (+)-(*S*)-1,2,3,10-tetramethoxy-9-hydroxyaporphine (**197**) via an Ullmann reaction was reported in 1969, confirming structural assignment and stereochemistry [444,445].

Adiantifoline produced a brief hypotensive response (−25 mm Hg) when administered intravenously to rabbits in a dosage of 1.0 mg/kg but failed to exhibit antimicrobial activity against a standard set of microorganisms [446].

Adiantifoline has also been isolated from *Thalictrum minus* L. [230], *T. minus* L. var. *elatum* Jacq. [230], *T. minus* L. race B [143,446], and *T. minus* var. *microphyllum* Boiss. [146].

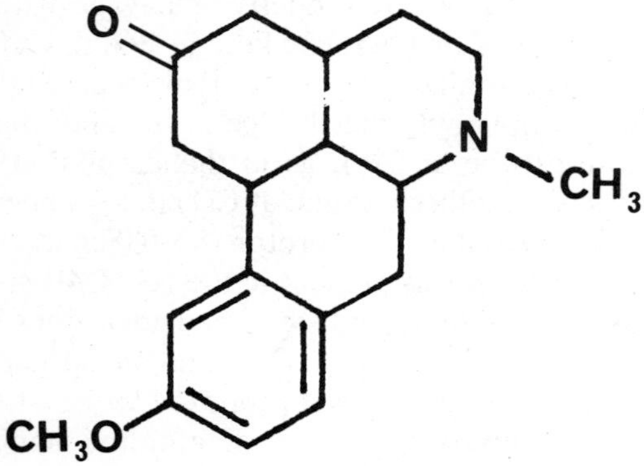

195

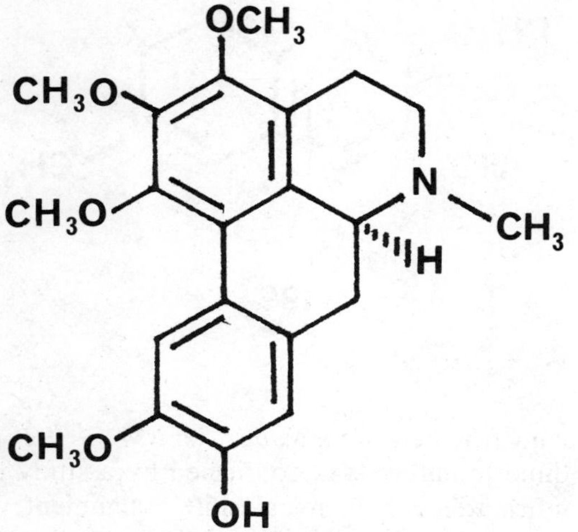

197

3.1.2. Bursanine. Tabular review [437, alkaloid No. 30], UV [437,447], ^{1}H NMR [437,447], MS [437,447], CD [437,447].

Bursanine (**198**), $C_{40}H_{46}N_2O_9$ (698.3200), $[\alpha]_D^{25}$ +117° (c 0.17, MeOH), was isolated from *Thalictrum minus* L. var. *microphyllum* Boiss. in 1982 [447]. The UV spectrum was typical of an aporphinoid system and showed maxima at 209 nm (log ϵ 4.81), 221 nm (sh) (4.75), 283 nm (log ϵ 4.34), 304 nm (sh) (log ϵ 4.24), and 314 nm (log ϵ 4.19), and the mass spectrum was characterized by the lack of a prominent parent ion [M^+, m/z 698 (0.1%)] but the presence

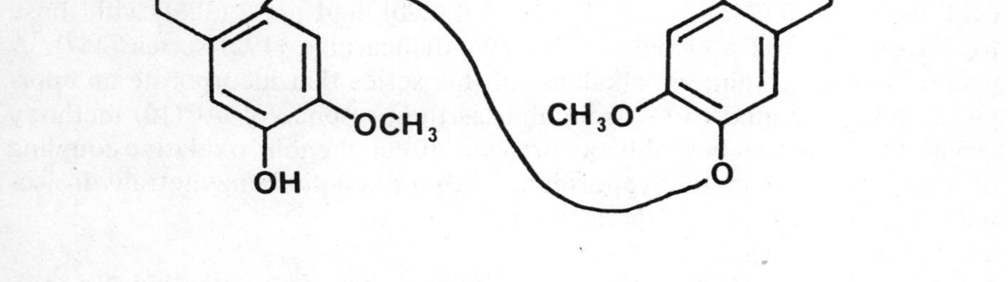

198

199

of several significant fragment ions at m/z 506 ($M^+ - 192$) and 192 (100%)
(**199**) [447]. The dimeric nature was confirmed by a study of the ^1H NMR
spectrum and confirmation of chemical shift assignment was via nuclear
Overhauser enhancement studies. The ^1H NMR spectrum revealed the presence of two *N*-methyl groups as singlets at δ 2.38 [3H, N(2′)] and 2.49 [3H,
N(6)]; six methoxy groups as six singlets at δ 3.75 [3H, C(1)], 3.79 [3H,
C(12′)], 3.80 [3H, C(6′)], 3.88 [3H, C(3)], 3.96 [3H, C(2)], and 3.98 [3H,
C(10)]; and six aromatic protons as four singlets at δ 6.44 [1H, H(8′)], 6.50
[2H, H(5′) and H(8)], 6.59 [1H, H(11′)], 6.82 [1H, H(14′)], and 8.03 [1H,
H(11)]. The low-field aromatic proton at δ 8.03, the mass spectral fragment
ion at m/z 192 (**199**), the UV spectrum, and the ^1H NMR spectrum were
characteristic of an adiantifoline (**190**)-like analog containing an identical
aporphine monomer as adiantifoline but with a diphenolic benzylisoquinoline
monomer, the latter containing one phenolic hydroxy group in the isoquinoline portion of that system. The final assignment of structure was via a
consideration of relevant NOE studies, which showed a 5% NOE of H(11′)
upon irradiation of H(8) and 3% NOE of the C(12′) methoxy group on irradiation of H(11′). These results fixed the position of the methoxy at C(12′)
and thus the hydroxy at C(13′). Furthermore, irradiation of the H(5′) proton
produced a 3% NOE of the C(6′) methoxy group, placing that group at C(6′)
and thus the hydroxy group at C(7′). Although not undertaken, methylation
of bursanine should afford adiantifoline (**190**). Finally, the absolute configuration of bursanine was assigned via a comparison of its CD spectrum [− 11.0
(214 nm), + 49.0 (241), − 4.9 (275), − 2.6 (290), and − 5.2 (306)] with those
for closely related alkaloids of the (+)-thalicarpine (**192**) series [447]. A
general feature of dimeric alkaloids of this series that incorporate an aporphine moiety related to (+)-reticuline is the presence of a C(10) methoxy
group. Dimerization probably occurs via a direct phenolic oxidative coupling
of a 9-hydroxy-10-methoxyaporphine with a phenolic benzyltetrahydroisoquinoline of the (+)-reticuline type [447].

3.1.3. *O*-Desmethyladiantifoline.   Tabular review [436, alkaloid No. 14],
IR [230,436], ^1H NMR [230,436].

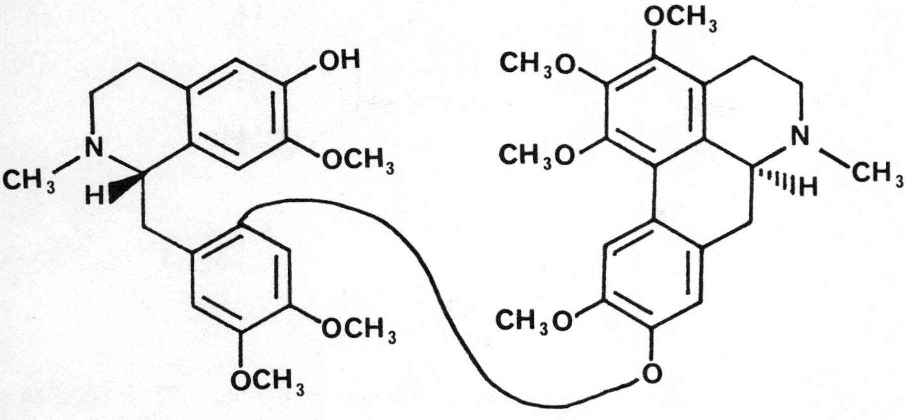

200

O-Desmethyladiantifoline (**200**), $C_{41}H_{48}N_2O_9$ (712.3360), m.p. 125–126°
(Et$_2$O), $[\alpha]_D^{22}$ +18° (c 0.4, CHCl$_3$), was first isolated from *Thalictrum minus*
L. var. *elatum* Jacq. (more recently designated *T. minus* L. ssp. *majus*) in
1970 [230]. Although no UV nor mass spectra were reported, the IR spectrum
suggested the phenolic nature of the alkaloid [230,436]. The ^{1}H NMR spectrum indicated the presence of two *N*-methyl groups as singlets at δ 2.45

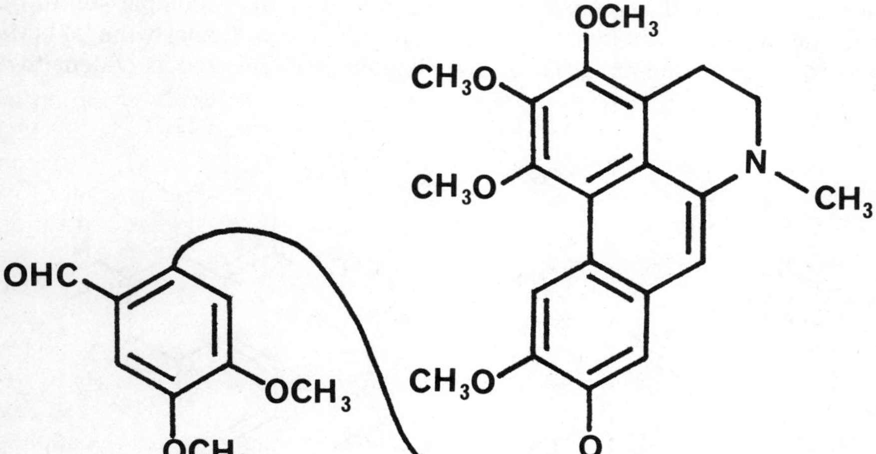

201

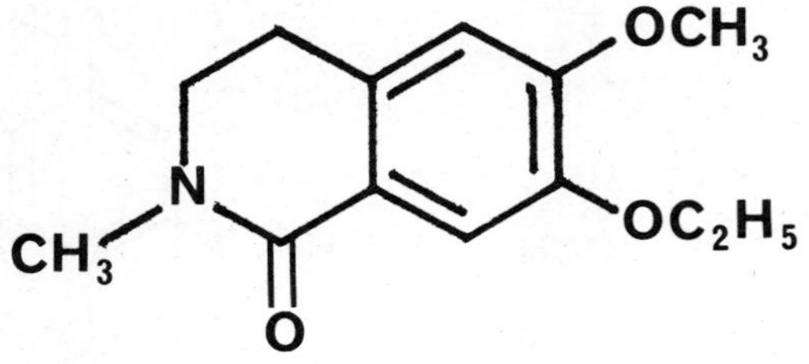

202

and 2.50; seven methoxy groups as six singlets at δ 3.56, 3.77, 3.80, 3.83, 3.90, and 3.96 (6H); and six aromatic protons as five singlets at δ 5.78, 6.43, 6.50, 6.60 (2H), and 8.05 [230,436]. Methylation of *O*-desmethyladiantifoline with ethereal CH_2N_2 afforded adiantifoline (**190**) thereby fixing the skeletal structure and both the positions of oxygenation and termination of the biphenyl ether linkage. Treatment of *O*-desmethyladiantifoline with diazoethane afforded the *O*-ethyl ether of the alkaloid, which was then oxidized with potassium permanganate in acetone to give the aldehyde **201** and what was then identified as 1-oxo-2-methyl-6-methoxy-7-ethoxy-1,2,3,4-tetrahydroisoquinoline (**202**). On the basis of these data, *O*-desmethyladiantifoline was characterized as 7′-demethyladiantifoline (**203**) [230]. However, in 1978 Liao et al. reported the isolation and identification of thaliadanine (**203**), a new alkaloid from *Thalictrum minus* L. race B [446]. A comparison of the specific rotation (although in different solvents) and particularly the ¹H NMR spectra of thaliadanine (**203**) and the alkaloid characterized as *O*-demethyl-

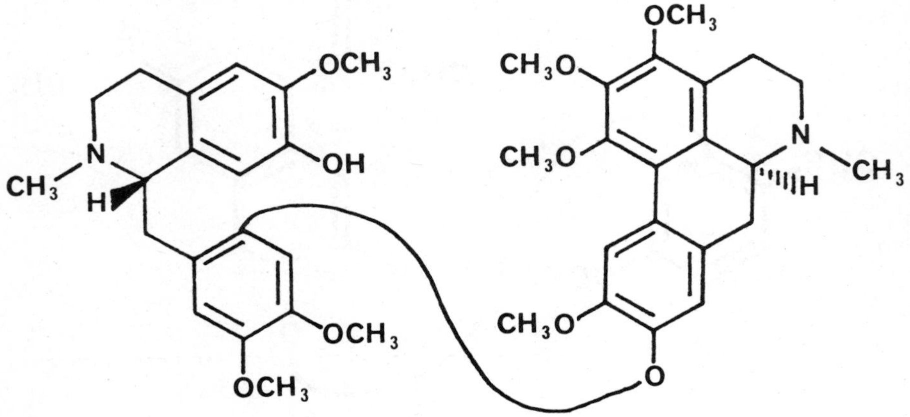

203

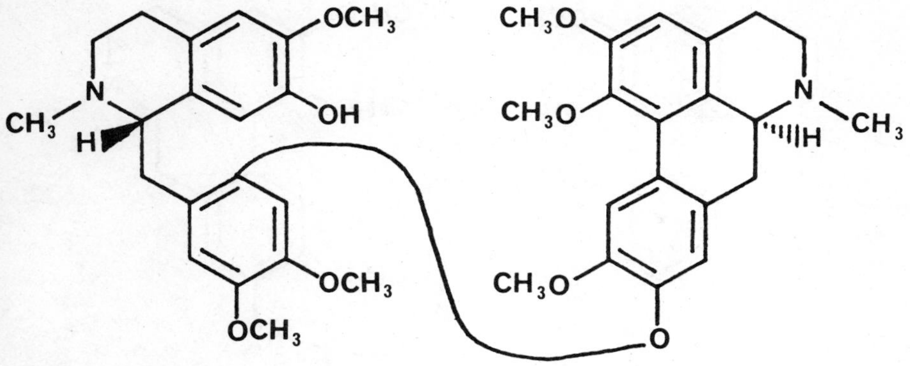

204

adiantifoline and assigned as **203** [230] showed them to be clearly different. The high-field methoxy singlet at δ 3.56 must be assigned to the C(7′) rather than the C(6′) position based on clearly established precedent [436,437,446]. Furthermore, the H(8′) high-field aromatic proton at δ 5.78 is characteristic for being located ortho to a methoxy in this dimeric system [436,437,446]. On this basis, the earlier isolated [230] O-desmethyladiantifoline was assigned as **200** some 8 years later [446]. In retrospect, the real point of contention in these studies is that both groups claim that 1-oxo-2-methyl-6-methoxy-7-ethoxy-1,2,3,4-tetrahydroisoquinoline was the isoquinolone produced by potassium permanganate oxidation of the respective O-ethyl aklaloids. Liao et al. compared their oxidation product with product prepared via ethylation of authentic thalifoline (**103**) and found identity via a number of parameters (TLC, UV, IR, ¹H NMR, MS, and m.p.) [446]. In addition, the literature m.p. (121–122°) was in agreement with that of the experimental oxidation product (120.5°) [446]. Although the isoquinolone of Mollov et al. had m.p. 118–120° and was directly compared (m.p., IR) with the isoquinolone oxidation product prepared from the O-ethyl ether of thalmelatine (**204**), one cannot escape the conclusion that the assignment of Liao et al. [446] is more valid because of the aforementioned ¹H NMR spectral characteristics.

O-Desmethyladiantifoline (**200**) was just recently reisolated from T. minus L. ssp. majus [299].

3.1.4. Faberidine. UV [471], IR [471], ¹H NMR [471], MS [471], CD [471]. Faberidine (**205**), $C_{40}H_{46}N_2O_8$ (682.3254), $[\alpha]_D^{22}$ + 106° (c 0.675, MeOH), was isolated as an amorphous solid from Thalictrum faberi Ulbr. in 1984 [471]. The UV spectrum was characterized by three maxima at 281 nm (log ε 4.30), 302 nm (log ε 4.14), and 313 nm (sh) (log ε 4.06) and was typical of a 1,2,9,10-

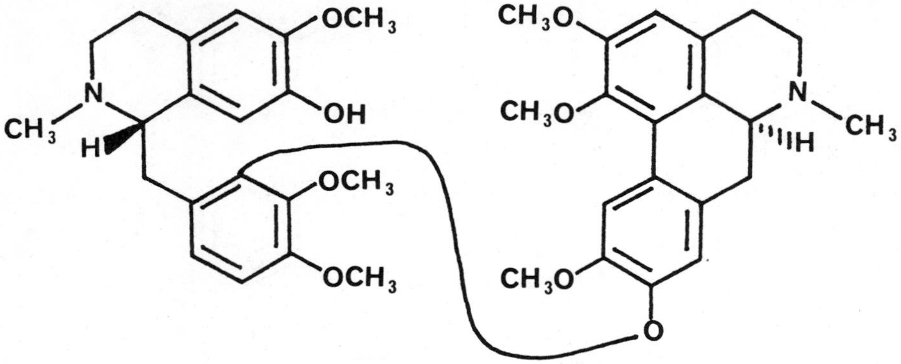

205

oxygenated aporphine. The mass spectrum showed the molecular ion at M^+ m/z 682 with other diagnostic fragment ions at m/z 490 (M-192), 489, 340 (M-342) (**206**), 324 (M-358) (**207**), and 192 (100%) (**199**). The ^{1}H NMR spectrum was very informative and demonstrated the presence of two N-methyl groups as singlets at δ 2.28 and 2.44; six methoxy groups as five singlets at δ 3.66, 3.78 (6H), 3.86, 3.88, and 3.99; and five aromatic protons as four singlets at δ 6.39 (2H), 6.45, 6.53, and 8.11 [H(11)], plus two additional aromatic protons as an AB quartet at δ 6.72 [H(13′)], and 6.86 [H(14′)] (J = 8.5 Hz). These spectral data were characteristic of an aporphine–benzylisoquinoline dimer containing three methoxy groups in the aporphine half of the alkaloid. In

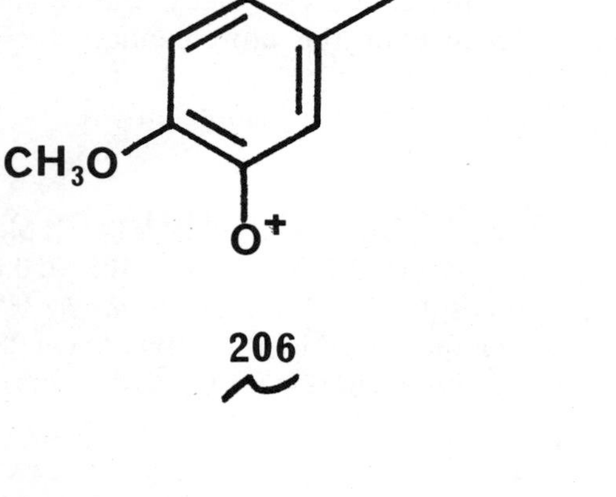

206

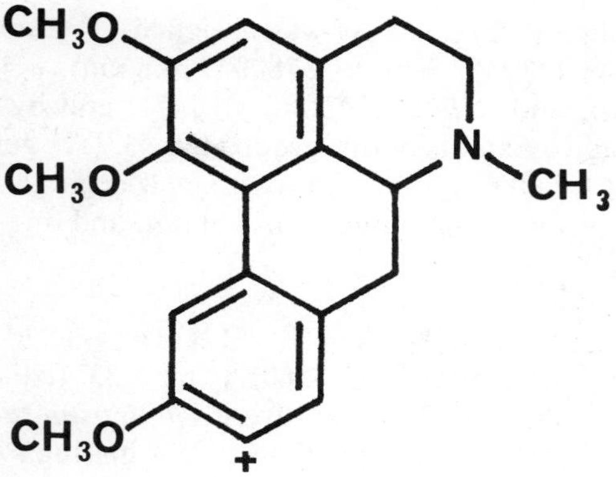

207

addition, there were two methoxy groups in the benzyl portion plus one methoxy group and one phenolic hydroxyl group in the isoquinoline portion of the benzylisoquinoline half of the dimer. The presence of an aromatic proton at δ 6.39 is characteristic of a H(8′) proton adjacent to a C(7′) hydroxy group, because the chemical shift of the same proton is around δ 6.20 if it is adjacent to a C(7′) methoxy group [469]. Furthermore, the absence of a methoxy signal near δ 3.60 also signals the presence of a phenolic hydroxy at C(7′), rather than a methoxy group at that position [469]. Finally, the presence of the AB quartet with δ$_A$ 6.72 and δ$_B$ 6.86 (J = 8.5 Hz) is characteristic of the H(13′) and H(14′) protons of fetidine (**208**)-like dimers [469]. These spectral data and the resulting correlations prompted the assignment of the structure of faberidine as **205**. The stereochemistry of the asymmetric

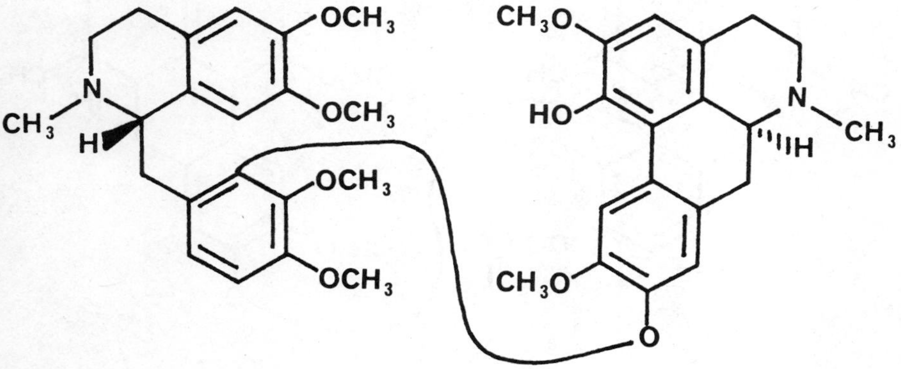

208

centers of faberidine (**205**) as (*S*,*S*) was assigned via a consideration of its CD spectrum [$-44,133$ (198.6 nm), $+7800$ (210.8 nm), $+45,833$ (241.0 nm), -6333 (277.8 nm), and -3700 (302.2 nm)] [471] and by comparison with other closely related alkaloids of this group [436,437]. In general, these bases are characterized by CD spectra with two small negative troughs near 275 and 305 nm, one positive maximum near 240 nm, and one negative tail near 210 nm [436,437].

3.1.5.   Faberonine.   UV [471], IR [471], ^{1}H NMR [471], MS [471], CD [471]. Faberonine (**209**), $C_{41}H_{48}N_2O_9$ (712.3360), $[\alpha]_D^{24}$ $+83°$ (c 0.498, MeOH), was isolated as an amorphous yellow solid from *Thalictrum faberi* Ulbr. in 1984 [471]. The UV spectrum showed maxima at 281 nm (log ϵ 4.33), 301 nm (log ϵ 4.18), and 312 nm (sh) (log ϵ 4.12) with no change in strong base. The mass spectrum revealed a parent ion at M$^+$ *m/z* 712 (0.1%) with other important fragment ions at *m/z* 520 (M-192), 519, 370 (M-342) (**210**), 354 (M-358) (**211**), and 192 (100%) (**199**). The ^{1}H NMR spectrum indicated the presence of two *N*-methyl groups as singlets at δ 2.30 and 2.43; seven methoxy groups as six singlets at δ 3.70, 3.78 (6H), 3.86, 3.88, 3.96, and 4.01; and six aromatic protons, four of which were as singlets at δ 6.39 (2H), 6.45, and 7.99 [H(11)] and two of which were part of an AB quartet (*J* = 8.5 Hz) at δ 6.72 [H(13')] and 6.89 [H(14')]. These spectral data were consistent for a faberidine (**205**)-like alkaloid containing one extra methoxy group in the aporphine moiety, most likely at C(3). As with faberidine (**205**), similar logic was employed in the deduction of structure. Thus the presence of the H(8') proton at δ 6.39 instead of around δ 6.20 and the absence of a methoxy signal at around δ 3.60 were diagnostic of the presence of a C(7) hydroxy group. Furthermore, the ortho-coupled protons at H(13') and H(14') also indicated faberonine

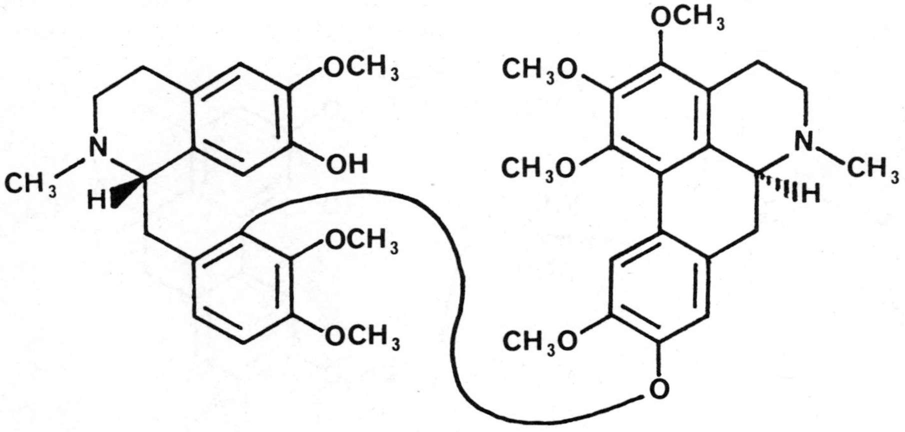

209

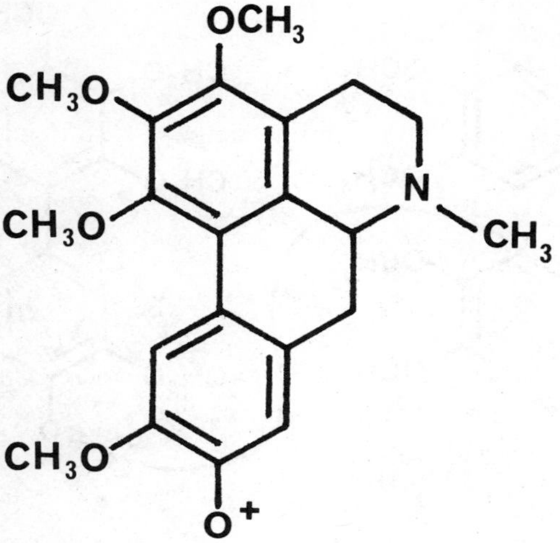

210

was a fetidine (**208**)-type alkaloid and was really 7'-desmethylhuangshanine (**209**). The CD spectrum of faberonine (**209**) [−38,967 (198.2 nm), +6,200 (213.4), +47,000 (242.6), −5,833 (277.8), and −4,900 (306.6)] was consistent with those of faberidine (**205**), fetidine (**208**), and huangshanine (**212**) and thus allowed the stereochemical assignment of faberonine (**209**) as S,S [471].

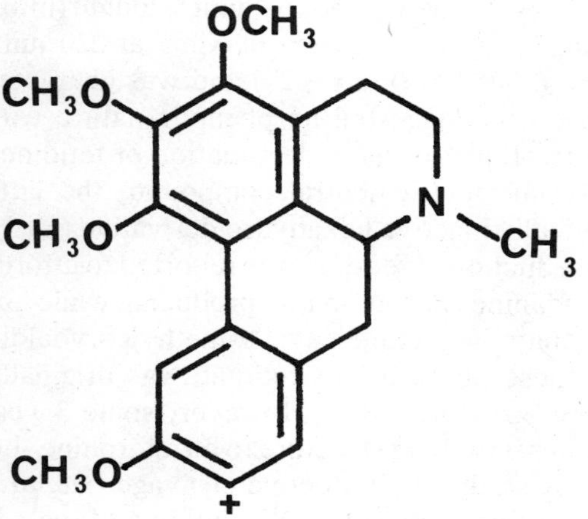

211

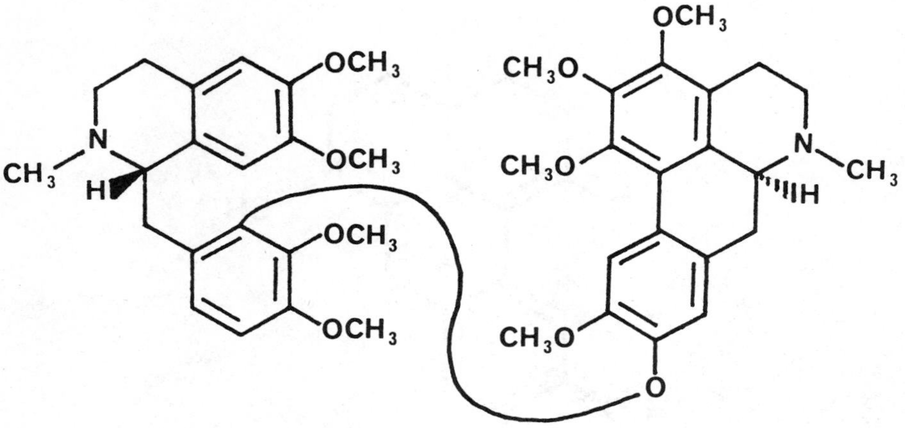

212

3.1.6. Fetidine (Foetidine).

Tabular review [436, alkaloid No. 19], UV [436,448,449], IR [436,449], ^{1}H NMR [436,450,451], MS [436,452], ORD [453].

Fetidine (208), $C_{40}H_{46}N_2O_8$ (682.3254), m.p. 132–135° (EtOAc) [448,449], $[\alpha]_D^{15}$ +121.4° (c 2.09, MeOH) [448,449], was first isolated from *Thalictrum foetidum* L. in 1963 [448]. A number of salts of fetidine were prepared which included the dihydrochloride (m.p. 228–230° dec.; $[\alpha]_D^{20}$ −30.9° [c 2.52, H$_2$O)], the dinitrate (m.p. 200° dec.), the dihydrobromide (m.p. 225–230°), the sulfate (m.p. 215–218° dec.), and the dimethiodide (m.p. 205–210°) [448]. The UV spectrum of fetidine showed maxima at 220 nm (log ε 4.80), 280 nm (log ε 4.36), and 305 nm (log ε 4.24) and was characteristic of an aporphine; the IR spectrum suggested its phenolic nature with a peak at 3400 cm^{-1} [449]. The twofold Hofmann degradation of fetidine dimethoiodide afforded trimethylamine and a neutral compound, the latter of which was reduced to an optically inactive hexahydro derivative ($C_{39}H_{48}O_8$), m.p. 126–127° [448]. Birch reduction of fetidine was reported to afford (+)-laudanosine (213) and (+)-laudanine (55) as major products, while oxidation with potassium permanganate in acetone gave *N*-methylcorydaldine (88) [448,449]. On the basis of these observations fetidine was originally assigned a bisbenzylisoquinoline structure [448]. However, some 3 years later a subsequent investigation revised the structure to an aporphine–benzylisoquinoline dimer [449]. In this study, the reductive cleavage (Na/liq. NH$_3$) of fetidine was repeated and a third product (+)-isoboldine (4) was isolated. Because treatment of (+)-laudanosine (213) with sodium in liquid ammonia afforded (+)-laudanine (55), it was proposed that the latter was formed from the former during the Birch reduction of fetidine. Although the terminus of the diphenyl ether linkage onto the benzyl portion of laudanosine had yet to be

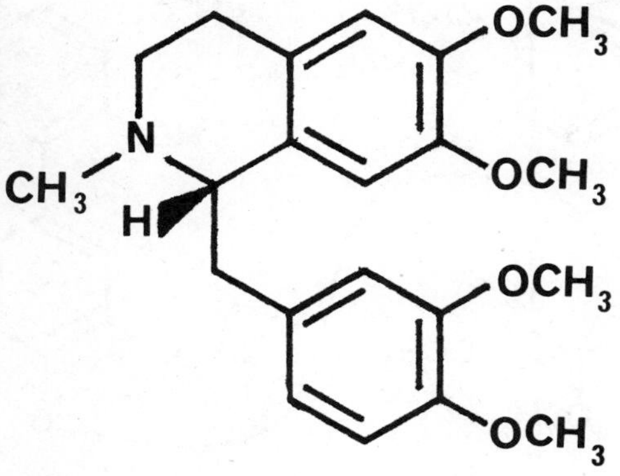

213

established, the structure of fetidine was nevertheless proposed as **214** [449]. In 1972 a definitve structure for fetidine was proposed [451]. Ullmann condensation of (+)-isoboldine (**4**) with (+)-6′-bromolaudanosine (**196**) afforded 1-O-demethylthalicarpine (**215**) whose spectral properties (UV, ¹H NMR) differed from those of fetidine. Furthermore, neither fetidine nor base **215** showed UV absorption in strongly basic solution characteristic of a 9-hydroxyaporphine. These observations precluded not only structure **215** for fetidine but also the possibility that the aporphine terminus of the alkaloid was at C(1) of the isoboldine monomer [451]. The ¹H NMR spectrum (220

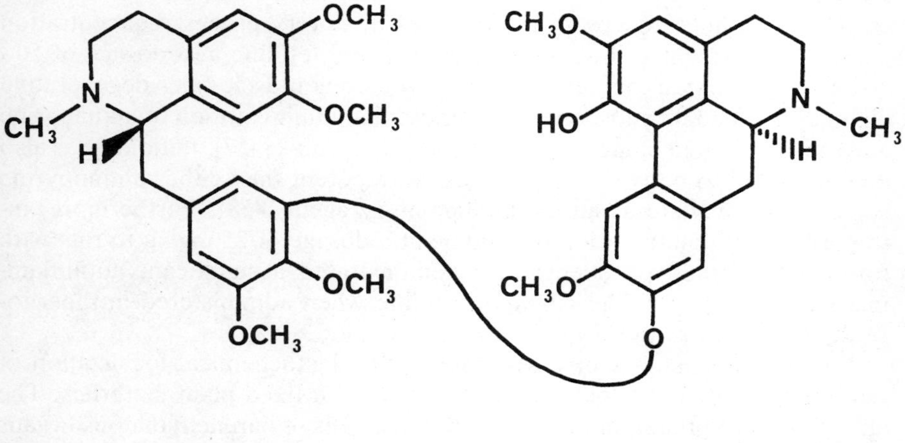

214

215

MHz) of fetidine was scrutinized in the aromatic region and the following observations and assignments were made: δ 6.14 [1H, s, H(8′)], 6.42 (1H, s), 6.52 (2H, s), 6.75 (1H), and 6.81 (1H) as an AB quartet (J = 8.5 Hz) [H(13′) and H(14′) or vice versa] and 8.12 [1H, s, H(11′)]. Therefore, because the observed quartet can be accounted for only by the presence of adjacent aromatic protons, fetidine was finally assigned as **208**. Fetidine (**208**) was the first example of a dimeric alkaloid of this series to contain 10′,11′,12′ trioxygenation [451].

Fetidine was found to exhibit central nervous system depressant activity [454] and hypotensive activity [455]. Administration of fetidine hydrochloride to rats intraperitoneally on a once-daily dosage of 10–30 mg/kg for 8 days resulted in conditioned reflex suppression and an improvement in differentiation. Prolonged administration of 5–10 mg/kg doses produced related effects, including a decrease in degree of inhibitory process concentration [456]. Pretreatment of rats with intraperitoneal fetidine at a dosage of 50–70 mg/kg prevented the development myocardial muscle fiber degenerative changes induced by subsequent intramuscular administration of epinephrine and thus prevented epinephrine-induced myocarditis [457]. Fetidine was also demonstrated to be two to four times more potent than either aminopyrine or sodium salicylate as an antiinflammatory agent [458]. Furthermore, intraperitoneal administration of fetidine at a dosage of 25 mg/kg to rats with formaldehyde-induced edema also demonstrated a significant antiinflammatory action [459]. The LD_{100} of fetidine when administered intraperitoneally was 230 mg/kg [459].

Luminescent microscopy was used in the histochemical localization of fetidine and other alkaloids of *T. foetidum* L. in fixed plant materials. The alkaloids were predominantly located in the cells of parenchymatous organs [460]. The quantitative determination of fetidine in extracts from raw plant material of *T. foetidum* L. was determined via chromatopolarography [461].

The alkaloid was also assayed in tablets via spectrophotometry [461]. The distribution coefficient of fetidine between various organic solvents/solvent mixtures and both pH 8 and pH 10 solutions was determined [462].

Fetidine has been repeatedly isolated from extracts of *T. foetidum* L. [192,455,463–466] but no other *Thalictrum* species to date.

3.1.7. Huangshanine.

Tabular review [437, alkaloid No. 33], UV [437,467], ^{1}H NMR [437,467], MS [437,467], CD [437,467].

Huangshanine (212), $C_{42}H_{50}N_2O_9$, (726.3516), $[\alpha]_D^{25}$ +120.6° (c 0.43, MeOH), was obtained as an amorphous yellow powder from *Thalictrum faberi* Ulbr. in 1983 [467]. The UV spectrum was typical of a 1,2,3,9,10-oxygenated aporphine and showed maxima at 281 nm (log ϵ 4.41), 302 nm (log ϵ 4.20), and 312 nm (sh) (log ϵ 4.13); the mass spectrum showed characteristic fragment ions for a benzylisoquinoline–aporphine dimer at m/z 725 (M-1), 520 (M-206), 370 (M-benzylisoquinoline half not containing the diphenyl ether oxygenation), 354 (M-benzylisoquinoline half containing diphenyl ether oxygenation), and 206 (100%) (191). These spectral data were consistent for an aporphine–benzylisoquinoline dimer containing two methoxy groups in both the benzyl and the isoquinoline portion of the benzylisoquinoline and four methoxy groups in the aporphine half of the alkaloid. The ^{1}H NMR spectrum was most informative and demonstrated the presence of two N-methyl groups as singlets at δ 2.36 [N(2')] and 2.45 [N(6)]; eight methoxy groups as seven singlets at δ 3.61 [C(7')], 3.75 [C(1)], 3.81, 3.83, 3.88 (6H), 3.97, and 4.00; and four aromatic protons as four singlets at δ 6.16 [H(8')], 6.41, 6.53, and 8.03 [H(11)]. In addition, an AB quartet (J = 8.5 Hz) at δ 6.73 and δ 6.76 was suggestive of the presence of *ortho*-coupled

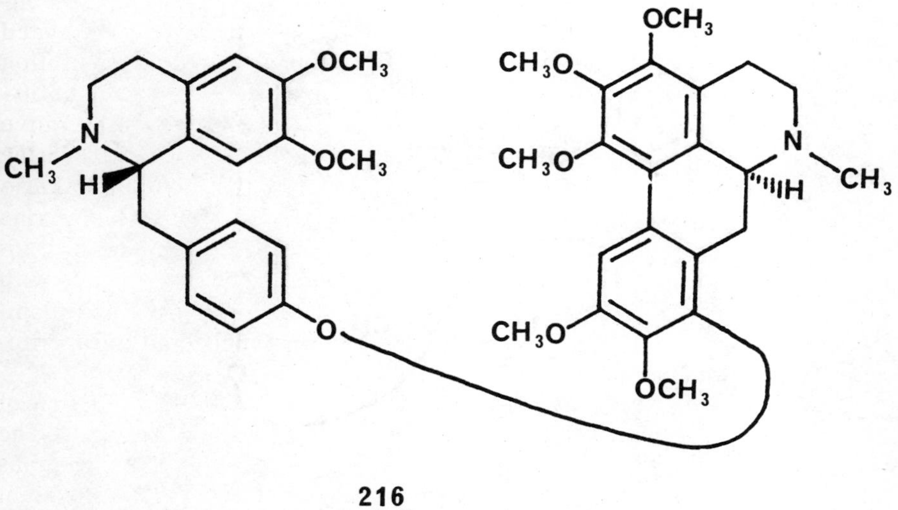

216

H(13') and H(14') protons and suggested placement of this alkaloid in the fetidine (208) series. A consideration of the CD curve [+29.2 (243 nm), −4.9 (277 nm), and −4.8 (301 nm)] and its similarity to that of (+)-thalifaberine (216) suggested an identical absolute configuration for huangshanine, which was assigned structure 212 [467]. Huangshanine (212) is thus the 7',11"-dimethoxy ether of iznikine (218) [447].

Huangshanine (212) has been isolated only from *T. faberi* Ulbr. to date and was shown to be cytotoxic as measured against the Walker 256 carcinoma [447].

3.1.8. Istanbulamine. Tabular review [437, alkaloid No. 34], UV [437,447], ^{1}H NMR [437,447], MS [437,447], CD [437,447].

Istanbulamine (217), $C_{39}H_{44}N_2O_8$ (668.3097), $[\alpha]_D^{25}$ +60° (c 0.09, MeOH), was isolated from *Thalictrum minus* L. var. *microphyllum* Boiss. in 1982 [447]. The UV spectrum reflected its 1,2,3,9,10-oxygenated aporphine character and showed maxima at 205 nm (log ε 4.87), 225 nm (sh) (log ε 4.75), 270 nm (sh) (log ε 4.20), 282 nm (log ε 4.34), 304 nm (sh) (log ε 4.15), and 313 nm (log ε 4.12). The mass spectrum failed to display a prominent parent ion which was found at M$^+$ m/z 668 (0.2%) but did show characteristic fragment ions at m/z 476 (4%) (M-192) and 192 (100%) (199). The ^{1}H NMR spectrum revealed the dimeric nature of the alkaloid and showed the presence of two *N*-methyl groups as singlets at δ 2.43 [3H, N(2')] and 2.51 [3H, N(6)]; five methoxy groups as five singlets at δ 3.76 [C(1)], 3.81 [C(6')], 3.90 [C(3)], 3.93 [C(10)] and 3.96 [C(2)]; four aromatic protons as four singlets at δ 6.44 [H(8')], 6.49 [H(5')], 6.84 [H(8)] and 8.03 [H(11)], and three more aromatic protons as part of an ABX system at δ 6.80 (1H, dd, J = 2.14,

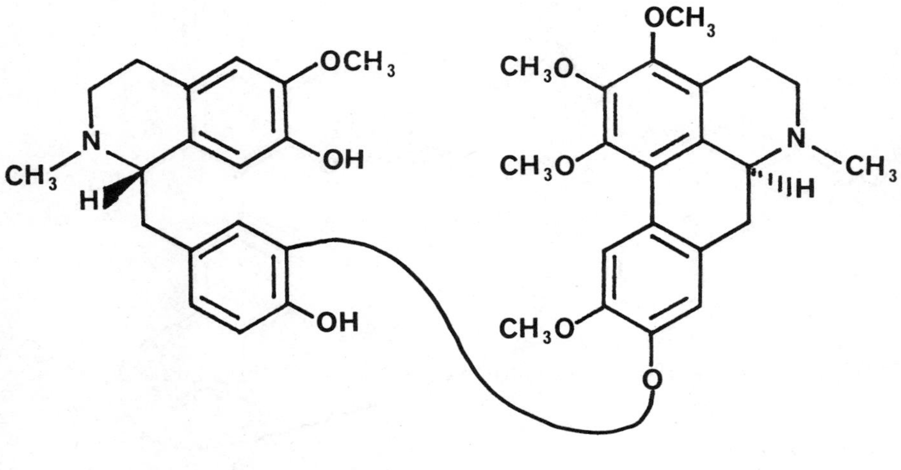

217

8.24 Hz) [H(14′)], 6.86 (1H, d, J = 2.14 Hz) [H(10′)], and 6.89 (1H, d, J = 8.24 Hz) [H(13′)]. The UV spectrum, the low-field proton singlet at δ 8.03, the aromatic ABX system, and the mass spectral fragment ion at m/z 192 all indicated an aporphine–benzylisoquinoline dimer containing four methoxy groups in the aporphine ring with an aromatic proton at C(11) and a diphenolic benzylisoquinoline half containing one phenolic group and one methoxy group in the isoquinoline portion. Confirmation of chemical shift assignments was achieved via a NOE study which demonstrated an 8% relaxation enhancement of the C(2) methoxy signal upon irradiation of the C(1) methoxy, which verified the chemical shift of the former. In addition, irradiation of the proton at C(11) produced a 3% NOE on the C(10) methoxy, verifying the chemical shift assignment of that group and demonstrating that C(10) cannot be the terminus of the diaryl ether on the aporphine side. Other NOE studies allowed for the placement of isoquinoline methoxy at C(6′). These data clearly indicated that istanbulamine was a new type of alkaloid in this group in that it possessed a diaryl ether terminus at C(11′), rather that at the more commonly encountered C(10′), and ultimately allowed for the assignment of istanbulamine as **217**. An inspection of the CD spectrum suggested that the absolute configuration of istanbulamine (**217**) [− 16.0 (210 nm), + 61.2 (243), − 10.9 (275), − 8.7 (284), and − 15.3 (306)] was the same as alkaloids of the (+)-thalicarpine (**192**) series. Like bursanine (**198**), which is also a dimer whose aporphine moiety is related to (+)-reticuline (**57**) and contains a methoxy group at C(10), istanbulamine is also probably biosynthesized via a direct phenolic oxidative coupling of a 9-hydroxy-10-methoxyaporphine with a (+)-N-methylcoclaurine (**52**)-type alkaloid [447].

3.1.9. Iznikine. Tabular review [437, alkaloid No. 32], UV [437,447], ¹H NMR [437,447], MS [437,447], CD [437,447].

Iznikine (**218**), $C_{40}H_{46}N_2O_9$ (698.3200), $[\alpha]_D^{25}$ + 76° (c 0.068, MeOH), was isolated from *Thalictrum minus* L. var. *microphyllum* Boiss. in 1982 [447]. The UV spectrum was characteristic of a 1,2,3,9,10-oxygenated aporphine with maxima at 208 nm (log ε 4.68), 222 nm (sh) (log ε 4.54), 281 nm (log ε 4.17), 301 nm (sh) (log ε 4.03), and 312 nm (log ε 3.98) [447]. The mass spectrum displayed only a weak M-1 ion at m/z 697 (0.3%) with two other characteristic fragment ions at m/z 506 (M-192) (2%) and 192 (100%) (**199**) [447]. Analogously to (+)-bursanine (**198**) and (+)-istanbulamine (**217**), the ¹H NMR spectrum with related NOE studies was instrumental in the structural assignment of iznikine (**218**). The ¹H NMR spectrum of this dimeric alkaloid was characterized by two N-methyl groups as singlets at δ 2.35 [N(2′)] and 2.47 [N(6)]; six methoxy groups as singlets at δ 3.72 [C(1)], 3.81 [C(6′)], 3.88 [C(3)], 3.92 [C(12′)], 3.96 [C(2)], and 4.02 [C(10)]; and six aromatic protons as five singlets at δ 6.37 [H(8′)], 6.44 [H(8)], 6.50 [H(5′)], 6.75 (2H) [H(12′), H(13′)], and 8.06 [H(11)]. These spectral data were characteristic of an aporphine–benzylisoquinoline dimer containing an identical aporphine portion to that found in (+)-bursanine (**198**) and (+)-istanbulam-

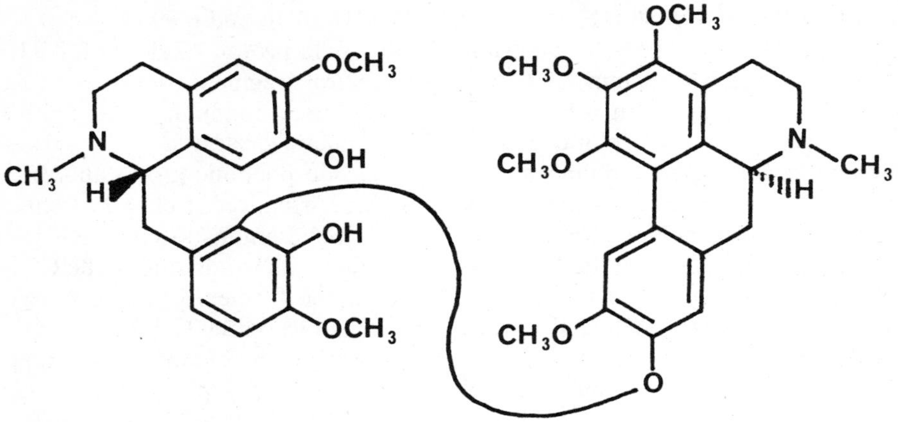

218

ine (**217**) and a benzylisoquinoline portion with one methoxy and one phenolic hydroxy group in both the isoquinoline half and the benzyl half of the molecule. Key NOE effects were observed via irradiation of the two proton aromatic singlet [C(13′) and C(14′)] at δ 6.75 with the resultant 3% NOE of the methoxy signal at δ 3.92. This allowed placement of the methoxy group at C(12′) and thus fixed the phenolic group at C(11′). Furthermore, irradiation of the H(5) proton at δ 6.50 produced a 4% enhancement of the methoxy signal at δ 3.81, thereby fixing the isoquinoline methoxy at C(6′). Also, a small but significant NOE was observed on the methylene multiplet at δ 2.79–2.84 upon irradiation of the aromatic proton at δ 6.44, thus fixing the proton at C(8). Finally, the absolute configuration of (+)-iznikine (**218**) was derived via a comparison of the shape of its curve [−3.4 (220 nm), +27.5 (241 nm), −5.3 (273 nm), −1.3 (285 nm), and −6.6 (305 nm)] with (+)-bursanine (**198**) and (+)-istanbulamine (**217**) [447]. As with these latter two alkaloids, the biosynthesis of iznikine probably proceeds via a direct phenolic oxidative coupling of a 9-hydroxy-10-methoxyaporphine with a (+)-reticuline (**57**)-type phenolic benzyltetrahydroisoquinoline [447].

3.1.10. *N*-2′-Noradiantifoline. Tabular review [437, alkaloid No. 31], UV [437,468], ^{1}H NMR [437,468], MS [437,468], CD [437,468].
N-2′-Noradiantifoline (**219**), $C_{41}H_{48}N_2O_9$ (712.3360), $[\alpha]_D^{25}$ +39° (c 0.082, MeOH), was isolated from *Thalictrum minus* L. var. *microphyllum* Boiss. as an amorphous solid in 1982 [468]. The UV spectrum was typical of a 1,2,3,9,10-oxygenated aporphine with maxima at 208 nm (log ε 4.77), 220 nm (sh) (log ε 4.69), 280 nm (log ε 4.31), 296 nm (sh) (log ε 4.24), 302 nm (sh) (log ε 4.17), and 314 nm (log ε 4.10). The mass spectrum showed a very weak parent ion at M$^+$ *m/z* 712 (0.3%) with other characteristic fragment

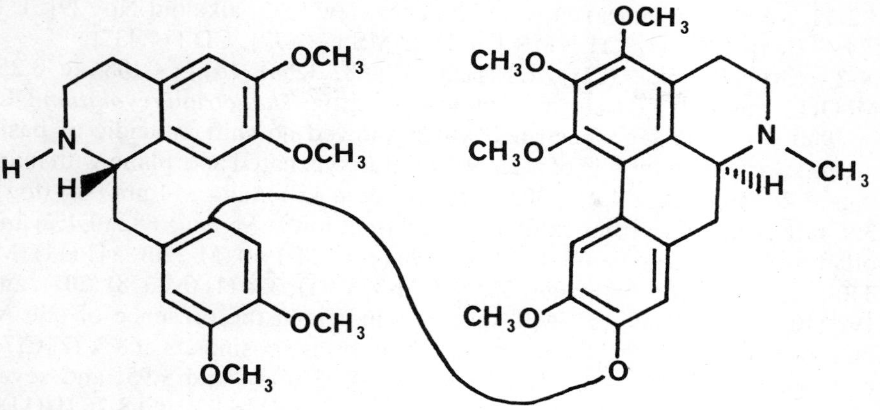

219

ions at m/z 520 (6%) (M-192), 519 (7), 192 (100%) (**220**), and 177 (**220**—CH$_3$).
The ^{1}H NMR spectrum indicated the presence of one methyl group at δ 2.51
(s); eight methoxy groups as seven singlets at δ 3.57 [C(7')], 3.60 [C(1)],
3.74, 3.78, 3.81, 3.90 (6H), and 3.96; and six aromatic protons as singlets at
δ 6.21 [H(8')], 6.47 [H(5')], 6.55 [H(8) or H(11')], 6.56 [H(11') or H(8)], 6.73
[H(14')], and 8.05 [H(11)]. These data were characteristic of an adiantifoline
(**190**) analog, lacking only one N-methyl group. The position of the secondary
nitrogen in the benzylisoquinoline portion was established by the presence
of the base at m/z 192 (**220**). Apparently a lack of a sufficient quantity of the
alkaloid precluded N-methylation, which would have afforded adiantifoline
(**190**). Nevertheless, it was appropriate that the new alkaloid was characterized as (+)-N-2'-noradiantifoline (**219**) and that the stereochemical configuration was assigned via a comparison of its CD spectrum [− 16 (213 nm),
+ 38 (246 nm), − 5.9 (272 nm), − 4.0 (287 nm), and − 5.3 (304 nm)] with that
of adiantifoline (**190**) [468]. This is only the second aporphine–benzylisoquinoline dimer that is not a ditertiary (NCH$_3$) amine. (+)-Northalicarpine
(**221**) was the first example of a dimer in this group bearing a secondary
nitrogen in the benzylisoquinoline portion of the molecule [74].

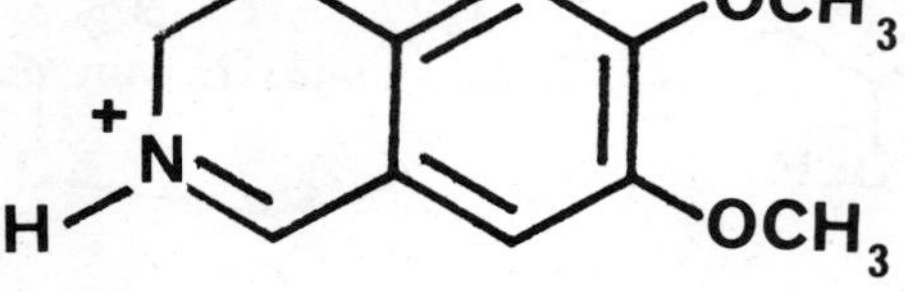

220

3.1.11.   *N*-2′-Northalicarpine.   Tabular review [437, alkaloid No. 29], UV [74,437], IR [74,437], ^{1}H NMR [74,437], MS [74,437], CD [74,437].

N-2′-Northalicarpine (**221**), $C_{40}H_{46}N_2O_8$ (682.3254), $[\alpha]_D^{21}$ + 108° (c 0.25, MeOH), was isolated as an amorphous solid from *Thalictrum revolutum* DC. in 1980 [74]. The UV spectrum, which showed no shift in acidic or basic media, was characteristic of a 1,2,9,10-tetraoxygenated aporphine with maxima at 282 nm (log ϵ 4.21), 303 nm (sh) (log ϵ 4.08), and 314 nm (sh) (log ϵ 3.97); the mass spectrum exhibited the parent ion at M^+ *m/z* 682 (0.4%) and other diagnostic fragment ions at *m/z* 490 (4%) (M-192) (M-220), 341 (11) (M-340 + 1), 340 (2) (M-342) (**206**), 325 (2) (M-358 + 1), 324 (1) (M-358) (**207**), and 192 (100) (**220**). The ^{1}H NMR spectrum indicated the presence of one *N*-methyl group at δ 2.50; seven methoxy groups as six singlets at δ 3.72 [C(7′) or C(1)], 3.76 [C(1) or C(7′)], 3.83, 3.86, 3.93 (6H), and 3.95; and seven aromatic protons as four singlets at δ 6.61, 6.66 (4H), 6.92, and 8.26 [H(11)]. These spectral data were indicative of a benzylisoquinoline–aporphine dimer very similar to thalicarpine (**192**) but with the replacement of the tertiary *N*-methyl group on the benzylisoquinoline moiety with a secondary nitrogen [74]. Treatment of *N*-2′-northalicarpine (**221**) with formaldehyde/formic acid afforded thalicarpine (**192**), thus fixing the skeletal structure, positions of oxygenation, and termini of the diphenyl ether linkage in the nor-alkaloid. Although assignment of the secondary amino group to the benzylisoquinoline half of the dimer might have been made via a consideration of the intense mass spectral fragment ion at *m/z* 192 (**220**), confirmation of this was achieved via sodium–liquid ammonia cleavage of the dimer, which afforded 6′-hydroxynorlaudanosine (**222**) as the phenolic product. The assignment of stereochemistry of *N*-2′-northalicarpine (**221**) was made via a consideration of its CD spectrum ([θ]$_{220}$ 0, [θ]$_{238}$ + 171,000, [θ]$_{256}$ 0, [θ]$_{276}$ − 25,000, [θ]$_{290}$ − 8000, [θ]$_{304}$ − 16,000, and [θ]$_{330}$ 0) and confirmed when the CD spectrum

221

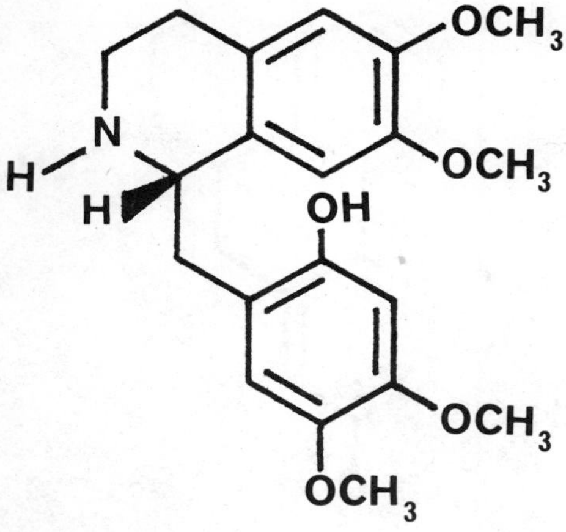

$$\underset{\sim}{\underline{2\,22}}$$

of its *N*-methylation product was identical with that of thalicarpine (**192**) [74].

3.1.12. Pennsylvanamine.

Tabular review [436, alkaloid No. 2], UV [436,469], ¹H NMR [436,469], MS [436,469].

Pennsylvanamine (**223**), $C_{39}H_{44}N_2O_8$ (668.3097), m.p. 128–129° (Me₂CO–Et₂O) or 107–108° (Et₂O) [469]; $[\alpha]_D^{25}$ +119° (c 0.94, MeOH) [469], was isolated from *Thalictrum polygamum* Muhl. in 1974 [469,470]. The alkaloid exhibited maxima in the UV spectrum at 276 nm (sh) (log ε 4.07), 284 nm (log ε 4.17), 297 nm (sh) (log ε 4.11), and 312 nm (sh) (log ε 4.06), with a

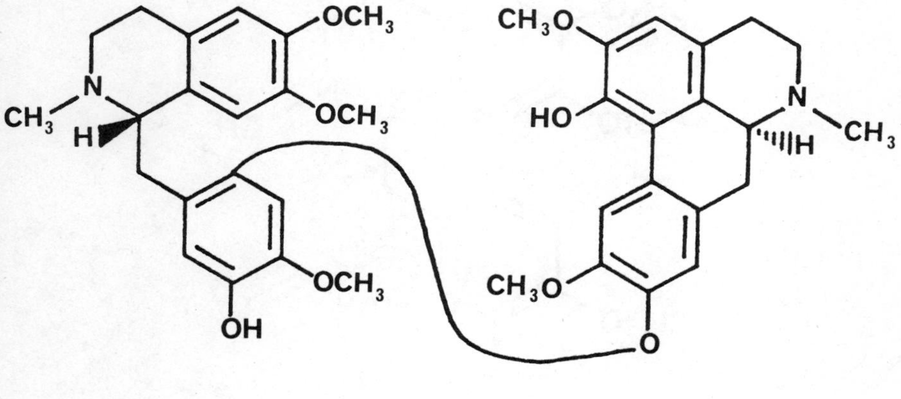

$$\underset{\sim}{\underline{223}}$$

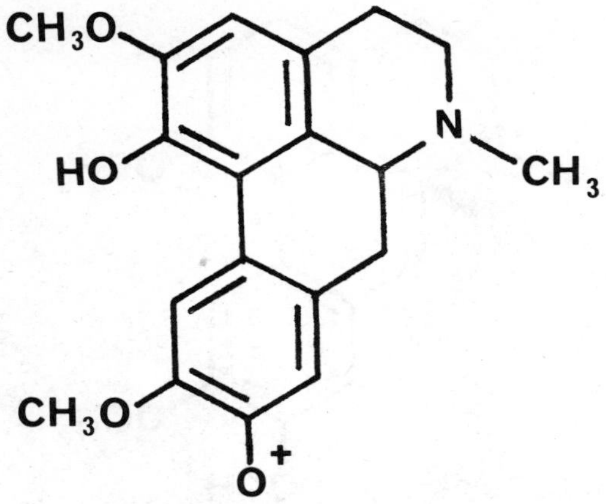

224

distinct bathochromic shift in strong base characteristic of a C(1) phenolic
1,2,9,10-tetraoxygenated aporphine. The principal mass spectral fragment
ions were at M^+ m/z 668, 466 (M-206), 326 (M-342) (**224**), 325 (M-342 + 1),
309 (M-358 + 1) (**225** + 1), and 206 (100%) (**191**). The ^{1}H NMR spectrum indicated the presence of two *N*-methyl groups as singlets as δ 2.47 and 2.53;
five methoxy groups as singlets at δ 3.60 [C(7′)], 3.80 [C(12′)], 3.83 [C(6′)],
3.92 [C(2)], and 3.94 [C(10)]; and seven aromatic protons as five singlets at
δ 6.25 [H(8′)], 6.55 (2H), 6.58 (2H), 6.80, and 8.20 [H(11)]. These spectral
data were characteristic of a benzylisoquinoline–aporphine dimer containing
two phenolic hydroxy groups, one of which was on the benzyl ring and the

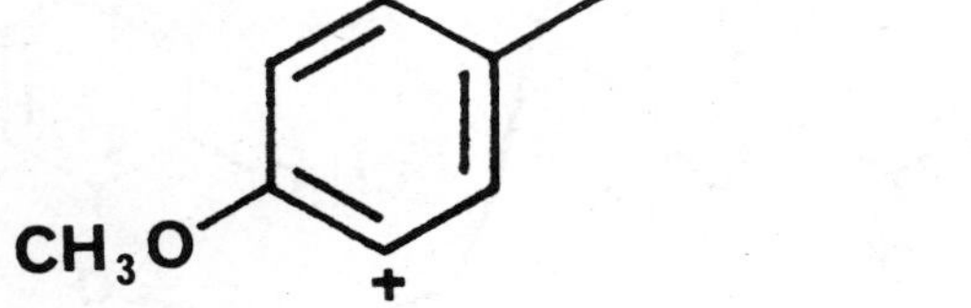

225

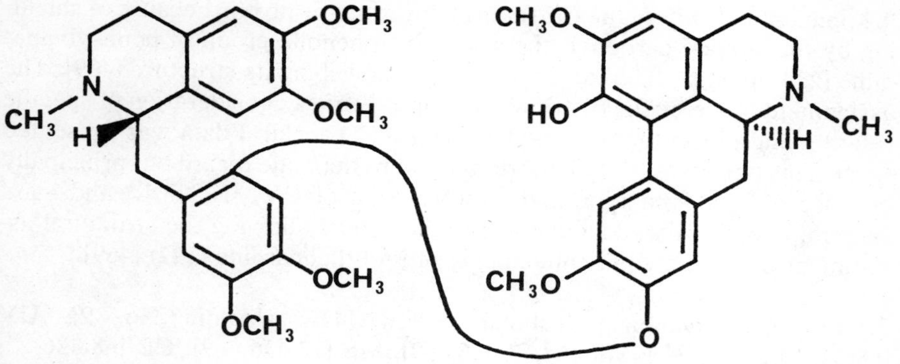

226

second at the C(1) of the aporphine (because there was no higher field methoxy around δ 3.70). Methylation of pennsylvanamine with ethereal CH₂N₂ afforded at 1:1 mixture of (+)-thalictropine (**226**) and (+)-thalicarpine (**192**), thus establishing the skeletal structure, the positions of oxygenation, and the termini of the diphenyl ether linkage, plus strongly suggesting the presence of a C(1) phenolic group on the aporphine. Acetylation of pennsylvanamine with acetic anhydride and pyridine afforded an *O-O*-diacetyl ester, m.p. 147–148° (Et₂O), in which the signal for the aporphine C(11) proton was shifted upfield to δ 7.63, which was consistent for the presence of a C(1) acetate function. There was no other upfield shift for any other proton in the spectrum. Because the only positions available for the remaining phenolic hydroxy group of pennsylvanamine were at C(12′) and C(13′) and because acetylation of the C(12′) group in thalidoxine (**227**) produced a 0.1–

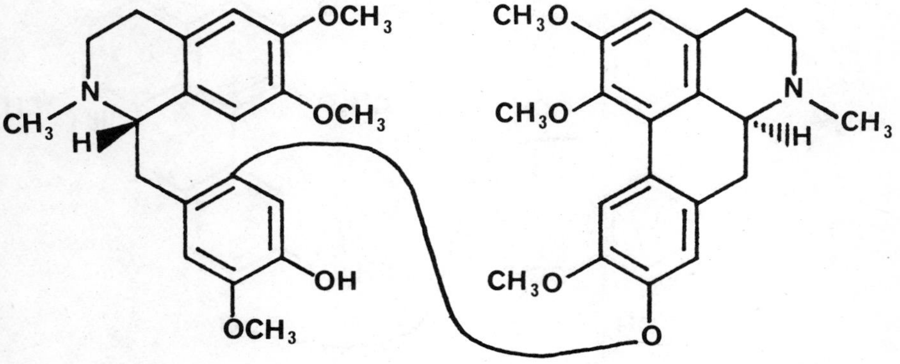

227

0.2 ppm upfield shift of the C(8) aporphine aromatic proton because of shielding by the acetate carbonyl, the remaining phenolic group of pennsylvanamine (**223**) must be fixed at C(13'), thus establishing its structure [469]. The assignment of stereochemistry was apparently via a consideration of specific rotation and biosynthetic precedent, as no CD spectral data was presented [469]. This paper was very important in more than one regard but principally because of its compilation and interpretation of UV, ^{1}H NMR, and mass spectral data for six alkaloids of this class, thus allowing the structural assignment of any new phenolic analog of (+)-thalicarpine (**192**) [469].

3.1.13. Pennsylvanine. Tabular review [436, alkaloid No. 9], UV [68,72,436,496], ^{1}H NMR [68,72,436,469], MS [72,436,469], CD [68,436]. Pennsylvanine (**228**), $C_{40}H_{46}N_2O_8$ (682.3254), m.p. 112–113° (Et$_2$O) [469], $[\alpha]_D^{25}$ + 131° (c 0.7, MeOH) [469], was first isolated from *Thalictrum polygamum* Muhl. in 1974 [469,470]. The UV spectrum showed maxima at 284 nm (log ϵ 4.26), 304 nm (log ϵ 4.18), and 320 nm (sh) (log ϵ 4.05) with both a hyperchromic and bathochromic in strong alkali to 284 nm (log ϵ 4.32), 311 nm (log ϵ 4.29), and 320 nm (sh) (log ϵ 4.15) and was characteristic of a 1,2,9,10-oxygenated aporphine. The principal fragment ions in the mass spectrum were found at M$^+$ *m/z* 682, 476 (M-206), 340 (M-324) (**206**), 324 (M-358) (**207**), 206 (100%) (**191**) and were identical with those in the mass spectrum of the monophenolic benzylisoquinoline–aporphine dimer thalidoxine (**227**). The ^{1}H NMR spectrum of pennsylvanine indicated the presence of two *N*-methyl groups as singlets at δ 2.46 and 2.50; six methoxy groups as singlets at δ 3.58 [C(7')], 3.71 [C(1)], 3.79 [C(12')], 3.82 [C(6')], 3.90 [C(2)], and 3.92 [C(10)]; and six aromatic protons as singlets at δ 6.52, 6.56, 6.59, 6.62, 6.76, and 8.18 [C(11)]. The spectral data were indicative of a benzylisoquinoline–aporphine dimer of the thalicarpine (**192**) type containing two methoxy groups in the isoquinoline ring, one methoxy and one

228

phenolic hydroxy group in the benzyl portion of the benzylisoquinoline, and three methoxy groups in the aporphine half of the alkaloid. Methylation of pennsylvanine with ethereal CH_2N_2 yielded thalicarpine (**192**), which fixed the skeletal structure, positions of oxygenation, and diphenyl ether termini of pennsylvanine. It only remained to assign the single phenolic hydroxy group of pennsylvanine to either C(12′) or C(13′). Acetylation of pennsylvanine with acetic anhydride–pyridine afforded a monoacetate, m.p. 137–138° (Et$_2$O). The ^{1}H NMR spectrum of this monoacetate did not exhibit a 0.1–0.2 ppm upfield shift of the C(8) aporphine proton (due to shielding by the acetate carbonyl), as was the case with acetate of thalidoxine (**227**). Therefore, the phenolic group of pennsylvanine (**228**) was assigned to the C(13′) position, thus completing structural assignment [469]. An identical stereochemistry with that of thalicarpine (**192**) was apparently assigned via a consideration of spectral data and specific rotation [469], although the CD spectrum ([θ]$_{237}$ +235,000, [θ]$_{276}$ −8900, and [θ]$_{308}$ −6500) was reported later in a separate isolation [68].

Pennsylvanine (**228**) was found to inhibit the growth of *Mycobacterium smegmatis* at a dosage of 100 μg/ml in a standard antimicrobial screen [101].

Pennsylvanine has also been isolated from *Thalictrum dioicum* L. [72] and *T. revolutum* DC. [68,101] but has not been found outside of the genus *Thalictrum* to date.

3.1.14. Revolutopine. Tabular review [436, alkaloid No. 20], UV [68,436,472], IR [68,436,472], ^{1}H NMR [68,436,472], MS [68,436,472], CD [68,436,472].

Revolutopine (**229**), $C_{39}H_{44}N_2O_8$ (668.3097), $[\alpha]_D^{25}$ +126° (c 0.1, MeOH), was isolated as an amorphous solid from *Thalictrum revolutum* DC. in 1977 [68,472]. The UV spectrum showed maxima at 281 nm (log ε 4.40), 302 nm (log ε 4.24), and 314 nm (sh) (log ε 4.15), with a bathochromic and hyper-

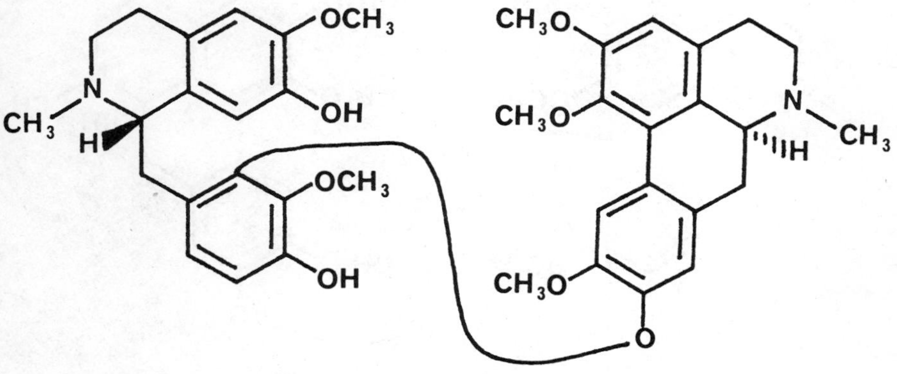

229

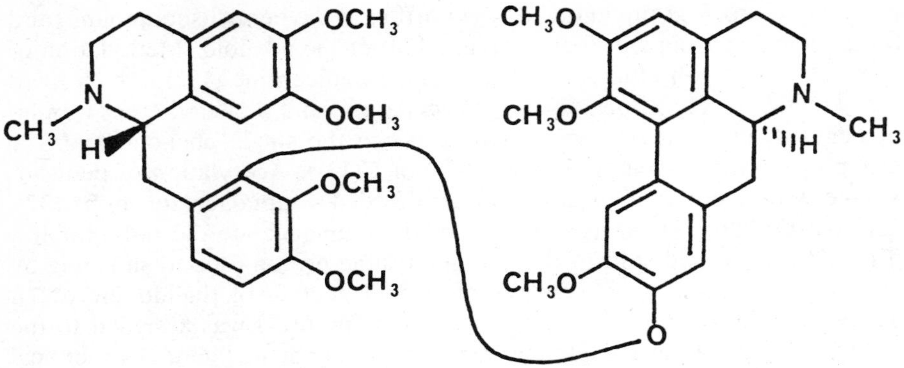

230

chromic shift in strong alkali. The ^{1}H NMR spectrum indicated the presence of two *N*-methyl groups as singlets at δ 2.29 and 2.43; five methoxy groups as singlets at δ 3.65, 3.75, 3.83, 3.86, and 3.96; and seven aromatic protons as five singlets at δ 6.33, 6.46 (2H), 6.56, 6.67 (2H), and 8.16 [H(11)] with a broad, D$_2$O-exchanged singlet at δ 5.23 for phenolic hydroxy groups. The mass spectrum showed the parent ion at M$^+$ *m/z* 668 (1.5%) with other diagnostic fragment ions at *m/z* 476 (M-192), 340 (M-342) (**206**), 324 (M-358) (**207**), and 192 (100%) (**199**). These spectral data were consistent for an aporphine–benzylisoquinoline dimer bearing three methoxy groups in the 1,2,9,10-oxygenated aporphine portion and two methoxy groups and two phenolic groups in the benzylisoquinoline portion. In addition, one methoxy and one phenolic hydroxy group in the benzylisoquinoline portion were in the isoquinoline ring. Treatment of revolutopine with ethereal CH$_2$N$_2$ af-

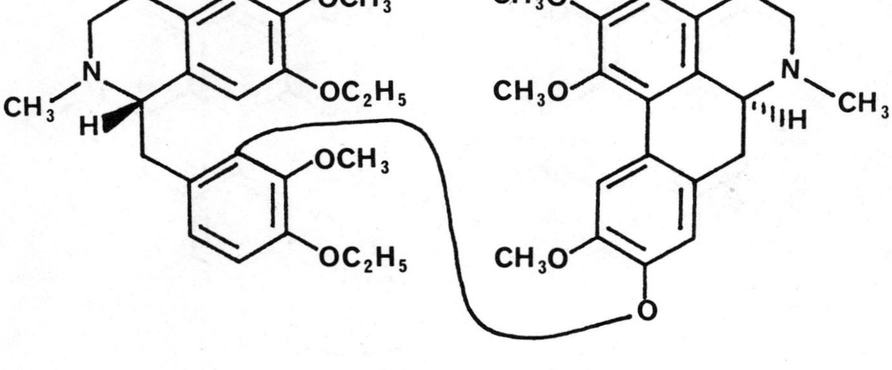

231

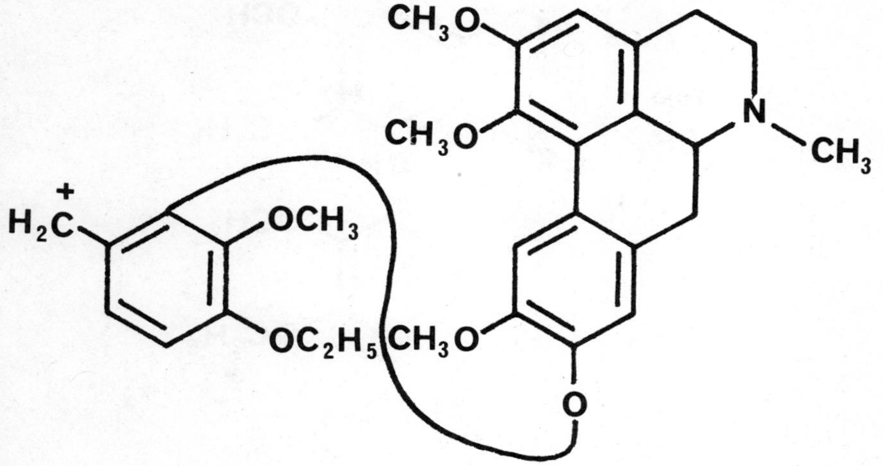

$$\underset{\sim}{\underline{232}}$$

forded *O*-methylfetidine (thalirevolutine) (**230**), thus establishing the skeletal structure, positions of oxygenation, and termini of the diphenyl ether bridge. Placement of the isoquinoline ring phenolic group at C(7′) was proposed because of the absence of a high-field methoxy resonance (about δ 3.58) characteristic of a C(7′) methoxy group and the presence of the H(8′) proton at δ 6.33, downfield from that expected for this proton ortho to a methoxy group (about δ 6.23) [469]. The second phenolic group of the alkaloid was placed at the C(12′) position because of a negative Gibbs' indophenol test, which requires a *para*-substituted phenol. Confirmation of this placement in the benzyl ring was obtained by a consideration of the mass spectral fragment ions of *O-O*-diethylrevolutopine (**231**) (prepared via treatment of revoluto-pine with diazoethane) at M$^+$ *m/z* 724 (0.7%), 505 (M-220) (**232**), and 220 (100%) (**233**). Confirmation of the specific placement at C(12′) was obtained when *O-O*-diethylrevolutopine was cleaved with sodium in liquid ammonia to yield the phenolic cleavage product (**234**), which gave a positive Gibbs'

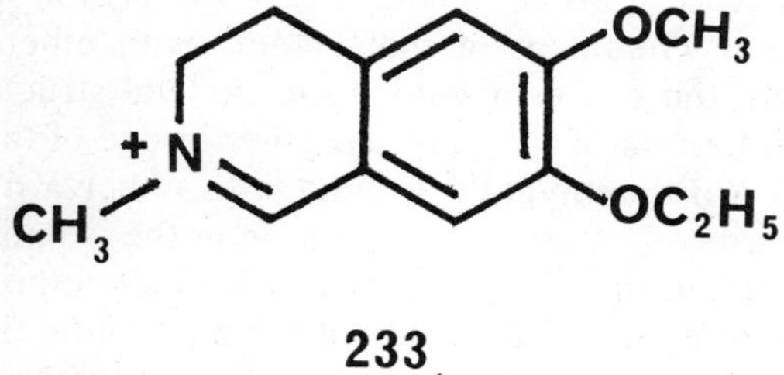

$$\underset{\sim}{\underline{233}}$$

234

indophenol test. The assignment of configuration as (*S*,*S*) was determined via a comparison of the CD spectrum of revolutopine (**229**) ([θ]$_{240}$ +43,000, [θ]$_{275}$ −5,300, and [θ]$_{305}$ −4,700) with that of (+)-fetidine (**208**) [68,472].

Revolutopine was found to lack antimicrobial activity screened against a series of standard microorganisms [68].

3.1.15. Thaliadanine. Tabular review [436, alkaloid No. 15], UV [146,436,446], IR [146,436,446], ¹H NMR [146,436,446], MS [146,436,446], CD [436,446].

Thaliadanine (**203**), C$_{41}$H$_{48}$N$_2$O$_9$ (712.3360), [α]$_D^{26}$ +81° (c 0.41, MeOH), was isolated as an amorphous solid from *Thalictrum minus* L. race B in 1978 [446]. The UV spectrum showed maxima at 281 nm (log ε 4.33), 302 nm (log ε 4.18), and 312 nm (log ε 4.11), with no shift in acid or base, and was characteristic of a 1,2,3,9,10-oxygenated aporphine. The mass spectrum was characterized by the presence of a very small parent ion at *m/z* 712 (0.1%) but other significant fragment ions at *m/z* 520 (13%) (M-192), 370 (7) (**210**), 354 (3) (**211**), and 192 (100) (**199**). The ¹H NMR spectrum (90 MHz) indicated the presence of two *N*-methyl groups as singlets at δ 2.38 and 2.48; seven methoxy groups as five singlets at δ 3.76, 3.79 (6H), 3.81, 3.89, and 3.96 (6H); six aromatic protons as five singlets at δ 6.44, 6.51 (2H), 6.60, 6.68, and 8.05 [H(11)]; with a broad singlet at δ 4.9 (OH) (60 MHz) which was D$_2$O exchangeable. Treatment of thaliadanine with ethereal CH$_2$N$_2$ gave adiantifoline (**190**), thereby establishing the skeletal structure, positions of oxygenation, and termini of the diphenyl ether bridge of thaliadanine. Thus thaliadanine was a desmethyladiantifoline with either a 6′-methoxy-7′-hydroxy or a 6′-hydroxy-7′-methoxy orientation in the isoquinoline portion of the benzylisoquinoline ring. The lack of a high-field methoxy at about δ 3.6 and a high-field aromatic proton at about δ 5.8 preclude the latter arrangement and strongly suggest that thaliadanine is best represented as **203**. Confirmation of structure was via a consideration of the potassium permanganate oxidation products of O-ethylthiadanine [prepared via treatment of thaliadanine (**203**) with ethereal diazoethane], which were dehydrothaliadine (**193**) [also obtained by a similar oxidation of adiantifoline (**190**)] and 6-methoxy-7-ethoxy-N-methyl-1-oxo-1,2,3,4-tetrahydroisoquinoline (**202**) [also obtained by ethylation of thalifoline (**103**) [446]. Finally, the assignment of (S,S) stereochemistry was made via a comparison of the CD spectrum ([θ]$_{241}$ + 166,000, [θ]$_{276}$ − 17,600, and [θ]$_{308}$ − 15,900) of thaliadanine (**203**) with similar alkaloids [436,437,446].

Thaliadanine (**203**) was found to exhibit hypotensive activity in a dosage of 0.1–1.0 mg/kg in normotensive rabbits and to possess weak in vitro antimicrobial activity against *Mycobacterium smegmatis* at a concentration of 100 μg/ml [446].

Thaliadanine (**203**) has been isolated from only one other *Thalictrum* species, that being *T. minus* L. var. *microphyllum* Boiss. [146].

3.1.16. Thalicarpine. Tabular review [436, alkaloid No. 10], UV [129,436,475], IR [269,436,475], ^{1}H NMR [129,436,475], ^{13}C NMR [477], MS [14,436,520].

Thalicarpine (**192**), $C_{41}H_{48}O_8N_2$ (696.3410), m.p. 160–161° (EtOAc) [129]; m.p. 105–106° (Et$_2$O) [269]; [α]$_D^{25}$ + 133° (c 0.83, MeOH); [α]$_D^{25}$ + 89° (c 0.88, CHCl$_3$), was first isolated from *Thalictrum dasycarpum* Fisch. and Lall as colorless crystals in 1963 [129]. The UV spectrum displayed maxima at 282 nm (log ε 4.23) and 302 (4.11) and the ^{1}H NMR spectrum showed signals at δ 2.45 (s) and 2.48 (s) for two N-methyl groups; δ 3.58 (s) [C(7′)], 3.71 (s) [C(1)], 3.80 (s) (6H), 3.83 (s), 3.91 (s), and 3.95 (s) [C(10)] for seven methoxy groups; and 6.21 (s) [H(8′)], 6.53 (s), 6.56 (s), 6.60 (s), 6.62 (s), 6.67 (s) [H(11′)], and 8.19 (s) [H(11)] for seven aromatic protons [129,436]. Hofmann degradation produced a methine and a des-N-methine of thalicarpine [129]. Early physicochemical studies, including elemental analyses, nonaqueous titration, Hofmann degradation reactions, Birch reduction, and synthesis led to the proposal of an aporphine–benzylisoquinoline dimeric structure (**235**) for thalicarpine [473,474]. Reductive cleavage of the alkaloid with sodium in liquid ammonia yielded (−)-6-hydroxylaudanosine (**194**), (+)-2,10-dimethoxyaporphine (**236**), and (+)-keto fragment **195** and synthesis of compound **235** achieved via Ullmann condensation of (−)-(R)-6′-bromolaudanosine (**196**) [(R)isomer] with (+)-(S)-isocorydine (**21**) [473,474]. Shortly thereafter spectral studies of thalicarpine and structurally related analogs indicated these compounds possessed properties inconsistent with structure **235** [475,476]. In particular, the presence of a low-field singlet at δ 8.23 in the spectrum of thalicarpine was strongly suggestive of the deshielded H(11) proton in the aporphine series, thus precluding the terminus of the diphenyl ether bridge of the dimer at C(11) in the aporphine moiety [15–17]. Furthermore, the coupled H(8) and H(9) protons, which would have likely been observed as an AB quartet, were not present in the ^{1}H NMR spectrum.

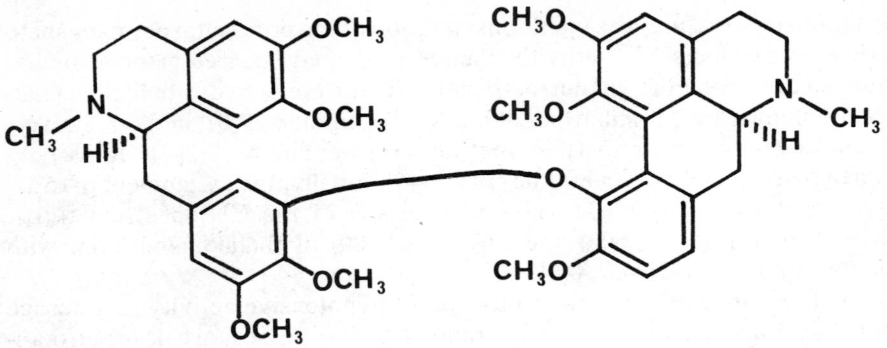

235

Finally, the UV spectrum of thalicarpine was consistent with those of C(11) unsubstituted aporphines [15–17]. A reexamination of the spectral properties of the Ullmann condensation product **235** with those of natural thalicarpine showed distinct differences [475,476]. Ullmann condensation of both the (−)-(R)- and (+)-(S)-isomers of 6'-bromolaudanosine (**196**) with (+)-(S)-N-methyllaurotetanine (**5**) produced a mixture of diastereomeric products, with the synthetic condensation product **192** being identical with natural thalicarpine [475,476]. Some seven years later, thalicarpine was also synthesized via condensation of (+)-(S)-hernandaline (**237**) with the appropriate Reissert compound (1-cyano-2-benzoyl-6,7-dimethoxy-1,2-dihydroisoquinoline) followed by reduction and N-methylation [478].

After the structural elucidation of thalicarpine was clarified, a number of interesting chemical reactions were noted. Oxidation of thalicarpine with

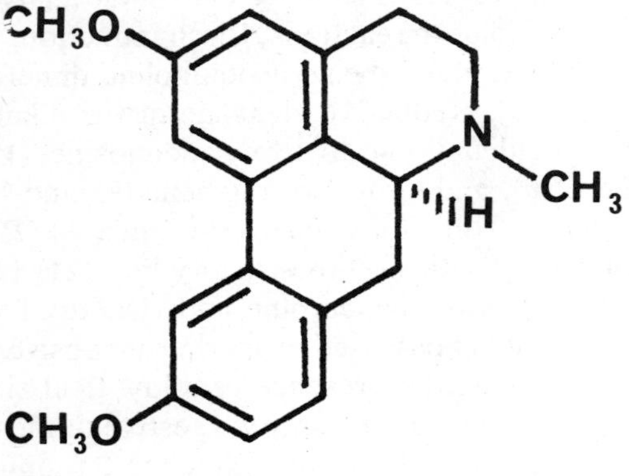

236

potassium permanganate in acetone afforded 1-oxo-2-methyl-6,7-dimethox-ytetrahydroisoquinoline (*N*-methylcorydaldine) (**88**) and 6a,7-dehydroher-nandaline (**238**), the latter of which upon reduction with Adams catalyst followed by cleavage with sodium in liquid ammonia gave 2,10-dimethox-yaporphine (**236**) [479,480]. A series of bisquaternary salts and methines of thalicarpine was also prepared [481]. A regioselective *O*-demethylation of thalicarpine (**192**) with sodium benzylselenolate in refluxing DMF yielded 1-*O*-demethylthalicarpine (**215**) (51% yield) and 1,12′-*O*-*O*-didemethylthali-carpine (**239**) (22% yield) [482]. An interesting microbial transformation uti-lizing *Streptomyces punipalus* (NRRL 3529) produced (+)-hernandalinol (**240**) from thalicarpine (**192**). It is most probable that hernandalinol (**240**) was formed in a two-step process involving cleavage of the isoquinoline ring from thalicarpine followed by reduction of the intermediate hernandaline (**237**) to hernandalinol (**240**) [483].

The biosynthesis of thalicarpine was reported in some detail in a series of papers appearing in 1981 and 1982. The first paper describes the incor-poration of (±)-1-^{14}C-reticuline (**57**) into both the aporphine and the ben-zylisoquinoline halves of thalicarpine (**192**) when administered to intact plants of *Thalictrum minus* L. var. *elatum* [484]. Shortly thereafter, a more detailed study utilizing (±)-1-^{14}C-reticuline (**57**) and (+)-8-^{3}H-isoboldine (**4**) demonstrated that these alkaloids were the benzylisoquinoline and aporphine precursors, respectively, of the dimeric thalicarpine in the intact *Thalictrum*

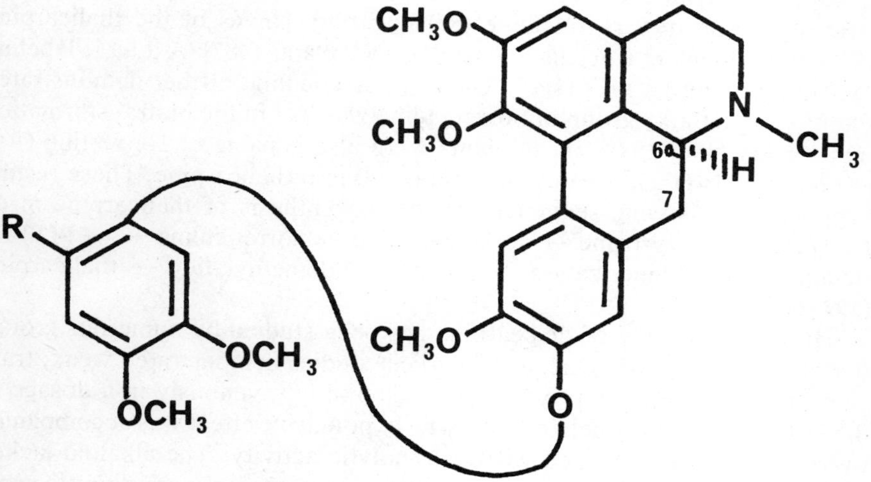

237 R= CHO
238 R= CHO; 6a,7-DEHYDRO-
240 R= CH$_2$OH

239

minus [485]. A separate study revealed the incorporation of 3-methoxy-4-hydroxyphenethylamine-1-^{14}C into intact *Thalictrum minus* plants exceeding the incorporation of either dopamine-1-^{14}C or 3-hydroxy-4-methoxyphene-thylamine-1-^{14}C, suggesting that the substitution pattern suitable for the formation of the aporphine structure of thalicarpine was obtained not later than at the phenethylamine stage [486]. A separate study also confirmed the specific utilization of ($\pm$)-reticuline (**57**) into both halves of the thalicarpine (**192**) dimer in intact *Cocculus laurifolius* DC. plants [487]. A double labeling experiment with ($\pm$)-[1-^{3}H,4'-^{14}O-CH$_3$]-norreticuline further demonstrated that the 4'-methoxy group of norreticuline was lost in the biotransformation into thalicarpine (**192**). These plants were also capable of converting ($+$)-(*S*)-boldine (**241**) and ($+$)-(*S*)-isoboldine (**4**) into thalicarpine. These results support the following sequence for the biosynthesis of thalicarpine in *C. laurifolius* DC.: tyrosine → norlaudanosine → norreticuline → ($+$)-(*S*)-reticuline (**57**) → dimerization → selective *O*-demethylation → thalicarpine (**192**) [487].

The pharmacology of thalicarpine has been studied by numerous groups over the last 20 years. One of the earliest studies demonstrated weak, transient hypotensive activity when administered intravenously in a dosage of 0.5–5.0 mg/kg to anesthetized cats. The hypotensive effect was accompanied by respiratory toxicity and weak adrenolytic activity. The alkaloid lacked significant antiinflammatory, anticoagulant, hypoglycemic, or diuretic properties [129]. A subsequent study in anesthetized dogs showed that the alkaloid produced moderate pressor activity of long duration when administered intravenously in a dosage of 2 mg/kg. This activity was also sometimes accompanied by a mild tachycardia and does not appear to involve a neural pathway but may be due to a direct effect on the heart or the vascular smooth muscle. Doses of 10 mg/kg produced an intense, long-lasting, noncholinergic

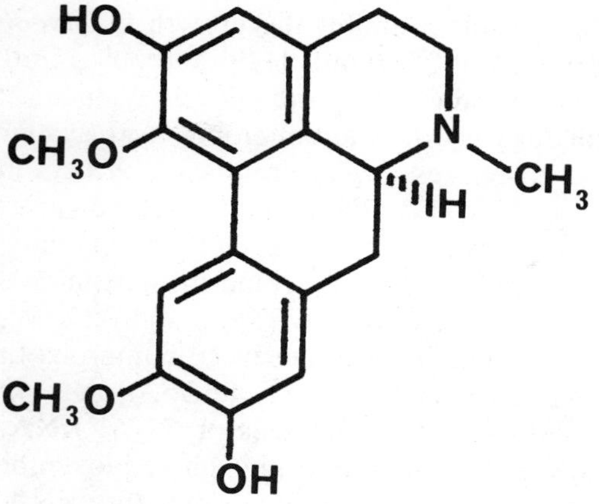

241

hypotension. Bath concentrations of 10–40 mcg/mL of thalicarpine produced prompt relaxation of the musculature of the isolated ileum and uterus. In general, there was a nonspecific inhibition of smooth muscle activity with the ability to antagonize the spasmogenic effects of various drugs [488]. Thalicarpine from *Thalictrum revolutum* DC. was also found to possess hypotensive activity in rabbits [101], and that from *T. dasycarpum* had the same effect in cats [129]. A detailed study of the hypotensive effects of thalicarpine in rhesus monkeys demonstrated an increase in pulse pressure accompanied by various degrees of hypotension on intravenous administration of 2.5–20.0-mg/kg doses. Carotid artery blood flow and heart rate increase after doses of 2.5–10.0 mg/kg but these effects were blocked by pretreatment with hexamethonium or (±)-propranolol. At a dosage of 5.0 or 10.0 mg/kg, there was decreased peripheral resistance and increased cardiac output. All dosages of thalicarpine from 2.5 to 20.0 mg/kg depressed the force of contraction, heart rate, and coronary perfusion pressure in the isolated, blood-perfused canine heart. Thalicarpine-induced hypotension appeared to be due to a nonspecific vasodilation and myocardial depression [489]. In a separate investigation, intravenous administration of thalicarpine in a dosage of 10 mg/kg to anesthetized rhesus monkeys produced a depressed mean arterial pressure and force of contraction but increased carotid artery blood flow and heart rate. The alkaloid produced an immediate negative inotropic and chronotropic response in isolated, blood-perfused dog hearts. Thalicarpine induced a persistent decrease in coronary perfusion pressure. The aporphine portion of thalicarpine was held responsible for many of the cardiovascular actions, but some appear to require the intact dimeric structure [490].

Thalicarpine was found to inhibit the growth of *Mycobacterium smegmatis* at a concentration of 25 μg/mL [149] and 100 μg/mL [101] and both *Staphylococcus aureus* and *Candida albicans* at a concentration of 1000 μg/mL [101] in a standard in vitro antimicrobial test system [101,149]. The alkaloid was found to possess weak antitussive effects in dogs when administered intravenously [112]. Early studies demonstrated the cytotoxic effect of the alkaloid to monolayer KB cell culture at a concentration of 100 μg/mL [262]. Incorporation of thalicarpine in liposomes was found to enhance its cytotoxicity significantly in Walker S and TLX-5 cell cultures [491]. This in vitro cytotoxic activity was observed numerous times in KB cells, L1210 cells, human embryonal fibroblasts, and SAg cells [492–494]. In L1210 cells, the alkaloid inhibited the synthesis of DNA, RNA, and protein, as well as the early steps of nucleotide triphosphate biosynthesis [492]. Thymidine incorporation into L1210 DNA and uridine into RNA was inhibited by thalicarpine (100 μM) within 20 min. Washing the cells to remove the drug resulted in a partially reversible inhibition of DNA synthesis and an almost total reversal on inhibition of RNA synthesis [492]. Thalicarpine was found to inhibit the growth of a diploid culture of embryonic human lung cells, with a concentration of ≥20 μg/mL totally inhibiting growth. There was an increase in the mitotic index as a result of the arrest of metaphase [495]. The alkaloid also appeared to be a weak mutagen in long-term human embryonal lung cell cultures at a dosage of 1–15 mg/mL of culture medium. There was a blockage of metaphase, with some tetraploid metaphases being observed as well as frequent breaks at the chromosomes. Finally, the alkaloid inhibited DNA formation in the same cultures [496]. Thalicarpine inhibited the uptake of thymidine and uridine in HeLa (human cervical carcinoma) cells. A 50% inhibition of this uptake was achieved via a 5 μM solution of the alkaloid but this concentration of alkaloid did not inhibit DNA, RNA, or protein synthesis [497]. Rodent hepatic microsomal aniline hydroxylase activity was inhibited by thalicarpine. However, pretreatment of rodents with the alkaloid (75 mg/kg × 3) does not affect this activity nor does it affect microsomal cytochrome *P*-450, cytochrome b_5, NADPH cytochrome *c* reductase, RNA, or nuclear DNA. Thalicarpine pretreatment does prevent the methylcholanthrene-induced increase in aniline hydroxylase, microsomal RNA, and nuclear DNA but doesn't affect the induction of cytochromes *P*-450 and b_5 [498]. In another study, it was observed that thalicarpine inhibited the in vitro biosynthesis of DNA, RNA, and proteins in S180 cells. Oxidation of glucose [^{14}C] to $^{14}CO_2$ was not affected by levels of the alkaloid up to 100 μg/mL in vitro but incorporation of labeled acetate into lipids was inhibited at that concentration. Inhibition of thymidine incorporation into DNA was observed in vivo after treatment with the alkaloid (30–120 mg/kg) but the synthesis of RNA and protein was not inhibited. Gel filtration and dialysis experiments showed that thalicarpine associated with both DNA and RNA and was bound by polyguanylic and polyadenylic acids but not polycytidilic acid [499].

Intraperitoneal injections of thalicarpine in a dosage range of 100–250 mg/kg prolonged the life of mice with Ehrlich ascites tumors and lymphoma NK/Ly [500]. Thalicarpine was administered intraperitoneally, intramuscularly, or subcutaneously to tumorous mice in a single dose of 100–250 mg/kg, which was repeated up to 11 times. The alkaloid had a pronounced antitumor effect on Ehrlich ascitic tumor, NK/Ly lymphoma, sarcoma 37, sarcoma 180, and Lewis lung carcinoma. Either single or intermittent alkaloid administration produced a better therapeutic effect on numerous tumors than chronic treatment [501]. The toxicology of thalicarpine in various animals was observed and the alkaloid was observed to produce convulsions, apnea, hypotension, and cardiac arrest on acute administration. Renal and hepatic toxicities were observed on chronic administration. The LD_{50} in the rat was determined to be 3000 mg/m^2 when the drug was administered as a 2 hr infusion [502]. The high toxic dose in the rhesus monkey was 1080 mg/m^2 and the low toxic dose was 525 mg/m^2 [503]. A Phase I initial clinical trial of thalicarpine was carried out in 1975. Twenty-eight patients received 33 courses of single-dose intravenous therapy in the range of 200–1900 mg/m^2. At the maximum tolerable dosage of 1400 mg/m^2, adverse effects included arm pain, central nervous system depression, nausea and vomiting, both hypotension and hypertension, and ventricular arrhythmias with electrocardiographic changes. The same type of adverse effects were observed at the maximum tolerable dose (1100 mg/m^2) for weekly intravenous administration. No objective tumor response was observed in any of the patients in the protocol with only acute toxic effects being noted. Renal, hepatic, and hematopoietic toxicities were not observed nor were any toxicities noted for up to 4 months post-treatment [504]. The plasma decay and urinary excretion of ^{3}H-thalicarpine were studied in 19 patients at intravenous doses of 300–1900 mg/m^2. The plasma decay was triexponential with a terminal phase half-life of 198–1386 hr in five patients. Only small amounts of unchanged alkaloid appeared in the cerebrospinal fluid, with renal excretion being slow and erratic (about 20% of the dose excreted in the urine of nine patients within 285 hr). These data suggest extensive tissue localization of the alkaloid with the danger of drug accumulation being very real if the alkaloid is administered at short dose intervals [505]. An abbreviated phase II clinical trial of thalicarpine was reported in 1980. In this study, 14 previously treated patients with advanced malignant diseases were treated with the alkaloid at a dose of 1100 mg/m^2 weekly via a constant 2 hr intravenous infusion. There were no complete nor even partial tumor responses but toxicities were apparent. These adverse effects commonly included nausea, arm pain, lethargy, and electrocardiographic changes, and less frequently, vomiting, diarrhea, urticaria, chills, mydriasis, pain distant from the infusion site, tachycardia, and hypotension. No renal, hepatic, nor hematologic toxicities were observed [506]. Single- and multiple-dose acute lethality and toxicity of intravenous thalicarpine in mice were finally reported in 1982 [507]. Since the Division of Cancer Treatment Decision Network

Review of August 1, 1979 resulted in the closure of the existing IND for thalicarpine [508], it is unlikely that thalicarpine or its related alkaloids will find a place in clinical oncology in the United States.

Thalicarpine has also been isolated From *Thalictrum alpinium* L. [70], *T. dasycarpum* Fisch. and Lall. [509], *T. dioicum* L. [72,130], *T. fendleri* Engelm. ex Gray [62], *T. flavum* L. [269], *T. foetidum* L. [451], *T. foliolosum* DC. [38,66 (HPLC)], *T. minus* L. [66,484–486,510,511,512 (quantitative analysis by MS)], *T. minus* L. var. *elatum* Jacq. [513], *T. polygamum* Muhl. [149,514(tlc)], and *T. revolutum* DC. [101,104,105,509].

3.1.17. Thalictrogamine.

Tabular review [436, alkaloid No. 1], UV [68,72,436,515], IR [436,515], ^{1}H NMR [68,72,436,515], MS [68,72,436,515], CD [68,436].

Thalictrogamine (**242**), $C_{39}H_{44}N_2O_8$ (668.3097), $[\alpha]_D^{25}$ + 135° (c 0.2, MeOH), was isolated as an amorphous solid from *Thalictrum polygamum* Muhl. in 1973 [515]. The alkaloid was very labile to aerial oxidation and turned green upon standing. The UV spectrum displayed maxima at 230 (sh) (log ϵ 4.39), 277 nm (log ϵ 4.11), 298 nm (sh) (log ϵ 3.98), and 307 nm (sh) (log ϵ 3.82) with a bathochromic shift in alkali. The mass spectrum showed the molecular ion at M$^+$ *m/z* 668 with diagnostic fragment ions at *m/z* 476 (M-192), 326 (M-342) (**224**), 309 (M-358-1) (**225**-1), and 192 (100%) (**199**). These spectral data were characteristic of a diphenolic aporphine–benzylisoquinoline containing one phenolic group in the aporphine moiety and one phenolic group in the isoquinoline portion of the benzylisoquinoline ring. Acetylation of thalictrogamine produced a diacetyl ester, m.p. 147–148° (Et$_2$O), whose mass spectral fragment ions at M$^+$ *m/z* 752, 518 (M-234), 368 (**243**), and 234 (100%) (**244**) further confirmed the relative locations of the two phenolic groups in the dimer. The ^{1}H NMR spectrum indicated the presence of two

242

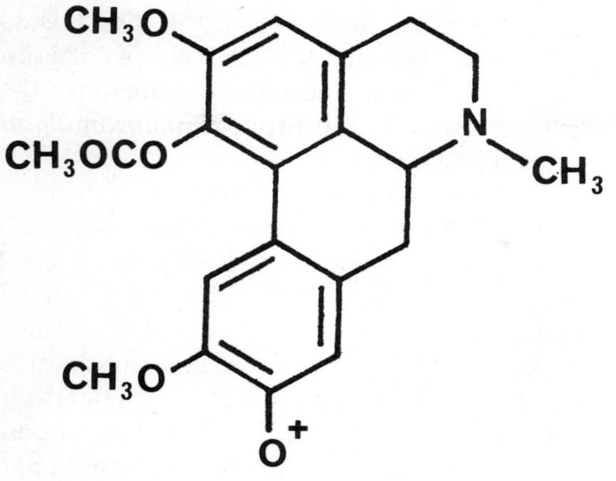

243

N-methyl groups as singlets at δ 2.49 and 2.53; five methoxy groups as four singlets at δ 3.79 [C(13′)], 3.83 (6H) [C(6′) and C(12′)], 3.92 [C(2)], and 3.95 [C(10)]; and seven aromatic protons at δ 6.40 [C(8′)], 6.51, 6.57 (3H), 6.78 [H(11′)], and 8.18 [C(11)]. Methylation of thalictrogamine with CH_2N_2 afforded both thalictropine (**226**) and thalicarpine (**192**) and thus established the skeletal structure, positions of oxygenation, and termini of the diphenyl ether bridge of thalictrogamine. A consideration of molecular models of aporphine–benzylisoquinoline dimers of the thalicarpine (**192**) class demonstrated that the presence of a C(7′) phenolic hydroxy group results in conformational changes in the molecule because of hydrogen bonding of the hydroxyl hydrogen with electron-rich groups [such as the C(1) oxygen or the aporphine nitrogen]. As a result, the C(8) aromatic proton is located at a lower field (near δ 6.4) than usual (near δ 6.2). Furthermore, O-acetylation of the C(7′) hydroxy group obviates this hydrogen bonding potential and the C(8′) proton is once again back upfield at the anticipated position [as in thalictropine (**226**) and thalicarpine (**192**)]. Additionally, acetylation also produced an upfield shift of the C(11) proton at δ 8.18 to δ 7.60, thus denoting

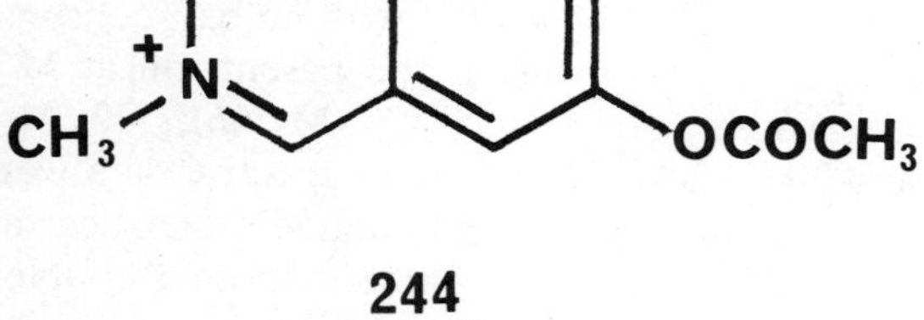

244

the presence of an acetoxy group at C(1) [515]. The CD spectrum ([θ]$_{238}$ +186,000, [θ]$_{275}$ −19,300, and [θ]$_{310}$ −28,200) of thalictrogamine (**242**) strongly suggests the (*S,S*) stereochemical assignment for the alkaloid [68].

Thalictrogamine (**242**) was subsequently isolated from *Thalictrum dioicum* L. [72] and *T. revolutum* DC. [68].

3.1.18. Thalictropine.   Tabular review [436, alkaloid No. 3], UV [72,436,515], IR [436,515], ^{1}H NMR [72,436,515], MS [72,436,515].

Thalictropine (**226**), $C_{40}H_{46}N_2O_8$ (682.3254), m.p. 167° (MeOH), $[\alpha]_D^{25}$ +120° (c 0.3, MeOH), was isolated as white needles from *Thalictrum polygamum* Muhl. in 1973 [515]. The UV spectrum was characterized by maxima at 225 nm (log ε 4.46), 278 nm (log ε 4.12), 298 nm (sh) (log ε 3.88), and 310 nm (log ε 3.70) with a bathochromic shift in strong alkali. The mass spectrum displayed a parent ion at M$^+$ *m/z* 682 and other important fragment ions at *m/z* 476 (M-206), 326 (M-356) (**224**), 310 (M-372) (**225**), and 206 (100%) (**191**). These spectral data were characteristic of a monophenolic aporphine–benzylisoquinoline dimer containing the phenolic group in the aporphine moiety. The ^{1}H NMR spectrum indicated the presence of two *N*-methyl groups as singlets at δ 2.47 and 2.50; six methoxy groups as five singlets at δ 3.58 [C(7')], 3.78 (6H) [C(12') and C(13')], 3.82 [C(6')], 3.88 [C(2)] and 3.92 [C(10)]; and seven aromatic protons as five singlets at δ 6.20 [H(8')], 6.55 (3H), 6.59, 6.67 [H(11')], and 8.18 [H(11)]. The alkaloid was noticeably labile to aerial oxidation and the *O*-acetyl ester, m.p. 182–183° (MeOH), was prepared via treatment with acetic anhydride–pyridine. The ^{1}H NMR spectrum of *O*-acetylthalictropine displayed a singlet at δ 7.60 for the H(11) aromatic proton. This upfield shift of the H(11) proton denoted the presence of C(1) acetoxy group, which located the single phenolic group of thalictropine at C(1). Finally, a direct comparison of the ^{1}H NMR spectrum of thalictropine (**226**) with that of synthetic 1-*O*-demethylthalicarpine confirmed the structural assignment [515].

Thalictropine (**226**) was subsequently isolated from only one other *Thalictrum* species, *T. dioicum* L. [72].

3.1.19. Thalidoxine.   Tabular review [436, alkaloid No. 8], UV [436,516], IR [436,516], ^{1}H NMR [436,516], MS [436,516].

Thalidoxine (**227**), $C_{40}H_{46}N_2O_8$ (682.3254), $[\alpha]_D^{25}$ +113° (c 0.2, MeOH), was isolated as an amorphous solid from *Thalictrum dioicum* L. in 1973 [516]. The UV spectrum showed maxima at 275 nm (log ε 4.23), 296 nm (sh) (log ε 4.08), and 310 nm (sh) (log ε 4.02) with no bathochromic shift in basic media. The mass spectrum displayed the parent ion at M$^+$ *m/z* 682 with other important fragment ions at *m/z* 476 (M-206), 340 (M-342) (**206**), 324 (M-358) (**207**), and 206 (100%) (**191**). These spectral data were suggestive of a monophenolic aporphine–benzylisoquinoline dimer bearing a 1,2,9,10-tetraoxygenated, trimethoxylated aporphine monomer coupled with a trimethoxylated benzylisoquinoline moiety, the latter containing the phenolic group in the benzyl portion. The ^{1}H NMR spectrum was characterized by the presence of two *N*-methyl groups as singlets at δ 2.47 and 2.48; six methoxy groups as singlets at δ 3.57 [C(7′)], 3.70 [C(1)], 3.75 [C(13′)], 3.78 [C(6′)], 3.88 [C(2)], and 3.90 [C(10)]; and seven aromatic protons as five singlets at δ 6.23 [H(8′)], 6.50 (3H), 6.57, 6.77 [H(11′)], and 8.15 [H(11)]. Methylation of thalidoxine with ethereal CH_2N_2 yielded thalicarpine (192) thereby establishing the skeletal structure, positions of oxygenation, positions of the termini of the diphenyl ether linkage, and stereochemistry of the asymmetric centers. Acetylation of thalidoxine with acetic anhydride–pyridine produced a monoacetate, m.p. 128–129° (CHCl₃). The most salient feature of the ^{1}H NMR spectrum of this acetate was the presence of a singlet at δ 6.40, which corresponded to one of the proton signals originally found at δ 6.50 (3H) or 6.57 (1H) in the spectrum of thalidoxine. Inspection of molecular models demonstrated that an acetate carbonyl at C(12′) could be located in close proximity to the protons at both the H(11′) and H(8) positions, thus exerting a shielding effect of 0.1–0.2 ppm on these protons. On the other hand, placement of the acetyl ester at C(13′) would not have permitted intramolecular shielding to have occurred. Thus thalidoxine (227) was characterized as 12′-*O*-demethylthalicarpine [516].

A subsequent investigation of *T. dioicum* L. in 1978 failed to demonstrate the presence of thalidoxine [72].

3.1.20. Thalifabatine. UV [471], IR [471], ^{1}H NMR [471], MS [471], CD [471].

Thalifabatine (247), $C_{41}H_{48}N_2O_9$ (712.3360), $[\alpha]_D^{22}$ +60.5° (c 0.154, MeOH), was isolated as a yellow amorphous solid from *Thalictrum faberi* Ulbr. in

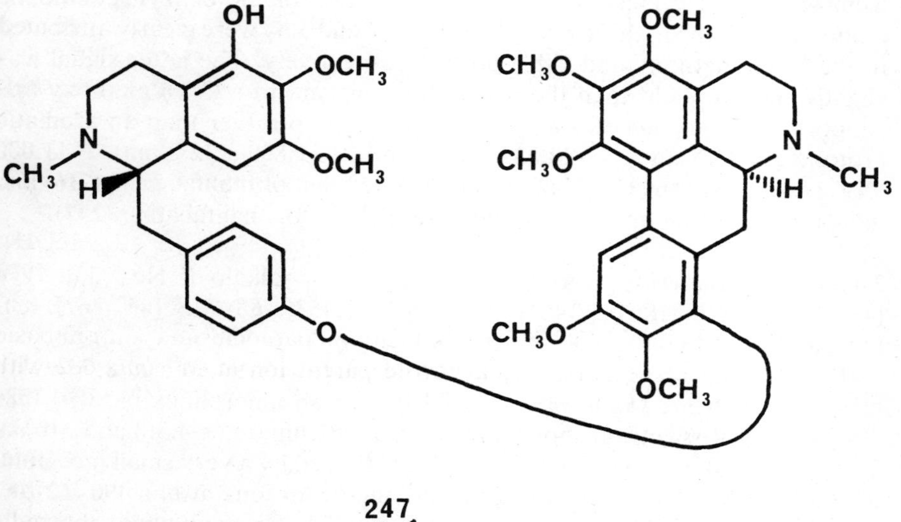

247

248

1984 [471]. The UV spectrum of the alkaloid was characterized by two maxima at 281 nm (log ϵ 4.26) and 311 nm (sh) (log ϵ 3.93) with no shift in strong alkali. The IR spectrum showed a strong absorption at 3530 cm^{-1}. The ^{1}H NMR spectrum indicated the presence of two *N*-methyl groups as singlets at 2.32 [N(6)] and 2.51 [N(2')]; seven methoxy groups as six singlets at δ 3.53 [C(7')], 3.79 (6H) [C(1) and C(9)], 3.82 [C(6')], 3.87 [C(3)], 3.90 [C(2) or C(10)], and 3.93 [C(10) or C(2)]; and six aromatic protons, as two singlets at δ 5.64 [H(8')] and 7.85 [H(11)] plus a pair of doublets at δ 6.74 (2H, J = 8.5 Hz) and 6.95 (2H, J = 8.5 Hz). The mass spectrum exhibited a very weak parent ion at M$^+$ m/z 712 (0.1%) with other characteristic fragment ions at m/z 490 (M$^+$ − 222), 489, and 222 (100%) (**248**). These spectral data were indicative of a 1,2,3,9,10-pentaoxygenated aporphine–monomer and two methoxy groups plus one hydroxy group in the isoquinoline portion of the benzylisoquinoline monomer. The A$_2$B$_2$ protons in the ^{1}H NMR spectrum were clearly located in the benzyl portion of the benzylisoquinoline monomer, and the high-field singlets at δ 3.53 and 5.64 were clearly attributed to the C(7') methoxy and C(8') proton, respectively. The latter signal was slightly more upfield than the usual H(8) adjacent to a C(7') methoxy and supported the presence of a C(5') hydroxy group rather than an aromatic proton. The CD curve [−24,333 (211.6 nm), +51,800 (242.0 nm), −11,000 (281.2 nm), and −6033 (305.2 nm)] resembled that of thalifaberine (**216**) and indicated the same absolute configuration (*S*,*S*) for thalifabatine (**247**).

3.1.21. Thalifaberine. Tabular review [437, alkaloid No. 35], UV [437,467], ^{1}H NMR [437,467], EI–MS [437,467], CI–MS [437,467], CD [437,467].

Thalifaberine (**216**), C$_{41}$H$_{49}$N$_2$O$_8$ (696.3410), [α]$_D^{25}$ +94.6° (c 0.38, MeOH), was isolated from *Thalictrum faberi* Ulbr. as an amorphous base in 1983 [467]. The UV spectrum showed maxima at 282 nm (log ϵ 4.36) and 310 nm (sh) (log ϵ 3.98), and the EI–MS was characterized by a very small molecular ion at M$^+$ m/z 696 (<0.1%) with other significant ions at m/z 490 (M-206) and 206 (100%) (**191**). On the other hand, the CI–MS (isobutane) showed a single high-mass ion at m/z 697 (MH$^+$). The ^{1}H NMR spectrum indicated the presence of two N-methyl groups as singlets at δ 2.34 [N(2')] and 2.53 [N(6)]; seven methoxy groups as six singlets at δ 3.60 [C(7')], 3.80 (6H), 3.83, 3.91, 3.94, and 3.97; and seven aromatic protons as three singlets at δ 6.01 [C(8')], 6.57 [C(5')], and 7.91 [C(11)] plus an AA'BB' system at δ 6.81 [C(11') and C(13')] and 7.00 [C(10') and C(14')] (J = 8.5 Hz). Oxidation of thalifaberine with potassium permanganate afforded a 6a,7-dehydroaporphine (**245**), with UV maxima at 257 nm (log ε 5.13), 270 nm (log ε 5.13), and 335 nm (log ε 4.49) and M$^+$ at m/z 503 (C$_{29}$H$_{29}$NO$_7$). The ^{1}H NMR spectrum of the oxidation product showed the presence of one N-methyl group at δ 2.90 and five methoxy groups at δ 3.92, 3.99, 4.05, 4.09, and 4.12. The aromatic protons were observed as two singlets at 6.60 [H(7)] and 9.05 [H(11)] and one AA'BB' system at δ 7.10 and 7.80, as confirmed via double irradiation. Extensive NOE experiments revealed clear enhancements between the H(11) and C(10) methoxy, the H(11) and C(1) methoxy, the H(7) and the N-methyl, and the H(7) and H(11') or H(13'). In particular, the NOE between H(7) and H(11')/H(13'), in addition to the lack of NOE between H(7) and any single methoxy group, led to the conclusion that the diphenyl ether terminus in the aporphine ring was at C(8) and not C(9), as is the case in many other dimers of this series. Finally the absolute configuration of thalifaberine (**216**) was determined via a consideration of its CD spectrum

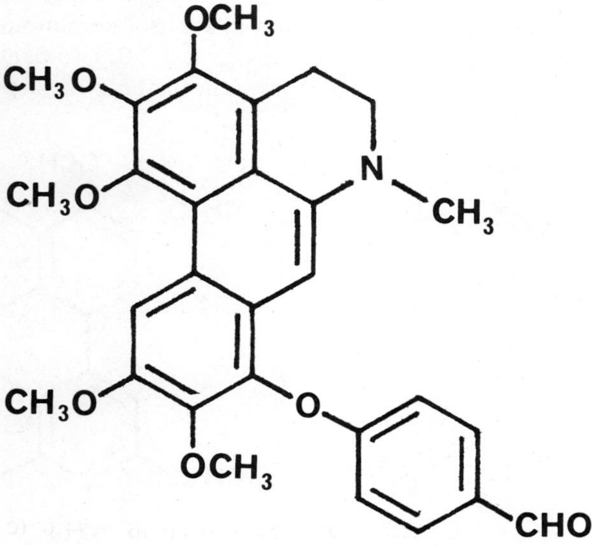

245

([θ]$_{299}$ −8,316, [θ]$_{278}$ −10,471, and [θ]$_{241}$ +98,779) which resembled those of alkaloids of the (+)-thalicarpine (**192**) series [467]. Thalifaberine (**216**) and its closely related companion alkaloid thalifabine (**246**) are the first benzyl-isoquinoline–aporphine dimers with a 1,2,3,8,9,10-oxygenation in the aporphine system and a C(8) to C(12′) diphenyl ether linkage between the aporphine and the benzylisoquinoline systems [436,437,467].

Thalifaberine was found to exhibit cytotoxicity as measured via the Walker 256 carcinoma cell system [467].

3.1.22.   Thalifabine.   Tabular review [437, alkaloid No. 36], UV [437,467], ¹H NMR [437,467], MS [437,467], CD [437,467].

Thalifabine (**246**), C$_{41}$H$_{46}$N$_2$O$_9$ (710.3191), [α]$_D^{25}$ +78.3° (c 0.53, MeOH), was isolated from *Thalictrum faberi* Ulbr. as a yellow amorphous solid in 1983 [467]. The UV spectrum was characterized by maxima at 282 nm (log ε 4.30) and 310 nm (sh) (log ε 4.03) and the mass spectrum displayed prominent fragment ions at m/z 490 (M-220) and 220 (100%) (**249**) with the molecular ion at m/z 710. The ¹H NMR spectrum indicated the presence of two N-methyl groups as singlets at δ 2.35 [N(2′)] and 2.52 [N(6)]; six methoxy groups as five singlets at δ 3.64 [C(1)], 3.81 (6H), 3.91, 3.95, and 3.98; one methylenedioxy group as a singlet at δ 5.95; and six aromatic protons as two singlets at δ 5.78 [H(8′)] and 7.92 [H(11)] and an A$_2$B$_2$ quartet at δ 6.82 [d, H(11′) and H(13′)] and 7.01 [d, H(10′) and H(14′)]. These spectral data were suggestive of a benzylisoquinoline–aporphine dimer containing one methylenedioxy and one methoxy group in the isoquinoline portion of the benzylisoquinoline; five methoxy groups in a 1,2,3,9,10-pentaoxygenated aporphine molecule; and an AA′BB′ system in the benzyl portion of the

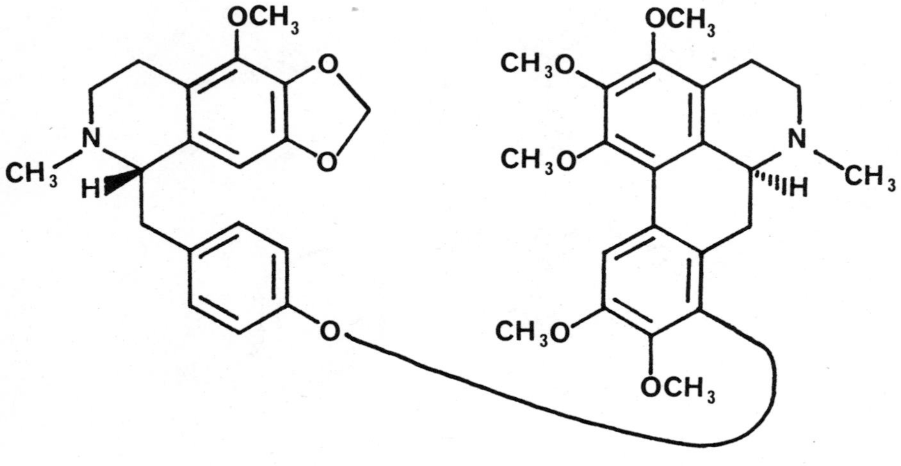

246

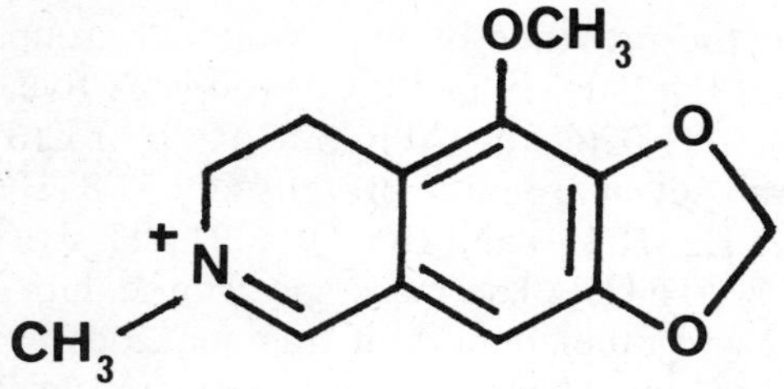

249

benzylisoquinoline moiety. Assignment of structure and stereochemistry of thalifabine (**246**) was by a comparison of data, including CD ([θ]$_{295}$ −27,720, [θ]$_{280}$ −37,181, and [θ]$_{242}$ +290,179) to that obtained for the alkaloid thalifaberine (**216**) [467]. Thalifabine (**246**) and thalifaberine (**216**) are unique in that they are the first benzylisoquinoline–aporphine dimers containing a 1,2,3,8,9,10-hexaoxygenated aporphine with a diphenyl ether linkage between the C(12′) of the benzylisoquinoline and the C(8) of the aporphine [437,467].

3.1.23. Thalifarapine. UV [471], ^{1}H NMR [471], MS [471], CD [471].
Thalifarapine (**250**), $C_{40}H_{46}N_2O_8$ (682.3254); [α]$_D^{24}$ +98.6° (c 0.422, MeOH), was isolated as an amorphous solid from *Thalictrum faberi* Ulbr. in 1984 [471]. The UV spectrum was characterized by maxima at 283 nm (log ε 4.33) and 310 nm (sh) (log ε 4.10) with a bathochromic shift in alkali to 293 nm (log ε 4.19), 320 nm (log ε 4.26), and 3.28 nm (log ε 4.26). The ^{1}H NMR

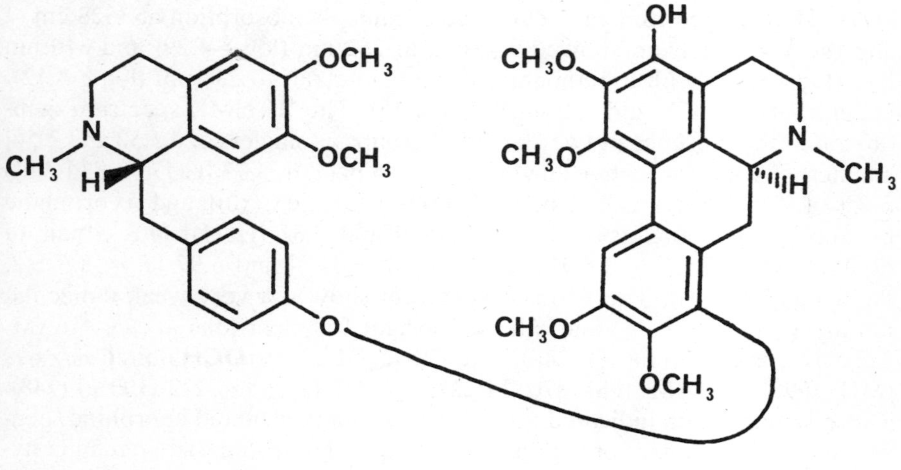

250

spectrum indicated the presence of two N-methyl groups as singlets at δ 2.30 [N(6)] and 2.50 [N(2')]; six methoxy groups as five singlets at δ 3.55 [C(7')], 3.76 [C(1)], 3.81 [C(6') or C(9)], 3.87 [C(9) or C(6')], and 3.94 (6H) [C(2) and C(10)]; and seven aromatic protons at δ 5.98 [H(8')], 6.51 [H(5')], 6.77 [2H, d, J = 8.5 Hz, H(11') and H(13')], 6.93 [2H, d, J = 8.5 Hz, H(10') and H(14')] and 7.84 [H(11)]. The mass spectrum exhibited a weak parent ion at M^+ m/z 682 and other important fragment ions at m/z 476 (M-206), 475, and 206 (100%) (**191**). These spectral data were typical of a 1,2,3,9,10-pentaoxygenated aporphine–benzylisoquinoline dimer containing four methoxy and one phenolic hydroxy groups in the aporphine ring and a dimethoxylated isoquinoline portion within the benzylisoquinoline ring. These data further indicated that thalifarapine was an O-desmethyl analog of thalifaberine (**216**), with the phenolic group being located at C(3) or C(9) because of the bathochromic and hyperchromic shift in alkali in the UV spectrum [71]. A comparison of the ^{1}H NMR spectrum with that of thalifasine (**251**) demonstrated the presence of all six methoxy groups with almost totally identical chemical shifts. This correlation, added to the biosynthetic precedent for C(3) hydroxylation in this system, suggested that thalifarapine be assigned as **250**. The CD spectrum of thalifarapine (**250**) [Δε/nm − 17,167 (216.6), +63,933 (243.0), −8400 (277.6), −7733 (304.8)] showed a close similarity to that of thalifaberine (**216**) and other alkaloids of this plant [471] plus other dimers of the thalicarpine (**192**) series [436,437], and the absolute configuration was thus assigned as (S,S) [471].

3.1.24. Thalifasine.

UV [471], IR [471], ^{1}H NMR [471], EI–MS [471], FD–MS [471], CD [471].

Thalifasine (**251**), $C_{40}H_{46}N_2O_9$ (698.3200), $[\alpha]_D^{25}$ +67.9° (c 0.80, MeOH), was isolated as an amorphous yellow solid from *Thalictrum faberi* Ulbr. in 1984 [471]. The IR spectrum gave evidence of phenolic absorption at 3528 cm^{-1} and the UV spectrum showed maxima at 282 nm (log ε 4.29) and 310 nm (sh) (log ε 4.05) with a bathochromic shift in alkali to 238 nm (log ε 4.12), 312 nm (log ε 4.17), and 330 nm (log ε 4.15). The ^{1}H NMR spectrum demonstrated the presence of two N-methyl groups as singlets of δ 2.30 and 2.50; six methoxy groups as five singlets at δ 3.55 [C(7')], 3.77 [C(1)], 3.83 [C(6') or C(9)], 3.88 [C(9) or C(6')], and 3.95 (6H) [C(2) and C(10)]; and six aromatic protons as two singlets at δ 5.67 [H(8')] and 7.87 [H(11)] plus a pair of doublets at 6.81 [J = 8.5 Hz, H(11') and H(13')] and 6.97 [J = 8.5 Hz, H(10') and H(14')]. The EI-mass spectrum showed a very weak molecular ion at m/z 698 (~0.1%) and other important fragment ions at m/z 476 (M-222), 475, and 222 (100%) (**248**). The FD–mass spectrum exhibited m/z 699 (MH) (69%), 698 (M) (68), 476 (M-222) (5), 475 (22), and 222 (100%) (**248**). These spectral data indicated a 1,2,3,9,10-pentaoxygenated aporphine–benzylisoquinoline dimer containing four methoxy groups and one phenolic hydroxy group in the aporphine moiety plus a dimethoxylated, monohydroxylated isoquinoline portion within the benzylisoquinoline monomer. The

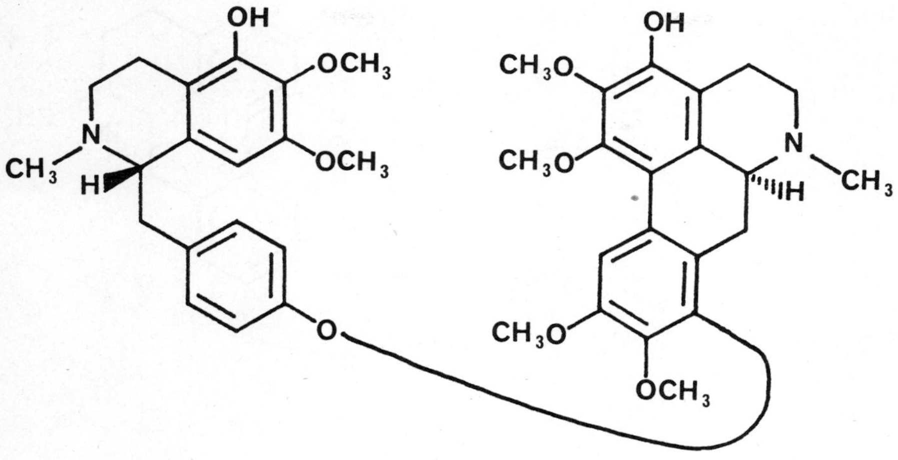

251

characteristic UV bathochromic plus hyperchromic shift in strong alkali suggested that the phenolic group in the aporphine half of the molecule should
be at the C(3) or C(9) position [71]. Preparation of the diacetyl ester (Ac₂O/
pyridine) and examination of the ¹H NMR spectrum revealed a downfield
shift of the H(8′) proton by 0.29 ppm to δ 5.96 but no downfield shift of the
H(11) proton (δ 7.83), thus indicating that one methoxy group rather than a
phenolic hydroxy group was attached to the C(9) position of the same aromatic ring as H(11). Hence the molecular structure of thalifasine (**251**) was
established and a consideration of its CD spectrum [Δε/nm −21,933 (216.4),
+61,167 (243.2), −7,833 (285.0), and 6,600 (298.4)] suggested the same absolute configuration (*S,S*) as thalifaberine (**216**) [471].

3.1.25. Thalilutidine. Tabular review [436, alkaloid No. 4], UV [436,517],
IR [436,517], ¹H NMR [436,517], MS [436,517], CD [436,517].
Thalilutidine (**252**), C₄₀H₄₆N₂O₈ (682.3254), [α]²⁰_D +74.2° (c 0.11, MeOH),
was isolated as an amorphous solid from *Thalictrum revolutum* DC. in 1977
[517]. The UV spectrum was characterized by maxima at 280 nm (log ε 4.38)
and 304 nm (sh) (log ε 4.21) with a bathochromic shift in base. The ¹H NMR
spectrum indicated the presence of two *N*-methyl groups as singlets at δ
2.47 and 2.48; six methoxy groups as five singlets at δ 3.55 [C(7′)], 3.70
[C(1)], 3.78 [C(12′) and C(13′)], 3.88 [C(2)], and 3.92 [C(10)]; and seven
aromatic protons at δ 6.13 [H(8′)], 8.17 [H(11)], and 6.5–6.7 (5H). A D₂O-
exchangeable hydroxy group was also observed at δ 5.1. The mass spectrum
showed the parent ion at M⁺ *m/z* 682 (1%), 490 (M-192) (6), 340 (8) (**206**),
324 (3) (**207**), and 192 (100) (**253**). These spectral data were indicative of a
1,2,9,10-tetraoxygenated aporphine–benzylisoquinoline dimer bearing one

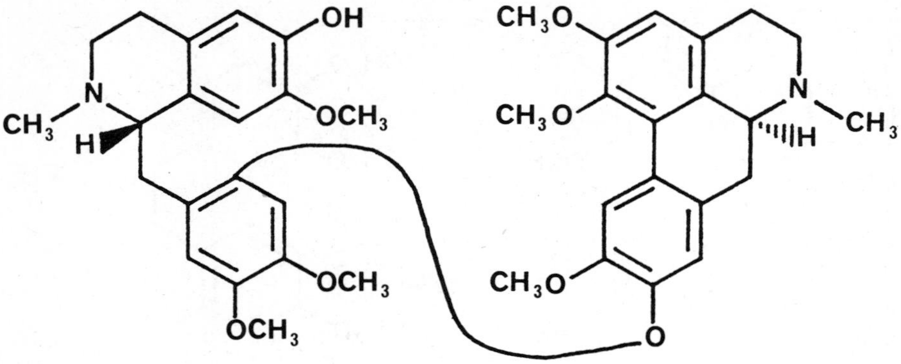

252

phenolic group and one methoxy group in the isoquinoline portion and two methoxy groups in the benzyl portion of the benzylisoquinoline monomer. Methylation of thalilutidine with ethereal CH_2N_2 afforded thalicarpine (**192**), thus establishing the skeletal structure, positions of oxygenation, and di-phenyl ether termini of thalilutidine. The presence of a characteristic high-field methoxy signal (δ 3.55) and high-field aromatic proton (δ 6.13) were characteristic of a C(7') methoxy and a H(8) proton in this dimeric system [436,437] and mandated the placement of the phenolic hydroxy group at C(6'). In addition, a direct comparison of thalilutidine (**252**) with thalmelatine (**204**) [the C(7') phenolic analogue of thalicarpine] demonstrated their dif-ference. The CD spectrum of thalilutidine (**252**) was reminiscent of other dimers with the (S,S) absolute configuration ($[\theta]_{238}$ + 180,000, $[\theta]_{274}$ − 16,100, and $[\theta]_{306}$ − 13,900) and the methylation product possessed the same CD curve as thalicarpine (**192**) (S,S).

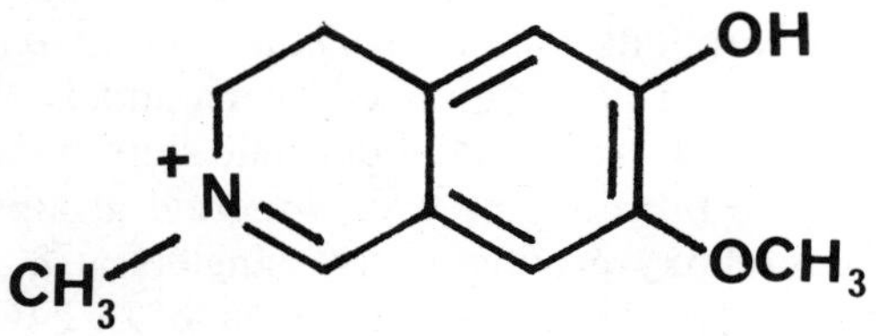

253

3.1.26.   Thalilutine.   Tabular review [436, alkaloid No. 13], UV [436,517], IR [436,517], ^{1}H NMR [436,517], MS [436,517], CD [436,517].

Thalilutine (**254**), $C_{41}H_{48}N_2O_9$ (712.3360), $[\alpha]_D^{20}$ +92° (c 0.175, MeOH), was isolated as an amorphous solid from *Thalictrum revolutum* DC. in 1977 [517]. The UV spectrum possessed maxima at 282 nm (log ϵ 4.40), 303 nm (sh) (log ϵ 4.28), and 312 nm (sh) (log ϵ 4.23) with a bathochromic shift in strong alkali. The ^{1}H NMR spectrum demonstrated the presence of two *N*-methyl groups as singlets at δ 2.48 and 2.49; seven methoxy groups as six singlets at δ 3.58 [C(7′)], 3.76 [C(1)], 3.78 (6H) [C(12′) and C(13′)], 3.83 [C(6′)], 3.94 [C(2)] and 3.98 [C(10)]; and six aromatic protons as singlets at δ 6.21 [H(8′)], 6.53, 6.56, 6.58, 6.63, and 8.00 [H(11)]. A D_2O-exchangeable hydroxy group was also observed at δ 5.65. The mass spectrum was characterized by a parent ion at M$^+$ *m/z* 712 (0.01%) and other important fragment ions at *m/z* 506 (0.4%) (M-206), 356 (0.5) (**255**), 340 (2.5) (**256**), and 206 (100) (**191**). These spectral data were suggestive of a 1,2,3,9,10-pentaoxygenated monophenolic aporphine–benzylisoquinoline dimer containing two methoxy groups each in the isoquinoline portion and the benzyl portion of the benzylisoquinoline monomeric half of the molecule. Methylation of thalilutine ([θ]$_{239}$ +198,000, [θ]$_{276}$ −19,300, and [θ]$_{307}$ −15,900) afforded adiantifoline (**190**), thus establishing the positions of oxygenation, positions of diphenyl ether termini, and absolute configuration of thalilutine (**254**). Of the four possible positions for the phenolic hydroxy group in the aporphine portion of thalilutine, the C(10) position was excluded because of the characteristic methoxy signal at δ 3.98 [436,437]. Furthermore, acetylation of thalilutine produced a monoacetate which was characterized by exhibiting chemical shifts for the aromatic protons with minimal differences from those in either thalilutine or adiantifoline

254

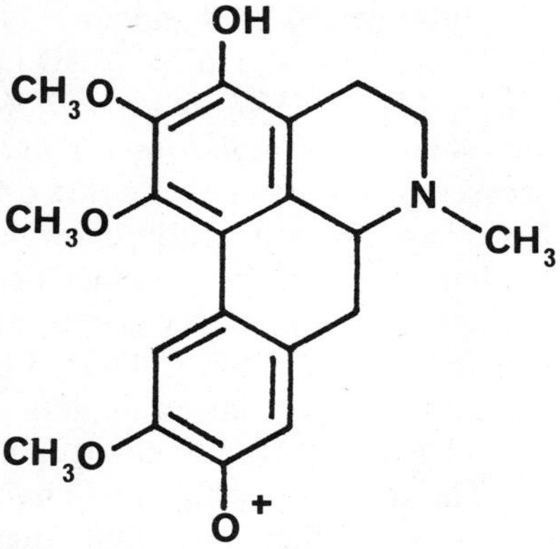

255

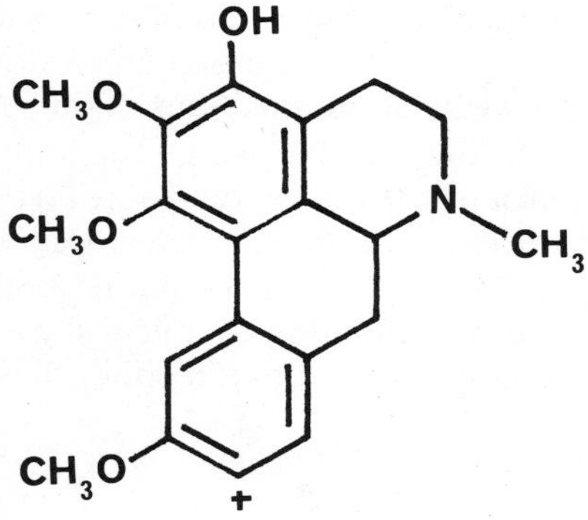

256

(190). Hence the C(1) position was excluded as the possible phenolic site because thalictropine (226) and thalictrogamine (242), both alkaloids of this class bearing a C(1) hydroxy, show large upfield shifts (0.58 ppm) of the H(11) proton on acetylation of a C(1) hydroxy [515]. This narrowed the choice to either the C(2) or the C(3) position and although a direct correlation could not be made, the C(3) position was chosen on biogenetic considerations. Hence thalilutine (254) was postulated to be 3-demethyladiantifoline [517].

3.1.27. Thalipine. Tabular review [436, alkaloid No. 5], UV [68,436,472,518], IR [68,472], ^{1}H NMR [68,436,472,518], MS [68,436,472,518], CD [68,436,472,518].

$$\underset{\sim}{\textbf{257}}$$

Thalipine (**257**), $C_{39}H_{44}N_2O_8$ (668.3097), $[\alpha]_D$ +59° (c 0.48; EtOH) [518], $[\alpha]_D^{25}$ +141° (c 0.19, MeOH) [472], was isolated as an amorphous solid from both *Thalictrum polygamum* Muhl. [518] and *T. revolutum* DC. [472] in 1977. The UV spectrum displayed maxima at 282 nm (log ε 4.49), 303 nm (sh) (log ε 4.34), and 316 nm (sh) (log ε 4.19) with a hyperchromic shift in strong alkali. The ^1H NMR spectrum indicated the presence of two N-methyl groups as singlets at δ 2.34 and 2.47; five methoxy groups as four singlets at δ 3.68 [C(1)], 3.79 (6H) [C(6') and C(12')], 3.88 [C(2)], and 3.96 [C(10)]; and seven aromatic protons as five singlets at δ 6.44 [H(8')], 6.49 (2H), 6.59 (2H), 6.81 [H(11')], and 8.17 [H(11)]. The mass spectrum displayed a very weak parent ion at M$^+$ *m/z* 668 (0.9%) with other important fragment ions at *m/z* 476 (3%) (M-192), 340 (38) (**206**), 324 (10) (**207**), and 192 (100) (**199**). These spectral data were characteristic of a 1,2,9,10-tetraoxygenated aporphine–benzylisoquinoline dimer containing one phenolic hydroxy group and one methoxy group in both the isoquinoline portion and the benzyl portion of the benzylisoquinoline monomer. The absence of a high-field methoxy signal (around δ 3.58) coupled with the presence of the H(8') proton at a downfield position (δ 6.35) suggested the placement of one phenolic group at C(7') [436,437]. Complete methylation of thalipine with CH_2N_2 produced thalicarpine (**192**), thereby establishing the positions of oxygenation, termini of the diphenyl ether linkage, and stereochemistry of the asymmetric centers (*S,S*) [472]. Partial methylation of thalipine with CH_2N_2 produced thalmelatine (**204**) and pennsylvanine (**228**) [472], thus firmly establishing the structure of thalipine (**257**). In a similar manner, short-term methylation of thalipine with CH_2N_2 produced (+)-thalmelatine (**204**) and (+)-thalicarpine (**192**), and a controlled acetylation afforded both a mono- and a diacetate. Treatment of the crude mixture of these acetyl esters with CH_2N_2 yielded thalipine diacetate and (+)-pennsylvanine acetate, further confirming the structural assignment of thalipine (**257**) [518].

Thalipine was found to exhibit *in vitro* antimicrobial activity against *Mycobacterium smegmatis* at a concentration of 100 μg/mL [68].

Thalipine was detected in *Thalictrum minus* and quantitatively determined by HPLC [66]. The alkaloid was also reisolated from *Thalictrum revolutum* DC. [68,105].

3.1.28. Thalirevoline.   Tabular review [436, alkaloid No. 21], UV [68,436,517], IR [68,436,517], ^{1}H NMR [68,436,517], MS [68,436,517], CD [68,436,517].

Thalirevoline (**258**), $C_{40}H_{46}N_2O_8$ (682.3254), m.p. 123–125° (Et$_2$O), $[\alpha]_D^{20}$ +95° (c 0.1, MeOH), was isolated as colorless crystals from *Thalictrum revolutum* DC. in 1977 [517]. The alkaloid exhibited UV maxima at 270 nm (sh) (log ε 4.31), 280 nm (log ε 4.40), 301 nm (log ε 4.24), and 310 nm (sh) (log ε 4.18) with a bathochromic and hyperchromic shift under alkaline conditions. The ^{1}H NMR spectrum was characterized by the presence of two *N*-methyl groups at δ 2.37 and 2.43; six methoxy groups at δ 3.56 [C(7′)], 3.67 [C(1)], 3.80, 3.87 (6H), and 3.96 [C(10)]; and seven aromatic protons at 6.14 [H(8′)], 6.46, 6.52, 6.57, 8.17 [H(11)], and 6.56 plus 6.68 (AB quartet, J = 8.5 Hz) [C(13) and C(14′)/vice versa]. The mass spectrum exhibited a weak parent ion at M$^+$ *m/z* 682 (0.2%) with other important fragment ions at *m/z* 476 (4%) (M-206), 340 (13) (**206**), 324 (5) (**207**), and 206 (100%) (**191**). These spectral data indicated a 1,2,9,10-tetraoxygenated aporphine–benzylisoquinoline dimer bearing two methoxy groups in the isoquinoline portion plus one methoxy group and one phenolic hydroxy group in the benzyl portion of the benzylisoquinoline monomer. Furthermore, the AB quartet suggested that thalirevoline was a fetidine (**208**)-like analog [436]. Methylation of thalirevoline with CH$_2$N$_2$ afforded thalirevolutine (**230**), thus confirming the positions of oxygenation, termini of the diphenyl ether bridge,

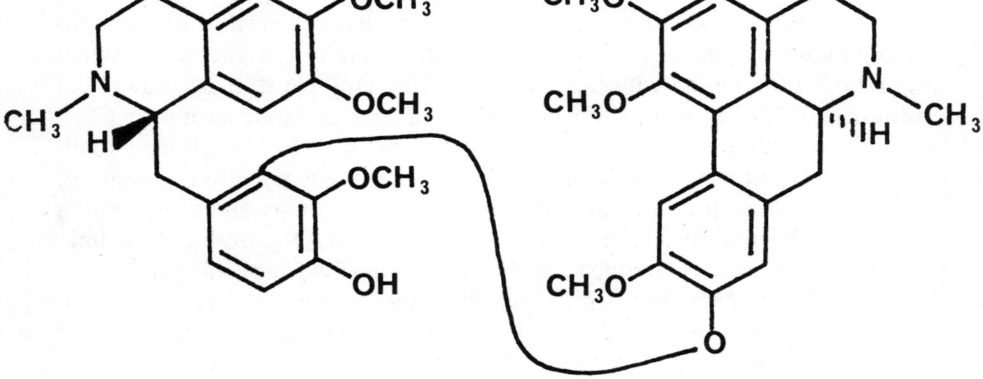

258

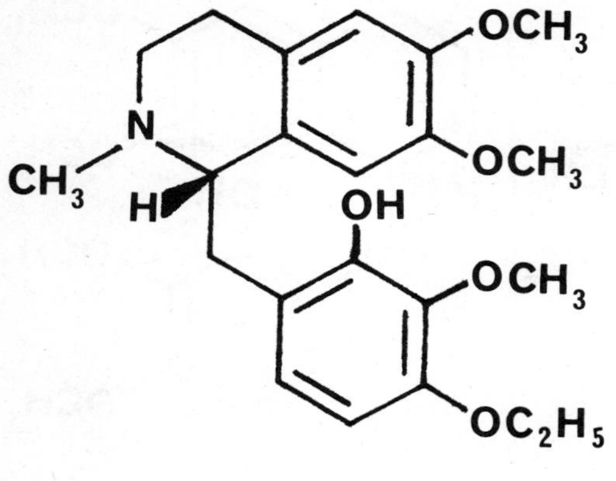

$$\underset{\sim}{\underline{259}}$$

and stereochemistry (*S*,*S*) of thalirevoline (CD [θ]₂₄₀ + 122,000, [θ]₂₇₇ − 12,800, and [θ]₃₀₀ − 11,700). Treatment of thalirevoline with diazoethane afforded the *O*-ethyl ether, which upon reductive cleavage with sodium in liquid ammonia gave 1-(2-hydroxy-3-methoxy-4-ethoxybenzyl)-2-methyl-6,7-dimethoxytetrahydroisoquinoline (**259**) plus 2,10-dimethoxyaporphine (**236**) [the hindered C(1) methoxy being lost in the cleavage]. Placement of the phenolic hydroxy group of thalirevoline at C(12′) was directed by a negative Gibbs test. Hence thalirevoline (**258**) was assigned as C(12′) de-methylthalirevolutine [517].

Thalirevoline (**258**) was shown to possess hypotensive activity in nor-motensive rabbits and to inhibit the in vitro growth of *Mycobacterium smeg-matis* [517].

Thalirevoline (**258**) was also reisolated from *Thalictrum revolutum* DC. [68,105] but has not been reported in any other *Thalictrum* species to date.

3.1.29. Thalirevolutine (*O-Methylfetidine*). Tabular review [436, alkaloid No. 22], UV [436,517], ¹H NMR [436,517], MS [436,517], CD [436,517].

Thalirevolutine (**230**), $C_{41}H_{48}N_2O_8$ (696.3410), m.p. 105–108° (Et₂O), $[\alpha]_D^{20}$ + 134° (c 0.1, MeOH), was isolated from *Thalictrum revolutum* DC. in 1977 [517]. The UV spectrum was characterized by maxima at 270 nm (sh) (log ε 4.31), 280 nm (log ε 4.38), 302 nm (log ε 4.21), and 315 nm (sh) (log ε 4.10) with no shift in either acidic or basic medium. The ¹H NMR spectrum indicated the presence of two *N*-methyl groups as singlets at δ 2.36 and 2.43; seven methoxy groups as six singlets at δ 3.59 [C(7′)], 3.66 [C(1)], 3.78, 3.80, 3.87 (6H), and 3.96 [C(10)]; and seven aromatic protons as six singlets at δ 6.18 [C(8′)], 6.42, 6.52, 6.57, 6.75 (2H), and 8.16 [H(11)]. The mass spectrum was characterized by a weak parent ion at M⁺ *m/z* 696 (1%) with other significant fragment ions at *m/z* 490 (7%) (M-206), 355 (12) (M-340-H), 340

$$\underline{\mathbf{260}}$$

(10) (**206**), 324 (7) (**207**), and 206 (100) (**191**). These spectral data were consistent for a 1,2,9,10-tetraoxygenated aporphine–benzylisoquinoline dimer containing four methoxy groups in the benzylisoquinoline portion, two of these groups in the isoquinoline half and two in the benzyl half. Oxidation of thalirevolutine with potassium permanganate in acetone afforded *N*-methylcorydaldine (**88**), thereby establishing the 6,7-dimethoxy oxygenation of the tetrahydroisoquinoline portion of the molecule. Reductive cleavage of thalirevolutine with sodium in liquid ammonia gave 1-(2-hydroxy-3,4-dimethoxybenzyl)-2-methyl-6,7-dimethoxytetrahydroisoquinoline (**260**) and (+)-*S*-2,10-dimethoxyaporphine (**236**), the latter of which was also obtained via a similar cleavage of thalicarpine (**192**) [473,474] and originates from the 1,2,9,10-tetraoxygenated aprophine half of the dimer. Finally, a direct comparison of thalirevolutine (CD $[\theta]_{240}$ +240,000, $[\theta]_{277}$ −26,000, and $[\theta]_{300}$ −20,200) with *O*-methylfetidine [prepared via diazomethylation of fetidine (**208**)] demonstrated their identity and confirmed the structure and stereochemisty of thalirevolutine (**230**) [517].

Thalirevolutine (**230**) was found to produce a hypotensive response in normotensive rabbits [517].

3.1.30.   Thalmelatidine.   Tabular review [436, alkaloid No. 18], UV [146], IR [146,270,436], ¹H NMR [146,270,436], MS [146].

Thalmelatidine (**101**), $C_{42}H_{48}N_2O_{10}$ (740.3309), m.p. 120–122° (abs EtOH), $[\alpha]_D^{19}$ +47° (c 1.0, CHCl₃), was first isolated from *Thalictrum minus* L. var. *elatum* Jacq. in 1970 [519]. The ¹H NMR spectrum suggested the presence of two *N*-methyl groups as singlets at δ 2.45 and 2.50; seven methoxy groups as five singlets at δ 3.62, 3.78, 3.80 (6H), 3.90, and 3.96 (6H); one methylenedioxy group at δ 5.94 and five aromatic protons as four singlets at 5.86 [H(8′)], 6.41 (2H), 6.50, and 8.04 [H(11)] [270]. Oxidation of thalmelatidine with potassium permanganate in acetone afforded what was then assigned as thalflavine (1-oxo-2-methyl-5-methoxy-6,7-methylenedioxy-1,2,3,4-tetrahydroisoquinoline) (**100a**) plus aldehyde **193**, the same aldehyde obtained on similar oxidation of adiantifoline (**190**) [270,443], thus apparently establishing the structure of thalmelatidine (**101a**) [270]. A recent communication has resulted in the structural revision of thalmelatidine as **101** [519a]. This revision was based on the results of extensive ¹H NMR studies, particularly double-irradiation techniques and NOE experiments. The oxidation product, thalflavine, must thus be assigned as 1-oxo-2-methyl-5,6-methylenedioxy-7-methoxy-1,2,3,4-tetrahydroisoquinoline (**100**). The (*S,S*) configuration was assigned to thalmelatidine by a comparison of the sign and magnitude of its specific rotation with those of other alkaloids of this series [270].

Thalmelatidine was reisolated from *T. minus* L. var. *elatum* Jacq. (more properly called *T. minus* L. ssp. *majus*) in 1984 [299]. The alkaloid has also been isolated from *T. minus* var. *microphyllum* Boiss., from which both its UV spectrum (λ_{max} 234, 282, 303, 315 nm) and its mass spectrum [M⁺ *m/z* 740 (<1%), 520 (3), 370 (1), 354 (1), and 220 (100)] were reported in the literature for the first time [146].

3.1.31. Thalmelatine. Tabular review [436, alkaloid No. 6], UV [72,101], IR [101], ¹H NMR [72,101,104,516], MS [72,520].

Thalmelatine (**204**), $C_{40}H_{46}N_2O_8$ (682.3254), m.p. 131–135° (Et₂O) [513], 110–112° (Et₂O) [101,104], 120–123° (abs. EtOH) [513]; $[\alpha]_D^{21}$ +110° (c 1.0, EtOH) [513], $[\alpha]_D^{25}$ +110° (c 0.4 MeOH) [72], was first isolated from *Thalictrum minus* var. *elatum* Jacq. in 1964 [513,521]. Methylation with CH₂N₂ afforded thalicarpine (**192**), thereby establishing the skeletal structure and positions of oxygenation of thalmelatine [513]. Ethylation with diazoethane yielded *O*-ethylthalmelatine (**261**) with upon reductive cleavage with sodium in liquid ammonia gave 2,10-dimethoxyaporphine (**236**) as one of the prod-

261

$$262$$

ucts. This product was also obtained on reductive cleavage of thalicarpine [473,474]. Oxidation of thalicarpine with potassium permanganate in acetone produced *N*-methylcorydaldine (**88**) and similar oxidation of *O*-ethylthalmelatine gave 1-oxo-2-methyl-6-methoxy-7-ethoxytetrahydroisoquinoline (**202**). In addition, the second product of each oxidation, although not fully characterized, was identical and was an optically inactive, yellow, crystalline substance, m.p. 153–155°. The structure of thalmelatine as a 7'-demethyl-thalicarpine was assigned based on these findings [513]. However, because thalicarpine had been mistakenly assigned as **235** at that time [owing to the placement of the terminus of the diphenyl ether at the C(11) instead of the C(9) position in the aporphine ring], thalmelatine was likewise incorrectly assigned as **262** [513]. Revision of the structure of thalicarpine to **192** in 1965 naturally led to the correction of thalmelatine as **204** [475]. A complete study of the oxidation products with potassium permanganate in acetone of thalicarpine (**192**) and *O*-ethylthalmelatine (**261**), appeared shortly thereafter and identified the aldehyde **238** (later named 6a,7-dehydrohernandaline) as a common oxidative product of both alkaloids in addition to *N*-methylcorydaldine (**88**) and 1-oxo-2-methyl-6-methoxy-7-ethoxytetrahydroisoquinoline (**202**), respectively [479]. Reduction of **238** with Adams catalyst and hydrogen afforded a tetrahydro derivative, alcohol **240**, which, upon reductive cleavage with sodium in liquid ammonia, gave 2,10-dimethoxyaporphine (**236**) [479]. A series of quarternary salts and methines of thalmelatine (**204**) were subsequently prepared and their physical properties noted [481]. The mass spectral fragmentation of thalmelatine was studied in detail and reported in 1970. In general, the mass spectrum was very similar to that of thalicarpine and was characterized by a very weak parent ion followed by facile cleavage of the double benzylic bond of the benzylisoquinoline portion to yield a stable ion at *m/z* 192 (100%) (**199**) plus various smaller aporphine-derived fragment ions [520]. Ultimately, a more complete description of the spectral properties of thalmelatine appeared in the literature. The UV spectrum of the alkaloid was typical of this group with two maxima at 287 nm (log ϵ 4.18) and 302 nm (log ϵ 4.05) [101]. The ^{1}H NMR spectrum showed the N-methyl groups as two singlets δ 2.42 and 2.48; the methoxy groups as four singlets at δ 3.72 [C(1)], 3.79 (9H) [C(6'), C(12'), and C(13')], 3.88 [C(2)], and 3.95 [C(10)]; and the aromatic protons as six singlets at 6.43 [H(8')], 6.52, 6.55, 6.60 (2H), 6.68 [H(11')], and 8.18 [H(11)] [516].

Thalmelatine (**204**) and O-ethylthalmelatine (**261**) were both found to possess cytotoxic activity as measured via inhibition of the growth of monolayer KB cell cultures in a concentration of 100 µg/mL [262]. In addition, the alkaloid inhibited the growth of *Mycobacterium smegmatis* in a standardized in vitro test [101].

Thalmelatine (**204**) has been isolated from several other *Thalictrum* species including *T. dioicum* L. [72], *T. minus* L. [66 (HPLC quantitation), 510] and *T. revolutum* DC. [101,104,105].

3.1.32. Thalmineline. Tabular review [436, alkaloid No. 17], UV [436,522], IR [436,522], ^{1}H NMR [436,522], MS [436,522]. Thalmineline (**263**), $C_{42}H_{50}N_2O_{10}$ (742.3465), m.p. 96–98° (Et$_2$O–heptane), 108–110° (EtOH); [α]$_D^{24}$ +22° (c 0.9, MeOH) was isolated from *Thalictrum minus* L. var. *elatum* Jacq. in 1970 [522]. The UV spectrum of the alkaloid was reported as possessing a single maximum at 283 nm (log ϵ 5.46). The absence of a second peak or shoulder at approximately 300 nm and the extraordinarily high extinction coefficient are not characteristic of alkaloids of this class. The ^{1}H NMR spectrum indicated the presence of two N-methyl groups as singlets at δ 2.45 and 2.50; eight methoxy groups as six singlets at δ 3.47 [C(7')], 3.70 [C(1)], 3.73, 3.76, 3.83, and 3.89 (but with no mention

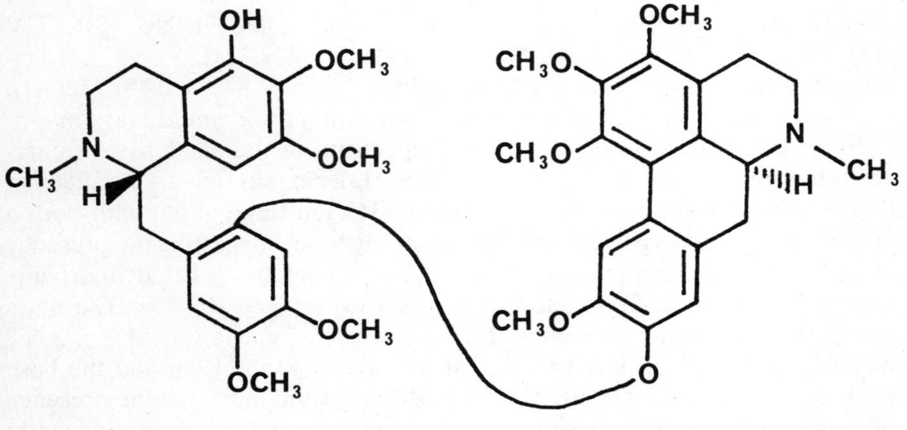

263

of which signals represented six protons/two methoxy groups); and five aromatic protons as singlets at δ 5.71 [H(8')], 6.43, 6.48, 6.66 [H(8)] and 7.98 [H(11)]. The mass spectrum was characterized by a weak molecular ion at M⁺ *m/z* 742 (less than 1%), an intense ion that is presumably the base beak at *m/z* 222 (100%) (**248**) and a cluster of fragment ions at *m/z* 519, 520, and 521, the last three of which were also observed in the mass spectrum of adiantifoline (**190**) [443]. The spectral data were indicative of an aporphine–benzylisoquinoline alkaloid of the adiantifoline (**190**) type, with one extra hydroxy group in the isoquinoline portion of the benzylisoquinoline ring. Thalmineline exhibited a positive test with ferric chloride and coupled with diazotized *p*-nitroaniline, the latter of which is indicative of a phenol with an unsubstituted position ortho or para to the phenolic hydroxy group. The presence of a high-field aromatic proton at δ 5.71 indicates a H(8') proton and thus suggests 5',6',7' trioxygenation in the remainder of the ring. Thalicarpine (**192**), adiantifoline (**190**), *O*-desmethyladiantifoline (**200**), thalilutine (**254**), pennsylvananine (**223**), thalictropine (**226**), thalilutidine (**252**), thalidoxine (**227**), and pennsylvamine (**228**) all are alkaloids of thalicarpine (**192**) series which possess a C(7') methoxy group whose chemical shift is between δ 3.55 and 3.60 [436]. If it can be assumed that the high-field methoxy signal of thalmineline at δ 3.47 could be assigned to the C(7') position, this would leave C(5') or C(6') as the position for the hydroxy group. On the basis of the coupling reaction with diazotized *p*-nitroaniline, location of the phenolic hydroxy group at C(5') was the more likely possibility, with the resultant assignment of structure of thalmineline (**263**). The stereochemistry of the alkaloid was probably assigned via comparison with other alkaloids in the series, because no CD spectrum was determined [522].

Thalmineline (**263**) was subsequently isolated from one other *Thalictrum* species, *T. minus* L. [523].

3.1.33. Uskudaramine.

Tabular review [437, alkaloid No. 53], UV [437,524], ¹H NMR [437,524], MS [437,524], CD [437,524].
Uskudaramine (**264**), $C_{39}H_{44}N_2O_8$ (668.3097), $[\alpha]_D^{25}$ +84° (c 0.15, MeOH), was isolated as an amorphous solid from *Thalictrum minus* var. *microphyllum* Boiss. in 1982 [524]. The UV spectrum of the alkaloid was characterized by maxima at 209 nm (log ε 4.78), 221 nm (sh) (log ε 4.73), 286 nm (log ε 4.35), 300 nm (sh) (log ε 4.18), and 312 nm (log ε 4.06) with both a bathochromic and hyperchromic shift in strong base, suggesting the presence of a 1,2,9,10-tetraoxygenated or a 1,2,3,9,10-pentaoxygenated aporphine moiety with a phenolic function at either C(3) or C(9) [15–17]. The mass spectrum showed the presence of a very weak parent ion at M⁺ *m/z* 668 (0.1%), another small fragment ion at *m/z* 476 (3%) (M-192), and the base peak at *m/z* 192 (100%) (**199**). The ¹H NMR spectrum indicated the presence of two *N*-methyl groups as singlets at 2.34 [N(2')] and 2.57 [N(6)]; five methoxy groups as four singlets at δ 3.73 [C(1)], 3.78 [C(6')], 3.90 [C(3)], and 3.97 (6H) [C(2) and C(10)]; and six aromatic protons as three singlets at δ

264

6.28 [H(8')], 6.53 [H(5')], and 7.99 [H(11)] plus one doublet at δ 6.78 (*J* = 2.4 Hz) [H(10')], one doublet at 6.89 (*J* = 8.5 Hz) [H(13')], and one doublet of doublets at 7.26 (*J* = 2.4, 8.5 Hz) [H(14')]. Proof of the presence of a C(9) phenolic hydroxy group was obtained via acetylation (Ac$_2$O/pyridine) of uskudaramine to yield a triacetyl ester whose ^{1}H NMR spectrum showed a downfield shift of the H(11) proton from δ 7.99 to δ 8.11. A detailed ^{1}H NMR NOE study subsequently established the structure of uskudaramine (**264**). Significant ^{1}H NMR NOE findings included the irradiation of the H(11) proton (δ 7.99) with the subsequent 2% NOE on the C(1) methoxy (δ 3.73) and 4% NOE on the C(10) methoxy (δ 3.97). Separate irradiations of all the other aromatic protons confirmed their presence in the benzylisoquinoline moiety. Furthermore, irradiation of the H(8') proton (δ 6.28) produced a 1% dipole–dipole relaxation enhancement of the H(10') proton (δ 6.78) and a 2% enhancement of the H(1') proton (δ 3.67), and irradiation of the H(10') proton caused a 1% enhancement of the H(8') proton (δ 6.28) [524]. The CD spectrum of uskudaramine (**264**) (Δε(nm) −32(212), +48(244), −3.9(280), −4.8(297) [524]) with its Cotton effect maximum at 244 nm plus a negative trough at 212 nm is very similar to istanbulamine (**217**) and generally resembles alkaloids of the (+)-thalicarpine (**192**) (*S,S*) series [436,437].

Uskudaramine (**264**) is the first aporphine–benzylisoquinoline dimer to be joined via a carbon-to-carbon biphenyl linkage rather than the usual biphenyl ether linkage [436,437]. (+)-Uskudaramine (**264**), (+)-istanbulamine (**217**), (+)-thalicarpine (**192**), (+)-adiantifoline (**190**), (+)-*N*-2'-noradiantifoline (**219**), and (+)-thaliadanine (**203**) are all found in *T. minus* var. *microphyllum* Boiss. and all bear a C(10) methoxy group. These alkaloids are biosynthesized by a direct oxidative coupling of a totally formed 1,2,9,10-tetraoxygenated or 1,2,3,9,10-pentaoxygenated aporphine derived from (+)-

reticuline with an appropriate benzyltetrahydroisoquinoline [(+)-*N*-methylcoclaurine for (+)-uskudaramine (**264**) and (+)-istanbulamine (**217**); (+)-reticuline for (+)-thalicarpine (**192**), (+)-adiantifoline (**190**), (+)-*N*-2'-noradiantifoline (**219**), and (+)-thaliadanine (**203**)]. The process is in contrast to the biosynthesis or the dimeric aporphine–benzylisoquinoline alkaloids of the Berberidaceae, which are formed from a proaporphine–benzylisoquinoline intermediate that in turn was derived from the condensation of two *N*-methylcoclaurine momomers [436,437,524].

3.2. Aporphine–Benzyltetrahydroisoquinoline Dimeric Alkaloid Subgroups

An inspection of the aporphine–benzyltetrahydroisoquinoline dimeric alkaloids of the genus *Thalictrum* demonstrates the following:

1. These dimers may belong to three different structural subgroups, represented by general formulas A, B, and C (Table 1).

Table 1.   Aporphine–Benzyltetrahydroisoquinoline Dimeric Alkaloidal Subgroups (R Substituents May Be H, OH, or OCH₃)

Subgroup A

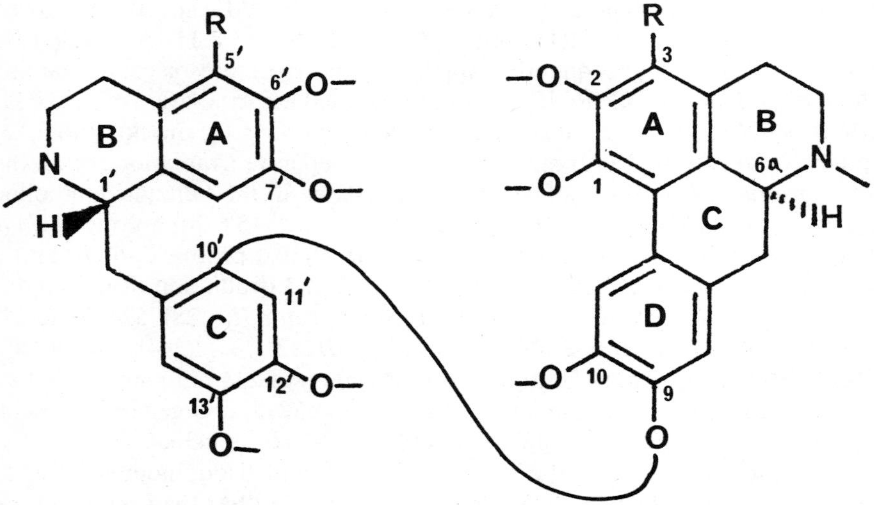

Adiantifoline, bursanine, *O*-desmethyladiantifoline, *O*-desmethylthalicarpine, 1,12'-*O,O*-didemethylthalicarpine, *N*-2'-noradiantifoline, *N*-2'-northalicarpine, pennsylvanamine, pennsylvanine, thaliadanine, thalicarpine, thalictrogramine, thalictropine, thalidoxine, thalilutidine, thalilutine, thalipine, thalmelatidine, thalmelatine, thalmineline

The specific rotations of the alkaloids in this subgroup range from +18 to +135°.

Table 1. (*continued*)

Subgroup B

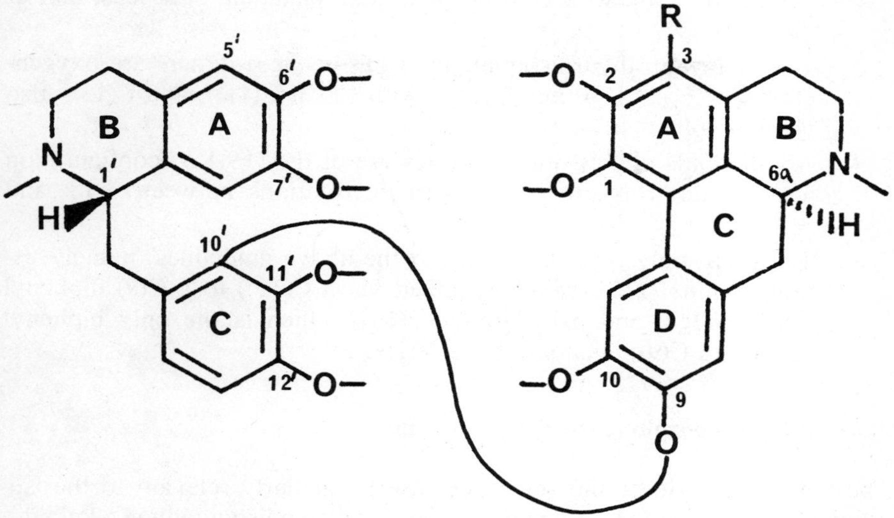

Faberidine, faberonine, fetidine, huangshanine, iznikine, revolutopine, thalirevoline, thalirevolutine
The specific rotations of the alkaloids in this subgroup range from $+76$ to $+229°$.

Subgroup C

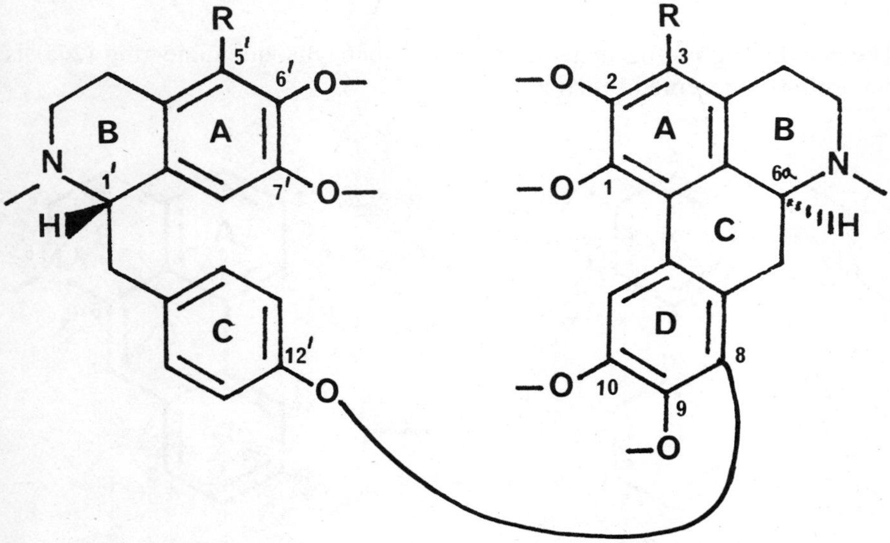

Thalifabatine, thalifaberine, thalifabine, thalifarapine, thalifasine
The specific rotations of the alkaloids in this subgroup range from $+61$ to $+99°$.

2. The aporphine rings of these dimers are oxygenated at C(1), C(2), C(9), and C(10), with C(3) being an additional site in at least half of the alkaloids.

3. The benzyltetrahydroisoquinoline rings of these dimers are oxygenated at C(6'), C(7'), and C(12'), with C(5'), C(11'), and C(13') also being involved sites.

4. All alkaloids of this dimeric series are of the (1*S*,1'*S*) configuration and are dextrorotatory with specific rotations between + 18° and + 229°.

5. There are only two exceptions to the above guidelines, namely, istanbulamine (**217**), which is linked via a C(11') to a C(9) diphenyl ether bridge, and uskudaramine (**264**) which is the only biphenyl [C(11') to C(9)] alkaloid of the series.

3.3. Dehydroaporphine–Benzylisoquinolines

The following reviews and references are particularly relevant to the isolation and identification of dehydroaprophine–benzylisoquinoline alkaloids:

1. Tabular reviews [10,18,19,436,437]
2. Annual reviews [22,24–30,194]
3. Other reviews [10–14]
4. Book chapters [14,438,441]

The numbering of the dehydroaporphine–benzylisoquinoline ring (**265**) is according to accepted practice [436].

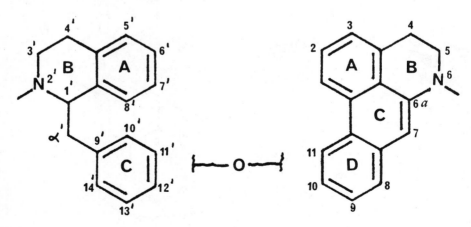

265

3.3.1. Dehydrohuangshanine.   UV [471], ¹H NMR [471], EI–MS [471], CI–MS [471], CD [471].

Dehydrohuangshanine (**266**), $C_{42}H_{48}N_2O_9$ (724.3360), $[\alpha]_D^{23}$ +42.2° (c 0.17, MeOH), was isolated as an amorphous yellow solid from *Thalictrum faberi* Ulbr. in 1984 [471]. The UV spectrum showed maxima at 257 nm (log ε 4.54), 267 nm (sh) (log ε 4.51), 275 nm (sh) (log ε 4.49), and 332 nm (log ε 3.85) and was characteristic of an alkaloid containing a 6a,7-dehydroaporphine moiety [15–17]. The mass spectrum was characterized by a weak molecular ion at M^+ m/z 724 (0.2%) with other diagnostic fragment ions at m/z 519 (4%) (M-206+H), 518 (2) (M-206), 517 (4) (M-206−H), and 206 (100) (**191**) and was suggestive of a dehydro derivative of huangshanine (**212**). The ¹H NMR spectrum indicated the presence of two *N*-methyl groups as singlets at δ 2.39 [N(2′)] and 2.95 [N(6)]; eight methoxy groups as singlets at δ 3.55 [C(7′)], 3.78, 3.79, 3.90, 3.93, 3.99, 4.06, and 4.09; and seven aromatic protons as five singlets at δ 6.06 [H(8′)], 6.38, 6.49, 6.72, and 9.10 [H(11)] plus an AB quartet at δ 6.75 and 6.82 (J = 8.5 Hz) [H(13′) and H(14′)]. These spectral data were strongly suggestive of fetidine (**208**)-type dehydroaporphine–benzylisoquinoline dimeric alkaloid, particularly the low-field *N*-methyl signal (δ 2.95), the very-low-field H(11) proton (δ 9.10), and the AB quartet for ortho-coupled protons [H(13′) and H(14′)] [436]. The structure of dehydrohuangshanine (**266**) was assigned via a consideration of these spectral data plus a consideration of its CD curve [Δε(nm) −21.3(202), +18.4(213), +7.04(233), and +1.6 (290)] in relation to dehydrothalifaberine (**267**), an alkaloid of this series with the (*S*) configuration at the C(1′) asymmetric center [471].

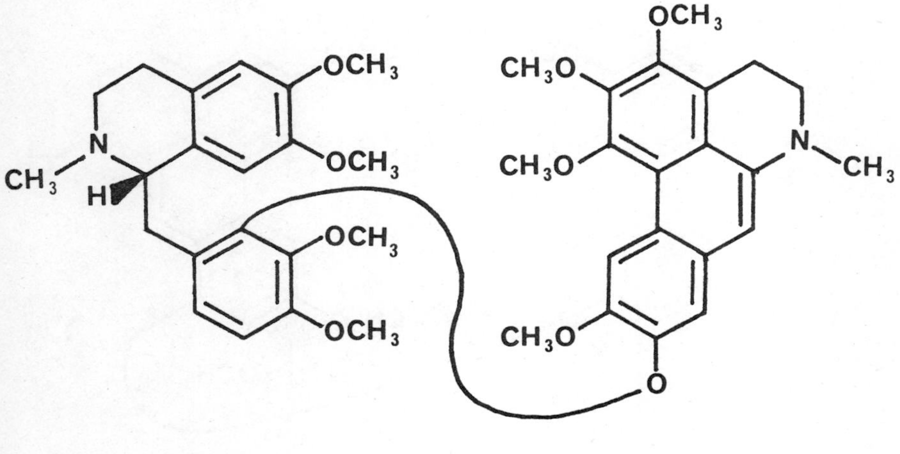

266

3.3.2. Dehydrothalifaberine.   UV [471], ^{1}H NMR [471], MS [471], CD [471].

Dehydrothalifaberine (267), $C_{41}H_{46}N_2O_8$ (694.3254), $[\alpha]_D^{24}$ +95.9° (c 0.143, MeOH) was isolated as a yellow amorphous solid from *Thalictrum faberi* Ulbr. in 1984 [471]. The UV spectrum was characterized by three maxima at 256 nm (log ε 4.56), 272 nm (log ε 4.56), and 332 nm (log ε 4.06), which were characteristic of a 6a,7-dehydroaporphine chromophore [436]. The mass spectrum showed a very small parent ion at M^+ *m/z* 694 (0.2%) with other important fragment ions at *m/z* 488 (M-206), 487, and 206 (100%) (191). The ^{1}H NMR spectrum indicated the presence of two *N*-methyl groups as singlets at δ 2.57 [N(2′)] and 2.92 [N(6)]; seven methoxy groups as seven singlets at δ 3.46 [C(7′)], 3.85, 3.94, 3.99, 4.04, 4.07, and 4.11; and seven aromatic protons as four singlets at δ 5.97 [H(8′)], 6.59 [H(5′)], 6.80 [H(7)], and 9.05 [H(11)] plus one AA′BB′ quartet at δ 6.90 and 6.99 (*J* = 8.5 Hz) [H(11′), H(13′), H(10′), H(14′)]. These spectral data were characteristic of a dehydroaporphine–benzylisoquinoline alkaloid of the thalifaberine (216)-type. Of particular significance were the low-field signals at δ 9.05 [H(11)] and δ 2.92 [N(6)] and the aromatic proton at δ 6.80 [H(7)] [436]. Confirmation of the structure was achieved via DDQ oxidation of thalifaberine (216) to afford dehydrothalifaberine (267), which was identical with the natural product. The CD spectrum of dehydrothalifaberine (267) [Δε(nm) −26(196), +12,733(212.6), +13,000(231.4), +4,689(290.0), +0.840(334.4), and +0.170(385.0)] indicated the (S) configuration at the sole asymmetric center [C(1′)] of the alkaloid. Finally, the premise that dehydrothalifaberine (267) was possibly of artifactual origin was dispelled when an ethanolic solution of thalifaberine (216) was stirred at room temperature for 4 days and then refluxed for 24 hr with no resultant oxidation [471].

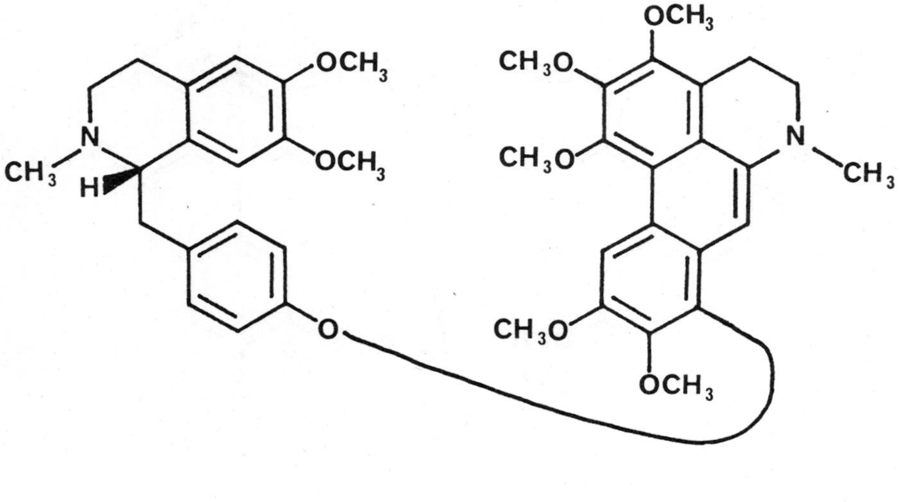

267

3.3.3. Dehydrothalicarpine.

Tabular review [436, alkaloid No. 12], UV [282,436,525], IR [282,436], ^{1}H NMR [282,436,525], MS [436,520].

Dehydrothalicarpine (**268**), $C_{41}H_{46}N_2O_8$ (694.3254), m.p. 180–182° (MeOH–Et$_2$O) [525], 182–183° (EtOAc) [282], $[\alpha]_D^{22}$ +54° (cl, CHCl$_3$) [525], was first isolated from *Thalictrum minus* L. var. *elatum* Jacq. in 1966 [525]. The UV spectrum of the alkaloid showed two maxima at 268 nm (log ε 4.82) and 331 (4.34). The ^{1}H NMR spectrum was highly informative and indicated the presence of two *N*-methyl groups as singlets at δ 2.49 [N(2')] and 3.02 [N(6)]; seven methoxy groups as six singlets at δ 3.49 [C(7')], 3.72, 3.74 (6H), 3.88, 3.96, and 3.99; and eight aromatic protons as seven singlets at δ 6.27 [H(8'), 6.41, 6.52, 6.77 (2H), 6.88, 7.04, and 9.38 [H(11)]. Both the UV and the ^{1}H NMR spectrum were similar to those of thalicarpine (**192**) but indicated a higher degree of aromatization and suggested the presence of a phenanthrene moiety within dehydrothalicarpine. Hydrogenation of dehydrothalicarpine in acetic acid resulted in the consumption of 1 mole of hydrogen and afforded thalicarpine (**192**) [525]. Furthermore, reductive cleavage of dehydrothali-carpine with sodium in liquid ammonia afforded (−)-6'-hydroxylaudanosine (**194**) and (+)-2,10-dimethoxyaporphine (**236**) [525], identical products obtained by like cleavage of thalicarpine (**192**) [473,474]. Although dehydroth-alicarpine can be generated via aerial oxidation of benzene, methanol, or chloroform solutions of thalicarpine over long periods of time at room temperature, the alkaloid is isolable from a freshly isolated alkaloid mixture obtained under mild conditions [525].

Dehydrothalicarpine (**268**) was found to lack cytotoxicity when evaluated against monolayer KB cell cultures [262].

Dehydrothalicarpine was subsequently isolated from *Thalictrum dasycarpum* Fisch. and Lall. in 1960 [282]. Catalytic hydrogenation (HOAc/PtO$_2$) of dehydrothalicarpine afforded thalicarpine (**192**) and oxi-

268

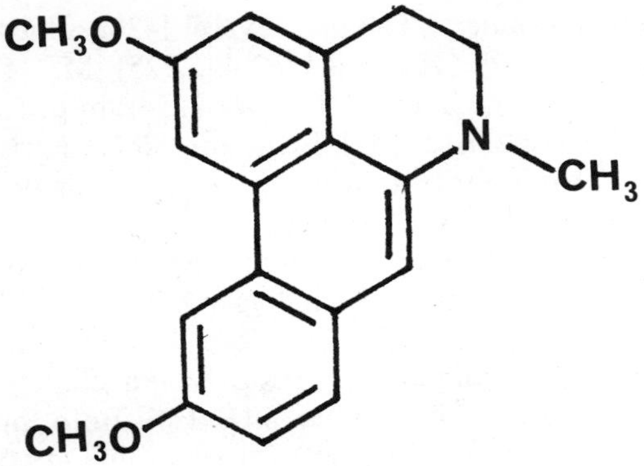

269

dation of thalicarpine with DDQ-produced dehydrothalicarpine (**268**) [282]. Finally, reductive cleavage sith sodium in liquid ammonia afforded 6′-hydroxylaudanosine (**194**), 2,10-dimethoxydehydroaporphine (**269**), 2,10-dimethoxydibenzo[de,g]quinolin-7-one (**270**), and 1,2,10-trimethoxydibenzo[de,g]quinolin-7-one (**271**), the latter two products being formed via oxidation of the appropriate dehydroaporphine precursors during work-up [282].

Some 12 years later, the structure of cleavage product **271** was revised to 2,10-dimethoxy-4,5-dioxodehydroaporphine (**272**) [526]. Dehydrothalicarpine may also be prepared via oxidation (I_2/NaOAc) of thalicarpine (**192**) in refluxing dioxan [527].

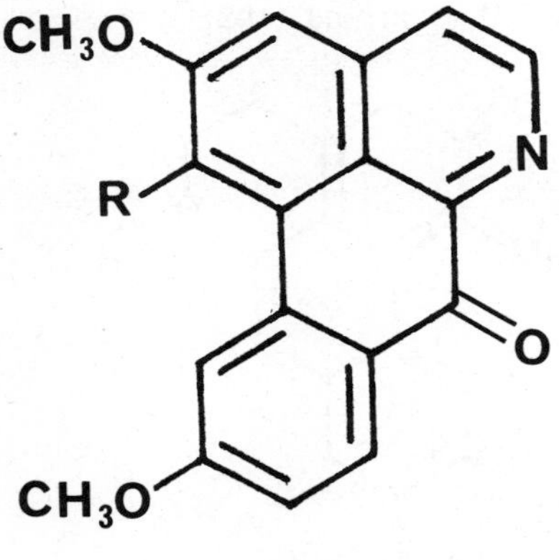

270 R=H
271 R=OCH₃

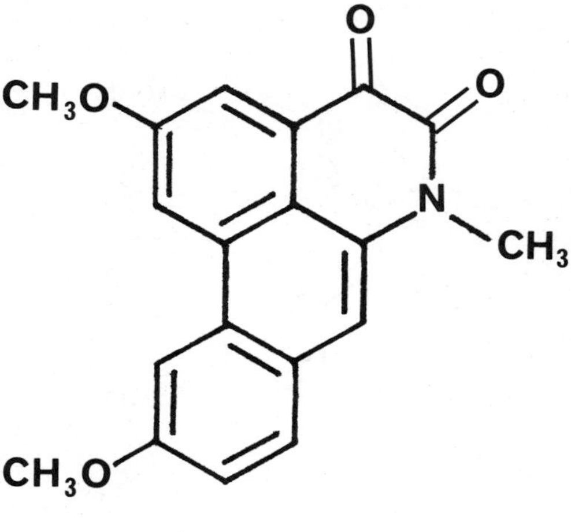

272

3.4. Aporphine–Pavines

The following reviews and references are particularly relevant to the isolation and identification of aporphine–pavine dimeric alkaloids:

1. Tabular reviews [436]
2. Annual reviews [25]
3. Other reviews [13]
4. Book chapters [439,528]

The numbering of the aporphine–pavine ring (**273**) is according to accepted practice [436].

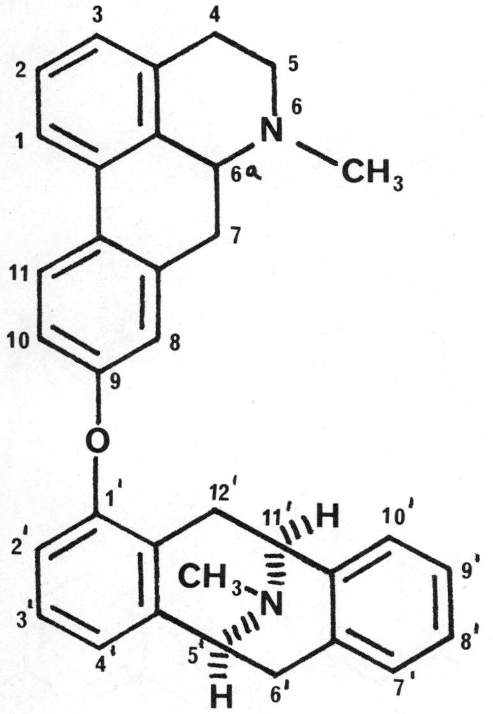

273

3.4.1. Pennsylpavine. Tabular review [436, alkaloid No. 27], UV [436,470], ^{1}H NMR [436,470], MS [436,470], CD [470].

Pennsylpavine (**274**), $C_{40}H_{44}N_2O_8$ (680.3097), m.p. 122–123° (Et$_2$O), $[\alpha]_D^{25}$ $-174°$ (c 0.6, MeOH), was isolated from *Thalictrum polygamum* Muhl. in 1974 [470]. The UV spectrum showed maxima at 230 nm (log ε 4.62), 280 nm (sh) (log ε 4.38), 288 nm (log ε 4.40), 308 nm (sh) (log ε 4.23), and 320 nm (sh) (log ε 4.15) and was characteristic of a 1,2,9,10-tetraoxygenated aporphine superimposed on that of a pavine or isopavine (288 nm) [15–17,237]. The mass spectrum showed the parent ion at M$^+$ m/z 680 (36%) with other important fragment ions at m/z 529 (22%) (**275**), 475 (5) (**276**), 355 (28) (**277**), 340 (20) (**206**), and 204 (100) (**81**). The spectrum was indicative of a trimethoxylated aporphine-trimethoxylated, monophenolic pavine joined through a diaryl ether oxygen with fragment ions **277** (m/z 355) and **206** (m/z 340) corresponding to cleavages on either side of the oxygen bridge. The fragment ion **275** (m/z 529) corresponds to cleavage through the pavine ring with retention of the pavine nigrogen, whereas fragment ions **276** (m/z 475) and **81** (m/z 204) indicated a like cleavage but loss of the pavine nitrogen

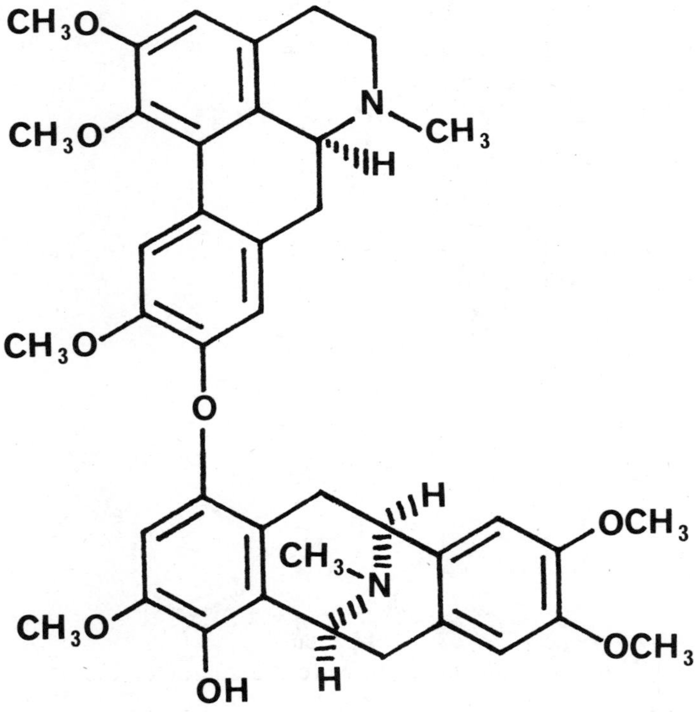

274

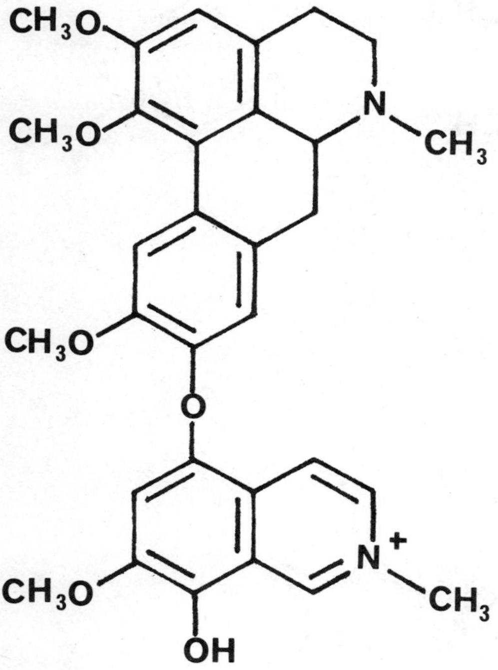

275

as a separate fragment ion. Less intense fragment ions were found at m/z 649, 648, and 637 for losses of CH_3NH_2, $CH_3\overset{+}{N}H_3$, and $CH_3\overset{+}{=}\overset{+}{N}CH_2$, respectively, from the parent ion. Because the mass spectra of both aporphines and isopavines are characterized by losses of $CH_3\overset{+}{N}=CH_2$ from the parent ion but not by losses of either CH_3NH_2 or $CH_3\overset{+}{N}H_3$, the fragment ions at m/z 649 and 648 suggested the presence of a pavine moiety within the dimeric structure of pennsylpavine. The ^1H NMR spectrum indicated the presence of two N-methyl groups as singlets at δ 2.50 [N(6)] and 2.57 [N′(CH₃)]; six methoxy groups as five singlets at δ 3.71 [C(1)], 3.76 (6H), 3.78, 3.88, and 3.91 [C(10)]; and six aromatic protons as five singlets at 6.23 [H(9′)], 6.48 (2H), 6.52, 6.60, and 8.15 [H(11)]. Methine protons were observed as double doublets at δ 4.06 [H(12′)] (J = 6 Hz) and 4.50 [H(6′)] (J = 6 Hz). The distinctive features of this spectrum were the characteristic methoxy resonances [C(1) and C(10)] and the low-field aromatic proton [H(11)] for the aporphine moiety plus the pair of one proton doublets [H(6′)] and [H(12′)], characteristic of the bridgehead protons of the unsymmetrically substituted pavines munitagine (**278**) and platycerine (**279**). Acetylation of pennsylpavine (**274**) afforded a monoacetate, m.p. 203–204° (Et₂O), in whose ^1H NMR

276

277

$$\underset{\sim}{278} \quad R_1 = R_2 = H$$

$$\underset{\sim}{279} \quad R_1 = H, R_2 = CH_3$$

spectrum was observed a downfield shift (δ 6.23 to δ 6.43) of only one aromatic proton signal [H(9′)] which was indicative of a meta orientation of that proton to the acetyl ester function. In addition, the position of the methine proton [H(6′)] (δ 4.50) shifted upfield to δ 4.35, indicating the proximal relationship of the phenolic hydroxy group of pennsylpavine (274) and the bridgehead proton [H(6′)]. This relationship was also noted between platycerine (279) and its acetate. Thus the molecular structure of pennsylpavine (274) was established. The absolute configuration was determined by the aromatic chirality method, with the strong positive extremum at 242 nm and the strong negative extremum at 209 nm being associated with the ¹B transitions of the aporphine and pavine moieties, respectively, which led to the ultimate assignment of configuration [470]. The concurrent isolation of the aporphine–benzylisoquinoline alkaloid pennsylvanine suggests that pennsylvanine (228) may indeed be the biogenetic precursor of pennsylpavine (274).

3.4.2. Pennsylpavoline.

Tabular review [436, alkaloid No. 28], UV [436,470], ¹H NMR [436,470], MS [436,470], CD [436,470].

Pennsylpavoline (280), $C_{39}H_{42}N_2O_8$ (666.2941), m.p. 145–146° (Et$_2$O), $[\alpha]_D^{25}$ −245° (c 0.66, MeOH) was isolated, along with pennsylpavine (274), from *Thalictrum polygamum* Muhl. in 1974 [470]. The UV spectrum was characterized by maxima at 230 nm (log ε 4.47), 280 nm (sh) (log ε 4.06), 288 nm (log ε 4.13), 306 nm (sh) (log ε 4.01), and 320 nm (sh) (log ε 3.96) and was very similar to that of its companion alkaloid pennsylpavine (274) [470]. The mass spectrum was also very similar to that of pennsylpavine (274) but with certain key fragment ions being 14 a.m.u. lower. The parent ion was found at M⁺ m/z 666 (10%) with other important fragment ions being found at m/z 515 (14%) (275a), 461 (18) (276a), 355 (25) (277), 326 (22) (224), and 204 (100) (81). A similar interpretation of the mehanism of fragmentation

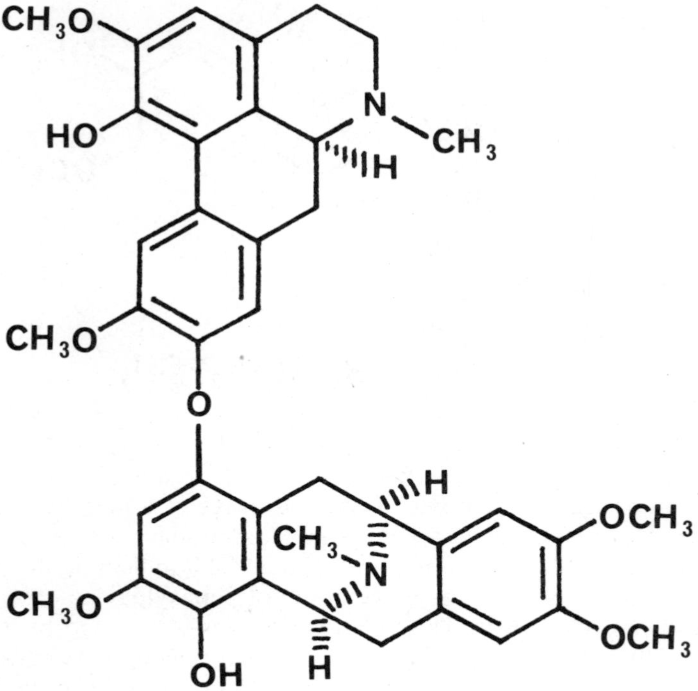

$$\underset{\sim}{\underline{\textbf{280}}}$$

and the nature of these fragment ions as compared to pennsylpavine (**274**) led to the speculation that pennsylpavoline was a desmethylpennsylpavine, with the absence of this methyl group from the aporphine ring. The ^{1}H NMR spectra of pennsylpavine (**274**) and pennsylpavoline (**280**) were very similar to the spectrum of the latter, indicating the presence of two *N*-methyl groups as singlets at δ 2.48 [N(6)] and 2.55 (N'—CH$_3$); five methoxy groups as three singlets at δ 3.75, 3.78 (6H), and 3.91 (6H); and six aromatic protons as five singlets at δ 6.26 [H(9')], 6.45, 6.49, 6.55 (2H), and 8.14 [H(11)]. Methine protons were observed as double doublets at δ 4.01 (J = 6 Hz) [H(12')] and δ 4.43 (J = 6 Hz) [H(6')]. The major difference in the ^{1}H NMR spectrum of pennsylpavine (**274**) and pennsylpavoline (**280**) was the absence of the relatively high-field methoxy singlet at about δ 3.7 in the spectrum of the latter, thus suggesting the absence of a C(1) methoxy group. Confirmation of the structure of pennsylpavoline (**280**) was obtained via acetylation of the alkaloid to a diacetate, m.p. 188–189° (Et$_2$O), whose ^{1}H NMR spectrum showed an upfield shift of the C–(6) bridgehead proton doublet (δ 4.43 to δ

275 a

276 a

4.21) and a similar upfield shift of the H(11) proton (δ 8.14 to δ 7.60) due to the presence of the acetoxy function at [C(1)]. The CD curve of pennsyl-pavoline (**280**) was superimposable with that of pennsylpavine (**274**), which permitted the assignment of absolute configuration of the former, as the aromatic chirality method had been utilized for the latter [470]. Pennsyl-vanamine (**223**), an aporphine–benzylisoquinoline alkaloid also isolated from *Thalictrum polygamum* [469,470], could well assume the role of biogenetic precursor of pennsylpavoline (**280**) [470]. Pennsylpavoline has been isolated only from *T. polygamum* Muhl. to date and this plant is the only natural source of aporphine–pavine dimers at this time.

3.5. Bisbenzylisoquinolines

The following reviews and references are particularly relevant to the iso-lation and identification of bisbenzylisoquinoline alkaloids.

1. Tabular reviews [10,18,19,529,530]
2. Annual reviews [20–22,24,194–201]
3. Other reviews [10–14]
4. Book chapters [14,531–537]

The numbering of the bisbenzylisoquinoline ring (**281**) is according to ac-cepted practice [529].

3.5.1. Aromoline (Thalicrine).   Tabular review [529, alkaloid No. 31, 530], UV [538], IR [538], ^{1}H NMR [171,538], MS [538], CD [538].

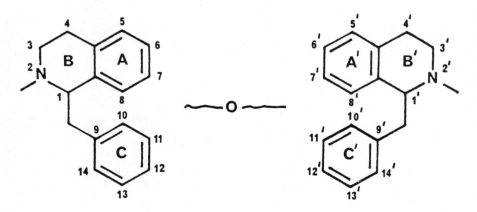

281

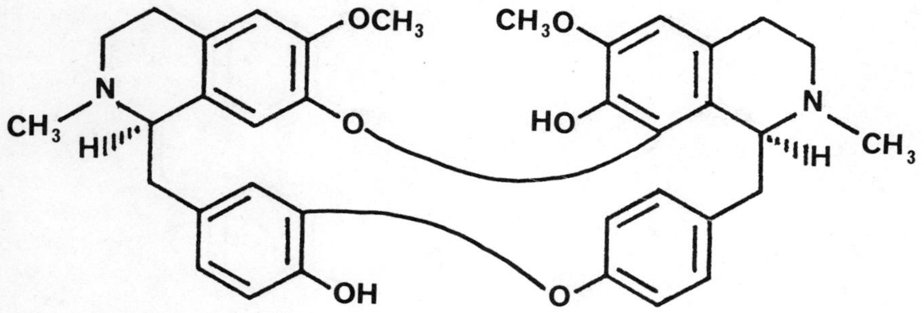

282

Aromoline (**282**), $C_{36}H_{38}N_2O_6$ (594.2730), m.p. 178–180° (Et$_2$O) [538], $[\alpha]_D^{25}$ +318° (c 0.06) [538], was first isolated from *Daphnandra aromatica* (Monimiaceae) in 1949 [539].

The alkaloid was found to inhibit the growth of *Mycobacterium smegmatis* at a concentration of 1000 μg/ml in a routine in vitro analysis [538].

The phenolic alkaloid thalicrine was isolated from *Thalictrum thunbergii* DC. in 1962 and assigned the dimeric structure **283** on the basis of oxidative and reductive cleavage studies [540–542]. However, in 1964, a detailed structural reinvestigation resulted in the revision of its structure to that previously established for aromoline (**282**) [543]. Hence the name thalicrine was dropped and aromoline was thus first isolated from a *Thalictrum* species, *T. thunbergii* DC., in 1962 [540–542]. Aromoline was subsequently isolated from *Thalictrum lucidum* L. [538], *T. minus* L. [181], *T. minus* L. var. *microphyllum* Boiss. [544], and *T. rugosum* Ait. [234].

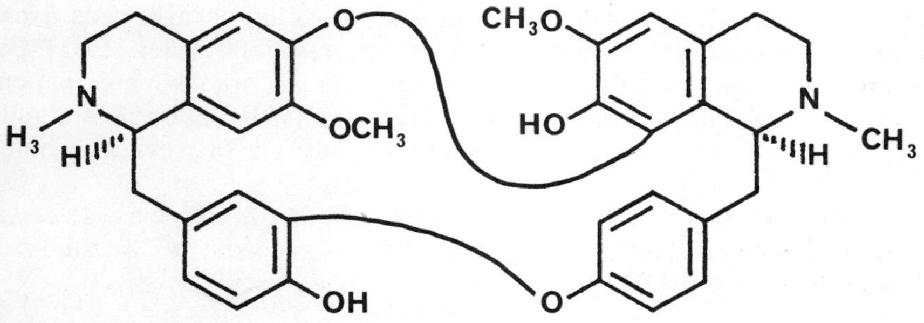

283

284

3.5.2. Berbamine.

Tabular review [529, alkaloid No. 57, 530], UV [529], ^{1}H NMR [171,529], ^{13}C NMR [530,545], MS [529,546], ORD [529,547], fluorescence spectra [530,548], phosphorescence spectrum [530,548].

Berbamine (**284**), $C_{37}H_{40}N_2O_6$ (608.2886), m.p. 161–163° [549], $[\alpha]_D^{21}$ +109° (c 1.0, CHCl₃) [549], was first isolated from several *Berberis* species (Berberidaceae) in the last decade of the nineteenth century [531].

Berbamine methiodide exhibited a curariform effect in rabbits as determined by the head-drop test [550]. Berbamine was found to exhibit a weak in vitro cytotoxicity against HeLa–S₃ cells (ED₅₀ of >10 µg/mL). In addition, the alkaloid did not inhibit the in vivo growth of Ehrlich ascites tumor or Sarcoma-180 solid tumor [551]. The LD₅₀ of berbamine in mice was determined to be 75 mg/kg and the alkaloid was not observed to produce hemolysis of rabbit erythrocytes at a dose of 100 µg/mL [551]. Berbamine exhibited weak in vitro antimicrobial effects against the Gram-positive organisms *Staphylococcus aureus* (250 µg/mL), *S. epidermidis* (125 µg/mL), *Sarcina lutea* (500 µg/mL), *Bacillus subtilis* (500 µg/mL), and *B. anthracis* (500 µg/mL), and weak to negligible effects against the Gram-positive rods *Escherichia coli* (1000 µg/mL) and *Klebsiella pneumoniae* (500 µg/mL) [551]. Berbamine was found to be an inducer of leukocytosis in leukopenic patients [552]. The effect of berbamine on blood pressure [553] and its general pharmacology and toxicology [554] were the subject of several Chinese publications.

Berbamine (**284**) was first isolated from a *Thalictrum* species, *T. pedunculatum* Edgew. in 1964 [352]. The alkaloid was subsequently isolated from *T. foetidum* L. in 1967 [549].

3.5.3. *N*-Desmethylthalidasine.

Tabular review [530, alkaloid No. 196], UV [349,530], IR [349], ^{1}H NMR [349,530], MS [349,530].

N-Desmethylthalidasine (**285**), $C_{38}H_{42}N_2O_7$ (638.2992), m.p. 137–139° [349]; $[\alpha]_D$ −86.9° (c 0.414, MeOH) [349], was isolated as a yellow, amorphous

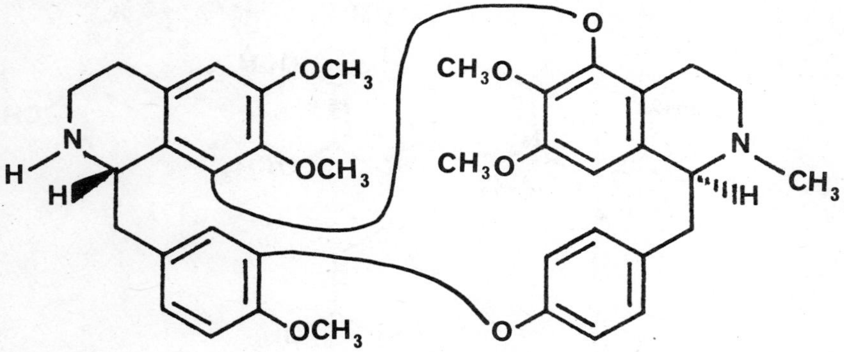

285

powder from *Thalictrum faberi* Ulbr. in 1980 [555]. The UV spectrum showed two maxima at 275 nm (log ϵ 3.78) and 283 nm (log ϵ 3.78) and the ^{1}H NMR spectrum indicated the presence of one N-methyl group as a singlet at δ 2.62; five methoxy groups as singlets at δ 3.25, 3.46, 3.75, 3.86, and 3.89; and nine aromatic protons as a multiplet from δ 6.20 to 7.75 [349]. The mass spectrum showed the molecular ion at M$^+$ *m/z* 638 (638.292) with other important fragment ions at *m/z* 623, 607, 411, 397, 380, 365, 206.5, 206, 204, 183.5, and 183 [349]. Treatment of the alkaloid with formaldehyde–sodium borohydride afforded thalidasine (**286**), thus establishing the skeletal structure, positions of oxygenation, and termini of the two diphenyl ether bridges. Reductive cleavage of N-desmethylthalidasine (**285**) with sodium in liquid

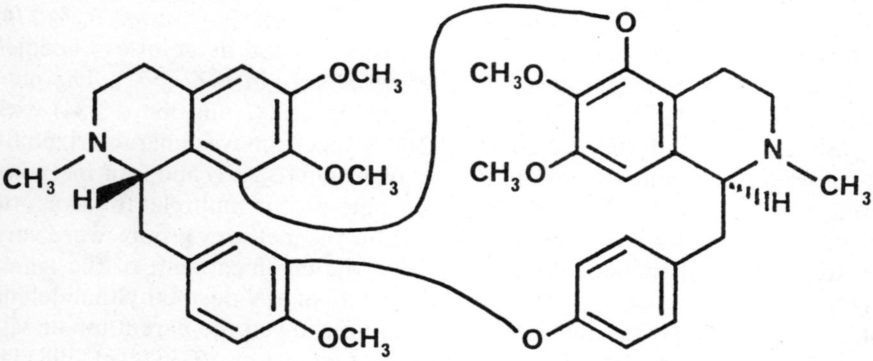

286

287

ammonia produced 2-methyl-5,4'-dihydroxy-6,7-dimethoxy-1-benzyltetrah-ydroisoquinoline (**287**), thus confirming the position of the single *N*-methyl function of *N*-desmethylthalidasine in the right-hand ring and establishing its structure as **285** [349]. The absolute configuration of *N*-desmethylthali-dasine (**285**) as (*S*,*S*) was determined by a comparison of its ORD curve with that of thalidasine (**286**) and by a similar comparison of the ORD curve of its *N*-methylation product with that of thalidasine [349].

N-Desmethylthalidasine was reported to possess anticancer activity in various animals [349].

3.5.4. *N*-Desmethylthalidezine. Tabular review [529, alkaloid No. 80], UV [301,529], IR [301], [superscript]1H NMR [301,529], MS [301,529], CD [301,529].

N-Desmethylthalidezine (**288**), $C_{37}H_{40}N_2O_7$ (624.2836), m.p. 173–174° (MeOH), $[\alpha]_D^{25}$ +280° (C 0.14, MeOH), was isolated as colorless needles from an extract of *Thalictrum podocarpum* Humb. in 1977 [301]. The phe-nolic alkaloid exhibited a single UV maximum at 282 nm (log ϵ 3.84) with a bathochromic shift in alkali. The [superscript]1H NMR spectrum was characterized by five three-proton singlets for one *N*-methyl group (δ 2.61) and four methoxy groups (δ 3.28, 3.34, 3.75, and 3.92) and a nine-proton multiplet for aromatic protons (δ 5.97–7.42). The chemical shifts for the methoxy groups were very close to those of thalidezine (**289**), as was the chemical shift of the single *N*-methyl group, thus suggesting the possibility of a *N*-desmethylthalidezine structure. The high-resolution mass spectrum showed the parent ion at M[superscript]+ m/z 624 (19%) with other significant fragment ions at m/z 623 (13%), 398 (11) (**290**), 208 (9) (**291**), 199 (398/2) (21) (**292**), and 192 (100) (**199**). The fragment ions at m/z 398 (**290**) and 199 (**292**) represented the facile double benzylic cleavage of dimeric alkaloids of this type and support a thalidezine-like struc-ture. Treatment of the alkaloid with formaldehyde–sodium borohydride af-

288

289

290

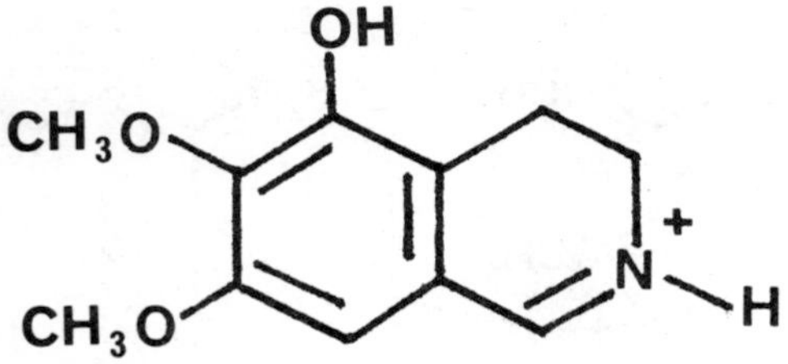

291

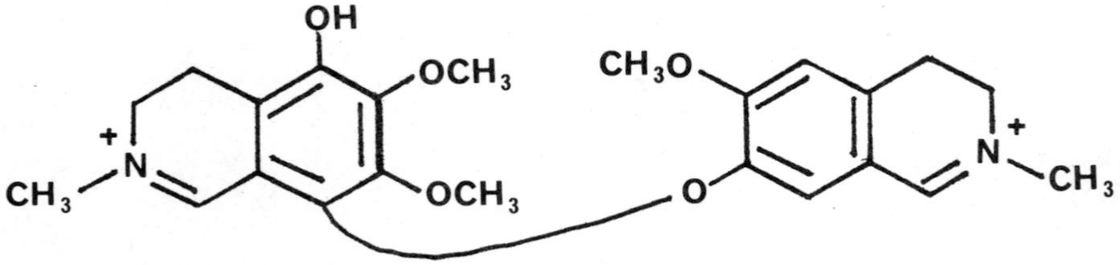

292

forded thalidezine (**289**), thus confirming the skeletal structure, positions of oxygenation, termini of the diphenyl ether linkages, and stereochemistry of the asymmetric centers. The presence of fragment ions at m/z 208 (**291**) and 192 (**199**) served to locate the presence of the secondary nitrogen atom in the trioxygenated isoquinoline ring and thus established the structure of *N*-desmethythalidezine (**288**) [301].

$\quad$ *N*-Desmethylthalidezine (**288**) was found to inhibit the in vitro growth of *Staphylococcus aureus* and *Candida albicans* at a concentration of 1000 μg/ mL plus *Mycobacterium smegmatis* at 100 μg/mL [301]. In addition, the alkaloid produced a hypotensive response in normotensive dogs and rabbits [301].

3.5.5. *N*-Desmethylthalistyline.
Tabular review [529, alkaloid No. 16, 530], UV [137, 529], ^{1}H NMR [137,529], MS [137,529], CD [137,529]. *N*-Desmethylthalistyline (**293**), $C_{40}H_{46}N_2O_8$ (682.3254), $[\alpha]_D^{25}$ +151° (c 0.2, MeOH) [137], (CD $[\theta]_{226}$ +87,800 and $[\theta]_{284}$ +6,820) [137], was first isolated as an amorphous solid from *Thalictrum longistylum* DC. [137] and *T. podocarpum* Humb. [301] in 1976 [556]. The alkaloid was characterized by a single UV maximum at 282 nm (log ε 3.81) [137]. The ^{1}H NMR spectrum indicated the presence of two *N*-methyl groups as singlets at δ 2.47 and 2.50; five methoxy groups as four singlets at δ 3.60, 3.63, 3.77, and 3.82 (6H); one methylenedioxy group as a singlet at δ 5.90; and nine aromatic protons as

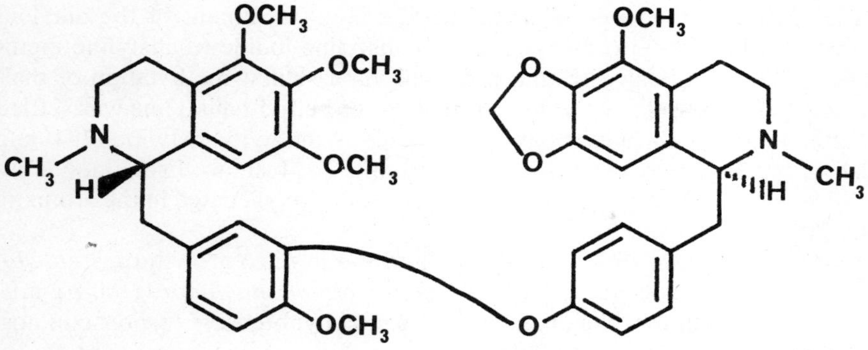

293

two singlets at δ 5.75 (1H) [H(8′)] and 5.95 (1H) [H(8)] plus a multiplet at δ 6.51–7.10 (7H) [137]. The mass spectrum was characterized by a very-low-intensity parent ion at M⁺ m/z 682 (0.6%) with other significant fragment ions at m/z 236 (94%) (**66**), 221 (17) (**66**—CH₃), 220 (100) (**249**), 206 (14) (**66**—2CH₃), 205 (6) (**249**—CH₃), 204 (7), and 192 (4) [137]. The very small parent ion and the intense fragment ions at m/z 236 (**66**) and 220 (**249**) were characteristic of a bisbenzylisoquinoline alkaloid containing three methoxy groups in one isoquinoline ring and one methylenedioxy group plus one methoxy group in the other isoquinoline ring with the dimer linked by a single tail-to-tail oxygen bridge [546]. Furthermore, the high-field singlet protons at δ 5.75 and 5.95 were indicative of the H(8′) and H(8) protons,

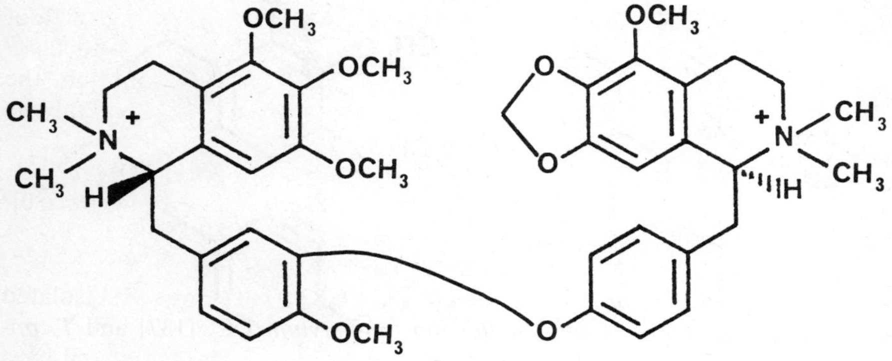

294

respectively, in this type of alkaloid [529,530]. Treatment of the alkaloid with methyl iodide afforded N-methylthalistyline iodide (thalistyline methiodide) (**294**) identical with that prepared via a similar methylation of thalistyline (**102**) [137]. Thus the structure of N-desmethylthalistyline was settled as **292**, which, with its companion alkaloids N-methylthalistyline (**294**) and thalistyline (**102**), represent the first examples of bisbenzylisoquinoline alkaloids in which both isoquinoline portions are trioxygenated in the aromatic ring [556].

N-Desmethylthalistyline (**293**) inhibited the in vitro growth of *Staphylococcus aureus* (1000 µg/mL) and *Mycobacterium smegmatis* (100 µg/mL) [137,301]. In addition, the alkaloid produced a hypotensive response in normotensive dogs and rabbits [301].

3.5.6. N-Desmethylthalrugosidine.

Tabular review [530, alkaloid No. 197], UV [70,530], IR [70,530], ^{1}H NMR [70,530], MS [70,530], CD [70,530].

N-Desmethylthalrugosidine (**295**), $C_{37}H_{40}N_2O_7$ (624.2836), m.p. 205–206° (MeOH), $[\alpha]_D^{21}$ −57° (c 0.23, MeOH), was isolated as colorless needles from *Thalictrum alpinum* L. in 1980 [70]. The UV spectrum showed maxima at 278 nm (log ε 3.90) and 283 nm (log ε 3.91) with no bathochromic shift in either acidic or alkaline medium. The ^{1}H NMR spectrum indicated the presence of one N-methyl group at δ 2.62(s), four methoxy groups as singlets at δ 3.52, 3.77, 3.88, and 3.92, and nine aromatic protons δ 6.2–7.7 (m). The mass spectrum showed an intense molecular ion at M^{+} *m/z* 624 (85%) with other diagnostic fragment ions at 623 (17%), 398 (33) (**296**), 397 (100) (**296**—H), 383 (6) (**296**—CH$_3$), 222 (4) (**248**), 206 (8) (**191**), 199 (55) (**297**), and 178 (4) (**298**). These spectral data and the CD curve ([θ]$_{225}$ −21,000, [θ]$_{232}$ 0, [θ]$_{242}$ +67,000, [θ]$_{258}$ +3600, [θ]$_{270}$ +6400, [θ]$_{273}$ 0, [θ]$_{285}$ −24,000, and

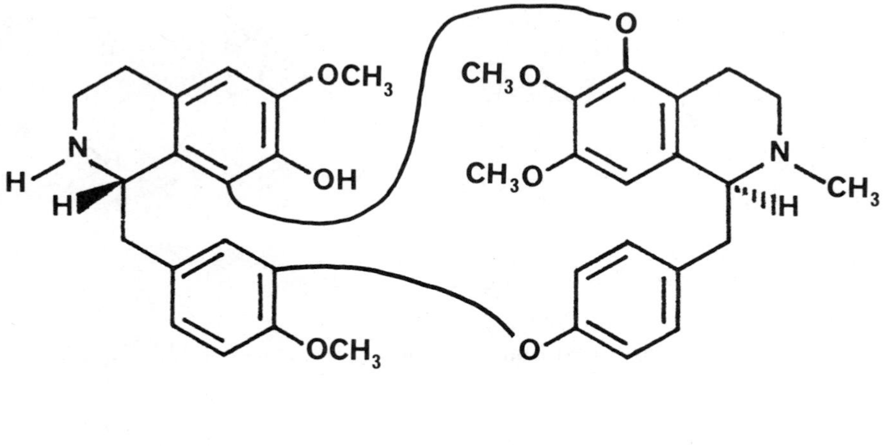

295

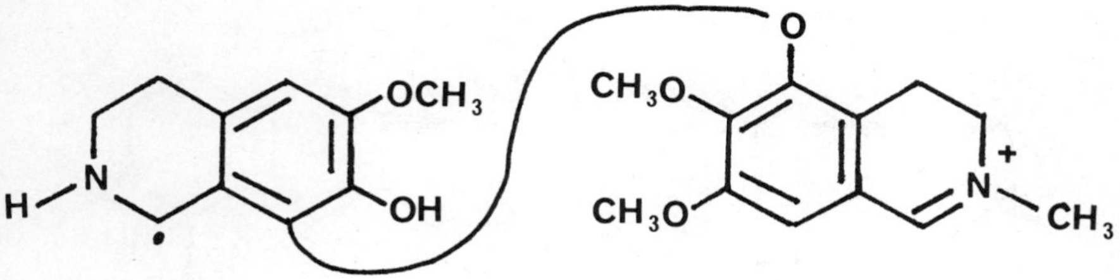

296

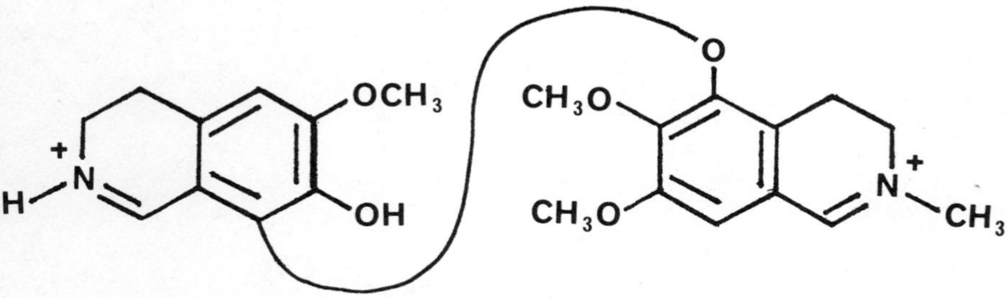

297

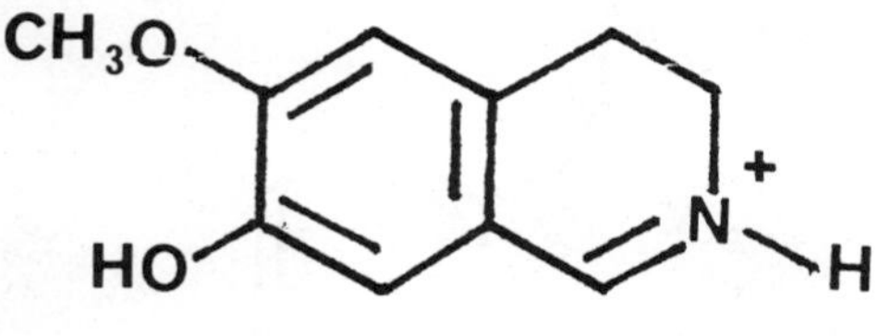

298

[θ]$_{300}$ 0) were indicative of a thalidasine (**286**)-like structure with a secondary nitrogen and a monophenolic hydroxy in the isoquinoline portion of the dimer [529,546]. Treatment of N-desmethylthalrugosidine with formaldehyde–sodium borohydride afforded thalrugosidine (**299**), thus establishing the skeletal structure, positions of oxygenation, positions of termini of the diphenyl ether bridges, and stereochemistry of N-desmethylthalrugosidine. Reductive cleavage of O-ethyl-N-desmethylthalrugosidine (prepared by ethylation of the alkaloid with diazoethane) gave (+)-(S)-4′,6-dimethoxy-7-ethoxybenzyltetrahydroisoquinoline (**300**) and (+)-(S)-2-methyl-4′,5-dihydroxy-6,7-di-

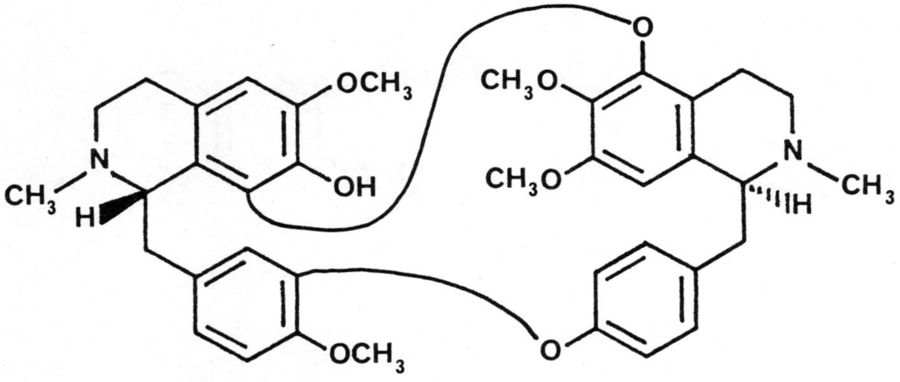

299

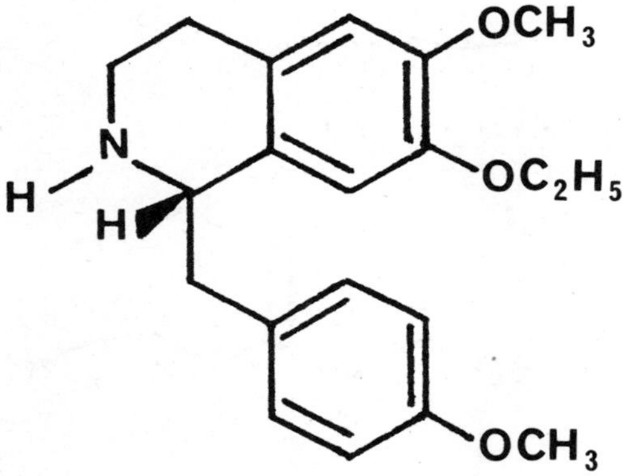

300

methoxytetrahydroisoquinoline (**287**), thereby locating the secondary nitrogen atom in the left ring [N(2)] of the dimer and establishing the structure of *N*-desmethylthalrugosidine (**295**).

3.5.7. Dihydrothalictrinine.   Tabular review [530, alkaloid No. 198], UV [530,557], IR [530,557], ^{1}H NMR [530,557], MS [530,557], CD [530,557]. Dihydrothalictrinine (**301**), $C_{38}H_{38}N_2O_9$ (666.2577), m.p. 194–197°, $[\alpha]_D^{22}$ −125° (c 0.13, MeOH), was isolated as colorless needles from *Thalictrum*

rochebrunianum Franc. and Sav. in 1979 [557]. The UV spectrum was characterized by a series of maxima at 238 nm (log ϵ 4.81), 249 nm (sh) (log ϵ
4.73), 285 nm (sh) (log ϵ 4.05), 299 nm (sh) (log ϵ 3.95), and 327 nm (log ϵ
3.62) with a bathochromic shift in acidic medium. The ^{1}H NMR spectrum
suggested the presence of one *N*-methyl group as a singlet at δ 2.49, five
methoxy groups as singlets at δ 3.45, 3.70, 3.79, 3.86, and 3.91; and 10
aromatic protons as signals at δ 6.13 (s) [H(8′)], 6.46 (s) (1H), 7.02 (s) (2H),
with a split ABXY pattern at δ 6.47 (br) (1H), 6.73 (dd, J = 2.5, 8.3 Hz)
(1H), ~7.1 (br) (1H), and 7.82 (dd, J = 2.2, 8.0 Hz) (1H) plus an AB quartet
at δ 7.48 (J = 5.7 Hz) (1H) and 8.40 (J = 5.7 Hz) (1H). In addition, a D_2O-
exchangeable signal was observed at δ 12.05 (OH). The mass spectrum was
characterized by the base peak and molecular ion both being observed at
M^+ m/z 666 (100%) with other fragment ions at m/z 651 (21), 635 (7), 513
(3), 409 (1), 332 (13), 325 (3), 188 (12), 142 (38), 129 (48), and 112 (13). These
spectral data were suggestive of a thalictrinine (**302**)-like alkaloid in which
the carbonyl of thalictrinine has been replaced by a secondary alcohol. This
is particularly true because the IR spectrum of dihydrothalictrinine was devoid of a carbonyl absorption but did have a peak at 3280 cm^{-1} characteristic
of hydroxyl stretching. In addition, the two peaks of the split ABXY pattern
were characteristic of protons *ortho* to a benzylic carbon that bears an alcohol function. Reduction (NaBH$_4$) of thalictrinine (**302**) produced a dihydro
derivative identical with naturally occurring dihydrothalictrinine (CD [θ]$_{217}$
+222,000, [θ]$_{231}$ 0, [θ]$_{245}$ −266,000, [θ]$_{270}$ −70,000 (sh), and [θ]$_{290}$ 0), suggesting that steric hindrance was a significant and controlling factor. Indeed,
examination of a space-filling molecular model demonstrated that the least
hindered approach for the hydride attack would produce the alcohol with

301

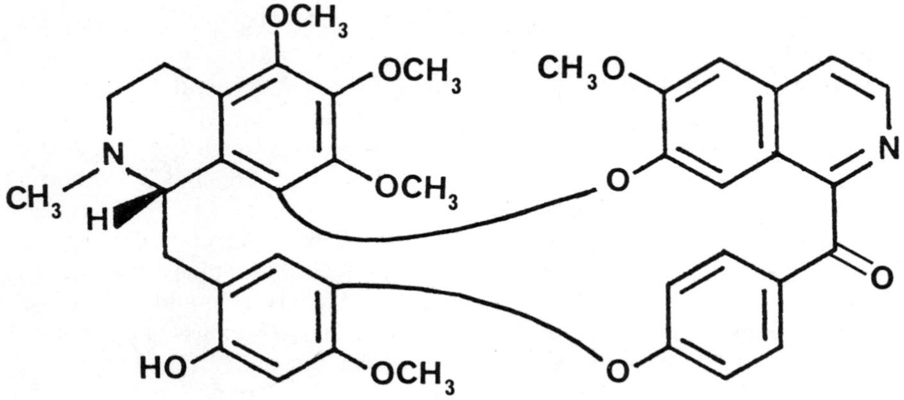

302

the (S) configuration. Hence the structure of dihydrothalictrinine (**301**), the first benbenzylisoquinoline alkaloid bearing a benzylic hydroxy group, was established [557].

3.5.8. Hernandezine (Thalicsimine, Thaliximine).   Tabular review [529, alkaloid No. 81, 530], UV [529,558], IR [558], ^{1}H NMR [301,529], ^{13}C NMR [559], MS [529,546,560,561], ORD [453,547], CD [562], fluorescence spectra [530,548], phosphorescence spectrum [530,548].
Hernandezine (**303**), $C_{39}H_{44}N_2O_7$ (652.3149), m.p. 192–193° (hexane) [558], 157–158° (MeOH) [558], 122–124° (Me$_2$CO) [558], 158–159° (Et$_2$O) [558];

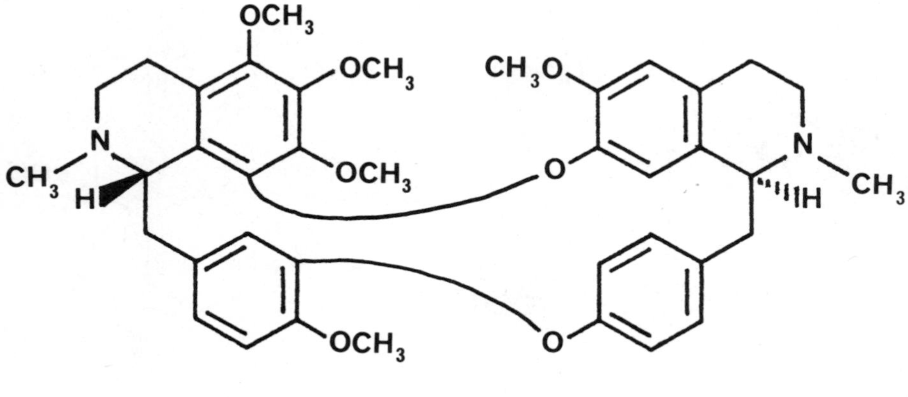

303

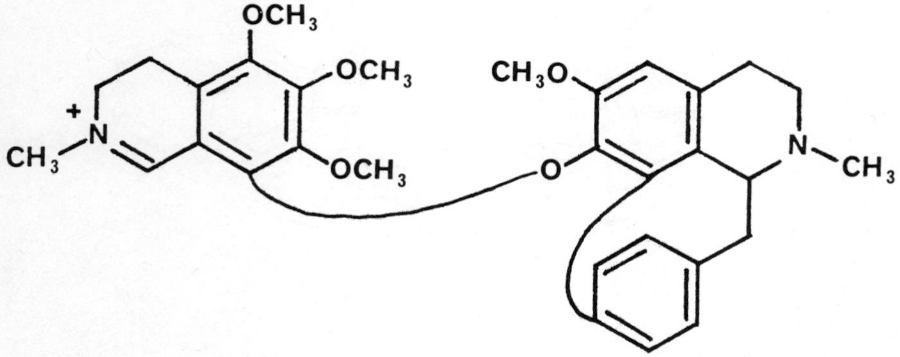

304

$[\alpha]_D^{20}$ +250° (c 0.2, CHCl$_3$) [558] was first isolated from *Thalictrum hernan-dezii* Tausch in 1962 [558]. The UV spectrum was characterized by a single maximum at 283 nm (log ϵ 3.90) [558]. The ^{1}H NMR spectrum indicated the presence of two *N*-methyl groups as singlets at δ 2.30 and 2.63; five methoxy groups as singlets at δ 3.24, 3.34, 3.79, 3.83, and 3.91; and nine aromatic protons as a multiplet from δ 6.02 to 7.50 [558]. The mass spectrum was characterized by an intense parent ion at M$^+$ *m/z* 652 (62%) and other characteristic fragment ions at *m/z* 651 (31%), 637 (12), 621 (5), 515 (3) (**304**), 461 (16) (**305**), 446 (2) (**305**—CH$_3$), 426 (17) (**306**), 213 (100) (**307**), 206 (3)

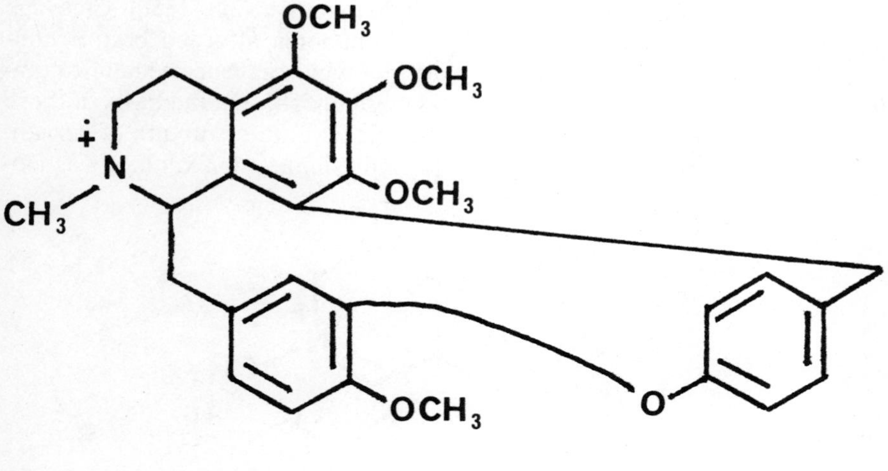

305

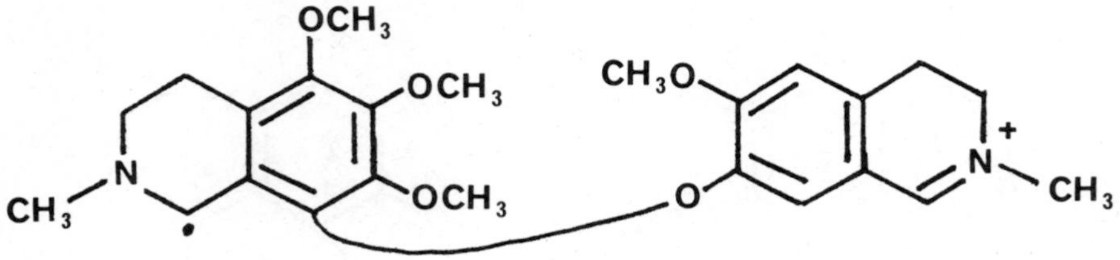

306

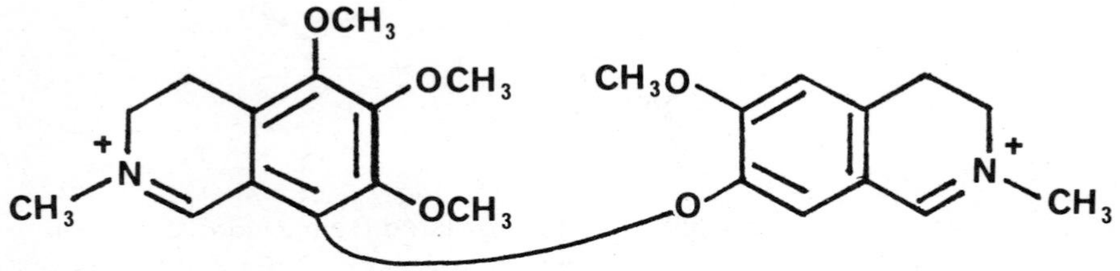

307

(**307**—CH$_3$), 190 (13) [**307**—(CH$_3$)$_2$O], 175 (8) (**308**), and 174 (27) (**308**—H) [560,561]. Oxidation of hernandezine with potassium permanganate afforded 5,4'-dicarboxy-2-methoxydiphenyl ether (**309**), and alkaline fission gave *p*-hydroxybenzoic acid [558]. Hernandezinemethine (m.p. 189–191°) was prepared via classical Hofmann degradation, and oxidation (CrO$_3$) of this methine furnished 5,4'-dicarboxy-2-methoxydiphenyl ether (**309**) [558]. Cleavage of hernandezine with potassium in liquid ammonia afforded both a nonphenolic and a phenolic fraction, the latter of which gave a crystalline compound, m.p. 138–139°, which was not identified [558]. On the basis of these degradative studies and certain spectroscopic data, the structure of hernandezine was first proposed as **310** [558]. Thalicsimine (thaliximine) was iso-

308

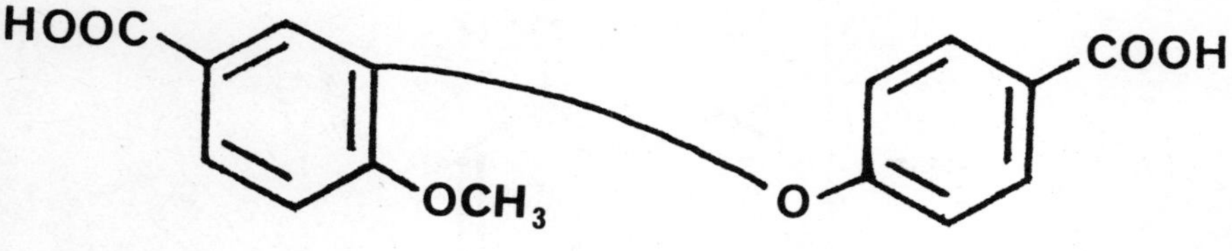

309

lated from *Thalictrum alpinum* L. in the late 1950s or early 1960s [563] and was shown to be identical with hernandezine in 1963 [564]. Reductive cleavage of hernandezine with sodium in liquid ammonia was reported to afford *N*-methylcoclaurine (**52**) and another phenolic base [565,566]. However, in 1966, a detailed reinvestigation of the mass spectrum and ¹H NMR spectral data of hernandezine and the cleavage reaction products led to a structural revision of hernandezine to **303** [567]. In particular, the high-field aromatic proton in the ¹H NMR spectrum of hernandezine indicated a C(8) proton that is shielded by the benzyl group of the benzyltetrahydroisoquinoline ring [567,568]. Furthermore, the mass spectral fragment ion at *m/z* 461 (M-191) (**303**) suggests that the lost C- and D-ring fragment contains only one methoxy group and not two, as would be consistent for structure **310**. Finally, repetition of the metal–ammonia cleavage of hernandezine afforded the trimethoxy base **311** and *N*-methylcoclaurine (**52**) [567]. Because a similar cleavage of *O*-methylthalifendlerine (tetrahydrotakatonine) (**62**) also produced base **311** [567] and because it is well known that hindered aromatic methoxy groups tend to cleave under these conditions [35], the revision of the structure of hernandezine (**303**) was complete [567]. Hernandezine (**303**)

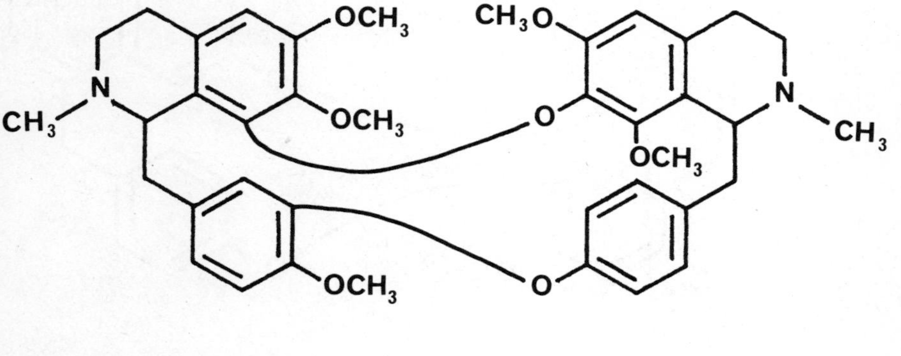

310

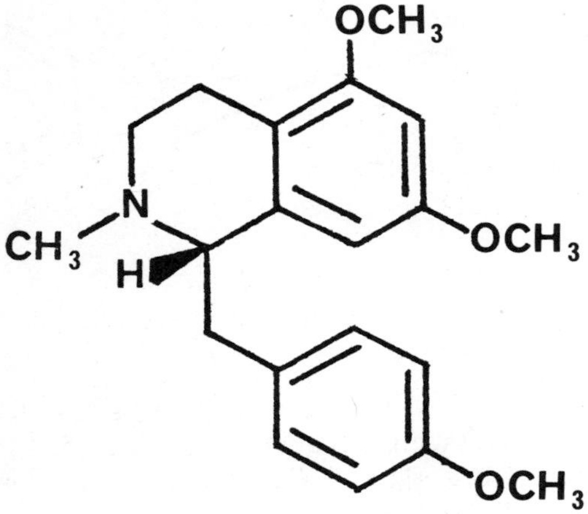

311

and its imino isomer thalsimine (**312**) were the first bisbenzylisoquinoline alkaloids that were oxygenated at C(5). Ceric ammonium nitrate was found to be an excellent one-electron oxidant of bisbenzylisoquinoline alkaloids and treatment of hernandezine with this reagent, followed by suitable work-up, afforded a diamine (**313**) and a diol (**314**) in 96 and 94% yields, respectively [569].

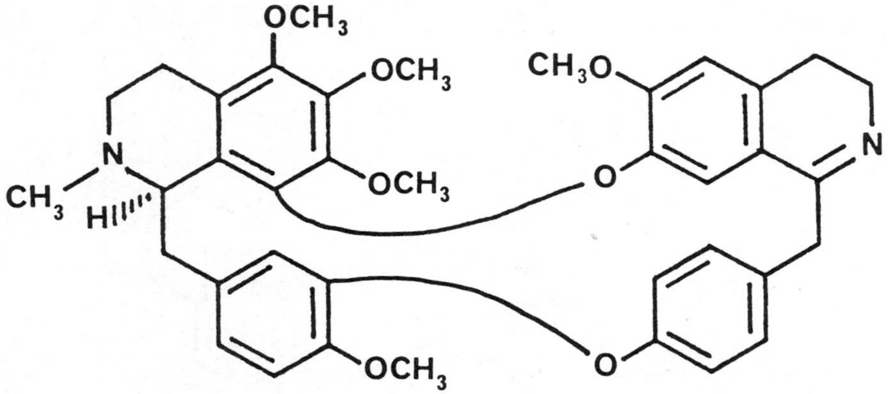

312

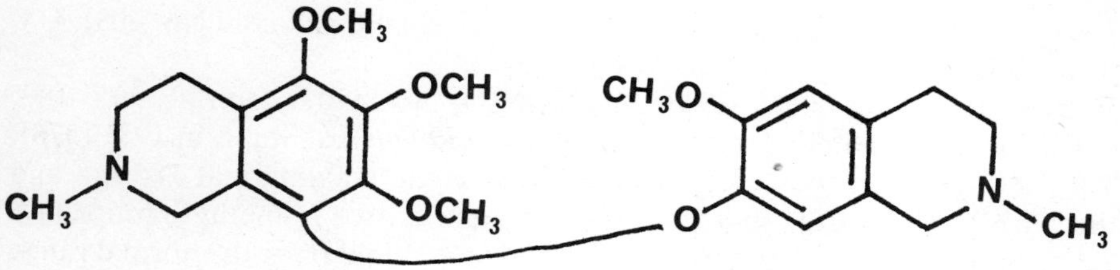

313

Early pharmacological studies on hernandezine demonstrated that an intravenous injection of 1–3 mg/kg to cats caused a temporary hypotensive response, with a dose of 10 mg/kg producing a sharp decrease in blood pressure, leading to death. The LD_{50} of the alkaloid in mice was 282 mg/kg [570]. Hernandezine and was also found to exhibit antiinflammatory activity at a dosage of 10–50 mg/kg [571] and was found to be two to four times more active against experimentally induced inflammation than aminopyrine or sodium salicylate [458]. The antiinflammatory effect of hernandezine on adrenalectomized rats was similar to that of intact ones [571]. The alkaloid possessed a hypothermic effect in rabbits when administered intraperitoneally [571]. Furthermore, the alkaloid inhibited conditioned avoidance reactions and motor-conditioned reflexes associated with movement and eating in rats at an intravenous dosage of both 50 and 100 mg/kg [572]. Hernandezine was observed to produce only weakly inhibitory effects on the growth of lymphoma NK/ly, alveolar hepatoma PC-1, or Pliss lymphosarcoma in rodents at a dosage of 30–160 mg/kg [573]. The alkaloid also produced a hypotensive response in normotensive dogs and rabbits [301], as well as inhibiting the in vitro growth of *Staphylococcus aureus* (100 μg/mL), *Mycobacterium smegmatis* (25 μg/mL), and *Candida albicans* (50 μg/mL) [301].

Hernandezine (**303**) has also been isolated from *Thalictrum fendleri* Engelm. [62], *T. podocarpum* Humb. [301], *T. rochebrunianum* Franc. and Sav. [121], *T. simplex* [153,570,574], and *T. sultanabadense* Stapf [575,576].

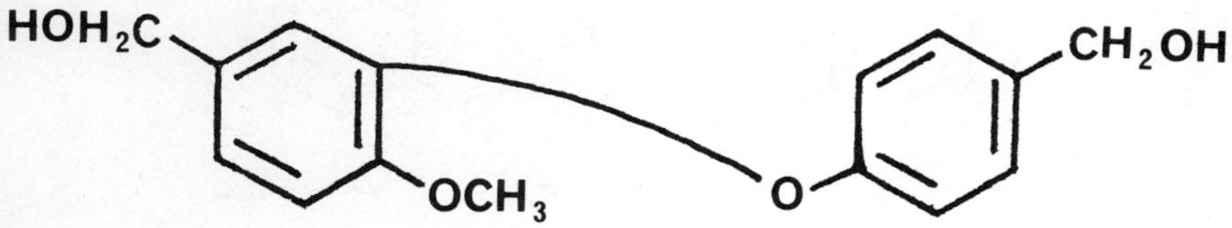

314

3.5.9.   Hernandezine N-Oxide.   Tabular review [430, alkaloid No. 203], UV [430,576], ¹H NMR [430,576], MS [430,576].

Hernandezine N-oxide (**315**), $C_{39}H_{44}N_2O_8$ (668.3097), m.p. 178–180° (CHCl₃), was isolated from *Thalictrum sultanabadense* Stapf. in 1981 [576]. The UV spectrum was characterized by a single maximum at 285 nm and the ¹H NMR spectrum indicated the presence of two N-methyl groups at δ 2.31 and 3.65 (the latter being considerably downfield from the normal range of δ 2.2–2.6); five methoxy groups at δ 3.15, 3.27, 3.71 (6H), and 3.81; and nine aromatic protons from δ 5.97 to 7.15. The mass spectrum showed a parent ion at M⁺ *m/z* 668 (15%) with other intense ions at *m/z* 652 (100%) (M—16), 461 (25), 460 (21), 425 (34), 424 (31), and 411 (62). These spectral data were suggestive of a bisbenzylisoquinoline alkaloid of the hernandezine (**303**)-type but containing a single N-oxide. This was supported by the prominent M—16 ion in the mass spectrum, the low-field N-methyl group (N-oxide) in the ¹H NMR spectrum, and the lower solubility in organic solvents, usually associated with N-oxides. Reduction (Zn/H₂SO₄) of the alkaloid afforded hernandezine, thus establishing skeletal structure, positions of oxygenation, and termini of the diphenyl ether bridges. The assignment of the N-oxide function to the N(2′) nitrogen was apparently via a comparison of the chemical shifts of the N(2) N-methyl groups (~δ2.2–2.3) and the N(2′) N-methyl groups (~δ2.5–2.6) of alkaloids of this and related series [529] and the generation of the appropriate signal upon reduction of hernandezine N-oxide (**315**). Although no specific rotation nor CD data was reported for the N-oxide or for its reduction products, the assumption is that the N-oxide is of the (S,S) configuration [576].

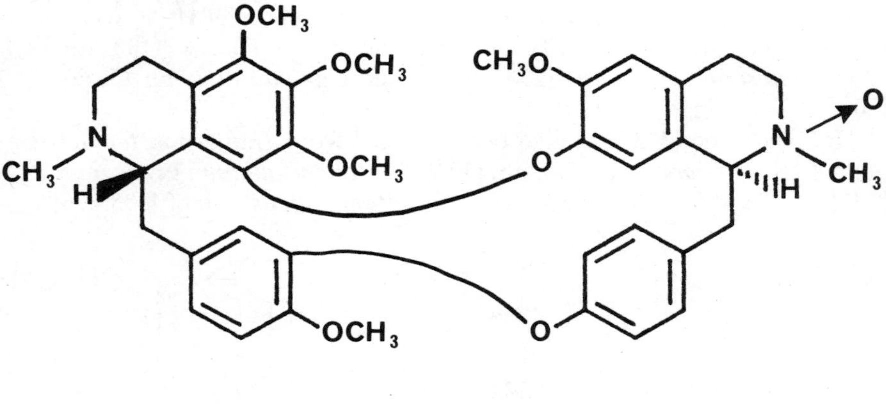

315

3.5.10. **Homoaromoline (Homothalicrine, *O*-Methylaromoline, Thalrugosa-mine).** Tabular review [529, alkaloids Nos. 42, 52], UV [139,529,577], IR [139], ^{1}H NMR [139,529,577], MS [529,577], CD [529,577].

Homoaromoline (**316**), $C_{37}H_{40}N_2O_6$ (608.2886), m.p. 142–143° (Et$_2$O) [139], 122–125° [577]; $[\alpha]_D^{30}$ +280° (MeOH) [577], was first isolated from *Thalictrum thunbergii* DC. in 1962 and assigned the dimeric structure **317** on the basis of various degradative studies and its production on CH_2N_2 methylation of thalicrine, which had been assigned as **283** [540–542]. However, in 1964 a detailed structural reinvestigation resulted in the revision of structure of thalicrine to that previously established for aromoline (**282**) and thus hom-othalicrine was reassigned as homoaromoline (**316**) [543]. In 1972, the al-

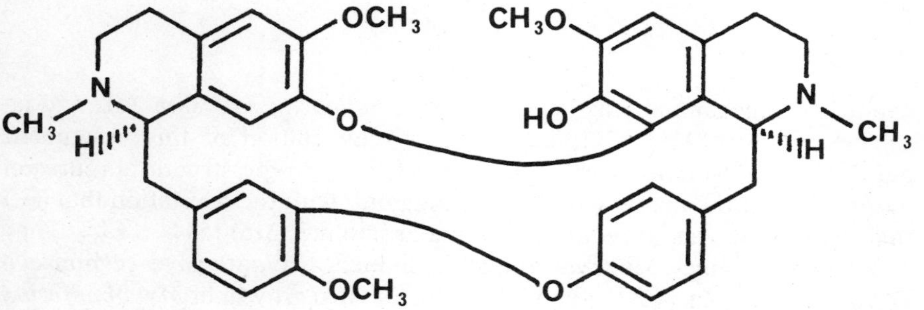

316

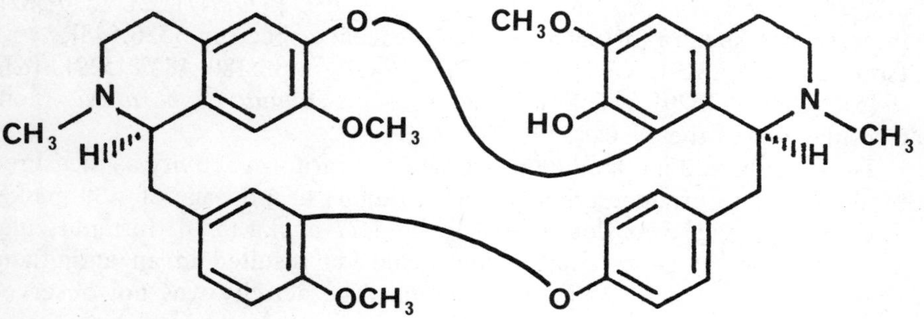

317

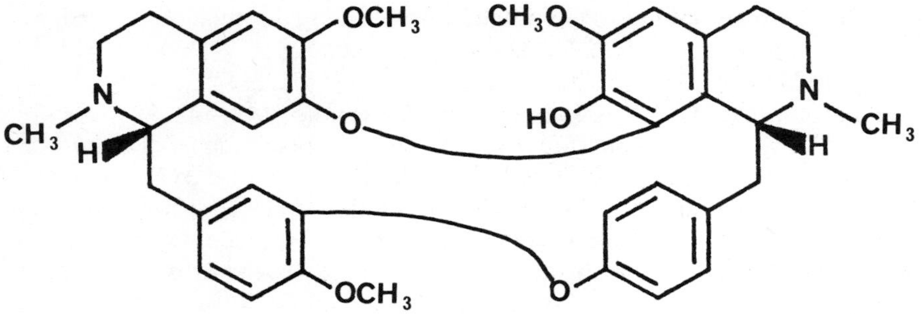

318

kaloid thalrugosamine was isolated from *Thalictrum rugosum* Ait. and assigned structure **318** [577]. However, a reevaluation of this assignment published in 1984 demonstrated an inconsistency in the structural representation (but not the actual work) of the alkaloid, with the revelation that (+)-thalrugosamine was in reality (+)-homoaromoline (**316**) [544].

Homoaromoline (**316**) was found to induce a hypotensive response in normotensive dogs [139] and to inhibit the in vitro growth of *Mycobacterium smegmatis* [577] at a concentration of 100 μg/mL [139] and *Candida albicans* at 1000 μg/mL [139].

Homoaromoline (**316**) has also been isolated from *Thalictrum lucidum* L. [139] and *T. minus* L. var. *microphyllum* Boiss. [544].

3.5.11. Isotetrandrine. Tabular review [529, alkaloid No. 62, 530], UV [529], ^{1}H NMR [171,529], MS [529,546,578], ORD [529,547], CD [530,562], fluorescence spectra [530,548], phosphorescence spectrum [530,548]. Isotetrandrine (**319**), $C_{38}H_{42}N_2O_6$ (622.3043), m.p. 180–182° [529], $[\alpha]_D$ +151° (CHCl$_3$) [529], was first isolated from *Stephania cepharantha* (Menispermaceae) in the 1930s [579].

Isotetrandrine dimethiodide possessed a curariform activity as measured by its ability to produce a head drop in rabbits at a dosage of 1.39 mg/kg. The lethal dosage was determined to be 1.77 mg/kg [550]. Intramuscular administration of isotetrandrine hydrochloride resulted in an antiinflammatory effect on induced rat paw edema. The activity was not observed when the drug was administered orally and may be related to adrenal stimulation [580].

The biosynthesis of isotetrandrine in *Cocculus laurifolius* DC. (Menispermaceae) was studied utilizing ^{3}H- and ^{14}C-labeled (±)-coclaurine, (±)-*N*-methylcoclaurine (**52**), didehydro-*N*-methylcoclaurinium iodide, (+)-(*S*)-*N*-methylcoclaurine, and (−)-(*R*)-*N*-methylcoclaurine. The study supported

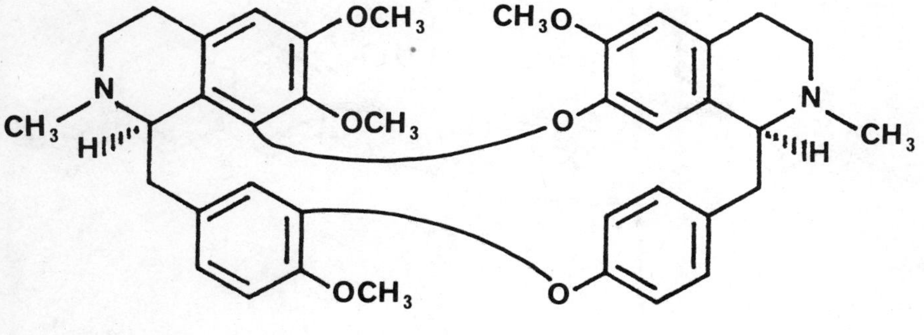

319

the following sequence for the biosynthesis of isotetrandrine (**319**): coclaurine → (+)-(S)-N-methylcoclaurine + (−)-(R)-N-methylcoclaurine → inter-
and intramolecular oxidative coupling → isotetrandrine [581].

Isotetrandrine (**319**) has been isolated from only one species of *Thalictrum*, *T. foetidum* L. [549].

3.5.12. Isothalidezine. Tabular review [529, alkaloid No. 82], UV
[301,529], IR [301], ^{1}H NMR [301,529], MS [301,529], CD [301,529].
Isothalidezine (**320**), $C_{38}H_{42}N_2O_7$ (638.2992), m.p. 136–138°, $[\alpha]_D^{25}$ −70° (c
0.13, MeOH), was isolated as a white crystalline residue from *Thalictrum
podocarpum* Humb. in 1977 [301]. The UV spectrum showed a single maximum at 282 nm (log ε 3.99) and the IR spectrum showed hydroxy absorption.

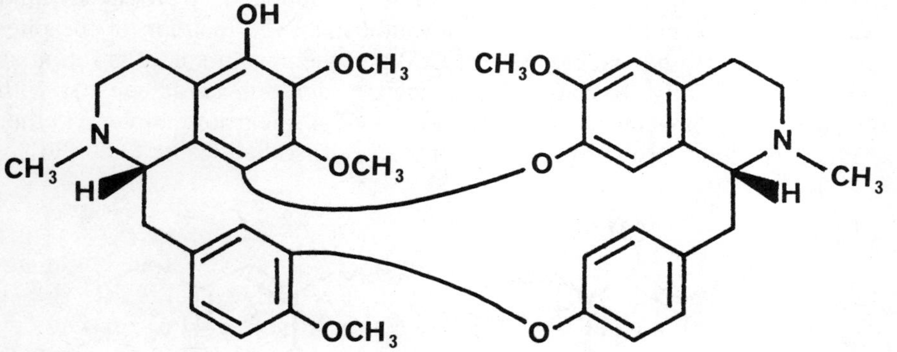

320

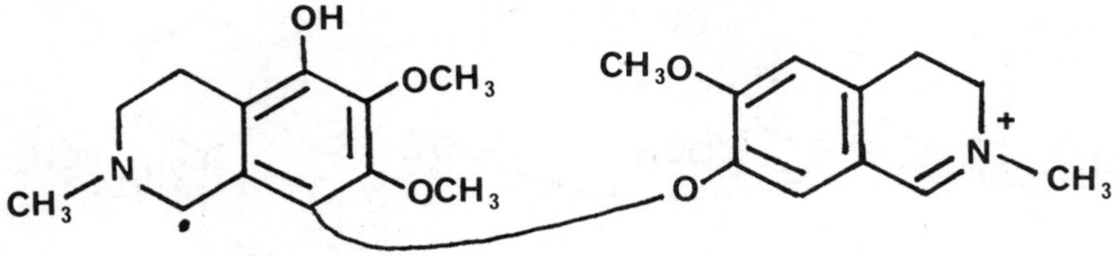

321

The ^{1}H NMR spectrum indicated the presence of two *N*-methyl groups as singlets at δ 2.23 and 2.54, four methoxy groups as singlets at δ 3.18, 3.62, 3.75, and 3.88; a D_2O-exchangeable phenolic group at δ 4.85; and nine aromatic protons from δ 5.99 to 7.33. The mass spectrum exhibited the parent ion at M$^+$ *m/z* 638 (47%) with other characteristic fragment ions at *m/z* 637 (18%), 412 (23) (**321**), 411 (80) (**321**—H), 397 (38) (**321**—CH$_3$), 222 (11) (**248**), 206 (86) (**322**), and 192 (100) (**199**). These spectral data were similar to those obtained for thalidezine (**289**), except for a different CD spectrum ([θ]$_{230}$ −63,800, [θ]$_{247}$ +37,000, and [θ]$_{285}$ −21,700) for isothalidezine [301]. Treatment of isothalidezine with ethereal CH_2N_2 afforded a new compound, iso-hernandezine (**323**), which showed mass spectral fragment ions at *m/z* 426 (**306**) and 236 (**66**), suggesting the presence of the phenolic group of isothalidezine in ring B and establishing both skeletal structure and positions of oxygenation for isothalidezine. Treatment of isothalidezine with diazoethane produced an *O*-ethyl ether which subsequently underwent cleavage with sodium in liquid ammonia to afford (*S*)-1-(4'-methoxybenzyl)-2-methyl-5-ethoxy-7-methoxytetrahydroisoquinoline (**324**) (minor nonphenolic product), (*S*)-1-(4'-methoxybenzyl-2-methyl-5-ethoxy-6,7-dimethoxytetrahydroisoquinoline (**325**) (major nonphenolic product), and (*R*)-*N*-methylcoclaurine (**52**), thereby establishing the location of the phenolic group of isothalidezine (**320**) at C(5) and the absolute configuration as (*S,R*) [301]. Finally, oxidation of an acetone solution isothalidezine with potassium permanganate afforded 2-methoxy-5,4'-dicarboxy biphenyl ether (**309**), consistent with the proposed structure of isothalidezine (**320**) [301].

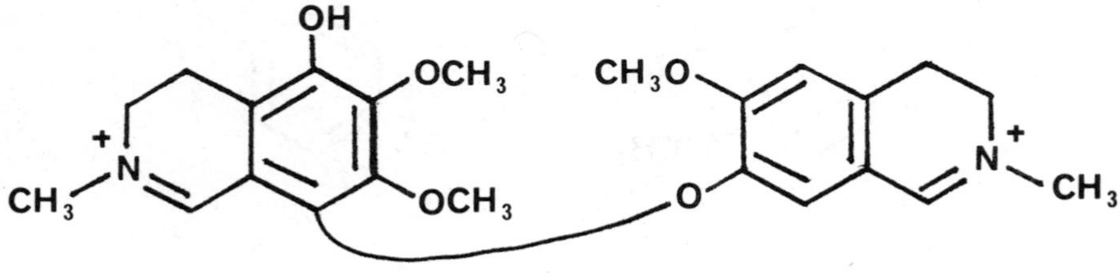

322

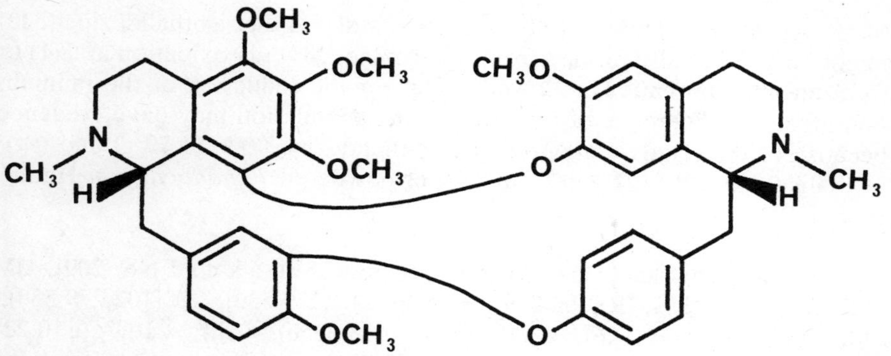

323

In a recent paper, Guinaudeau et al. described four empirical rules for correlation of the structures of bisbenzylisoquinoline *Thalictrum* alkaloids with their stereochemistry at C(1) and C(1') [544]. (−)-Isothalidezine (**320**) was cited as the sole exception, out of 58 alkaloids, to the rule that stated that the right-hand benzylisoquinoline moiety of the dimer is characterized

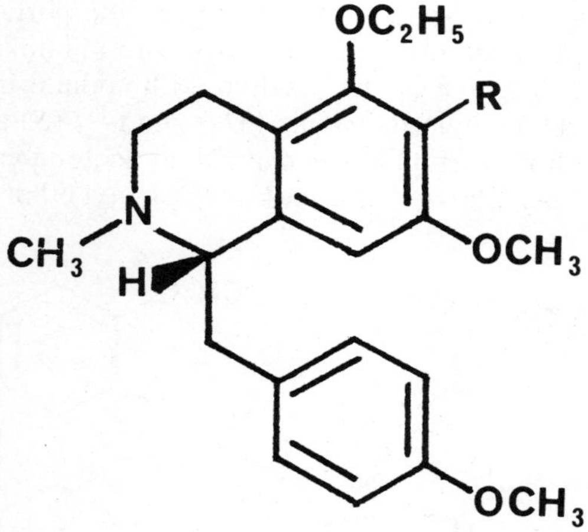

324 R = H

325 R = OCH$_3$

by the (*S*) configuration at C(1′). It was postulated that isothalidezine (**320**) might be artifactually formed from thalidezine (**289**) via oxidation at C(1) to form an iminium cation, followed by enzymatic reduction of the iminium bond from the beta side of the dimer. This postulation may have credence because of the relative amounts of (+)-thalidezine (**289**) (1.79 g) and (−)-isothalidezine (**320**) (92 mg) from 7 kg of powdered *T. podocarpum* Humb. roots [301,544].

3.5.13. *O*-Methylthalibrine. Tabular review [530, alkaloid No. 209], UV [102,279,530], ^{1}H NMR [102,279,530], MS [102,279,530], CD [102,279,530]. *O*-Methylthalibrine (**326**), $C_{39}H_{46}N_2O_6$ (638.3356), $[\alpha]_D^{20}$ +109° (c 0.22, MeOH) [102]; $[\alpha]_D$ +82° (c 0.36, CHCl$_3$) [582], was isolated as an amorphous solid from *Thalictrum minus* L. race B in 1980 [102]. The UV spectrum was characterized by maxima at 280 nm (log ε 4.02) and 285 nm (sh) (log ε 4.01); the CD spectrum (MeOH) exhibited $[\theta]_{228}$ +75,500, $[\theta]_{250}$ 0, $[\theta]_{270}$ 0, $[\theta]_{287}$ +15,500, and $[\theta]_{300}$ 0 [102]. The ^{1}H NMR spectrum indicated the presence of two *N*-methyl groups as singlets at δ 2.49 and 2.53; five methoxy groups as singlets at δ 3.60, 3.63, 3.78, 3.80, and 3.83; and 11 aromatic protons as four singlets at δ 6.10 [H(8)], 6.16 [H(8′)], 6.53 [H(5)], 6.56 [H(5′)], and a multiplet from δ 6.6 to 7.2 (7H) [102]. The mass spectrum was characterized by a very weak parent ion at M$^+$ *m/z* 638 (0.03%) with other important fragment ions at *m/z* 206 (100%) (**191**), 191 (9) (**191**—CH$_3$), and 190 (9) [102]. These spectral data were characteristic of a bisbenzylisoquinoline alkaloid containing four methoxy groups in the isoquinoline portion with unsubstituted C(8) and C(8′) positions and with one tail-to-tail linked diphenyl ether bridge [529,530]. Cleavage of the alkaloid with sodium in liquid ammonia afforded (+)-(*S*)-*O*-methylarmepavine (**327**) and (+)-(*S*)-armepavine (**328**), and oxidation with potassium permanganate in acetone gave *N*-methylcorydaldine (**88**) and 2-methoxy-4′,5-dicarboxybiphenyl ether (**309**), thus firmly

326

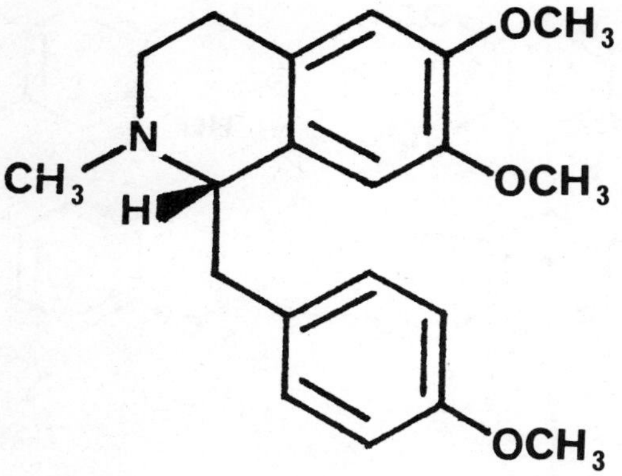

327

establishing the structure of *O*-methylthalibrine (**326**) [102]. Finally, a direct comparison of the natural material with a synthetic product prepared via methylation of thalibrine (**329**) [582] confirmed the assignment [102].

O-Methylthalibrine (**326**) was observed to inhibit the in vitro growth of *Mycobacterium smegmatis* at a concentration of 100 μg/mL and *Candida albicans* at 500 μg/mL [102].

O-Methylthalbrine has been isolated from only one other *Thalictrum* species, *T. faberi* Ulbr. [279].

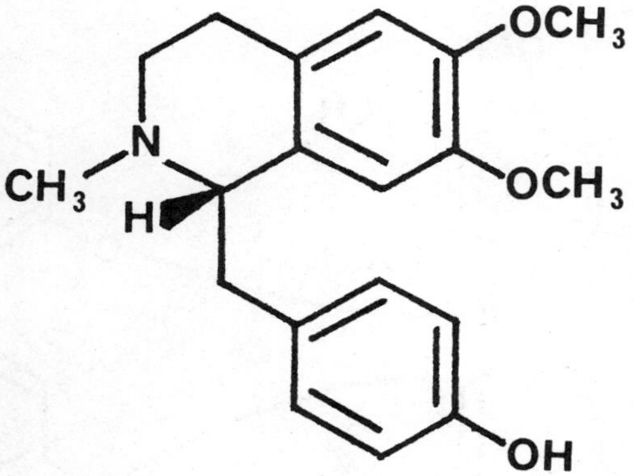

328

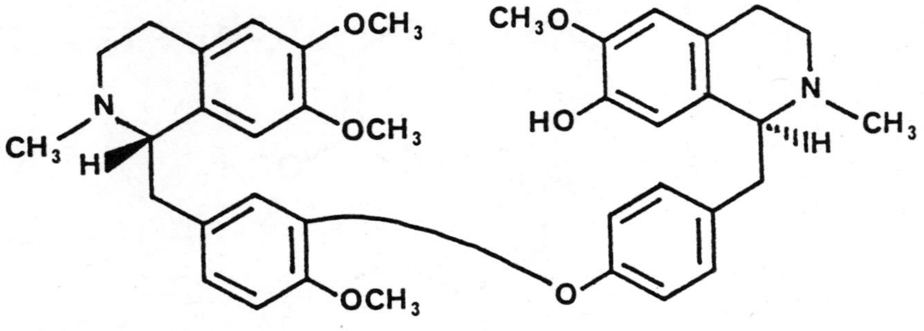

329

3.5.14. *O*-Methylthalibrunimine.

Tabular review [530, alkaloid No. 210], UV [530,583], ¹H NMR [530,583], MS [530,583].

O-Methylthalibrunimine (**330**) (incorrectly spelled as *O*-methylthalibruna-mine [583]), $C_{39}H_{42}N_2O_8$ (666.2941), m.p. 183–185° (CHCl₃, $[\alpha]_D$ −103.7° (c 0.5, CHCl₃), was isolated from *Thalictrum rochebrunianum* Franc. and Sav. in 1978 [583]. The UV spectrum was characterized by maxima at 240 nm (sh) (log ε 4.46), 282 nm (log ε 4.01), and 305 nm (sh) (log ε 3.92) and the ¹H NMR spectrum showed signals for one *N*-methyl group (δ 2.52); six methoxy groups (δ 3.45, 3.74, 3.81, 3.88, and 3.92, one of which must for six protons but was unspecified); a benzylic methylene (δ 4.39) and seven aromatic protons [δ 6.02 − 7.60 (m)] (eight aromatic protons are actually present). The mass spectrum showed a molecular ion at M⁺ *m/z* 665 (65%) (this is impossible because the alkaloid contains two nitrogen atoms and

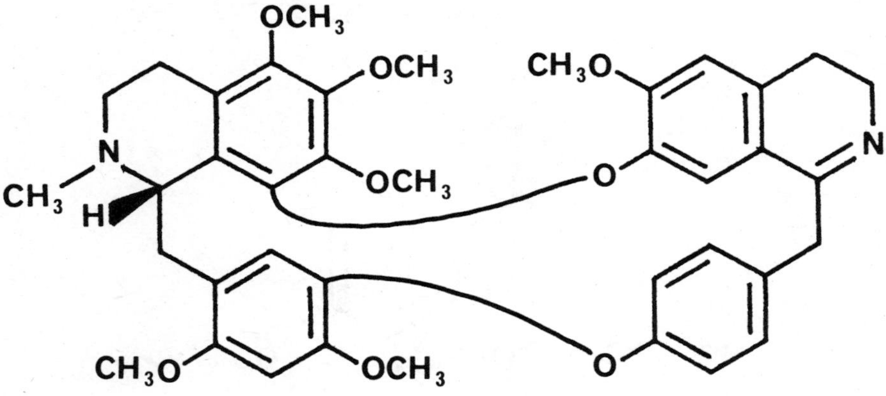

330

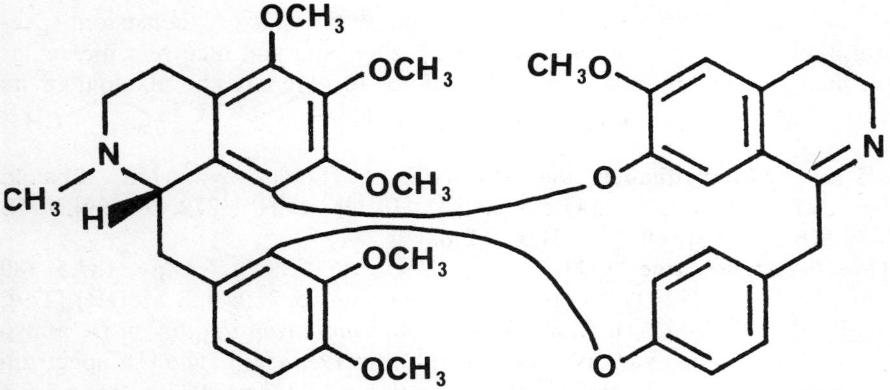

331

must therefore have an even molecular ion/molecular weight) with other fragment ions at 664 (100), 650 (54), 635 (26), 620 (3), 605 (13), 410 (13), 395 (3), 377 (20), 363 (20), 336 (33), 234 (6), and 205 (13) [583]. Reduction of the alkaloid with sodium borohydride produced an oily dihydro derivative which was treated with formaldehyde–sodium borohydride to afford a crystalline compound m.p. 130–135°, whose ¹H NMR spectrum showed the presence of an additional N-methyl group at δ 2.49. O-Methylthalibrunimine was assigned as **331** but because the structure of thalibrunimine had been revised to **332**, O-methylthalibrunimine should be assigned as **330**. The structural

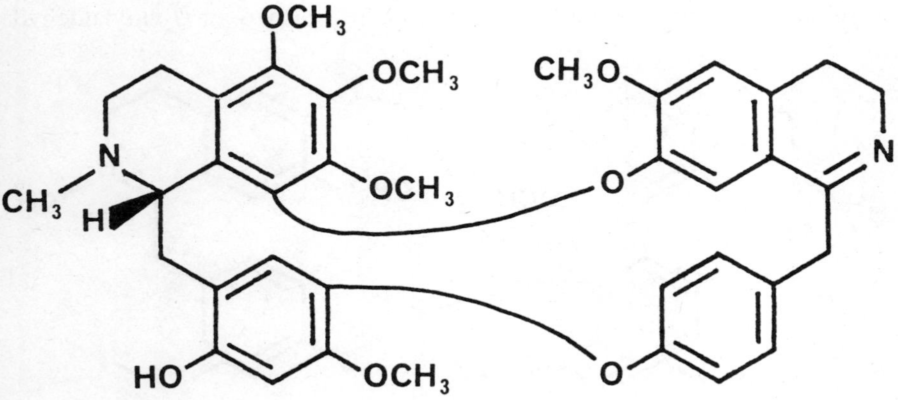

332

assignment presented in this paper is in doubt because of inconsistent spectral data (shoulder at 305 nm in the UV spectrum and incorrect molecular ion in the mass spectrum) and the failure to compare the alkaloid or its derivatives with authentic samples.

3.5.15.   O-Methylthalicberine (Thalmidine).   Tabular review [529, alkaloid No. 95, 530], UV [145,279,529], ` ¹H NMR [145,279,450,529], MS [529,560,584,585], ORD [453], CD [562].

O-Methylthalicberine (**333**), $C_{38}H_{42}N_2O_6$ (622.3043), m.p. 187.5–189 (MeOH) [145], 172–174° (Et₂O) [114]; $[\alpha]_D^{25}$ + 175.5° (c 0.2, MeOH) [279]; $[\alpha]_D^{19}$ + 244.6° [586], was first isolated from *Thalictrum thunbergii* DC. (also known as *T. minus* var. *hypoleucum* [142]) in 1959 [586]. The UV spectrum was characterized by maxima at 278 nm (log ε 3.90) and 282 nm (log ε 3.90) and the ¹H NMR spectrum showed the presence of two *N*-methyl groups as singlets at δ 2.09 and 2.57; four methoxy groups as singlets at δ 3.64, 3.75, 3.86, and 3.88; and 10 aromatic protons as a singlet at δ 6.06 (1H) [H(8′)] and multiplet from δ 6.34 to 7.16 [279]. The mass spectrum showed the parent ion at M⁺ *m/z* 622 (52%) with other important fragment ions at *m/z* 621 (26), 607 (6), 591 (2), 396 (100), 198 (24), 175 (5), 174 (10), 90 (2), and 89 (2) [560]. A methyl methine was produced via Hofmann degradation and this product was oxidized with potassium permanganate to 2-methoxy-5,4′-dicarboxybiphenyl ether (**309**) [586]. Reductive cleavage of *O*-methylthalicberbrine with sodium in liquid ammonia gave (+)-*N,O,O*-trimethylcoclaurine (*O*-methylarmepavine) (**327**) and (+)-*N*-methylisococlaurine (**334**) (the latter of which methylated to the former) [587]. In a subsequent study, ozonlysis of the Hofmann-generated methyl methine produced 4-methoxy-3,4′-oxydibenzaldehyde, which on oxidation furnished the corresponding acid (**309**). The mother liquor from which the aldehyde was removed yielded an oily diaminodialdehyde, the methiodide of which furnished an *O*-vinylaldehyde

333

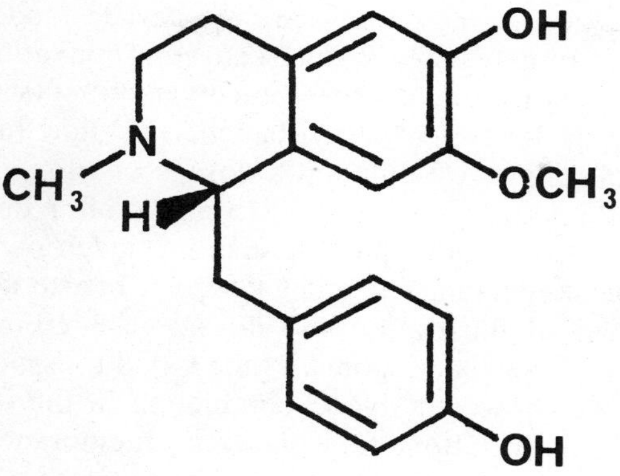

334

on treatment with alkali. Catalytic reduction of the *O*-vinyl aldehyde yielded an ethyl aldehyde, which on Clemmensen reduction afforded 2-methoxy-4-methyl-5-(ethylphenyl)-2′-methyl-3′-ethyl-5′,6′-dimethoxyphenyl ether (**335**) [588]. A later study again confirmed that (+)-*N*-methylisococlaurine (**334**) was the phenolic fragment obtained via Birch reduction of *O*-methylthalicberine [589]. These studies allowed the unambiguious assignment of structure of *O*-methylthalicberine (**333**) [586–589], which was subsequently confirmed by total synthesis [590,591].

O-Methylthalicberine (**333**) produced restlessness, dyspnea, and paralysis of the extremities when administered to mice. The LD_{50} in these experiments was 310 mg/kg. In rabbits, a dose of 10–25 mg/kg produced death. Suppression of the orientation reaction and antagonism of the stimulant action of phenamine in mice was produced at dosages of 50 mg/kg or less of the

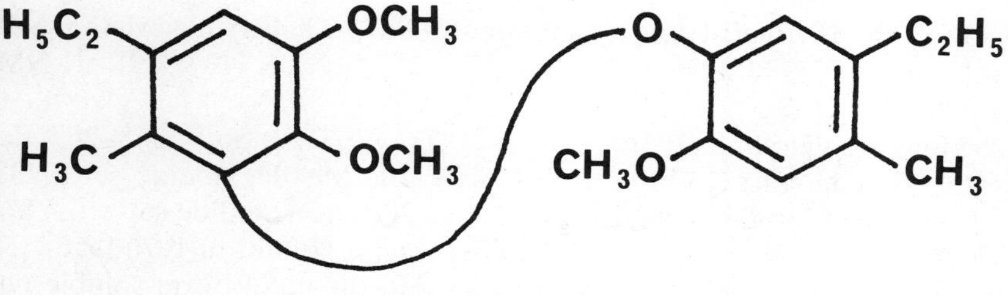

335

alkaloid. A dosage of 5 mg/kg decreased the arterial blood pressure in cats by more than 50%. Finally, the alkaloid caused a relaxation of an isolated segment of rabbit intestine and decreased acetylcholine- or barium chloride-induced contractions [592]. *O*-Methylthalicberine dimethiodide produced skeletal muscle relaxation in rabbits at a dosage of 10 mg/kg. The LD_{50} of the alkaloid dimethiodides in mice was 115 mg/kg after intraperitoneal administration. The alkaloid dimethiodide also produced partial or complete relaxation of feline skeletal muscle and a decrease in arterial blood pressure after intraperitoneal administration in a dosage of 3–10 mg/kg. The dimethiodide salt did not alter the response of the blood pressure on injection of acetylcholine. It did, however, block conduction in the intracardial vagal ganglia and induced a relaxation of the nictitating membrane. The quaternary salt relaxed the muscular contractions of isolated intestine and decreased acetylcholine- or barium chloride-induced intestinal contractions [593]. *O*-Methythalicberine exhibited antiinflammatory activity in the experimentally induced inflammed mouse paw, exceeding both aminopyrine and sodium salicylate in activity. This activity was apparently related to an alkaloid-induced decreased permeability in blood vessels [594]. The alkaloid possessed weak inhibitory effects on the growth of lymphoma NK/Ly, alveolar hepatoma PC-1, or Pliss lymphosarcoma in a dosage of 30–100 mg/kg [573]. *O*-Methylthalicberine produced a hypotensive response in rabbits [101] and decreased the blood pressure of normotensive dogs by 32 mm Hg at a dosage of 1.0 mg/kg and by 97 mm Hg at 2.0 mg/kg [139]. The alkaloid possessed weak in vitro antimicrobial effects against *Mycobacterium smegmatis* and *Candida albicans*, both at 1000 μg/mL [139].

The alkaloid thalmidine, first isolated from *Thalictrum minus* L. in 1950 [84] and later in 1956 [595], was shown to be identical with *O*-methylthalicberine (**333**) [596,597] in 1965 after earlier structural studies were inconclusive [564,598].

O-Methylthalicberine (**333**) has also been isolated from *Thalictrum faberi* Ulbr. [279], *T. lucidum* L. [139], *T. minus* L. [65,114,181,272], *T. minus* L. var. *adiantifolium* Hort. [145], *T. minus* L. var. *microphyllum* Boiss. [146,544], and *T. revolutum* DC. [101,105].

3.5.16.   *N*-Methylthalistyline (Methothalistyline, Thalistyline Metho Salt).

Tabular review [529, alkaloid No. 17], UV [137,529], IR [137,529], ^{1}H NMR [137,529], CD [137,529].

N-Methylthalistyline (**294**), $C_{42}H_{52}N_2^{2+}O_8$ (712.3724), m.p. 265–267° dec. (MeOH) (iodide salt) [137]; $[\alpha]_D^{21} +125°$ (c 0.1, MeOH) (iodide salt) [137], was first isolated in the form of colorless needles as its iodide salt from both *Thalictrum longistylum* DC. and *T. podocarpum* Humb. in 1976 [556]. The bisquaternary alkaloid was isolated from both the chloroform-soluble nonquaternary alkaloid fraction and the quaternary alkaloid fraction of the extract and showed UV maxima at 276 nm (log ε 3.89) and 283 nm (log ε 3.87) and a simple CD spectrum ($[\theta]_{226} +134,000$ and $[\theta]_{280} +13,500$) [137]. The

^{1}H NMR spectrum (TFA) indicated the presence of two quaternary *N*-methyl groups as singlets at δ 3.27 (6H) and 3.53 (6H); five methoxy groups as singlets at δ 3.63, 3.73, 3.98, 4.03, and 4.06; one methylenedioxy group as a singlet at δ 6.08; and nine aromatic protons as two singlets at δ 5.75 [H(8)] and 5.88 [H(8′)] with a multiplet of seven protons from δ 6.73 to 7.22 [137]. The structural similarities to the alkaloid thalistyline (**102**), which co-occurs along with this alkaloid in the same plants [137,301,556], prompted consideration of a bisquaternary structure, which was confirmed by preparation of *N*-methylthalistyline (**294**) iodide via reaction of thalistyline (**102**) in acetone with methyl iodide [137]. The isolation of *N*-methylthalistyline (**294**), along with *N*-desmethylthalistyline (**293**) and thalistyline (**102**), represented the first examples of dimeric benzylisoquinoline alkaloids bearing trioxygenation in each of the isoquinoline rings [556].

N-Methylthalistyline (**294**) was found to produce a hypotensive response in normotensive dogs and rabbits [137,301] and to inhibit the in vitro growth of *Mycobacterium smegmatis* at a concentration of 100 μg/mL and *Staphylococcus aureus* at 1000 μg/mL [137,301].

3.5.17. *O*-Methylthalmethine.

Tabular review [529, alkaloid No. 96, 530], UV [529,599], IR [599], ^{1}H NMR [529,599], MS [529,585].
O-Methylthalmethine (**336**), $C_{37}H_{38}N_2O_6$ (606.2730), m.p. 245–246° (C_6H_6), $[\alpha]_D^{21}$ +237° (c 1, $CHCl_3$), was first isolated from a Bulgarian *Thalictrum minus* L. variety as colorless needles in 1965 [310,599]. The UV spectrum was characterized by maxima at 280 nm (log ε 4.13) and 314 nm (log ε 3.87), and the IR spectrum showed a band at 1633 cm^{-1} (imino) that increased to 1657 cm^{-1} in the spectrum of the HCl salt (m.p. 198–200° [MeOH]) [599]. The ^{1}H NMR spectrum indicated the presence of one *N*-methyl group at δ

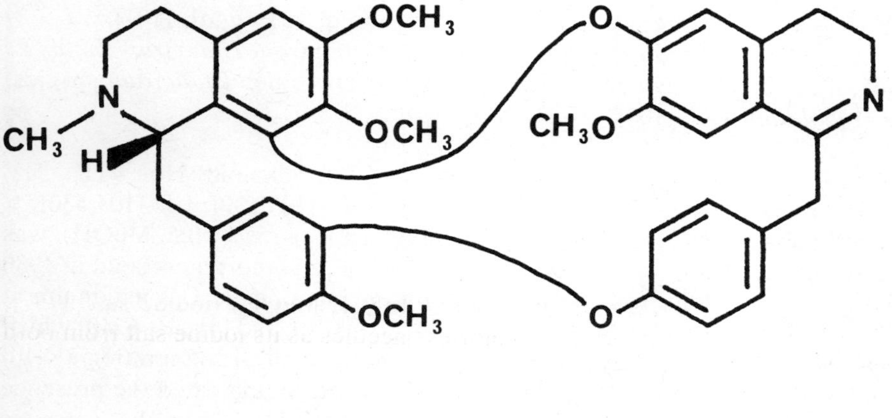

336

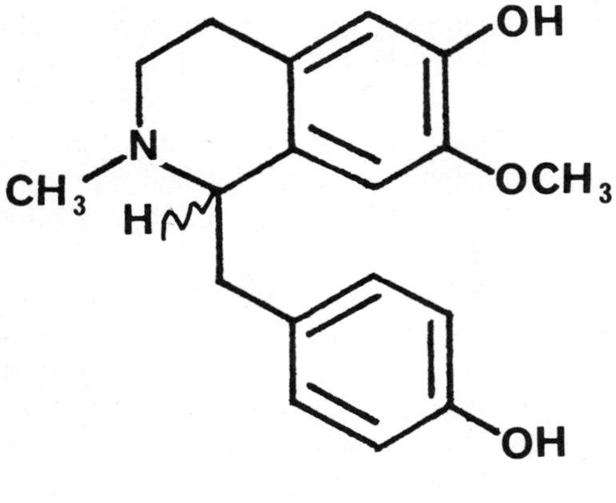

337

1.94; four methoxy groups at δ 3.70, 3.84, 3.89, and 3.91; and a methylene group at δ 4.19 (2H, J = 0.2 Hz) [599]. Reduction of the alkaloid with either Adams catalyst or sodium borohydride afforded a dihydro derivative, m.p. 278–280° (C_6H_6), which on treatment with formaldehyde–formic acid yielded *O*-methylthalicberine (**333**). Finally, treatment of *O*-methylthalmethine with sodium in liquid ammonia also furnished the same dihydro derivative whereas a like treatment with a greater quantity of reagents produced (+)-*N,O,O*-trimethylcoclaurine (**327**) and (±)-isococlaurine (**337**), thus establishing the structure of *O*-methylthalmethine (**336**) [599].

O-Methylthalmethine (**336**) was observed to inhibit the in vitro growth of *Mycobacterium smegmatis* at a concentration of 100 μg/mL [101].

O-Methylthalmethine (**336**) has been reisolated from *Thalictrum minus* L. [65,272,510] and has been isolated from only one other *Thalictrum* species, *T. revolutum* DC. [101].

3.5.18. Neothalibrine.   Tabular review [530, alkaloid No. 211], UV [105,530], IR [105,530], ¹H NMR [105,530], MS [105,530], CD [105,530]. Neothalibrine (**338**), $C_{38}H_{44}N_2O_6$ (624.3199), $[\alpha]_D^{27}$ +155° (c 0.5, MeOH), was first isolated from *Thalictrum revolutum* DC. as an amorphous solid in 1980 [105]. The alkaloid was characterized by having a single UV maximum at 284 nm (log ε 4.10) (CD $[\theta]_{231}$ +29,600, $[\theta]_{250}$ −1250, and $[\theta]_{288}$ +6240) which shifted bathochromically on addition of strong alkali to 285 nm (log ε 4.10) and 310 nm (sh) (log ε 3.68). The ¹H NMR spectrum indicated the presence of two *N*-methyl groups as singlets at δ 2.43 and 2.51; four methoxy groups as three singlets at δ 3.59, 3.78 (6H), and 3.82; a hydroxy group as a broad singlet at δ 5.17 (D_2O-exchangeable); and 11 aromatic protons as four singlets

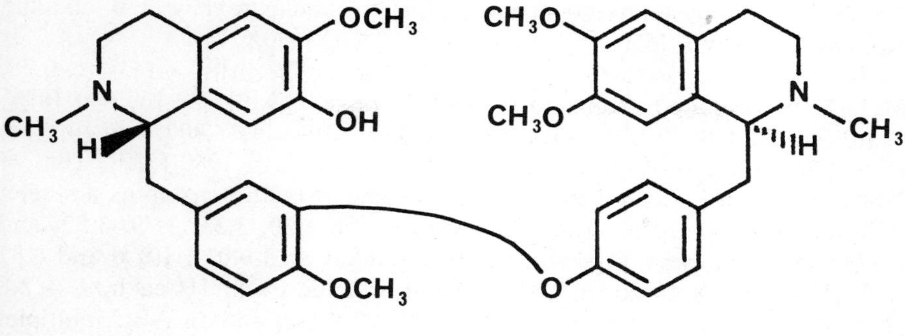

338

at δ 6.09 (1H) [H(8′)], 6.38 (1H) [H(8)], 6.46 (1H), 6.56 (1H); an AA′BB′ quartet at δ 6.78 and 6.98 (J_{AB} = 8.8 Hz); and an ABC multiplet between δ 6.6 and 6.9. The mass spectrum showed a very weak parent ion at M^+ m/z 624 (0.1%) and other important fragment ions at m/z 418 (0.3%) (M—206), 206 (100) (**191**), and 192 (80) (**199**) [105]. These spectral data were consistent with a thalibrine (**329**)-like structure [582] and methylation of neothalibrine with ethereal CH_2N_2 afforded O-methylthalibrine (**326**), thus defining the skeletal structure, positions of oxygenation, and termini of the diphenyl ether linkage of neothalibrine [105]. Ethylation of neothalibrine with ethereal dia-zoethane produced O-ethylneothalibrine (**339**), which underwent reductive cleavage with sodium in liquid ammonia to afford (+)-(S)-2-methyl-4′,6-dimethoxy-7-(ethoxybenzyl)tetrahydroisoquinoline (**428**) and (+)-S-arme-pavine (**328**) and thus established the structure of neothalibrine (**338**) [105].

Neothalibrine (**338**) was subsequently isolated from *Thalictrum alpinum* L. [70] and *T. rugosum* Ait. [234].

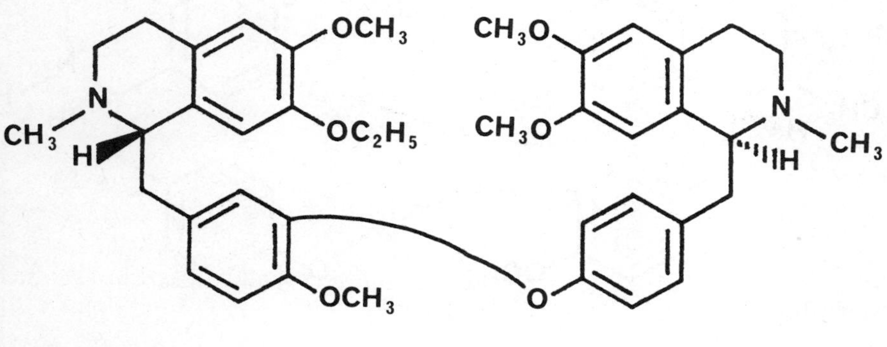

339

3.5.19. *N'*-Norhernandezine (Thalisamine).

Tabular review [530, alkaloid No. 212], ^{1}H NMR [530,557], MS [530,557], CD [530,557].

N'-Norhernandezine (340), $C_{38}H_{42}N_2O_7$ (638.2992), $[\alpha]_D^{22}$ + 143° (c 0.28, MeOH), was isolated as an amorphous substance (CD $[\theta]_{218}$ + 169,000, $[\theta]_{241}$ 0, $[\theta]_{247}$ − 31,400, $[\theta]_{261}$ 0, $[\theta]_{266}$ + 4,000 (sh), $[\theta]_{287}$ + 17,300, and $[\theta]_{320}$ 0) from *Thalictrum rochebrunianum* Franc. and Sav. in 1980 [557]. The ^{1}H NMR spectrum indicated the presence of one *N*-methyl group as a singlet at δ 2.30; five methoxy groups as singlets at δ 3.30, 3.35, 3.79, 3.82, and 3.93; and nine aromatic protons as two singlets at δ 6.01 [H(8')] and 6.87 [H(5')], and ABXY pattern at δ 6.36 and 6.81 (dd each, 1H each, *J* = 2,8 Hz) and 7.14, 7.36 (dd each, 1H each, *J* = 2,8 Hz) and an ABC multiplet from δ 6.5 to 6.9 [557]. The mass spectrum showed a parent ion at M$^+$ *m/z* 638 (9%) with other important fragment ions at *m/z* 623 (4%), 607 (2), 501 (1), 460 (13), 425 (15), 411 (34), 397 (22), and 206 (100) [557]. Reduction of thalsimine (312) with sodium borohydride afforded norhernandezine (340) and epinorhernandezine (341), the former of which was identical with the naturally occurring alkaloid and thus confirmed the structure and stereochemistry of *N'*-norhernandezine (340) [557]. It was not until 1984 that Guinaudeau et al. called attention to the close resemblance between *N'*-norhernandezine (340) and thalisamine, an alkaloid isolated from *Thalictrum simplex* L. in 1967 and assigned as 342 [600]. Thalisamine, which was isolated as a crystalline substance, m.p. 191–194° (Et$_2$O), with a single UV maximum at 284 nm (log ε 4.60), was assigned structure 342 after consideration of its IR, ^{1}H NMR, and mass spectral data plus its conversion to hernandezine on methylation (formaldehyde–formic acid) [600]. Because the *N*-methyl signal in the ^{1}H NMR spectrum of the closely related (+)-*N*-desmethyl-thalidezine (288) is at δ 2.61 and because the *N*-methyl signals of hernan-

340

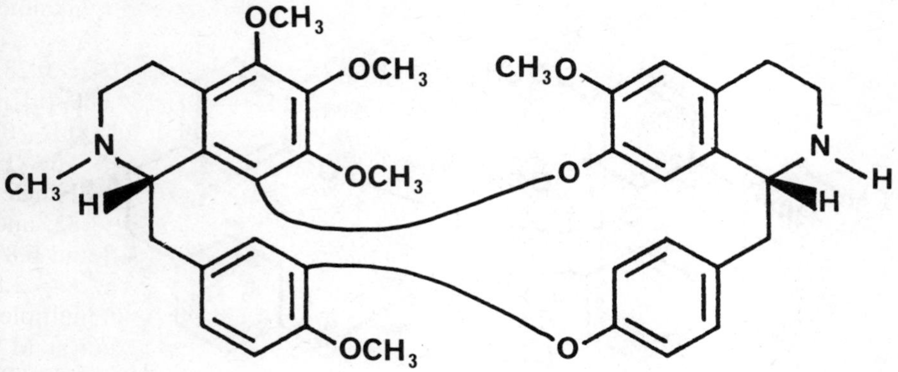

341

dezine are well separated at δ 2.30 and 2.63 [558], thalisamine and *N'*-nor-hernandezine must be assumed to be identical [544]. Although a negative specific rotation $[\alpha]_D^{24}$ −138° (c 0.194, CHCl₃) is reported for thalisamine [600], the methylation product was identified as hernandezine but with no report of the specific rotation of this methylation product [600]. One is left with the conclusion that either there is an error in the sign of the specific rotation of thalisamine or that thalisamine is the (*R,R*) isomer represented by structure **343**. However, because hernandezine (**303**) (*S,S*) and thalide-zine (**289**) (*S,S*) were also isolated from this mixture [600] and because no

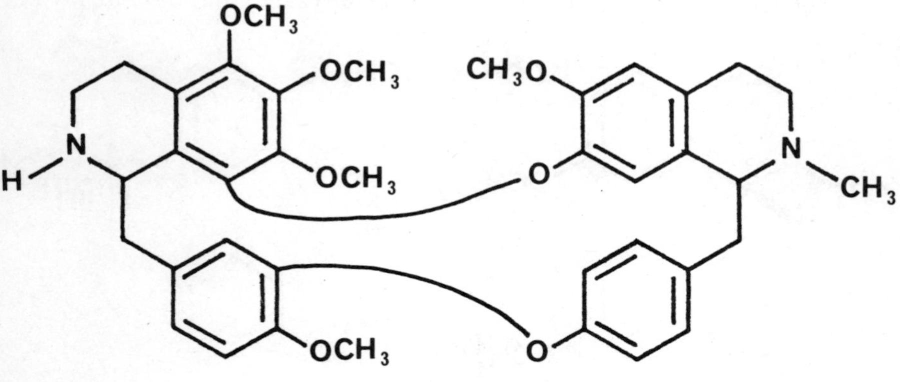

342

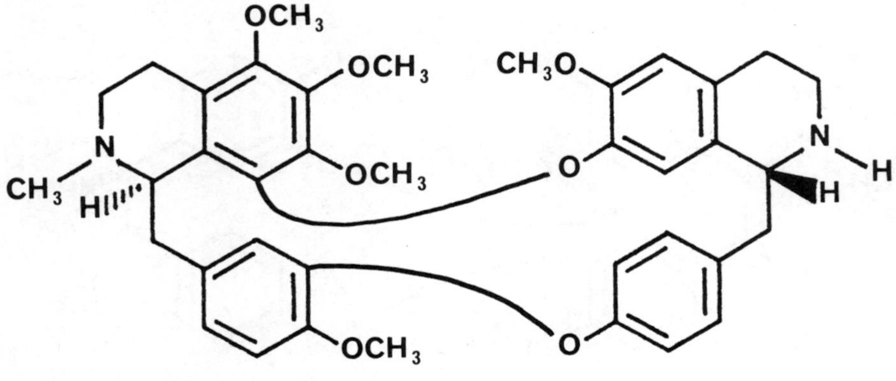

343

alkaloids of the (*R*,*R*) configuration have been reported for this series [529,530], it is more likely that thalisamine is, in reality, (+)-*N'*-norhernandezine. Finally, alkaloids of the (*R*,*R*) configuration would be in great variation to the empirical rules that govern *Thalictrum* bisbenzylisoquinoline alkaloid formation [544].

3.5.20. Northalibrine. Tabular review [529, alkaloid No. 13], UV [529,582], ¹H NMR [529,582], MS [529,582].
Northalibrine (**344**), $C_{37}H_{42}N_2O_6$ (610.3043), $[\alpha]_D$ +47° (c 0.2, CHCl₃), was first isolated from *Thalictrum rochebrunianum* Franc. and Sav. in 1976 as

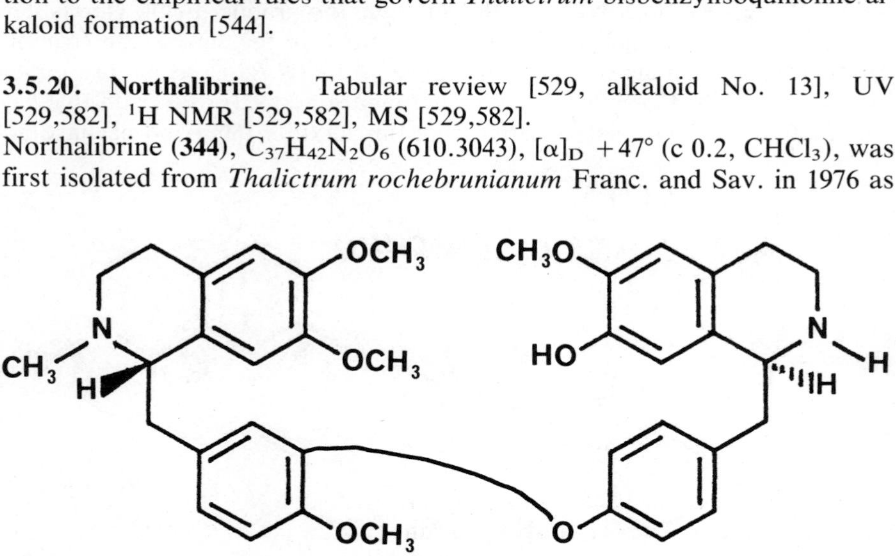

344

345

an amorphous solid [582]. The UV spectrum of the alkaloid was characterized by a single maximum at 284 nm (log ϵ 3.70) and the ^{1}H NMR spectrum
suggested the presence of one *N*-methyl group at δ 2.46 (3H, s) and four
methoxy groups at δ 3.60 (3H, s), 3.71 (3H, s), and 3.73 (6H, s). The mass
spectrum showed the base peak at *m/z* 206 (100%) (**191**) and a diagnostic
fragment ion at *m/z* 178 (23%) (**298**). These spectral data were consistent for
a thalibrine (**329**)-like alkaloid containing a secondary nitrogen atom at N(2′).
Treatment of northalibrine with formaldehyde/sodium borohydride produced thalibrine (**329**) whereas reductive cleavage with sodium in liquid ammonia afforded (+)-(*S*)-*O*-methylarmepavine (**327**) and (+)-(*S*)-coclaurine
(**345**), thereby establishing the structure of northalibrine (**344**) [582].

3.5.21. *N*′-Northalibrunine (*N*-2′-Northalibrunine, 2′-Northalibrunine).

Tabular review [530, alkaloid No. 214], UV [529,583], ^{1}H NMR
[529,557,583], MS [529,557,583], CD [529,557].
N′-Northalibrunine (**346**), $C_{38}H_{42}N_2O_8$ (654.2941), m.p. 158–161° [557],
$[\alpha]_D^{20}$ +79° (c 0.16, MeOH) [557], CD ($[\theta]_{220}$ +180,000, $[\theta]_{236}$ 0, $[\theta]_{245}$
−14,000, $[\theta]_{273}$ −36,000, $[\theta]_{282}$ 0, $[\theta]_{294}$ +53,000, and $[\theta]_{320}$ 0) [557] was first
isolated from *Thalictrum rochebrunianum* Franc. and Sav. in 1978 [583] and
shortly thereafter from the same plant [557]. The UV spectrum was characterized by maxima at 226 nm (log ϵ 4.41), 236 nm (log ϵ 4.56), and 284 nm
(log ϵ 4.24) [583]. The ^{1}H NMR spectrum indicated the presence of one *N*-
methyl group as a singlet at δ 2.47; five methoxy groups as singlets at δ 3.23,
3.35, 3.77, 3.83, and 3.89; and eight aromatic protons as four singlets for
four protons at δ 5.92 [H(8′)], 6.39, 6.48, 6.53, and four other protons as an
ABXY pattern with δ_{AB} δ 7.1–7.4, δ_x 6.4–6.7, and δ_y 6.1–6.3 ($J_{AB} \approx J_{xy} \approx$
8 Hz) [557]. The mass spectrum showed the molecular ion at M$^+$ *m/z* 654

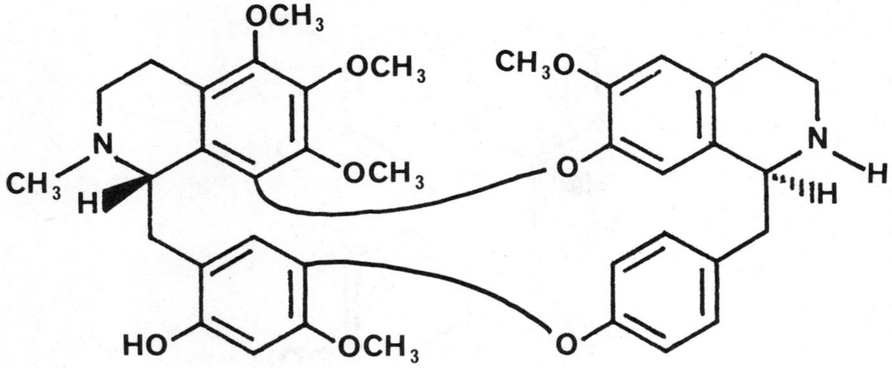

OCH3
OCH3
OCH3
CH3O
N
CH3
H
O
N
H
H
HO
OCH3
O

346

OCH3
OCH3
+
N
OCH3
CH3
HO
OCH3
O

347

OCH3
OCH3
CH3O
N
CH3
OCH3
O
+
N
H

348

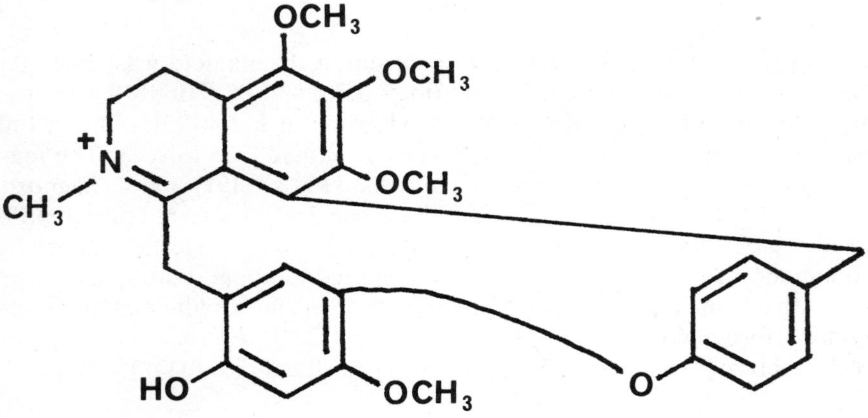

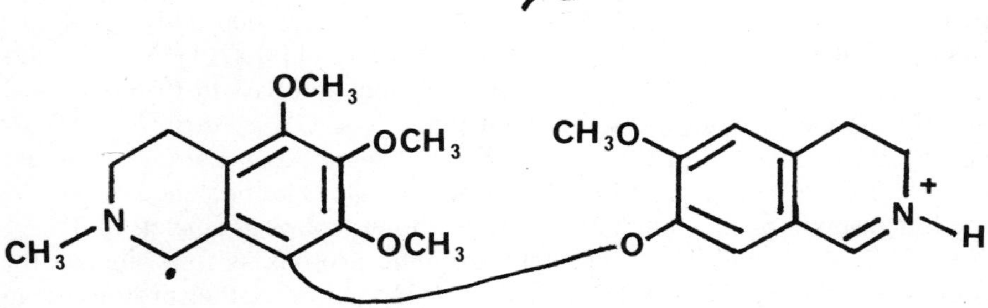

(100%) with other important fragment ions at m/z 632 (12), 476 (7) (**347**), 411 (40) (**348**—H), 397 (27), 222 (12), 206 (37) (**349**), and 177 (23) (**350**) [557]. The spectral data were characteristic of a thalibrunine (**351**)-like structure bearing a secondary nitrogen atom in the right-hand ring. A comparison of the physical properties of the natural alkaloid with those of one of the products of sodium borohydride reduction of thalibrunimine (**332**) [601,602] established their identity and thus the structure of N'-Northalibrunine (**346**) [557].

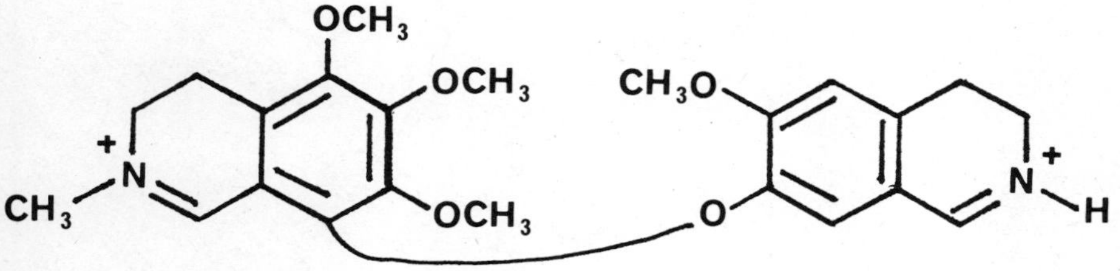

349

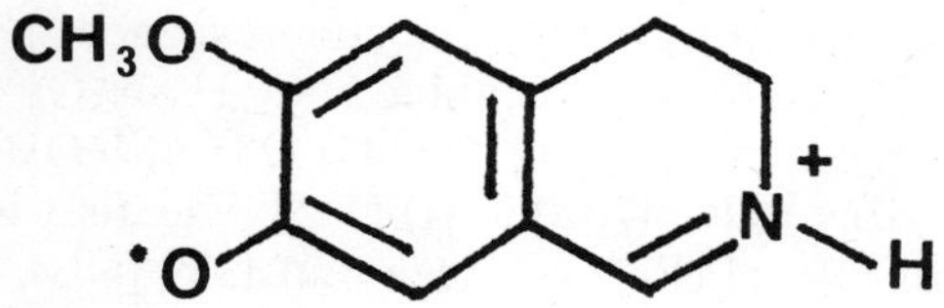

350

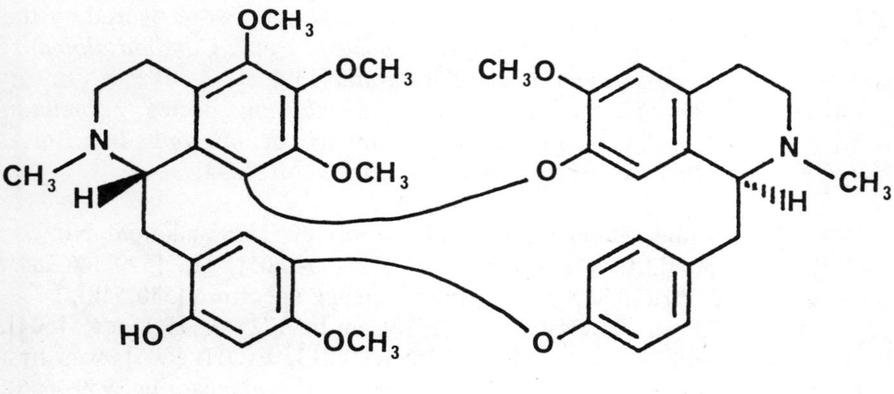

351

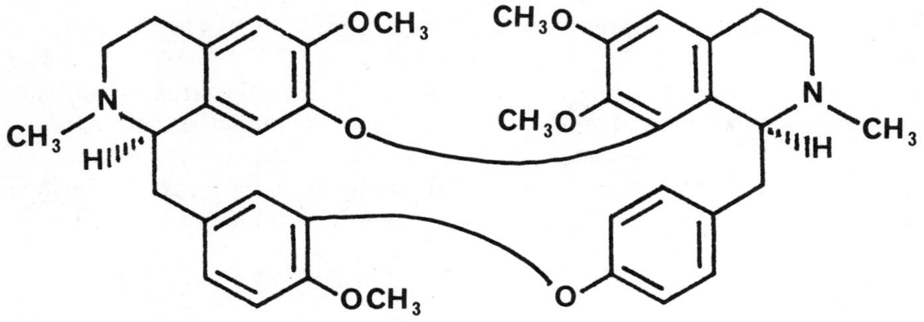

$$352$$

3.5.22 Obaberine (*O*-Methyloxyacanthine).

Tabular review [529, alkaloid No. 46, 530], UV [139,529], ^{1}H NMR [139,171,529], MS [529,546], ORD [547], fluorescence spectra [529,548], phosphorescence spectrum [529,548]. Obaberine (**352**) (*O*-Methyloxyacanthine), $C_{38}H_{42}N_2O_6$ (622.3043), m.p. 138–140° (petroleum ether–C_6H_6 [1:1]) [139] (255° [MeOH] dihydrochloride salt [215]), $[\alpha]_D^{26}$ +169° (c 0.147, MeOH) [215], was first isolated from *Berberis tschonoskyana* Regel (Berberidaceae) in 1959 [603].

Obaberine was noted to produce a hypotensive response in normotensive dogs [139] and rabbits [215]. Administration of an intravenous dosage of 1.0 mg/kg produced no hypotensive response in dogs but at 2.0 and 4.0-mg/kg dosage levels, there was a 40 mm Hg decrease in mean blood pressure [139]. In rabbits, a similar administration of both 1.0 and 4.0 mg/kg produced a 2–3 min decrease in mean blood pressure of 45 mm Hg [215]. Obaberine was also observed to possess a weak antimicrobial activity as measured by the in vitro inhibition of growth of *Staphylococcus aureus, Candida albicans,* and *Mycobacterium smegmatis* at 1000 µg/mL [139].

Obaberine (**352**) was first isolated from a *Thalictrum* species, *T. lucidum* L. in 1976 [139], and subsequently from *T. minus* var. *microphyllum* Boiss. [146], *T. minus* L. race B [215], and *T. rugosum* Ait. [234].

3.5.23. Obamegine (Stepholine).

Tabular review [529, alkaloid No. 71, 530], UV [529,604], IR [604], ^{1}H NMR [529,604,605], MS [529,546,584], fluorescence spectra [530,548], phosphorescence spectrum [530,548]. Obamegine (**353**), $C_{36}H_{38}N_2O_6$ (594.2730), m.p. 172° (C_6H_6) dec. [604], $[\alpha]_D^{30}$ +241° (CHCl$_3$) [604], $[\alpha]_D^{19}$ +225° (c 0.013, EtOH) [605], was first isolated from *Berberis tschonoskyana* Regel (Berberidaceae) in 1959 [606]. The first isolation from a *Thalictrum* species, *T. rugosum* Ait., was reported in 1966 [604].

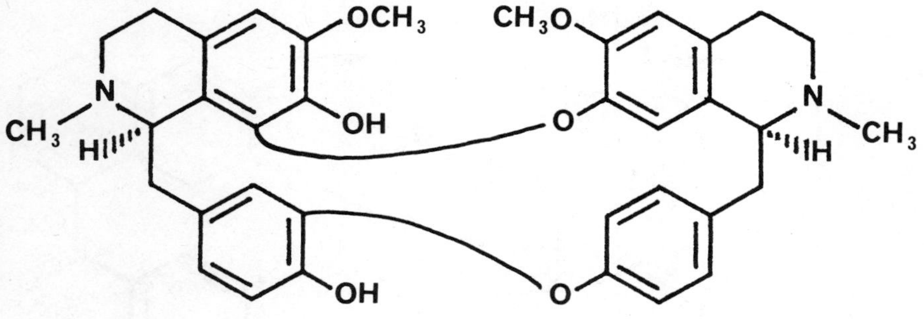

353

Obamegine was screened for antimicrobial activity [152,353] and found to inhibit the growth of *Staphylococcus aureus*, *Escherichia coli*, *Salmonella gallinarum*, *Klebsiella pneumoniae*, and *Candida albicans* at a concentration of 100 μg/mL [139] in addition to *Mycobacterium smegmatis* at 50 μg/mL [139]. The hypotensive effect of obamegine in normotensive dogs was first described in 1976 when intravenous dosages of 0.5, 1.0, and 2.0 mg/kg produced a decrease in mean blood pressure of 58, 65, and 75 mm Hg, respectively [139]. In subsequent work [606], an intravenous dose of 4 mg/kg produced a decrease in mean arterial pressure slightly greater than 100 mm Hg. The onset of action was 5 sec after injection, with the peak effect occurring at 1 min and 30 sec and lasting 1 min and 45 sec. The total duration of hypotensive action was about 30 min. The alkaloid produced bradycardia during the hypotensive response but had no inhibitory respiratory effects. Tachyphylaxis to the hypotensive effect was noted; dosages of 0.5–4 mg/kg were not lethal to the test animals (dogs) [607]. Obamegine was found to produce inconsistent results when administered as a hypertensive-blocking agent prior to the administration of dopamine and norepinephrine [607]. Obamegine was observed to antagonize the phenylephrine-induced contractions of rabbit aortic strips and thus possesses α-adrenoreceptor blocking activity [607]. The alkaloid had no effect on canine respiration nor did it inhibit neuromuscular transmission at concentrations of $1–30 \times 10^{-6}$ *M* [607].

Obamegine (**353**) has also been isolated from *Thalictrum lucidum* L. [139] and *T. minus* L. var. *microphyllum* Boiss. [544], as well as having been reisolated from *T. rugosum* Ait. [152,353].

3.5.24. Oxothalibrunimine. Tabular review [530, alkaloid No. 215], UV [530,557], IR [530,557], ^{1}H NMR [530,557], ^{13}C NMR [530,557], MS [530,557], CD [530,557].

354

Oxothalibrunimine (**354**), $C_{38}H_{38}N_2O_9$ (666.2577), m.p. 198–200 ° (Me₂CO), [α]$_D^{22}$ −70° (c 0.25, MeOH); CD ([θ]$_{222}$ +28,000, [θ]$_{222}$ 0, [θ]$_{243}$ −156,000, [θ]$_{270}$ 0, [θ]$_{278}$ −6,700, [θ]$_{285}$ 0, [θ]$_{299}$ +33,000, [θ]$_{320}$ +12,000, [θ]$_{342}$ 0, [θ]$_{365}$ −9,900, [θ]$_{390}$ 0); was isolated as yellow prisms from *Thalictrum rochebrunianum* Franc. and Sav. in 1980 [557]. The UV spectrum was characterized by a series of undulating shoulders at 220 nm (log ε 4.34), 240 nm (sh) (log ε 4.10), 270 nm (sh) (log ε 3.86), and 330 nm (sh) (log ε 3.40) with a bathochromic shift in acidic media. The IR spectrum displayed bands at 1680 and 1625 cm^{-1} for carbonyl and imine absorption, respectively, which was supported by the appropriate ^{13}C NMR signals at δ 192.2 (carbonyl) and 165.0 (imine). The ^{1}H NMR spectrum indicated the presence of one *N*-methyl group as a singlet at δ 2.43; five methoxy groups as singlets at δ 3.35, 3.47, 3.79, 3.84, and 3.91; a hydroxy group as a broad singlet at δ 12.86; and eight aromatic protons as four singlets at δ 5.95 [H(8′)], 6.42 (1H), 6.52 (1H), and 6.62 (1H) plus an ABXY pattern with split doublets for four protons at δ 6.78 (*J* = 2.2, 8.6 Hz), 7.05 (*J* = 1.9, 8.6 Hz), 7.41 (*J* = 2.2, 8.3 Hz), and 8.23 (*J* = 1.9, 8.3 Hz). The mass spectrum exhibited the parent ion and base peak at M$^+$ *m/z* 666 with other important fragment ions at *m/z* 651 (37%), 649 (29) (M—OH), 638 (M—CO), 635 (16) (M—OCH₃), 410 (2), 409 (6), and 333 (10) (1/2M^{2+}). These spectral data were characteristic of a thalibrunimine (**332**)-like structure containing a carbonyl function. Aerial oxidation of thalibrunimine (**332**) via refluxing a benzene solution of the alkaloid for 6 hr afforded oxothalibrunimine (**354**) which was identical with the isolated alkaloid. Because it was not indicated that a spot corresponding in R_f to oxothalibrunimine (**354**) was present in the TLC of a fresh plant extract and because aerial oxidation of a solution of thalibrunimine (**332**) produced oxothalibrunimine, it is possible that oxothalibrunimine may be an artifact. Is it not possible that N'-northalibrunine (**346**) could be oxidized to thalibrunimine (**332**) and that to oxothalibrunimine (**354**) via aerial oxidation of solutions of the alkaloids in organic solvents?

3.5.25. Oxyacanthine. Tabular review [529, alkaloid No. 48, 530], UV [529,608], IR [608], ^{1}H NMR [171,529,608], MS [529,546,584], ORD [529,547], fluorescence spectra [530,548], phosphorescence spectrum [530,548].

Oxyacanthine (**355**), $C_{37}H_{40}N_2O_6$ (608.2886), m.p. 215–216° (Et_2O) [139]; 212–214° (petroleum ether) [608]; $[\alpha]_D^{29} + 285.6°$ (c 0.5, $CHCl_3$) [608], was first isolated from the roots of several *Berberis* species (Berberidaceae) in the nineteenth century [609]. The first isolation of this alkaloid from the genus *Thalictrum* occurred in 1976 when oxyacanthine was found in *T. lucidum* L. [139].

The alkaloid was found to induce a sympatholytic effect in dogs [610] and to product vasodilation of canine vessels [611]. Administration of the alkaloid to normotensive dogs in an intravenous dosage of 1.0 and 2.0 mg/kg failed to produce a hypotensive response [139]. Oxyacanthine was found to inhibit the in vitro growth of *Staphylococcus aureus*, *Mycobacterium smegmatis*, and *Candida albicans* at a concentration of 1000 µg/ml [139] and to possess an antibacterial activity [612,613].

Oxyacanthine (**355**) has not been isolated from any other species of *Thalictrum* to date.

3.5.26. Thalabadensine. Tabular review [529, alkaloid 106a, 530], UV [530,575], ^{1}H NMR [530,575], MS [530,575].

Thalabadensine (**356**), $C_{36}H_{38}N_2O_6$ (594.2730), was isolated as a yellow amorphous base from *Thalictrum sultanabadense* Stapf. in 1978 [575]. The

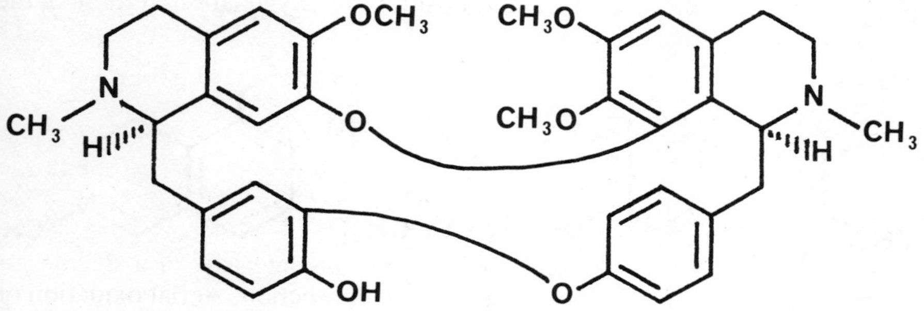

355

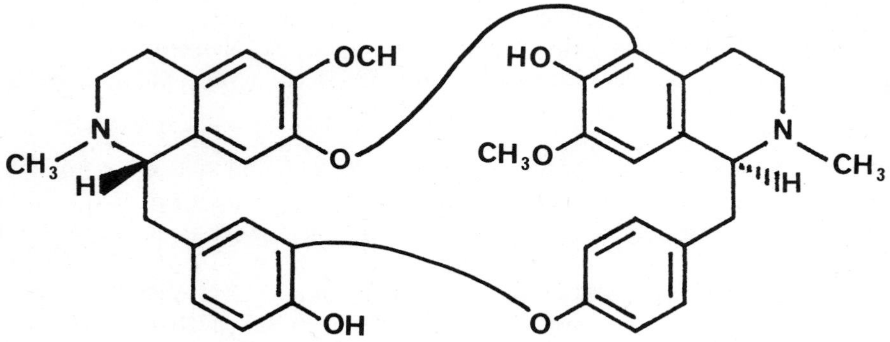

356

UV spectrum of the alkaloid showed a single maximum at 286 nm (no extinction coefficient reported) although neither the specific rotation nor CD spectrum was reported [575]. The ^{1}H NMR spectrum indicated the presence of two N-methyl groups as singlets at δ 2.17 and 2.56; two methoxy groups or singlets at δ 3.81 and 3.87; two hydroxy groups as a broad singlet at δ 4.63; and 10 aromatic protons from δ 5.95 to 6.76, two of which were a two-proton singlet at δ 5.95 [H(8) and H(8')]. The mass spectrum displayed the parent ion at M$^+$ *m/z* 594 and an intense fragment ion at *m/z* 381 (357—H). These data were consistent with a thalmine (358)-like structure bearing two methoxy groups and one phenolic hydroxy group in the top half of the dimer and one phenolic hydroxy group in the bottom half of the molecule. Methylation (CH$_2$N$_2$) of thalabadensine afforded *O*-methylthalmine (359) thereby establishing the molecular structure, positions of oxygenation, termini of the

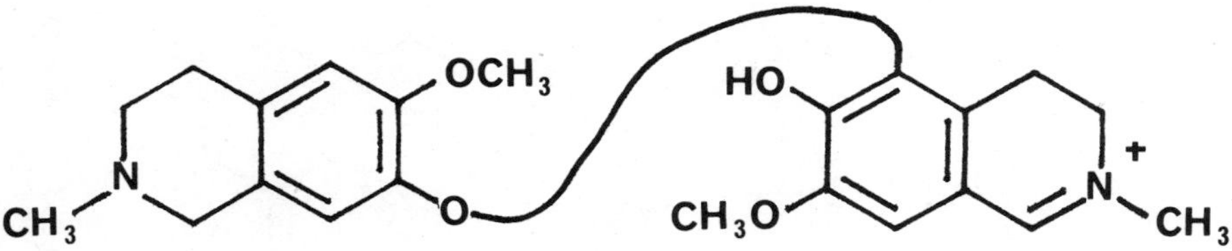

357

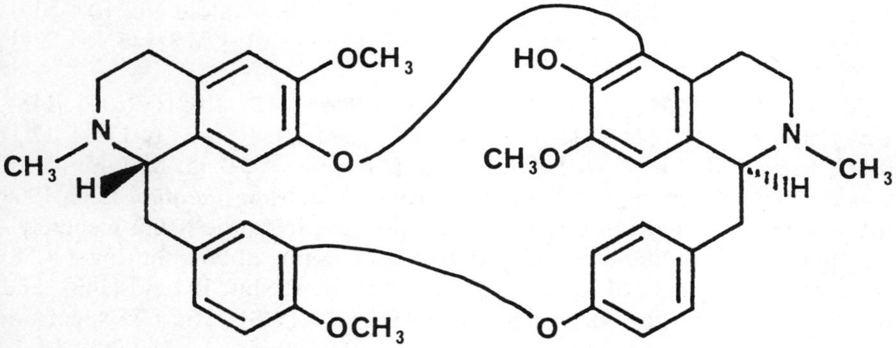

358

diphenyl ether bridges, and presumably, the stereochemistry (data not reported) [575]. The position of the hydroxy group in the top ring of thalabadensine was established to be C(6'), because on methylation signals at δ 3.61 and 3.84 appeared, the former of which is characteristic of a C(6) methhoxy group in this series [529,530]. The second hydroxy group must therefore reside in the bottom ring, as demonstrated by the methylation. On the basis of these spectral data and the product of methylation, the structure of thalabadensine as **356** was considered most probable [575].

Thalabadensine was reisolated from *Thalictrum sultanabadense* Stapf. in 1981 [576] and isolated from *T. minus* L. in 1983 [181].

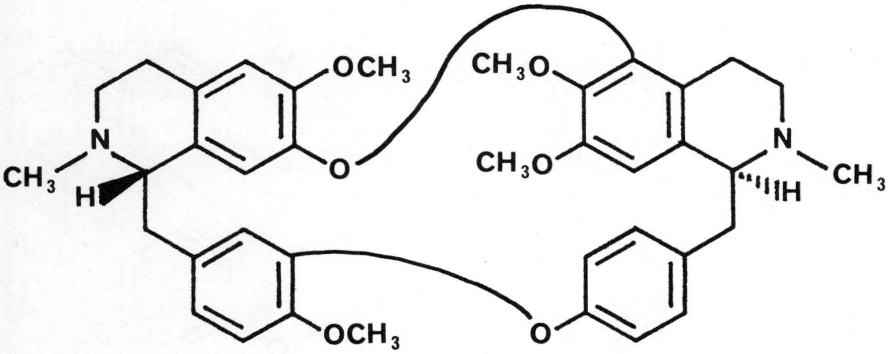

359

3.5.27. Thalfine (Thalphine).   Tabular review [529, alkaloid No. 102, 530], UV[143,529,615], IR[143,616], ^{1}H NMR[143,529,615,616], MS[143,215,529], CD [215,530,562].

Thalfine (**360**), $C_{38}H_{36}N_2O_8$ (648.2472), m.p. 141–142° (EtOH) [615], 148–149° (Et$_2$O–MeOH) [143], 150–151° (MeOH) [215]; $[\alpha]_D^{15}$ +69° (c 1.0, EtOH) [615]; $[\alpha]_D^{25}$ +18.3 (c 0.49, EtOH) [143]; $[\alpha]_D^{26}$ +85° (c 0.35, MeOH) [215], was first isolated as prismatic crystals from *Thalictrum foetidum* L. in 1968 [615]. The UV spectrum was distinctly different from the usual bisbenzyl-tetrahydroisoquinoline alkaloid and showed maxima at 260 nm (log ϵ 4.58) and 348 nm (log ϵ 3.86) [615] with a bathochromic shift in acid [143], and the IR spectrum showed a signal at 1640 cm^{-1} [143]. The CD spectrum showed a multiphasic curve with $[\theta]_{208}$ +134,000, $[\theta]_{233}$ +67,000, $[\theta]_{263}$ +82,000, $[\theta]_{289}$ −28,000, $[\theta]_{388}$ +690, and $[\theta]_{360}$ −1390 [215]. The ^{1}H NMR spectrum indicated the presence of one *N*-methyl group at δ 2.20; four methoxy groups at δ 3.40 [C(7)], 3.50 [C(6)], 3.61 [C(5′)], and 3.76 [C(13)]; one methylenedioxy group at δ 6.04; and 10 aromatic protons, the most high field being at δ 5.93 [H(8)] [616]. The mass spectrum showed the molecular ion at M$^+$ *m/z* 648 (100%) with other significant fragment ions at *m/z* 647 (31%), 633 (83) (M—CH$_3$), 617 (21) (M—OCH$_3$), 442 (7) (**361**+CH$_3$), 421 (7) (**362**—H), 324 (49) (M$^+$/2), 220 (12) (**363**—H), and 204 (21) (**364**—H) [215]. These spectral data were consistent with a benzyltetrahydroisoquinoline–benzylisoquinoline dimeric structure. Hofmann degradation (2X) of thalfine dimethiodide produced trimethylamine but the remaining product remained nitrogenous, further suggesting the presence of a benzylisoquinoline moiety in the structure of thalfine [616]. Oxidation of thalifine with potassium permanganate in acetone afforded 2-methoxy-4,5′-dicarboxybiphenyl ether

360

360a

361

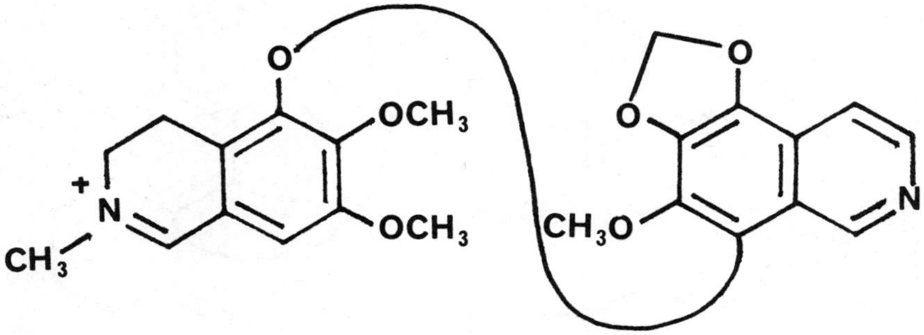

362

(**309**), confirming the dimeric nature of thalfine and establishing the placement of one methoxy group and the termini of one diphenyl ether bridge
[616]. Reduction (Zn/H_2SO_4) of thalfine dimethiodide afforded *N*-methyl-
tetrahydrothalfine methiodide which on *N*-demethylation(ethanolamine)
gave *N*-methyltetrahydrothalfine (**365**) (λ_{max} 283 nm), a substance with an
IR spectrum identical to that of thalfinine (**366**) [616]. Reductive cleavage
of thalfine with sodium in liquid ammonia afforded laudanidine (**55**) (stereochemistry unspecified) plus *O*-methylarmepavine (**327**) (stereochemistry
unspecified) and prompted the assignment of thalfine as **360a** (stereochemistry not defined) [616]. In a subsequent study, the stereochemistry of the
asymmetric center in thalfine was assigned as (*S*) [215]. This was demonstrated by reduction of thalfine with zinc–hydrochloric acid followed by *N*-
methylation. The resulting diastereoisomers were separated and identified

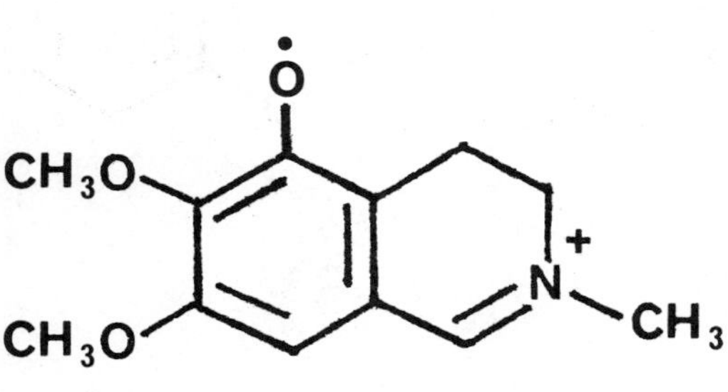

363

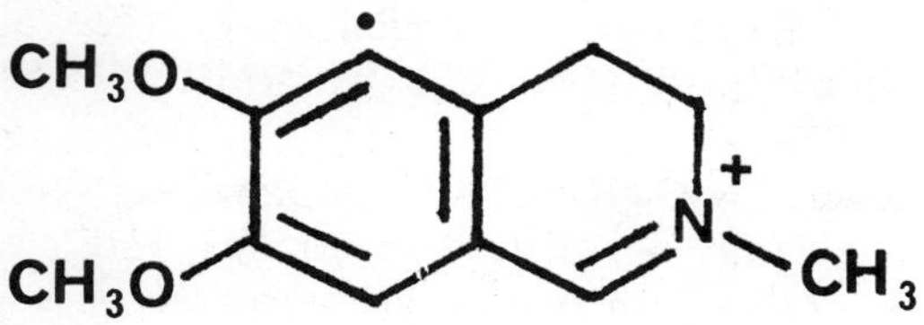

364

as thalfinine (**366**) and epithalfinine (**366**) [with opposite stereochemistry at C(1′)]. Reductive cleavage of thalfinine with sodium and liquid ammonia afforded (+)-(S)-O-methylarmepavine (**327**) as the major nonphenolic product, thereby confirming the (S) stereochemistry at the C(1) carbon in thalfine [215]. A recent communication revised the structure of thalfine to **360** [519a]. This was based on extensive ^{1}H NMR studies, particularly double irradiation techniques, and NOE experiments.

Administration of thalfine to normotensive rabbits in intravenous dosages of 0.1, 0.2, and 1.0 mg/kg failed to produce a hypotensive effect [215]. Thalfine exhibited an in vitro antimicrobial effect against *Mycobacterium smegmatis* at a minimum inhibitory concentration of 100 µg/mL [215].

Thalfine has been reisolated from *Thalictrum foetidum* L. [192] and isolated from *T. minus* L. race B [143,215].

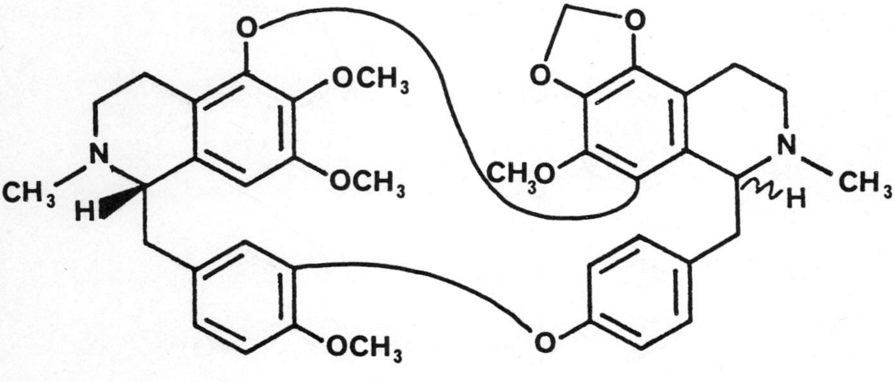

365

3.5.28.   Thalfinine (Thalphinine).   Tabular review [529, alkaloid No. 103, 530], UV [215,529,615], IR [529,615], ^{1}H NMR [215,529,615,616], MS [215,529], CD [215,529].

Thalfinine (**366**), $C_{39}H_{42}N_2O_8$ (666.2941), m.p. 117–118° [615]; $[\alpha]_D^{16}$ +115° (c 0.95, EtOH) [615]; $[a]_D^{26}$ +141° (c 0.25, MeOH) [215], was first isolated from *Thalictrum foetidum* L. in 1968 as an amorphous base [615]. The perchlorate salt, m.p. 234–235° dec., $[\alpha]_D^{21}$ +135° [c 1.16, EtOH–H$_2$O(2:1)], and the hydrochloride salt, m.p. 223–226 dec. (H$_2$O), were both crystalline [615], as well as the hydriodide, m.p. 234–236° dec. (MeOH–H$_2$O) [215]. The UV spectrum was characterized by a single maximum at 282 nm (log ϵ 3.76) [615] and the CD spectrum showed $[\theta]_{207}$ +195,000, $[\theta]_{227}$ (sh) +57,100, $[\theta]_{245}$ (min) +10,900, $[\theta]_{255}$ +36,200, and $[\theta]_{288}$ −12,800 [215]. The ^{1}H NMR spectrum indicated the presence of two *N*-methyl groups as singlets at δ 2.35 [N(2)] and 2.60 [N(2′)]; four methoxy groups as singlets at δ 3.42 [C(7)], 3.49 [C(6)], 3.72 [C(5′)], and 3.86 [C(12)]; one methylenedioxy group at δ 5.86 (s); eight aromatic protons as singlets at δ 6.02 [H(8)], 6.42 (1H), 6.77 (2H), and an AA′BB′ system at δ 6.69 and 7.16 (q, J9, 4 Hz); and one methine proton [H(1) or H(1′) at δ 4.44] [215]. The mass spectrum showed the parent ion at M$^+$ *m/z* 666 (95%) with other important fragment ions at *m/z* 440 (14%) (**367**), 220 (100) (**368**), 204 (16) (**364**—H) [215]. On the basis of spectral data and its relationship to what was then assigned as thalfine (**360a**), the structure of thalfinine was proposed as **366a** [615,616]. Reductive cleavage of thalfinine with sodium in liquid ammonia afforded (+)-(*S*)-*O*-methylarmepavine (**327**) as the major nonphenolic product with no pure phenolic compounds being isolated, although two were detected by TLC. Hence the asymmetric center at C(1) (the trimethoxylated benzylisoquinoline portion) was assigned the

366

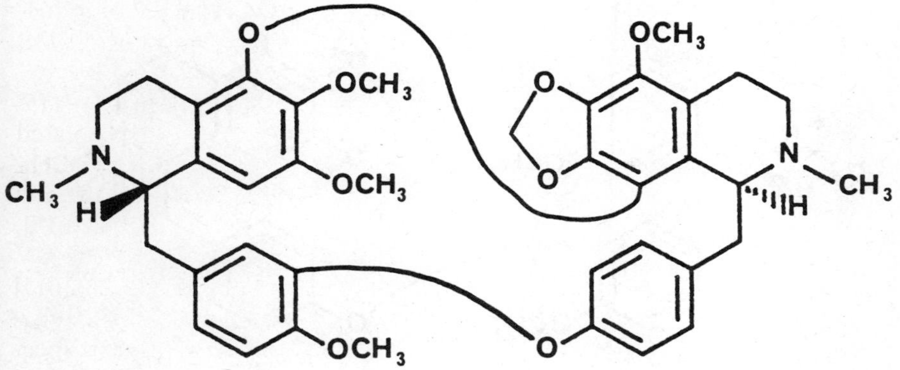

366a

(S) configuration [215]. The assignment of the stereochemistry at C(1′) (the methylenedioxy-containing portion) remained undefined but was tentatively suggested to likewise be (S). This was proposed to be likely because the alkaloid thalirabine (**369**) from the same plant is of the (S,S) configuration and both thalfinine and thalirabine could be visualized as sharing a common

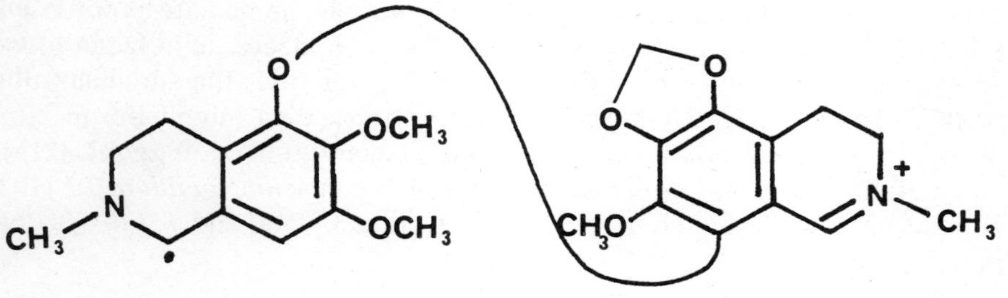

367

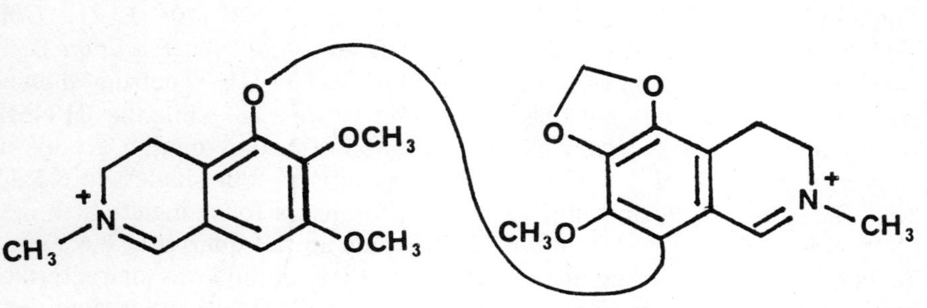

368

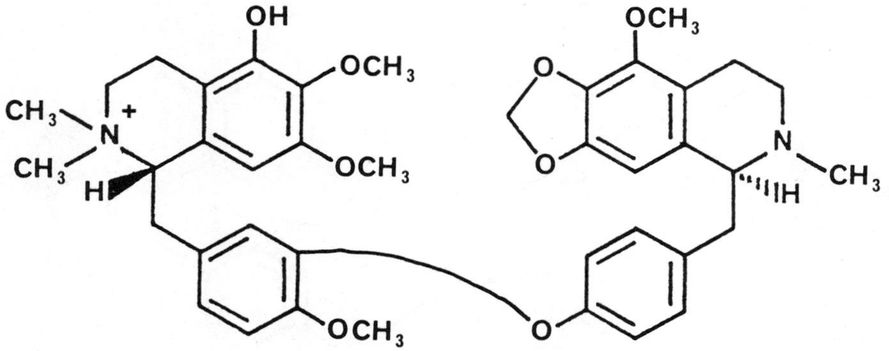

369

biosynthetic pathway [215]. A recent communication described the structural reassignment of thalfine as **360** [519a]. Thus thalfinine must also be reassigned as **366** based on detailed ^{1}H NMR experiments and the conversion of thalfine to thalfinine via reduction [519a].

Thalfinine (**366**) was administered to normotensive rabbits in an intravenous dosage of 0.1, 0.2, 0.4, and 1.0 mg/kg with no immediate hypotensive effects being observed. However, a 10–25 mm Hg decrease in mean blood pressure that lasted 0.5 min was observed 2–3 min after the administration of the alkaloid [215]. Thalfinine was also observed to inhibit the in vitro growth of *Mycobacterium smegmatis* at a concentration of 50 μg/mL [215].

Thalfinine (**366**) has been reisolated from *Thalictrum foetidum* L. [192] and has also been isolated from *T. minus* L. race B [215] and *T. faberi* Ulbr. [471,555].

3.5.29. Thalfoetidine. Tabular review [529, alkaloid No. 99], UV [529,617], IR [617], ^{1}H NMR [529,617], ORD [453,529,618], MS [529,560]. Thalfoetidine (**370**), $C_{38}H_{42}N_2O_7$ (638.2992), m.p. 168–170° (Et$_2$O), $[\alpha]_D^{21}$ −88.6° (c 1.0, CHCl$_3$), was first isolated from *Thalictrum foetidum* L. in 1966 in the form of colorless prisms [549,617]. The UV spectrum showed no maxima at 275 nm (log ε 3.88) and 285 nm (log ε 3.88) while the ^{1}H NMR spectrum was characterized by the presence of two *N*-methyl groups as singlets at δ 2.32 and 2.70; four methoxy groups as four singlets at δ 3.32, 3.51, 3.77, and 3.89; and nine aromatic protons as four singlets at δ 6.33 (2H), 6.40 (1H), 6.47 (2H), and 6.75 (2H) plus an AB quartet at δ 6.93 and 7.57 ($J = 9$ Hz) [617]. The mass spectrum of thalfoetidine was characterized by the parent ion and base peak at M$^+$ *m/z* 638 (100%) with other significant fragment ions at *m/z* 637 (46%) (M—1), 623 (9) (M—CH$_3$), 607 (6)

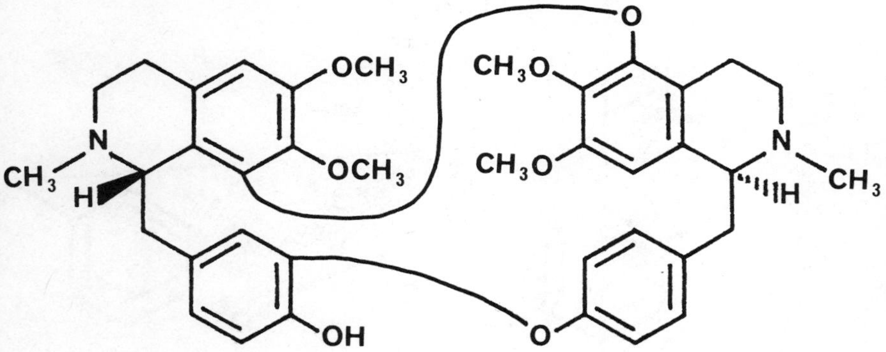

$$\underset{\sim}{\textbf{370}}$$

(M—OCH$_3$), 515 (2) (**371**), 417 (63) (**372**), 402 (57) (**372**—CH$_3$), 213 (67) (**373**), 206 (18) (**373**—CH$_3$), 190 (69) (**373**—(CH$_3$)$_2$O), 175 (7), 174 (5), 90 (4), and 89 (2) [560]. Treatment of thalfoetidine with CH$_2$N$_2$ produced O-methyl-thalfoetidine (**286**) which on reductive cleavage with sodium in liquid ammonia gave (+)-N-O,O-trimethylcoclaurine (**327**) as the nonphenolic base [617]. A similar cleavage of O-ethylthalfoetidine (**374**) afforded (+)-O-ethy-larmepavine (**375**) and a phenolic product characterized as the diphenol **376**. The structure of thalfoetidine was thus proposed as **377** on the basis of the spectral data and the products of the reductive cleavage [617]. In 1969, the structure of thalfoetidine was revised to **370** by interrelation with thalidasine

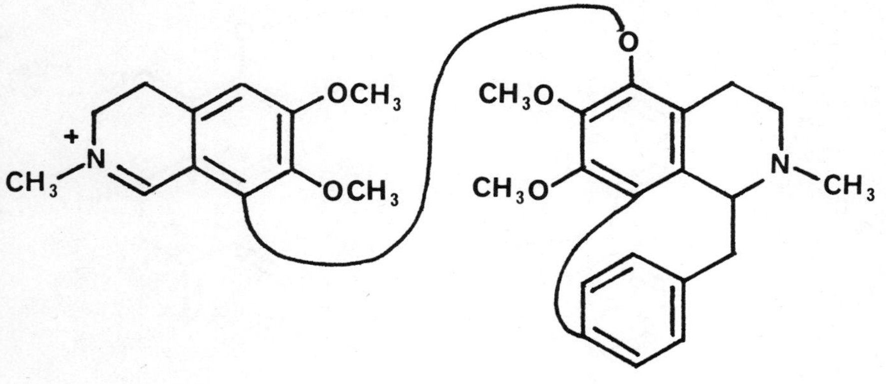

$$\underset{\sim}{\textbf{371}}$$

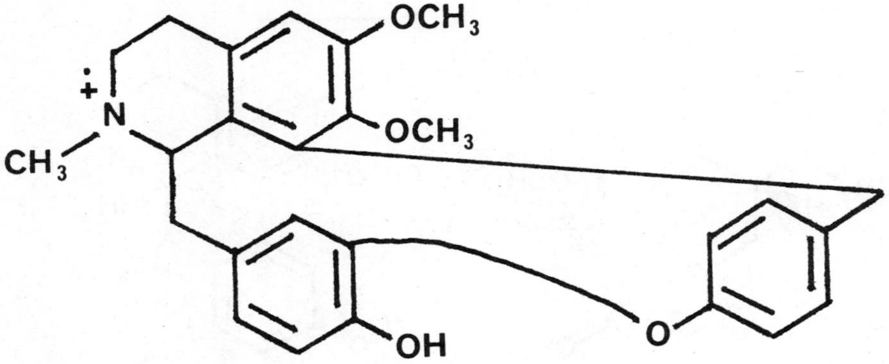

OCH3
OCH3
N
CH3
OH
O

372

OCH3
OCH3
N
CH3
O
CH3O
CH3O
N
CH3

373

OCH3
OCH3
N
CH3
H
O
CH3O
CH3O
N
CH3
H
OC2H5
O

374

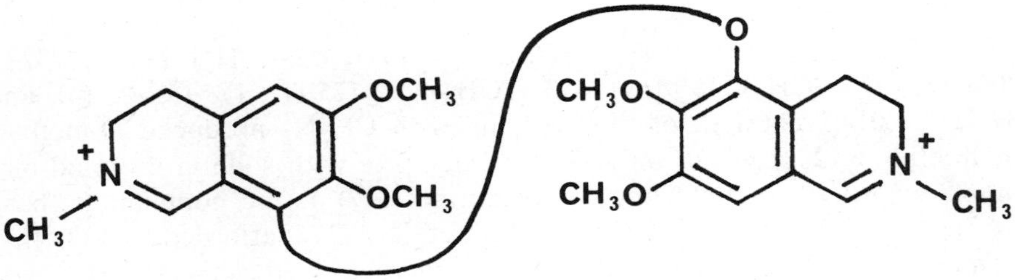

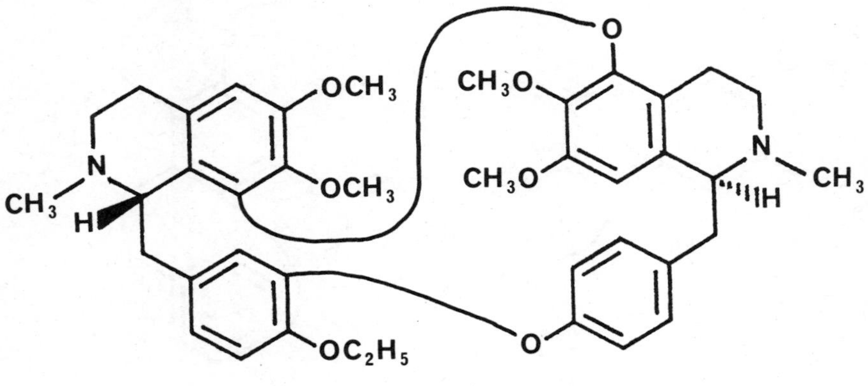

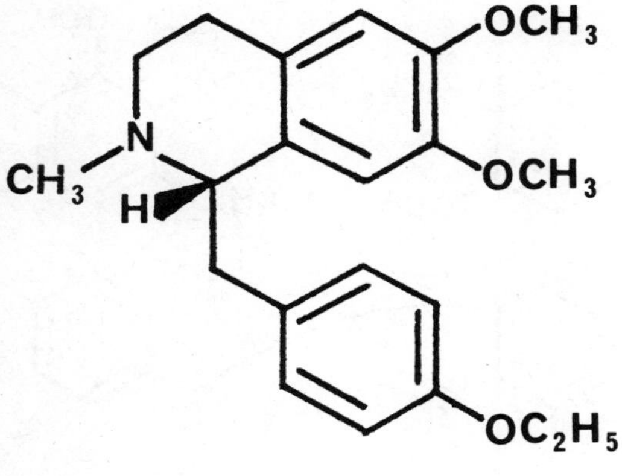

375

(**286**) [231]. Methylation of thalfoetidine produced *O*-methylthalfoetidine, which was identical with thalidasine (**286**), whose structure had been unequivocally assigned from spectral and degradative evidence [231]. Shortly thereafter, the reductive cleavage of both *O*-methylthalfoetidine (**286**) and *O*-ethylthalfoetidine (**374**) with sodium–liquid ammonia was repeated with the reisolation of the nonphenolic cleavage products (+)-*N,O,O*-trimethyl-

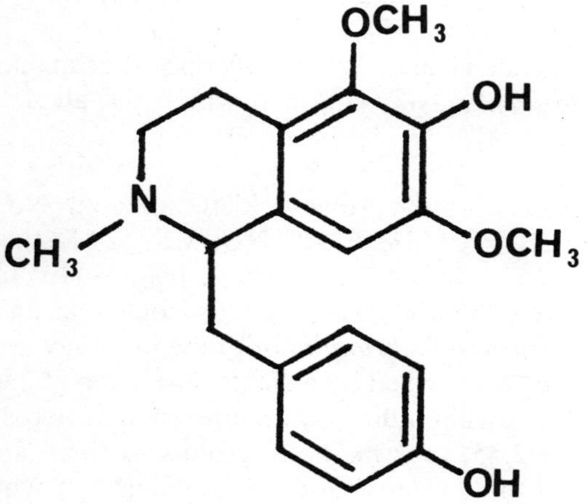

376

$$\underset{\displaystyle\mathbf{\widetilde{377}}}{}$$

coclaurine (**327**) and (+)-*O*-ethylarmepavine (**375**), respectively. In addition, (+)-1-(4-hydroxybenzyl)-2-methyl-5-hydroxy-6,7-dimethoxy-1,2,3,4-tetrahydroisoquinoline (**287**) was identified as the phenolic cleavage product from both reactions, thereby again confirming the structure of thalfoetidine (**370**).

Administration of thalfoetidine to mice and its evaluation on their conditioned reflexes indicated that the alkaloid did not depress high nervous system activity [454]. Thalfoetidine (called thalictrinine in this reference) was also known to be two to four times more effective than either aminopyrine or sodium salicylate against experimentally induced inflammation [458].

Thalfoetidine has been isolated from only one other species of *Thalictrum*, *T. longipedunculatum* E. Nik. [134], in which the alkaloid was formerly called thalictrinine [134].

3.5.30. Thalibrine. Tabular review [529, alkaloid No. 14, 530], UV [137,529,582], IR [582], ^{1}H NMR [529,582], MS [529,582].

Thalibrine (**329**), $C_{38}H_{44}N_2O_6$ (624.3199), $[\alpha]_D$ + 110° (c 0.135, CHCl$_3$) [582], $[\alpha]_D^{25}$ + 120° (c 0.26, MeOH) [137], was isolated as an amorphous substance from *Thalictrum rochebrunianum* Franc. and Sav. in 1976 [582]. The UV spectrum was characterized by a single maximum at 284 nm (log ϵ 3.90) and the ^{1}H NMR spectrum indicated the presence of two superimposed *N*-methyl groups at δ 2.45, four methoxy groups as three singlets at δ 3.60, 3.75 (6H), and 3.77; and 11 aromatic protons, eight of which were readily discernible as four singlets at δ 6.15 [H(8)], 6.31 [(8')], 6.50 [H(5)], and 6.53 [H(5')] and two pairs of doublets centered at δ 6.80 (2H, *J* = 8.5 Hz) and 7.03 (2H, *J* = 8.5 Hz) [582]. The mass spectrum did not show an observable parent ion but did show base peaks at *m/z* 206 (**191**) and 192 (**199**) and was characteristic of a tail-to-tail linked bisbenzylisoquinoline alkaloid [546].

Methylation (CH_2N_2) of thalibrine afforded O-methylthalibrine (**326**), which, except for the opposite sign of its rotation, possessed identical data to those observed for O-methyldauricine (**326**, R,R) suggesting the (S,S) stereochemistry for O-methylthalibrine. Treatment of thalibrine with CD_2N_2 in dioxane–D_2O afforded O-trideuteriomethylthalibrine and a comparison of the ^{1}H NMR spectra of thalibrine and its trideuteriomethyl ether demonstrated that the new methyl group of the latter was a shielded methoxy group at δ 3.58. This corresponds to the C(7) or C(7') methoxy group in a dauricine-type alkaloid [171], hence suggesting the presence of a phenolic hydroxy group at one of those positions in thalibrine [582]. Reductive cleavage of this trideuteriomethyl ether with sodium in liquid ammonia yielded ($+$)-(S)-O-methyarmepavine (**327**) and ($+$)-(S)-7-trideuteriomethyl-N-methylcoclaurine [**52**—OCD_3 at C(7)], thereby confirming the structure of thalibrine (**329**) [582].

Thalibrine was observed to inhibit the in vitro growth of *Staphylococcus aureus*, *Mycobacterium smegmatis* and *Candida albicans* at 1000 μg/mL [137].

Thalibrine has been isolated from only one other *Thalictrum* species to date, that being *T. longistylum* DC. [137].

3.5.31. Thalibrunimine. Tabular review [529, alkaloid No. 112, 530], UV [529,601], ^{1}H NMR [529,601], ^{13}C NMR [559], MS [529,601].

Thalibrunimine (**332**), $C_{38}H_{40}N_2O_8$ (652.2785), m.p. 198–200° (MeOH), $[\alpha]_D$ $+28°$ (c 0.19, $CHCl_3$), was first isolated from *Thalictrum rochebruniaum* Franc. and Sav. in 1976 [601]. The UV spectrum showed maxima at 241 nm (sh) (log ε 4.48), 283 nm (log ε 4.02), and 300 nm (sh) (log ε 3.91), and the ^{1}H NMR spectrum (C_5D_5N) indicated the presence of one N-methyl group at δ 2.38; five methoxy groups at δ 3.21, 3.55, 3.79, 3.83, and 3.88; one low-

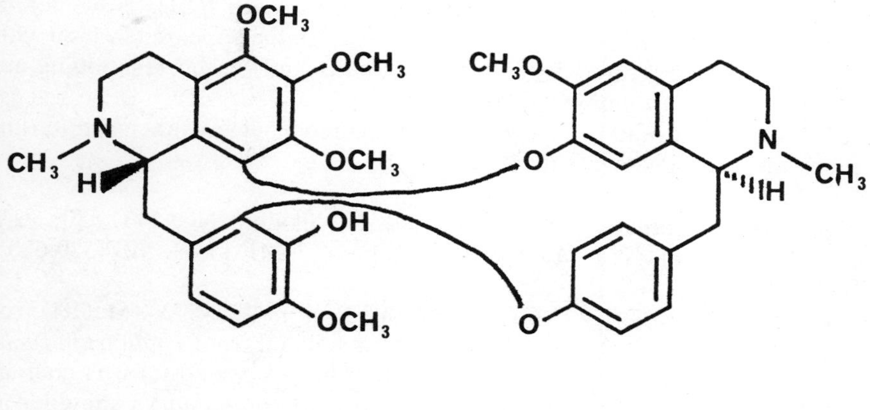

378

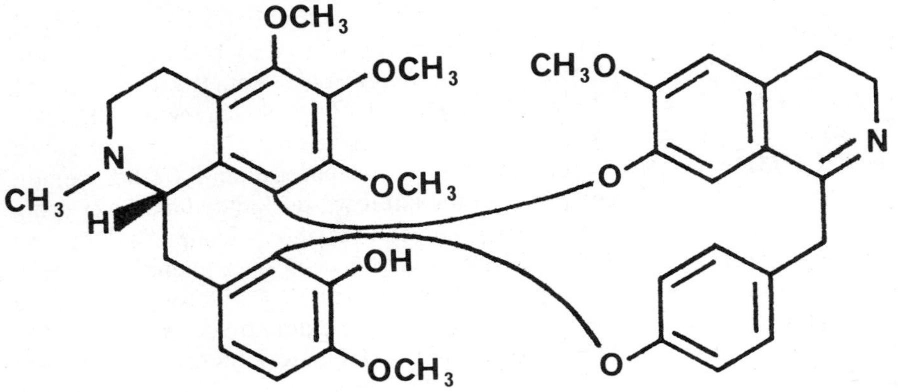

379

field benzylic methylene group at δ 4.40 (s); and eight aromatic protons from δ 6.42 to 7.58 (m) [601]. The mass spectrum was characterized by the molecular ion and base peak at M$^+$ *m/z* 652 with an intense M—1 ion at *m/z* 651 (85%), and was consistent with the spectra of other iminobisbenzylisoquinoline alkaloids [546,601]. Reduction of thalibrunimine with sodium borohydride produced a mixture (4:1) of two dihydro derivates that were very difficult to separate. Separation of the mixture resulted only in the isolation of the major constituent, N(2′)-northalibrunine (**346**), which on treatment with formaldehyde–sodium borohydride afforded thalibrunine (**351**) [601]. Because the structure of thalibrunine had been assigned as **378** at that time, thalibrunimine was first assigned as **379** [601]. However, a detailed reinvestigation of the structure of thalibrunine appeared in 1980 with the revision of structure of thalibrunine to **351** and the corresponding assignment of thalibrunimine as **332** [619].

To date, thalibrunimine has not been reisolated from any other *Thalictrum* species or other higher plant.

3.5.32. Thalibrunine.   Tabular review [529, alkaloid No. 113, 530], UV [121,529], IR [121], ^{1}H NMR [529,619,620], ^{13}C NMR [559], MS [529,620], CD [529,619,620].

Thalibrunine (**351**), $C_{39}H_{44}N_2O_8$ (668.3097), m.p. 172–173° (MeOH) (rosettes), $[\alpha]_D^{28}$ +160° (c 0.9, MeOH), was first isolated from *Thalictrum rochebrunianum* Franc. and Sav. in 1966 [121]. The UV spectrum was characterized by a single maximum at 281–282 nm (log ε 3.93) and a shoulder at 240–242 nm and displayed no bathochromic shift in strong alkali [121]. It was not until 1974 that a proposal of the structure of thalibrunine appeared

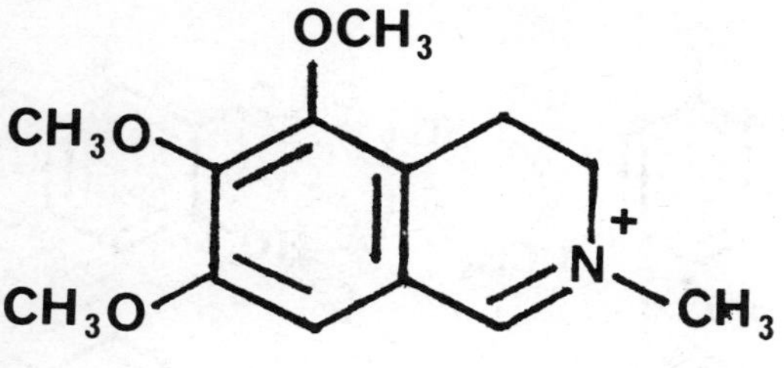

380

in the literature [620]. The ^{1}H NMR spectrum (CDCl$_3$) indicated the presence of two *N*-methyl groups at δ 2.45 and 2.58; five methoxy groups at δ 3.16, 3.36, 3.77, 3.82, and 3.89; and eight aromatic protons, six of which were present as an unresolved multiplet with two singlets at δ 5.90 [H(8′)] and 6.39 [H(5′)] [620]. The mass spectrum showed the parent ion at M$^+$ *m/z* 668 (55%) with other diagnostic fragment ions at *m/z* 561 (very weak) (M—loss of ring F), 515 (very weak) (**304**), 476 (weak) (**347**), 425 (53) (**306**—H), 234 (5) (**380**—H), 213 (100) (**307**), and 192 (70) (**381**+H) [620]. These spectral data suggested the presence of both a methoxy group and a hydroxy group in ring E of the dimer [561,620]. Reductive cleavage of thalibrunine with sodium in liquid ammonia gave a complex mixture of products which apparently was almost totally phenolic in nature. Two compounds isolable from this mixture were dihydrothalibrunine (**382**) and (+)-(*S*)-*N*-methylcoclaurine (**52**) [620]. Photooxidative cleavage of thalibrunine followed by zinc–hydrochloric acid reduction produced another complex mixture of products from which the diamine **313** was isolated [620]. The same diamine was subsequently obtained via oxidative cleavage with ceric ammonium nitrate [569].

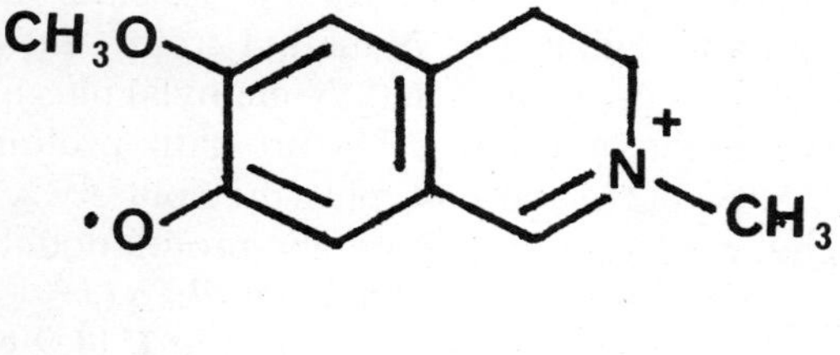

381

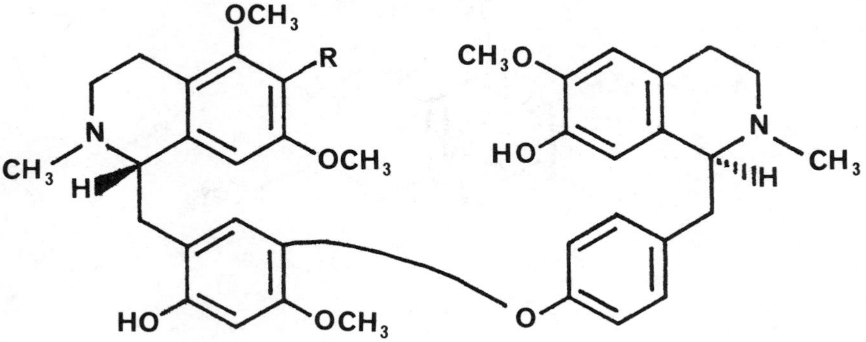

382 R = OCH$_3$

385 R = H

Attempts to derivatize the alkaloid with either Ac$_2$O or CH$_2$N$_2$ failed due to the highly hindered nature of the phenolic hydroxy group. Warm alkali produced a 12 nm bathochromic shift of the 282 nm UV maximum and the alkaloid exhibited a positive Gibbs reaction, characteristic of an unsubstituted position para to a phenolic hydroxy group [620]. On the assumption that the two ^{1}H NMR singlets were at the C(5') and C(8') positions, phenolic ring E was proposed to have a 1,2,3,4-tetrasubstitution pattern and thus the structure of thalibrunine was assigned as 378, the only structure under these conditions that would accommodate a hindered phenolic group [620]. The (S,S) stereochemistry was assigned via a comparison of its curve with that of hernandezine (303), the latter being of known (S,S) configuration [562,620]. In 1979, a detailed reinvestigation of the structure of thalibrunine was published which presented evidence of the revision of structure to 351 [619]. The ^{1}H NMR spectrum was obtained in acetone-d$_6$ under pulsed-signal Fourier transform conditions which resulted in a clearer (almost total first-order patterns) aromatic region. The N-methyl groups and methoxy groups were seen as singlets at δ 2.46 and 2.56 (2 N-methyls) plus δ 3.15, 3.38, 3.73, 3.79, and 3.82 (five methoxy groups). The aromatic protons were found as four singlets at δ 5.89, 6.37, and 6.46 plus the split ABXY pattern of the disubstituted phenyl ring (ring F), each a one-proton doublet of doublets at δ 6.16 (J = 2.0, 8.3 Hz), 6.36 (J = 2.4, 8.3 Hz), 7.16 (J = 2.4, 8.1 Hz), and 7.37 (J = 2.0, 8.1 Hz). Finally, a broad singlet at δ 11.9 was observed for a hydrogen-bonded phenolic hydroxy group. There were no signals for the outer less intense peaks of a typical AB quartet expected for the ortho protons in ring E of structure 378. These data would require para protons in ring E, for which six structures (two each for 17-, 18-, and 19-membered

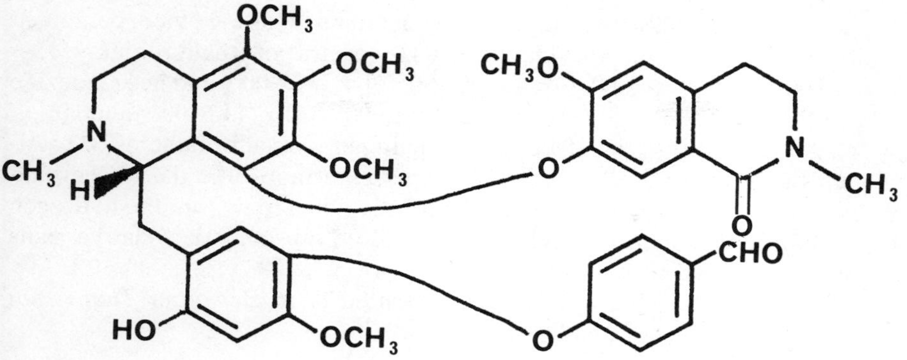

383

rings) were possible. The Gibbs test was performed again but a negative result was obtained, obviating the earlier evidence that led to the proposal of thalibrunine as **378**. An examination of Dreiding models demonstrated that an intramolecular hydrogen bond could be formed with the electron pair of the tertiary nitrogen atom to yield an unstrained seven-membered ring only when the phenolic hydroxyl was located ortho to the benzylic carbon. This situation would require the diphenyl ether to be placed para to the phenolic hydroxyl with structure **351** being the most likely possibility for thalibrunine. Thalibrunine acetate was prepared (Ac$_2$O/pyridine), but exposure to hydroxylic solvents was avoided to avoid rapid hydrolysis. Oxidation of thalibrunine acetate with potassium permanganate in acetone afforded secothalibrunine aldehydolactam (**383**), with more vigorous conditions producing complex mixtures. Oxidation of thalibrunine acetate with ceric ammonium nitrate and modified work-up yielded the diphenyl ether dialdehyde **384** and the diamine **313**. The reductive cleavage of thalibrunine was repeated with an additional product, characterized as 6-de-

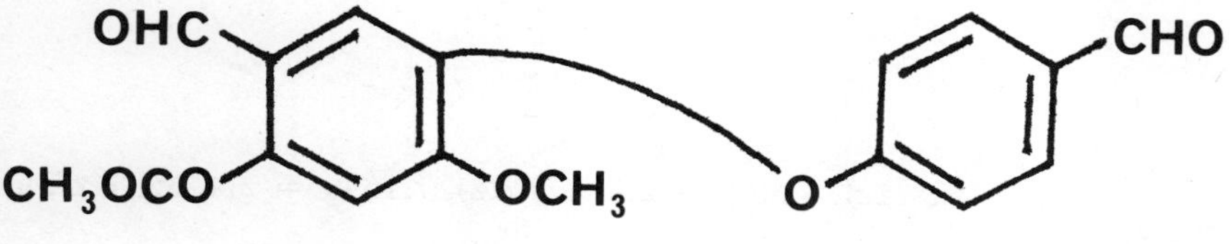

384

methoxydihydrothalibrunine (**385**), being obtained. This evidence, combined with the similar, but not identical CD spectra of thalibrunine ($[\theta]_{222}$ $+244,000$, $[\theta]_{245} -124,000$, $[\theta]_{273} -28,000$, $[\theta]_{293} +62,000$) and hernandezine (S,S)-(**303**) ($[\theta]_{272} +300,000$, $[\theta]_{246} -52,000$, $[\theta]_{265} +6700$, and $[\theta]_{286}$ $+28,000$) allowed for the structural assignment of thalibrunine as **351**. Although thalibrunine acetate had a CD spectrum more like that of hernandezine, protonation of thalibrunine apparently failed to disrupt the hydrogen bonded structure of the diphenyl ether and did not alter the relative signs of the maxima [619].

To date, thalibrunine has not been isolated from any other *Thalictrum* species or other higher plant.

3.5.33. Thalicberine.

Tabular review [529, alkaloid No. 97, 530], UV [529,538], IR [538], ^{1}H NMR [529,538], MS [529,585], CD [529,538].

Thalicberine (**386**), $C_{37}H_{40}N_2O_6$ (608.2886), m.p. 160–161° (Et$_2$O) (needles) [538,586], $[\alpha]_D^{19} +231.2°$ [586], $[\alpha]_D^{25} +210°$ (c 0.02) [538]; oxalate m.p. 239–240° (prisms) [586]; dihydrobromide·3H$_2$O m.p. 272–273° (dec.) (needles) [586]; was first isolated from *Thalictrum thunbergii* DC. (also known as *T. minus* L. var. *hypoleucum* [142]) in 1959 [586]. The UV spectrum displayed a single maximum at 282 nm (log ε 3.81) [538] and the CD spectrum showed a curve characterized by three peak/troughs ($[\theta]_{214} +249,000$, $[\theta]_{250} -12,200$, and $[\theta]_{285} +39,500$) [538]. The ^{1}H NMR spectrum indicated the presence of two *N*-methyl groups as singlets at δ 2.10 and 2.58; three methoxy groups as singlets at δ 3.66, 3.77, and 3.88; and 10 aromatic protons as a multiplet at δ 6.07–7.20 [538]. Methylation with CH$_2$N$_2$ produced *O*-methylthalicberine (**333**) which underwent Hofmann degradation to afford a methylmethine. Oxidation of this methylmethine with potassium permanganate produced 2-methoxy-5,4′-dicarboxybiphenyl ether (**309**) [586]. Treatment of thalicberine with ethereal diazoethane afforded *O*-ethylthalicberine (**387**) as an oil

386

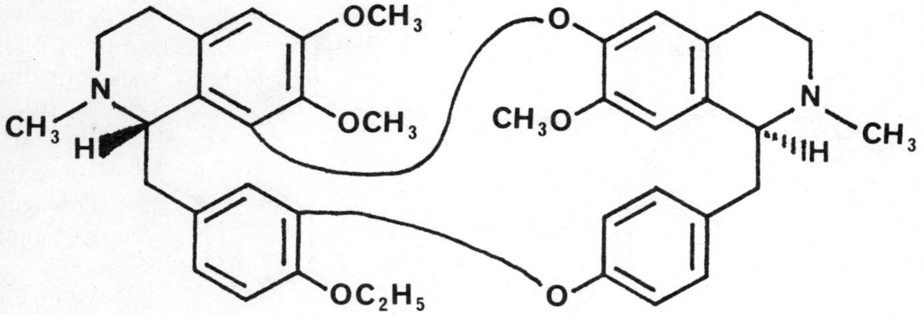

387

([α]$_D^{29}$ +208.7°), which underwent reductive cleavage with sodium in liquid ammonia to yield (+)-*O*-ethylarmepavine (**375**) and (+)-*N*-methylisoco-claurine (**334**) [587]. An identical reductive cleavage of thalicberine gave a phenolic base fraction (but no nonphenolic bases) which was ethylated with diazoethane and subsequently recleaved with sodium in liquid ammonia to afford (+)-*O*-ethylarmepavine (**375**) and (+)-1-(4-hydroxybenzyl)-2-methyl-6-ethoxy-7-methoxy-1,2,3,4,-tetrahydroisoquinoline (**388**) (characterized as its *O*-methyl-ether) [621]. Finally, another reductive cleavage of thalicberine with sodium in liquid ammonia followed and resulted in separation of the phenolic product. Methylation of this product yielded **389**, the optical an-

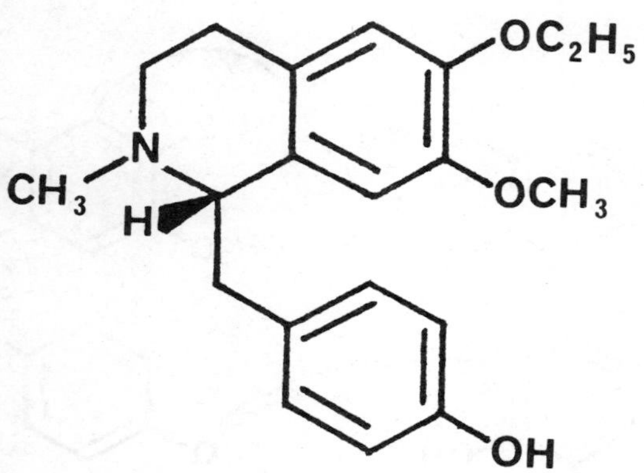

388

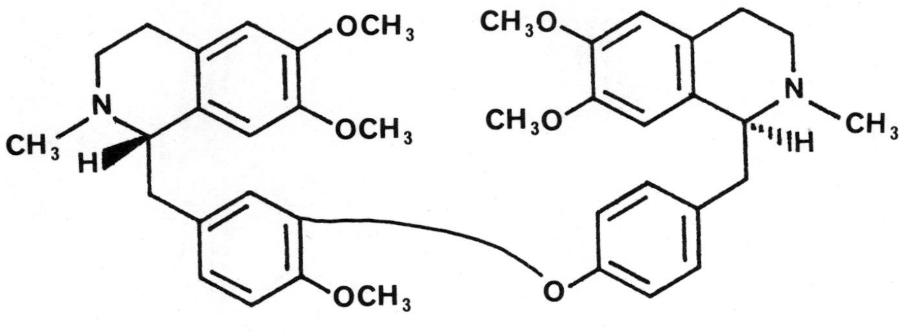

389

tipode of *O*-methyldauricine (*R,R*) [622]. On the basis of the above cited evidence [586,587,621,622], thalicberine was assigned as **386**.

Thalicberine (**386**) has also been isolated from *Thalictrum lucidum* L. [538], *T. minus* L. [65,272], and *T. minus* L. var. *microphyllum* Boiss. [544].

3.5.34. Thalictine. Tabular review [529, alkaloid No. 107], UV [529,623], IR [623], ^{1}H NMR [529,623], MS [529,623].

Thalictine (**390**), $C_{37}H_{40}N_2O_6$ (608.2886), m.p. 226–228° (MeOH) (nitrate salt), $[\alpha]_D^{28}$ − 15.8° (c 1.203, CHCl$_3$), was isolated from *Thalictrum thunbergii* DC. (also known as *T. minus* L. var. *hypoleucum* [142]) in 1975 [623]. The free base, which was amorphous and resisted crystallization, showed a single

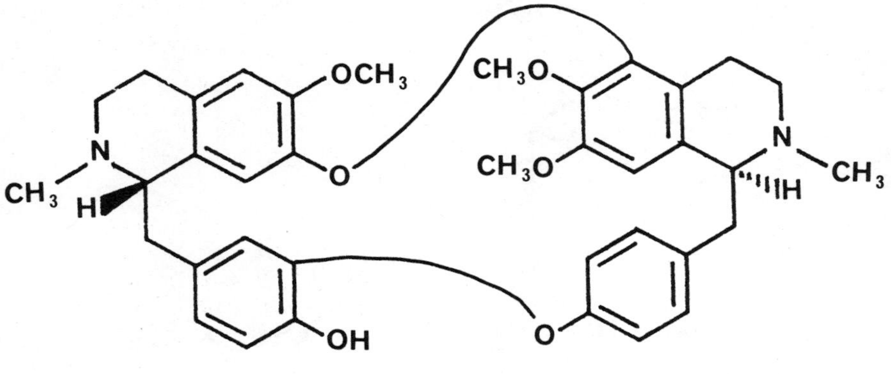

390

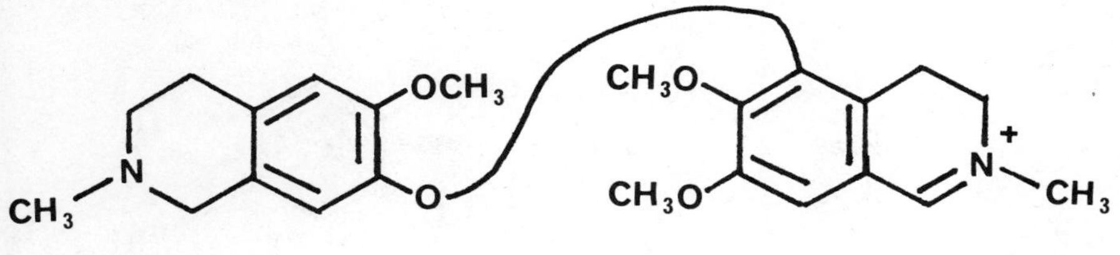

391

UV maximum at 284 nm. The alkaloid was insoluble in dilute aqueous alkali but gave a positive test (color) with ammonium phosphomolybdate. The compound was unstable to air and rapidly oxidized. The ^{1}H NMR spectrum suggested the presence of two N-methyl groups at δ 2.19 [N(2')] and 2.62 [N(2)]; three methoxy groups at δ 3.62 [C(6')], 3.82 [C(7')], and 3.86 [C(6)]; and 10 aromatic protons from δ 5.84 to 6.90, two of which appeared as singlets at δ 5.84 [H(8)] and 6.01 [H(8')]. The mass spectrum displayed the molecular ion at M$^+$ m/z 608 (~60%) with other significant fragment ions at m/z 396 (**391**), 395 (**391**—H), 381 (**391**—CH$_3$), 198 (100%) (**392**), and 175 [**392**—(CH$_3$)$_2$O]. These spectral data are characteristic of a bisbenzyliso-quinoline alkaloid containing three methoxy groups in a 7–5' ether-linked isoquinoline portion with cryptophenolic hydroxy group in the benzyl por-tion of the dimer [614]. Treatment of thalictine with ethereal diazoethane afforded O-ethylthalictine (**393**), which on Hofmann degradation furnished a methylmethine. Oxidation of this methine with potassium permanganate in acetone afforded 4-ethoxy-3,4'-oxydibenzoic acid (**394**), thus confirming the presence of the phenolic hydroxy group in the lower (benzyl) portion of the alkaloid. Reductive cleavage of O-ethylthalictine (**393**) with sodium in liquid ammonia afforded (+)-(S)-1-(4-ethoxybenzyl)-2-methyl-6-methoxy-7-hydroxy-1,2,3,4-tetrahydroisoquinoline (**395**) [identified by conversion to (+)-(S)-O-ethylarmepavine (**375**) by treatment with CH$_2$N$_2$] and (+)-(S)-armepavine (**328**). The same cleavage of O-methylthalictine with sodium in liquid ammonia yielded (+)-(S)-1-(4-methoxybenzyl)-2-methyl-6-methoxy-7-hydroxy-1,2,3,4-tetrahydroisoquinoline [characterized as its O-ethyl ether

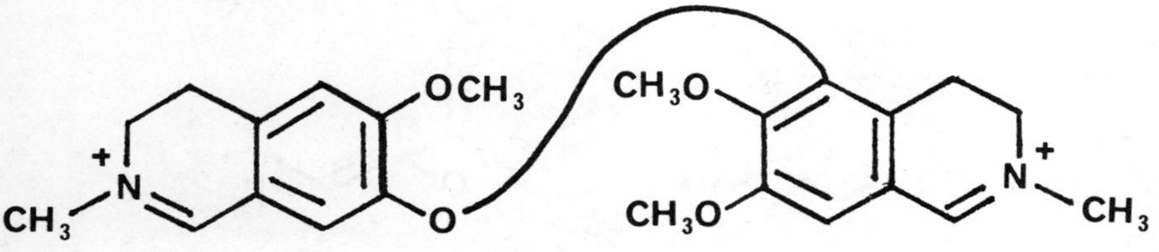

392

OCH3
CH3O
CH3O
N
N
CH3
CH3
H
H
O
O
OC2H5

393

HOOC
COOH
OC2H5
O

394

OCH3
N
CH3
H
OH
OC2H5

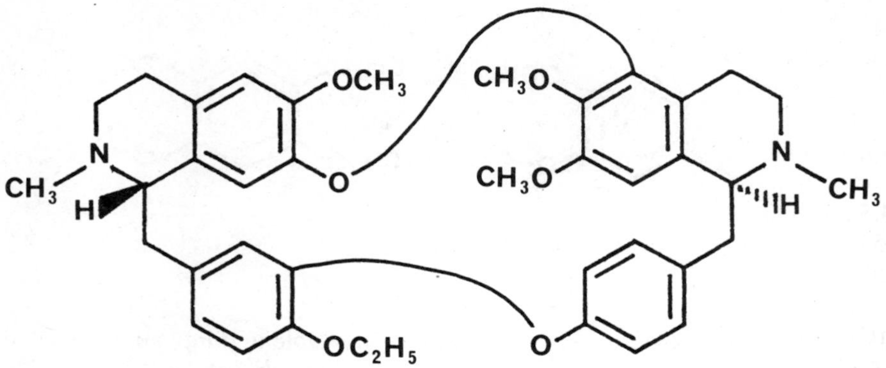

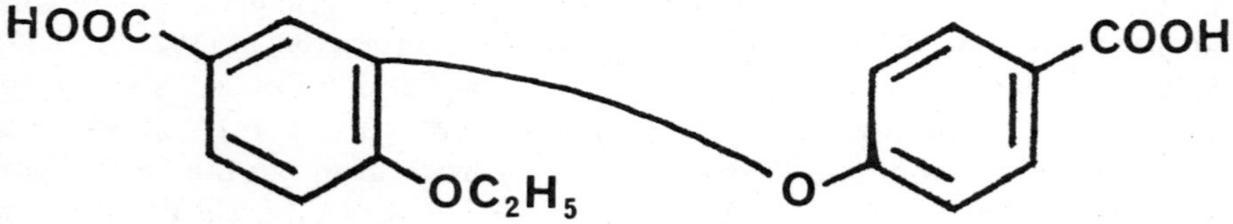

395

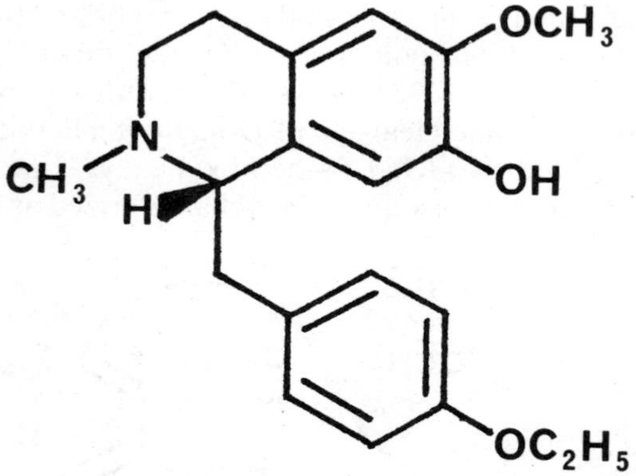

(428)] and (+)-(S)-armepavine (328). On the basis of the spectral data and the cleavage products, thalictine was characterized as 390, a new alkaloid of the thalmine (358) series [623].

To date, thalictine has not been isolated from any other species of *Thalictrum* or other higher plant.

3.5.35. Thalictrinine.
Tabular review [530, alkaloid No. 220], UV [530,557], IR [530,557], ¹H NMR [530,557], ¹³C NMR [530,557], MS [530,557], CD [530,557].

Thalictrinine (302), $C_{38}H_{36}N_2O_9$ (664.2421), m.p. 199–201° dec. [$(CH_3)_2CO$], $[\alpha]_D^{22}$ −255° (c 0.24, MeOH), was isolated as colorless rhombic crystals from *Thalictrum rochebrunianum* Franc. and Sav. in 1980 [557]. The UV spectrum was characterized by a series of undulating maxima at 205 nm (sh) (log ε 4.79), 236 nm (log ε 4.62), 251 nm (sh) (log ε 4.50), 285 nm (sh) (log ε 4.01), 301 nm (sh) (log ε 3.84), and 330 nm (log ε 3.73), which simplified to 282 nm (sh) (log ε 4.13) and 340 nm (log ε 3.64) in 0.1 N methanolic hydrochloric acid. The IR spectrum possessed a carbonyl band at 1675 cm⁻¹ which was also observed in the ¹³C NMR spectrum at δ 194.3. The CD spectrum exhibited several extrema at $[\theta]_{230}$ +115,000 (end), $[\theta]_{241}$ 0, $[\theta]_{254}$ −112,000, $[\theta]_{275}$ −76,000 (sh), $[\theta]_{310}$ 0, $[\theta]_{355}$ −35,000, and $[\theta]_{395}$ 0. The ¹H NMR spectrum indicated the presence of one N-methyl group as a singlet at δ 2.47; five methoxy groups as singlets at δ 3.28, 3.61, 3.79, 3.86, and 3.90; and 10 aromatic protons as four singlets at δ 6.05 [H(8′)], 6.51, 6.84, 7.02; an ABXY system with split doublets at ~δ6.8 and ~6.9 (J = 2, 8 Hz); a double doublet at δ 7.49 and 8.37 (J = 1.9, 8.3 Hz); and an AB quartet at δ 7.64 [H(4′)] and 8.62 [H(3′)] (2d, 5.1 Hz). The presence of a hydroxy group was demonstrated by the appearance of a broad singlet at δ 12.80 which disappeared on addition of D_2O. The mass spectrum showed the parent ion and base peak at M⁺ m/z 664 with other prominent ions at m/z 649 (37) (M—CH₃) and 332 (15) (1/2 M⁺²). These spectral data were suggestive of a benzyltetrahydroisoquinoline–benzylisoquinoline dimer similar to oxothalibrunimine (354) and to dihydrothalictrinine (301) and most likely being 3′,4′-dehydrooxothalibrunimine (302). The accuracy of this proposal was confirmed by the preparation of thalictrinine via oxidation of thalibrunimine (332) with palladium on carbon as catalyst [557]. Thalictrinine appears to be the first example of a dimeric alkaloid of this series bearing a carbonyl group on the benzyl carbon of one half of the bisbenzylisoquinoline moiety [557].

To date, thalictrinine has not been isolated from any other *Thalictrum* species or other plant.

Note that some confusion may result in the literature with reference to the alkaloid called "thalictrinine," because an alkaloid once designated "thalictrinine" [624,625] was later shown to be identical with thalfoetidine (370) [134]. Hence thalictrinine in current *Thalictrum* alkaloid nomenclature should now always refer to dimer 302.

3.5.36.   Thalidasine.   Tabular review [529, alkaloid No. 100, 530], UV [529,626], ^{1}H NMR [529,626], MS [231,529,585,627].

Thalidasine (**286**), $C_{39}H_{44}N_2O_7$ (652.3149), m.p. 105–107° (amorphous) [231], 178° (picrate) (EtOH) [321], 144–145° (oxalate) (EtOH) [627], 146–148° (oxalate) [627, footnote #11], 160–162° (oxalate) [626], 182–183° (meth-iodide) [626]; $[\alpha]_D^{27}$ −70° (c 0.89, MeOH) [231,626]; $[\alpha]_D^{30}$ −71.6° (c 1.10, CHCl$_3$) [627] was first isolated in 1967 from *Thalictrum dasycarpum* Fisch. and Lall. [626]. The UV spectrum was characterized by two maxima at 275 nm (log ε 3.66) and 282 nm (log ε 3.66) and the ^{1}H NMR spectrum indicated the presence of two *N*-methyl groups as singlets at δ 2.25 and 2.62; five methoxy groups as singlets at δ 3.27, 3.50, 3.75, 3.87, and 3.91; and nine aromatic protons as a multiplet at δ 6.30–7.54 [231,626]. The mass spectrum showed the molecular ion at M$^+$ *m/z* 652 with other significant fragment ions at *m/z* 637 (M—CH$_3$), 621 (M—OCH$_3$), 425 (**396**—H), 411 (**396**—CH$_3$), 394, 379, 213 (100%) (**373**), 204, and 190 [231]. Oxidation of thalidasine with potassium permanganate in aqueous media at pH 6 afforded 5,4′-dicarboxy-2-methoxydiphenyl ether (**309**), and reductive cleavage with sodium in liquid ammonia gave (+)-(*S*)-*O*-methylarmepavine (**327**) and (+)-(*S*)-1-(4-hydrox-ybenzyl)-2-methyl-5-hydroxy-6,7-dimethoxy-1,2,3,4-tetrahydroisoquinoline (**287**) as major products and (+)-(*S*)-armepavine (**328**) and (+)-(*S*)-(4-meth-oxybenzyl)-2-methyl-6,7-dimethoxy-8-hydroxy-1,2,3,4-tetrahydroisoquino-line (**397**) as minor products [231]. An analysis of the spectral data and the degradative reactions permitted the rational assignment of structure of thal-idasine as **286**. Thalidasine (**286**) was the first bisbenzylisoquinoline alkaloid to possess 8,5′ termini of the diphenyl ether bridge in the isoquinoline por-tions of the dimer and as such was the first asymmetric alkaloid of this series containing a 20-membered ring [529,626]. Although thalfoetidine (**370**) (12-demethylthalidasine) was isolated 1 year prior to thalidasine (**286**), it was not until the structure of thalidasine was established that the structure of thalfoetidine was finally clarified [231,626].

Thalidasine (**286**) was found to inhibit the growth of *Staphylococcus au-reus* (100 µg/mL) [139,152], *Streptococcus faecalis* (100 µg/mL) [152], *Esch-*

396

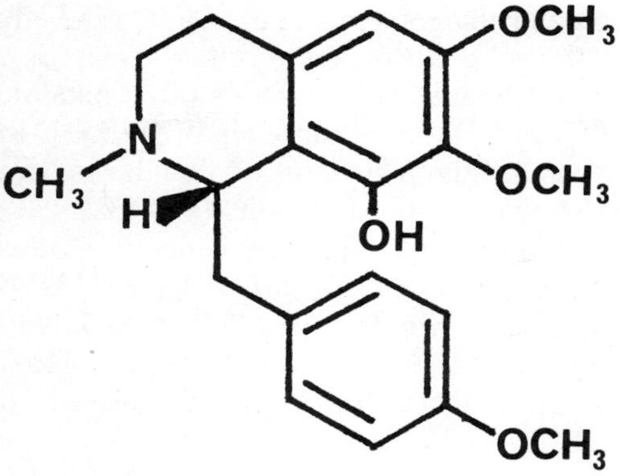

$$\underset{\sim}{\underline{397}}$$

erichia coli (100 μg/mL) [139,152], *Klebsiella pneumoniae* (100 μg/mL) [139,152], *Pseudomonas aeruginosa* (100 μg/mL) [152], *Pasteurella multocida* (100 μg/mL) [152], *Salmonella typhimurium* (100 μg/mL) [152], *Salmonella gallinarum* [139], *Shigella dysenteriae* (100 μg/mL) [152], *Mycobacterium smegmatis* (200 μg/mL [152] and 25 μg/mL [139,152]), *Proteus vulgaris* (100 μg/mL) [152], *Proteus mirabilis* (100 μg/mL) [152], and *Candida albicans* (100 μg/mL) [139]. Other less specific antimicrobial effects were also reported [353,550]. Intravenous administration of thalidasine in dosages of 1.0 and 2.0 mg/kg failed to produce a hypotensive response in normotensive dogs [139]. However, similar administration of the alkaloid in dosages of 1.0, 2.0, and 4.0 mg/kg to normotensive rabbits produced a mean decrease in blood pressure of 2, 11, and 18 mm Hg, respectively [101,215]. The action of *O*-methylthalfoetidine (thalidasine) on conditioned reflexes in mice was evaluated with the result that the alkaloid failed to depress higher nervous activity [454]. Thalidasine was first reported to possess antitumor activity in 1967 when the alkaloid showed significant inhibitory activity carcinosarcoma 256 in rats at a dosage of 200 mg/kg [231,626]. Subsequent studies confirmed this activity and attempted to define the structural requirements for such biological activity [628,629]. Intraperitoneal injections of thalidasine in mice in a dosage range of 70–100 mg/kg day suppressed about 50% Ehrlich ascites tumor growth, about 50–80% of Lewis lung cancer growth and about 25% of S-180 sarcoma growth [630,631]. The alkaloid had no effect on the growth of hepatic or uterine tumors [631]. The intraperitoneal LD_{50} was 520 mg/kg [630,631] but decreased to 300 mg/kg in animals undergoing repeated treatment [630]. The intravenous LD_{50} was 120 mg/kg with toxicity and accompanying pathological changes appearing in mice receiving a dosage greater than 300 mg/kg day [631]. The effective therapeutic dosage was 70–100 mg/kg [630]. The crude extract of *Thalictrum faberi* Ulbr., which contains thalidasine as well as other alkaloids, has been used in the treatment of stomach cancer in China [349,555]. Recently, intraperitoneal and intravenous administration of thalidasine polyphase liposome preparations have shown promising results [632,633].

Thalidasine has also been isolated from numerous other *Thalictrum* species including *T. alpinum* L. [70], *T. faberi* Ulbr. [349,555], *T. foliolosum* DC. [38], *T. longipedunculatum* E. Nik. [297], *T. lucidum* L. [139], *T. minus* L. [114], *T. minus* L. race B [215], *T. revolutum* DC. [101], and *T. rugosum* Ait. [114,152,353,627], as well as having been reisolated from *T. dasycarpum* Fisch. and Lall. [282].

3.5.37. Thalidezine.   Tabular review [529, alkaloid No. 83, 530], UV [79,301,529], IR [301], ^{1}H NMR [79,301,529], MS [79,529,576], CD [301,529].

Thalidezine (**289**), $C_{38}H_{42}N_2O_7$ (638.2992), m.p. 158–159° $((CH_3)_2CO)$ [79], 145–147° $[(CH_3)_2CO]$ (colorless needles) [301]; $[\alpha]_D^{25}$ +235° (CHCl$_3$) [79]; $[\alpha]_D^{25}$ +224° (c 0.12, MeOH) [301], was first isolated from *Thalictrum fendleri* Engelm. ex Gray in 1967 [62,79]. The UV spectrum showed a single maximum at 283 nm (log ε 4.02) with a bathochromic shift in alkaline medium [79], and the CD spectrum showed extrema at $[\theta]_{216}$ +318,000, $[\theta]_{248}$ −46,600, and $[\theta]_{288}$ +38,200 [301]. The ^{1}H NMR spectrum indicated the presence of two *N*-methyl groups as singlets at δ 2.33 and 2.64; four methoxy groups as singlets at δ 3.27, 3.37, 3.78, and 3.92; and nine aromatic protons as a singlet at δ 6.02 (1H) and a multiplet from δ 6.12 to 7.5 (8H) [79]. A broad peak at δ 5.00 which was D$_2$O-exchangeable indicated the presence of a phenolic hydroxy group [301]. The mass spectrum showed the molecular

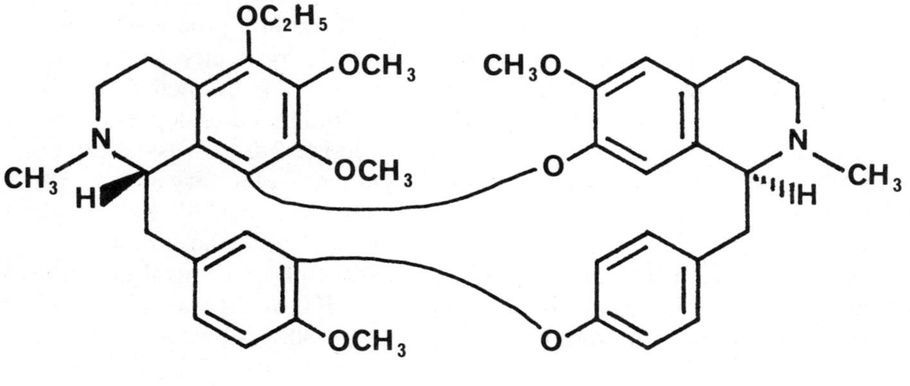

398

ion at M^+ m/z 638 with other significant fragment ions at m/z 411 (**321**—H) and 192 (**199**) [79]. These data were characteristic of an O-demethylhernandezine containing the phenolic hydroxy group in ring B [558,560,561], and this was confirmed by methylation (CH_2N_2) to afford hernandezine (**303**) [79]. Ethylation of thalidezine with diazoethane produced O-ethylthalidezine (**398**) which on reductive cleavage with sodium in liquid ammonia yielded (+)-(S)-1-(4-methoxybenzyl)-2-methyl-5-ethoxy-6,7-dimethoxytetrahydroisoquinoline (**325**) as the major nonphenolic cleavage product and 1-(4-methoxybenzyl)-2-methyl-5-ethoxy-7-methoxy-1,2,3,4-tetrahydroisoquinoline (**324**) as the minor nonphenolic cleavage product [79]. This permitted thalidezine to be characterized as 5-demethylhernandezine and to be assigned as **289** [79].

Thalidezine (**289**) was found to inhibit the in vitro microbial growth of *Klebsiella pneumoniae*, *Mycobacterium smegmatis*, and *Candida albicans*, each at a concentration of 100 µg/mL of alkaloid [301].

Thalidezine has also been isolated from *Thalictrum podocarpum* Humb. [301], *T. simplex* L. [600], and *T. sultanabadense* Stapf. [576].

3.5.38. Thaligosidine. Tabular review [529, alkaloid No. 100a, 530], UV [529,634], IR [529,634], ¹H NMR [529,634], MS [529,634], CD [529,634].
Thaligosidine (**399**), $C_{37}H_{40}N_2O_7$ (624.2836), m.p. 175–177° (C_6H_6) (colorless crystals), $[\alpha]_D^{20}$ −45° (c 0.26, MeOH), was isolated from *Thalictrum rugosum* Ait. in 1978 [634]. The UV spectrum of the alkaloid showed two maxima at 275 nm (log ϵ 3.72) and 283 nm (log ϵ 3.72) with a small bathochromic shift and the appearance of a shoulder at 310 nm (log ϵ 3.22) in dilute alkali [634]. The ¹H NMR spectrum was characterized by six singlets at δ 2.25 (NCH₃), 2.66 (NCH₃), 3.49 (OCH₃), 3.75 (OCH₃), 3.86 (OCH₃), and 5.6 (br) (2OH)

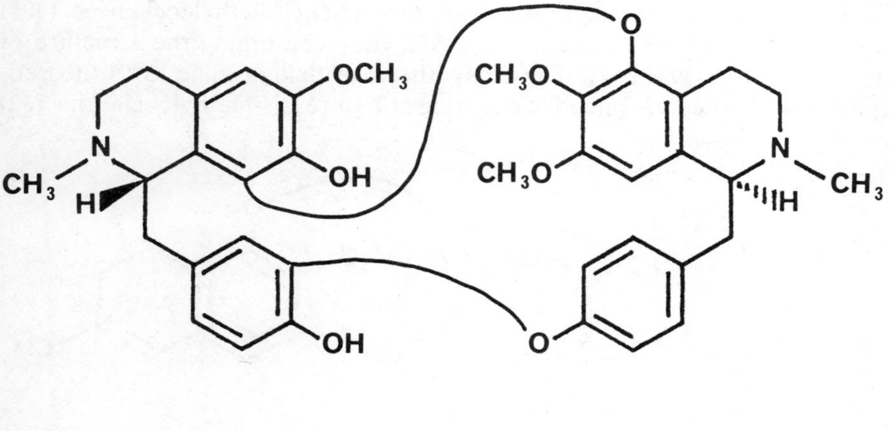

399

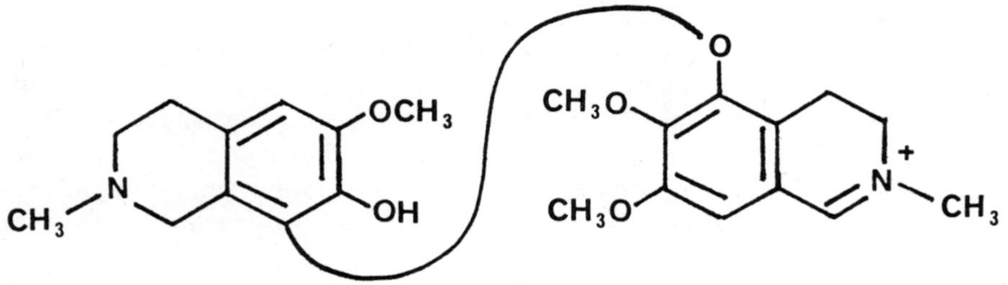

400

(D₂O-exchangeable) and a broad multiplet at δ 6.2–7.7 (9 ArH). The mass spectrum showed the molecular ion at M⁺ 624 (40%) with other diagnostic fragment ions at *m/z* 412 (7%) (**400**), 411 (20) (**400**—H), 206 (23) (**401** and **191**), and 192 (100) (**199**) [634]. These spectral data were characteristic of a thalidasine (**286**)-like structure bearing one phenolic hydroxy group in both the isoquinoline (top) portion and the benzyl (bottom) portion of the dimer [231,560,585,626]. Methylation (CH₂N₂) of thaligosidine ([θ]₂₂₄ −31,800, [θ]₂₄₂ +5490, [θ]₂₆₈ +6700, and [θ]₂₈₇ −19,000) afforded thalidasine (**286**), thereby fixing the skeletal structure, positions of oxygenation, termini of the diphenyl ether linkages, and stereochemistry of thaligosidine (**399**) [634]. Treatment of thaligosidine with ethereal diazoethane afforded *O,O*-diethyl-thaligosidine (**402**), m.p. 198–200° (MeOH), whose mass spectrum showed the parent ion at M⁺ *m/z* 680 (100%) with other diagnostic fragment ions at *m/z* 440 (7) (**403**), 439 (22) (**403**—H), 220 (37) (**404** and/or **232**), 206 (4) (**191**), and 190 (10) (**404** and/or **232**—[CH₂CH₃ and —H]), thereby confirming the location of one phenolic group in the isoquinoline (top) portion of the alkaloid [634]. Finally, reductive cleavage of *O,O*-diethylthaligosidine with sodium in liquid ammonia yielded (+)-(*S*)-*N*-methyl-*O,O*-diethylcoclaurine (**405**) and (+)-(*S*)-5-hydroxyamrepavine (**287**), thus confirming the structure of thaligosidine as **399** [634]. The biosynthesis of thaligosidine (**399**) theoretically may proceed via phenolic coupling of (+)-(*S*)-*N*-methylcoclaurine (**52**)

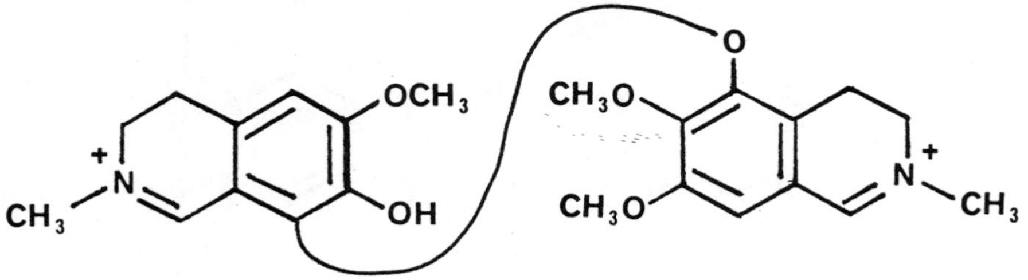

401

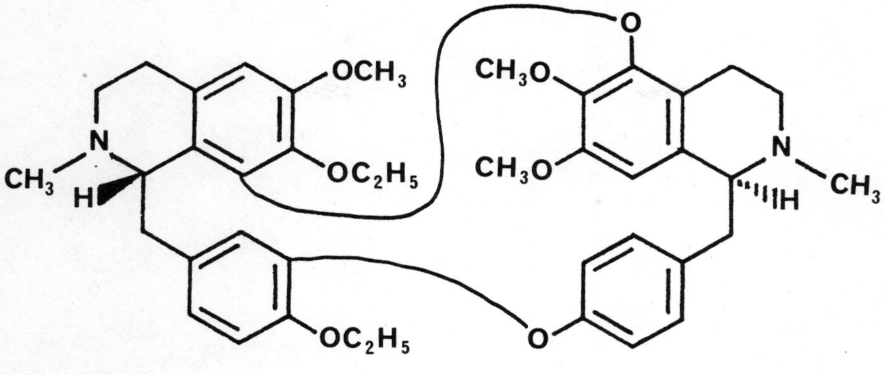

402

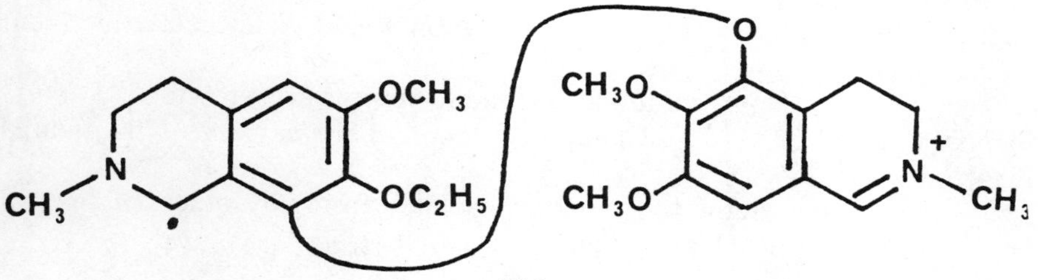

403

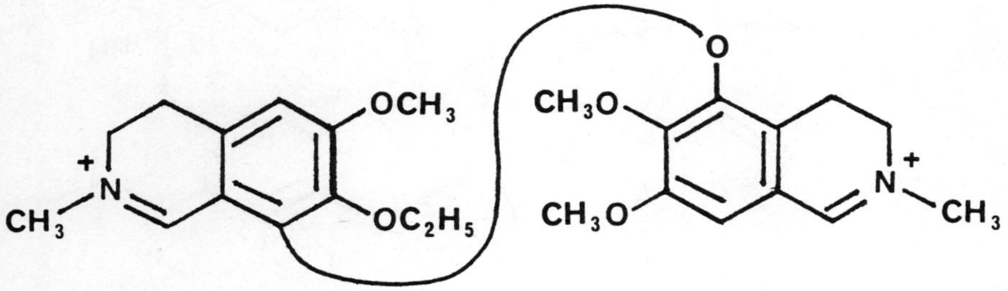

404

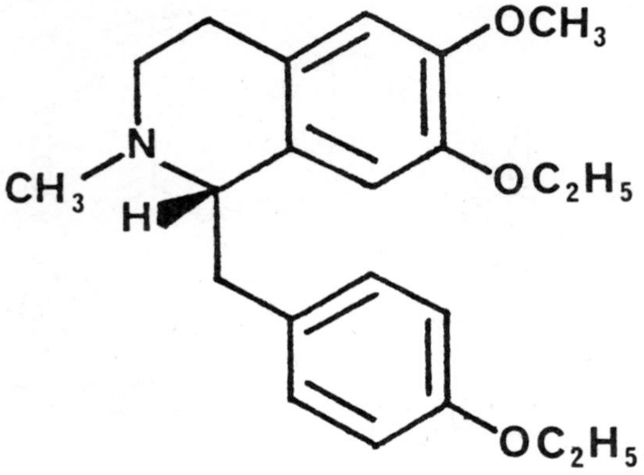

405

and (+)-(S)-5-demethylthalifendlerine (**287**) to afford the hypothetical tail-to-tail linked intermediate alkaloid of structure **406**. Intramolecular coupling between the 5'-hydroxy and the C(8) position would thus produce thaligosidine (**399**), which on biological methylation would yield thalidasine (**286**) [634].

Thaligosidine (**399**) was found to inhibit the in vitro growth of *Staphylococcus aureus* (1000 μg/mL) and *Mycobacterium smegmatis* (100 μg/mL) [234].

To date, thaligosidine has not been isolated from any other *Thalictrum* species nor any other higher plant.

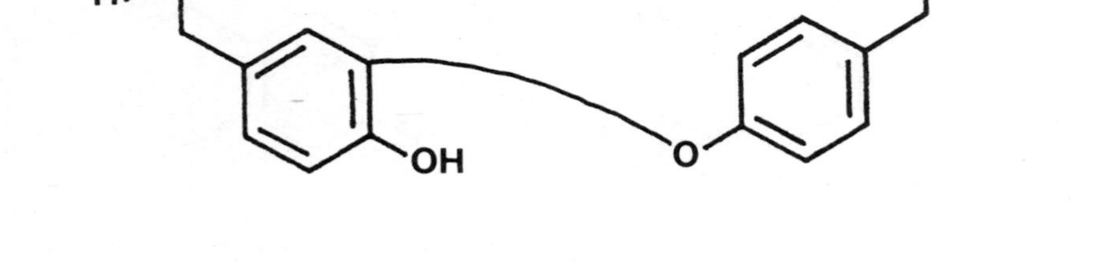

406

3.5.39. Thaligosinine. Tabular review [529, alkaloid No. 52b, 530], UV [529,634], IR [529,634], ^{1}H NMR [529,634], MS [529,634], CD [529,634]. Thaligosinine (**407**), $C_{38}H_{42}N_2O_7$ (638.2992), m.p. 233–234.5° dec. (Et$_2$O), $[\alpha]_D^{21}$ −58.5° (c 0.316, MeOH), was first isolated from *Thalictrum rugosum* Ait. in 1978 [634]. The UV spectrum of the alkaloid showed a single maximum at 282 nm (log ϵ 3.90) with a bathochromic shift in strong alkali to 284 nm (log ϵ 4.11) and 307 nm (sh) (log ϵ 3.53) while the CD spectrum showed extrema at $[\theta]_{230}$ +100,000, $[\theta]_{242}$ −48,000, and $[\theta]_{275}$ −13,000 [634]. The ^{1}H NMR spectrum indicated the presence of two *N*-methyl groups as singlets at δ 2.51 and 2.56; four methoxy groups as singlets at δ 3.04, 3.40, 3.80, and 3.84; one hydroxy group at around δ 5.0 (br, D$_2$O-exchangeable); and nine aromatic protons as two singlets at δ 6.36 [H(8′)] and 6.47 [H(5′)] plus a multiplet at δ 6.5–7.5 (7H) [634]. The mass spectrum showed the molecular ion at M$^+$ *m/z* 638 (100%) with other characteristic fragment ions at *m/z* 426 (15%) (**306**), 425 (37) (**306**—H), 411 (34) (**306**—CH$_3$), 236 (2) (66), 213 (97) (**307**), 192 (31) (**381**+H), 191 (6) (**381**), and 190 (9) (**381**—H) [634]. These spectral data were characteristic of a thalrugosaminine (**408**)-like alkaloid bearing a phenolic hydroxyl group in the benzyl (bottom) portion of the dimer [98,560]. Treatment of thaligosinine with ethereal CH$_2$N$_2$ produced thalrugosaminine (**408**), thereby establishing the skeletal structure, oxygenation pattern, positions of diphenyl ether termini, and stereochemistry of thaligosinine. The presence of the mass spectral fragment ion at *m/z* 426 corresponding to **306** and other related ions fixed the positions of the phenolic group of thaligosinine at C(12′), thereby establishing the structure of thaligosinine (**407**) as 12′-demethylthalrugosaminine [634]. A biosynthetic intramolecular coupling of (+)-(*S*)-*N*-methylcoclaurine (**52**) and (+)-(*S*)-5-demethylthalifendlerine (**287**) would lead to the hypothetical alkaloid **406**.

407

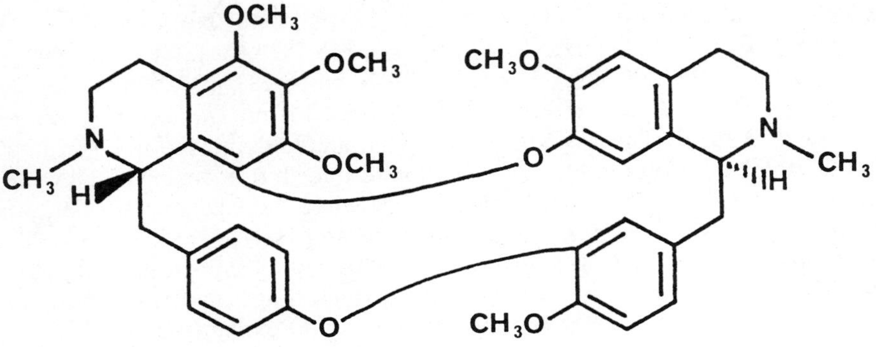

408

Phenolic coupling between the C(7) phenol and the C(8') position, followed by *O*-methylation, would provide the genesis of thaligosinine (**407**) [634].

Thaligosinine inhibited the in vitro growth of *Staphylococcus aureus* and *Mycobacterium smegmatis* at a concentration of 1000 μg/mL [234].

Thaligosinine has not been isolated from any other *Thalictrum* species nor other higher plant to date.

3.5.40. Thaligrisine.   UV [544], ¹H NMR [544], MS [544], CD [544].
Thaligrisine (**409**), $C_{37}H_{42}N_2O_6$ (610.3043), $[\alpha]_D^{25}$ +57° (c 0.13, MeOH), was isolated from *Thalictrum minus* L. var. *microphyllum* Boiss. in 1984 [544]. The UV spectrum of the alkaloid was characterized by peaks at 226 nm (sh) (log ε 4.48) and 284 nm (log ε 3.98); the CD spectrum showed Δε(nm) positive

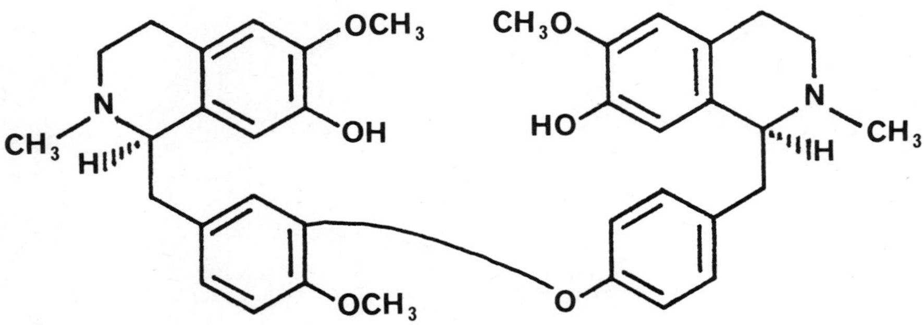

409

tail at 215 nm, +2(228), +15(244), +2.5(sh)(254), 0(300). The mass spectrum failed to exhibit a molecular ion but did show a very small M—1 ion at m/z 609 (0.2%) with the base peak at m/z 192 (100%) (**199**). These data were characteristic of a thalibrine (**329**)-like alkaloid with two isoquinoline portions (top portion), each bearing one methoxy and one hydroxy group. The ^{1}H NMR spectrum was highly informative and indicated the presence of two *N*-methyl groups as singlets at δ 2.44 and 2.51; three methoxy groups as singlets at δ 3.83 (6H) and 3.84 (3H); and 11 aromatic protons as four singlets at δ 6.26 [H(8) or H(8′)], 6.35 [H(8′) or H(8)], 6.45 [H(5) or H(5′)], 6.54 [H(5′) or H(5)], 6.58 (br) [H(10)]; a double doublet at δ 6.81 (J = 8.5 Hz) [H(11′) and H(13′)] and 7.03 (J = 8.5 Hz) [H(10′) and H(14′)]; and a doublet at δ 6.85 (J = 8.2 Hz) [H(14)] and 6.86 (J = 8.2 Hz) [H(14′)]. Thaligrisine was subsequently assigned as **409** and is the enantiomer of (−)-grisabine [544].

Thaligrisine has not been isolated from any other *Thalictrum* species nor higher plant to date.

3.5.41. Thaliphylline. UV [544], ^{1}H NMR [544], MS [544], CD [544]. Thaliphylline (**410**), $C_{37}H_{40}N_2O_6$ (608.2886), $[\alpha]_D^{25}$ +198° (c 0.12, MeOH), was isolated from *Thalictrum minus* L. var. *microphyllum* Boiss. in 1984 [544]. The alkaloid showed UV maxima at 211 nm (log ε 4.71), 279 nm (log ε 3.89), and 290 nm (sh) (log ε 3.75); the CD spectrum was rich in peaks and exhibited Δε(nm) +61(214), 0(246), −4.2(250), 0(270), +21(286), and 0(300). The ^{1}H NMR spectrum demonstrated the presence of two *N*-methyl groups as singlets at δ 2.10 [N(2)] and 2.57 [N(2′)]; three methoxy groups as singlets at δ 3.65 [H(7′)] and 3.90 (6H) [H(6) and H(12)]; and 10 aromatic protons as three singlets at δ 6.07 [H(8′)], 6.12 [H(5′)], and 6.55 [H(5)], a series of double doublets at δ 6.74 [H(10′)], 6.78 [H(11′)], 7.01 [H(13′)], and 7.23 [H(14′)] (J = 2.0 Hz, 8.2 Hz), plus δ 6.34 (d) (J = 2.2 Hz) [H(10)], 6.71

410

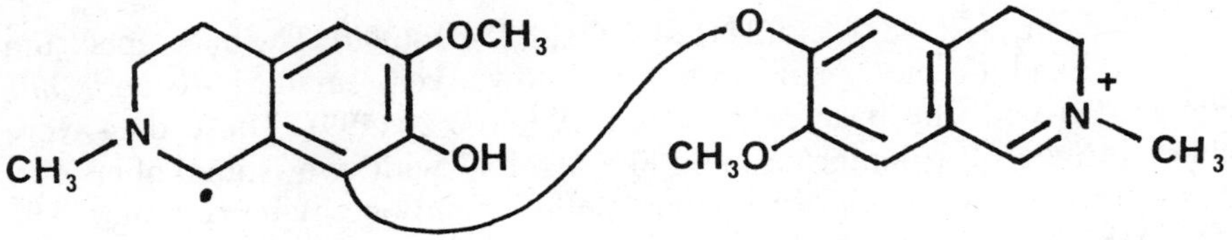

411

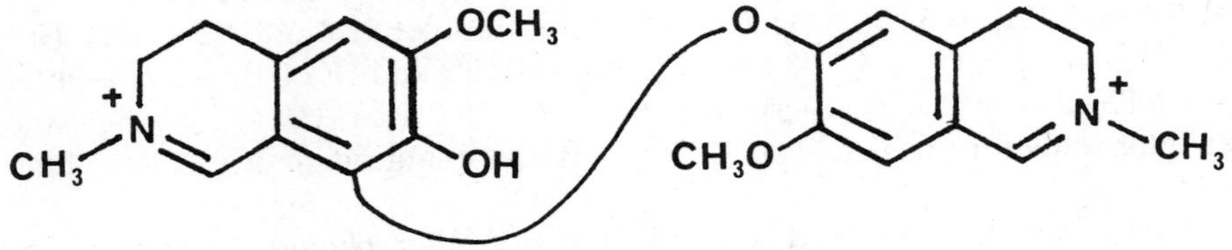

412

(dd) (J = 2.2 Hz, 8.2 Hz) [H(14)], and 6.82 (d) (J = 8.2 Hz) [H(13)]. The mass spectrum showed the molecular ion at M$^+$ m/z 608 (38%), 607 (24), 593 (5) (M—CH$_3$), 577 (2) (M—OCH$_3$), 381 (100) (**411**—H), 367 (16) (**411**—CH$_3$), 204 (3), 192 (21), 191 (89) (**412**), 190 (22) (**412**—H), 176 (33), and 174 (48). These spectral data were consistent with an *O*-methylthalic-berine (**333**)-like structure with a phenolic group in the top (isoquinoline) portion of the dimer [529,585]. Methylation of thaliphylline with CH$_2$N$_2$ af-forded *O*-methylthalicberine (**333**), which underwent reductive cleavage with

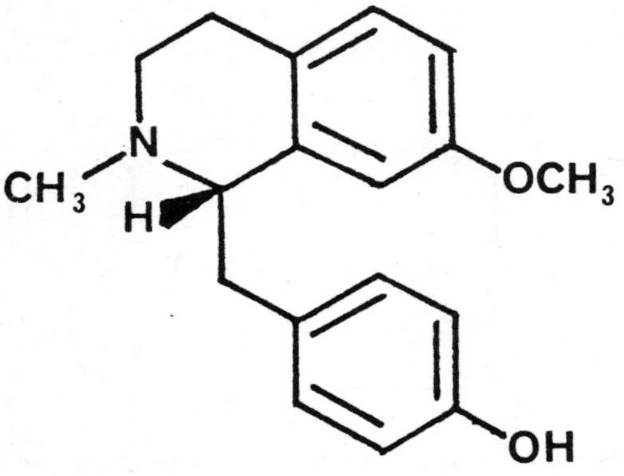

413

sodium in liquid ammonia to yield (+)-(S)-O-methylarmepavine (**327**) and (+)-(S)-N-methylisococlaurine (**334**) as major products and (+)-1-(4-hydroxybenzyl)-2-methyl-7-methoxy-1,2,3,4-tetrahydroisoquinoline (**413**) as a minor product. These data established the skeletal structure, positions of oxygenation, termini of the diphenyl ether linkages, and stereochemistry of thaliphylline (**410**). A comparison of the ^{1}H NMR spectra of thalicberine (**386**) and O-methylthalicberine (**333**) with that of thaliphylline showed that the δ 3.77–3.78 singlet absorption associated with a C(7) methoxy group of **386** and **333** was missing in the spectrum of thaliphylline and thus the single phenolic group of thaliphylline was placed at C(7), confirming the structure as **410** [544].

Thaliphylline (**410**) has not been isolated from any other *Thalictrum* species or other plants to this time.

3.5.42. Thalirabine (5-O-Desmethylthalistyline). Tabular review [529, alkaloid No. 17a, 530], UV [215,529], IR [215], ^{1}H NMR [215,529], MS [215,529], CD [215,529].

Thalirabine (**369**), $C_{40}H_{47}N_2^+O_8$ (683.3332), m.p. 131–132° (hydroxide), 205–206° dec. (MeOH–H$_2$O) (iodide), 176–177° dec. (hydriodide), 198–200° dec. (MeOH) (methiodide); $[\alpha]_D^{26}$ +142° (c 0.548, MeOH) (hydroxide), $[\alpha]_D^{26}$ +106° (c 0.013, MeOH) [215] was isolated as its amorphous hydroxide from *Thalictrum minus* L. race B in 1978 [215]. The UV spectrum was characterized by the presence of maxima at 207 nm (log ε 4.99), 276 nm (log ε 3.82), and 283 nm (log ε 3.80) and the CD spectrum (methiodide) showed $[\theta]_{225}$ +135,000 and $[\theta]_{278}$ +8,100 [215]. The ^{1}H NMR spectrum (CDCl$_3$) indicated the presence of one tertiary N-methyl group as a singlet at δ 2.55 and quaternary N-methyl group as two singlets at δ 3.45 and 3.78; four methoxy groups as two singlets at δ 3.65 (3H) and 3.78 (9H); one methylenedioxy group as a singlet at δ 5.85; and nine aromatic protons as two singlets at δ 5.43 [H(8)] and 5.71 [H(8')] plus a multiplet at δ 6.2–7.3 (7H) [215]. The ^{1}H NMR spectrum was also determined in CD$_3$NO$_2$ as well as that of the hydriodide salt in pyridine-d$_5$, both of which aided in the resolution of the N-methyl and O-methyl peaks into separate regions, as well as sharpening the signals. The EI–MS showed no significant high mass ions but showed two major fragment ions at m/z 222 (57%) (**248**) and 220 (100) (**249**) and the CI–MS showed a prominent ion at M$^+$ m/z 683 (26%) (thermal Hofmann + H) and 669 (100) (M—CH$_3$ + H). Various tests (Zn–HCl, Fe^{2+}– NH$_4$OH, Fe^{2+}–H$_2$O, triphenylphosphine) for the presence of a N-oxide were negative, but other tests suggested the presence of a methylenedioxy group and a *para*-unsubstituted phenol. Methylation (CH$_2$N$_2$) of thalirabine afforded thalistyline (**102**) thereby establishing the skeletal structure, portions of oxygenation, termini of the diphenyl ether linkage, and stereochemistry of thalirabine [215]. Because the high-resolution mass spectrum of thalistyline (O-methylthalirabine) showed the presence of a fragment ion at m/z 236 (**66**) and the disappearance of the ion at m/z 222 (**248**), the single phenolic

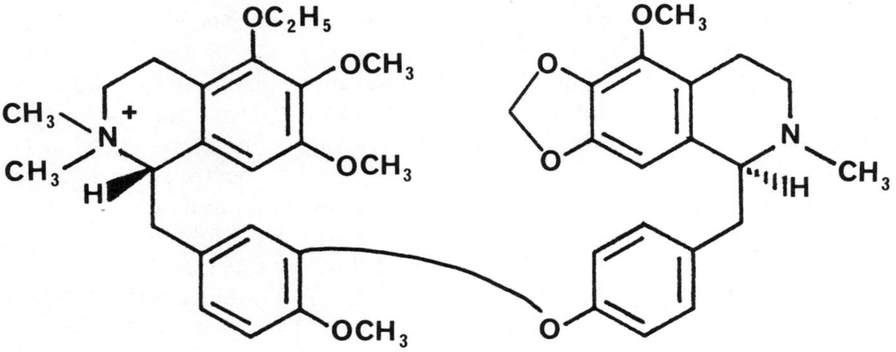

414

group of thalirabine must be in ring B. Furthermore, because the Gibbs test for para-unsubstituted phenols was positive, the phenolic group must be placed at C(5) and thalirabine assigned as **369** [215]. Treatment of thalirabine with ethereal diazoethane produced *O*-ethylthalirabine (**414**), which upon reductive cleavage with sodium in liquid ammonia afforded amines **415** and

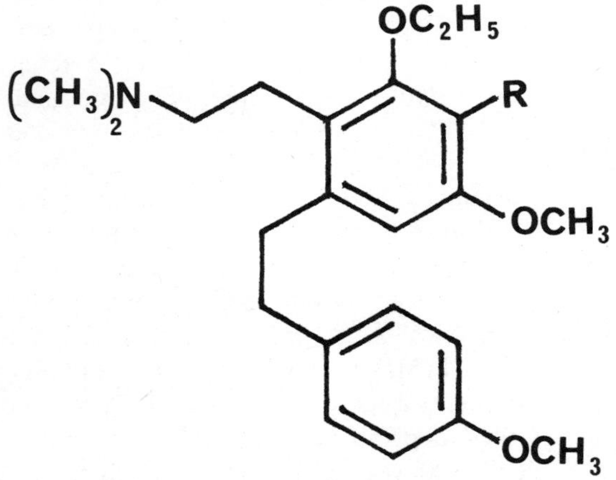

415 R = OCH$_3$

416 R = H

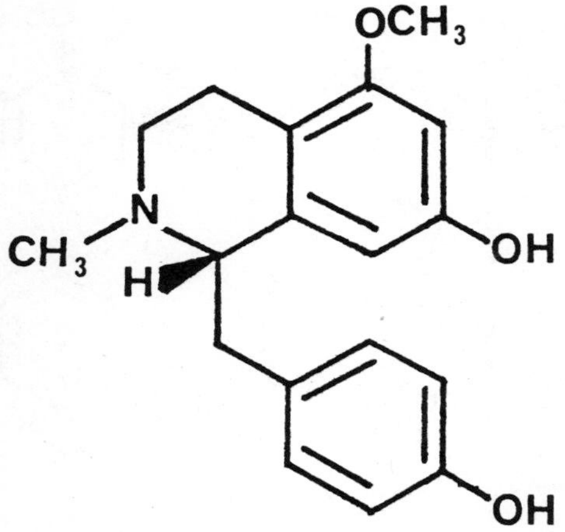

$$\underline{417}$$

416 (both produced by Emde-type reductive degradation) and $(+)$-(S)-1-(4-hydroxybenzyl)-2-methyl-5-methoxy-7-hydroxy-1,2,3,4-tetrahydroisoquinoline (**417**), further confirming the structure of thalirabine (**369**) [215].

Thalirabine (**369**) was found to produce a hypotensive response (13–28 mm Hg) in normotensive rabbits when administered in an intravenous dosage of 0.1–1.0 mg/kg. There was an initial decrease in blood pressure followed by a drop to a fatal level within 4 min. A gradual bradycardia was also observed [215]. Thalirabine was also found to inhibit the in vitro growth of *Mycobacterium smegmatis* at 100 μg/mL [215].

Thalirabine, only the second monoquaternary bisbenzylisoquinoline alkaloid to be isolated from *Thalictrum* species (the first being thalistyline (**102**) [556]), has not been isolated from other *Thalictrum* species or other plants to date.

3.5.43. Thaliracebine. Tabular review [529, alkaloid No. 14a, 530], UV [215,529], IR [215], ^{1}H NMR [215,529], MS [215,529], CD [215,529].
Thaliracebine (**418**), $C_{39}H_{44}N_2O_7$ (652.3149), m.p. 83–84° (amorphous), $[\alpha]_D^{26}$ $+121°$ (c 0.28, MeOH), was first isolated from *Thalictrum minus* L. race B in 1978 [215]. The UV spectrum was characterized by a single maximum at 278 nm (log ϵ 3.90) with no shift in acidic or alkaline medium, and the CD spectrum showed extrema at $[\theta]_{238}$ $+103,000$, $[\theta]_{289}$ $+11,100$, and $[\theta]_{310}$ -510 [215]. The ^{1}H NMR spectrum suggested the presence of two N-methyl groups as a singlet at δ 2.48 (6H); four methoxy groups as three singlets at δ 3.62 (6H), 3.76, and 3.78; one methylenedioxy group as a singlet

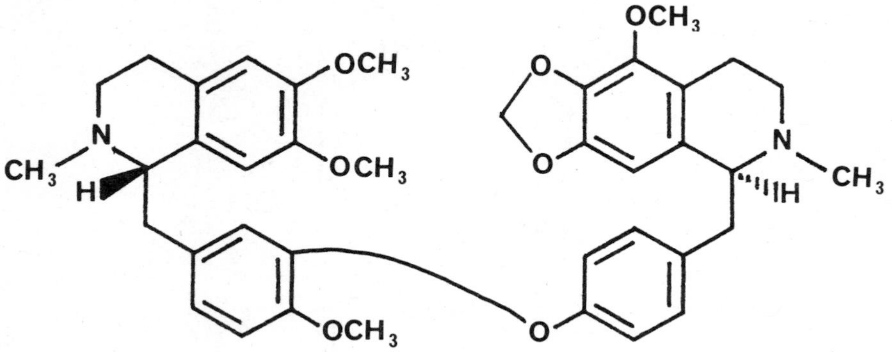

418

at δ 5.87; and 10 aromatic protons as three singlets at δ 5.77 [H(8)], 6.15 [H(8′)], and 6.52 [H(5)] plus a multiplet at δ 6.6–7.2 (7H) [215]. The mass spectrum showed an extremely weak molecular ion at M$^+$ *m/z* 652 (0.05%) with other intense fragment ions at *m/z* 220 (95%) (**249**) and 206 (100) (**191**) [215]. These spectral data were characteristic of a thalibrine (**329**)-like alkaloid in which one isoquinoline ring bears two methoxy groups and the other bears one methoxy group and one methylenedioxy group [529,546]. Reductive cleavage of thaliracebine with sodium in liquid ammonia produced (+)-(*S*)-*O*-methylarmepavine (**327**) plus (+)-(*S*)-1-(4-hydroxybenzyl)-2-methyl-5-methoxy-7-hydroxy-1,2,3,4-tetrahydroisoquinoline (**417**) and thus established the structures of the two benzylisoquinoline parts of the dimer, including their stereochemistry [215]. Oxidation of thaliracebine with potassium permanganate afforded *N*-methylcorydaldine (**88**), 2-methyl-5-methoxy-6,7-methylenedioxy-1-oxo-1,2,3,4-tetrahydroisoquinoline (thalflavine) (**100**), and 2-methoxy-4′,5-dicarboxydiphenyl ether (**309**), thereby confirming the structure of the isoquinoline rings and the termini of the diphenyl ether of thaliracebine (**418**) [215].

The intravenous administration of thalracebine (**418**) to normotensive rabbits in a dosage range of 0.1–1.0 mg/kg produced a delayed (1–2 min) hypotensive response of 8–22 mm Hg. In addition, the alkaloid was found to inhibit the in vitro growth of *Mycobacterium smegmatis* at a concentration of 100 μg/mL [215].

Thaliracebine (**418**) has been isolated from only one other *Thalictrum* species, that being *T. faberi* Ulbr. [555].

3.5.44. Thalirugidine. Tabular review [529, alkaloid No. 17b, 530] UV [268,529,634], IR [268,529,634], ^{1}H NMR [268,529,634], MS [268,529,634], CD [529,634].

Thalirugidine (**419**), $C_{39}H_{46}N_2O_8$ (670.3254), $[\alpha]_D^{20}$ +112° (c 0.19, MeOH), was isolated as a white amorphous solid from *Thalictrum rugosum* Ait. in 1978 [634]. The UV spectrum showed a single maximum at 278 nm (log ϵ 3.82) with a bathochromic shift in alkali while the CD spectrum was characterized by several extrema ($[\theta]_{230}$ +98,500 and $[\theta]_{280}$ +7,200) [634]. The ^1H NMR spectrum suggested the presence of two *N*-methyl groups as singlets at δ 2.48 and 2.51; five methoxy groups as singlets at δ 3.61, 3.63, 3.81 (6H), and 3.85; two hydroxy groups as a broad singlet at δ 5.1 (D_2O-exchangeable); and nine aromatic protons as two singlets at δ 5.76 [H(8)] and 5.79 [H(8')] plus a multiplet at δ 6.6–7.2 (7H). The mass spectrum displayed a weak parent ion at M^+ *m/z* 670 (1%) with other important fragment ions at *m/z* 222 (100%) (**248**), 221 (3) (**248**—H), 220 (3) (**248**—2H), 207 (3) (**248**—CH$_3$), 206 (8) (**248**—CH$_3$—H), 192 (8) (**248**—2H—CO), and 178 (2) (**248**—CH$_3$—H—CO) [634]. These data were indicative of a thalibrine (**329**)-like alkaloid bearing two methoxy groups and one phenolic group in each of the upper isoquinoline rings and one methoxy group in one of the lower benzyl rings [529,546]. Methylation (CH_2N_2) of thalirugidine produced an *O,O*-dimethyl derivative (**420**) [amorphous, $[\alpha]_D^{20}$ +53 ° (c 0.21, MeOH)], and diazoethane ethylation afforded *O,O*-diethylthalirugidine (**421**) [amorphous, $[\alpha]_D^{20}$ +44° (c 0.7, MeOH)]. Reductive cleavage of the *O,O*-diethyl ether **421** with sodium in liquid ammonia yielded (+)-(*S*)-1-(4-methoxybenzyl)-2-methyl-5-ethoxy-6,7-dimethoxy-1,2,3,4-tetrahydroisoquinoline (**325**) and (+)-(*S*)-1-(4-hydroxybenzyl)-2-methyl-5-ethoxy-6,7-dimethoxy-1,2,3,4-tetrahydroisoquinoline (**422**), thus identifying the two monomeric units that constitute the dimer, fixing the position of the phenolic hydroxy groups, and establishing the (*S,S*) stereochemistry for the dimer. Oxidation of an acetone solution of *O,O*-diethylthalirugidine (**421**) with potassium permanganate afforded 1-oxo-2-methyl-5-ethoxy-6,7-dimethoxy-1,2,3,4-tetrahydroisoquino-

419

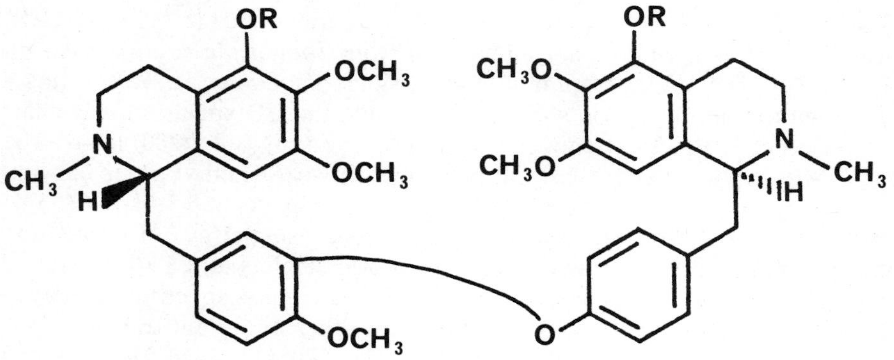

420   R = CH$_3$

421   R = C$_2$H$_5$

line (423) and 2-methoxy-4',5-dicarboxybiphenyl ether (309), thus furnishing corroborating evidence for the structural assignment of thalirugidine as 419 [634]. Thalirugidine (419) is a plausible biosynthetic product of the coupling of two molecules of (+)-(*S*)-5-demethylthalifendlerine (287) followed by methylation at the C(12) phenol [634].

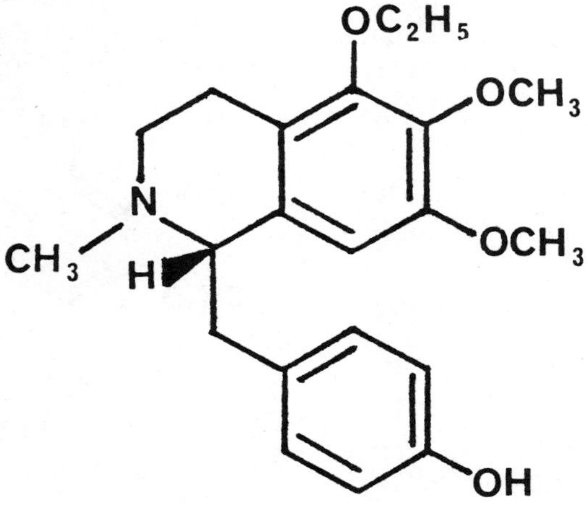

422

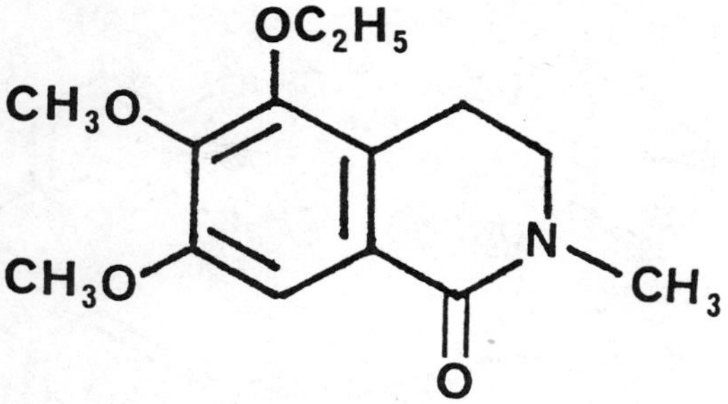

423

Thalirugidine was found to inhibit the in vitro growth of *Staphylococcus aureus* and *Mycobacterium smegmatis* at a concentration of 1000 µg/ml [234].

Thalirugidine (419) has also been isolated from *Thalictrum foliolosum* DC. [268].

3.5.45. Thalirugine. Tabular review [529, alkaloid No. 14b, 530], UV [529,634], IR [634], ^{1}H NMR [529,634], MS [529,634], CD [529,634].
Thalirugine (424), $C_{38}H_{44}N_2O_7$ (640.3149), $[\alpha]_D^{20}$ +92° (c 0.25, MeOH), was isolated as an amorphous solid from *Thalictrum rugosum* Ait. in 1978 [634]. The UV spectrum was characterized by a single maximum at 280 nm (log ε 3.81) with additional peaks at 285 nm (log ε 3.87) and 310 nm (sh) (log ε 3.45) in dilute alkali. The CD spectrum showed extrema at $[\theta]_{226}$ +78,000, $[\theta]_{248}$ −3,100, and $[\theta]_{282}$ +6,400 [634]. The ^{1}H NMR spectrum indicated the pres-

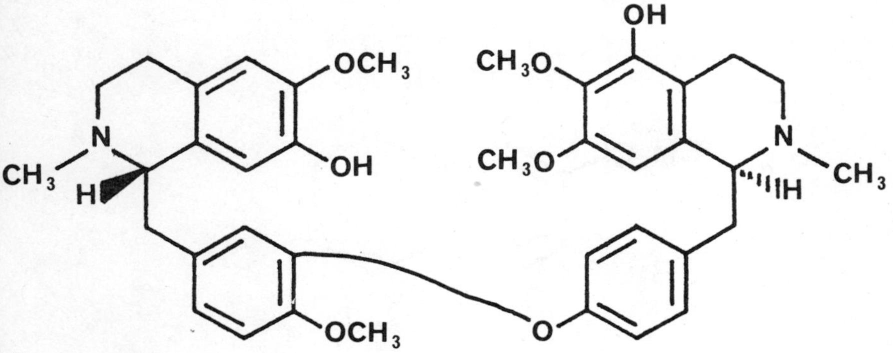

424

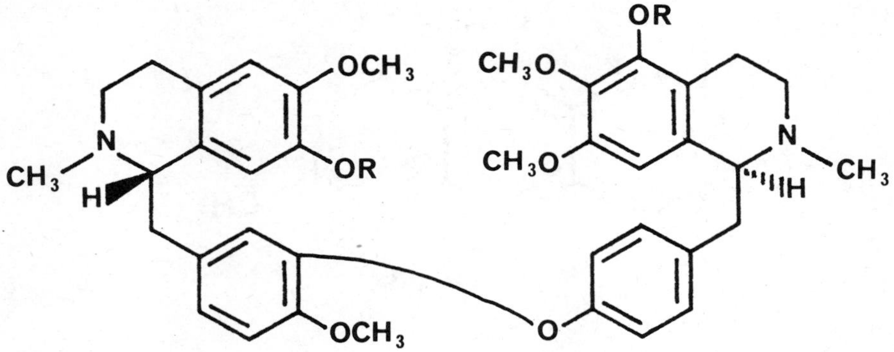

425 R = CH₃

426 R = C₂H₅

ence of two N-methyl groups as singlets at δ 2.43 and 2.49; four methoxy groups as singlets at δ 3.58, 3.78 (6H), and 3.83; a phenolic functionality as a broad, D₂O-exchangeable singlet at δ 5.50; and 10 aromatic protons as three singlets at δ 5.73 [H(8′)], 6.38 [H(8)], and 6.47 [H(5)] plus a multiplet from δ 6.6 to 7.2 (7H) [634]. The mass spectrum showed a very weak molecular ion at M⁺ *m/z* 640 (0.01%) with other important fragment ions at *m/z* 222 (100%) (**248**), 207 (35) (**248**—CH₃), and 192 (83) (**199**) [634]. These spectral data were characteristic of a tail-to-tail linked bisbenzylisoquinoline alkaloid containing one methoxy group plus one phenolic hydroxy group in one isoquinoline ring and two methoxy groups plus one phenolic hydroxy

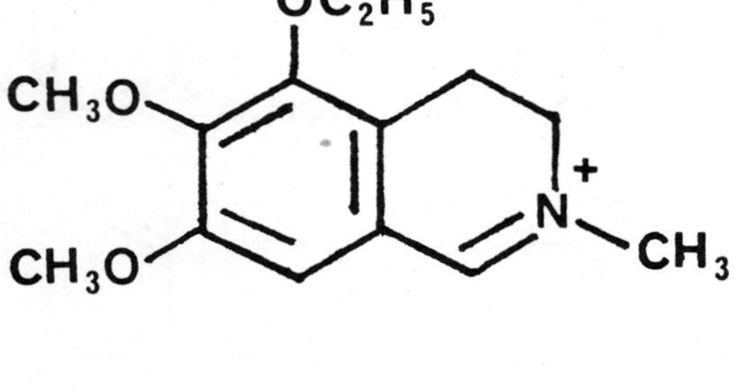

427

group in the other isoquinoline ring [529,546]. Methylation of thalirugine with CH_2N_2 produced an O,O-dimethyl derivative (425) and diazoethane ethylation afforded an O,O-diethyl derivative (426). Intense mass spectral fragment ions at m/z 236 (100%) (66) and 206 (91) (191) for the O,O-dimethyl derivative (425) plus 250 (45%) (427) and 220 (63) (232) for the O,O-diethyl derivative (426) supported the presence of the two phenolic hydroxy groups in the isoquinoline rings [546,634]. Reductive cleavage of O,O-diethylthalirugine (426) with sodium in liquid ammonia afforded (+)-(S)-1-(4-methoxybenzyl)-2-methyl-6-methoxy-7-ethoxy-1,2,3,4-tetrahydroisoquinoline (428) plus (+)-(S)-1-(4-hydroxybenzyl)-2-methyl-5-ethoxy-6,7-dimethoxy-1,2,3,4-tetrahydroisoquinoline (422). Oxidation of O,O-diethylthalirugine (426) with potassium permanganate (acetone) yielded 1-oxo-2-methyl-6-methoxy-7-ethoxy-1,2,3,4-tetrahydroisoquinoline (202), 1-oxo-2-methyl-5-ethoxy-6,7-dimethoxy-1,2,3,4-tetrahydroisoquinoline (423), and 2-methoxy-4'-5-dicarboxybiphenyl ether (309). The metal–ammonia cleavage and oxidative products permitted the unequivocal assignment of structure of thalirugine as 424 [634]. Thalirugine may be considered as a monomethyl derivative of the hypothetical dimeric alkaloid 406, formed by the biosynthetic phenolic coupling of (+)-(S)-N-methylcoclaurine (52) and (+)-(S)-5-demethylthalifendlerine (287) [634].

Thalirugine (424) was noted to inhibit the in vitro growth of *Staphylococcus aureus* and *Mycobacterium smegmatis* at a concentration of 1000 μg/mL [234].

Thalirugine has been isolated from only one other *Thalictrum* species to date, that being *T. minus* L. var. *microphyllum* Boiss. [544].

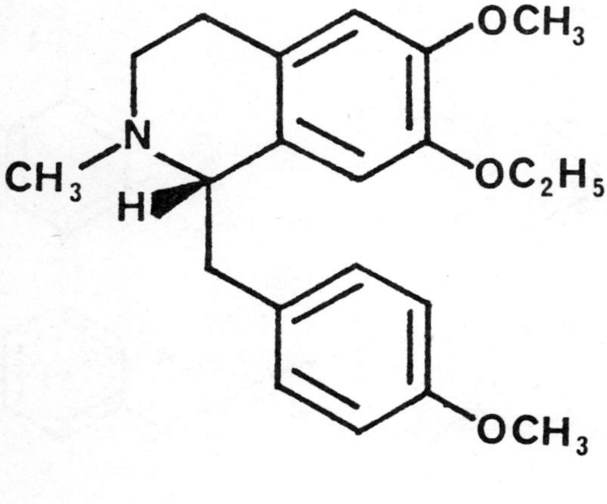

428

3.5.46. Thaliruginine. Tabular review [529, alkaloid No. 14c], UV [529,634], IR [634], ^{1}H NMR [529,634], MS [529,634], CD [529,634].

Thalilruginine (**429**), $C_{39}H_{46}N_2O_7$ (654.3305), $[\alpha]_D^{20}$ +104° (c 0.16, MeOH), was isolated as a white amorphous solid from *Thalictrum rugosum* Ait. in 1978 [634]. The alkaloid possessed a single UV maximum at 281 nm (log ϵ 3.90) with an additional absorption at 309 nm (sh) (log ϵ 3.09) in dilute alkali. The CD spectrum displayed extrema at $[\theta]_{230}$ +93,000, $[\theta]_{252}$ −2,100, and $[\theta]_{287}$ +14,500. The ^{1}H NMR spectrum suggested the presence of two N-methyl groups as singlets at δ 2.48 and 2.50; five methoxy groups as singlets at δ 3.57, 3.61, 3.78, 3.80, and 3.85; hydroxy functionality as a broad D_2O-exchangeable singlet at δ 5.4; and 10 aromatic protons as three singlets at δ 5.71 [H(8′)], 6.11 [H(8)], and 6.53 [H(5)] plus a multiplet at δ 6.6–7.2 (7H). The mass spectrum displayed a weak molecular ion at M$^+$ m/z 654 (0.8%) with other diagnostic fragment ions at m/z 222 (68%) (**248**), 206 (100) (**191**), and 192 (26). These spectral data were characteristic of a tail-to-tail linked bisbenzylisoquinoline alkaloid bearing two methoxy groups in one isoquinoline ring, two methoxy groups plus one hydroxy group in the other isoquinoline ring, plus one methoxy group in the bottom biphenyl ether system [529,546]. Diazomethane methylation of thaliruginine produced an O-methyl ether that was identical to O,O-dimethylthalirugine (**425**). Ethylation of thaliruginine with ethereal diazoethane afforded O-ethylthaliruginine (**430**), which on reductive cleavage with sodium in liquid ammonia yielded (+)-(S)-O-methylarmepavine (**327**) and (+)-(S)-1-(4-hydroxybenzyl)-2-methyl-5-ethoxy-6,7-dimethoxy-1,2,3,4-tetrahydroisoquinoline (**422**), thereby establishing the structure and stereochemistry of thaliruginine as **429** [634]. Thaliruginine (**429**) may be visualized as a dimethyl derivative of the hypothetical but biosynthetically plausible alkaloid **406**. This hypothetical alkaloid could

429

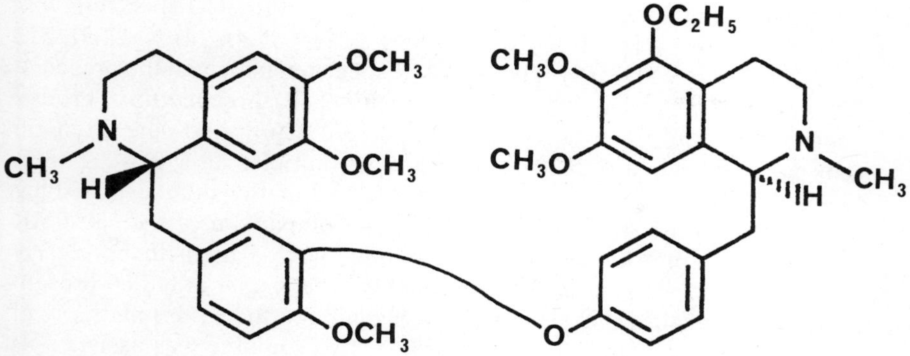

$$\underset{\displaystyle\sim}{\mathbf{430}}$$

be formed via the biosynthetic phenolic coupling of $(+)$-(S)-N-methylco-claurine (52) and $(+)$-(S)-5-demethylthalifendlerine (287) [634].

Thaliruginine (429) has not been isolated from any other *Thalictrum* species or higher plant since this report.

3.5.47. Thalisopidine. Tabular review [529, alkaloid No. 53, 530], UV [174,529], ^{1}H NMR [174,529,635], CD [530,562].
Thalisopidine (431), $C_{37}H_{40}N_2O_7$ (624.2836), m.p. 215–216° $((CH_3)_2CO)$, $[\alpha]_D^{19}$ $-9°$ (c 0.96, EtOH), was first isolated from *Thalictrum isopyroides* C.A.M. in 1968 [174]. The UV spectrum was characterized by the presence

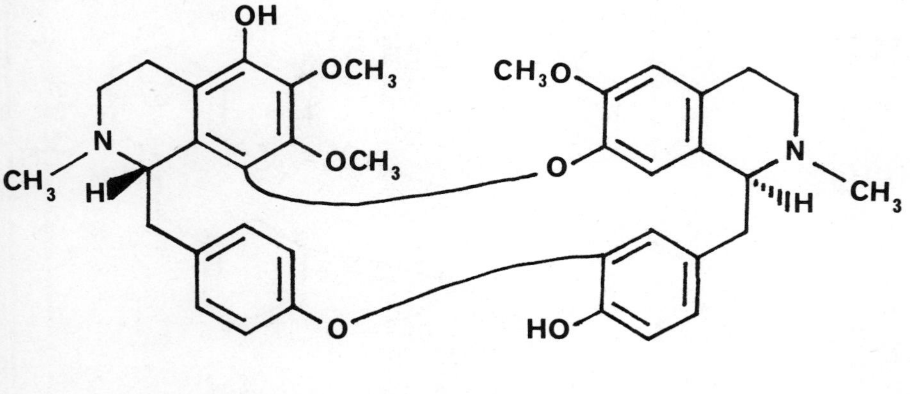

$$\underset{\displaystyle\sim}{\mathbf{431}}$$

of a single maximum at 285 nm (log ϵ 4.04) [174] and the CD spectrum was rich in Cotton effects: [nm ($\Delta\epsilon$)] 209 (-69.7), 224 ($+34.4$), 241 (-22.9), 272 (-6.9), and 290 ($+3.51$) [562]. The ^{1}H NMR spectrum indicated the presence of two *N*-methyl groups as singlets at δ 2.44 and 2.49; three methoxy groups as singlets at δ 2.96 [C(7)], 3.30 [C(6′)], and 3.70 [C(6)]; and nine aromatic protons as a singlet at δ 6.30 [H(8′)] and a multiplet at δ 6.4–7.2 (8H) [174,635]. Methylation of thalisopidine afforded a dimethyl ether establishing the biphenolic nature of the alkaloid [635]. A comparison of the ^{1}H NMR spectra of thalisopine (432) and thalisopidine showed them to be clearly related, with the spectrum of the latter lacking the singlet at δ 3.86 present in the former. Because this singlet was assignable to the C(12′) methoxy of thalisopine and was missing in thalisopidine, thalisopidine was assigned as 431 [635]. Apparently the assignment of stereochemistry as (*S*,*S*) was later determined and made on the basis of the relationship of thalisopidine (431) to thalisopine (432) [562].

Thalisopidine has not been isolated from any other *Thalictrum* species nor other higher plant to date.

3.5.48. Thalisopine (Thaligosine).
Tabular review [529, alkaloid No. 54, 530] UV [529,634], IR [174,529,634], ^{1}H NMR [174,529,634], MS [174,529,634], ORD [453,529], CD [529,562,634].

Thalisopine (432), $C_{38}H_{42}N_2O_7$ (638.2992), m.p. 151–153° [H$_2$O–MeOH(1:3)] (needles) [636], m.p. 143–145° (Et$_2$O) [634], 235–237° (EtOH) (hydriodide salt) [636]; $[\alpha]_D^{20}$ $-104.9°$ (Me$_2$CO), $[\alpha]_D^{20}$ $-71.02°$ (CHCl$_3$), $[\alpha]_D$ $-109°$ (c 0.17, MeOH), was first isolated from *Thalictrum isopyroides* C.A.M. in 1961 [636]. The UV spectrum was characterized by a single maximum at 282 nm (log ϵ 3.86) with the addition of a shoulder at 305 nm (log ϵ 3.64) in strong alkali [634]. The ^{1}H NMR spectrum indicated the presence

432

433

of two *N*-methyl groups as singlets at δ 2.43 and 2.48; four methoxy groups as singlets at δ 3.00 [C(7)], 3.29 [C(6′)], 3.70 [C(6)], and 3.86 [C(12′)]; one hydroxy group at δ 5.1; and nine aromatic protons at δ 6.31 (s) [H(8′)] and from δ 6.38 to 7.06 (8H) [174]. The mass spectrum was characterized by a parent ion at M$^+$ *m/z* 638 (11%) with other diagnostic fragment ions at *m/z* 412 (89%) (**321**), 397 (38) (**321**—CH$_3$), 221 (18), 206 (100) (**322**), 183 (17) [**322**—(CH$_3$)$_2$O], 174 (18), 173 (29), 172 (89), 90 (9), and 89 (20) [174]. Consecutive Hofmann degradations of thalisopine produced a des-*N*-methine and oxidation with potassium permanganate afforded 2-methoxy-4,5′-dicarboxybiphenyl ether (**309**) [636]. In a subsequent study, reductive cleavage of thalisopine with sodium in liquid ammonia afforded two products that were characterized as (+)-1-(4-methoxybenzyl)-6-methoxy-2-methyl-1,2,3,4-tetrahydroisoquinoline (**433**) and armepavine (**328**). Oxidation of **433** with potassium permanganate (Me$_2$CO) produced anisic acid (*p*-methoxybenzoic acid) and a weak base, the latter of which on further oxidation also yielded anisic acid. On the basis of these results and a comparison of the nonphenolic cleavage product with related isoquinolines, thalisopine was initially assigned as **434** [637]. Some 5 years later, a detailed ^{1}H NMR and mass spectral analysis of thalisopine prompted the proposal that thalisopine was best represented by structure **432** [174]. It was not until 1978 that the structure of thalisopine as (**432**) was settled with certainty, however. Extracts of *Thalictrum rugosum* Ait. furnished an alkaloid that was named thaligosine (CD[θ]$_{225}$ +16,700, [θ]$_{240}$ −7,980, [θ]$_{272}$ −1,200, [θ]$_{287}$ +3,200) [634]. Methylation (CH$_2$N$_2$) of thaligosine produced an *O*-methyl ether that was identical with thalrugosaminine (**408**). Diazoethane ethylation furnished an *O*-ethyl ether (**435**) that on reductive cleavage with sodium in liquid am-

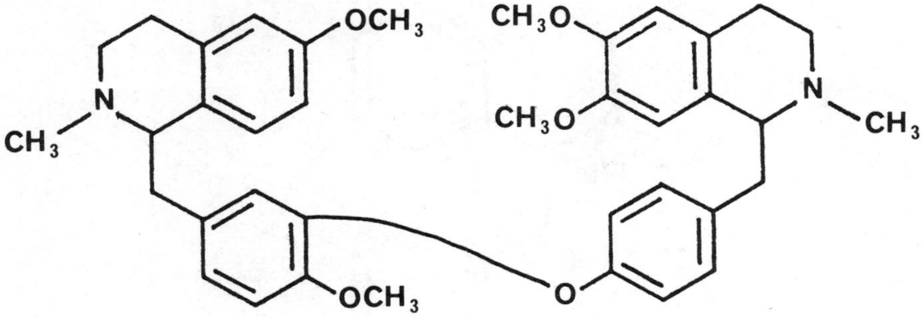

434

monia yielded (+)-(S)-1-(4-hydroxybenzyl)-2-methyl-5-ethoxy-6,7-dime-
thoxy-1,2,3,4-tetrahydroisoquinoline (**422**) plus (+)-(S)-1-(4-methoxyben-
zyl)-2-methyl-6-methoxy-7-hydroxy-1,2,3,4-tetrahydroisoquinoline   (**436**)
and permitted the unequivocal assignment of structure of thaligosine as **432**
[634]. However, because structure **432** had been previously assigned to thalisopine [174] in 1968 and since m.p., specific rotation, and other data reported for the two alkaloids agree closely, thalisopine should take preference
with regard to literature nomenclature.

Administration of thalisopine to rabbits, frogs, and albino mice produced
a state of decreased mobility and an apparent tranquilization. The intravenous LD_{50} in mice is 70 mg/kg and in rabbits is 21 mg/kg; the subcutaneous
LD_{50} in mice is 397 mg/kg. The alkaloid produced an antagonism to the

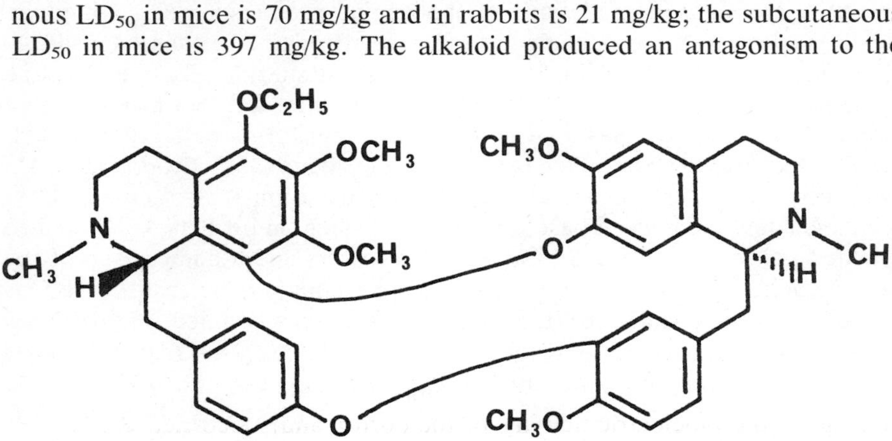

435

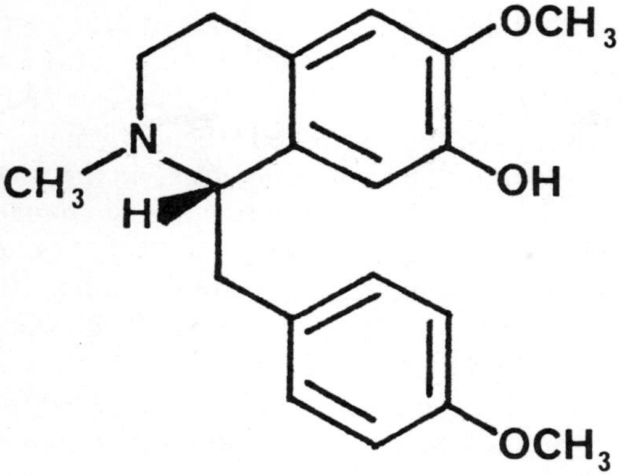

$$\underset{\displaystyle \sim}{436}$$

effects of camphor but only mildly so to those of caffeine. Intraperitoneal administration (50 mg/kg) protected against the convulsions of a lethal dose of strychnine [638]. An intraperitoneal dose (100 mg/kg) of thalisopine in white mice prevented strychnine nitrate (1.36 mg/kg LD_{90})-induced lethality. The same dosage of thalisopine also prevented pentamethylenetetrazole (76.5 mg/kg 90% toxic dose)-induced convulsions and markedly reduced the convulsions caused by vincanine (16.9 mg/kg 90% toxic dose). The action of thalisopine is greater than that of chloral hydrate but less than that of phenobarbital [639]. Parenteral administration of thalisopine to rats (61–122 mg/kg) or rabbits (2.16 and 4.32 mg/kg) did not produce symptoms of intoxication or structural alterations in internal organs. Rodents receiving 245 mg/kg orally or 17.6 mg/kg parenterally showed irregular hyperemia of the brain and liver, edema of the brain and spinal cells (white matter), and protein dystrophy of parenchymal cells. Intravenous administration of 26 mg/kg to rabbits produced death within 3 min, with arterial spasms, congestive venous plethora, hypoxia, pulmonary edema, and hyperemia of the brain, liver, and kidney. Rodents receiving 480 mg/kg died within 3–7 days, and those receiving 612 mg/kg died on the third day. Pronounced ventral edema and hepatic micronecrosis were noted [640]. Intravenous administration of the alkaloid in a 3.5 or 10-mg/kg dosage suppressed the effects of corazol and decreased the bioelectric activity of the cortex and subcortical formations [641]. Administration of thalisopine to rats in oral dosages of 61 and 122 mg/kg and to rabbits in intravenous doses of 2.16 and 4.32 mg/kg produced no organic changes. An intravenous dose of 25.3 mg/kg to rabbits produced organ edema and hyperemia and death within 3–5 min. Oral administration of 480 mg/kg to rats caused death in 3–7 days and a dose of 612 mg/kg

produced death on the third day [642]. Administration of thalisopine dihy-
driodide in an intravenous dosage of 1–30 mg/kg produced hypotension in
narcotized dogs, cats, and rabbits. There was no effect on feline respiration
but a brief arrest in rabbits and increased frequency and amplitude of respiratory movements in dogs. There was no effect on the tone and magnitude
of contraction of the second eyelid of cats nor any change on the pressor
effect of epinephrine, hypotensive action of acetylcholine, or stimulation of
the coronary vagal branch. A bilateral vagotomy did not effect the intensity
of the hypotensive action in cats. The alkaloid did not antagonize the respiratory stimulation or hypertension induced by cytisine. The alkaloid (10–
100 mg/kg) administered intravenously to frogs produced negative chronotropic and inotropic actions on frog heart contractions and a negative chronotropic action on the isolated frog heart. The alkaloid decreased the tone of
isolated sections of rodent and rabbit small intestine and relieved barium
chloride- and acetylcholine-induced spasms [643]. Administration of thalisopine dihydroiodie at a dosage of 45 mg/kg prevented weak epileptic shock
in 50% of animals but was ineffective in animals with major electroshock.
The alkaloid enhanced the anticonvulsant effect of diphenine in mice and
produced sedation plus elongation of hexenal- or chloral hydrate-induced
sleep [644]. Thalisopine depressed higher nervous activity and conditioned
reflexes in mice [454] and exerted antiarrhythmic effects in various animal
models against nicotinamide or acontine-induced arrhythmias. The alkaloid
prevented the loss of potassium ion from myocardial cells and prevented
fibrillation (10 mg/kg intravenously) induced by atrial electrical stimulation
[418]. Intravenous injections of the dihydrochloride salt in dosages of 5–10
mg/kg to dogs, cats, or rodents with induced (aconitine or epinephrine or
K-strophanthin or $CaCl_2$) myocardial arrhythmias produced a significant antiarrhythmic effect. This effect was equal to or greater than that produced
by quinidine (10–15 mg/kg) or procainamide (10–20 mg/kg) [644a]. Thalisopine (thaligosine) inhibited the growth of *Staphylococcus aureus* and *Mycobacterium smegmatis* at a concentration of 100 µg/mL and *Klebsiella
pneumoniae* and *Candida albicans* at a concentration of 1000 µg/mL in a
standardized in vitro test [234].

Thalisopine has also been isolated from *Thalictrum faberi* Ulbr. [279], *T.
foliolosum* DC. [268], and *T. minus* L. var. *microphyllum* Boiss. [146,544].

3.5.49. Thalistine.   Tabular review [530, alkaloid No. 221], UV [102,530],
IR [102,530], ^{1}H NMR [102,530], MS [102,530], CD [102,530].
Thalistine (**437**), $C_{39}H_{44}N_2O_8$ (668.3097), $[\alpha]_D^{20}$ + 104° (c 0.35, MeOH), was
isolated from *Thalictrum minus* L. race B as an amorphous residue in 1980
[102]. The alkaloid showed a single UV maximum at 278 nm (log ε 3.90)
with Cotton effects in the CD spectrum at $[\theta]_{226}$ +64,000 and $[\theta]_{290}$ − 1,530
[102]. The ^{1}H NMR spectrum indicated the presence of two *N*-methyl groups
at δ 2.47 and 2.50; four methoxy groups as singlets at δ 3.60, 3.63, and 3.78
(6H); one methylenedioxy group as a singlet at δ 5.88; a hydroxy group as

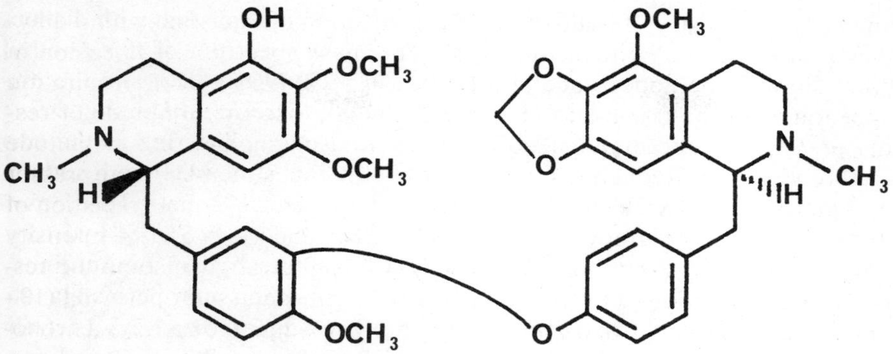

437

a broad, D$_2$O-exchangeable singlet at δ 5.8; and nine aromatic protons as a singlet at δ 5.76 (2H) [H(8) and H(8′)] and a multiplet from δ 6.4 to 7.2 (7H). The mass spectrum displayed a weak parent ion at M$^+$ *m/z* 668 (5%), with other diagonstic fragment ions at *m/z* 236 (15%), 222 (100) (**248**), 221 (82), 220 (91) (**249**), 205 (33), 204 (31), 192 (50), and 176 (10). These spectral data were indicative of a tail-to-tail linked dimeric benzylisoquinoline alkaloid containing two methoxy groups and one hydroxy group in one part of the isoquinoline (head) portion plus one methoxy group and one methylenedioxy group in the other isoquinoline ring [529,546]. Methylation of thalistine with CH$_2$N$_2$ produced on *O*-methyl ether that was identical to *N*-desmethylthal-istyline (**293**), thus fixing the skeletal structure, positions of oxygenation,

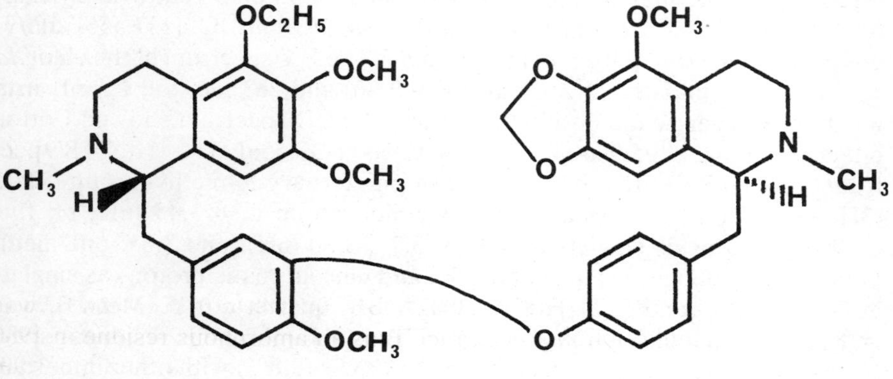

438

and stereochemistry of thalistine [102]. Ethylation of thalistine with diazoethane afforded *O*-ethylthalistine (**438**). The mass spectrum of this *O*-ethyl ether showed the appearance of an ion at m/z 250 (96%) (**427**) and the disappearance of the fragment ion at m/z 222 (**248**), thereby further suggesting the presence of the monophenol in the same isoquinoline ring as the two methoxy groups. Reductive cleavage of *O*-ethylthalistine (**438**) with sodium in liquid ammonia yielded (+)-(*S*)-1-(4-methoxybenzyl)-2-methyl-5-ethoxy-7-methoxy-1,2,3,4-tetrahydroisoquinoline (**324**) and (+)-(*S*)-1-(4-hydroxybenzyl)-2-methyl-5-methoxy-7-hydroxy-1,2,3,4-tetrahydroisoquinoline (**417**). Finally, oxidation of *O*-ethylthalistine with potassium permanganate in acetone afforded 1-oxo-2-methyl-5-ethoxy-6,7-dimethoxy-1,2,3,4-tetrahydroisoquinoline (**423**), 1-oxo-2-methyl-5-methoxy-6,7-methylenedioxy-1,2,3,4-tetrahydroisoquinoline (**100**), and 2-methoxy-4′,5-dicarboxybiphenyl ether (**309**), thereby establishing the structure of thalistine as **437** [102]. The genesis of cleavage products **324** and **417** can easily be rationalized because of ample precedents that demonstrate that trioxygenated tetrahydroisoquinolines do not survive intact on reductive cleavage [137,215].

Thalistine exhibited in vitro antimicrobial activity against *Staphylococcus aureus* (1000 μg/mL) and *Mycobacterium smegmatis* (100 μg/mL) [102].

To date, thalistine has not been isolated from any other *Thalictrum* species nor other higher plant.

3.5.50. Thalistyline.

Tabular review [529, alkaloid No. 18, 530], UV[137,529,556], IR [137], ^{1}H NMR [137,529,556], MS [137,529,556], CD [137,529,556].

Thalistyline (**102**), $C_{41}H_{49}N_2^+O_8$ (697.3489), m.p. 150–153° (chloride) [137,556], 220–223° dec. (iodide) [137,556], m.p. 266–268° dec. (methiodide) [137,556]; $[\alpha]_D^{25}$ +146° (c 0.1, MeOH) (chloride) [137,556]; $[\alpha]_D^{21}$ +125° (c 0.1, MeOH) (methiodide) [137]; was isolated as its pale yellow microcrystalline chloride salt from both *Thalictrum longistylum* DC. [137,556] and *T. podocarpum* Humb. [301,556] in 1976 [556]. The UV spectrum of the chloride salt displayed maxima at 276 nm (log ε 3.86) and 283 nm (log ε 3.84) with no shift in either acidic or alkaline media; the CD spectrum showed Cotton effects at $[\theta]_{225}$ +105,000 and $[\theta]_{284}$ +12,500 [137,556]. The ^{1}H NMR spectrum (CDCl₃) indicated the presence of one tertiary *N*-methyl group δ 2.48 (3H, s) and one quaternary dimethylamino group at δ 3.45 (6H, s); five methoxy groups as singlets at δ 3.63, 3.77, 3.80 (6H), and 3.85; one methylenedioxy group as a singlet at δ 5.89; and nine aromatic protons as singlets at δ 5.70 [H(8)] and 5.77 [H(8′)], an AA′BB′ quartet at δ 6.63 and 6.98 (*J* = 8 Hz) and a multiplet from δ 6.39 to 7.44 (3H) [137,556]. The mass spectrum displayed a weak molecular at M⁺ m/z 697 (0.8%) with other important fragment ions at m/z 682 (2%) (M—CH₃), 236 (100) (**66**), 221 (16) (**66**—CH₃), 220 (88) (**249**), and 205 (2) (**249**—CH₃) [137,556]. These spectral data were indicative of a monoquaternary–monotertiary bisbenzylisoquinoline alkaloid linked via a single tail-to-tail ether bridge and containing three methoxy

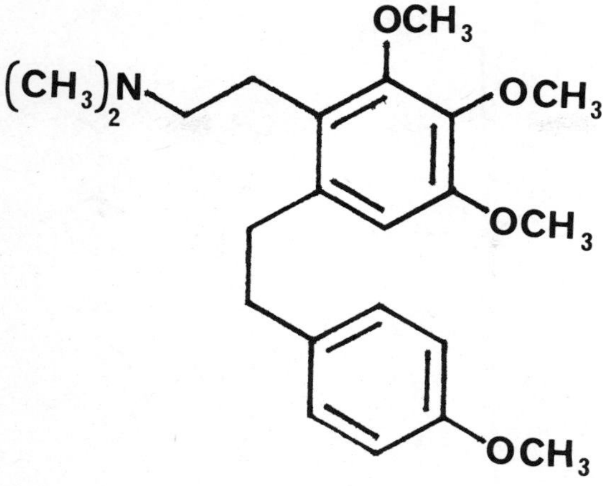

$$\underset{\displaystyle \sim}{\underline{439}}$$

groups in one half of the isoquinoline head and one methoxy group plus one methylenedioxy group in the other half of the isoquinoline head [529,546]. Reductive cleavage with sodium in liquid ammonia afforded the optically inactive amine **439** and ($+$)-(S)-1-(4-hydroxybenzyl)-2-methyl-5-methoxy-7-hydroxy-1,2,3,4-tetrahydroisoquinoline (**417**), and oxidation with potassium permanganate (acetone) yielded 1-oxo-2-methyl-5-methoxy-6,7-methylene-dioxy-1,2,3,4-tetrahydroisoquinoline (**100**) plus 2-methoxy-4',5-dicarboxy-biphenyl ether (**309**) [137,556]. The CD curve of thalistyline chloride showed two positive maxima and was similar to that of thalibrine (**329**) of established (S,S) configuration but opposite to another structurally similar alkaloid, dauricine (**440**), of established (R,R) configuration [556]. These data led to the assignment of structure of thalistyline as **102**. The dimethylaminoethyl portion of amine **439** obtained via reductive cleavage was the quaternary nitrogen center in the original alkaloid, and the two phenolic groups in cleavage product **417** arose from methylenedioxy fragmentation and biphenyl ether cleavage, respectively [556]. Thalistyline is characterized by unusually high water and chloroform solubility and thus was isolated from fractions of greatly differing polarity [137]. Thalistyline (**102**), *N*-methylthalistyline (**294**), and *N*-desmethylthalistyline (**293**) constitute the first examples of dimeric benzylisoquinoline alkaloids in which both isoquinoline rings are trioxygenated [556].

Thalistyline (**102**) was observed to inhibit the in vitro growth of *Staphylococcus aureus* and *Mycobacterium smegmatis* at a concentration of 50

440

μg/mL [137,301]. Intravenous administration of thalistyline chloride (0.1–0.4 mg/kg) to normotensive dogs decreased the blood pressure up to 40 mm Hg. However, doses greater than 0.4 mg/kg produced respiratory depression and death. Tachyphylaxis to the cardiovascular effects of thalistyline was apparent. Administration of dopamine (50–100 μg/kg) and norepinephrine (1 or 2 μg/kg) intravenously prior to and during the administration of thalistyline resulted in inconsistent blocking effects by thalistyline. Thalistyline produced competitive antagonism of the contractile responses of rabbit aortic strips (α-adrenoreceptors) as visualized via a Schild plot, thus demonstrating the α-adrenoreceptor blocking properties of the alkaloid. Thalistyline (>0.4 mg/kg) produced respiratory arrest in dogs as measured by the inhibition of neuromuscular transmission via isolated rat left hemidiaphragm and phrenic nerve preparations. The blocking effect of thalistyline (if compared at the 50% level, IC$_{50}$) was one-quarter that of (+)-tubocurarine. This neuromuscular inhibition was rapidly reversed upon washing the preparation. Addition of the cholinesterase inhibitor physostigmine sulfate to the hemidiaphragm preparations during thalistyline blockade reversed the inhibitory effect of thalistyline on neuromuscular transmission. When the neuromediated stimulation of the hemidiaphragm was fully blocked by incubation with (+)-tubocurarine, thalistyline was added to the bath. Direct muscle stimulation via electrodes was then undertaken with reductions in contraction by thalistyline being small and insignificant. Finally, drugs that interact at nicotinic cholinergic receptors either as agonists or antagonists can compete with α-bungarotoxin (a low-molecular-weight protein) in binding to the receptor, such as a skeletal muscle. It was observed that thalistyline was unable to protect the total population of functional receptors on the motor end plate completely and, instead, protected a fraction of the total population of nicotinic cholinergic receptors subject to irreversible blockade by α-bungarotoxin [607].

Thalistyline has not been isolated from other *Thalictrum* species or other higher plants to date.

3.5.51. Thalmethine. Tabular review [529, alkaloid No. 98, 530], UV [645], IR [645], ^{1}H NMR [645], MS [520,529,585,645].

Thalmethine (**441**), $C_{36}H_{36}N_2O_6$ (592.2573), m.p. 275–277° (MeOH) [599], 265–268° (MeOH) [645], 275–277° (C_6H_6) [272]; $[\alpha]_D^{21}$ +200° (c 1, CHCl$_3$), was first isolated from *Thalictrum minus* L. in 1965 [599]. The UV spectrum was characterized by two maxima at 283 nm (log ϵ 3.73) and 315 (log ϵ 3.41); the IR spectrum showed a band at 1630 cm^{-1} (imine) [645]. The ^{1}H NMR spectrum indicated the presence of one *N*-methyl group at δ 1.87 and three methoxy groups at δ 3.60, 3.77, and 3.87 [645]. The mass spectrum showed the molecular ion and base peak at M$^+$ *m/z* 592 (100%) with other significant fragment ions at *m/z* 591 (93%) (M—H), 577 (6) (M—CH$_3$), 544 [M—H—(CH$_3$)$_2$O] (M—47), and 469 (M—133) (**442**) [585]. These spectral data are characteristic of a benzytetrahydroisoquinoline–benzyl-3,4-dihydrosioquinoline dimer, which was confirmed by methylation (CH$_2$N$_2$) of thalmethine to the *O*-methylthalmethine (**336**), the latter of established structure and stereochemistry [599]. This established the skeletal structure, positions of oxygenation and termini of the diphenyl ether linkages, and stereochemistry of thalmethine. Reduction of thalmethine with sodium borohydride afforded a dihydro derivative which on treatment with formaldehyde–formic acid yielded thalicberine (**386**) [599]. This established the position of the phenolic group of thalmethine at C(12) and fixed the structure of thalmethine as **441** [599].

Thalmethine has been isolated from *Thalictrum minus* L. on numerous occasions [65,181,272,310,510,646] but has not been isolated from other species of *Thalictrum* nor other higher plants.

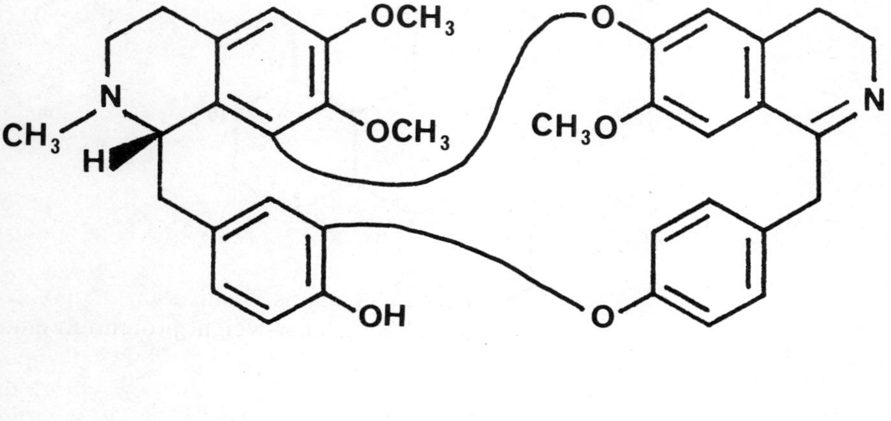

441

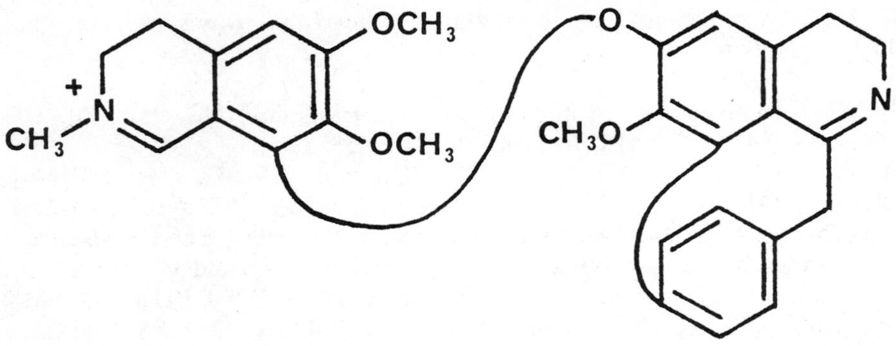

442

3.5.52.   Thalmine.

Tabular review [529, alkaloid No. 108, 530], ^{1}H NMR [450,529,614], MS [529,560,585,614], ORD [453], CD [562].

Thalmine (**358**), $C_{37}H_{40}N_2O_6$ (608.2886), m.p. 253° (EtOH–CHCl$_3$) [84], 147–157° dec. (HCl salt) [84], 238–241° dec. (HClO$_4$ salt) [84], 250 ° dec. (CH$_3$I salt) [84]; [α]$_D$ − 64.5° [84]; was first isolated from *Thalictrum minus* L. in 1950 [84]. Early structural studies, including zinc dust distillation, Hofmann degradation, and potassium permanganate oxidation, resulted in the proposal of a triphenylidine structure, specifically 3,5-dimethoxy-*N*-methyl-11-hydroxyhexahydrotriphenylidene (**443**), for thalmine [647]. The

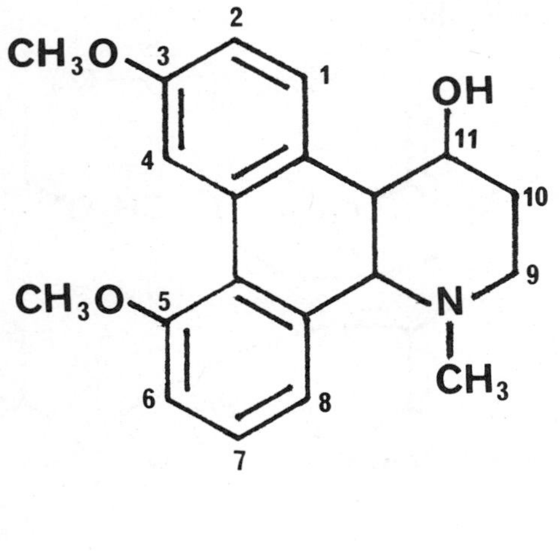

443

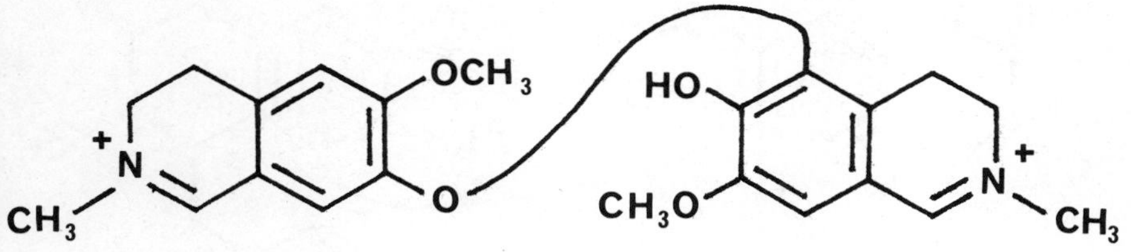

444

^{1}H NMR spectrum of thalmine indicated the presence of two N-methyl groups as singlets at δ 2.22 and 2.64; three methoxy groups as a singlet at δ 3.93 (9H); and 10 aromatic protons, two of which appear as a singlet at δ 6.06 (2H) [H(8) and H(8′)] [614]. The mass spectrum showed the molecular ion at M$^+$ m/z 608 (40%) with other important fragment ions m/z 607 (23%) (M—1), 593 (5) (M—CH$_3$), 577 (5) (M—OCH$_3$), 382 (70) (**357**), 191 (100) (**444**), 183 (10) (**444**—CH$_3$), 175 (15), 174 (40), 90 (10), and 89 (5) [560]. The CD spectrum showed Cotton effects at [$\lambda_{max}(\Delta\epsilon)$] 207($-28.9$), 242($+8.23$), 275($-3.31$), and 292($+10.8$) [562]. Oxidation of thalmine with potassium permanganate afforded 2-methoxy-4′,5-dicarboxybiphenyl ether (**309**) [564], and reductive cleavage with sodium in liquid ammonia gave 1-(4-methoxy-benzyl)-2-methyl-6-methoxy-1,2,3,4-tetrahydroisoquinoline (**433**) [596]. An identical cleavage of O-ethylthalmine (**445**) yielded 1-(4-methoxybenzyl)-2-methyl-6-methoxy-7-hydroxy-1,2,3,4-tetrahydroisoquinoline (**436**) and 1-(4-

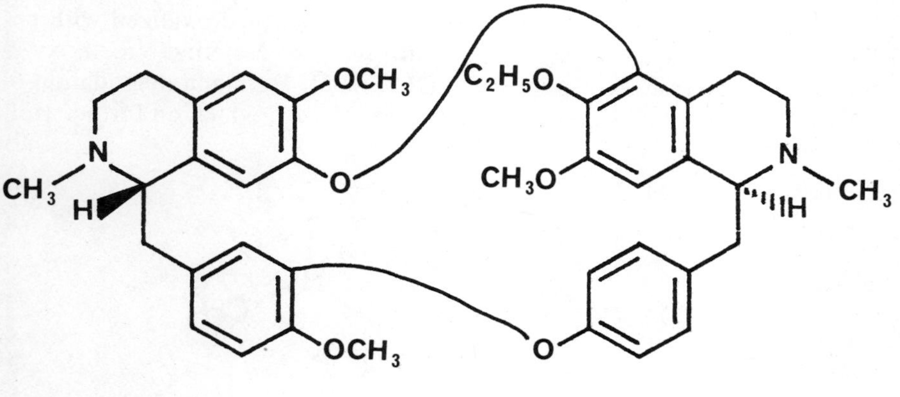

445

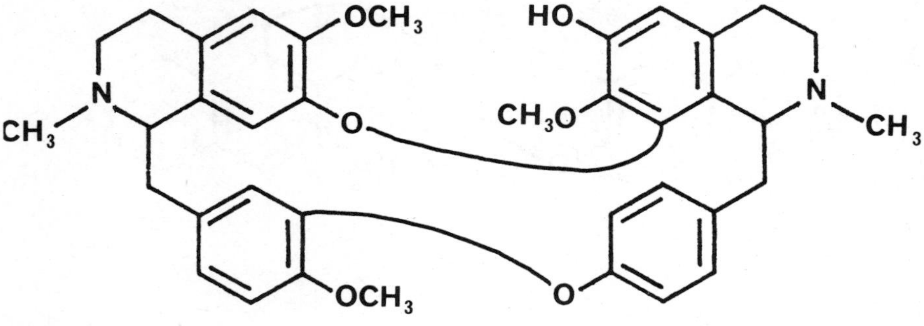

446

hydroxybenzyl)-2-methyl-6-ethoxy-7-methoxy-1,2,3,4-tetrahydroisoquino-line (**388**) but with the erroneous assignment of thalmine as **446** [596]. In a later study, reductive cleavage of thalmine with sodium in liquid ammonia again afforded amine **433** as the nonphenolic product, whereas the same cleavage of *O*-methylthalmine (**359**) gave armepavine (**328**) as the nonphenolic product [597]. Oxidation of *O*-methylthalmine (**359**) with potassium permanganate in acetone furnished a dimeric tetrahydroisoquinoline ($C_{23}H_{26}O_6N_2$), which demonstrated that thalmine possessed a second ether bridge, this one in the top half of the dimer. A similar oxidation of amine **388** yielded 1-oxo-2-methyl-6-ethoxy-7-methoxy-1,2,3,4-tetrahydroisoqui-noline (**447**). A further product of the reductive cleavage of *O*-ethylthalmine (**445**) was quaternized with methyl iodide, ethylated with ethyl iodide and sodium ethoxide, dequaternized with ethanolamine, and oxidized with potassium permanganate in acetone to afford 1-oxo-2-methyl-6-methoxy-7-ethoxy-1,2,3,4-tetrahydroisoquinoline (**202**) [597]. Hofmann degradation of *O*-methylthalmine (**359**) gave a mixture of des-bases, which on further Hof-

447

mann degradation produced trimethylamine and a nitrogen-free product. Ozonolysis of the mixture of the above-cited des-bases followed by catalytic hydrogenation of the ozonide produced a diaminodialdehyde. This dialdehyde, on Hofmann degradation, furnished a divinyldialdehyde different from the one prepared from O-methyloxyacanthine (obaberine) (**352**). Furthermore, the diaminodialdehydes prepared from the ozonolysis of deoxymethylthalmine and deshernandezine were different. These degradations permitted the assignment of thalmine as **358** [597].

Thalmine was found to inhibit the in vivo growth of Ehrlich's carcinoma cells [648] and rodent ascites lymphoma NK/ly [573]. The alkaloid produced restlessness, dyspnea, and paralysis of the extremities in mice with an LD_{50} of 535 mg/kg [529]. Doses of ≤ 50 mg/kg suppressed the orientation reaction and antagonized the stimulant action of phenamine in mice [592]. A dose of 5 mg/kg decreased feline arterial blood pressure by greater than 50%, and rabbits died after intravenous injections of 20–25 mg/kg [592]. The alkaloid also produced relaxation of an isolated segment of rabbit intestine and decreased acetylcholine of barium chloride-induced contractures [592]. Thalmine dimethiodide produced a skeletal muscle relaxation in rabbits at a dose of about 3 mg/kg [593]. The LD_{50} in mice was 18.7 mg/kg after intraperitoneal administration [593]. A dosage range of 3–10 mg/kg decreased or totally abolished neurostimulated feline skeletal muscle contractions. Although the alkaloid did not alter the acetylcholine-induced (parenteral) hypotensive response, it did elicit relaxation of the nictitating membrane and produced a block of conduction in the intracardial vagal ganglia. Finally, the alkaloid relaxed the contractions of isolated intestinal musculature and decreased acetylcholine or barium chloride–induced contractions [593]. Thalmine was observed to possess antiinflammatory effects in experimentally induced mouse paw edema which exceeded those of both aminopyrine and sodium salicylate. This activity was related to the apparent ability of the alkaloid to decrease blood vessel permeability [458,594].

Although thalmine has been reported as being isolated from five species of *Thalictrum*, these species were not named [595]. Hence *Thalictrum minus* L. [84,181] remains as the single species that is an unequivocal source of thalmine.

3.5.53. Thalmirabine. Tabular review [530, alkaloid No. 222], UV [102,530], IR [102,530], ^{1}H NMR [102,530], MS [102,530], CD [102,530]. Thalmirabine (**448**), $C_{39}H_{44}N_2O_8$ (668.3097), $[\alpha]_D^{20} + 116°$ (c 0.2, MeOH), was isolated from *Thalictrum minus* L. race B in 1980 as an amorphous residue [102]. The UV spectrum of the alkaloid showed maxima at 280 nm (log ϵ 3.95) and 314 nm (sh) (log ϵ 3.34), and the CD spectrum displayed Cotton effects at $[\theta]_{230} + 65,100$, $[\theta]_{269} 0$, $[\theta]_{285} - 11,000$, and $[\theta]_{300} 0$ [102]. The ^{1}H NMR spectrum suggested the presence of two N-methyl groups as singlets at δ 2.36 and 2.60; five methoxy groups as singlets at δ 3.38, 3.42, 3.72, 3.80, and 3.86; a hydroxy group as a broad singlet at δ 5.20 (D_2O-exchangeable);

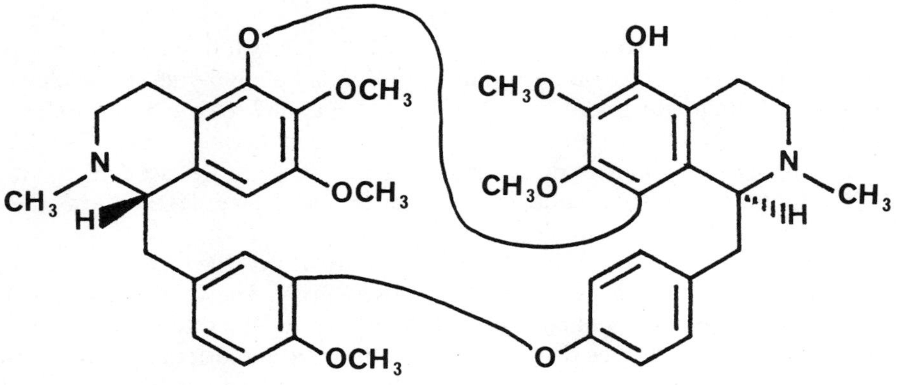

448

and eight aromatic protons as a singlet at δ 6.00 (1H) [H(8)] plus a multiplet from δ 6.4 to 7.3 (7H). The mass spectrum showed a strong molecular ion at M$^+$ *m/z* 668 (37%) with other significant fragment ions at *m/z* 442 (4%) (**449**), 222 (56) (**450** + H), 221 (100) (**450** or **451**), and 206 (**364** + H). These spectral data were suggestive of a head-to-head and tail-to-tail ether-linked dimeric alkaloid bearing two methoxy groups and one phenolic hydroxy group in one part of the head (one isoquinoline ring) plus two methoxy groups in the other part of the head (other isoquinoline ring) and one methoxy group in the tail (benzyl) portion [529,546]. Treatment of thalmirabine with ethereal diazoethane produced an *O*-ethyl ether (**452**) which on reductive cleavage with sodium in liquid ammonia gave (+)-(*S*)-*O*-methylarmepavine (**327**), (+)-(*S*)-1-(4-methoxybenzyl)-2-methyl-5-hydroxy-6,7-dimethoxy-1,2,3,4-tetrahydroisoquinoline (**452a**), and (+)-(*S*)-1-(4-hydroxybenzyl)-2-methyl-5-ethoxy-6,7-dimethoxy-1,2,3,4-tetrahydroisoquinoline (**422**). Finally, oxidation (KMnO$_4$–Me$_2$CO) of thalmirabine produced 2-methoxy-4′5-dicarbox-

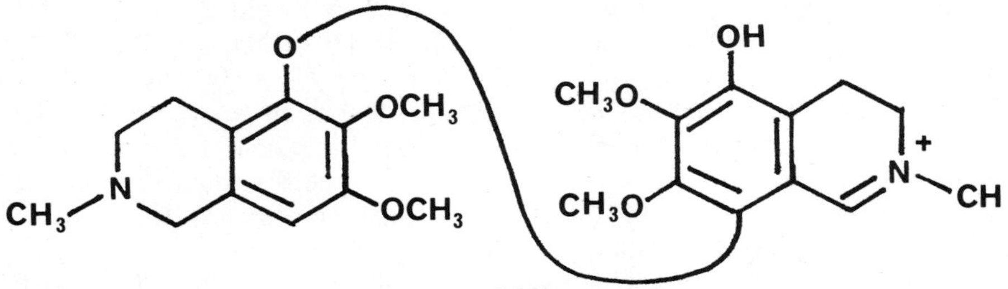

449

O
OCH3
OCH3
N
CH3
OH
CH3O
CH3O
N
CH

450

OH
CH3O
CH3O
N
CH3

451

O
OCH3
OCH3
N
CH3
H
OC2H5
CH3O
CH3O
N
CH3
H
OCH3
O

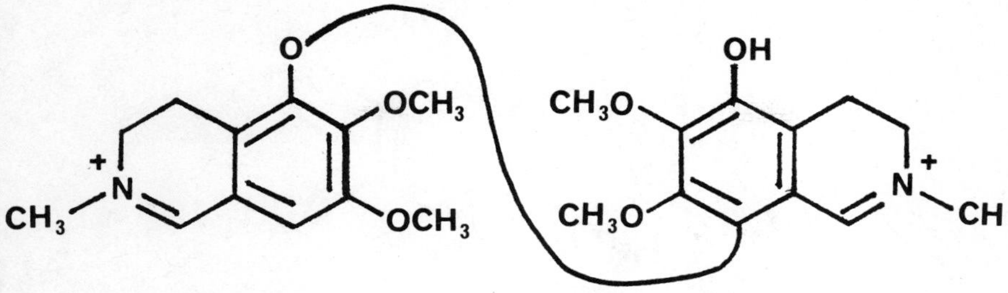

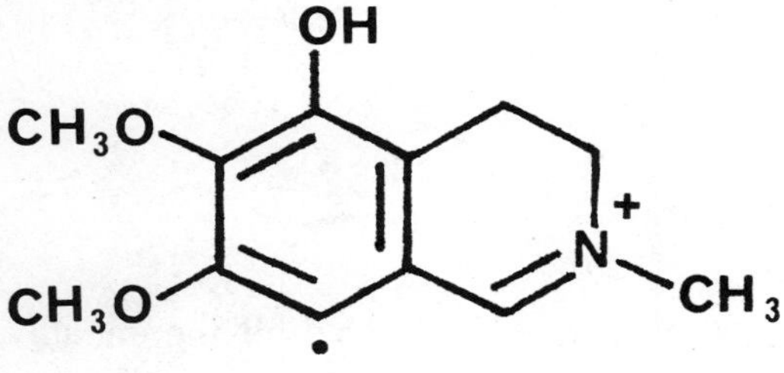

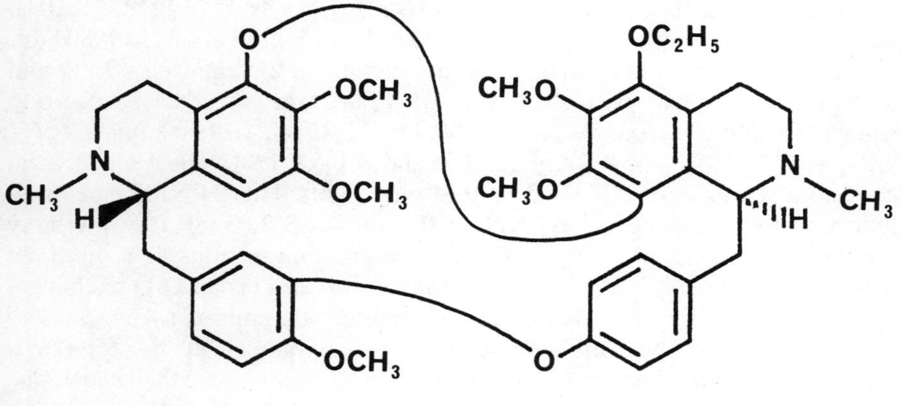

452

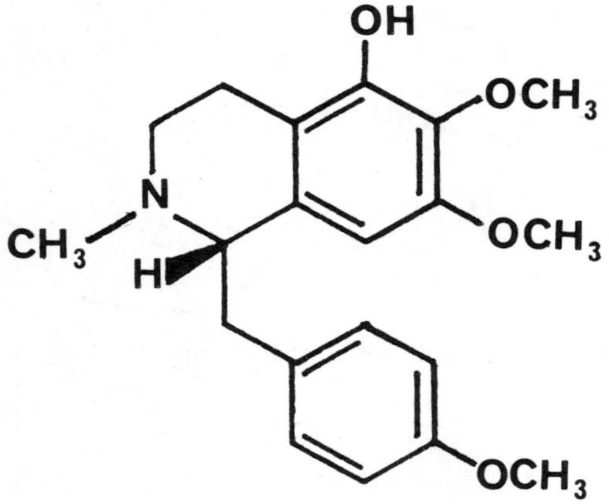

452 a

ybiphenyl ether (**309**). These data allowed for the unequivocal assignment of structure of thalmirabine as **448**.

Thalmirabine (**448**) was found to inhibit the in vitro growth of *Mycobacterium smegmatis* at a concentration of 100 μg/mL [102].

Thalmirabine has not been isolated from any other *Thalictrum* species or other higher plant to date.

3.5.54. Thalpindione.   Tabular review [530, alkaloid No. 223], UV [70,530], IR [70,530], ^{1}H NMR [70,530], MS [70,530], CD [70,530].
Thalpindione (**453**), $C_{37}H_{36}N_2O_9$ (652.2421), $[\alpha]_D^{21}$ −41.5° (c 0.29, MeOH), was isolated as an amorphous base from *Thalictrum alpinum* L. in 1980 [70]. The UV spectrum was characterized by maxima at 275 nm (log ε 3.78) and 283 nm (sh) (log ε 3.77) with change in 0.01 *N* NaOH or HCl; the CD spectrum showed Cotton effects at $[\theta]_{215}$ −80,000, $[\theta]_{227}$ 0, $[\theta]_{240}$ +78,000, $[\theta]_{258}$ +560, $[\theta]_{265}$ +2,230, $[\theta]_{270}$ 0, $[\theta]_{285}$ −22,900, and $[\theta]_{305}$ 0 [70]. The IR spectrum showed a strong carbonyl band at 1663 cm^{-1} [70]. The ^{1}H NMR spectrum indicated the presence of one *N*-methyl group at δ 2.63 (s); four methoxy groups at δ 3.47, 3.80, 3.89, and 3.90; two methine protons as a multiplet between δ 4.4 and 4.7; a phenolic hydroxy group as a broad D$_2$O-exchangeable singlet at δ 5.2; and nine aromatic protons as a multiplet from δ 6.1 to 7.7 [70]. The mass spectrum revealed the molecular ion at M$^+$ *m/z* 652 (26%) [70]. These spectra data were consistent with a monophenolic thalrugosinone (**454**)-like structure, and this was confirmed by methylation (CH$_2$N$_2$) of thalpindione to thalrugosinone (**454**). This established the molecular structure, positions of oxygenation, number and positions of termini of diphenyl ether bridges, and stereochemistry for thalpindione [70]. The CD spectrum of thal-

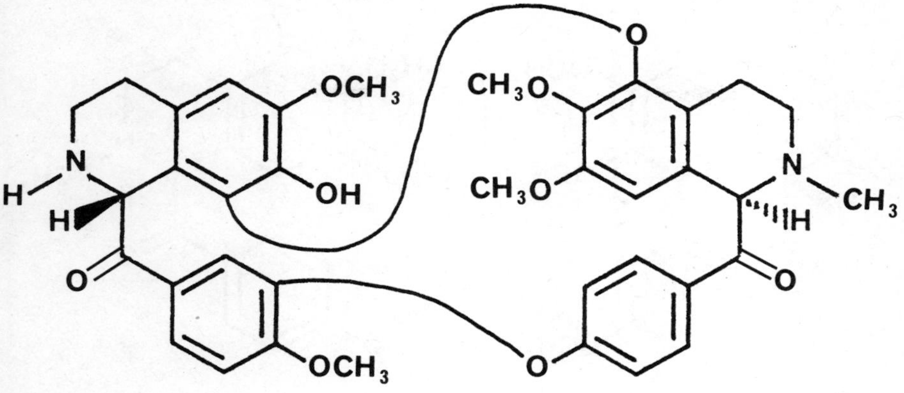

453

pindione (**453**), thalrugosinone (**454**), thalidasine (**286**), and thalrugosidine (**299**) are very similar, which suggests a similar conformation for these alkaloids. Thus the N-methyl and O-methyl ^{1}H NMR chemical shifts would be expected to be relatively constant and may be utilized for assignment of location of these functional groups. The N-methyl and methoxy resonances for these four alkaloids deviate at the most by 0.05 ppm and are located

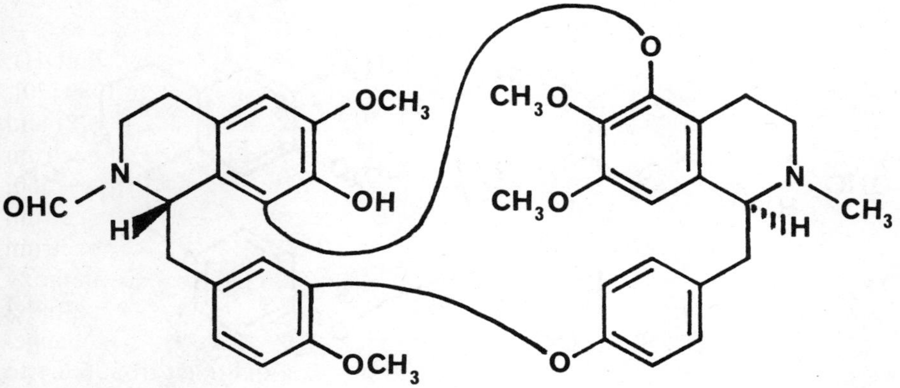

453a

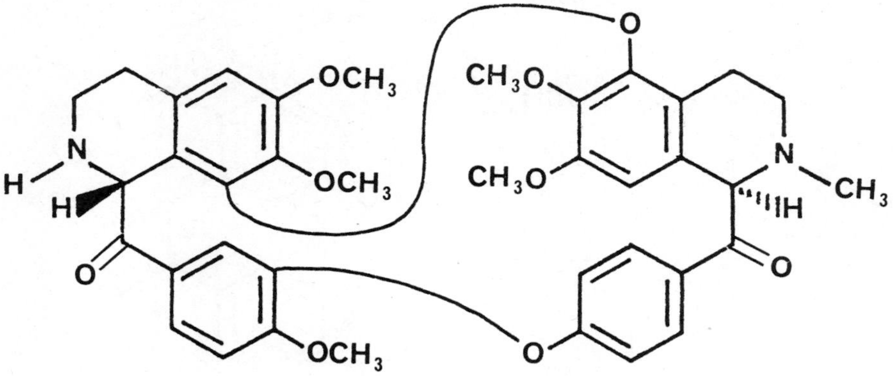

454

within ± 0.02 ppm on the average. The peak at δ 3.33 [C(7)] in thalrugosinone is missing in thalpindione but the other positions are relatively constant. Thus the phenolic group of thalpindione was placed at C(7) with the assignment of structure of thalpindione as **453**. A recent communication has revised the structure of thalrugosinone to **454a** based on chemical reactions plus high-resolution ^{1}H NMR (NOE) and mass spectral data [648a]. Consequently, thalpindione must be assigned structure **453a**.

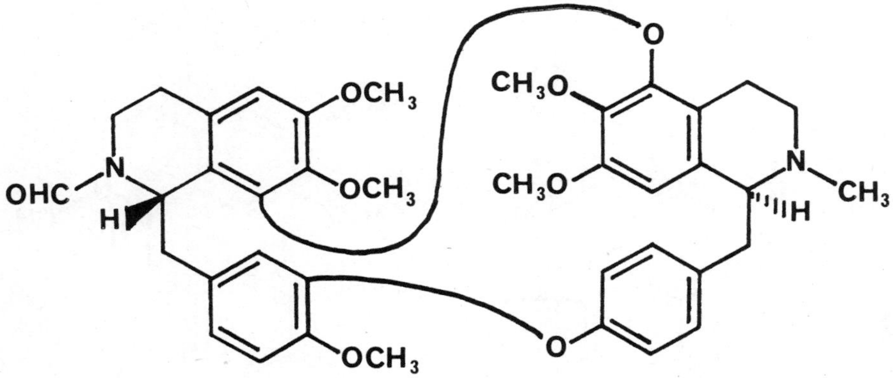

454a

Thalpindione has not been isolated from any other *Thalictrum* species or other higher plant to date.

3.5.55. Thalrugosaminine (*O*-Methylthalisopine). Tabular review [529, alkaloid No. 55, 530], UV [98,101,529], IR [268], ^{1}H NMR [98,101,529], MS [98,101,529], CD [98,101,529].

Thalrugosaminine (**408**), $C_{39}H_{44}N_2O_7$ (652.3149), m.p. 90–95° [98], 92–97° [268], 103–105° [101], $[\alpha]_D^{25}$ −91° [98], $[\alpha]_D^{25}$ −90.4° (c 0.104, MeOH) [101], $[\alpha]_D^{26}$ −103° (c 0.60, MeOH) [268], was isolated as an amorphous residue from *Thalictrum rugosum* Ait. in 1976 [98]. The UV spectrum was characterized by maxima at 205 nm (log ε 4.89), 227 nm (sh) (log ε 4.51), and 282 ng (log ε 3.94), and the CD spectrum had Cotton effects at $[\theta]_{225}$ +55,200, $[\theta]_{240}$ −46,400, $[\theta]_{272}$ −14,400, and $[\theta]_{288}$ +1,300 [101]. The ^{1}H NMR spectrum indicated the presence of two *N*-methyl groups at δ 2.51 and 2.55; five methoxy groups at δ 3.08, 3.39, 3.78, 3.82, and 3.94; and nine aromatic protons between δ 6.39 and 7.35 (m) [101]. The mass spectrum displayed the molecular ion and base peak at M$^+$ *m/z* 652 (100%) with other important fragment ions at *m/z* 651 (40%) (M—H), 426 (26) (**306**), 425 (87) (**306**—H), 411 (62) (**306**—CH$_3$), 213 (33) (**307**), and 190 (18) [101]. These spectral data were indicative of a dimeric benzylisoquinoline alkaloid with two diphenyl ether bridges of the thalisopine (**432**)-type [529,546]. Reductive cleavage with sodium in liquid ammonia afforded (+)-(*S*)-1-(4-methoxybenzyl)-2-methyl-6-methoxy-7-hydroxy-1,2,3,4-tetrahydroisoquinoline (**436**) and another unidentified product [98]. Repeated cleavage of thalrugosaminine again afforded fragment **436** but also identified the other cleavage product as (+)-thalifendlerine ((+)-(*S*)-1-(4-hydroxybenzyl)-2-methyl-5,6,7-trimethoxy-1,2,3,4-tetrahydroisoquinoline) (**65**) [101]. Finally, oxidation of thalrugosaminine with potassium permanganate in acetone afforded 2-methoxy-4′,5-dicarboxybiphenyl ether (**309**) [101]. These data permitted the unequivocal assignment of thalrugosaminine as **408**. A compound tentatively identified as *O*-methylthalisopine by a TLC comparison with the methyl ether of thalisopine (**432**) was described in 1968 [174]. However, this compound was not unequivocally identified and hence the name thalrugosaminine should prevail.

Thalrugosaminine (**408**) inhibited the in vitro growth of *Mycobacterium smegmatis* at a concentration of 50 μg/mL [101] and 100 μg/mL [98].

Thalrugosaminine (**408**) has also been isolated from *Thalictrum alpinum* L. [70], *T. foliolosum* DC. [268], *T. minus* L. race B [215], and *T. revolutum* DC. [101].

3.5.56. Thalrugosidine. Tabular review [529, alkaloid No. 101, 530], UV [152,353,529], IR [152,345], ^{1}H NMR [152,353,529], MS [152,353,529], CD [152,268,353].

Thalrugosidine (**299**), $C_{38}H_{42}N_2O_7$ (638.2992), m.p. 172–174° (MeOH) [152,353], 176–178° (EtOH) [268], $[\alpha]_D^{30}$ −185° (MeOH) [152,353], was first isolated from *Thalictrum rugosum* Ait. in 1971 [152,353]. The UV spectrum was characterized by two maxima at 275 nm (log ϵ 3.99) and 282 nm (log ϵ 3.99), and the CD spectrum had a series of Cotton effects at $[\theta]_{226}$ $-3,430$, $[\theta]_{241}$ $+9,800$, $[\theta]_{248}$ $+17,150$, $[\theta]_{267}$ 5,880, and $[\theta]_{286}$ $-20,335$ [152,353]. The ^{1}H NMR spectrum indicated the presence of two *N*-methyl groups as singlets at δ 2.25 and 2.60; four methoxy groups as singlets at δ 3.51, 3.75, 3.85, and 3.87; and nine aromatic protons as a multiplet form δ 6.3 to 7.65 [152,353]. The mass spectrum displayed the molecular ion at M$^+$ *m/z* 638 (64%) with other diagnostic fragment ions at *m/z* 623 (11) (M—CH$_3$), 608 (6) (M—2CH$_3$), 417 (2) (**455**), 416 (2) (**455**—H), 412 (15) (**400**), 411 (48) (**400**—H), and 206 (100) (**401**) [152,268]. These spectral data were suggestive of a thalidasine (**286**)-like alkaloid bearing a single phenolic hydroxy group in the top half (isoquinoline half) of the dimer [529,585]. Treatment of thalrugosidine with ethereal CH$_2$N$_2$ for 7 days afforded thalidasine (**286**), thereby determining the skeletal structure, positions of oxygenation, number and termini of the diphenyl ether linkages, and stereochemistry [152,353]. To establish the position of the single phenolic group, *O*-ethylthalrugosidine (**456**) (thalrugosidine + ethereal diazoethane, 8 days) was reductively cleaved with sodium in liquid ammonia to (+)-(*S*)-1-(4-methoxybenzyl)-2-methyl-6-methoxy-7-ethoxy-1,2,3,4-tetrahydroisoquinoline (**428**) and (+)-(*S*)-1-(4-hydroxyben-zyl)-2-methyl-5-hydroxy-6,7-dimethoxy-1,2,3,4-tetrahydroisoquinoline (**287**). This established the position of the phenolic group at C(7), supporting its cryptophenolic nature, and allowed the unambiguous assignment of structure of thalrugosidine as **299** [152,353].

Thalrugosidine was found to inhibit the in vitro growth of the following microbes at a mininum inhibitory concentration of 100 µg/mL: *Staphylococcus aureus*, *Streptococcus faecalis*, *Escherichia coli*, *Klebsiella pneu-*

455

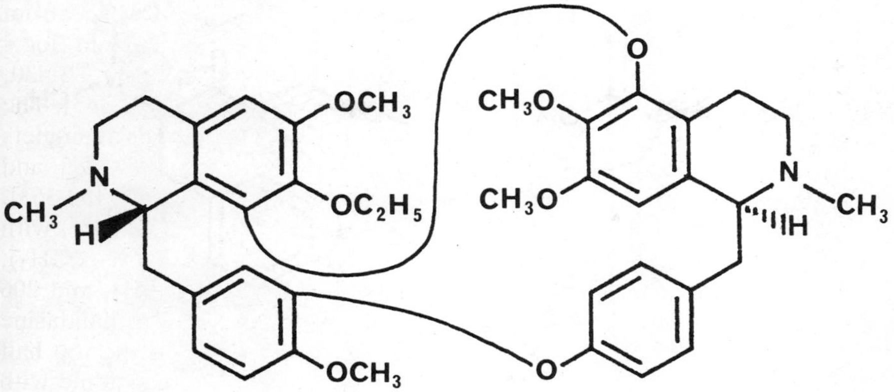

456

moniae, *Pseudomonas aeruginosa, Pasteurella multocida, Salmonella ty-phimurium, Shigella dysenteriae, Proteus vulgaris, Proteus mirabilis*, and *Mycobacterium smegmatis* [152].

Thalrugosidine has also been isolated from *Thalictrum alpinum* L. [70], *T. faberi* Ulbr. [279], and *T. foliolosum* DC. [38,268].

3.5.57. Thalrugosine (Thaligine, Isofangchinoline). Tabular review [529, alkaloid No. 79, 530], UV [152,353,529], IR [152], ¹H NMR [152,353,529], ORD [649], CD [152], MS [152,353,529,649].
Thalrugosine (**457**), $C_{37}H_{40}N_2O_6$ (608.2886), m.p. 212–214° (Et$_2$O) [152], 216–218° (Et$_2$O) [139], 153° (MeOH) [649], $[\alpha]_D^{30}$ +128° (MeOH) [152],

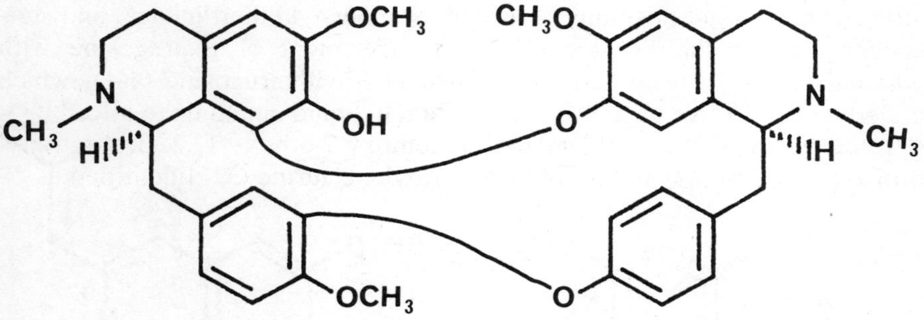

457

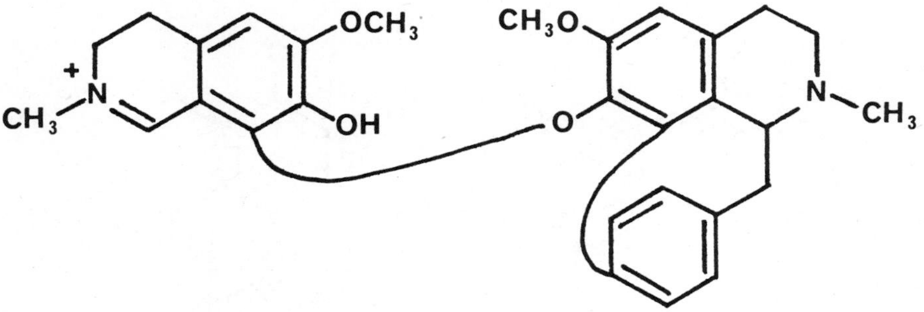

458

$[\alpha]_D^{25}$ +87° (c 0.33, MeOH), $[\alpha]_D^{21}$ +163° (c 1.16, MeOH) [650], $[\alpha]_D^{26}$ +264° (c 1.45, CHCl₃) [650], was first isolated from *Thalictrum rugosum* Ait. in 1971 [152,353]. The UV spectrum was characterized by a single maximum at 283 nm (log ε 3.78) and the CD spectrum showed Cotton effects at $[\theta]_{227}$ −7,600, $[\theta]_{268}$ −4,100, and $[\theta]_{287}$ +3,900 [152]. The ¹H NMR spectrum indicated the presence of two *N*-methyl groups as singlets at δ 2.31 and 2.50; three methoxy groups as singlets at δ 3.77, 3.90, and 3.91; and 10 aromatic protons as a multiplet from δ 6.10 to 7.43 [152,353]. The mass spectrum displayed the molecular ion at M⁺ *m/z* 608 (70%) with other characteristic fragment ions at *m/z* 471 (2) (**458**), 417 (9) (**455**), 382 (27) (**459**), 381 (19) (**459**—H), 367 (38) (**459**—CH₃), 191 (100) (**460**), 174 (28), and 168 (21) [**460**—(CH₃)₂O] [152,649]. These spectral data were suggestive of a monophenolic analog of isotetrandrine (**319**) in which the phenolic group is located in the top (isoquinoline) half of the alkaloid [171,546]. Methylation (CH₂N₂) of thalrugosine proceeded slowly over a period of 7 days to afford isotetrandrine (**319**), thereby fixing the skeletal structure, positions of oxygenation, numbers and positions of termini of diphenyl ether linkage, and stereochemistry of thalrugosine [152,353]. Treatment of thalrugosine with ethereal diazoethane for 8 days yielded *O*-ethylthalrugosine (**461**), which underwent reductive cleavage with sodium in liquid ammonia to afford (−)-(*R*)-1- (4-methoxybenzyl) -2-methyl-6-methoxy-7-ethoxy-1, 2, 3, 4-tetrahydroisoquinoline (**462**) and (+)-(*S*)-*N*-methylcoclaurine (**52**) [identified as *O*-

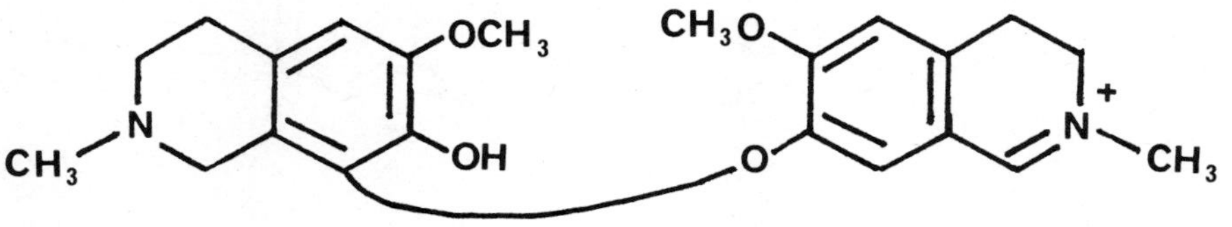

459

460

461

462

methylarmepavine (**327**)] [152,353]. These data permitted the unequivocal assignment of structure of thalrugosine as **457**. In 1973, an alkaloid thought to be new and named thaligine was isolated from *Thalictrum polygamum* Muhl. [649]. Furthermore, a second apparently new alkaloid was isolated from *Cyclea barbata* (Menispermaceae) in 1974 and designated isofangchinoline [651]. The latter two alkaloids were subsequently shown to be identical with thalrugosine and hence the alkaloid should be designated as thalrugosine in the literature.

Thalrugosine was observed to inhibit the in vitro growth of the following organsims at a minimum inhibitory concentration of 100 µg/mL: *Staphylococcus aureus. Streptococcus faecalis, Escherichia coli, Klebsiella pneumoniae, Salmonella gallinarum, Salmonella typhimurium, Pseudomonas aeruginosa, Pasturella multocida, Shigella dysenteriae, Proteus vulgaris, Proteus mirabilis, Mycobacterium smegmatis,* and *Candida albicans* [139,152]. Intravenous administration of thalrugosine in a dosage of 0.5 mg/kg to normotensive dogs produced a 10 mm Hg decrease in mean blood pressure and 1.0 mg/kg dose induced a hypotensive response of 52 mm Hg [139].

Thalrugosine (**457**) has been reisolated from *Thalictrum rugosum* Ait. [634] and has also been isolated from *Thalictrum lucidum* L. [139], *T. minus* L. var. *microphyllum* Boiss. [146], *T. minus* L. race B [102], *T. polygamum* Muhl. [649], and *T. sachalinense* Lecoyer. [67].

3.5.58. Thalrugosinone. Tabular review [530, alkaloid No. 224], UV [234,530], IR [234,530], ^{1}H NMR [234,530], MS [234,530], CD [234,530].

Thalrugosinone (**454**), $C_{38}H_{38}N_2O_9$ (666.2577), $[\alpha]_D^{22}$ $-46.4°$ (c 0.125, MeOH), was isolated as an amorphous substance from *Thalictrum rugosum* Ait. in 1980 [234]. The alkaloid, which darkened on exposure to air, was characterized by a UV spectrum with maxima at 274 nm (log ε 3.89) and 283 nm (sh) (log ε 3.86) and a CD spectrum with Cotton effects at $[\theta]_{220}$ $-48,000$ (end abs.), $[\theta]_{230}$ 0, $[\theta]_{242}$ $+67,000$, $[\theta]_{260}$ 0, $[\theta]_{268}$ $+7,500$, $[\theta]_{274}$ 0, $[\theta]_{286}$ $-25,000$, and $[\theta]_{300}$ 0 [234]. The IR spectrum showed a strong carbonyl band at 1660 cm^{-1} and the ^{1}H NMR spectrum indicated the presence of one

463

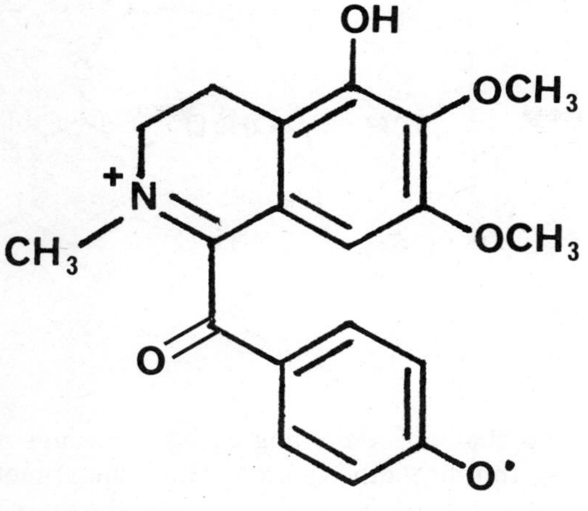

464

N-methyl group as a singlet at δ 2.64; five methoxy groups as singlets at δ 3.33, 3.48, 3.78, 3.88, and 3.90; two methine protons as a broad multiplet from δ 4.4 to 4.75; and nine aromatic protons as a multiplet from δ 6.1 to 7.7 [234]. The mass spectrum showed the molecular ion at M^+ m/z 666 (56%) and other important fragment ions at m/z 412 (16%) (**463**), 341 (12) (**464**), 325 (100) (**465**), 221 (11) (**363**), 207 (8) (**465**), 206 (10) (**466**), 205 (6) (**364**), and 191 (6) (**467**) [234]. These spectral data were indicative of a dimeric benzylte-

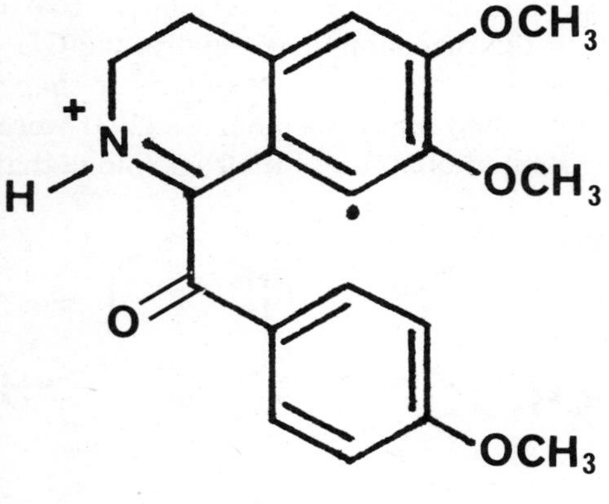

465

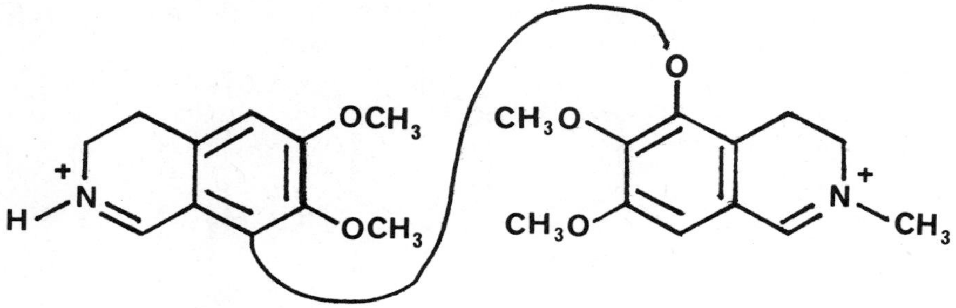

466

trahydroisoquinoline alkaloid containing carbonyl groups at both of the α-benzylic positions with four methoxy groups and one N-methyl group in the upper (isoquinoline) portion of the dimer. Because the mass spectal fragment ion at m/z 412 corresponding to head-to-head linked isoquinoline portions requires four methoxy groups, thalrugosinone could be placed into one of three groups of dimeric alkaloids; namely, those related to thalidasine (**286**), thalrugosaminine (**408**), or hernandezine (**303**). A close examination of the mass spectrum excluded a hernandezine (**303**)-like structure but couldn't differentiate between a thalidasine (**286**)-like or thalrugosaminine (**408**)-like structure. Thalidasine (**286**) and thalrugosaminine (**408**) have both been isolated from *Thalictrum rugosum* Ait., with thalidasine being the major alkaloid [98,101,114,152,353,627]. A comparison of the ¹H NMR chemical shifts for the N-methyl and methoxy groups of thalrugosinone with thalidasine (**286**) [627] and thalrugosaminine (**408**) [98] demonstrated a very small difference from the thalidasine positions but a much greater difference for the thalrugosaminine positions. In addition, the multiplicity pattern of the aromatic protons of thalrugosinone was similar to that of thalidasine and different from that of thalrugosaminine. Finally, the CD spectral curves of thalrugosinone and thalidasine (**286**) were of similar shape whereas those of both thalrugosaminine (**408**) and hernandezine (**303**) were clearly different. Hence although the limited quantity of thalrugosinone that was isolated pre-

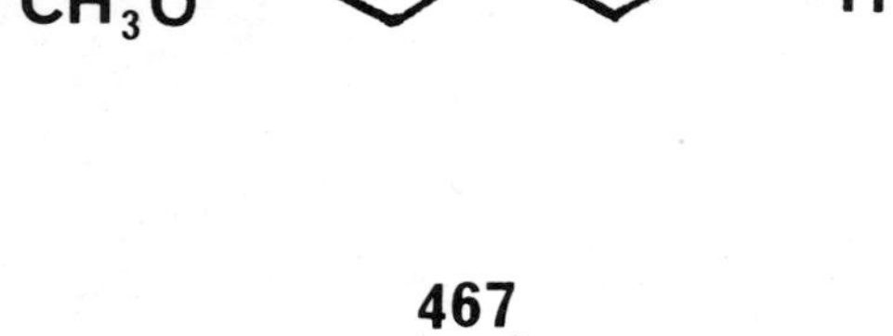

467

cluded a direct chemical characterization, thalrugosinone was assigned as **454** based on the above-cited spectral evidence [234]. The spectral comparisons were based on the assumption that changing the benzylic sp^3 carbons (methylene) of the usual bisbenzyltetrahydroisoquinoline alkaloids to sp^2 carbons (carbonyl) of thalrugosinone would not significantly change the conformation of the alkaloid and that the ^{1}H NMR chemical shifts would be minimally affected. Both the ^{1}H NMR spectra and the CD spectra appear to support this assumption. Thalrugosinone (**454**) is the first reported naturally occurring dioxobisbenzyltetrahydroisoquinoline alkaloid. A recent communication has revised the structure of thalrugosinone to **454a** based on chemical reactions plus high-resolution ^{1}H NMR (NOE) and mass spectral data [648a].

3.5.59. Thalsimidine (Thalcimidine). Tabular review [529, alkaloid No. 85, 530], UV [529,652], IR [652], MS [529,653], ORD [453,529], CD [530,562]. Thalsimidine (**468**), $C_{37}H_{38}N_2O_7$ (622.2679), m.p. 195° (EtOH), $[\alpha]_D^{14}$ +48° (c 1.10, CHCl$_3$), was isolated from *Thalictrum simplex* L. in 1968 [652]. The UV spectrum displayed maxima at 280 nm (log ϵ 4.12) and 312 nm (log ϵ 3.76) and the IR spectrum showed bands at 3490 and 1630 cm^{-1} [652]. The mass spectrum showed the molecular ion and base peak at M$^+$ m/z 622 (100%) and other significant fragment ions at m/z 621 (60%) (M—H), 607 (56) (M—CH$_3$), 591 (20) (M—OCH$_3$), 485 (10) (**469**), 311 (28) (M^{++}/2), 221 (8) (**451**), 190 (10) (**451**—OCH$_3$), and 175 (6) (**470**) [653]. The CD spectrum showed numerous Cotton effects [$\lambda_{max}(\Delta\epsilon)$]: 200(−41.0), 220(+25.4), 251(−1.86), 274(+4.30), 290(−2.80), and 323(−0.57) [562]. Thalsimidine was insoluble in aqueous alkaline solutions and in Claisen's cryptophenol

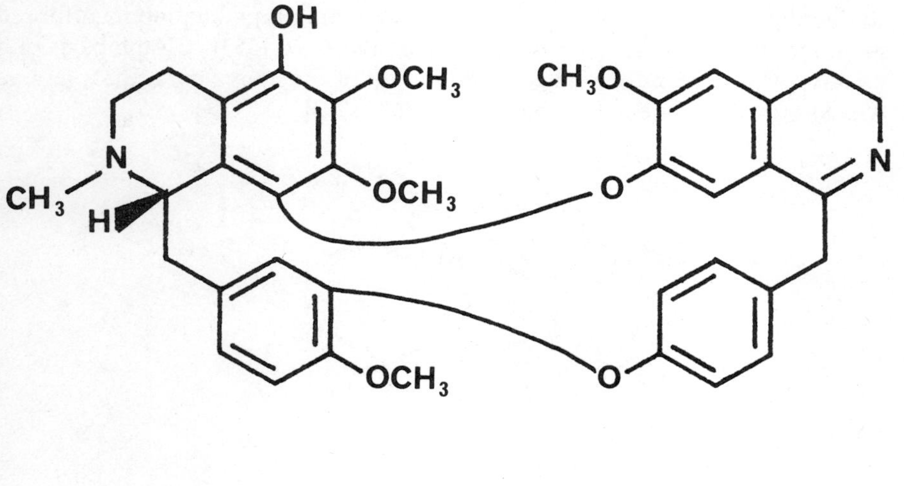

468

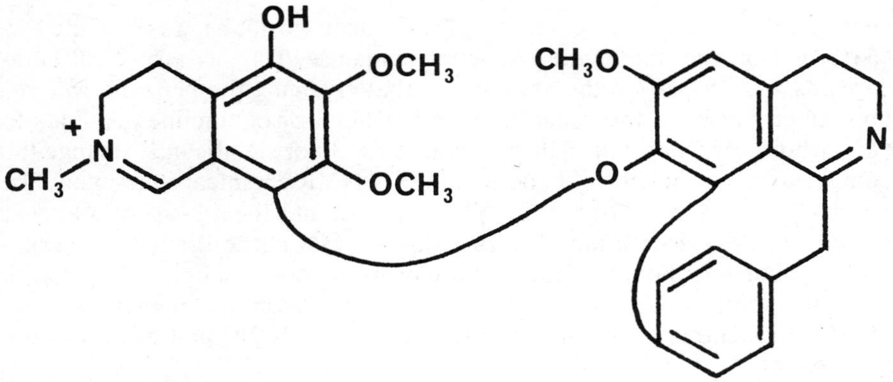

469

reagent but did form a dihydro derivative (471) [λ_{max} 285 nm (log ϵ 3.94)] on reduction with Adams catalyst [652]. Acetylation (Ac_2O/pyridine) of dihydrothalsimidine (471) produced a diacetyl derivative (472) with an IR spectrum displaying intense carbonyl absorptions at 1778 cm^{-1} (phenolic ester) and 1650 cm^{-1} (amide) [652]. These data were suggestive of a phenolic analog of thalsimine (312) in which the phenolic group is present in ring B [529,546]. This was confirmed upon methylation (CH_2N_2) of thalsimidine to afford thalsimine (312), thus fixing the oxygenation pattern, number and termini of diphenyl ether bridges, and stereochemistry of thalsimidine [653]. Reductive cleavage of thalsimidine with sodium in liquid ammonia afforded monomeric compounds that were not characterized [653]. On the basis of a positive Millon reaction and a consideration of the above data, thalsimidine was assigned as 5-demethylthalsimine (468) [653].

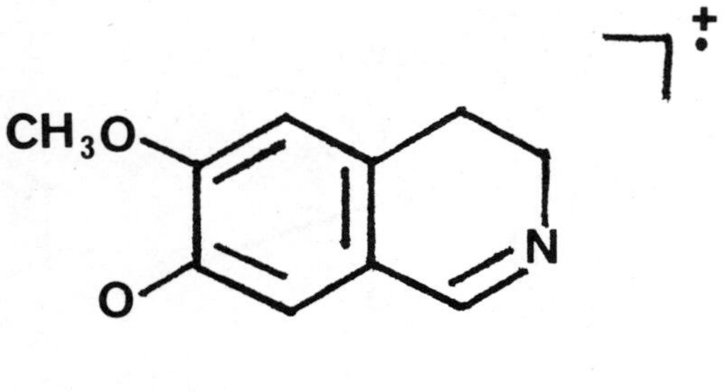

470

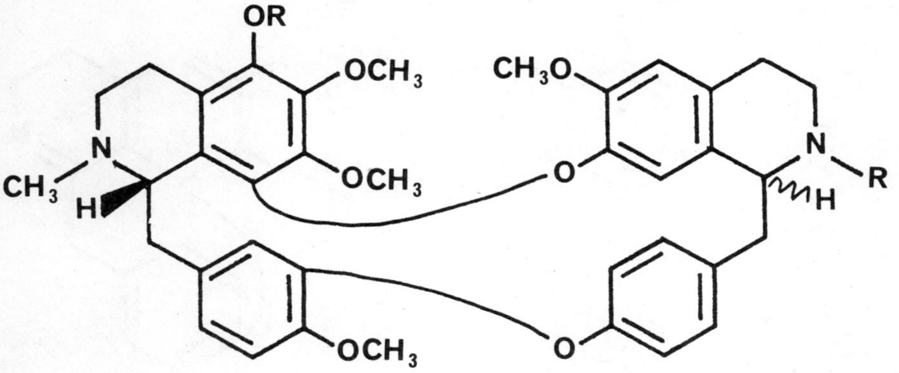

$\underset{\sim}{471}$ R=H

$\underset{\sim}{472}$ R=COCH$_3$

Thalsimidine (**468**) has not been isolated from any other *Thalictrum* species nor any other higher plant to date.

3.5.60. Thalsimine. Tabular review [529, alkaloid No. 86, 530], UV [529,565], IR [565], ^{1}H NMR [450,529,601], MS [520,529,560], ORD [453], CD [530,562].

Thalsimine (**312**), C$_{38}$H$_{40}$N$_2$O$_7$ (636.2836), m.p. 140–142° [625], 128–130° [574], 165–167° (Et$_2$O) [114], 149–150° [601], 233–235° dec. (HCl salt) [625], 209–211° dec. (picrate) [625], 198–218° dec. (nitrate) [654]; [α]$_D^{20}$ +27.45° (c 6.72, CHCl$_3$) [625], [α]$_D^{22}$ +20.5° (CHCl$_3$) [574], [α]$_D$ +22.6° (c 0.7, CHCl$_3$), was first isolated from *Thalictrum simplex* L. in 1960 and assigned the empirical formula C$_{19}$H$_{21}$O$_4$N [625]. The UV spectrum of thalsimine showed two maxima at 278 nm (log ε 4.02) and 310 nm (log ε 4.76) and the IR spectrum displayed a triplet in the 1550–1630 cm^{-1} region [565]. The CD spectrum was characterized by a series of Cotton effects at [λ$_{max}$(Δε)] 206(−23.0), 217(+11.4), 240(+16.8), 269(+4.95), 289(−2.58), and 318(−0.91) [562]. The ^{1}H NMR spectrum in C$_5$D$_5$N at 23°C indicated a 1:1 mixture of isomers, showing two *N*-methyl groups as singlets at δ 2.28 and 2.33 and 10 methoxy groups as singlets at δ 3.39, 3.46, 3.52, 3.63, 3.78, 3.83 (6H), 3.88, 3.91, and 3.93. However, on heating to 95°C, the chemical shifts reverted to singlets at δ 2.29 (*N*CH$_3$) plus δ 3.47, 3.54, 3.82 (6H), and 3.89 (*O*CH$_3$), suggesting that at room temperature thalsimine is a 1:1 mixture of two stable conformers [601]. The mass spectrum displayed the molecular ion and base peak at M$^+$ m/z 636 (100%) with other diagnostic fragment ions at m/z 635 (64%) (M—H),

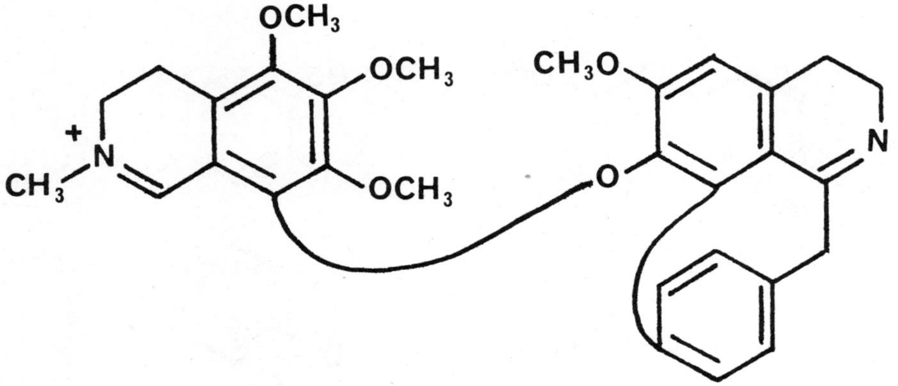

473

621 (54) (M—CH$_3$), 605 (17) (M—OCH$_3$), 499 (11) (**473**), 205 (13) (**474**), 175 (5) (**470**), 174 (7) (**470**—H), 90 (15), and 89 (6) [560]. Early structure elucidation studies demonstrated the presence of two tertiary nitrogen atoms (one of which was a *N*-methyl group), five methoxy groups, and two ether oxygen atoms [655]. Oxidation of thalsimine (KMnO$_4$–Me$_2$CO) gave 2-methoxy-4,5-dicarboxybiphenyl ether (**309**), and reduction (Zn + H$_2$SO$_4$ or NaBH$_4$) afforded dihydrothalsimine (**475**) [655]. *N*-methyldihydrothalsimine (**476**) underwent a twofold Hofmann degradation to afford a compound that on catalytic hydrogenation gave an octahydro derivative, identical with the corresponding derivative of hernandezine (**303**) [655]. Reductive cleavage of

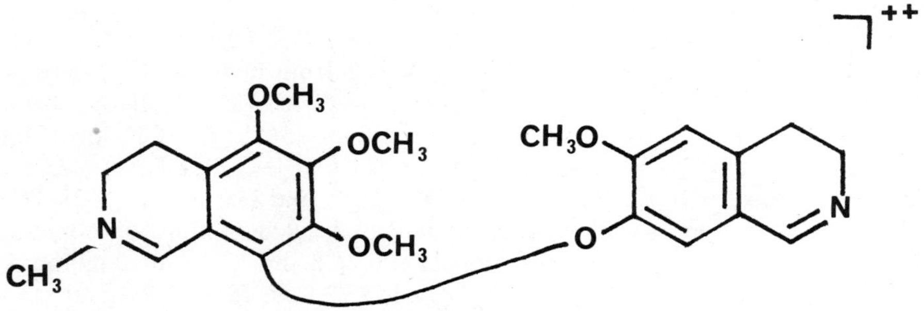

474

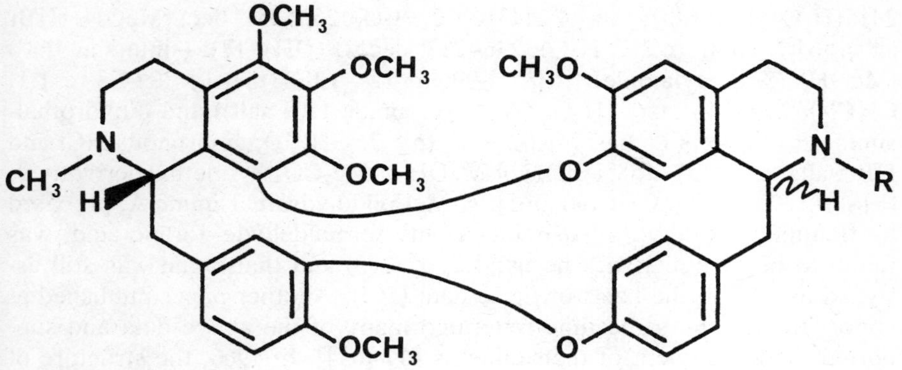

$\underset{\sim}{475}$ R=H

$\underset{\sim}{476}$ R=CH$_3$

either thalsimine or dihydrothalsimine with sodium in liquid ammonia gave coclaurine (**345**) [655]. On the basis of these studies, thalsimine was first assigned as **477** [655]. In a separate study, reductive cleavage of thalsimine and dihydrothalsimine (**475**) with sodium in liquid ammonia afforded *O*-methylarmepavine (**327**) and coclaurine (**345**) [565]. Reduction of thalsimine with zinc–sulfuric acid yielded two diastereomeric isomers, dihydrothalsimine A (softening at 115° with m.p. 130–132° (Me$_2$CO), $[\alpha]_D^{15}$ +241° (CHCl$_3$), 237–

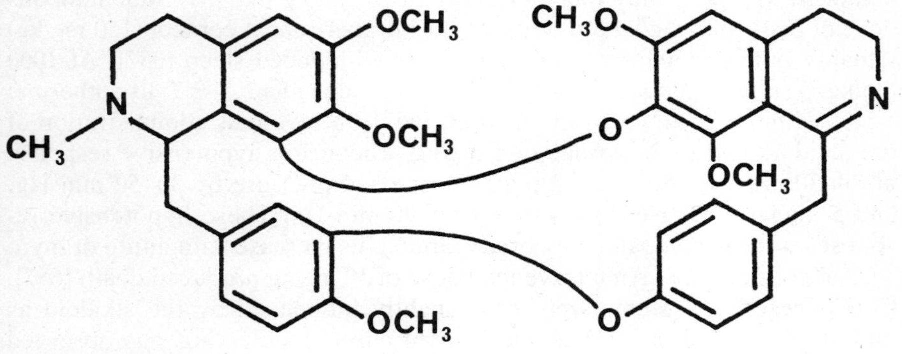

$\underset{\sim}{477}$

241° (H_2O) (HCl salt), $[\alpha]_D^{25}$ +211° (c 1.2, H_2O), 250–251° dec. (MeOH) (HBr salt), $[\alpha]_D^{29}$ + 188° (c 2.2, H_2O), 236–237° dec. (EtOH), 171° (sinters at 168°) (MeOH) (*N*-acetylamide), $[\alpha]_D^{28}$ +90° (c 1.1, EtOH), $[\alpha]_D^{28}$ +143° (c 1.1, $CHCl_3$), 239–242° dec. (H_2O) (*N*-acetyl amide HCl salt)) and dihydrothalsimine B (amorphous base, $[\alpha]_D$ −42° (c 5.2, $CHCl_3$), amorphous HCl and HBr salts, m.p. 157–158° (EtOH or MeOH or Me_2CO) (*N*-methyl derivative), $[\alpha]_D^{27}$ − 127° (c 1.29, $CHCl_3$)) [565]. *N*-Methyldihydrothalsimine A, prepared by treatment of dihydrothalsimine A with formaldehyde–formic acid, was found to be identical with hernandezine (**303**) and thalsimine was still assigned as **477** on the basis of these data [565]. Another paper published at approximately the same time reiterated many of the above data and supported the assignment of thalsimine as **477** [654]. In 1966, the structure of hernandezine was revised to **303** on the basis of spectral interpretations and sodium–liquid ammonia reductive cleavage products [567]. It followed that thalsimine must be therefore represented as **312** [567] and that the principal errors in its earlier structural assignment as **477** resulted from the erroneous assumption that the C(8′) methoxy of **477** was lost during reductive cleavage plus the postulation that *O*-methylarmepavine (**327**) and not (+)-(*S*)-1-(4-methoxybenzyl)-2-methyl-5,7-dimethoxy-1,2,3,4-tetrahydroisoquinoline (**311**) was obtained as a second cleavage product. A study of the polarographic behavior of thalsimine in an aqueous alcoholic solution of tetraethylammonium hydroxide using a dropping mercury electrode established that dihydrothalsimine was the major product of the electrode reaction. In addition, a chromatopolarographic method for the quantitation of thalsimine in the epigeal part of *Thalictrum simplex* L. was developed [656]. Thalsimine (**312**) along with its closely related analog hernandezine (**303**) were the first dimeric benzylisoquinoline alkaloids bearing C(5) oxygenation.

Thalsimine produced a sedative effect in mice followed by death when administered in an intravenous dosage of 71 mg/kg [657]. A subcutaneous dose of 500–1000 mg/kg did not prevent intraperitoneal corazol (150 mg/kg) lethality but did (500 mg/kg) prolong hexenal-induced sleep [657]. At 1000 mg/kg, a subcutaneous dose of the alkaloid produced a 2.5–2.7° hypothermic effect within 2 hr and 5.5–6.0° in 18 hr [657]. Intravenous administration of the alkaloid to cats in dosage of 1 mg/kg produced a hypotensive response of 20–30 mm Hg (arterial), 2 mg/kg decreased pressure by 30–50 mm Hg, and 5 mg/kg decreased pressure by 70–90 mm Hg. These hypotensive responses were accompanied by bradycardia plus increased amplitude of myocardial contractions. An intravenous dose of 10 mg/kg produced death [657]. Cholinoreactive systems were only slightly influenced by the alkaloid as measured in atropinized cats and a slight adrenolytic action was observed with a dosage of 5 mg/kg. Cholinomimetic properties were minimal as determined with feline intestinal and uterine musculature with spasmolysis observed with a 10^{-4} dilution of thalsimine [657]. In another study, the LD_{50} of thalsimine in mice upon intramuscular injection was 502 mg/kg and it was 262 mg/kg for dihydrothalsimine [658]. Both alkaloids inhibited motor activity in mice and doses of 2.5, 5, and 10 mg/kg of thalsimine produced sedation in rats. A dosage of 15–20 mg/kg of thalsimine increased motor activity in rats [658]. Neither alkaloid had any effect on the somnifacient intensity of chloral hydrate, hexenal, or luminal but prolonged the length of sleep by about 30% [658]. Intraperitoneal administration of thalsimine and dihydrothalsimine to rats 3 days before and 3 days after formalin-induced aseptic inflammation induced an antiinflammatory and antipyretic response in both intact and adrenalectomized animals [659]. Thalsimine (10 mg/kg) and (+)-dihydrothalsimine (5 mg/kg) decreased the vascular permeability of rabbits by 2.2 and 2 times, respectively, and at 10–20 mg/kg decreased body temperature by 1–1.6° [659]. Intraperitoneal administration of both alkaloids at 25–30 mg/kg increased the pain sensitivity threshold of rats by 1.5–2 times [659]. The antiiflammatory activity of these two alkaloids was 2–4 times greater than aminopyrine or sodium salicylate [458,659]. Apparently, the ether oxygen atoms are not involved in the antiinflammatory effect but acetylation of the nitrogen atom of dihydrothalsimine greatly reduced its effect [458]. Thalsimine and dihydrothalsimine inhibited the conditioned avoidance reactions and motor-conditioned reflexes associated with movement and eating in rodents. In addition, the alkaloids temporarily reduced the time for dogs to run through a labyrinth [572]. Administration of thalsimine in a dosage of 60–150 mg/kg to mice or rats produced only weak inhibition of the growth of lymphoma NK/Ly, alveolar hepatoma PC-1, or Pliss lymphosarcoma [573]. Finally, thalsimine was observed to possess weak antitussive activity in nerve-stimulated coughing in dogs [112].

Thalsimine has been isolated repeatedly from *Thalictrum simplex* L. [574,600,660] and has also been isolated from *T. rochebrunianum* Franc. and Sav. [601] and *T. rugosum* Ait. [114,661].

Note: Because the structure of thalflavine is now best represented as **100**, not **100a** (see discussions of thalflavine and thalmelatidine), it is not unlikely that the structures of thalistyline (**102**), *N*-desmethylthalistyline (**293**), *N*-methylthalistyline (**294**), thalirabine (**369**), thaliracebine (**418**), and thalistine (**437**) may be revised in the future to reflect a 5′,6′-methylenedioxy-7′-methoxy substitution rather than a 5′-methoxy-6′,7′-methylenedioxy substitution. These revisions are possible because of the past comparison of methylation products of one or more of the above alkaloids with a thalistyline analog or by a similar comparison of oxidation products with that of flavine.

3.6. Evolutionary Pattern and Chemosystematics of the Thalictrum Alkaloids

There are two papers that eloquently address the evalutionary pattern and chemosystematics of the *Thalictrum* alkaloids [668,669]. The first paper addresses the chemosystematics of the *Thalictrum minus* L. complex [668] which is widely distributed in Europe, the Caucasus, Siberia, and South-western Asia and is characterized by cytologic as well as morphologic variability. This paper lists the alkaloids found in *Thalictrum* species via their chemical nuclei and also cross-references particular alkaloids and alkaloid types via section and subsection. Finally, it lists the alkaloids of the *Thalictrum minus* complex and their habitat and presents some correlations between the alkaloid pattern of the *T. minus* complex and the genus as a whole. The evolutionary pattern of the complex is evidently connected with adaptive irradiation, hybridization, and polyploidy. Hexaploids are of the most frequent occurrence with decaploids also being quite common. Diploids and dodecaploids occur far less frequently. The biogenesis of alkaloids in hexaploids frequently extends to the bisbenzylisoquinolines whereas in decaploids the aporphine–benzylisoquinoline dimers prevail. These differences may be due to different genomes and to the involvement of different putative parents in the process of hybridization and polyploidization [668]. The second paper contains a table showing the chromosome numbers and alkaloid content in 17 species of *Thalictrum* [669]. In all these species, quaternary protoberberine salts (primarily berberine) and/or aporphines were isolated. Furthermore, aporphine–benzylisoquinoline and bisbenzylisoquinoline dimeric alkaloids are common to most of these species. On the other hand, benzylisoquinoline monomeric alkaloids occur only infrequently, lending credence to the propensity of these bases to dimerization. Aporphines that are present are postulated to be so because of certain structural features that are not conducive to dimerization. Isoquinolones are believed to result from oxidation of a parent monomer or dimer. Polyploid species of *Thalictrum* are generally richer in both monomeric and dimer alkaloids than diploid species, whereas within the diploids, protoberberines prevail [669].

3.7. Enzymatic Control of Stereochemistry Among the Thalictrum Bisbenzylisoquinoline Alkaloids

Bisbenzylisoquinoline alkaloids have long been known to be biosynthesized via an in vivo condensation of two tetrahydrobenzylisoquinoline monomers through phenolic oxidative coupling. This coupling may initially occur in a tail-to-tail fashion (benzyl group linking to benzyl group) or a head-to-tail fashion (isoquinoline group linking to a benzyl group) and be subsequently followed by a second and/or third oxidative coupling, thus producing numerous alkaloids with different permutations in oxidative linkage. Guinaudeau et al. have recently considered the general pathway for the biogenesis of the bisbenzylisoquinolines of *Thalictrum minus* L. *microphyllum* Boiss. and, perhaps more importantly, proposed four rules that appear to govern the formation of these alkaloids in *Thalictrum* species [544]. In *T. minus* L. var. *microphyllum* Boiss., thaligrisine (409) is the obvious precursor to obamegine (353), aromoline (282), and homoaromoline (316), whereas thalirugine (424) is the likely precursor of thalisopine (thaligosine) (432). Phenolic analogs of thalirugine (**424**) lacking the C(5′) hydroxyl group are the possible precursors of thaliphylline (**410**), *O*-methylthalicberine (**333**), and thalicberine (**386**). When the *Thalictrum* bisbenzylisoquinolines were considered as a group, these alkaloids were first uniformly drawn with their two lower aromatic rings (tail portions) in a position such that the phenolic hydroxyl or methoxyl group in ring C at C(12) would be in the lower left ring, which then fixed the termini of the diphenyl ether bridge at C(11) and C(12′). The following four rules were proposed to govern the formation of the *Thalictrum* bisbenzylisoquinolines:

1. The dimers may belong to any of seven different structural subgroups represented by general formulas A to G (Table 2).
2. When a benzylisoquinoline moiety is oxygenated at C(5) or C(5′), it has the (*S*) configuration.
3. The right-hand benzylisoquinoline moiety incorporates the (*S*) configuration at C(1′).
4. The left-hand benzylisoquinoline moiety has the (*S*) configuration at C(1), except in subgroups A, B, and C, where it may be (*R*).

As an extension of rules 3 and 4, bisbenzylisoquinolines whose occurrence is restricted to the genus *Thalictrum* or to its chemotaxonomically close

Table 2. Bisbenzylisoquinoline Alkaloidal Subgroups [544] (R Substituents May Be H, OH, or OCH₃)

Subgroup A

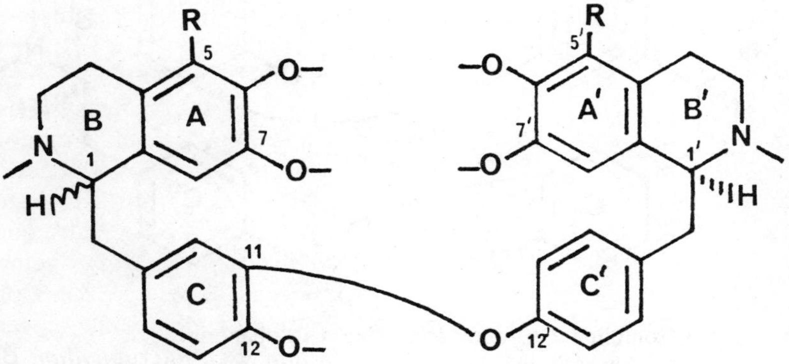

(1*R*,1′*S*): (+)-Thaligrisine
(1*S*,1′*S*): (+)-*O*-Methylthalibrine, (+)-*N*-methylthalistyline, (+)-neothalibrine,
         (+)-northalibrine, (+)-thalibrine, (+)-thalirabine, (+)-thaliracebine,
         (+)-thalirugidine, (+)-thalirugine, (+)-thaliruginine, (+)-thalistine,
         (+)-thalistyline

Table 2. (*continued*)

Subgroup B

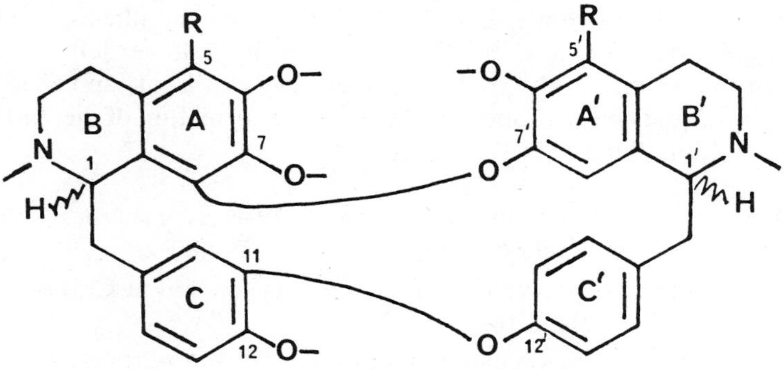

(1*R*,1'*S*): (+)-Berbamine, (+)-isotetrandrine, (+)-obamegine, (+)-thalrugosine
(1*S*,1'*S*): (+)-*N*-Desmethylthalidezine, (+)-hernandezine, (+)-hernandezine
       N-Oxide, (+)-*N*'-norhernandezine, (+)-*N*'-northalibrunine,
       (+)-thalibrunine, (+)-thalidezine
(1*S*,1'*R*): (−)-Isothalidezine

Subgroup C

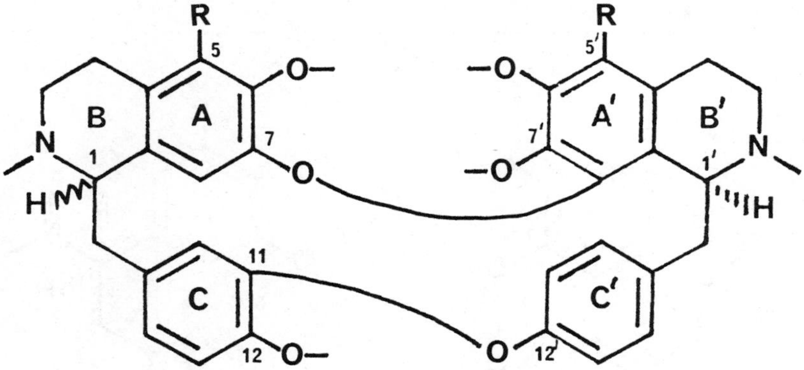

(1*R*,1'*S*): (+)-Aromoline, (+)-homoaromoline, (+)-obaberine,
       (+)-oxyacanthine
(1*S*,1'*S*): (−)-Thalisopidine, (−)-thalisopine [(−)-thaligosine],
       (−)-thalrugosaminine

Table 2.   (*continued*)

Subgroup D

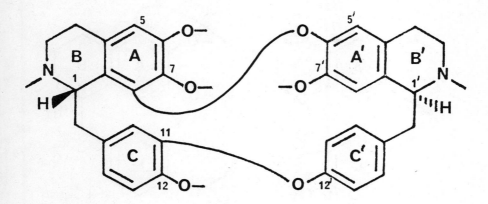

(1*S*,1'*S*): (+)-*O*-Methylthalicberine, (+)-thalicberine, (+)-thaliphylline

Subgroup E

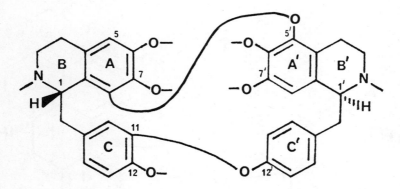

(1*S*,1'*S*): (−)-*N*-Desmethylthalidasine, (−)-*N*-desmethylthalrugosidine,
(−)-thalfoetidine, (−)-thalidasine, (−)-thaligosidine,
(−)-thalpindione, (−)-thalrugosidine, (−)-thalrugosinone

Table 2. (*continued*)

Subgroup F

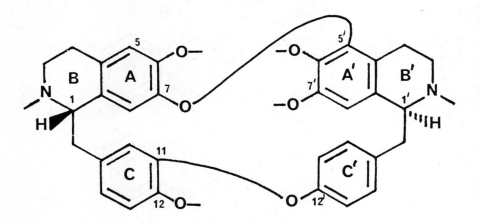

1S,1'S: (−)-Thalabadensine, (−)-thalictine, (−)-thalmine

Subgroup G

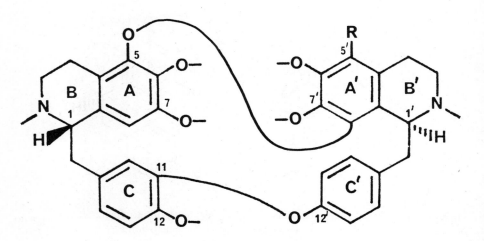

(1S,1'S): (+)-Thalfinine, (+)-thalmirabine

relative the *Hernandia* genus (Hernandiaceae) possess the $(1S,1'S)$ configuration. However, *Thalictrum* bisbenzylisoquinolines of $(1R,1'S)$ configuration are also found in other botanical families and include some fairly common alkaloids [e.g., isotetrandrine (**319**), obaberine (**352**), oxyacanthine (**355**)] of this type. It was also noted that when an imine function (either as

a single $-\overset{|}{C}=N-$ bond or as part of an isoquinoline system) was present in a *Thalictrum* bisbenzylisoquinoline, the imine was found on the right side of the dimer whereas the left side was of the (S) configuration.

The following alkaloids are those with imine functions. Subgroup B: (+)-thalsimine (**312**), (+)-thalsimidine (**468**), (+)-thalibrunimine (**332**), (−)-dihydrothalictrinine (**301**), (−)-thalictrinine (**302**), (−)-oxothalibrunimine (**354**), and (−)-*O*-methylthalbrunimine (**330**). Subgroup D: (+)-thalmethine (**441**) and (+)-*O*-methylthalmethine (**336**). Subgroup G: (+)-thalfine (thalphine) (**360**). It was via a consideration of these proposed rules that the structures of thalrugosamine (**408**) and (+)-thalisamine (**340**) were revised [544]. (−)-Isothalidezine (**320**) was cited as the sole exception, out of 58 alkaloids, to the rule that states that the right-hand benzylisoquinoline moiety of the dimer is characterized by the (S) configuration at C(1'). It was postulated that isothalidezine (**320**) might be artifactually formed from thalidezine (**289**) via oxidation at C(1) to form an iminium cation, followed by enzymatic reduction of the iminium bond from the beta side of the dimer. The postulation may have credence because of the relative amounts of (+)-thalidezine (**289**) (1.79 g) and (−)-isothalidezine (**320**) (92 mg) from 7 kg of powdered *T. podocarpum* Humb. roots [301,544].

4. MISCELLANEOUS ALKALOIDS

4.1. Revolutinone

UV [105], IR [105], ^{1}H NMR [105], MS [105], CD [105].
Revolutinone (**478**), $C_{38}H_{40}N_2O_8$ (652.2785), $[\alpha]_D^{28} - 10°$ (c 0.5, MeOH), was isolated from *Thalictrum revolutum* DC. as an amorphous solid in 1980 [105]. The UV spectrum was characterized by a series of maxima at 205 nm (sh) (log ϵ 5.13), 250 nm (log ϵ 4.78), 258 nm (log ϵ 4.76), 272 nm (log ϵ 4.69), 280 nm (sh) (log ϵ 4.66), and 301 nm (sh) (log ϵ 4.31) and the CD spectrum displayed Cotton effects at $[\theta]_{230} + 26,000$, $[\theta]_{260} - 14,000$, and $[\theta]_{396} - 2600$ [105]. The IR spectrum showed strong carbonyl absorption bands at 1644 (tertiary lactam) and 1695 cm^{-1} (aryl aldehyde) [105]. The ^{1}H NMR spectrum indicated the presence of two *N*-methyl groups as singlets at δ 2.27 and 3.09 (lactam); four methoxy groups as singlets at δ 3.67, 3.73, 3.85, and 3.86; 10 aromatic protons as singlets at δ 6.27 [H(5)], 6.56 [H(5')], 7.63 [H(8')], plus a multiplet at δ 6.79–7.04 (3H, ABC pattern) and a quartet at δ 6.92 and 7.77

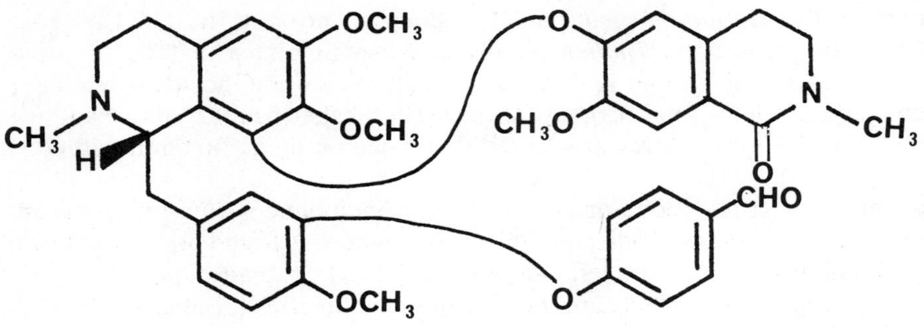

478

(J = 8.9 Hz); and one aldehydic proton as a singlet at δ 9.89 [105]. The mass spectrum showed a molecular ion at M$^+$ m/z 652 (0.15%) with other important fragment ions at m/z 411 (100%) (**479**), 241 (4) (**480**), and 221 (3) (**481**) [105]. These spectral data were characteristic of a secobisbenzyliso-quinoline aldehydo-lactam similar to baluchistanamine (**482**) [662] but derived from *O*-methylthalicberine (**333**) rather than from oxyacanthine (**355**). Oxidation of *O*-methylthalicberine (**333**) with potassium permanganate in acetone afforded revolutinone (**478**), thereby confirming the structure and establishing the stereochemistry as (*S*) [105]. Stirring of an ethanolic solution of *O*-methylthalicberine (**333**) at room temperature for 24 hr followed by refluxing for 2.5 hr produced no change, which tended to preclude the notion that revolutinone was of artifactual origin [105]. To date, revolutinone has not been isolated from any other *Thalictrum* species or other higher plant.

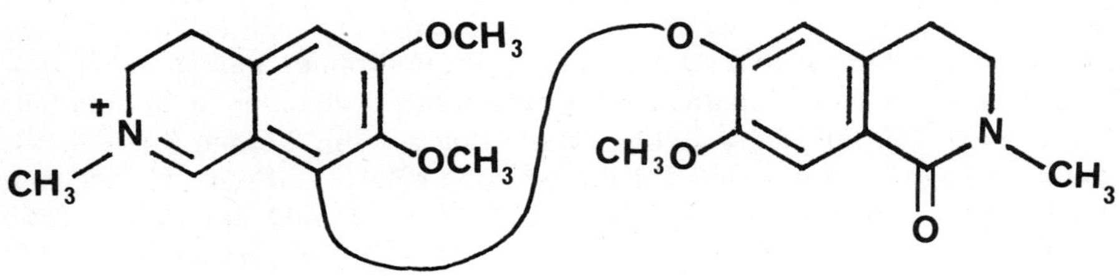

479

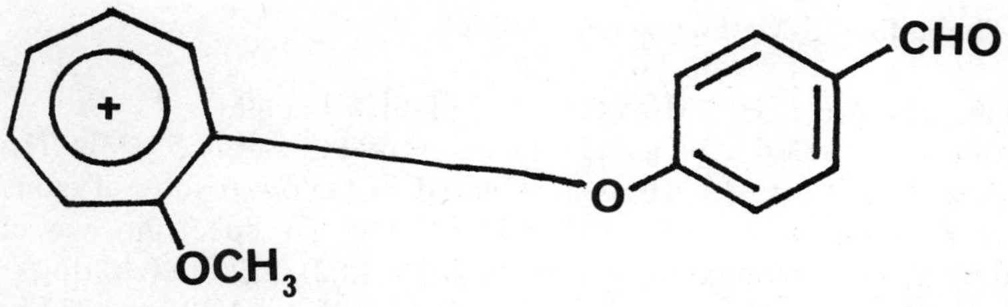

480

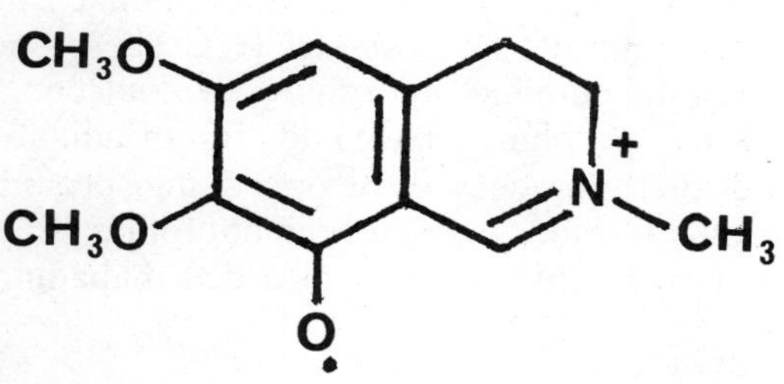

481

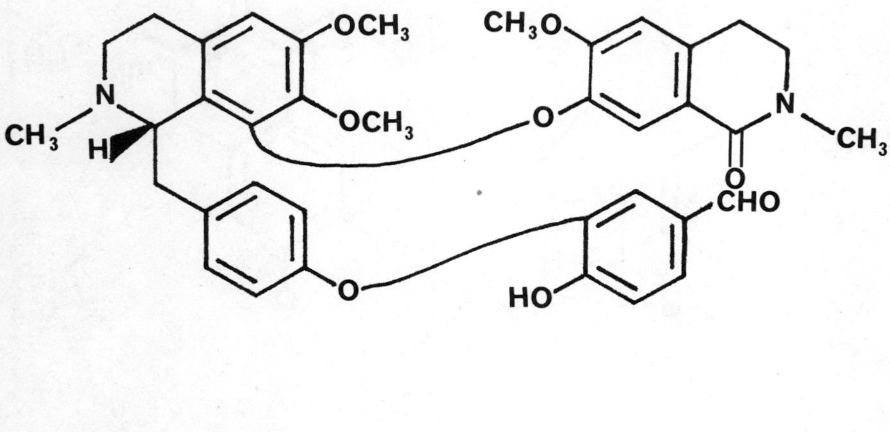

482

4.2. Thaliadine (3-Methoxyhernandaline)

UV [446], IR [446], ^{1}H NMR [446], MS [446], CD [446].

Thaliadine (**483**), $C_{30}H_{33}NO_8$ (535.2206), m.p. 143.5–144.5° (MeOH); $[\alpha]_D^{26}$ 0° (c 0.22, CHCl$_3$ or MeOH), was isolated as yellow rosette crystals from *Thalictrum minus* L. race B in 1978 [446]. The UV spectrum was characterized by a series of maxima at 220 nm (log ϵ 4.62), 237 nm (sh) (log ϵ 4.48), 277 nm (log ϵ 4.50), 300 nm (log ϵ 4.30), 312 nm (log ϵ 4.30) and 337 nm (sh) (log ϵ 4.01) and the IR spectrum showed a carbonyl absorption at 1675 cm^{-1}. The ^{1}H NMR spectrum suggested the presence of one *N*-methyl group at δ 2.50 (s); six methoxy groups as singlets at δ 3.79, 3.81, 3.91 (6H), 3.93, and 3.96; four aromatic protons as singlets at δ 6.45, 6.77, 7.40, and 8.08 [H(11)]; and one aromatic aldehydic proton at δ 10.37. The mass spectrum showed the parent ion at M$^+$ *m/z* 535 (100%) with other intense fragment ions at *m/z* 520 (45) (M—CH$_3$) and 504 (38) (M—OCH$_3$). These spectral data were suggestive of a hernandaline-like aporphine alkaloid containing one extra methoxy group in the aporphine ring. Oxidation of adiantifoline (**190**) with potassium permanganate in acetone for only 5 min produced a mixture of four oxidation products plus unreacted adiantifoline. Careful chromatography of the oxidation product mixture afforded thaliadine (**483**) as one of

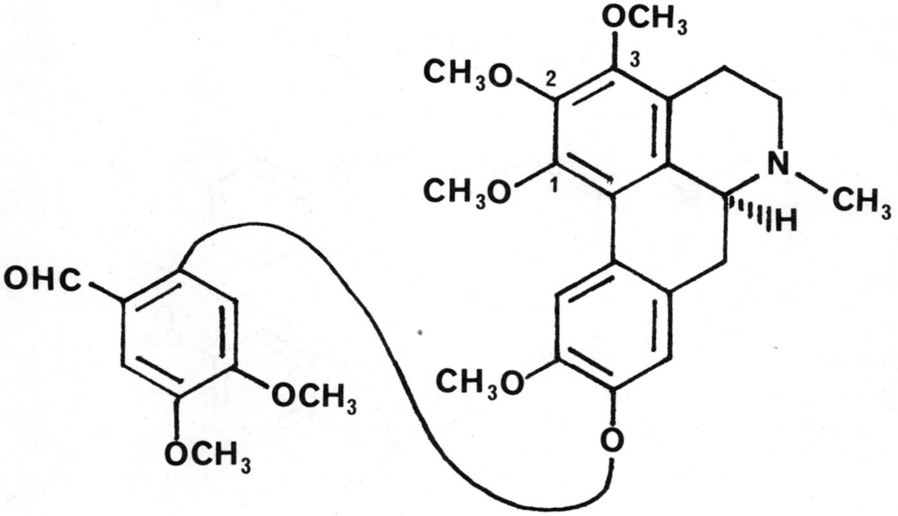

483

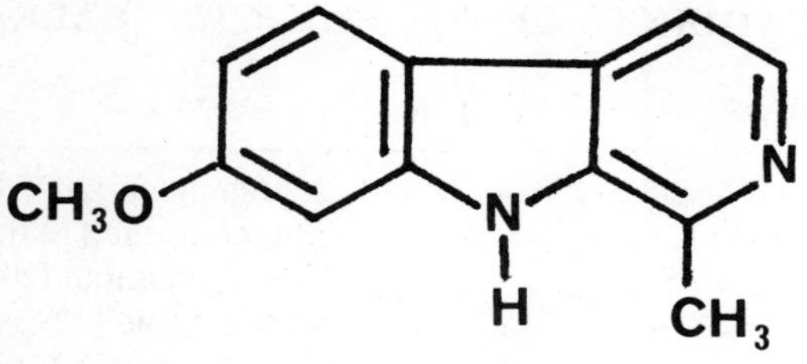

484

the oxidation products. The CD spectrum of the isolated thaliadine (**483**) ($[\theta]_{240}$ +214,000, $[\theta]_{278}$ −37,500, and $[\theta]_{300}$ −23,600) was similar to those observed for aporphine–benzylisoquinoline dimers of the (*S*,*S*) configuration and hence the asymmetric center was assigned the (*S*) stereochemistry [446].

Thaliadine (**483**) was observed to produce a hypotensive effect in normotensive dogs when administered in a dosage of 0.1–1.0 mg/kg but to be devoid of in vitro antimicrobial activity [446].

Thaliadine (**483**) has been isolated from only one other *Thalictrum* species, *T. minus* L. ssp. *majus* (also known as *T. minus* ssp. *elatum*) [299].

4.3. Harmine

Harmine (**484**), $C_{13}H_{12}N_2O$ (212.0950), m.p. 252° (CHCl$_3$–MeOH [1:1]) [192], is a β-carboline alkaloid first isolated from *Peganum harmala* (Rutaceae) well over 100 years ago [667]. This indole alkaloid was recently isolated from *Thalictrum foetidum* L. and identified by spectral data and m.p. comparisons with an authentic sample [192]. This is the first report of a tryptophan-derived alkaloid as a constituent of *Thalictrum* species and is a clear exception to the chemotaxonomic character of the genus.

ACKNOWLEDGMENTS The author would like to acknowledge the tireless efforts of Ms. Sharyn Harvey in typing this manuscript; Ms. Deirdre Gallagher and Ms. Judy Cherevka for limited typing duties; Ms. Suzanne Walker and Ms. Nancy Hudson for extensive photocopying; and Dr. Paul Wang, Dr. Shing-Shing Wu, Ms. Mei-Chao Lin, and Dr. Jiang-Sheng Zhang for translation of several Chinese papers.

5. BOTANICAL SOURCES OF THE *THALICTRUM* ALKALOIDS

Table 3

T. acteaefolium Sieb. & Zucc.
Tetrahydroberberine [401]
T. alpinum L.
  Berberine [70]
  Columbamine [70]
  N-Desmethylthalrugosidine [70]
  Hernandezine (thalicsimine,
    thaliximine) [563]
  Isoboldine [70]
  Jatrorrhizine [70]
  Magnoflorine [70]
  N-Methyl-6,7-dimethoxy-1,2-
    dihydroisoquinolin-1-one [70]
  Neothalibrine [70]
  Noroxyhydrastinine [70]
  Oxyberberine (berlambine) [70]
  Palmatine [70]
  Thalicarpine [70]
  Thalidasine [70]
  Thalifendine [70]
  Thaliporphine (*O*-methylisoboldine)
    [70]
  Thalpindione [70]
  Thalrugosaminine [70]
  Thalrugosidine [70]
T. amurense Maxim.
  β-Allocryptopine [63]
  Thalictricine (thalictrisine) [63]
T. aquilegifolium L.
  Isocorydine [114]
T. baicalense Turcz.
  Baicalidine [61]
  Baicaline [61, 127]
  Berberine [61, 127]
  Glaucine [61]
  Magnoflorine [127]
T. calabricum Spreng.
  Magnoflorine [128]
T. contortum L.
  β-Allocryptopine [63]
T. dasycarpum Fisch. and Lall.
  Argemonine [282]
  Bisnorargemonine [207]
  Corypalline [207]
  Dehydrothalicarpine [282]

Laudanidine [207]
Magnoflorine [129]
Norargemonine [207]
Thalicarpine [129, 473, 509]
Thalidasine [282, 626]
Thalisopavine [207]
T. dasycarpum Fisch. and Lall. var.
  hypoglaucum (Rydb.) Boivin
  Berberine [122]
  Magnoflorine [122]
T. dioicum L.
  Berberine [72, 193]
  Berberrubine [72]
  Corydine [72, 106]
  Isocorypalmine [72]
    (tetrahydrocolumbamine) [72]
  Magnoflorine [130]
  N-Methylcoclaurine [72]
  O-Methylcorypalline [72]
  N-Methyllaurotetanine [72]
  Pallidine [72, 106, 130]
  Pennsylvanine [72]
  Thalicarpine [72, 130]
  Thalictrogamine [72]
  Thalictropine [72]
  Thalicminine [193]
  Thalidicine [130]
  Thalidine [72, 243]
  Thalidoxine [516]
  Thalifendine [72]
  Thalmelatine [72]
T. faberi Ulbr.
  Berberine [349]
  Dehydrohuangshanine [471]
  Dehydrothalifaberine [471]
  N-Desmethylthalidasine [349,
    555]
  Faberidine [471]
  Faberonine [471]
  Huangshanine [467, 471]
  O-Methylthalibrine [279]
  O-Methylthalicberine [279]
  Pallidine [279]
  Thalfinine [555]
  Thalidasine [349, 555]

Table 3. (*continued*)

Thalifabatine [471]
Thalifaberine [467, 471]
Thalfabine [467, 471]
Thalifarapine [471]
Thalifasine [471]
Thaliracebine [555]
Thalisopine (thaligosine) [279]
Thalrugosidine [279]
T. fauriei Hayata
  Dehydrodiscretine [131]
  Magnoflorine [131]
  Thalifaurine [131]
T. fendleri Engelm. ex Gray
  Berberine [62, 132]
  Glaucine [62]
  Hernandezine [62]
  Jatrorrhizine [62, 132]
  Magnoflorine [62, 132]
  N-Methylcorydaldine [232]
  N-Methylthalidaldine [232]
  Ocoteine [62]
  Preocoteine [62, 79]
  Tetrahydrothalifendine [232]
  Thaldimerine [62]
  Thalicarpine [62]
  Thalidastine [62, 366]
  Thalidezine [62, 79]
  Thalifendine [62, 132]
  Thalifendlerine [62, 132]
  Thaliporphine [62, 79]
  Veronamine [226, 232]
T. filamentosum Maxim.
  Glaucine [63]
  Thalicsimidine [63]
T. flavum L.
  Berberine (thalsine) [133, 296]
  Cryptopine [133]
  Magnoflorine [133]
  Thalflavidine [293]
  Thalflavine [269, 670]
  Thalicarpine [269]
  Thalicflavine [663]
  Thalicsine (thalixine) [296]
T. foetidum L.
  Berbamine [549]
  Berberine [134, 192]
  Corunnine [64, 192]

Fetidine [134, 192, 448, 449, 455, 460, 463–466]
Glaucine [64, 192]
Harmine [192]
Isoboldine [64, 192]
Isotetrandrine [549]
Magnoflorine [64, 134, 192]
Oxoglaucine [192]
Thalfine [192, 615, 616, 670]
Thalfinine [192, 615, 616, 670]
Thalfoetidine [549, 617]
Thalicarpine [451]
Thaliporphine (thalicmidine) [64, 192]
T. foliolosum DC.
  Berberine [38, 223, 347, 348]
  Columbamine [223]
  Dehydrodiscretamine [223]
  Jatrorrhizine [223, 348]
  Magnoflorine (thalictrine) [38, 135, 223, 347]
  Noroxyhydrastinine [268]
  Oxyberberine (berlambine) [268]
  Palmatine [38, 223, 348]
  Reticuline [38]
  Rugosinone [223]
  Tembetarine (*N*-methylreticuline) [223]
  Thalicarpine [38]
  Thalidasine [38]
  Thalidastine [223]
  Thalifendine [223]
  Thalirugidine [268]
  Thalisopine (thaligosine) [268]
  Thalrugosaminine [268]
  Thalrugosidine [38, 268]
  N,O,O-Trimethylsparsiflorine [38]
  Xanthoplanine [223]
T. glaucum Desf.
  Berberine [354]
  Magnoflorine [354]
T. hernandezii Tausch
  Hernandezine [558]
T. isopyroides C.A.M.
  Cabudine [183]
  Cryptopine [174, 432]

Table 3.   (*continued*)

Dehydrothalicmine
  (dehydroocoteine) [185]
Magnoflorine [134]
Ocoteine (thalicmine) [174]
N-Methyl-6,7-dimethoxy-1,2-
  dihydroisoquinolin-1-one [175]
Thalicminine [174]
Thalisopidine [174]
Thalisopine [174, 636, 637]
Thalisopynine (thalisopinine) [175]
T. javanicum B1.
  Berberine [136]
  Columbamine [136]
  Demethyleneberberine [136]
  Jatrorrhizine [136]
  Magnoflorine [136]
  Palmatine [136]
T. longepedunculatum E. Nik.
  Berberine [134]
  Magnoflorine [134]
  Thalfoetidine [134]
  Thalicsine (thalixine) [297]
  Thalidasine [297]
T. longistylum DC.
  Berberine [137]
  Columbamine [137]
  N-Desmethylthalistyline [137, 556]
  Jatrorrhizine [137]
  Magnoflorine [137]
  N-Methylthalistyline (thalistyline
    metho salt, methothalistyline)
    [137, 556]
  Oxyberberine (berlambine) [137]
  Palmatine [137]
  Thalibrine [137]
  Thalifendine [137]
  Thaliglucinone [137]
  Thalistyline [137, 556]
  8-Trichloromethyldihydroberberine
    [137]
T. lucidum L.
  Aromoline (thalicrine) [538]
  Berberine [138, 139]
  Columbamine [139]
  Homoaromoline (homothalicrine,
    O-methylaromoline,
    thalrugosamine) [139]

Jatrorrhizine [138, 139]
Magnoflorine [138, 139]
O-Methylthalicberine [139]
Obaberine [139]
Obamegine [139]
Oxyacanthine [139]
Oxyberberine (berlambine) [139]
Palmatine [138, 139]
Thalicberine [538]
Thalidasine [139]
Thalifendine [139]
Thaliglucinone [139]
Thalrugosine [139]
8-Trichloromethyldihydroberberine
  [139]
T. minus L.
  Adiantifoline [230]
  β-Allocryptopine (thalictrimine)
    [427–429]
  Argemonine [63, 65, 181]
  Aromoline (thalicrine) [181]
  Berberine [65, 66, 134, 140, 141,
    181, 272, 310]
  Bursanine [447]
  Corunnine [181]
  O-Desmethyladiantifoline [230]
  Eschscholtzidine [65]
  Glaucine [45, 65, 66]
  Istanbulamine [447]
  Iznikine [447]
  Jatrorrhizine [140]
  Magnoflorine [128, 140, 141, 181]
  N-Methylargemonine [181]
  N-Methylcanadine [65, 380]
  O-Methylthalicberine (thalmidine)
    [65, 84, 114, 181, 272, 595, 596,
    598]
  O-Methylthalmethine [65, 272, 310,
    510, 599]
  Ocoteine (thalicmine) [45, 66, 68,
    84, 95, 96, 181]
  Preocoteine *N*-oxide [96]
  Thalabadensine [181]
  Thalactamine [263, 272]
  Thalflavidine [511]
  Thalicarpine [66, 367, 484–486, 511,
    513, 665]

Table 3. (*continued*)

Thalicberine [65, 272]
Thalicmidine *N*-oxide
(thaliporphine *N*-oxide) [96]
Thalicminine [95, 96, 181]
Thalicsimidine [181]
Thalictrine [664]
Thalidasine [114]
Thalifendine [140]
Thalipine [66]
Thalipoline [65]
Thaliporphine (thalicmidine) [45,
84, 87, 95, 96]
Thalmelatidine [230, 670]
Thalmelatine [66, 510]
Thalmethine [65, 181, 272, 310, 510,
599, 645, 646]
Thalmine [84, 181, 595, 596, 647]
Thalmineline [523]
T. minus L. var. *adiantifolium* Hort.
Adiantifoline [145, 443]
Berberine [144, 145]
Magnoflorine [144, 145]
O-Methylthalicberine [145]
Noroxyhydrastinine [145, 264]
Thalifendine [145]
Thalifoline [145, 264]
Thaliglucinone [145] (see notation
in [149])
T. minus L. var. *elatum* Jacq. (*T.
minus* L. ssp. *majus*)
Adiantifoline [230]
Berberine [229, 262, 350, 351]
Dehydrothalicarpine [230, 262, 525]
Dehydrothalmelatine [230]
O-Demethyladiantifoline [230, 299]
Hernandaline [262]
Noroxyhydrastinine [230, 270]
1-Oxo-6,7-dimethoxy-*N*-
methyltetrahydroisoquinoline
[262]
Thaliadine [299]
Thalicarpine [230, 262]
Thalicthuberine [299]
Thalmelatidine [230, 270, 299]
Thalmelatine [230, 262, 513, 521]
Thalmeline [230, 270]
Thalmineline [522]

T. minus L. var. *microphyllum* Boiss.
Adiantifoline [146]
Aromoline (thalicrine) [544]
Berberine [146]
Bursanine [447]
Homoaromoline (homothalicrine,
O-methylaromoline,
thalrugosamine) [544]
Istanbulamine [447]
Iznikine [447]
Jatrorrhizine [146]
Magnoflorine [146]
O-Methylthalicberine [146, 544]
N(2')Noradiantifoline [468]
Obaberine [146]
Obamegine [544]
Palmatine [146]
Takatonine [217]
Thalactamine [217]
Thaliadanine [146]
Thalicberine [544]
Thaliglucinone [146]
Thaligrisine [544]
Thaliphylline [544]
Thalirugine [544]
Thalisopine (thaligosine) [146, 544]
Thalmelatidine [146]
Thalmicrinone [217]
Thalrugosine [146]
8-Trichloromethyldihydroberberine
[146]
Uskudaramine [524]
T. minus L. var. *minus*
Berberine [147, 148]
Magnoflorine [147, 148]
T. minus L. race B
Adiantifoline [143, 446]
Berberine [143]
Columbamine [102]
Jatrorrhizine [102]
Magnoflorine [143]
N-Methylcorydaldine [102]
O-Methylthalibrine [102]
Obaberine [215]
Oxyberberine (berlambine) [143]
Palmatine [143]
Reticuline [215]

Table 3.   (*continued*)

Thalfine [143, 215]
Thalfinine [215]
Thaliadanine [446]
Thaliadine [446]
Thalidasine [215]
Thalifendine [102]
Thaliglucinone [215]
Thalirabine [215]
Thaliracebine [215]
Thalistine [102]
Thalmirabine [102]
Thalphenine [102]
Thalrugosaminine [215]
Thalrugosine [102]
T. pedunculatum Edgew.
  Berbamine [352]
  Berberine [352]
T. petaloideum L.
  Magnoflorine [128]
T. podocarpum Humb.
  Berberine [301]
  Columbamine [301]
  N-Desmethylthalidezine [301]
  N-Desmethylthalistyline [301, 556]
  Hernandezine [301]
  Isothalidezine [301]
  Jatrorrhizine [301]
  Magnoflorine [301]
  N-Methylthalistyline
    (methothalistyline, thalistyline
    metho salt) [301, 556]
  Oxyberberine (berlambine) [301]
  Palmatine [301]
  Thalidezine [301]
  Thalifendine [301]
  Thaliglucinone [301]
  Thalistyline [301, 556]
  8-Trichloromethyldihydroberberine
    [301]
T. polygamum Muhl.
  Berberine [149, 212]
  Berberrubine [149]
  Bisnorthalphenine [103]
  Deoxythalidastine [149]
  Magnoflorine [149, 212]
  N-Methylnantenine [78]
  N-Methylpalaudinium [212]

N-Methylthaliglucine [103]
N-Methylthaliglucinone [103]
Pennsylpavine [470]
Pennsylpavoline [470]
Pennsylvanamine [469, 470]
Pennsylvanine [469, 470]
Thalicarpine [149]
Thalictrogamine [515]
Thalictropine [515]
Thalifendine [149, 212]
Thaliglucinone [149]
Thalphenine [97]
Thalphenine methine [97]
Thalipine [518]
Thalrugosine (thaligine) [649]
T. revolutum DC.
  β-Allocryptopine [104, 105]
  Argemonine [104, 105]
  Berberine [101, 105, 368]
  Bisnorthalphenine [104]
  Choline chloride [105]
  Columbamine [101]
  N-Desmethylthalphenine [104, 105]
  Deoxythalidastine [101]
  Eschscholtzidine [104, 105]
  Isonorargemonine [68]
  Jatrorrhizine [101, 368]
  Laudanidine [68]
  Magnoflorine [101, 368]
  N-Methylargemonine [105]
  N-Methylarmepavine [105]
  N-Methylcoclaurine [68, 105]
  N-Methyl-6,7-
    Dimethoxyisoquinolinium [105]
  N-Methyleschscholtzidine [105]
  N-Methyllaurotetanine [68, 74]
  O-Methylthalicberine [101, 105]
  O-Methylthalmethine [101]
  Neothalibrine [105]
  N(2′)Northalicarpine [74]
  Palmatine [101]
  Pennsylvanine [68, 101]
  Platycerine [68, 105]
  Reticuline [68]
  Revolutinone [105]
  Revolutopine [68, 472]
  Thalflavidine [74]

Table 3. (*continued*)

Thalicarpine [101, 104, 105, 509]
Thalictrogamine [68]
Thalidasine [101]
Thalifendine [101]
Thaliglucinone [101]
Thalilutidine [517]
Thalilutine [517]
Thalipine [68, 105, 472]
Thalirevoline [68, 105, 517]
Thalirevolutine [517]
Thalmelatine [101, 104, 105]
Thalphenine [101]
Thalrugosaminine [101]
T. rochebrunianum Franc. and Sav.
Berberine [121]
Dihydrothalictrinine [557]
Hernandezine [121]
Jatrorrhizine [121]
Magnoflorine [121]
O-Methylthalibrunimine [583]
N'-Norhernandezine [557]
Northalibrine [582]
N'-Northalibrunine [557, 583]
Oxothalibrunimine [557]
Thalibrine [582]
Thalibrunimine [557, 601, 619]
Thalibrunine [121, 557, 619, 620]
Thalictrinine [557]
Thaliglucinone [302] (see notation
in [149])
Thalsimine [601]
T. rugosum Ait. (*T. glaucum* Desf.)
Aromoline (thalicrine) [234]
Berberine [150–152, 353, 354]
Columbamine [98, 151]
Corypalline [234]
Deoxythalidastine [98]
Homoaromoline (homothalicrine,
O-methylaromoline,
thalrugosamine) [544, 577]
Jatrorrhizine [150, 151]
Magnoflorine [128, 144, 150–152]
6,7-Methylenedioxy-1,2-
dihydroisoquinolin-1-one [234]
Neothalibrine [234]
Noroxyhydrastinine [234]
Obaberine [234]

Obamegine [152, 353, 604]
Oxyberberine (berlambine) [234]
Protothalipine [98]
Protopine [234]
Rugosine [151, 666]
Rugosinone [234]
Thalicthuberine [234]
Thalidasine [114, 152, 353, 627]
Thalifendine [98]
Thaliglucine [114, 300]
Thaliglucinone [98, 152, 300] (see
notation in [149])
Thaligosidine [634]
Thaligosine [634]
Thaligosinine [634]
Thalirugidine [634]
Thalirugine [634]
Thaliruginine [634]
Thalphenine [98]
Thalrugosaminine [98]
Thalrugosidine [152, 353]
Thalrugosine [152, 353, 634]
Thalrugosinone [234]
Thalsimine [114, 661]
T. sachalinense Lecoyer.
Berberine [67]
Glaucine [67]
Magnoflorine [67]
N-Methylnantenine [67]
Thalrugosine [67]
T. simplex L.
β-Allocryptopine [153, 154]
Hernandezine [153, 570, 574]
Magnoflorine [153, 154]
N'-Norhernandezine (thalisamine)
[557, 600]
Ocoteine (thalicmine) [153]
Thalfoetidine [624, 625]
Thalicminine [153]
Thalicsimidine [153, 154, 177]
Thalicsine (thalixine) [153, 294,
295]
Thalictrisine [153, 435]
Thalidezine [600]
Thalsimidine [652, 653]
Thalsimine [574, 600, 625, 652, 654,
655, 657, 660]

Table 3. (*continued*)

T. squarrosum Steph. ex Willd.
  Magnoflorine [128]
T. strictum Ledeb.
  Argemonine [155, 160]
  Berberine [155]
  Magnoflorine [155]
  O-Methylcassyfiline
    (Northalicmine,
    O-methylcassythine,
    hexahydrothalicminine) [160, 161]
  2,3-Methylenedioxy-4,8,9-
    trimethoxypavinane [155, 161]
  Ocoteine (thalicmine) [155, 160,
    161]
  Preocoteine [155]
  Thalicminine [160, 161]
  Thalicsimidine [155]
  Thalicthuberine [155]
T. sultanabadense Stapf.
  Hernandezine [575, 576]
  Hernandezine *N*-oxide [576]
  Thalabadensine [575, 576]
  Thalidezine [576]
T. thunbergii DC. (*T. minus* L. var.
  hypoleucum).

Aromoline (thalicrine) [540–543]
Berberine [142, 157]
Columbamine [142]
Deoxythalidastine [142]
Homoaromoline (homothalicrine,
  O-methylaromoline,
  thalrugosamine) [540–543]
Jatrorrhizine [142]
Magnoflorine [123, 142, 156, 157]
O-Methylthalicberine [542, 586–
  589, 621, 622]
Palmatine [142]
Takatonine [157, 218]
Thalicberine [586–588, 621, 622]
Thalicthuberine [298]
Thalictine [623]
Thalidastine [142]
Thalifendine [142]
T. tuberiferum Maxim.
  Berberine [355]
T. uchiyamai Nakai
  Corypalline [249]
T. urbaini Hayata
  Isocorydine [118]
  Oconovine [118]

6. CALCULATED MOLECULAR WEIGHTS OF *THALICTRUM* ALKALOIDS

Table 4

189.0426 $C_{10}H_7NO_3$
  6,7-Methylenedioxy-1,2-
  dihydroisoquinolin-1-one
  (**106**)
191.0582 $C_{10}H_9NO_3$
  Noroxyhydrastinine (**96**)
193.1103 $C_{11}H_{15}NO_2$
  Corypalline (**85**)
204.1024 $C_{12}H_{14}N^+O_2$
  N-Methyl-6,7-
  dimethoxyisoquinolinium (**81**)
207.0895 $C_{11}H_{13}NO_3$
  Thalifoline (**103**)
207.1259 $C_{12}H_{17}NO_2$
  O-Methylcorypalline (**86**)

212.0950 $C_{13}H_{12}N_2O$
  Harmine (**484**)
219.0895 $C_{12}H_{13}NO_3$
  N-Methyl-6,7-dimethoxy-1,2-
  dihydroisoquinolin-1-one (**89**)
221.1052 $C_{12}H_{15}NO_3$
  N-Methylcorydaldine (**88**)
235.0845 $C_{12}H_{13}NO_4$
  Thalflavine (**100**)
249.1001 $C_{13}H_{15}NO_4$
  Thalactamine (**107**)
251.1158 $C_{13}H_{17}NO_4$
  N-Methylthalidaldine (**93**)
299.1521 $C_{18}H_{21}NO_3$
  N-Methylcoclaurine (**52**)

Table 4. (*continued*)

320.0923 $C_{19}H_{14}N^+O_4$
Deoxythalidastine (**150**)
322.1079 $C_{19}H_{16}N^+O_4$
Berberrubine (**136**)
Thalifaurine (**164**)
Thalifendine (**171**)
323.1158 $C_{19}H_{17}NO_4$
Bisnorthalphenine (**18**)
324.1236 $C_{19}H_{18}N^+O_4$
Dehydrodiscretamine (**138**)
Demethyleneberberine (**144**)
325.1314 $C_{19}H_{19}NO_4$
Tetrahydrothalifendine (**152**)
325.1677 $C_{20}H_{23}NO_3$
N,O,O-Trimethylsparsiflorine (**2**)
327.1471 $C_{19}H_{21}NO_4$
Bisnorargemonine (**112**)
Isoboldine (N-methyllaurelliptine)
(**4**)
Pallidine (**58**)
Thalidicine (**75**)
Thalidine (**76**)
328.1913 $C_{20}H_{26}N^+O_3$
N-Methylarmepavine (**51**)
329.1627 $C_{19}H_{23}NO_4$
Reticuline (**57**)
Thalmeline (**68**)
336.1236 $C_{20}H_{18}N^+O_4$
Berberine (**135**)
337.1314 $C_{20}H_{19}NO_4$
Cabudine (**42**)
N-Demethylthalphenine (**19**)
338.1028 $C_{19}H_{16}N^+O_5$
Thalidastine (**151**)
338.1392 $C_{20}H_{20}N^+O_4$
Columbamine (**137**)
Dehydrodiscretine (**139**)
Jatrorrhizine (**153**)
339.1471 $C_{20}H_{21}NO_4$
Eschscholtzidine (**113**)
Tetrahydroberberine (canadine)
(**173**)
340.1547 $C_{20}H_{22}N^+O_4$
N-Methylpalaudinium (**53**)
341.1627 $C_{20}H_{23}NO_4$
Corydine (**20**)
Isocorydine (**21**)

Isonorargemonine (**114**)
N-Methyllaurotetanine (**5**)
Norargemonine (**117**)
Platycerine (**118**)
Tetrahydrocolumbamine
(isocorypalmine) (**176**)
Thaliporphine (thalicmidine) (**8**)
Thalisopavine (**78**)
342.1705 $C_{20}H_{24}N^+O_4$
Magnoflorine (**22**)
343.1783 $C_{20}H_{25}NO_4$
Laudanidine (**50**)
Thalifendlerine (**65**)
344.1862 $C_{20}H_{26}N^+O_4$
Tembetarine (**64**)
351.1107 $C_{20}H_{17}NO_5$
Corunnine (**45**)
Oxoglaucine (**46**)
Oxyberberine (**154**)
351.1471 $C_{21}H_{21}NO_4$
Thaliglucine (**123**)
352.1549 $C_{21}H_{22}N^+O_4$
Palmatine (**148**)
Thalphenine (**15**)
353.0899 $C_{19}H_{15}NO_6$
Rugosinone (**71**)
353.1263 $C_{20}H_{19}NO_5$
Protopine (**180**)
Thalmicrinone (**73**)
353.1627 $C_{21}H_{23}NO_4$
Thalicthuberine (**130**)
354.1705 $C_{21}H_{24}N^+O_4$
N-Methylescheeholtzidine (**116**)
N-Methylnantenine (**6**)
N-Methyltetrahydroberberine
(N-methylcanadine) (**175**)
Takatonine (**61**)
355.1420 $C_{20}H_{21}NO_5$
Baicaline (**27**)
O-Methylcassyfiline (northalicmine,
O-methylcassythine,
hexahydrothalicminine) (**30**)
Thalictricine (thalictrisine) (**187**)
355.1783 $C_{21}H_{25}NO_4$
Argemonine (**111**)
Glaucine (**3**)

Table 4. (*continued*)

356.1861 $C_{21}H_{26}N^+O_4$
Xanthoplanine (**14**)
357.1576 $C_{20}H_{23}NO_5$
Thalicmidine *N*-oxide
(thaliporphine *N*-oxide) (**13**)
365.0899 $C_{20}H_{15}NO_6$
Thalicminine (**47**)
365.1263 $C_{21}H_{19}NO_5$
Thaliglucinone (**125**)
Thalicsine (thalixine) (**128**)
366.1705 $C_{22}H_{24}N^+O_4$
N-Methylthaliglucine (thaliglucine
metho salt) (**122**)
367.1420 $C_{21}H_{21}NO_5$
Dehydrothalicmine
(dehydroocoteine) (**43**)
369.1576 $C_{21}H_{23}NO_5$
Allocryptopine (β-allocryptopine)
(**178**)
Baicalidine (**26**)
Cryptopine (**179**)
2,3-Methylenedioxy-4,8,9-
trimethoxypavinane (**119**)
Ocoteine (thalicmine) (**31**)
370.2018 $C_{22}H_{28}N^+O_4$
N-Methylargemonine (**115**)
371.1733 $C_{21}H_{25}NO_5$
Oconovine (**40**)
Preocoteine (**34**)
Protothalipine (**181**)
Thalisopynine (thalisopinine) (**39**)
380.1498 $C_{22}H_{22}N^+O_5$
N-Methylthaliglucinone
(thaliglucinone metho salt) (**124**)
385.1889 $C_{22}H_{27}NO_5$
Thalicsimidine (purpureine) (**35**)
387.1682 $C_{21}H_{25}NO_6$
Preocoteine *N*-oxide (**38**)
395.1369 $C_{22}H_{21}NO_6$
Thalflavidine (**126**)
454.4826 $C_{21}H_{18}NO_4Cl_3$
8-Trichloromethyldihydroberberine
(**174**)
489.2363 $C_{26}H_{35}NO_8$
Veronamine (**69**)
535.2206 $C_{30}H_{33}NO_8$
Thaliadine (**483**)

592.2573 $C_{36}H_{36}N_2O_6$
Thalmethine (**441**)
594.2730 $C_{36}H_{38}N_2O_6$
Aromoline (thalicrine) (**282**)
Obamegine (**353**)
Thalabadensine (**356**)
606.2730 $C_{37}H_{38}N_2O_6$
O-Methylthalmethine (**336**)
608.2886 $C_{37}H_{40}N_2O_6$
Berbamine (**284**)
Homoaromoline (homothalicrine,
O-methylaromoline,
thalrugosamine) (**316**)
Oxyacanthine (**355**)
Thalicberine (**386**)
Thalictine (**390**)
Thaliphylline (**410**)
Thalmine (**358**)
Thalrugosine (**457**)
610.3043 $C_{37}H_{42}N_2O_6$
Northalibrine (**344**)
Thaligrisine (**409**)
622.2679 $C_{37}H_{38}N_2O_7$
Thalsimidine (**468**)
622.3043 $C_{38}H_{42}N_2O_6$
Isotetrandrine (**319**)
O-Methylthalicberine (**333**)
Obaberine (*O*-methyloxyacanthine)
(**352**)
624.2836 $C_{37}H_{40}N_2O_7$
N-Desmethylthalidezine (**288**)
N-Desmethylthalrugosidine (**295**)
Thaligosidine (**399**)
Thalisopidine (**431**)
624.3199 $C_{38}H_{44}N_2O_6$
Neothalibrine (**338**)
Thalibrine (**329**)
636.2836 $C_{38}H_{40}N_2O_7$
Thalsimine (**312**)
638.2992 $C_{38}H_{42}N_2O_7$
N-Desmethylthalidasine (**285**)
Isothalidezine (**320**)
N'-Norhernandezine (thalisamine)
(**340**)
Thalfoetidine (**370**)
Thalidezine (**289**)
Thaligosinine (**407**)

Table 4. (*continued*)

Thalisopine (thaligosine) (432)
Thalrugosidine (299)
638.3356 $C_{39}H_{46}N_2O_6$
O-Methylthalibrine (326)
640.3149 $C_{38}H_{44}N_2O_7$
Thalirugine (424)
648.2472 $C_{38}H_{36}N_2O_8$
Thalfine (360)
652.2421 $C_{37}H_{36}N_2O_9$
Thalpindione (453)
652.2785 $C_{38}H_{40}N_2O_8$
Revolutinone (478)
Thalibrunimine (332)
652.3149 $C_{39}H_{44}N_2O_7$
Hernandezine (303)
Thalidasine (286)
Thaliracebine (418)
Thalrugosaminine (408)
654.2941 $C_{38}H_{42}N_2O_8$
N'-Northalibrunine (346)
654.3305 $C_{39}H_{46}N_2O_7$
Thaliruginine (429)
664.2421 $C_{38}H_{36}N_2O_9$
Thalictrinine (302)
666.2577 $C_{38}H_{38}N_2O_9$
Dihydrothalictrinine (301)
Oxothalibrunimine (354)
Thalrugosinone (454)
666.2941 $C_{39}H_{42}N_2O_8$
Pennsylpavoline (280)
O-Methylthalibrunimine (330)
Thalfinine (366)
668.3097 $C_{39}H_{44}N_2O_8$
Hernandezine N-oxide (315)
Istanbulamine (217)
Pennsylvanamine (223)
Revolutopine (229)
Thalibrunine (351)
Thalictrogamine (242)
Thalipine (257)
Thalistine (437)
Thalmirabine (448)
Uskudaramine (264)
670.3254 $C_{39}H_{46}N_2O_8$
Thalirugidine (419)
680.3097 $C_{40}H_{44}N_2O_8$
Pennsylpavine (274)

682.3254 $C_{40}H_{46}N_2O_8$
N-Desmethylthalistyline (293)
Faberidine (205)
Fetidine (208)
N(2')-Northalicarpine (221)
Pennsylvanine (228)
Thalictropine (226)
Thalidoxine (227)
Thalifarapine (250)
Thalilutidine (252)
Thalirevoline (258)
Thalmelatine (204)
683.3332 $C_{40}H_{47}N_2^{2+}O_8$
Thalirabine (369)
694.3254 $C_{41}H_{46}N_2O_8$
Dehydrothalifaberine (267)
Dehydrothalicarpine (268)
696.3410 $C_{41}H_{48}N_2O_8$
Thalicarpine (192)
Thalifaberine (216)
Thalirevolutine (230)
697.3489 $C_{41}H_{49}N_2^+O_8$
Thalistyline (102)
698.3200 $C_{40}H_{46}N_2O_9$
Bursanine (198)
Iznikine (218)
Thalifasine (251)
710.3191 $C_{41}H_{46}N_2O_9$
Thalifabine (246)
712.3360 $C_{41}H_{48}N_2O_9$
O-Desmethyladiantifoline (200)
Faberonine (209)
N(2')-Noradiantifoline (219)
Thaliadanine (203)
Thalifabatine (247)
Thalilutine (254)
712.3724 $C_{42}H_{52}N_2^{2+}O_8$
N-Methylthalistyline (294)
724.3360 $C_{42}H_{48}N_2O_9$
Dehydrohuangshanine (266)
726.3516 $C_{42}H_{50}N_2O_9$
Adiantifoline (190)
Huangshanine (212)
740.3309 $C_{42}H_{48}N_2O_{10}$
Thalmelatidine (101)
742.3465 $C_{42}H_{50}N_2O_{10}$
Thalmineline (263)

REFERENCES

1. L. H. Bailey, *The Standard Cyclopedia of Horticulture*, Vol. 3, Macmillan, New York, 1943, p. 3326.
2. L. H. Bailey, *Manual of Cultivated Plants*, Macmillan, New York, 1949, p. 390.
3. J. C. Lecoyer, *Bull. Soc. Roy. Bot. Belgique* **23–24,** 146 (1885–1996).
4. J. Trelease, *Proc. Soc. Bost. Nat. Hist.* **23,** 293 (1886).
5. B. L. Robinson, *Synop. Fl. N. Am.*, I, Pt. 1, 17 (1895).
6. K. C. Davis, *Minn. Bot. Studies* **2,** 514 (1900).
7. B. Boivin, *Rhodora* **46,** 337 (1944).
8. H. A. Gleason, *The New Britton and Brown Illustrated Flora of the Northeastern United States and Adjacent Canada*, II, New York Botanical Garden, 1952, p. 158.
9. M. Tamura, *Acta Phytotax. Geobot.* **15,** 80 (1953).
10. P. L. Schiff, Jr., and R. W. Doskotch, *Lloydia* **33,** 403 (1970).
11. T. Tomimatsu, *Tokushima Daigaku Yakugabubu Kenkyu Nempo* **14,** 24 (1965).
12. T. Tomimatsu, *Tokushima Daigaku Yakugakubu Kenkyu Nempo* **16,** 40 (1967).
13. T. Tomimatsu, *Syoyakugaku Zasshi* **30,** 1 (1976).
14. N. M. Mollov, H. B. Dutschewska, and V. St. Georgiev, *Recent Developments in the Chemistry of Natural Carbon Compounds*, Vol. 4, R. Bognar, V. Brockner, and Cs. Szantay, Eds., Hungarian Academy of Sciences, Budapest, 1971, p. 195.
15. H. Guinaudean, M. Leboeuf, and A. Cavé, *Lloydia* **38,** 275 (1975).
16. H. Guinaudeau, M. Leboeuf, and A. Cavé, *J. Nat. Prod.* **42,** 325 (1979).
17. H. Guinaudeau, M. Leboeuf, and A. Cavé, *J. Nat. Prod.* **46,** 761 (1983).
18. T. Kametani, *The Chemistry of the Isoquinoline Alkaloids*, Hirokawa Publishing, Tokyo, and Elsevier, Amsterdam, 1969, 265 pp.
19. T. Kametani, *The Chemistry of the Isoquinoline Alkaloids*, Vol. 2, Hirokowa, Tokyo, and Elsevier, Amsterdam, 1974.
20. K. W. Bentley, *The Alkaloids*, Vol. 1, J. E. Saxton, Ed., The Chemical Society, London, 1971, p. 106.
21. H. O. Bernhard and V. A. Snieckus, *The Alkaloids*, Vol. 2, J. E. Saxton, Ed., The Chemical Society, London, 1972, p. 97.
22. M. Shamma, *The Alkaloids*, Vol. 3, J. E. Saxton, Ed., The Chemical Society, London, 1973, p. 116.
23. M. Shamma and S. S. Salgar, *The Alkaloids*, Vol. 4, J. E. Saxton, Ed., The Chemical Society, London, 1974, p. 197.
24. H. O. Bernhard and V. A. Snieckus, *The Alkaloids*, Vol. 5, J. E. Saxton, Ed., The Chemical Society, London, 1975, p. 111.
25. M. Shamma, *The Alkaloids*, Vol. 6, M. F. Grundon, Ed., The Chemical Society, London, 1976, p. 170.
26. M. Shamma, *The Alkaloids*, Vol. 7, M. F. Grundon, Ed., The Chemical Society, London, 1977, p. 152.
27. M. Shamma, *The Alkaloids*, Vol. 8, M. F. Grundon, Ed., The Chemical Society, London, 1978, p. 122.
28. M. Shamma, *The Alkaloids*, Vol. 9, M. F. Grundon, Ed., The Chemical Society, London, 1979, p. 126.
29. M. Shamma, *The Alkaloids*, Vol. 10, M. F. Grundon, Ed., The Royal Society of Chemistry, London, 1980, p. 126.

30. M. Shamma, *The Alkaloids*, Vol. 11, M. F. Grundon, Ed., The Royal Society of Chemistry, London, 1981, p. 117.

31. M. Shamma and H. Guinaudeau, *The Alkaloids*, Vol. 12, M. F. Grundon, The Royal Society of Chemistry, London, 1982, p. 135.

32. M. Shamma and W. A. Slusarchyk, *Chem. Rev.* **64,** 59 (1964).

33. M. P. Cava and A. Venkateswarlu, *Ann. Rep. Med. Chem.* **1969,** p. 331.

34. R. H. F. Manske, *The Alkaloids*, Vol. 4, R. H. F. Manske, Ed., Academic, New York, 1954, p. 119.

35. M. Shamma, *The Alkaloids*, Vol. 9, R. H. F. Manske, Ed., Academic, New York, 1976, p. 1.

36. M. Shamma, *The Isoquinoline Alkaloids*, Academic, New York, 1972, p. 194.

37. M. Shamma and J. L. Moniot, *Isoquinoline Alkaloids Research: 1972–1977*, Plenum, New York, 1978, p. 123.

38. D. S. Bhakuni and R. S. Singh, *J. Nat. Prod.* **45,** 252 (1982).

39. H. Hara, O. Hoshino, T. Ishige, and B. Umezawa, *Chem. Pharm. Bull.* **29,** 1083 (1981).

40. M. P. Dubey, R. C. Srimal, and B. N. Dhawan, *Indian J. Pharmacol.* **7,** 73 (1969).

41. L. M. Jackman, J. C. Trewella, J. L. Moniot, M. Shamma, R. L. Stephens, E. Wenkert, M. Leboeuf, and A. Cavé, *J. Nat. Prod.* **42,** 437 (1979).

42. L. Castedo, R. Riguera, and F. J. Sardina, *Anal. Quim.* **77,** 138 (1981).

43. A. H. Jackson and J. A. Martin, *J. Chem. Soc., C,* **1966,** 2181.

44. H. M. Fales, H. A. Lloyd, and G. W. A. Milne, *J. Am. Chem. Soc.* **92,** 1590 (1970).

45. S. Yunusov and N. N. Progressov, *Zhur. Obschei Khim.* **22,** 1047 (1952); *Chem. Abstr.* **47,** 8084 (1952).

46. Kh. S. Shakhabutdinova, I. K. Kamilov, and S. F. Fakhrutdinov, *Farmakol. Alkaloidov Glikozidov,* **1967,** 142; *Chem. Abstr.* **70,** 2219 (1969).

47. E. E. Aleshinskaya and V. V. Berezhinskaya, *Farmakol. Toksikol.* (Moscow) **29,** 611 (1966); *Chem. Abstr.* **66,** 17903 (1967).

48. N. Donev, *Tr Nauch-Issled. Khim.—Farm. Inst.* **5,** 92 (1966); *Chem. Abstr.* **67,** 9286 (1967).

49. O. Angelova and S. Zarkova, *Suvrem. Med.* **22,** 33 (1971); *Chem. Abstr.* **76,** 135836 (1972).

50. J. Berthe, B. Remandet, G. Mazue, and H. A. Tilson, *Neurobehav. Toxicol. Teratol.* **5,** 305 (1983).

51. V. V. Berezhinskaya, E. E. Aleshinskaya, and Y. A. Aleshina, *Farmakol. Toksikol.* (Moscow) **31,** 44 (1968); *Chem. Abstr.* **68,** 94521 (1968).

52. N. Donev, *Farmatsiya* (Sofia) **12,** 17 (1962); *Chem. Abstr.* **58,** 4941 (1963).

53. N. Donev, *Farmatsiya* (Sofia) **14,** 49 (1964); *Chem. Abstr.* **61,** 9928 (1964).

54. Y. Kase, M. Kawaguchi, K. Takahama, T. Miyata, I. Hirotsu, T. Hitoshi, and Y. Okano, *Arzneim.-Forsch.* **33,** 936 (1983).

55. Y. Kase, Y. Matsumoto, K. Takahama, T. Miyata, T. Hitoshi, I. Hirotsu, and Y. Okano, *Arzneim.-Forsch.* **33,** 947 (1983).

56. P. W. Erhardt and T. O. Soine, *J. Pharm. Sci.* **64,** 53 (1975).

57. J. Kovar, *Arch. Biochem. Biophys.* **221,** 271 (1983).

58. J. Ulrichová, D. Walterová, V. Preininger, and V. Simanek, *Planta Medica* **48,** 174 (1983).

59. M. I. Mironova, E. V. Arzamastsev, V. V. Bortnikova, L. V. Krepkova, and Yu. B. Kuznestov, *Farmakol. Toksikol.* (Moscow) **46,** 100 (1983); *Chem. Abstr.* **99,** 115844 (1983).

60. P. J. Davis, D. Wiese, and J. P. Rosazza, *J. Chem. Soc. Perkin I,* **1977,** 1.

61. S. Kh. Maekh, S. Yu. Yunusov, E. V. Boiko, and V. M. Starchenko, *Chem. Nat. Cpds.* **18,** 761 (1982).
62. M. Shamma and B. S. Dudock, *J. Pharm. Sci.* **57,** 262 (1968).
63. D. Umarov, S. Kh. Maekh, S. Yu. Yunusov, P. G. Govovoi, and E. V. Boiko, *Chem. Nat. Prod.* **12,** 706 (1976).
64. S. Mukhamedova, S. Kh. Maekh, and S. Yu. Yunusov, *Khim. Prir. Soedin* **1981,** 251; *Chem. Abstr.* **95,** 21334 (1981).
65. H. Dutschewska, B. Dimov, N. Mollov, and L. Evstatieva, *Planta Med.* **39,** 77 (1980).
66. H. Dutschewska, B. Dimov, V. Christov, B. Kuzmanov, and L. Evstatieva, *Planta Med.* **45,** 39 (1982).
67. D. Umarova, S. Kh.-Maekh, S. Yu. Yunusov, N. M. Zaitseva, S. A. Volkova, and P. G. Gorovoi, *Chem. Nat. Cpds.* **14,** 511 (1978).
68. J. Wu, J. L. Beal, W.-N Wu, and R. W. Doskotch, *J. Nat. Prod.* **40,** 593 (1977).
69. K. Wada and K. Munakata, *J. Agric. Food Chem.* **16,** 471 (1968).
70. W.-N. Wu, J. L. Beal, and R. W. Doskotch, *J. Nat. Prod.* **43,** 372 (1980).
71. M. Shamma, S. Y. Yao, B. R. Pai, and R. Charubala, *J. Org. Chem.* **36,** 3253 (1971).
72. M. Shamma and A. S. Rothenberg, *Lloydia* **41,** 169 (1978).
73. B. Ringdahl, R. P. K. Chan, J. C. Craig, M. P. Cava, and M. Shamma, *J. Nat. Prod.* **44,** 80 (1981).
74. W.-N. Wu, J. L. Beal, and R. W. Doskotch, *J. Nat. Prod.* **43,** 567 (1980).
75. E. H. Herman and D. P. Chadwick, *Pharmacology* **10,** 178 (1973).
76. Laboratoire R. Bellou, French Patent 2130107; *Chem. Abstr.* **78,** 136492 (1973).
77. J. R. Jatimliansky and E. M. Sivori, *Ann. Soc. Cient. Argentina* **187,** 49 (1969); *Chem. Abstr.* **72,** 974 (1970) and **74,** 94958 (1971).
78. M. Shamma and J. L. Moniot, *Heterocycles* **3,** 279 (1975).
79. M. Shamma and J. L. Shine, and B. S. Dudock, *Tetrahedron* **23,** 2887 (1967).
80. Z. F. Ismailov, M. R. Yagudaev, and S. Yu. Yunusov, *Chem. Nat. Cpds.* **4,** 175 (1968).
81. Z. F. Ismailov and S. Yu Yunusov, *Chem. Nat. Cpds.* **4,** 169 (1968).
82. Z. F. Ismailov, M. R. Yagudaev, and S. Yu. Yunusov, *Chem. Nat. Cpds.* **4,** 202 (1968).
83. M. Shamma, M. J. Hillman, R. Charubala, and B. R. Pai, *Indian J. Chem.* **7,** 1056 (1969).
84. S. Yunusov and N. N. Progressov, *Zhur. Obshchei Khim.* **20,** 1151 (1950); *Chem. Abstr.* **45,** 1608 (1950).
85. M. Shamma, *Experientia* **18,** 64 (1962).
86. P. S. Clezy, A. W. Nichol, and E. Gellert, *Experienta* **19,** 1 (1963).
87. Kh. G. Pulatova, Z. F. Ismailov, and S. Yu. Yunusov, *Chem. Nat. Cpds.* **3,** 57 (1967).
88. M. Shamma and W. A. Slusarchyk, *Tetrahedron Lett.* **1965,** 1509.
89. M. Shamma and W. A. Slusarchyk, *Tetrahedron* **23,** 2563 (1967).
90. T. Kametani, K. Fukumoto, and T. Nakano, *J. Heterocycl. Chem.* **9,** 1363 (1972).
91. T. Kametani, A. Ujiie, K. Takahashi, T. Nakano, T. Suzuki, and F. Fukumoto, *Chem. Pharm. Bull* **21,** 768 (1973).
92. T. Kametani, S. Shibuya, and S. Kano, *J. Chem. Soc. Perkin I* 1212 (1973).
93. S. V. Kessar, S. Bhatra, V. K. Nadir, and S. S. Gandhi, *Indian J. Chem.* **13,** 1109 (1975).
94. Kh. S. Shakhabutdinova, I. K. Kamilov, and S. F. Fakhrutdinov, *Med. Zh. Uzb.* **1967,** 36; *Chem. Abstr.* **68,** 20793 (1967).
95. Kh. G. Pulatova, Z. F. Ismailov, and S. Yu. Yunusov, *Chem. Nat. Cpds.* **2,** 349 (1966).
96. V. G. Khozhdaev, S. Kh. Maekh, and S. Yu. Yunusov, *Chem. Nat. Cpds.* **8,** 599 (1972).

97. M. Shamma, J. L. Moniot, S. Y. Yao, and J. A. Stanko, *J. Chem. Soc. Chem. Commun,* **1972,** 408.

98. W.-N. Wu, J. L. Beal, G. W. Clark, and L. A. Mitscher, *Lloydia* **39,** 65 (1976).

99. M. Shamma and D.-Y. Hwang, *Heterocycles* **1,** 31 (1973).

100. M. Shamma and D.-Y. Hwang, *Tetrahedron* **30,** 2279 (1974).

101. W.-N. Wu, J. L. Beal, and R. W. Doskotch, *Lloydia* **40,** 508 (1977).

102. W.-N. Wu, W.-T. Liao, Z. F. Mahmoud, J. L. Beal, and R. W. Doskotch, *J. Nat. Prod.* **43,** 472 (1980).

103. M. Shamma and J. L. Moniot, *Heterocycles* **2,** 427 (1974).

104. J. Wu, J. L. Beal, R. W. Doskotch, and W.-N. Wu, *Lloydia* **40,** 294 (1977).

105. J. Wu, J. L. Beal, W.-N. Wu, and R. W. Doskotch, *J. Nat. Prod.* **43,** 270 (1980).

106. M. Shamma and S. S. Salgar, *Phytochemistry* **12,** 1505 (1973).

107. V. Peters, J. L. Hartwell, A. J. Dalton, and M. J. Shear, *Cancer Res.* **6,** 490 (1946).

108. S. Fakhrutdinov and M. B. Sultanov, *Farmakol. Alkaloidov Ikh Proizvod.* **1972,** 118; *Chem. Abstr.* **80,** 91212 (1974).

109. V. V. Berezhinskaya, E. E. Aleshinskaya, and Y. A. Aleshkina, *Farmakol. Toksikol.* **31,** 44 (1968); *Chem. Abstr.* **68,** 94521 (1968).

110. P. N. Patil, A. M. Burkman, D. Yamaguchi, and S. Hetey, *J. Pharm. Pharmacol.* **25,** 221 (1973).

111. B. Borkowski, A. Desperak-Naciazek, K. Obojska, and Z. Szmal, *Diss. Pharm. Pharmacol.* **18,** 455 (1966).

112. I. Khamdamov, *Farmakol. Prir. Veschestv* **1978,** 29; *Chem. Abstr.* **91,** 32778 (1979).

113. K. S. Shamirzaeva and S. F. Fakhrutindov, *Farmakol. Alkaloidov Glikozidov* 141 (1971); V. Preininger, *The Alkaloids,* Vol. 15, R. H. F. Manske, Ed., Academic, New York, 1975, p. 227.

114. N. M. Mollov, P. Panov, L. N. Thuan, and L. Panova, *Compt. Rend. Akad. Bulg. Sci.* **23,** 181 (1970).

115. E. Moisset de Espanes, *Rev. Soc. Argent. Biol.* **31,** 241, 253 (1955); *C. R. Soc. Biol* **149,** 1789, 1791 (1955).

116. E. Moisset de Espanes, *Rev. Soc. Argent. Biol.* **32,** 108 (1956); *Chem. Abstr.* **51,** 5992 (1957).

117. K. Shakhabutdinova, S. F. Fakhrutdinov, and I. K. Kamilov, *Farmakol. Alkaloidov Glikozidov* 146 (1967); *Chem. Abstr.* **70,** 2220 (1969).

118. C.-H. Chen and J. Wu, *J. Taiwan Pharm. Assoc.* **28,** 121 (1976).

119. F. R. Stermitz, L. Castedo, and D. Dominguez, *J. Nat. Prod.* **43,** 140 (1980).

120. S. R. Hemingway, J. D. Phillipson, and R. Verpoorte, *J. Nat. Prod.* **44,** 67 (1981).

121. H. H. S. Fong, J. L. Beal, and M. P. Cava, *Lloydia* **29,** 94 (1966).

122. R. Hogg, J. L. Beal, and M. P. Cava, *Lloydia* **24,** 45 (1961).

123. E. Fujita and T. Tomimatsu, *Pharm. Bull.* (Tokyo) **4,** 489 (1956); *Chem. Abstr.* **51,** 13886 (1957).

124. R. W. Doskotch and J. E. Knapp, *Lloydia* **34,** 292 (1971).

125. S. F. Fakhrutdinov and I. K. Kamilov, *Farmakol. Alkaloidov Glikozidov* 149 (1967); *Chem. Abstr.* **70,** 2221 (1969).

126. S. F. Fakhrutdinov, *Farmakol. Alkaloidov Serdech. Glikozidov* 155 (1971); *Chem. Abstr.* **77,** 122094 (1972).

127. S. Kh. Maekh, S. Yu. Yunusov, E. V. Boiko, and V. M. Starchenko, *Chem. Nat. Cpds* **18,** 208 (1982).

128. M. Sobiczewska and B. Borkowski, *Acta Pol. Pharm.* **27**, 379 (1970); *Chem. Abstr.* **74**, 10299 (1971).

129. S. M. Kupchan, K. K. Chakravarti, and N. Yokoyama, *J. Pharm. Sci.* **52**, 985 (1963).

130. H. Ong and J. Beliveau, *Ann. Pharm. Fr.* **34**, 223 (1976).

131. C.-H. Chen, T.-M. Chen, and C. Lee, *J. Pharm. Sci.* **69**, 1061 (1980).

132. M. Shamma, M. A. Greenberg, and B. S. Dudock, *Tetrahedron Lett.* **1965**, 3595.

133. Z. F. Ismailov, K. L. Lutfullin, and S. Yu. Yunusov, *Chem. Nat. Cpds.* **4**, 173 (1968).

134. Kh. G. Pulatova, S. Abdizhabbarova, Z. F. Ismailov, and S. Yu. Yunusov, *Chem. Nat. Cpds.* **4**, 51 (1968).

135. K. W. Gopinath, T. R. Govindachari, S. Rajappa, and C. V. Ramadas, *J. Sci. Ind. Res.* (India) **18B**, 444 (1959).

136. S. Bahadur and A. K. Shukla, *J. Nat. Prod.* **46**, 454 (1983).

137. W.-N. Wu, J. L. Beal, R.-P. Leu, and R. W. Doskotch, *J. Nat. Prod.* **40**, 281 (1977).

138. T. Baytop and M. Berghmans, *Istanbul Univ. Eczacilik Fak. Mecm.* **11**, 58 (1975); *Chem. Abstr.* **84**, 40710 (1976).

139. W.-N. Wu, J. L. Beal, L. A. Mitscher, K. N. Salman, and P. Patil, *Lloydia* **39**, 204 (1976).

140. T. Kaniewska and B. Borkoneski, *Acta Pol. Pharm.* **28**, 503 (1971); *Chem. Abstr.* **76**, 124102 (1972).

141. I. Ciulei and P. A. Ionescu, *Farmacia* (Bucharest) **21**, 17 (1973); *Chem. Abstr.* **79**, 123633 (1973).

142. A. Ikuta and H. Itokawa, *Phytochemistry* **21**, 1419 (1982).

143. C. W. Geiselman, S. A. Gharbo, J. L. Beal, and R. W. Doskotch, *Lloydia* **35**, 296 (1972).

144. T. Tomimatsu, C. R. Gharbo, and J. L. Beal, *J. Pharm. Sci.* **54**, 1390 (1965).

145. R. W. Doskotch, P. L. Schiff, Jr., and J. L. Beal, *Lloydia* **32**, 29 (1969).

146. K. H. C. Baser, *Doga, Seri A,* **5**, 163 (1981); *Chem. Abstr.* **96**, 65701 (1982).

147. Y. Aynehchi, *Quart. J. Crude Drug Res.* **17**, 81 (1979).

148. Y. Aynehchi, *Pazhoohandeh (Tehran)* **23**, 165 (1979).

149. S. A. Gharbo, J. L. Beal, R. W. Doskotch, and L. A. Mitscher, *Lloydia* **36**, 349 (1973).

150. C. Tadeusz and B. Borkowski, *Acta Polon. Pharm.* **22**, 265 (1965); *Chem. Abstr.* **63**, 16658 (1965).

151. T. Cieszynski and B. Borkowski, *Acta Polon. Pharm.* **22**, 347 (1965); *Chem. Abstr.* **64**, 7036 (1966).

152. L. A. Mitscher, W.-N. Wu, R. W. Doskotch, and J. L. Beal, *Lloydia* **35**, 167 (1972).

153. Kh. S. Umarov, M. V. Telezhenetskaya, Z. F. Ismailov, and S. Yu. Yunusov, *Chem. Nat. Cpds.* **6**, 219 (1970).

154. Kh. S. Umarov, M. V. Telezhenetskaya, Z. F. Ismailov, and S. Yu. Yunusov, *Chem. Nat. Cpds.* **3**, 199 (1967).

155. S. Kh. Maekh, P. G. Gorovoi, and S. Yu. Yunusov, *Chem. Nat. Cpds.* **12**, 507 (1976).

156. E. Fujita and T. Tomimatsu, *Chem. Pharm. Bull.* (Tokyo) **6**, 107 (1958); *Chem. Abstr.* **53**, 5587 (1959).

157. T. Tomimatsu, M. Matsui, A. Uji, and Y. Kano, *Yakugaku Zasshi* **82**, 1560 (1962); *Chem. Abstr.* **58**, 11685 (1963).

158. M. P. Cava, K. V. Rao, B. Douglas, and J. A. Weisbach, *J. Org. Chem.* **33**, 2443 (1968).

159. S. R. Johns and J. A. Lamberton, *Aust. J. Chem.* **19**, 297 (1966).

160. P. G. Gorovoi, A. A. Ibragimov, S. Kh. Maekh, and S. Yu. Yunusov, *Chem. Nat. Cpds.* **11**, 568 (1975).

161. S. Kh. Maekh, S. Yu. Yunusov, and P. G. Gorovoi, *Chem. Nat. Cpds.* **12**, 110 (1976).

162. T. R. Govindachari, B. R. Pai, and G. Shanmugasundaram, *Tetrahedron* **20**, 2895 (1964).

163. Ya. V. Rashkes and M. R. Yagudaev, *Uzbeksk. Khim. Zh.* **7**, 62 (1963); *Chem. Abstr.* **59**, 8270 (1963).

164. A. J. Marsaioli, F. De A. M. Reis, A. F. Magalhaes, and E. A. Ruveda, *Phytochemistry* **18**, 165 (1979).

165. M. J. Vernengo, *Experientia* **17**, 420 (1961).

166. G. A. Iacobucci, *Cienc. Invest.* **7**, 48 (1951).

167. T. R. Govindachari, S. Rajadurai, C. V. Ramandas, and N. Viswanathan, *Chem. Ber.* **93**, 360 (1960); *Chem. Abstr.* **54**, 12182 (1960).

168. M. J. Vernengo, *Experientia* **19**, 294 (1963).

169. S. Goodwin, J. N. Shoolery, and L. F. Johnson, *Proc. Chem. Soc.* **1958**, 306.

170. M. J. Vernengo, *Experientia* **19**, 294 (1963).

171. I. R. C. Bick, J. Harley-Mason, N. Sheppard, and M. J. Vernengo, *J. Chem. Soc.* **1961**, 1896.

172. F. S. Sadritdinov and I. Khamdamov, *Farmakol. Toxsikol.* (Moscow) **38**, 490 (1975); *Biol. Abstr.* **61**, 44769 (1976).

173. A. G. Kurmukov, M. I. Aizikov, Kh. S. Akhmedkhodzhaeva, and M. B. Sultanov, *Rastit. Resur.* **9**, 341 (1973); *Biol. Abstr.* **59**, 16116 (1975).

174. Kh. G. Pulatova, S. Kh. Maekh, Z. F. Ismailov, and S. Yu. Yunusov, *Chem. Nat. Cpds.* **4**, 336 (1968).

175. S. Abduzhabbarova, S. Kh. Maekh, S. Yu. Yunusov, M. R. Yagudaev, and D. Kurbakov, *Chem. Nat. Cpds.* **14**, 400 (1978).

176. R. W. Doskotch, J. D. Phillipson, A. B. Ray, and J. L. Beal, *J. Org. Chem* **36**, 2409 (1971).

177. Z. F. Ismailov, M. V. Telezhenetskaya, and S. Yu. Yunusov, *Chem. Nat. Cpds.* **4**, 117 (1968).

178. R. W. Doskotch, J. D. Phillipson, A. B. Ray, and J. L. Beal, *Chem. Commun.* **1969**, 1083.

179. T. Kametani, K. Takahashi, T. Sugahara, M. Koizumi, and K. Fukumoto, *J. Chem. Soc.* (*C*) **1971**, 1032.

180. T. Kametani, K. Takahashi, and K. Fukumoto, *J. Chem. Soc.* (*C*) **1971**, 3617.

181. S. Mukhamedova, S. Kh. Maekh, and S. Yu. Yunusov, *Chem. Nat. Cpds.* **19**, 375 (1983).

182. M. P. Cava, Y. Watanabe, K. Bessho, and M. J. Mitchell, *Tetrahedron Lett.* **1968**, 2437.

183. M. Kurbanov, Kh. Sh. Khusainova, M. Khodzhimatov, A. E. Vezen, K. Kh. Khaidarov, and V. K. Burichenko, *Dokl. Akad. Nauk. Tadzh. SSR* **18**, 20 (1975); *Chem. Abstr.* **84**, 180440 (1976).

184. M. Kurbanov, Kh. Sh. Khusainova, M. Khodzhimatov, A. E. Vezen, K. Kh. Khaidarov, and V. K. Burichenko, *Otkrytiya, Izobret., Prom. Obraztsy, Tovarnye Znaki* **54**, 91 (1977); *Chem. Abstr.* **87**, 33690 (1977).

185. S. Kh. Maekh, V. G. Khodzhaev, and S. Yu. Yunusov, *Chem. Nat. Cpds.* **7**, 363 (1971).

186. M. Shamma and R. L. Castenson, *The Alkaloids*, Vol. 14, R. H. F. Manske, Ed., Academic, New York, 1973, p. 225.

187. M. Shamma, *The Isoquinoline Alkaloids*, Academic, New York, 1972, p. 245.

188. M. Shamma and J. L. Moniot, *Isoquinoline Alkaloids Research: 1972–1977*, Plenum, New York, 1978, p. 173.

189. I. Ribas, J. Sueiras, and L. Castedo, *Tetrahedron Lett.* **1971**, 3093.

190. A. J. Marsaioli, A. F. Magalhaes, E. A. Ruveda, and F. De A. M. Reis, *Phytochemistry* **19**, 995 (1980).

191. M. A. Buchanan and E. E. Dickey, *J. Org. Chem.* **25**, 1389 (1960).

192. S. Mukhamedova, S. Kh. Maekh, and S. Yu. Yunusov, *Chem. Nat. Cpds.* **19**, 376 (1983).

193. X. A. Dominguez, R. Franco O., G. Cano C., and S. Garcia y Socorro Tamez R., *Rev. Latinoam. Quim.* **12**, 61 (1981).

194. V. A. Snieckus, *The Alkaloids*, Vol. 4, J. E. Saxton, Ed., The Chemical Society, London, 1974, p. 128.

195. N. J. McCorkindale, *The Alkaloids*, Vol. 6, M. F. Grundon, Ed., The Chemical Society, London, 1976, p. 110.

196. N. J. McCorkindale, *The Alkaloids*, Vol. 7, M. F. Grundon, Ed., The Chemical Society, London, 1977, p. 92.

197. K. W. Bentley, *The Alkaloids*, Vol. 8, M. F. Grundon, Ed., The Chemical Society, London, 1978, p. 87.

198. K. W. Bentley, *The Alkaloids*, Vol. 9, M. F. Grundon, Ed., The Chemical Society, London, 1979, p. 89.

199. K. W. Bentley, *The Alkaloids*, Vol. 10, M. F. Grundon, Ed., The Royal Society of Chemistry, London, 1987, p. 84.

200. K. W. Bentley, *The Alkaloids*, Vol. 11, M. F. Grundon, Ed., The Royal Society of Chemistry, London, 1981, p. 78.

201. K. W. Bentley, *The Alkaloids*, Vol. 12, M. F. Grundon, Ed., The Royal Society of Chemistry, London, 1982, p. 94.

202. A. Burger, *The Alkaloids*, Vol. IV, R. H. F. Manske, Ed., Academic, New York, 1954, p. 29.

203. V. Deulofeu, J. Comin, and M. J. Vernengo, *The Alkaloids*, Vol. 10, R. H. F. Manske, Ed., Academic, New York, 1967, p. 402.

204. M. Shamma, *The Isoquinoline Alkaloids*, Academic, New York, 1972, p. 44.

205. M. Shamma and J. L. Moniot, *Isoquinoline Aklaloids Research: 1972–1977*, Plenum, New York, 1978, p. 27.

206. J. Cymerman Craig and S. K. Roy, *Tetrahedron* **21**, 401 (1965).

207. S. M. Kupchan and A. Yoshitake, *J. Org. Chem.* **34**, 1062 (1969).

208. V. Preininger, *The Alkaloids*, Vol. 15, R. H. F. Manske, Ed., Academic, New York, 1975, p. 207.

209. S. M. Albonico, J. Comin, A. M. Kuck, E. Sanchez, P. M. Scopes, R. J. Swan, and M. J. Vernego, *J. Chem. Soc. (C)* **1966**, 1340.

210. H. Ishi, H. Ohida, and J. Haginiwa, *Yakugaku Zasshi* **92**, 118 (1972); *Chem. Abstr.* **77**, 16530 (1972).

211. I. Kimura, M. Kimura, M. Yoshizaki, K. Yanada, S. Kadota, and T. Kikuchi, *Planta Med.* **48**, 43 (1983).

212. M. Shamma and J. L. Moniot, *J. Pharm. Sci.* **61**, 295 (1972).

213. K. W. Gopinath, T. R. Govindachari, B. R. Pai, and N. Viswanathan, *Chem. Ber.* **92**, 776 (1959).

214. M. Shamma and D. Hindenlang, *Carbon-13 NMR Shift Assignments of Amines and Alkaloids*, Plenum, New York, 1979, p. 119.

215. W.-T. Liao, J. L. Beal, W.-N. Wu, and R. W. Doskotch, *J. Nat. Prod.* **41**, 257 (1978).

216. T. Kametani, Y. Ohta, M. Takemura, M. Ihara, and F. Fukumoto, *Heterocycles* **6**, 415 (1977).

217. K. H. C. Baser, *J. Nat. Prod.* **45**, 704 (1982).

218. E. Fujita and T. Tomimatsu, *Yakugaku Zasshi* **79**, 1082 (1959); *Chem. Abstr.* **54**, 4643 (1960).

219. S. Kubota, T. Masui, E. Fujita, and S. M. Kupchan, *Tetrahedron Lett.* **1965,** 3599.

220. S. Kubota, T. Masui, E. Fujita, and S. M. Kupchan, *J. Org. Chem.* **31,** 516 (1966).

221. A. J. Birch, A. H. Jackson, and P. V. R. Shannon, *J. Chem. Soc. Perkin I,* **1974,** 2190.

222. Y. Ishida, T. Hiyama, and T. Tomimatsu, *Tokushima Daigaku Yakugaku Kenkyu Nempo* **19,** 17 (1970); *Chem. Abstr.* **77,** 28799 (1972).

223. S. K. Chattopadhyay, A. B. Ray, D. J. Slatkin, and P. L. Schiff, Jr., *Phytochemistry* **22,** 2607 (1983).

224. S. M. Albonico, A. M. Kuck, and V. Deulofeu, *Chem. Ind.* **1964,** 1580.

225. D. G. Patel, A. Tye, P. Patil, A. M. Burkman, and J. L. Beal, *Lloydia* **33,** 36 (1970).

226. M. Shamma, M. G. Kelley, and Sr. M. A. Podczasy, *Tetrahedron Lett.* **1969,** 4951.

227. J. C. Craig, M. Martin-Smith, S. K. Roy, and J. B. Stenlake, *Tetrahedron* **22,** 1335 (1966).

228. A. R. Battersby, I. R. C. Bick, W. Klyne, J. P. Jennings, P. M. Scopes, and M. J. Vernengo, *J. Chem. Soc.* **1965,** 2239.

229. T. R. Govindachari, N. Viswanathan, B. R. Pai, and S. Narayanaswami, *Indian J. Chem.* **6,** 4 (1968).

230. N. M. Mollov, P. P. Panov, L. N. Thuan, and L. N. Panova, *Compt. Rend. Acad. Bulg. Sci. (Dokl. Bolg. Akad. Nauk)* **23,** 1243 (1970).

231. S. M. Kupchan, T.-H. Yang, G. S. Vasilikiotis, M. H. Barnes, and M. L. King, *J. Org. Chem.* **34,** 3884 (1969).

232. M. Shamma and Sr. M. A. Podczasy, *Tetrahedron* **27,** 727 (1971).

233. R. A. Hahn, M. G. Kelly, M. Shamma, and J. L. Beal, *Arch. Int. Pharmacodyn.* **198,** 392 (1972).

234. W.-N. Wu, J. L. Beal, and R. W. Doskotch, *J. Nat. Prod.* **43,** 143 (1980).

235. H.-Y. Cheng and R. W. Doskotch, *J. Nat. Prod.* **43,** 151 (1980).

236. I. R. C. Bick, T. Sevenet, W. Sinchai, B. W. Skelton, and A. H. White, *Aust. J. Chem.* **34,** 195 (1981).

236a. S. Al-Khalil and P. L. Schiff, Jr., *J. Nat. Prod.* **48,** 989 (1985).

237. B. Gözler, M. S. Lantz, and M. Shamma, *J. Nat. Prod.* **46,** 293 (1983).

238. M. Shamma, *The Isoquinoline Alkaloids,* Academic, New York, 1972, p. 96.

239. M. Shamma and J. L. Moniot, *Isoquinoline Alkaloids Research: 1972–1977,* Plenum, New York, 1978, p. 61.

240. R. H. F. Manske, *The Alkaloids,* Vol. 10, R. H. F. Manske, Ed., Academic, New York, 1968, p. 467.

241. F. Santavy, *The Alkaloids,* Vol. 12, R. H. F. Manske, Ed., Academic, New York, 1970, p. 333.

242. F. Santavy, *The Alkaloids,* Vol. 17, R. H. F. Manske and R. G. A. Rodrigo, Eds., Academic, New York, 1979, p. 385.

243. M. Shamma, A. S. Rothenberg, S. S. Salgar, and G. S. Jayatilake, *J. Nat. Prod.* **39,** 395 (1976).

244. M. Shamma, J. L. Moniot, W. K. Chan, and K. Nakanishi, *Tetrahedron Lett.* **1971,** 3425.

245. L. Reti, *The Alkaloids,* Vol. 4, R. H. F. Manske, Ed., Academic, New York, 1954, p. 7.

246. R. H. F. Manske, *The Alkaloids,* Vol. 7, R. H. F. Manske and H. L. Holmes, Eds., Academic, New York, 1960, p. 423.

247. M. Shamma, *The Isoquinoline Alkaloids,* Academic, New York, 1972, p. 1.

248. M. Shamma and J. L. Moniot, *Isoquinoline Alkaloids Research: 1972–1977,* Plenum, New York, 1978, p. 1.

249. I. R. Iee and M. M. Lee, *Kor. J. Pharmacog.* **13**, 132 (1982).

250. R. H. F. Manske, *Can. J. Res.* **B15**, 159 (1937).

251. T.-H. Yang and C.-M. Chen, *J. Chinese Chem. Soc.* (*Taiwan*) **17**, 54 (1970); *Chem. Abstr.* **73**, 99072 (1970).

252. D. K. Dalling and D. M. Grant, *J. Am. Chem. Soc.* **89**, 6612 (1967).

253. R. Mata, C.-J Chang, and J. L. McLaughlin, *Phytochemistry* **22**, 1263 (1983).

254. J. L. Moniot and M. Shamma, *Heterocycles* **9**, 145 (1978).

255. B. D. Crane and M. Shamma, *J. Nat. Prod.* **45**, 377 (1982).

256. M. Shamma, *The Isoquinoline Alkaloids*, Academic, New York, 1972, p. 90.

257. M. Shamma and J. L. Moniot, *Isoquinoline Alkaloids Research: 1972–1977*, Plenum, New York, 1978, p. 57.

258. D. W. Hughes and D. B. MacLean, *The Alkaloids*, Vol. 17, R. H. F. Manske and R. G. Rodrigo, Eds., Academic, New York, 1981, p. 222.

259. M. P. Cava, K. Bessho, B. Douglas, S. Markey, and J. A. Weisbach, *Tetrahedron Lett.* **1966**, 4279.

260. M. P. Cava and K. T. Buck, *Tetrahedron* **25**, 2795 (1969).

261. S. M. Kupchan, S. Kubota, E. Fujita, S. Kobayashi, J. H. Block, and S. A. Telang, *J. Am. Chem. Soc.* **88**, 4212 (1966).

262. N. M. Mollov, H. B. Dutschewska, K. Siljanovska, and S. Stojcev, *Compt. Rend. Acad. Bulg. Sci.* **21**, 605 (1968).

263. N. M. Mollov and H. B. Dutschewska, *Tetrahedron Lett.* **1969**, 1951.

264. R. W. Doskotch, P. L. Schiff, Jr., and J. L. Beal, *Tetrahedron* **25**, 469 (1969).

265. I. R. C. Bick, T. Sevenet, W. Sinchai, B. W. Skeleton, and A. H. White, *Aust. J. Chem.* **34**, 195 (1981).

266. M. Shamma and D. M. Hindenlang, *Carbon-13 NMR Shift Assignment of Amines and Alkaloids*, Plenum, New York, 1979, p. 116.

267. V. H. Belgaonkar and R. N. Usgaonkar, *J. Het. Chem.* **15**, 257 (1978).

268. S. K. Chattopadhyay, A. B. Ray, D. J. Slatkin, J. E. Knapp, and P. L. Schiff, Jr., *J. Nat. Prod.* **44**, 45 (1981).

269. Kh. S. Umarov, Z. F. Ismailov, and S. Yu. Yunusov, *Chem. Nat. Cpds.* **6**, 452 (1970).

270. N. M. Mollov and L. N. Thuan, *Compt. Rend. Acad. Bulg. Sci.* **24**, 601 (1971).

271. F. Eloy and A. Deryckere, *Helv. Chim. Acta* **52**, 1755 (1969).

272. H. B. Dutschewska, A. V. Georgieva, N. M. Mollov, P. P. Panov, and N. K. Kotsev, *Compt. Rend. Acad. Bulg. Sci.* **24**, 467 (1971).

273. K. L. Stuart, *Chem. Rev.* **71**, 47 (1971).

274. H. L. Holmes, *The Alkaloids*, Vol. 2, R. H. F. Manske and H. L. Holmes, Eds., Academic, New York, 1952, p. 1.

275. H. L. Holmes and G. Stork, *The Alkaloids*, Vol. 2, R. H. F. Manske and H. L. Holmes, Eds., Academic, New York, 1952, p. 161.

276. G. Stork, *The Alkaloids*, Vol. 6, R. H. F. Manske, Ed., Academic, New York, 1960, p. 219.

277. T. Kametani, M. Ihara, and T. Honda, *Chem. Commun.* **1969**, 1301.

278. R. Ansa-Asamoah and G. A. Starmer, *Planta Med.* **1984**, 69.

279. H. Wagner, L.-Z. Lin, and O. Seligmann, *Planta Med.* **50**, 14 (1984).

280. M. Shamma, *The Isoquinoline Alkaloids*, Academic, New York, 1972, p. 96.

281. M. Shamma and J. L. Moniot, *Isoquinoline Alkaloids Research: 1972–1977*, Plenum, New York, 1978, p. 61.

282. S. M. Kupchan, T.-H. Yang, M. L. King, and R. T. Borchardt, *J. Org. Chem.* **33**, 1052 (1968).

283. T. O. Soine and O. Gisvold, *J. Am. Pharm. Assoc., Sci. Ed.* **33**, 185 (1944).

284. K. C. Rice, W. C. Ripka, J. Reden, and A. Brossi, *J. Org. Chem.* **45**, 601 (1980).

285. A. A. Genenah, T. O. Soine, and N. A. Shoath, *J. Pharm. Sci.* **64**, 62 (1975).

286. L. B. Kier and T. O. Soine, *J. Am. Pharm. Assoc., Sci. Ed.* **49**, 187 (1960).

286a. J. S. Glasby, *Encyclopedia of the Alkaloids*, Vol. 2, Plenum, New York, 1975, p. 1186.

287. R. H. F. Manske and K. H. Shin, *Can. J. Chem.* **44**, 1259 (1966).

288. F. R. Stermitz and K. D. McMurtrey, *J. Org. Chem.* **34**, 555 (1969).

289. J. Slavík and L. Slavíková, *Coll. Czech. Chem. Commun.* **28**, 1728 (1963).

290. M. Shamma, *The Isoquinoline Alkaloids*, Academic, New York, 1972, p. 260.

291. M. Shamma and J. L. Moniot, *Isoquinoline Alkaloids Research: 1972–1977*, Plenum, New York, 1978, p. 179.

292. J. L. Cashaw, S. Ruchirawat, Y. Nimit, and V. E. Davis, *Biomed. Mass Spec.* **11**, 63 (1984).

293. Kh. S. Umarov, Z. F. Ismailov, and S. Yu. Yunusov, *Chem. Nat. Cpds.* **9**, 660 (1973).

294. Z. F. Ismailov, S. Kh. Maekh, and S. Yu. Yunusov, *DAN UzSSR* No. 7, 32 (1959).

295. S. Yu. Yunusov and M. F. Telezhenetskaya, *DAN UzSSR* No. 5, 22 (1963).

296. S. T. Kholodkov, K. L. Lutfulin, and Z. F. Ismailov, *Dokl. Akad. Nauk Uz. SSR* **22**, 39 (1965); *Chem. Abstr.* **63**, 16670 (1965).

297. V. G. Khodzhaev, S. Kh. Maekh, and S. Yu. Yunusov, *Chem. Nat. Cpds.* **11**, 421 (1975).

298. E. Fujita and T. Tomimatsu, *Yakugaku Zasshi* **79**, 1252 (1959); *Chem. Abstr.* **54**, 4643 (1960).

299. A. Sidjimov and V. S. Christov, *J. Nat. Prod.* **47**, 387 (1984).

300. N. M. Mollov, L. N. Thuan, and P. P. Panov, *Compt. Rend. Acad. Bulg. Sci.* **24**, 1047 (1971).

301. W.-N. Wu, J. L. Beal, R.-P. Leu, and R. W. Doskotch, *Lloydia* **40**, 384 (1977).

302. R. L. Lyon, M. S. thesis, The Ohio State University, 1969, 102 pp.

303. R. H. F. Manske and W. R. Ashford, *The Alkaloids*, Vol. 4, R. H. F. Manske, Ed., Academic, New York, 1954, p. 77.

304. P. W. Jeffs, *The Alkaloids*, Vol. 9, R. H. F. Manske, Ed., Academic, New York, 1967, p. 41.

305. M. Shamma, M. J. Hillman, and C. D. Jones, *Chem. Rev.* **69**, 779 (1969).

306. M. P. Cava, T. A. Reed, and J. L. Beal, *Lloydia* **28**, 73 (1965).

307. K. Jewers, A. J. Manchanada, and P. N. Jenkins, *J. Chem. Soc. Perkin II*, **1972**, 1393.

308. M. Shamma and D. Hindenlang, *Carbon-13 NMR Shift Assignments of Amines and Alkaloids*, Plenum, New York, 1979, p. 134.

309. G. Habermehl, J. Schunck, and G. Schaden, *Annalen* **742**, 138 (1970).

310. N. M. Mollov, Kh. Duchevska, Kh. Kiryakov, B. Pjuskjulev, U. Georgiev, D. Jordanov, and P. Panov, *Compt. Rend. Acad Bulg. Sci.* **18**, 849 (1965); *Chem. Abstr.* **64**, 3957 (1966).

311. Y. Kondo, *Heterocycles* **4**, 197 (1976).

312. S. Uchizumi, *Nippon Yakurigaku Zasshi* **53**, 63 (1957); *Chem. Abstr.* **52**, 4016 (1958).

313. S. K. Kulkarni, P. C. Dandiya, and N. L. Varandani, *Jap. J. Pharmacol.* **22**, 11 (1972).

314. A. D. Turova, A. I. Leskov, and V. I. Bichevma, *Lekarstv. Sredstva iz Rast.* 303 (1962); *Chem. Abstr.* **58**, 2763 (1963).

315. S. M. Shanbhag, H. J. Kulkarni, and B. B. Gaitonde, *Jap. J. Pharmacol.* **20**, 482 (1970).

316. A. D. Turova, M. N. Konovalov, and A. I. Leskov, *Med. Prom. SSSR* **18**, 59 (1964); *Chem. Abstr.* **61**, 15242 (1964).

317. B. A. Vartazaryan and E. E. Koltochnik, *Sb. Nauchn. Tr. Vladirostoksk. Med. Inst.* **2**, 113 (1964); *Chem. Abstr.* **62**, 15304 (1965).

318. J. Maj, Z. Borzecki, and D. Chibowksi, *Diss. Pharm.* **17**, 437 (1965).

319. S. Suzuki, *Tohoku J. Exp. Med.* **36**, 134 (1939); *Chem. Abstr.* **33**, 9452 (1959).

320. C. S. Jang, *J. Pharmacol. Exp. Ther.* **71**, 178 (1941).

321. F. Honda, M. Oka, A. Akashi, and Y. Nishino, *Yakugaku Kenkyu* **32**, 836 (1960); *Chem. Abstr.* **55**, 10695 (1961).

322. T. Furuya, *Nippon Yakurigaku Zasshi* **55**, 1152 (1959); *Chem. Abstr.* **55**, 791 (1961).

323. T. Furuya, *Nippon Yakurigaku Zasshi* **55**, 1162 (1959); *Chem. Abstr.* **55**, 791 (1961).

324. M. Sabir and N. K. Bhide, *Indian J. Physiol. Pharmacol.* **15**, 111 (1971).

325. F. T. Schein and C. Hanna, *Arch. Int. Pharmacodyn.* **124**, 317 (1960).

326. M. B. Bhide, S. R. Chavan, and N. K. Dutta, *Indian J. Med. Res.* **57**, 2128 (1969).

327. G. L. Nardi and J. H. Seipel, *Surg. Forum Proc. 41st Congr.* 381 (1955); *Chem. Abstr.* **52**, 1490 (1958).

328. V. B. Schatz, B. C. O'Brien, W. M. Chadduck, A. M. Kanter, A. Burger, and W. R. Sandusky, *J. Med. Pharm. Chem.* **2**, 425 (1960).

329. H. Ogakurayama, I. Aramori, T. Murata, N. Sato, and S. Shibamoto, *Yakkyoku* **7**, 1089 (1956); *Chem. Abstr.* **51**, 7652 (1957).

330. H. Schmitz, *Z. Krebsforsch.* **57**, 405 (1951).

331. F. Honda, M. Oka, A. Akashi, and Y. Nisheno, *Yakugaku Kenkyu* **32**, 830 (1960); *Chem. Abstr.* **55**, 10695 (1951).

332. V. V. Berezhinskaya, E. E. Aleshinskaya, and E. A. Trutneva, *Farmakol. Toksikol.* (Moscow) **31**, 296 (1968); *Chem. Abstr.* **69**, 50676 (1968).

333. H. Schmitz, *Z. Krebsforsch.* **57**, 137 (1951).

334. I. Hilwig and H. Schmitz, *Naturwiss.* **38**, 336 (1951); *Chem. Abstr.* **46**, 3166 (1952).

335. K.-T. Liu, Y.-C. Chin, and H.-P. Lui, *Yao Hsueh Pao* **13**, 356 (1966); *Chem. Abstr.* **65**, 17532 (1966).

336. N. Tanaka, *Jap. J. Exp. Med.* **22**, 87 (1952).

337. S. Hasegawa and N. Tanaka, *Jap. J. Exp. Med.* **22**, 77 (1952).

338. J. Kudo, *Folia Pharmacol. J.* **49**, 255 (1953); *Chem. Abstr.* **48**, 7192 (1954).

339. J. Ulrichova, D. Walterova, V. Preininger, J. Slavik, J. Lenfeld, M. Cushman, and V. Simanek, *Planta Med.* **48**, 111 (1983).

340. M. H. Akhter, M. Sabir, and N. K. Bhide, *Indian J. Med. Res.* **65**, 133 (1977).

341. M. Sabir, M. H. Akhter, and N. K. Bhide, *Indian J. Physiol. Pharmacol.* **22**, 9 (1978); *Chem. Abstr.* **89**, 173608 (1978).

342. Z. Kowalewski, W. Kedzia, and I. Mirska, *Arch. Immunol. Therap. Exp.* **20**, 353 (1972).

343. T. Sawada, J. Yamahara, K. Goto, and M. Yamamura, *Shoyakugaku Zasshi* **25**, 74 (1971); *Chem. Abstr.* **77**, 122005 (1972).

344. K. C. Singhal, *Indian J. Exp. Biol.* **14**, 345 (1976).

345. A. I. Khadzhiolov, E. Tsvetkova, I. Vulkov, L. Gitsov, I. Cholakova, M. Marinov, and I. Tsvetkov, *Arch. Union Med. Balk.* **12**, 53 (1974); *Chem. Abstr.* **82**, 53611 (1975).

346. A. I. Khadzhiolov, E. Tsvetkova, E. M. Enchev, and K. G. Tsanev, *Dokl. Bolg. Akad. Nauk* **27**, 1301 (1974); *Chem. Abstr.* **82**, 71314 (1975).

347. S. K. Vashistha and S. Siddiqui, *J. Indian Chem. Soc.* **18**, 641 (1941); *Chem. Abstr.* **36**, 5478 (1942).

348. R. Chatterjee, M. P. Guha, and A. Chatterjee, *J. Indian Chem. Soc.* **29**, 371 (1952); *Chem. Abstr.* **47**, 11663 (1953).

349. L.-Z. Lin, Z.-Y. Fan, C.-Q. Song, C.-F. Du, and R.-S. Xu, *Hua Hsueh Hsueh Pao* **39**, 159 (1981); *Chem. Abstr.* **95**, 76882 (1981).

350. A. Gheorghiu, E. Ionescu-Matiu, and V. Calcandi, *Studii Cercetari Biochim.* **8**, 193 (1965); *Chem. Abstr.* **63**, 9745 (1965).

351. A. Gheorghiu, E. Ionescu-Matiu, and V. Calcandi, *Planta Med.* **15**, 179 (1967); *Chem. Abstr.* **67**, 76333 (1967).

352. B. K. Wali, V. Paul, and K. L. Handa, *Indian J. Pharm.* **26**, 69 (1964).

353. L. A. Mitscher, W.-N. Wu, R. W. Doskotch, and J. L. Beal, *J. Chem. Soc. D* **1971**, 589.

354. Z. Kowalewski, I. Frencel, and J. Schumacher, *Acta Polon. Pharm.* **23**, 305 (1966); *Chem. Abstr.* **66**, 17030 (1967).

355. H. J. Chi, *Yakhak Hoeji* **9**, 37 (1965); *Chem. Abstr.* **65**, 12563 (1966).

356. B. Zhu and F. A. Ahrens, *Am. J. Vet. Res.* **43**, 1594 (1982); *Chem. Abstr.* **97**, 192920 (1982).

357. A. K. Ghosh, M. M. Rakshit, and D. K. Ghosh, *Indian J. Med. Res.* 407 (1983); *Chem. Abstr.* **99**, 205663 (1983).

358. D. Hou, L. Li, Q. Wang, and G. Chen, *Nanjing Yaoxneyuan Xuebao* 30 (1983); *Chem. Abstr.* **99**, 115564 (1983).

359. R. J. Greco, A. M. Lefer, and P. R. Maroko, *IRCS Med. Sci: Libr. Compend.* **11**, 570 (1983); *Chem. Abstr.* **99**, 151843 (1983).

360. M. Sethi, *J. Pharm. Sci.* **72**, 538 (1983).

361. T. Y. Owen, S. Y. Wang, S. Y. Chang, F. L. Lu, C. L. Yang, and B. Hsu, *K'O Hsueh Tung Pao* **21**, 285 (1976); *Chem. Abstr.* **86**, 5660 (1976).

362. M. Cushman, F. W. Dekow, and L. B. Jacobsen, *J. Med. Chem.* **22**, 331 (1979).

363. K. I. Kuchkova, A. A. Semenov, and M. O. Broitman, *Chem. Nat. Cpds.* **12**, 751 (1976).

364. C. Tani, N. Nagakura, S. Saeki, and M. J. Kao, *Planta Med.* **41**, 403 (1981).

365. C.-H. Chen and T.-M Chen, *J. Taiwan Pharm. Assoc.* **35**, 1 (1983).

366. M. Shamma and B. S. Dudock, *Tetrahedron Lett.* **1965**, 3825.

367. H. F. Andrew and C. K. Bradsher, *Tetrahedron Lett.* **1966**, 3069.

368. D. W. Spiggle, M.S. thesis, The Ohio State University, 1960.

369. M. Shamma and D. Hindenlang, *Carbon-13 NMR Shift Assignments of Amines and Alkaloids*, Plenum, New York, 1979, p. 153.

370. H. Kondo and M. Tomita, *J. Pharm. Soc. Japan* **50**, 309 (1930); *Chem. Abstr.* **24**, 3512 (1930).

371. R. Chatterjee and A. Banerjee, *J. Indian Chem. Soc.* **30**, 705 (1953).

372. R. Chatterjee and P. C. Maiti, *J. Indian Chem. Soc.* **32**, 609 (1955).

373. M.-C. Chen and C.-Y. Chi, *Yao Hsueh Hsueh Pao* **12**, 185 (1965); *Chem. Abstr.* **63**, 3492 (1965).

374. M. M. Nijland, *Pharm. Weekblad.* **98**, 301 (1963).

375. M. H. AbuZarga and M. Shamma, *Tetrahedron Lett.* **21**, 3739 (1980).

376. G. Habermehl, J. Schunk, and G. Schaden, *Ann. Chem.* **742**, 138 (1970).

377. H. Suguna and B. R. Pai, *Coll. Czech. Chem. Commun.* **41**, 1219 (1976).

378. G. A. Miana, *Phytochemistry* **12**, 1822 (1973).

379. H. A. D. Jowett and F. L. Pyman, *J. Chem. Soc.* **103**, 290 (1913).

380. K. I. Kuchkuva, I. V. Terentèva, and G. V. Lazuŕevskii, *Chem. Nat. Cpds.* **3**, 118 (1967).

381. K.-ichi Yoshikawa, I. Morishima, J.-ichi Kunitomo, M. Ju-Ichi, and Y. Yoshida, *Chem. Lett.* **1975**, 961.

382. J. M. Calderwood and F. Fish, *Chem. Ind.* **1966**, 237.

383. M. Ohta, H. Tani, and S. Morozumi, *Chem. Pharm. Bull.* **12**, 1072 (1969).

384. F. C. Ohiri, R. Verpoorte, and A. Baerheim Svendsen, *Planta Med.* **49**, 162 (1983).

385. C.-Y. Chen and D. B. MacLean, *Can. J. Chem.* **46**, 2501 (1968).

386. D. W. Hughes, H. L. Holland, and D. B. MacLean, *Can. J. Chem.* **54**, 2252 (1976).

387. W. J. Richter and E. Brochmann-Hanssen, *Helv. Chim. Acta* **58**, 203 (1975).

388. G. Snatzke, J. Hrbek, Jr., L. Hruban, A. Horeau, and F. Santavy, *Tetrahedron* **26**, 5013 (1970).

389. T. Kametani, *The Chemistry of the Isoquinoline Alkaloids*, Hirokawa, Tokyo, and Elsevier, Amsterdam, 1969, p. 114.

390. R. H. F. Manske and W. R. Ashford, *The Alkaloids*, Vol. 4, R. H. F. Manske, Ed., Academic, New York, 1954, p. 91.

391. K. C. Chin, H.-Y. Chu, H.-T. Tang, and P. Hsu, *Sheng Li Hsueh Pao* **25**, 182 (1962); *Chem. Abstr.* **59**, 13249 (1963).

392. K. C. Chin, Y.-N. Wang, T.-Y. Pao, and P. Hsu, *Sheng Li Hsueh Pao* **28**, 72 (1965); *Chem. Abstr.* **63**, 12151 (1965).

393. F. Sadritdinov, *Med. Zh. Uzbekistana* 48 (1966); *Chem. Abstr.* **65**, 14306 (1966).

394. B. L. Danilevskii, N. T. Tulyaganov, and F. S. Sadritdinov, *Dokl. Akad. Nauk Uzb. SSR* **29**, 37 (1972); *Chem. Abstr.* **78**, 38001 (1973).

395. B. L. Danilevskii, N. Tulyaganov, and F. S. Sadritdinov, *Farmakol. Alkaloidov Ikh Proizvod.* 136 (1972); *Chem. Abstr.* **80**, 103861 (1974).

396. *Jpn. Kokai Tokkyo Koho* 80, 143, 914 (1980); *Chem. Abstr.* **94**, 96320 (1981).

397. F. S. Sadritdinov and Zh. Rezhepov, *Dokl. Akad. Nauk Uzb. SSR* 361 (1982); *Chem. Abstr.* **98**, 27772 (1983).

398. F. S. Sadritdinov and Zh. Rezhepov, *Dokl Akad. Nauk Uzb. SSR* 34 (1982); *Chem. Abstr.* **98**, 83240 (1983).

399. G. Jin, K. C. Kin, X. Wang, and J. Xu, *Shengli Xuebao* **35**, 112 (1983); *Chem. Abstr.* **99**, 33006 (1983).

400. X. Wang, G. Jin, K. C. Kin, L. Yu, and J. Li, *Zhongguo Yaoli Xuebao* **3**, 73 (1982); *Chem. Abstr.* **97**, 85241 (1982).

401. M. Matsui, T. Tomimatsu, and E. Fujita, *Yakugaku Zasshi* **82**, 308 (1962); *Chem. Abstr.* **57**, 3563 (1962).

402. R. H. F. Manske and W. R. Ashford, *The Alkaloids*, Vol. 4, R. H. F. Manske, Ed., Academic, New York, 1954, p. 96.

403. Y. Kitabatake, K. Ito, and M. Tajima, *Yakugaku Zasshi* **84**, 73 (1964); *Chem. Abstr.* **61**, 7575 (1964).

404. B. Ringdahl, R. P. K. Chan, J. Cymerman Craig, and R. H. F. Manske, *J. Nat. Prod.* **44**, 75 (1981).

405. H. Guinaudeau and M. Shamma, *J. Nat. Prod.* **45**, 237 (1982).

406. R. H. F. Manske, *The Alkaloids*, Vol. 4, R. H. F. Manske and H. L. Holmes, Eds., Academic, New York, 1954, p. 147.

407. M. Shamma, *The Isoquinoline Alkaloids*, Academic, New York, 1972, p. 269.

408. M. Shamma and J. L. Moniot, *Isoquinoline Alkaloids Research: 1972–1977*, Plenum, New York, 1978, p. 209.

409. M. Shamma, *The Isoquinoline Alkaloids*, Academic, New York, 1972, p. 345.

410. M. Shamma and J. L. Moniot, *Isoquinoline Alkaloids Research: 1972–1977*, Plenum, New York, 1978, p. 299.

411. K. Iwasa, M. Sugiura, and N. Takao, *J. Org. Chem.* **47**, 4275 (1982).

412. S. F. Hussain, B. Gozler, V. Fajardo, A. J. Freyer, and M. Shamma, *J. Nat. Prod.* **46**, 251 (1983).

413. L. Dolejs, V. Hanus, and J. Slavik, *Coll. Czech. Chem. Commun.* **29**, 2479 (1964).

414. E. L. McCawley, *The Alkaloids*, Vol. 5, R. H. F. Manske, Ed., Academic, New York, 1955, p. 79.

415. A. C. Taquini, *Am. Heart J.* **33**, 719 (1947).

416. D. Scherf, A. M. Silver, and L. D. Weinberg, *Ann. Int. Med.* **30**, 100 (1949).

417. P. I. Sizov and V. S. Yurasov, *Uch. Zap. Petrozavodsk. Gos. Univ.* **17**, 52 (1969); *Chem. Abstr.* **75**, 47154 (1971).

418. Z. S. Akbarov, Kh. U. Aliev, and M. B. Sultanov, *Farmakol. Prir. Veschestu* 11 (1978); *Chem. Abstr.* **91**, 32777 (1979).

419. Z. S. Akbarov, Kh. U. Aliev, and M. B. Sultanov, *Dokl. Akad. Nauk Uzbek. S.S.R.* **29**, 38 (1972); *Chem. Abstr.* **78**, 38003 (1973).

420. Kh. U. Aliev and M. B. Akbarov, *Farmacol. Alkaloidov Ikh Proizvod.* 133 (1972); *Chem. Abstr.* **80**, 103860 (1974).

421. P. I. Sizov, *Tr. Smolensk Med. Inst.* **26**, 288 (1968); *Biol. Abstr.* **51**, 32104 (1970).

422. P. I. Sizov, *Zdravookhr. Beloruss.* **15**, 44 (1969); *Chem. Abstr.* **72**, 119957 (1970).

423. P. I. Sizov, *Farmakol. Toksikol.* (Moscow) **33**, 40 (1970); *Chem. Abstr.* **72**, 119844 (1970).

424. P. I. Sizov, *Zdravookhr. Beloruss.* **16**, 17 (1970); *Chem. Abstr.* **73**, 108122 (1970).

425. V. S. Yasnetsov and P. I. Sizov, *Farmakol. Toksikol.* **35**, 201 (1972); *Biol. Abstr.* **55**, 10269 (1973).

426. N. S. Baksheev, I. N. Medvedeva, P. S. Bernatskii, G. T. Khmyz, N. R. Kokulenko, and O. F. Fortuna, *Ref. Zh. Otd. Vyp. Farmakol. Khimioter. Sredstva. Toksikol.* **54**, 321 (1967); *Biol. Abstr.* **48**, 107123 (1967).

427. N. A. Pakhareva and G. V. Lazur'evskii, *Uchenye Zapiski Vologodsk. Gosudarst. Pedagog. Inst.* **24**, 309 (1959); *Chem. Abstr.* **55**, 27394 (1961).

428. K. I. Kuchkova, G. V. Lazur'evskii, and I. V. Terent'eva, *Izv. Akad. Nauk Moldavsk. SSR* 98 (1962); *Chem. Abstr.* **62**, 9459 (1965).

429. K. I. Kuckkova and G. V. Lazur'evskii, *Izv. Akad. Nauk Mold. SSR, Ser. Khim. Biol.* 43 (1965); *Chem. Abstr.* **66**, 95252 (1967).

430. T. T. Nakashima and G. E. Maciel, *Org. Magn. Resonance* **5**, 9 (1973).

431. A. K. Reynolds, *The Alkaloids*, Vol. 5, R. H. F. Manske and H. L. Holmes, Eds., Academic, New York, 1955, p. 163.

432. Z. F. Ismailov, A. U. Rakhmatkariev, and S. Yu. Yunusov, *DAN UzSSR* 34 (1959); see references in ref. 133.

433. Kh. U. Aliev, *Farmakol. Alkaloidov Ikh Proizvod.* **1972**, 126; *Chem. Abstr.* **80**, 103859 (1974).

434. V. N. Burtsev, E. N. Dormidontov, and U. A. Salayev, *Kardiologiya* **18**, 76 (1978); M. F. Grundon, *The Alkaloids*, Vol. 10, The Royal Society of Chemistry, London, 1980, p. 106.

435. Kh. S. Umarov, Z. F. Ismailov, and S. Yu. Yunusov, *Chem. Nat. Cpds.* **4**, 280 (1968).

435a. *Jpn. Kokai Tokkyo JP* **82**, 112, 325; *Chem. Abstr.* **97**, 150739 (1982).

436. H. Guinaudeau, M. LeBoeuf, and A. Cavé, *J. Nat. Prod.* **42**, 133 (1979).

437. H. Guinaudeau, M. LeBoeuf, and A. Cavé, *J. Nat. Prod.* **47**, 565 (1984).

438. M. Curcumelli-Rodostamo, *The Alkaloids*, Vol. 13, R. H. F. Manske, Ed., Academic, New York, 1971, p. 303.

439. M. P. Cava, K. T. Buck, and K. L. Stuart, *The Alkaloids*, Vol. 16, R. H. F. Manske, Ed., Academic, New York, 1977, p. 249.

440. M. Shamma and V. St. Georgiev, *The Alkaloids*, Vol. 16, R. H. F. Manske, Ed., Academic, New York, 1977, p. 319.

441. M. Shamma, *The Isoquinoline Alkaloids*, Academic, New York, 1972, p. 232.

442. M. Shamma and J. L. Moniot, *Isoquinoline Alkaloids Research: 1972–1977*, Plenum, New York, 1978, p. 159.

443. R. W. Doskotch, P. L. Schiff, Jr., and J. L. Beal, *Tetrahedron Lett.* **1968**, 4999.

444. R. W. Doskotch, J. D. Phillipson, A. B. Ray, and J. L. Beal, *Chem. Commun.* **1969**, 1083.

445. R. W. Doskotch, J. D. Phillipson, A. B. Ray, and J. L. Beal, *J. Org. Chem.* **36**, 2409 (1971).

446. W.-T. Liao, J. L. Beal, W.-N. Wu, and R. W. Doskotch, *J. Nat. Prod.* **41**, 271 (1978).

447. H. Guinaudeau, A. J. Freyer, R. D. Minard, M. Shamma, and K. H. C. Baser, *Tetrahedron Lett.* **23**, 2523 (1982).

448. Dzh. Sargazakov, Z. F. Ismailov, and S. Yu. Yunusov, *Dokl. Akad Nauk Uz. SSR* **20**, 28 (1963); *Chem. Abstr.* **59**, 15336 (1963).

449. Z. F. Ismailov and S. Yu. Yunusov, *Chem. Nat. Cpds.* **2**, 35 (1966).

450. Z. F. Ismailov, M. R. Yagudaev, and S. Yu. Yunusov, *Chem. Nat. Cpds.* **4**, 226 (1968).

451. M. P. Cava and K. Wakisaka, *Tetrahedron Lett.* **1972**, 2309.

452. Z. F. Ismailov and S. Yu. Yunusov, *Chem. Nat. Prod.* **6**, 141 (1970).

453. G. P. Moiseeva, Z. F. Ismailov, and S. Yu. Yunusov, *Chem. Nat. Cpds.* **6**, 715 (1970).

454. I. Khamdamov, F. Sadritdinov, and M. B. Sultanov, *Farmakol. Alkaloidov Serdech. Glikozidov* **1971**, 135; *Chem. Abstr.* **77**, 122095 (1972).

455. R. I. Alimov, I. N. Zatorskaya, and T. T. Shakirov, *Uzb. Khim. Zh.* **17**, 47 (1973); *Chem. Abstr.* **80**, 149052 (1974).

456. I. Sh. Zabirov, *Ref. Zh. Otd. Uypusk. Farmakol. Khimioterap. Sredstva Toksikol.* 54.508 (No. 2) (1966); *Biol. Abstr.* **48**, 28876 (1967).

457. N. V. Gorbatenkova, I. Sh. Zabirov, and E. M. Serdyak, *Ref. Zh. Otd. Vypusk. Farmakol. Khimioterap. Sredstva Toksikol.* 54.513 (No. 2) (1966); *Biol. Abstr.* **48**, 28649 (1967).

458. F. Sadritdinov, *Farmakol. Alkaloidov Serdech. Glikozidov* **1971**, 122; *Chem. Abstr.* **78**, 79555 (1973).

459. F. Sadritdinov, *Farmakol. Alkaloidov Ikh Proizvod.* **1972**, 154; *Chem. Abstr.* **80**, 91211 (1974).

460. R. G. Medvedeva and Zh. S. Nuralieva, *Izv. Akad. Nauk Kaz. SSR Ser. Biol.,* **6**, 62 (1968); *Chem. Abstr.* **70**, 112250 (1969).

461. D. A. Rakhimova, E. K. Dobronravova, and T. T. Shakirov, *Chem. Nat. Prod.* **8**, 734 (1972).

462. R. I. Alimov, I. N. Zatorskaya, and T. T. Shakirov, *Chem. Nat. Cpds.* **9**, 537 (1973).

463. Zh. S. Nuralieva and P. K. Alimbaeva, *Fiziol. Aktiv. Soedin. Rast. Kirg.* **1970**, 99; *Chem. Abstr.* **76**, 17765 (1972).

464. I. N. Zatorskaya, R. Alimov, and T. T. Shakirov, *Chem. Nat. Cpds.* **8**, 633 (1972).

465. I. N. Zatorskaya, R. Alimov, and T. T. Shakirov, *Uzb. Khim. Zh.* **17**, 84 (1973); *Chem. Abstr.* **80**, 19461 (1974).

466. R. I. Alimov, I. N. Zatorskaya, and T. T. Shakirov, *Khim. Prir. Soedin.* **10**, 111 (1974); *Chem. Abstr.* **81**, 68420 (1974).

467. L.-Z. Lin, H. Wagner, and O. Seligmann, *Planta Med.* **49**, 55 (1983).

468. H. Guinaudeau, M. Shamma, and K. H. C. Baser, *J. Nat. Prod.* **45**, 505 (1982).

469. M. Shamma and J. L. Moniot, *Tetrahedron Lett.* **1974**, 2291.

470. M. Shamma and J. L. Moniot, *J. Am. Chem. Soc.* **69**, 3338 (1974).

471. H. Wagner, L.-Z. Lin, and O. Seligmann, *Tetrahedron* **40**, 2133 (1984).

472. J. Wu, J. L. Beal, W.-N. Wu, and R. W. Doskotch, *Heterocycles* **6**, 405 (1977).

473. S. M. Kupchan and N. Yokoyama, *J. Am. Chem. Soc.* **85**, 1361 (1963).

474. S. M. Kupchan and N. Yokoyama, *J. Am. Chem. Soc.* **86**, 2177 (1964).

475. M. Tomita, H. Furukawa, S.-T. Lu, and S. M. Kupchan, *Tetrahedron Lett.* **1965**, 4309.

476. M. Tomita, H. Furukawa, S.-T. Lu, and S. M. Kupchan, *Chem. Pharm. Bull.* **15**, 959 (1967).

477. T. A. Broadbent and E. G. Paul, *Heterocycles* **20**, 863 (1983).

478. S. M. Kupchan, A. J. Liepa, V. Kameswaran, and K. Sempuku, *J. Am. Chem. Soc.* **95**, 2995 (1973).

479. N. M. Mollov and Kh. B. Duchevska, *Tetrahedron Lett.* **1966**, 853.

480. H. B. Dutschewska and N. M. Mollov, *Chem. Ber.* **100**, 3135 (1967).

481. M. A. Haimova, N. M. Mollov, H. B. Dutschewska, I. R. Petrova, D. L. Kasseva, and N. G. Koitscheva, *Compt. Rend. Acad. Bulg. Sci.* **19**, 921 (1966); *Chem. Abstr.* **66**, 65689 (1967).

482. R. Ahmad, J. M. Saá, and M. P. Cava, *J. Org. Chem.* **42**, 1228 (1977).

483. T. Nabih, P. J. Davis, J. F. Caputo, and J. P. Rosazza, *J. Med. Chem.* **20**, 914 (1977).

484. N. L. Marekov and A. K. Sidjimov, *Tetrahedron Lett.* **22**, 2311 (1981).

485. A. K. Sidjimov and N. L. Marekov, *Phytochemistry* **21**, 871 (1982).

486. N. L. Marekov and A. K. Sidjimov, *Int. Conf. Chem. Biotechnol. Biol. Act. Nat. Prod. [Proc.] 1st*, **3**, 418 (1981); *Chem. Abstr.* **97**, 88762 (1982).

487. D. S. Bhakuni and S. Jain, *Tetrahedron* **38**, 729 (1982).

488. R. A. Hahn, J. W. Nelson, A. Tye, and J. L. Beal, *J. Pharm. Sci.* **55**, 466 (1966).

489. E. H. Herman and D. P. Chadwick, *Toxicol. Appl. Pharmacol.* **26**, 137 (1973).

490. E. H. Herman and D. P. Chadwick, *Pharmacology* **10**, 178 (1973).

491. D. K. Todorov and G. Deliconstantinos, *Experientia* **38**, 857 (1982).

492. L. M. Allen and P. J. Creaven, *Cancer Res.* **33**, 3112 (1973).

493. D. K. Todorov and M. S. Damyonova, *Compt. Rend. Bulg. Acad. Sci.* **28**, 709 (1975).

494. I. E. Broder and S. K. Carter, *Thalicarpine* (NSC-68075) Clinical Brochure, NCI, Bethesda, MD (1971).

495. V. Khadzhidekova, M. Vinarova, I. Bradvarova, and Z. Paskalev, *Onkologiya* (Sofia) **20**, 37 (1983); *Chem. Abstr.* **99**, 98918 (1983).

496. V. Khadzhidekova, B. Ivanov, M. Koleva, and A. Mincheva, *Onkologiya* (Sofia) **20**, 95 (1983); *Chem. Abstr.* **99**, 205708 (1983).

497. L.-L. H. Liao, *Proc. Natl. Sci. Counc. Repub. China* **4**, 285 (1980); *Biol. Abstr.* **71**, 54066 (1981).

498. P. J. Creaven, L. M. Allen, and C. P. Williams, *Xenobiotica* **4**, 255 (1974).

499. W. A. Creasey, *Biochem. Pharmacol.* **25**, 1887 (1976).

500. I. Mircheva and I. Stoichkov, *Probl. Onkol.* **5**, 17 (1977); *Chem. Abstr.* **90**, 348 (1979).

501. J. Mircheva and J. Stoychkov, *Biomed. Express.* (*Paris*) **25**, 280 (1976); *Biol. Abstr.* **63**, 52833 (1977).

502. P. E. Palm, M. S. Nuk, D. W. Yesair, et al., *U.S. Dept. of Commerce National Technical Information Service*, PB Rep-185765 (1969); *Chem. Abstr.* **72**, 109228 (1970).

503. P. E. Palm, M. S. Nick, and E. P. Arnold, *U.S. Dept. of Commerce National Technical Information Service*, PB Rep-201914 (1971); *Chem. Abstr.* **76**, 68093 (1972).

504. P. J. Creaven, M. H. Cohen, O. S. Selawry, F. Tejada, and L. E. Broder, *Cancer Chemother. Rep. Part I* **59**, 1001 (1975).

505. P. J. Creaven and L. M. Allen, *Cancer Treatment Rep.* **60**, 69 (1976).

506. J. T. Leimert, M. P. Corder, T. E. Elliott, and J. M. Lovett, *Cancer Treatment Rep.* **64**, 1389 (1980).

507. G. N. Rao, T. E. Palmer, M. W. Balk, G. W. Thompson, and S. M. Glaza, *Gov. Rep. Announce. Index (US)* **82**, 2088 (1982).

508. J. Henney, *Division of Cancer Treatment Bull.* **1979**, 4.

509. T. Tomimatsu, E. Vorperian, J. L. Beal, and M. P. Cava, *J. Pharm. Sci.* **54**, 1389 (1965).

510. N. M. Mollov and Kh. B. Duchevska, *Herba Hung.* **5**, 67 (1966); *Chem. Abstr.* **68**, 57371 (1968).

511. Kh. S. Umarov, Z. F. Ismailov, and Kh. B. Allayarov, *Izv. Akad. Nauk Turkem. SSR, Ser. Biol. Nauk* **1977**, 82; *Chem. Abstr.* **89**, 39379 (1978).

512. V. Christov, P. Demirev, N. Mollova, and V. Nenev, *Int. Chem. Biotechnol. Biol. Act. Nat. Prod. 1st,* **1981**, 367.

513. N. M. Mollov and H. B. Dutschewska, *Tetrahedron Lett.* **1964**, 2219.

514. I. Yankulov and L. Evstatieva, *Dokl. Bolg. Akad. Nauk* **29**, 1345 (1976); *Chem. Abstr.* **86**, 52298 (1977).

515. M. Shamma and J. L. Moniot, *Tetrahedron Lett.* **1973**, 775.

516. M. Shamma, S. S. Salgar, and J. L. Moniot, *Tetrahedron Lett. 1973*, 1859.

517. W.-N. Wu, J. L. Beal, and R. W. Doskotch, *Tetrahedron* **33**, 2919 (1977).

518. M. Shamma, J. L. Moniot, and P. Chinnasamy, *Heterocycles* **6**, 399 (1977).

519. N. M. Mollov, P. P. Panov, L. N. Thuan, and L. N. Panova, *Compt. Rend. Acad. Bulg. Sci.* **23**, 1243 (1970).

519a. M. O. A. ElSheikh, Ph.D. thesis, The Ohio State University, 1985; Jack L. Beal, College of Pharmacy, The Ohio State University, personal communication.

520. N. M. Mollov, V. St. Georgiev, and H. B. Dutschewska, *Compt. Rend. Acad. Bulg. Sci.* **23**, 383 (1970).

521. N. Mollov, Kh. Duchevska, and P. Panov, *Compt. Rend. Acad. Bulg. Sci.* **17**, 709 (1964); *Chem. Abstr.* **61**, 16351 (1964).

522. J. Reisch, H. Alfes, T. Kaniewska, and B. Borkowski, *Tetrahedron Lett.* **1970**, 2113.

523. T. Kaniewska and B. Borkowski, *Acta Pol. Pharm.* **28**, 413 (1971); *Chem. Abstr.* **76**, 43962 (1972).

524. H. Guinaudeau, A. J. Freyer, R. D. Minard, M. Shamma, and K. H. C. Baser, *J. Org. Chem.* **47**, 5407 (1982).

525. H. B. Dutschewska and N. M. Mollov, *Chem. Ind.* **1966**, 770.

526. J.-I. Kunitomo, Y. Murakami, and M. Akasu, *Yakugaku Zasshi* **100**, 337 (1980); *Biol. Abstr.* **70**, 46088 (1980).

527. M. P. Cava, A. Venkateswarlu, M. Srinivasan, and D. L. Edie, *Tetrahedron* **28**, 4299 (1972).

528. M. Shamma and J. L. Moniot, *Isoquinoline Alkaloids Research: 1972–1977*, Plenum, New York, 1978, p. 167.

529. K. P. Guha, B. Mukherjee, and R. Mukherjee, *J. Nat. Prod.* **42**, 1 (1979).

530. P. L. Schiff, Jr., *J. Nat. Prod.* **46**, 1 (1983).

531. M. Kulka, *The Alkaloids*, Vol. 4, R. H. F. Manske and H. L. Holmes, Eds., Academic, New York, 1954, p. 199.

532. M. Kulka, *The Alkaloids*, Vol. 7, R. H. F. Manske, Ed., Academic, New York, 1960, p. 439.

533. M. Curcumelli-Rodostamo and M. Kulka, *The Alkaloids*, Vol. 9, R. H. F. Manske, Ed., Academic, New York, 1967, p. 134.

534. M. Curcumelli-Rodostamo, *The Alkaloids*, Vol. 13, R. H. F. Manske, Ed., Academic, New York, 1971, p. 304.

535. M. P. Cava, K. T. Buck, and K. L. Stuart, *The Alkaloids*, Vol. 16, R. H. F. Manske, Ed., Academic, New York, 1977, p. 250.

536. M. Shamma, *The Isoquinoline Alkaloids*, Academic, New York, 1972, p. 115.

537. M. Shamma and J. L. Moniot, *Isoquinoline Alkaloids Research: 1972–1977*, Plenum, New York, 1978, p. 71.

538. W.-N. Wu, J. L. Beal, and R. W. Doskotch, *Lloydia* **39**, 378 (1976).

539. I. R. C. Bick, E. S. Ewen, and A. R. Todd, *J. Chem. Soc.* **1949**, 2767.

540. E. Fujita, T. Tomimatsu, and Y. Kano, *Yakugaku Zasshi* **82**, 311 (1962); *Chem. Abstr.* **58**, 3468 (1963).

541. T. Tomimatsu and Y. Kano, *Yakugaku Zasshi* **82**, 315 (1962); *Chem. Abstr.* **58**, 3468 (1963).

542. T. Tomimatsu and Y. Kano, *Yakugaku Zasshi* **82**, 320 (1962); *Chem. Abstr.* **58**, 3469 (1963).

543. E. Fujita, T. Tomimatsu, and Y. Kitamura, *Bull. Inst. Chem. Res., Kyoto Univ.* **42**, 235 (1964); *Chem. Abstr.* **62**, 5310 (1965).

544. H. Guinaudeau, A. J. Freyer, M. Shamma, and K. H. C. Baser, *Tetrahedron* **40**, 1975 (1984).

545. L. Koike, A. J. Marsaioli, E. A. Rúveda, and F. de A. M. Reis, *Tetrahedron Lett.* **1979**, 3765.

546. J. Baldas, I. R. C. Bick, T. Ibuka, R. S. Kapil, and Q. N. Porter, *J. Chem. Soc., Perkin I* **1972**, 592.

547. A. R. Battersby, I. R. C. Bick, W. Klyne, J. P. Jennings, P. M. Scopes, and M. J. Vernengo, *J. Chem. Soc.* **1965**, 2239.

548. E. P. Gibson and J. H. Turnbull, *J. Chem. Soc. Perkin 2*, **1980**, 1696.

549. N. M. Mollov, V. St. Georgiev, P. P. Panov, and D. Jordanov, *Compt. Rend. Acad. Bulg. Sci.* **20**, 333 (1967).

550. M. Okada and A. Asoda, *Ann. Rept. ITSUU Lab.* (*Tokyo*) **4**, 65 (1953); M. Kulka, *The Alkaloids*, Vol. 7, R. H. F. Manske, Ed., Academic, New York, 1960, p. 451.

551. H. Kuroda, S. Nakazawa, K. Katagiri, O. Shiratori, M. Kozuka, K. Fujitani, and M. Tomita, *Chem. Pharm. Bull.* **24**, 2413 (1976).

552. L. Guo-Sheng, C. Bi-Zhu, S. Wan-zhi, and X. Pei-gen, *Chih Wu Hsueh Pao* (*Acta Botanica Sinica*) **20**, 255 (1978); *Chem. Abstr.* **90**, 12202 (1979).

553. Z.-D. Zhou, C.-H Han, and P. Wang, *Yao Hsueh Hsueh Pao* **15**, 248 (1980); K. W. Bentley, *The Alkaloids*, Vol. 12, M. F. Grundon, Ed., The Royal Society of Chemistry, Burlington House, London, 1982, p. 103.

554. C. Liu, G. Liu, and P. Xiao, *Zhongcaoyao* **14**, 45 (1983); *Chem. Abstr.* **99**, 32628 (1983).

555. L.-C. Lin, C.-C. Sung, C.-Y. Fan, C.-F. Tu, M.-L. Chou, C.-C. Ma, and J.-S. Hsu, *Yao Hsueh T'ung Pao* **15**, 46 (1980); *Chem. Abstr.* **95**, 86198 (1981).

556. W.-N. Wu, J. L. Beal, and R. W. Doskotch, *Tetrahedron Lett.* **1976**, 3687.

557. J. Wu, J. L. Beal, and R. W. Doskotch, *J. Org. Chem.* **45**, 213 (1980).

558. J. Padilla and J. Herrán, *Tetrahedron* **18**, 427 (1962).

559. T. A. Broadbent and E. G. Paul, *Heterocycles* **20**, 899 (1983).

560. Z. F. Ismailov and S. Yu. Yunusov, *Chem. Nat. Cpds.* **4**, 220 (1968).

561. J. Baldas, Q. N. Porter, I. R. C. Bick, and M. J. Vernengo, *Tetrahedron Lett.* **1966**, 2059.

562. G. P. Moiseeva, S. Kh. Maekh, and S. Yu. Yunusov, *Chem. Nat. Cpds.* **15**, 723 (1979).

563. Z. F. Ismailov, D. Sargazakov, and S. Yu. Yunusov, *Dokl. Akad. Nauk Uz. SSR* 32 (1960); *Chem. Abstr.* **61**, 4700 (1964).

564. S. Yu. Yunusov and M. V. Telezhenetskaya, *Dokl. Akad. Nauk Uz. SSR* **20**, 22 (1963); *Chem. Abstr.* **61**, 14735 (1964).

565. S. Kh. Maekh and S. Yu. Yunusov, *Chem. Nat. Cpds.* **1**, 144 (1965).

566. S. Kh. Maekh and S. Yu. Yunusov, *Chem. Nat. Cpds.* **1**, 229 (1965).

567. M. Shamma, B. S. Dudock, M. P. Cava, K. V. Rao, D. R. Dalton, D. C. Dejongh, and S. R. Shrader, *Chem. Commun.* **1966**, 7.

568. D. R. Dalton, M. P. Cava, and K. T. Buck, *Tetrahedron Lett.* **1965**, 2687.

569. I. R. C. Bick, J. B. Bremner, M. P. Cava, and P. Wiriyachitra, *Aust. J. Chem.* **31**, 321 (1978).

570. D. K. Kasmaliev, *Godichn. Nauchn. Sessii, Kirgizsk. Inst. Kraev. Med., Akad. Med. Nauk SSSR, Frunze, Sb.* **1965**, 97; *Chem. Abstr.* **65**, 1261 (1966).

571. F. Sadritdinov, *Farmakol. Alkaloidov Serdechnykh Glikozidov* **1971**, 117; *Chem. Abstr.* **78**, 79629 (1973).

572. N. Tulyaganov and F. Sadritdinov, *Farmakol. Alkaloidov Serdechnykh Glikozidov* **1971**, 132; *Chem. Abstr.* **78**, 79631 (1973).

573. Sh. U. Ismailov and D. A. Asadov, *Farmakol. Alkaloidov Ikh Proizvod.* **1972**, 171; *Chem. Abstr.* **80**, 103857 (1974).

574. N. M. Mollov, V. St. Georgiev, D. Jordanov, and P. Panov, *Compt. Rend. Acad. Bulg. Sci.* **19**, 491 (1966); *Chem. Abstr.* **65**, 13780 (1966).

575. S. Abdizhabarova, S. Kh. Maekh, and S. Yu. Yunusov, *Chem. Nat. Cpds.* **14**, 114 (1978).

576. S. Mukhamedova, S. Kh. Maekh, and S. Yu. Yunusov, *Khim. Prir. Soedin.* **1981**, 250; *Chem. Abstr.* **95**, 58078 (1981).

577. L. A. Mitscher, W.-N. Wu, and J. L. Beal, *Experientia* **28**, 500 (1972).

578. M. Tomita, T. Kikuchi, K. Fujitani, A. Kato, H. Furukawa, Y. Aoyagi, M. Kitano, and T. Ibuka, *Tetrahedron Lett.* **1966**, 857.

579. M. Kulka, *The Alkaloids*, Vol. 4, R. H. F. Manske and H. L. Holmes, Eds., Academic, New York, 1954, p. 215.

580. J. Yamahara, T. Sawada, M. Kozuka, and H. Fugimara, *Shoyakugaku Zasshi* **28**, 83 (1974); *Chem. Abstr.* **83**, 53481 (1975).

581. D. S. Bhakuni, A. N. Singh, and S. Jain, *Tetrahedron* **36**, 2149 (1980).

582. J. M. Saá, M. J. Mitchell, M. P. Cava, and J. L. Beal, *Heterocycles* **4**, 753 (1976).

583. R. Ahmad, *Islamabad J. Sci.* **5**, 38 (1978).

584. D. C. DeJongh, S. R. Shrader, and M. P. Cava, *J. Am. Chem. Soc.* **88**, 1052 (1966).

585. J. Baldas, I. R. C. Bick, M. R. Falco, J. X. deVries, and Q. N. Porter, *J. Chem. Soc, Perkin I* **1972**, 597.

586. E. Fujita and T. Tomimatsu, *Yakugaku Zasshi* **79**, 1256 (1959); *Chem. Abstr.* **54**, 4643 (1960).

587. E. Fujita and T. Tomimatsu, *Yakugaku Zasshi* **79**, 1260 (1959); *Chem. Abstr.* **54**, 4644 (1960).

588. T. Tomimatsu and Y. Kano, *Yakugaku Zasshi* **83,** 153 (1963); *Chem. Abstr.* **59,** 3971 (1963).

589. E. Fujita, K. Fuji, and T. Suzuki, *Bull. Inst. Chem. Res., Kyoto Univ.* **43,** 449 (1965); *Chem. Abstr.* **65,** 7229 (1966).

590. E. Fujita and A. Sumi, *Chem. Pharm. Bull.* **18,** 2591 (1970).

591. E. Fujita, A. Sumi, and Y. Yoshimura, *Chem. Pharm. Bull.* **20,** 368 (1972).

592. S. A. Tursunova, Kh. I. Tashbaev, and M. B. Sultanov, *Farmakol. Alkaloidov Glikozidov* **1967,** 156; *Chem. Abstr.* **70,** 2223 (1969).

593. S. A. Tursunova, Kh. I. Tashbaev, and M. B. Sultanov, *Farmakol. Alkaloidov Glikozidov* **1967,** 160; *Chem. Abstr.* **70,** 2224 (1969).

594. F. Sadritdinov and M. B. Sultanov, *Farmakol. Alkaloidov Serdechnykh Glikozidov* 120 (1971); *Chem. Abstr.* **78,** 66916 (1973).

595. S. Yu. Yunusov and Z. F. Ismailov, *Dokl. Akad. Nauk Uzbek. S.S.R.* 17 (1956); *Chem. Abstr.* **52,** 12100 (1958).

596. M. V. Telezhenetskaya and S. Yu. Yunusov, *Dokl. Akad. Nauk SSR.* **162,** 254 (1965); *Chem. Abstr.* **63,** 5689 (1965).

597. M. V. Telezhenetskaya, Z. F. Ismailov, and S. Yu. Yunusov, *Khim. Prirodn. Soedin., Akad. Nauk Uz. SSR* **2,** 107 (1966); *Chem. Abstr.* **65,** 10629 (1966).

598. Z. F. Ismailov, Dzh. Sargazakov, and S. Yu. Yunusov, *Dokl. Akad Nauk Uz. SSR* 21 (1965); *Chem. Abstr.* **61,** 693 (1964).

599. N. M. Mollov, H. B. Dutschewska, and H. G. Kirjakov, *Chem. Ind.* **1965,** 1595.

600. N. M. Mollov and V. St. Georgiev, *Compt. Rend. Acad. Bulg. Sci.* **20,** 329 (1967).

601. J. M. Saá, M. V. Lakshmikantham, M. J. Mitchell, M. P. Cava, and J. L. Beal, *Tetrahedron Lett.* **1976,** 513.

602. J. Wu, J. L. Beal, and R. W. Doskotch, *J. Org. Chem.* **45,** 208 (1980).

603. M. Curcumelli-Rodostamo and M. Kulka, *The Alkaloids*, Vol. 9, R. H. F. Manske, Ed., Academic, New York, 1967, p. 148.

604. T. Tomimatsu and J. L. Beal, *J. Pharm. Sci.* **55,** 208 (1966).

605. T. R. Suess and F. R. Stermitz, *J. Nat. Prod.* **44,** 680 (1981).

606. M. Curcumelli-Rodostamo and M. Kulka, *The Alkaloids*, Vol. 9, R. H. F. Manske, Ed., Academic, New York, 1967, p. 154.

607. J. W. Banning, K. N. Salman, and P. N. Patil, *J. Nat. Prod.* **45,** 168 (1982).

608. J. E. Knapp, F. J. Hussein, J. L. Beal, R. W. Doskotch, and T. Tomimatsu, *J. Pharm. Sci.* **56,** 139 (1967).

609. M. Kulka, *The Alkaloids*, Vol. 4, R. H. F. Manske and H. L. Holmes, Eds., Academic, New York, 1954, p. 213.

610. R.-Hamet, *Compt. Rend.* **197,** 1354 (1933); *Chem. Abstr.* **28,** 1408 (1934).

611. R.-Hamet, *Compt. Rend. Soc. Biol.* **136,** 112 (1942); *Chem. Abstr.* **37,** 2812 (1943).

612. E. Andronescu, P. Petcu, T. Goina, and A. Radu, *Clujul Medical* **46,** 627 (1973); *Chem. Abstr.* **81,** 100062 (1974).

613. O. N. Tolkachev, S. A. Vichkanova, and L. V. Makarova, *Farmatsiya* (Moscow) **27,** 23 (1978); *Chem. Abstr.* **89,** 36464 (1978).

614. J. Baldas, Q. N. Porter, I. R. C. Bick, G. K. Douglas, M. R. Falco, J. X. deVries, and S. Yu. Yunusov, *Tetrahedron Lett.* **1968,** 6315.

615. S. Abdizhabbarova, Z. F. Ismailov, and S. Yu. Yunusov, *Chem. Nat. Cpds.* **4,** 281 (1968).

616. S. Abdizhabbarova, Z. F. Ismailov, and S. Yu. Yunusov, *Chem. Nat. Cpds.* **6,** 281 (1970).

617. N. M. Mollov and V. St. Georgiev, *Chem. Ind.* **1966,** 1178.

618. V. St. Georgiev and N. M. Mollov, *Phytochemistry* **10**, 2161 (1971).

619. J. Wu, J. L. Beal, and R. W. Doskotch, *J. Org. Chem.* **45**, 208 (1980).

620. M. P. Cava, J. M. Saa, M. V. Lakshmikantham, M. J. Mitchell, J. L. Beal, R. W. Doskotch, A. Ray, D. C. Dejongh, and S. R. Shrader, *Tetrahedron Lett.* **1974**, 4259.

621. T. Tomimatsu, *Yakugaku Zasshi* **79**, 1386 (1959); *Chem. Abstr.* **54**, 13163 (1960).

622. E. Fujita, T. Tomimatsu, and Y. Kano, *Yakugaku Zasshi* **80**, 1137 (1960); *Chem. Abstr.* **55**, 595 (1961).

623. T. Tomimatsu and M. Sasakawa, *Chem. Pharm. Bull.* **23**, 2279 (1975).

624. S. S. Norkina and N. A. Pakhareva, *Zhur. Obshchei Khim.* **20**, 1720 (1950); *Chem. Abstr.* **45**, 1306 (1951).

625. Z. F. Ismailov, S. Kh. Maekh, and S. Yu. Yunusov, *Dokl. Akad. Nauk Uzbek. S.S.R.* **1960**, 22; *Chem. Abstr.* **56**, 11646 (1962).

626. S. M. Kupchan, T.-H. Yang, G. S. Vasilikiotis, M. H. Barnes, and M. L. King, *J. Am. Chem. Soc.* **89**, 3076 (1967).

627. T. Tomimatsu, M. Hashimoto, and J. L. Beal, *Chem. Pharm. Bull.* **16**, 2070 (1968).

628. S. M. Kupchan and H. W. Altland, *J. Med. Chem.* **16**, 913 (1973).

629. H. Kuroda, S. Nakazawa, K. Katagiri, O. Shiratori, M. Kozuka, K. Fujitani, and M. Tomita, *Chem. Pharm. Bull.* **24**, 2413 (1976).

630. Z.-Q Ma, S.-M. Hsing, and H.-C. Chen, *Chung Tsao Yao* **11**, 217 (1980); *Chem. Abstr.* **94**, 76727 (1981).

631. C.-C. Ma, *Yao Hsueh T'ung Pao* **15**, 46 (1980); *Chem. Abstr.* **95**, 54810 (1981).

632. X. Gu, Z. Ma, H. Li, S. Sun, C. Yao, S. Xin, S. Pei, D. Ma, and J. Tao, *Zhongcaoyao* **13**, 13 (1982); *Chem. Abstr.* **97**, 203176 (1982).

633. X. Gu, Z. Ma, S. Xin, H. Li, S. Sun, S. Pei, X. Shen, M. Zhou, S. Tao, and D. Lin, *Zhangcaoyao* **13**, 15 (1982); *Chem. Abstr.* **97**, 222817 (1982).

634. W.-N. Wu, J. L. Beal, E. H. Fairchild, and R. W. Doskotch, *J. Org. Chem.* **43**, 580 (1978).

635. Kh. G. Pulatova, Z. F. Ismailov, and S. Yu. Yunusov, *Chem. Nat. Cpds.* **5**, 533 (1969).

636. Z. F. Ismailov, A. U. Rakhmatkariev, and S. Yu. Yunusov, *Uzbeksk. Khim. Zh. 1961*, 56; *Chem. Abstr.* **58**, 3469 (1963).

637. Z. F. Ismailov, A. U. Rakhmatkariev, and S. Yu. Yunusov, *Dokl. Akad. Nauk. Uz. SSR* **20**, 21 (1963); *Chem. Abstr.* **61**, 4407 (1964).

638. Kh. I. Tashbaev and I. K. Kamilov, *Vopr. Ispol'z Mineral'n i Rast. Syr'ya Srednei Azii, Akad. Nauk Uz. SSR, Otd. Geol.- Khim. Nauk* **1961**, 164; *Chem. Abstr.* **58**, 7278 (1963).

639. Kh. I. Tashbaev and M. B. Sultanov, *Farmakol. Allkaloidov, Akad. Nauk Uz SSR, Inst. Khim. Rast. Veshchestv 1962*, 210; *Chem. Abstr.* **61**, 8790 (1964).

640. Kh. M. Malikov, Kh. I. Tashbaev, and M. B. Sultanov, *Materialy l-oi (Pervoi) Nauchn.-Prakt. Konf. Patologoanatomov Uzbekistana* **1964**, 112; *Chem. Abstr.* **62**, 15295 (1965).

641. A. G. Kurmukov, Kh. I. Tashbaev, and M. B. Sultanov, *Dokl. Akad. Nauk Uz. SSR* **23**, 35 (1966); *Chem. Abstr.* **65**, 12731 (1966).

642. Kh. M. Malikov, Kh. I. Tashbaev, and M. B. Sultanov, *Farmakol. Alkaloidov, Akad. Nauk Uzb. SSR, Inst. Khim. Rast. Veshchestv* **1965**, 213; *Chem. Abstr.* **66**, 53948 (1967).

643. Kh. I. Tashbaev and M. B. Sultanov, *Farmakol. Farmakoter. Alkaloidov Glikozidov* **1966**, 12; *Chem. Abstr.* **67**, 1958 (1967).

644. Kh. I. Tashbaev and M. B. Sultanov, *Farmakol. Alkaloidov, Akad. Nauk Uzb. SSR, Inst. Khim. Rast. Veshchestv* **1965**, 197; *Chem. Abstr.* **67**, 10142 (1967).

644a. Z. S. Akbarov, Kh. U. Aliev, and M. B. Sultanov, *Farmakol. Alkaloidov Ikh Proizvod.* **1972**, 129; *Chem. Abstr.* **81**, 20880 (1074).

645. V. G. Khodzhaev and Kh. Allayarov, *Chem. Nat. Cpds.* **6**, 522 (1970).

646. H. B. Allayarov, V. G. Khodzhaev, and Z. F. Ismailov, *Izv. Akad. Nauk Turkin. SSR, Ser. Fiz.-Tekh., Khim. Geol. Nauk* 121 (1971); *Chem. Abstr.* **76**, 72686 (1972).

647. S. Yu. Yunusov and Z. F. Ismailov, *Zhur. Obshchei Khim.* **30**, 1721 (1960); *Chem. Abstr.* **55**, 3631 (1961).

648. B. Yu. Aizenman, M. O. Shvaiger, T. P. Mandrik, and A. M. Bredikhina, *Mikrobiol. Zh., Akad. Nauk Ukr. RSR* **25**, 52 (1963); *Chem. Abstr.* **60**, 3383 (1964).

648a. S. F. Hussain, H. Guinaudenu, A. J. Freyer, and M. Shamma, *J. Nat. Prod.*, submitted for publication.

649. M. Shamma and S. Y. Yao, *Experientia* **29**, 517 (1973).

650. J. S. K. Ayim, D. Dwuma-Badu, N. Y. Fiagbe, A. M. Ateya, D. J. Slatkin, J. E. Knapp, and P. L. Schiff, Jr., *Lloydia* **40**, 561 (1977).

651. C. Goepel, T. Yupraphat, P. Pachaly, and F. Zymalkowski, *Planta Med.* **26**, 94 (1974).

652. S. Kh. Maekh, Z. F. Ismailov, and S. Yu. Yunusov, *Chem. Nat. Cpds.* **4**, 119 (1968).

653. S. Kh. Maekh, Z. F. Ismailov, and S. Yu. Yunusov, *Chem. Nat. Cpds.* **4**, 335 (1968).

654. S. Kh. Maekh and S. Yu. Yunusov, *Izv. Akad. Nauk SSR, Ser. Khim. 1966*, 112; *Chem. Abstr.* **64**, 14233 (1966).

655. S. Kh. Maekh and S. Yu. Yunusov, *Dokl. Akad. Nauk Uz. SSR* **21**, 27 (1964); *Chem. Abstr.* **62**, 13191 (1965).

656. D. A. Rakhimova, E. K. Dobronravona, and T. T. Shakirov, *Chem. Nat. Cpds.* **11**, 398 (1975).

657. F. Sadritdinov and I. K. Kamilov, *Med Zh. Uzbekistana* **1964**, 56; *Chem. Abstr.* **62**, 16835 (1965).

658. S. A. Tursunova, Kh. I. Tashbaev, and M. B. Sultanov, *Farmakol. Farmakoter. Alkaloidov Glikozidov, Akad. Nauk Uzh. SSR Khim.-Tekhnol. Biol. Otd., 1966*, 16; *Chem. Abstr.* **67**, 42382 (1967).

659. F. S. Sadritdinov, *Farmakol. Toksikol.* (Moscow) **32**, 598 (1969); *Chem. Abstr.* **72**, 1988 (1970).

660. R. I. Alimov, I. N. Zatorskaya, T. T. Shakirov, V. P. Zakharov, and B. A. Yankovskii, *Deposited Doc.* VINITI 3271-74 (1974); *Chem. Abstr.* **87**, 206408 (1977).

661. N. M. Mollov, I. C. Ivanov, V. St. Georgiev, P. P. Panov, and N. Kotsev, *Planta Med.* **19**, 10 (1971).

662. M. Shamma, J. E. Foy, and G. A. Miana, *J. Am. Chem. Soc.* **96**, 7809 (1974).

663. S. A. Vichkanova, L. V. Makarova, and N. I. Gordeikina, *Fitontsidy Mater. Soveshch. 6th 1969*, 90; *Chem. Abstr.* **78**, 66905 (1973).

664. L. Evstatieva and I. Yankulov, *Fitologiya* **14**, 60 (1980); *Chem. Abstr.* **93**, 184908 (1980).

665. V. Khristov, P. Demirey, N. Mollova, and V. Nenov, *Int. Conf. Chem. Biotechol. Biol. Act. Nat. Prod. (Proc.) 1st*, **3**, 367 (1981); *Chem. Abstr.* **97**, 98419 (1982).

666. T. Cieszynski and B. Borkowski, *Acta Polon. Pharm.* **22**, 171 (1965); *Chem. Abstr.* **63**, 11918 (1965).

667. R. H. F. Manske, *The Alkaloids*, Vol. 7, R. H. F. Manske, Ed., Academic, New York, 1965, p. 47.

668. H. B. Dutschewska and B. A. Kuzmanov, *J. Nat. Prod.* **45**, 295 (1982).

669. B. Kuzmanov and H. Dutschewska, *J. Nat. Prod.* **45**, 766 (1982).

Chapter Five

Synthesis of Cephalotaxine Alkaloids

Tomas Hudlicky,* Lawrence D. Kwart, and Josephine W. Reed
Department of Chemistry
Virginia Polytechnic Institute and State University
Blacksburg, Virginia 24061

CONTENTS

1. INTRODUCTION

The *Cephalotaxus* alkaloids are structurally unique compounds isolated from several species of the genus *Cephalotaxus*. These evergreen shrubs are widely distributed around the world between 30° and 40° latitude and favor mildly elevated locations with high relative humidity [1]. Two of the eight known species, *C. harringtonia* and *C. manii*, are found in China, Japan, India, western Canada and the United States, parts of the eastern United

---

* NIH Research Career Development Award recipient, 1984–1989.

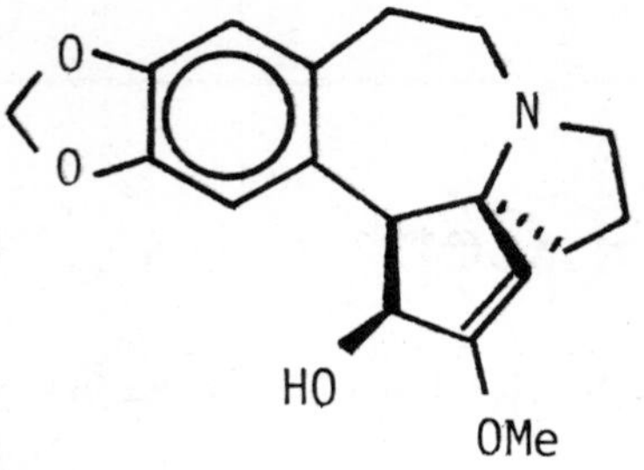

1

States, and also in the rolling regions surrounding Rio de la Plata in western Uruguay and northeastern Argentina [2].

The discovery of antileukemic activity exhibited by some of the esters of cephalotaxine (**1**), the major constituent of plant extracts, spurred vigorous research activity in China and in the United States. During the 1970s a total of 19 compounds containing the cephalotaxine nucleus were isolated from various species of *Cephalotaxus* (Table 1) and characterized. Of these about half are esters of cephalotaxine, and four possess significant antitumor properties.

As of this writing, several syntheses of cephalotaxine and the antitumor esters homoharringtonine and harringtonine have been attained, following the establishment of structures for most cephalotaxine alkaloids isolated to date and the proposition of plausible biogenic pathways for *Cephalotaxus* alkaloids. The current status of the field has been summarized in several excellent reviews that have appeared recently and that also allude to the preparation and possible medicinal use of unnatural derivatives of cephalotaxine [3–5]. This chapter concentrates on the advances made in the chemical synthesis of these interesting alkaloids.

Table 1.   Cephalotaxus Alkaloids

1 Cephalotaxine

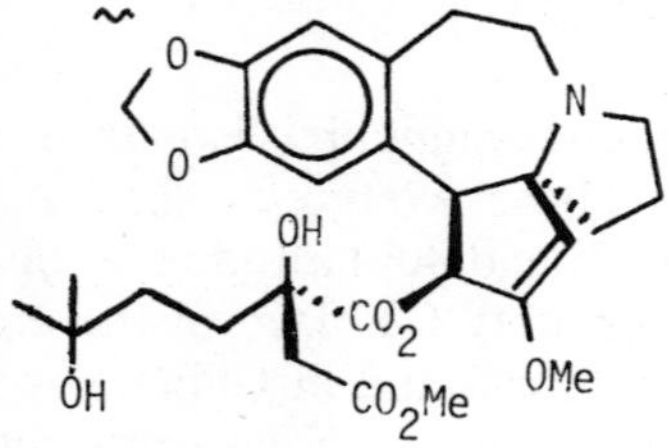

2 Harringtonine

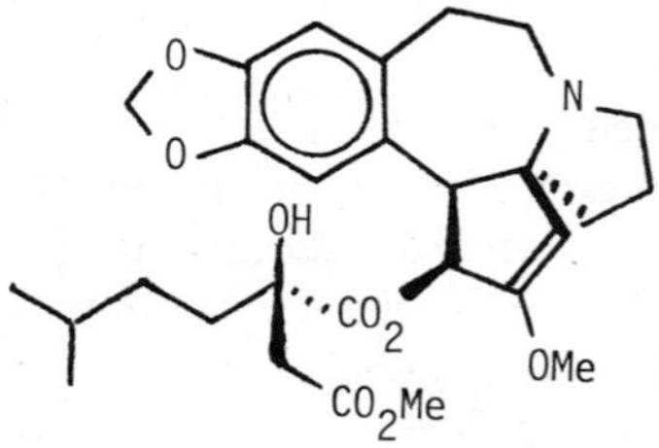

3 Deoxyharringtonine

Table 1. (*continued*)

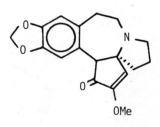

4 Homoharringtonine

5 Isoharringtonine

6 Acetylcephalotaxine
7 (+)-acetylcephalotaxine

8 R=OH Isoharringtonic Acid

9 R=H Deoxyharringtonic Acid

10 Epicephalotaxine

11 Cephalotaxinone

Table 1. (*continued*)

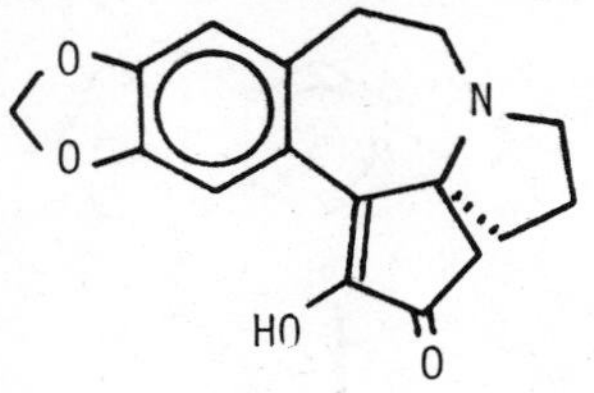

<u>12</u> Demethylcephalotaxinone

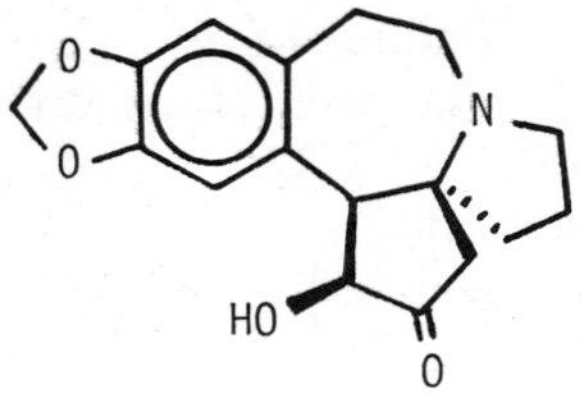

<u>13</u> Demethylcephalotaxine

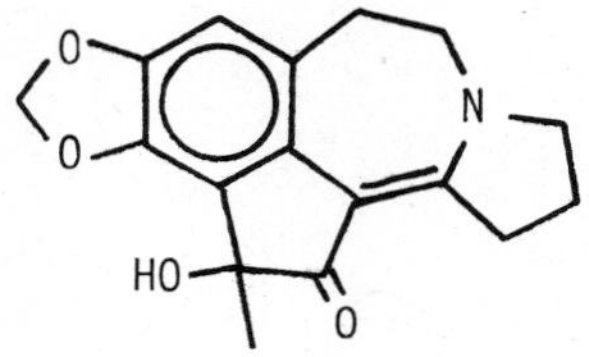

<u>14</u> Hainanensine

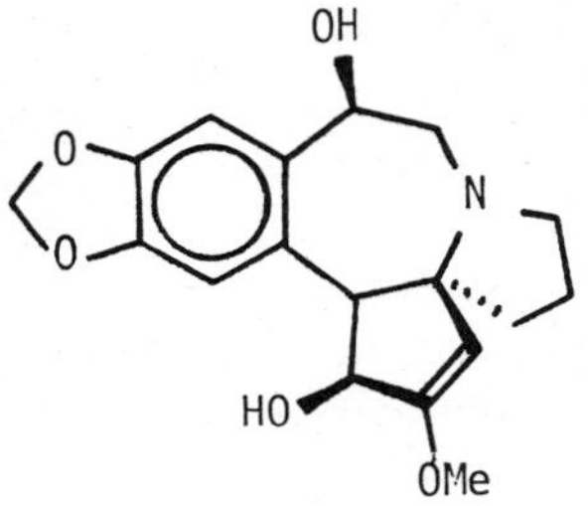

<u>15</u> 11-Hydroxycephalotaxine

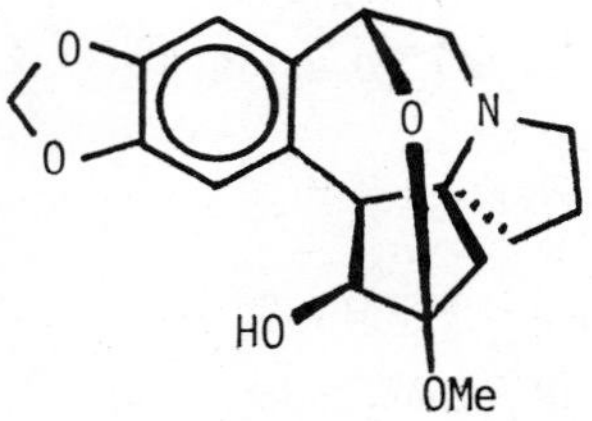

<u>16</u> Drupacine

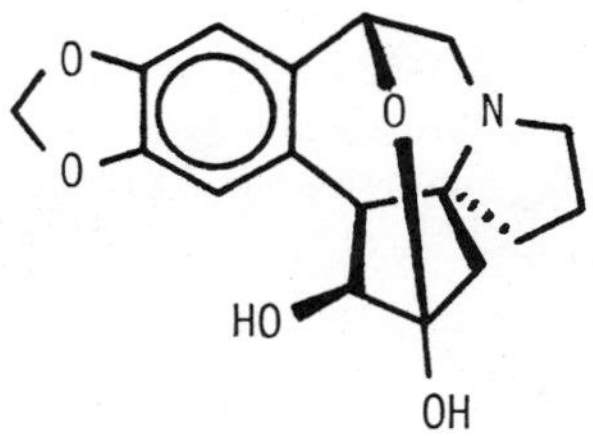

<u>17</u> Demethylneodrupacine

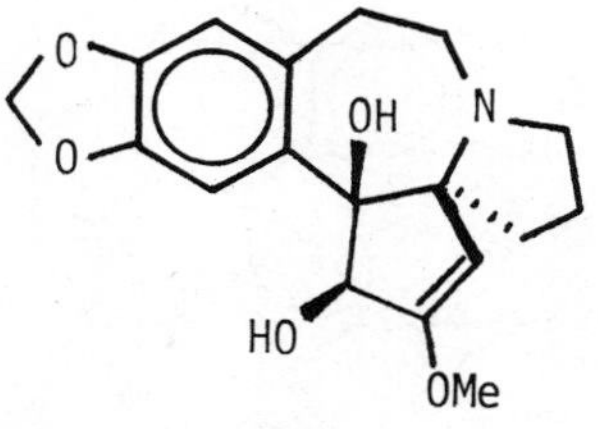

<u>18</u> 4-Hydroxycephalotaxine

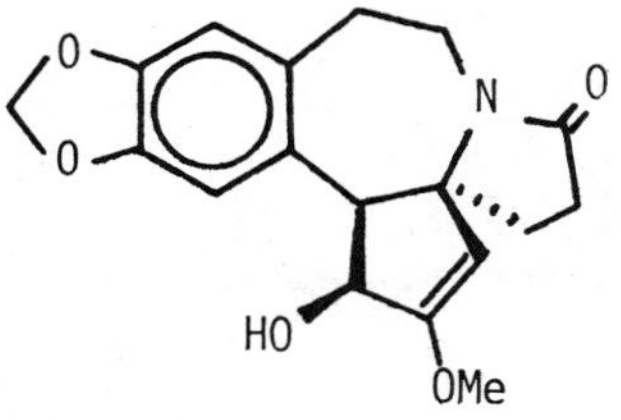

<u>19</u> Cephalotaxinamide

2. STRUCTURE ELUCIDATION

The structures of all *Cephalotaxus* alkaloids (except for hainanensine, **14**) contain the parent skeleton of cephalotaxine (**1**), the major constituent of the plant extracts. Its unusual skeleton suggests a biogenic relationship to the homoerythrina alkaloids, and, indeed, many homoerythrina alkaloids occur as minor constituents in the various species of *Cephalotaxus*.

Thirty years have passed since the presence of alkaloids in *Cephalotaxus* was first detected by Wall et al. [6]. Ten years after its initial detection, Paudler isolated cephalotaxine, and structure **20** was proposed by Paudler et al. and McKay [7]. This structure was subjected to revision by Powell et al., who suggested two alternatives, **21** and **1**, both of which conformed more precisely to the spectral characteristics exhibited by cephalotaxine [8]. This structural evolution culminated when formula **1** was established unambiguously by Abraham et al. through the X-ray analysis of cephalotaxine methiodide [9].

The structures of some of the esters, namely, harringtonine, isoharringtonine, and homoharringtonine, were established in 1970 [10]. Deoxyharringtonine was elucidated in 1972 by a group in Peoria, Illinois [11]. These studies also yielded the structures of side-chain acids obtained by alkaline hydrolysis of the esters. The unambiguous assignment of those carboxylates involved in esterification of cephalotaxine was eventually confirmed by syntheses, although the regiochemistry of the attachments was correctly advanced at the time of their isolation [11].

The absolute configuration of cephalotaxine was determined by Arora and co-workers in 1974 through X-ray crystallography of the *p*-bromobenzoate of **1** [12] and later of free cephalotaxine [13]. These studies established the configuration as (3*S*,4*S*,5*R*). In this context, the earlier work by Abraham et al. was performed on the methiodide of **1** [9]. Arora et al. reported the racemization of **22**, which seemed to have involved inversion at four chiral

20

21

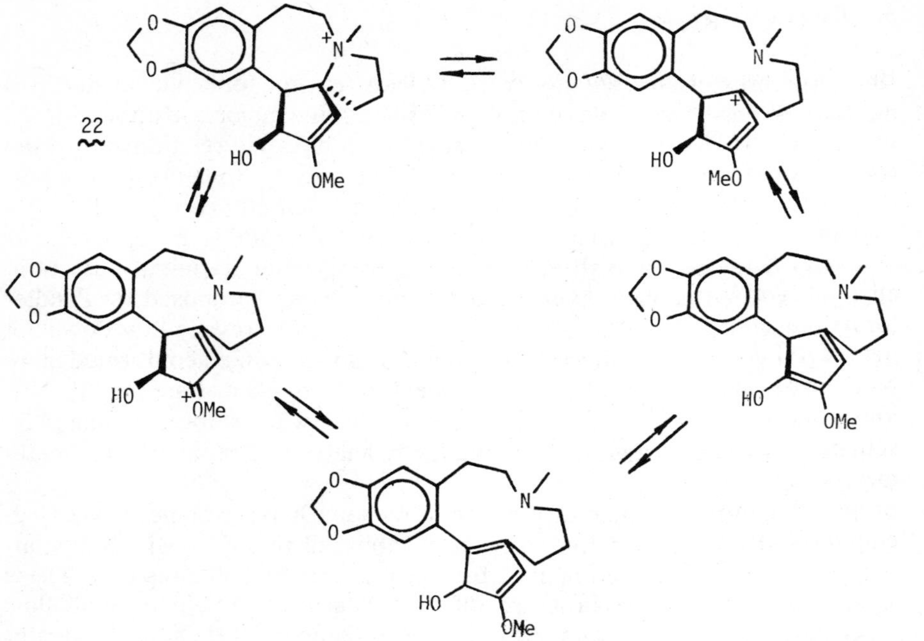

Scheme 1.

centers. This unusual event was rationalized by invoking a series of macrocyclic equilibrations portrayed in Scheme 1 and documented later in several synthetic approaches to cephalotaxine [27,29]. The absolute stereochemistry of the side-chain acids was determined by CD studies. The comparison of spectra of piscidic acid (23) with the diacid residue of isoharringtonine led Brandänge et al. to conclude that 24 has the (2R,3S) configuration [14]. Similarly, the configurations of 25 and 26 were assigned as (2R) because they displayed characteristics *opposite* to those of (S)-citramalic acid [15].

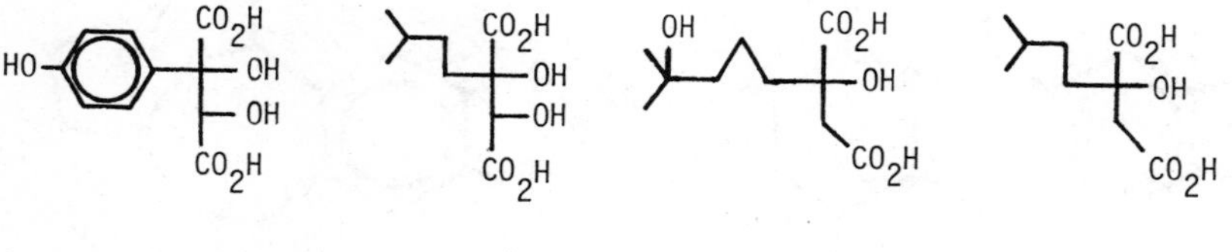

23   24   25   26

            (isoharringtonine) (homoharringtonine) (deoxyharringtonine)

3. BIOSYNTHETIC STUDIES

Because cephalotaxine functions as the biogenic precursor of its variously acylated congeners, the biosynthesis of the alkaloids should be considered in separate acyl and cephalotaxine routes [16,17]. Although both heterocyclic and acyl residues are derived from amino acids, the enzymology and exact chemical transformations of each biosynthetic step are not necessarily known.

The acyl portions of the *Cephalotaxus* alkaloids are derived from L-leucine. This amino acid is oxidatively deaminated, alkylated with acetyl CoA (or its biogenic equivalent), and ultimately decarboxylated to furnish **27**, the net result being methylene insertion into the oxidatively deaminated leucine. Not surprisingly, homoleucine can serve as a biogenic precursor of the various acyl residues. The steps in this process are outlined in Scheme 2. The scheme also indicates the carbon atoms of particular substrates and isolated metabolites that were traced radiometrically as part of the evidence for the proposed pathway. α-Keto acid **27** is then alkylated with acetyl CoA (or equivalent) to produce deoxyharringtonic diacid (**28**). Repetition of the de-

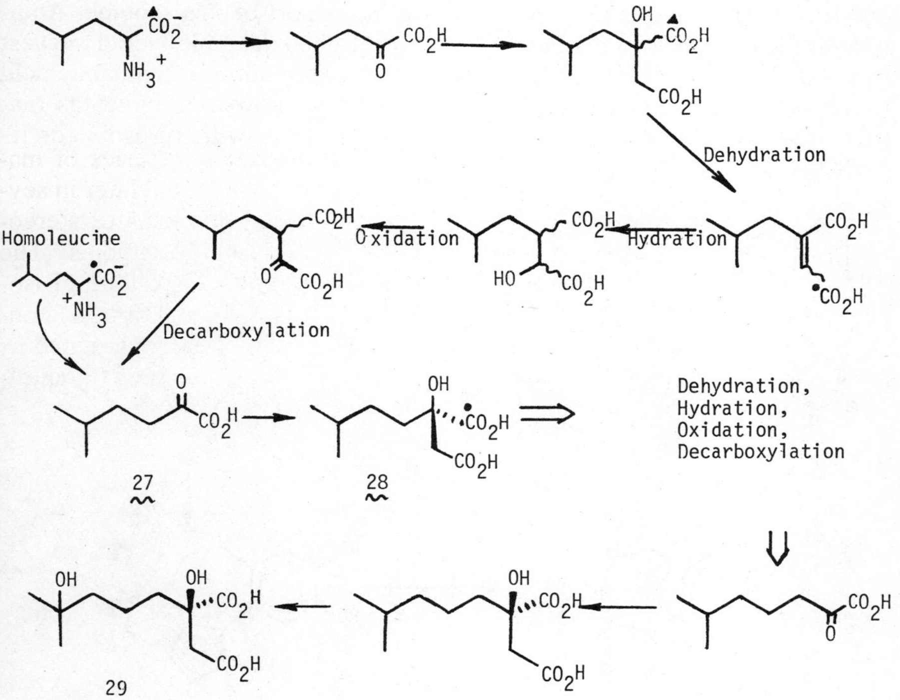

Scheme 2.

hydration, hydration, oxidation, and decarboxylation sequence results in an extension of the α-keto acid, whose alkylation with acetyl CoA and subsequent hydroxylation furnishes homoharringtonic diacid (29), as shown. Such a scheme is not intended to suggest that the *Cephalotaxus* diacids or their precursors are synthesized in vivo in unesterified form.

Because, in the experiments on side-chain biosynthesis, plant extractions were performed under conditions that effected ester hydrolysis, the presence of free acid pools was not verified. In vivo deacylation of alkaloids was suggested, but the relative roles of alkaloid versus free diacid or methyl ester–acid modifications were not ascertained. Deoxyharringtonine is believed to be converted to both harringtonine and isoharringtonine without significant deacylation. At what point the diacid residues or their precursors are either attached to the alkaloid nucleus or methylated is not clearly understood. Thus the exact sequence of events of side-chain attachment and modification leading to the wide spectrum of *Cephalotaxus* alkaloids is not known at this time.

Labeling experiments have indicated that the cephalotaxine nucleus is most likely synthesized from one molecule each of tyrosine and phenylalanine, as shown in Scheme 3 [17,18]. Again, carbon atoms are indicated in substrate and metabolite structures corresponding to the radioactive tracers, and their disposition is taken as evidence in support of this scheme. Additionally, loss of 50% of ^{3}H label from C(3) of phenylalanine would suggest that a dehydroamination process is operating. The failure of cinnamic acid to act as a biogenic equivalent of phenylalanine, however, indicates that such deamination may occur after its condensation with tyrosine (or its derivative).

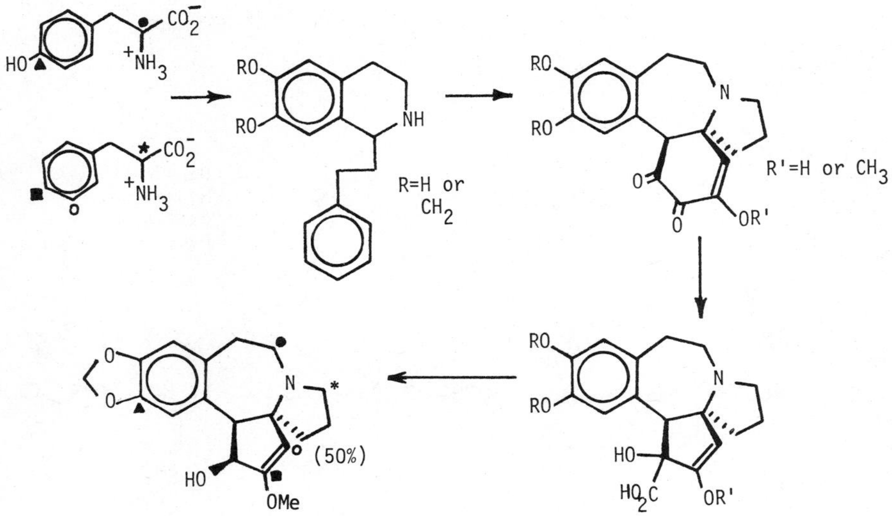

Scheme 3.

Interestingly, although both the aromatic amino acids incorporated into the cephalotaxine skeleton suffer severe oxidation, the phenylalanine and tyrosine moieties are not biogenically interchangeable. The exact sequence of events as portrayed in Scheme 3 is not known. Although dioxygenation of ring A would enhance the oxidative coupling with the phenylalanine-derived aromatic ring, and trioxygenation of the six-membered D ring precursor would likewise be helpful for the ring contraction–decarboxylation, the entities presented in the scheme are hypothetical.

4.   SYNTHESIS OF CEPHALOTAXUS ALKALOIDS

4.1.   Cephalotaxine

Of the 19 alkaloids listed in Table 1, only cephalotaxine (**1**) was prepared through total synthesis, with compounds **11** and **12** being intermediates in the synthesis of **1**. The esters harringtonine, deoxyharringtonine, homoharringtonine, and isoharringtonine were prepared by esterification methods from cephalotaxine of either natural or synthetic origin. The conversion of 11-hydroxycephalotaxine (**15**) to drupacine (**16**) is known [19,20], but a total synthesis of drupacine has not been reported even though several attempts were made to functionalize **1** oxidatively at the benzylic carbon [5,21].

The interest in the synthesis of *Cephalotaxus* alkaloids was spurred on by the discovery of the antitumor activity of several of its esters. In view of their relative unavailability from natural sources, and in view of the medicinal inactivity of cephalotaxine itself, the organic community turned to synthesis as means of attaining the esters.

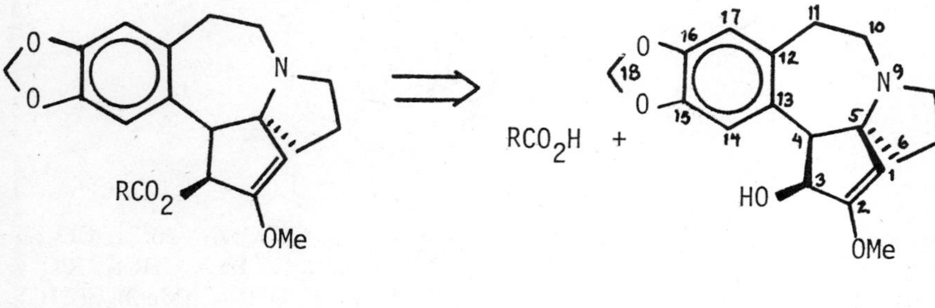

The attempts to prepare cephalotaxine esters serve well to illustrate the synthetic chemist's frequent misfortune—a series of judgmental errors associated with the perceived degree of difficulty of certain operations required to construct the target. At a glance, an ester of **1** would be reduced retrosynthetically to cephalotaxine and the side-chain acyl moiety. This step would appear straightforward to the planner who, having made this dissection, would decide that, of the two portions, the alkaloid nucleus was by far the more complex. Many attempts to carry out the synthesis of esters of type **30**, especially those possessing a large acyl group, proved uniformly unsuccessful [11,53].

On the other hand, some of the most logical concerns that every synthetic

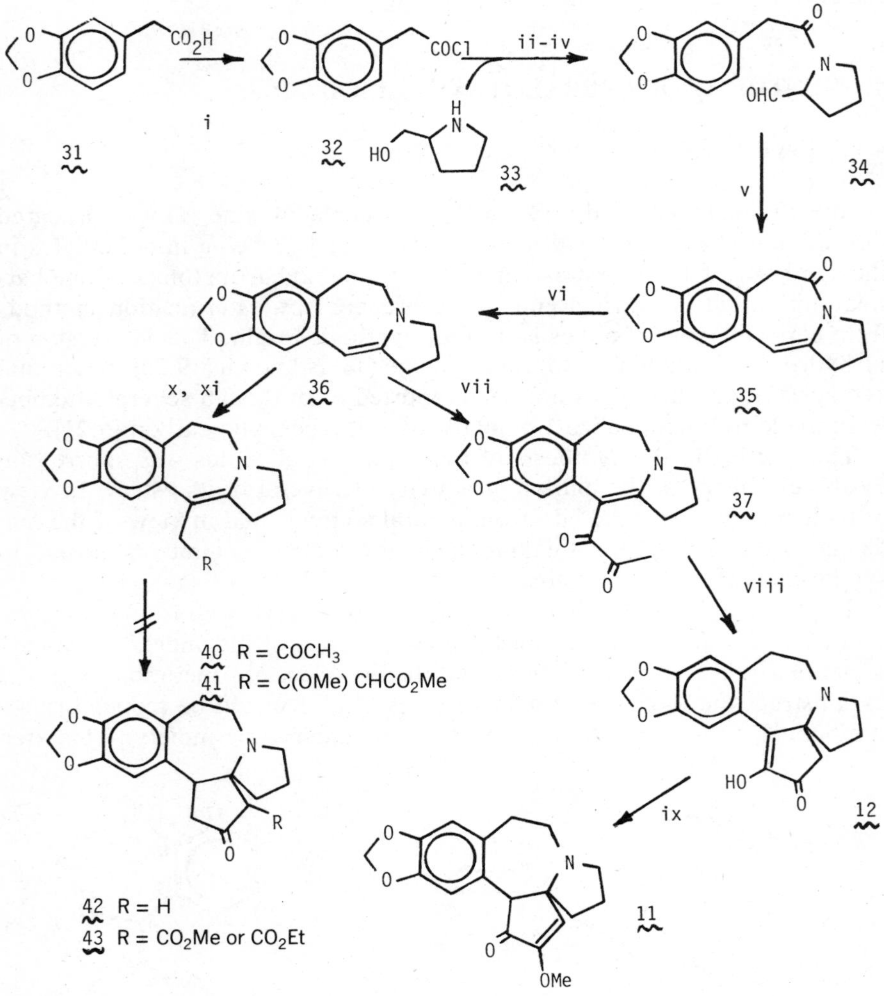

Scheme 4. Weinreb's synthesis. Reagents: i, $SOCl_2$; ii, CH_3CN, $-20°$, K_2CO_3; iii, K_2CO_3 (aq.), 82%; iv. DMSO, DCC, Cl_2CHCO_2H, 67%; v, BF_3, $CHCl_3$, RT; vi, $LiAlH_4$, THF; vii, $CH_3COCO_2CO_2Et$; viii, $Mg(OMe)_2$, MeOH; ix, $(MeO)_2C(CH_3)_2$, p-TsOH; x, propargyl bromide or methyl 4-bromo-3-methoxycrotonate; xi, Hg^{2+}, H_3O^+

chemist would assume to be problematic turned out to be inconsequential. Only through tedious experimentation did the organic community come to understand that the direct esterification of **1** was impossible, whereas concerns such as the stability of the enol ether or control of C(3) stereochemistry did not present major problems.

The synthetic approaches to the unique skeleton of **1** centered around and reflect the following concerns:

(a) formation of C(4)–C(13) bond
(b) quaternary center at C(5)
(c) enol ether moiety at C(1)–C(2)
(d) stereochemistry at C(3) and C(4)
(e) closures of benzazepines by classical methods

There are several reported total syntheses of cephalotaxine and a few approaches to the tetracylic intermediate lacking the cyclopentenyl ring. By and large, these approaches reflect the above concerns. Unfortunately for the organic community, only very few syntheses were subjected to "second-generation" preparations that would lead to improvement in both the yield and the design.

The first total synthesis was reported by Auerbach and Weinreb in 1972 [22,23]. In this approach the formation of the benzazepine ring system was performed early in the synthesis, and the construction of ring E as well as its functionalization was left until the final stages. In this way the data gathered during isolation of *Cephalotaxus* alkaloids could be used to full advantage because the intermediates **12** and **11** are, in fact, known natural products.

The synthesis commenced with 3,4-(methylenedioxy)phenylacetic acid, which was converted to its acid chloride and condensed with prolinol (Scheme 4). The small amount of *N,O*-dialkylated product could be removed by heating the crude mixture with aqueous base to effect selective ester hydrolysis. Oxidation of the resulting amide alcohol to **34** provided the required intermediate for cyclization. The closures of benzazepines by the application of Pictet–Spengler or similar reactions have been uniformly sluggish, leading instead to isoquinoline ring systems [24,25]:

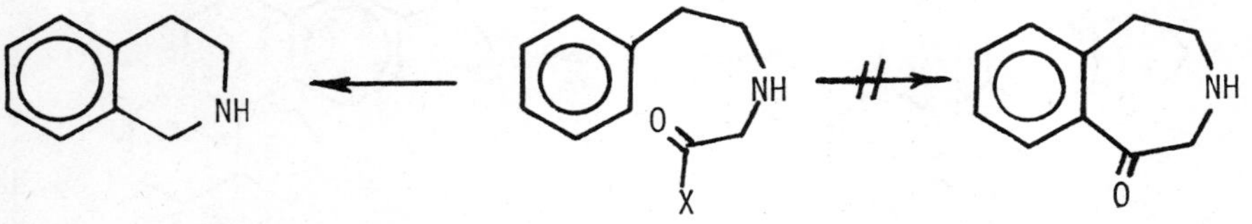

A solution to this problem appeared in the use of anilides as activating groups [26]; however, the acid-catalyzed closure of aldehyde **34** would appear to alleviate the problem of benzazepine ring closure. Enamide **35** was reduced to a somewhat labile tetracyclic enamine **36**, which was used immediately in the next steps. The attachment of ring E of cephalotaxine presented some initial problems. Alkylation of **36** with propargyl bromide and hydration of the acetylene gave ketone **40**, which did not yield to cyclization to afford **42** under various conditions of acid catalysis. Neither did the more activated keto ester prepared via crotonate **41** yield the cyclized keto ester **43**. The problem was eventually solved by the preparation of vinylogous amide **37** via the interaction of enamine **36** and a mixed anhydride prepared from pyruvic acid and ethyl chloroformate. This procedure was superior to other less direct attempts at the preparation of **37**, which was made available reproducibly in 73% yield [23]. Under improper conditions hainanensine (**14**) could be isolated from the reaction mixtures.

The failure of ketones and keto esters **40** and **41** to undergo the desired cyclization contrasted with the report that treatment of **1** with ethyl γ-bro-

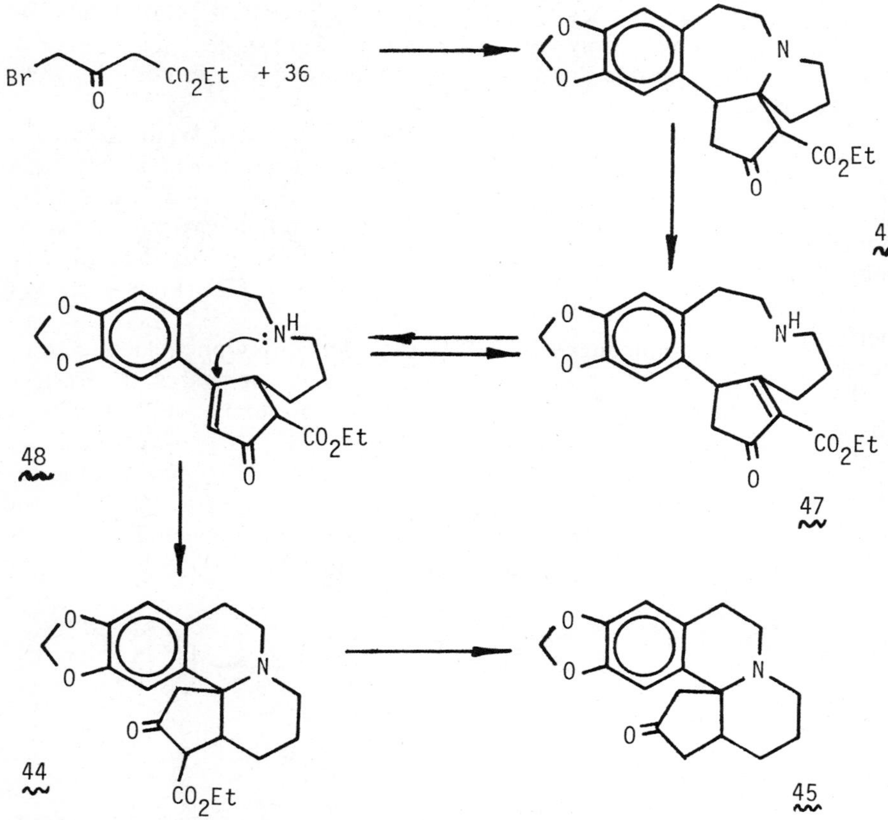

Scheme 5.

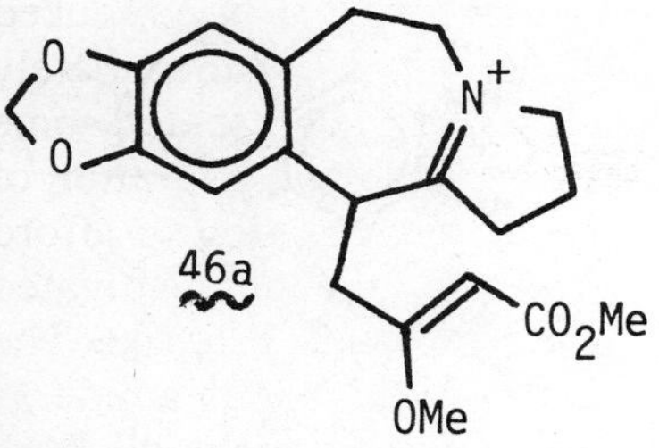

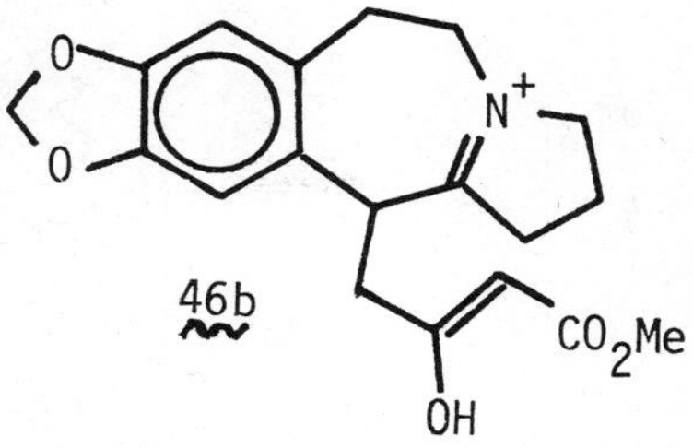

moacetoacetate (instead of ethyl 4-bromo-3-methoxycrotonate) led to the formation of keto ester **44**, which was initially thought to have the structure **43** (Scheme 5). Decarboxylation of this substance provided pentacyclic tetrahydroisoquinoline **45** [27]. The structure of this material was proved by its dioxygenation to a known compound [28].

It is interesting to ponder the subtleties of the two systems. The acid-catalyzed cyclization of **41** required the interaction of enol ether **46a** with an immonium ion, whereas a simple enol was the nucleophile in the case of **46b**. This difference may have been sufficient in the initial cyclization of **46b** to **44**. As Dolby rationalized, the initially formed pentacycle is susceptible to retro-Michael addition, which creates a stable chromophore **47** and would be favored under equilibrating conditions. Under these conditions, the creation of the regioisomeric enone **48** would set the stage for a new Michael reaction to produce **44**. These observations also lent credence to aformentioned proposed biosynthetic arguments relating to the existence of macrocyclic equilibria [29].

The successful cyclization of **37** to demethylcephalotaxinone (**12**) was performed with magnesium methoxide in methanol and was thought to proceed through a chelated enone of type **38**.

The resulting demethylcephalotaxinone was methylated with diazomethane exclusively to enol ether **39**, but under equilibrating conditions of *O*-methylation, it yielded cephalotaxinone (**11**) in high yield, reflecting the relative stability of the olefins in **39** and **11**. Reduction of **11** with sodium boro-

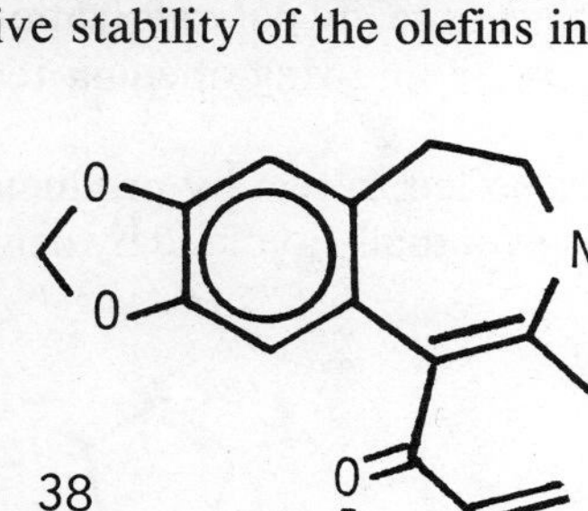

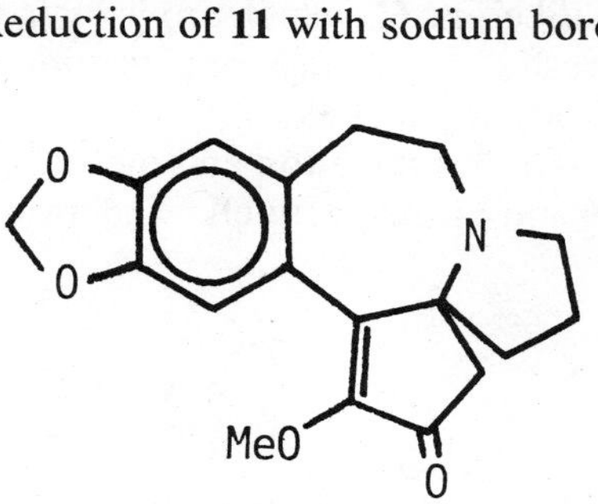

18

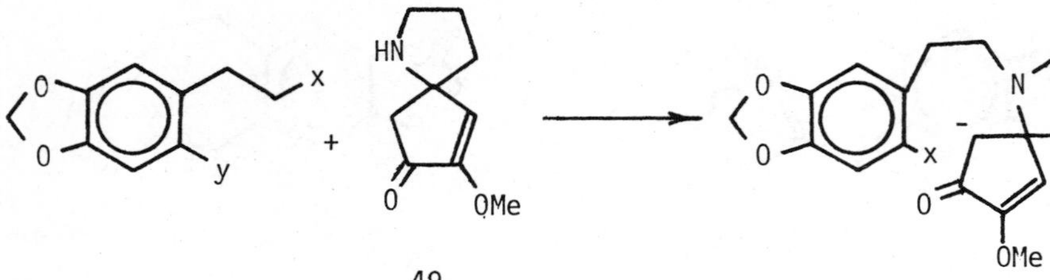

hydride gave cephalotaxine stereospecifically in 4.5% overall yield, which was later optimized by the Merck group to 10.4% [30].

The second cephalotaxine synthesis was reported by Semmelhack and co-workers [31,32]. Conceptually different from the Auerbach–Weinreb design, it is an exercise in aromatic nucleophilic substitution. The key element of the synthesis involved a separate construction of the spirocycle **49** and its union with tetrasubstituted phenethylamine to give **50**, which was predisposed to undergo the crucial C(4)–C(13) bond formation by an intramolecular nucleophilic substitution to give cephalotaxinone (**11**). The spirocycle **49** thus served as a double nucleophile, this property having been reinforced by appropriate functional consonance.

The synthesis began with 3,4-(methylenedioxy)-6-chlorophenylacetic acid (**51**) obtained in 55% yield from piperonal and converted in two steps to nitrobenzenesulfonate **52** (Scheme 6). The construction of spirocycle **49** commenced by conversion of 2-pyrrolidone to its imino ester and condensation with excess allylmagnesium bromide to afford amine **53**, which contained the necessary quaternary center of **1**. This substance was converted to diester **54** via oxidation of the olefin. The projected acyloin condensation to produce **55** or **56** met with initial failure until proper experimental conditions were established by using sodium/potassium alloy. This symmetrical intermediate was oxidized to **49** by treatment with bromine followed by fragmentation and O-methylation. Enone **49** was usually prepared from crude α-diketone **57**, which proved very labile and subject to ready polymerization, presumably because of its zwitterionic nature as **58** and ring-opening tendency to yield **59**.

The bromination and methylation sequence also led to two by-products, **60** and **61**, respectively. The formation of **60** was eventually avoided by using

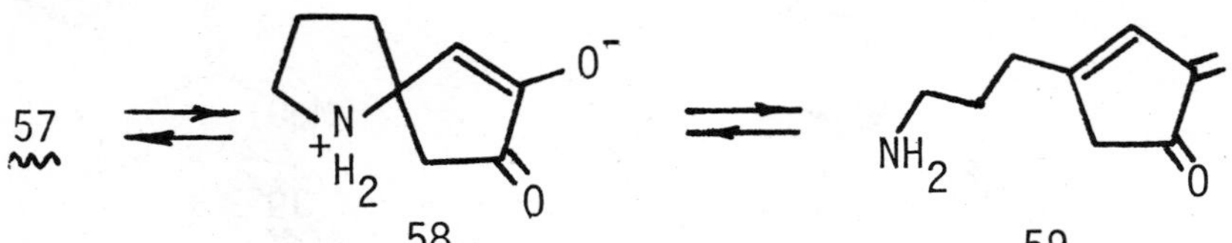

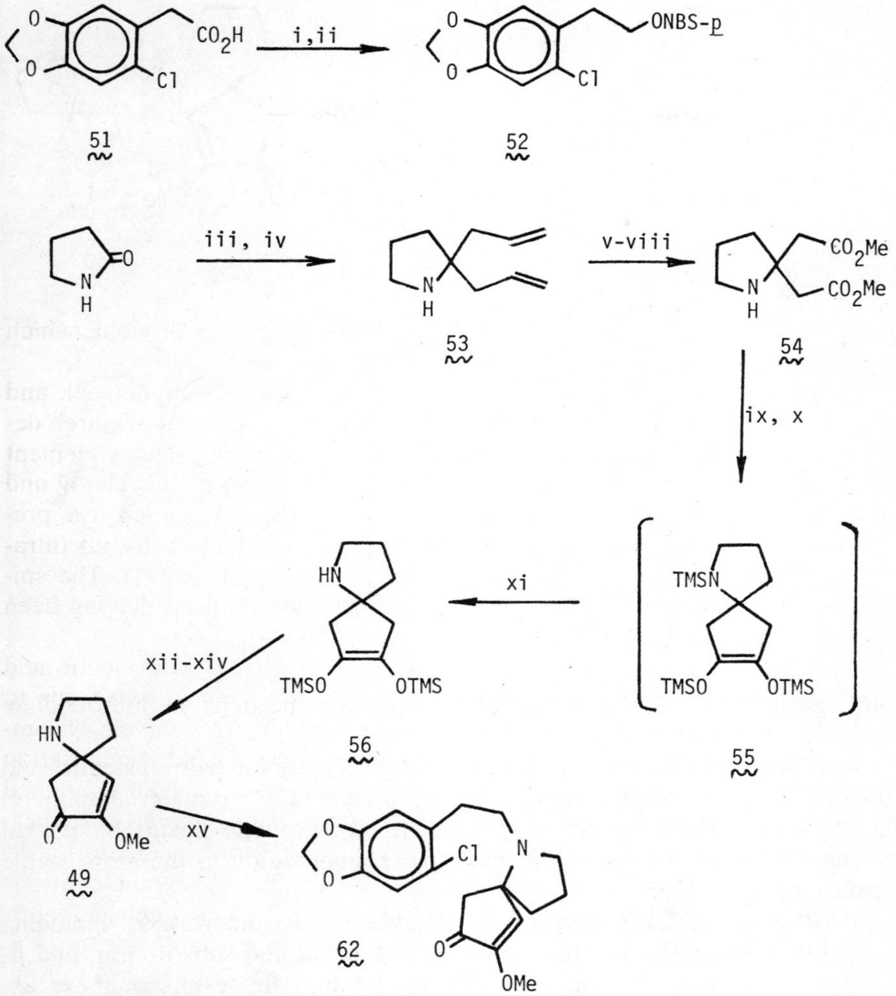

Scheme 6. Reagents: i, LiAlH$_4$; ii, *p*-NO$_2$C$_6$H$_4$SO$_2$Cl; iii, Et$_3$O$^+$BF$_4^-$; iv, allylmagnesium bromide, 3 equiv.; v, *t*-butoxycarbonyl azide; vi, O$_3$, MeOH; vii, Ag$_2$O; viii, MeOH, HCl; ix, Na/K (1:5); x, TMSCl; xi, moist ether; xii, Br$_2$, $-78°$; xiii, $-30°$; xiv, CH$_2$N$_2$, CH$_2$Cl$_2$; xv, **52**, (*i*-Pr)$_2$NEt.

stoichiometric quantities of bromine, and the production of **61** was minimized by the use of less polar solvents for the diazomethane methylation. Both these requirements would appear to be difficult to meet when crude preparations of **58** were used. The convergence of **52** and **49** was accomplished by treatment of these two substances with diisopropylethylamine in acetonitrile to yield the penultimate intermediate **62**. Treatment of this substance with two equivalents of potassium triphenylmethide in dimethoxyethane at

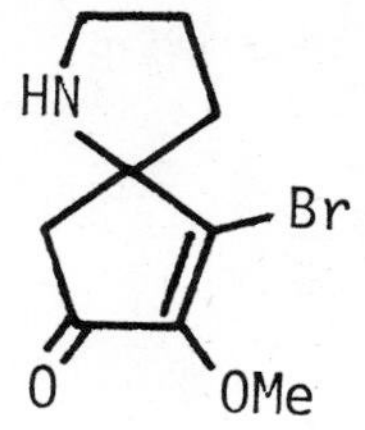

60

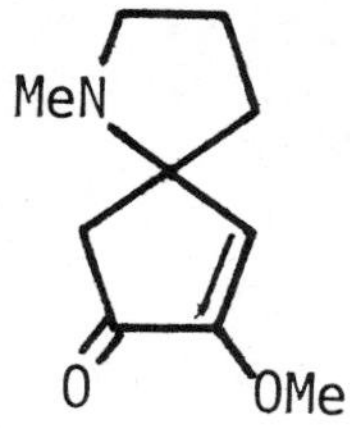

61

62 →

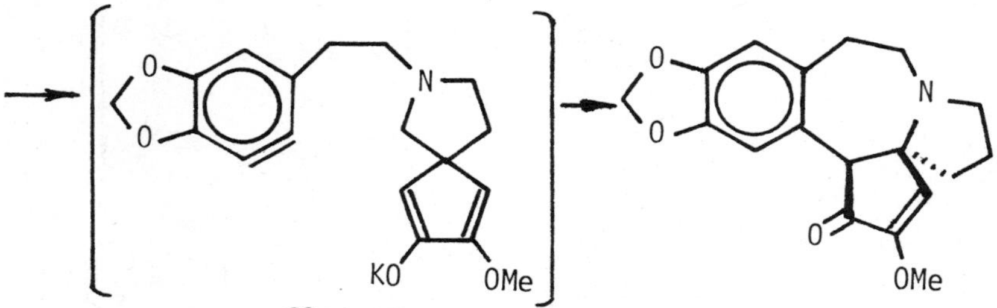

63

11

50°C gave 13–16% yield of cephalotaxinone via the benzyne intermediate **63**.

Substitution of bromine or iodine into the precursor haloarylacetic acid did not lead to substantial improvements in yield [31]. No material epimeric at C(4) was isolated; this attests perhaps to either kinetic closure having led to the correct stereochemistry or a rapid epimerization to the more stable natural configuration.

A detailed investigation of alternatives was undertaken, including nickel(0), copper(I), enolates, anion–radical-promoted substitution, and finally, photostimulated nucleophilic substitution. The results of these experiments consistently raised the yield of cephalotaxinone while decreasing the complexity of each successive reaction mixture. Table 2 summarizes the conditions and the yields of cephalotaxinone **11** [32,33]. This work indicates that the intermediate most likely present in the reaction mixtures was the

Table 2

Aryl Halide	Conditions	Yield of **11** (%)
Cl	$K^+C^-Ph_3$, DME, 50°	15
I	LDA, (biscyclooctadienyl) Ni(0)	35
I	$NaNH_2$, Na/K, NH_3	45
I	NH_3, t-BuOK, hv	94

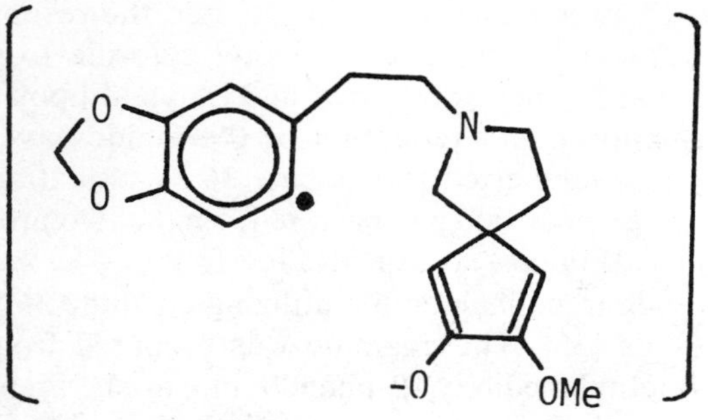

64

anion radical **64** generated from an aryl iodide by an electron-transfer process in accord with an $S_{RN}1$ mechanism of nucleophilic aromatic substitution.

The synthesis of **1** was completed by reduction of cephalotaxinone with diisobutylaluminum hydride. The two total syntheses presented above have been summarized in a review article [34].

The approach to cephalotaxine reported by Dolby et al. [27] and discussed above in the context of macrocyclic equilibria is conceptually similar to that of Auerbach and Weinreb in that the construction of **1** relied on the aforementioned alkylations and subsequent cyclization of **36**, which was, however, synthesized by quite a different route, as portrayed in Scheme 7. The first step consisted of a Vilsmeier–Haack condensation of N,N-dimethyl-piperonylamide and pyrrole leading to ketone **65**. The benzylic reduction

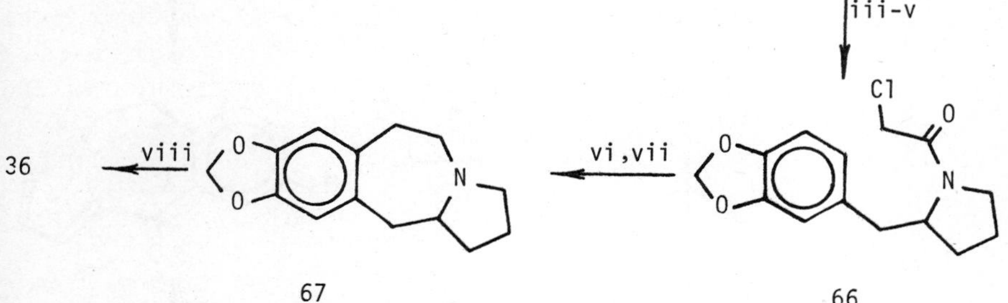

Scheme 7. Reagents: i, POCl₃; ii, H₂O, iii, NaBH₄; iv, H₂/Rh–Al₂O₃; v, ClCH₂COCl; vi, hν/40% EtOH/H₂O; vii, LiAlH₄; viii, Hg(OAc)₂.

was accomplished with sodium borohydride, and the resulting pyrrole was hydrogenated and acylated with chloroacetyl chloride to yield amide **66**. Cyclization of this substance proceeded in 25% yield upon irradiation with a Hanovia lamp. Subsequent reduction of the amide gave the tetracyclic amine **67**, which was converted to enamine **36** by oxidation with mercuric acetate. Such strategy provided a departure from the two previous syntheses in that the C(4)–C(13) bond was constructed first.

Another approach to cephalotaxine utilizing enamine **36** was reported by Weinstein and Craig [35]. The enamine was prepared from 2-carbethoxy-pyrrole and 3,4-(methylenedioxy)-β-phenethanol as shown in Scheme 8. The decision to use pyrrole-2-carboxylate rather than proline was necessitated by a ring expansion reaction that took place instead of the desired cyclization of proline-amide **68**.

None of the desired tetracycle **69** was obtained, and the tendency of **68** toward ring expansion to **71**, via aziridonium ion **70**, was attributed to the basicity of the proline nitrogen. Thus acid **73** was prepared by alkylation of the primary tosylate **72** and subsequent hydrolysis. Friedel–Crafts acylation of this material was accomplished by treatment with excess stannic chloride and trifluoroacetic anhydride in chloroform, and the resulting ketone **74** was hydrogenated to Dolby's saturated amine **67** with rhodium on charcoal at atmospheric pressure.

A unique approach to the same saturated amine **67** was undertaken by Tse and Snieckus [36]. The C(4)–C(13) bond was constructed by an intra-molecular photochemical cyclization of an enamide and an aryl halide

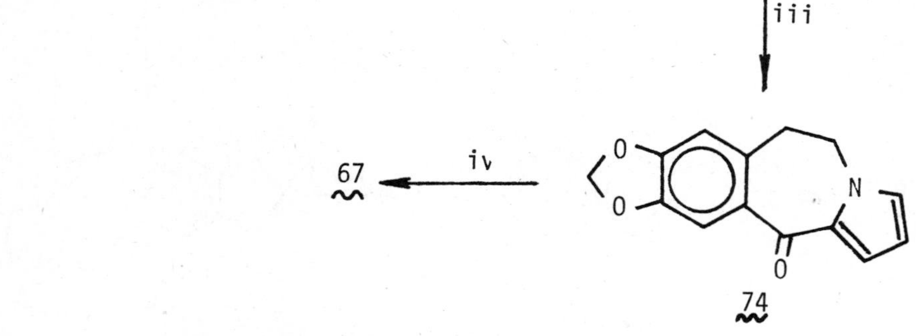

Scheme 8. Reagents: i, 2-carboethoxypyrrole, NaH, diglyme; ii, KOH, EtOH; iii, $(CF_3CO)_2O$, $SnCl_4$, $CHCl_3$; iv, Rh(C), H_2.

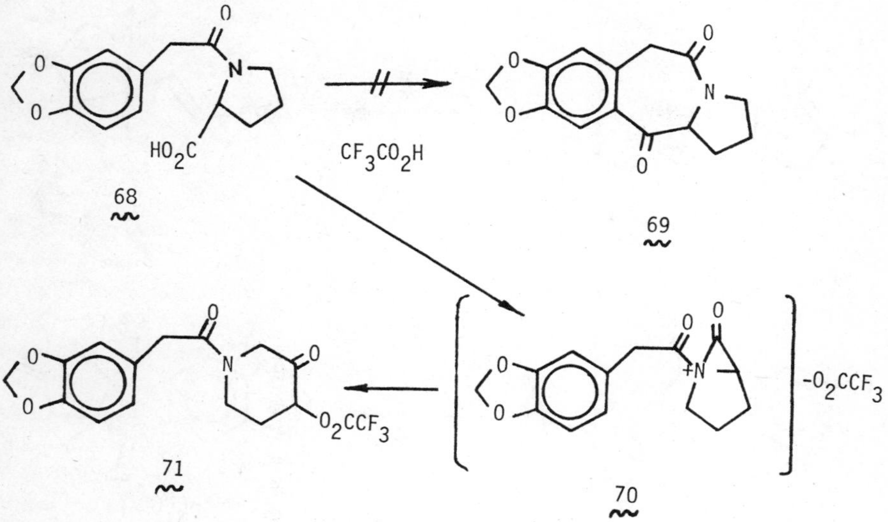

(Scheme 9). Imide **75**, prepared in two steps from 3,4-(methylenedioxy)-β-phenethylamine and maleic anhydride, was iodinated to afford **76**. Addition of methylmagnesium iodide followed by dehydration gave a labile enamide **77**, which was subjected to photocylization to afford unsaturated amide **78** in 40% yield. Hydrogenation and reduction of this material to saturated

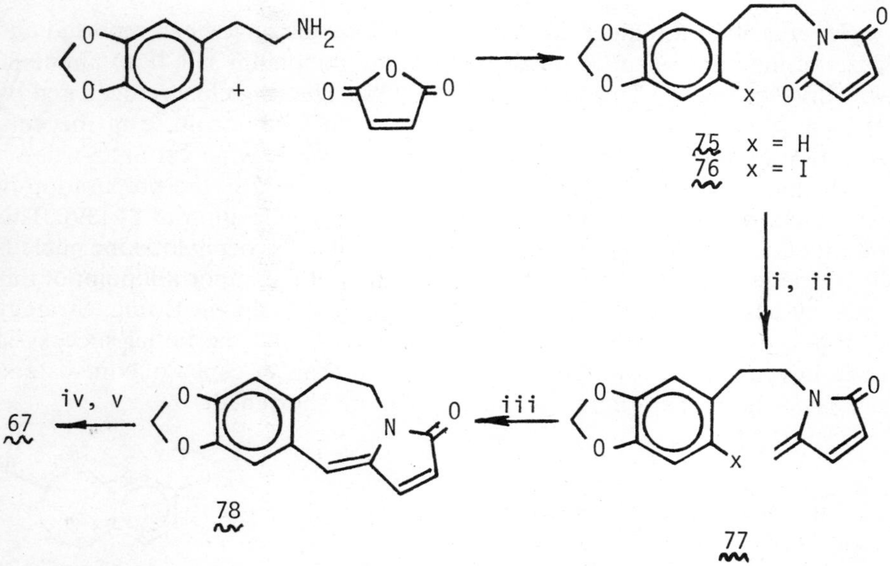

Scheme 9. Reagents: i, MeMgI; ii, TsOH/PhH; iii, PhH, Et₃N, *hν*; iv, H₂; v, LiAlH₄.

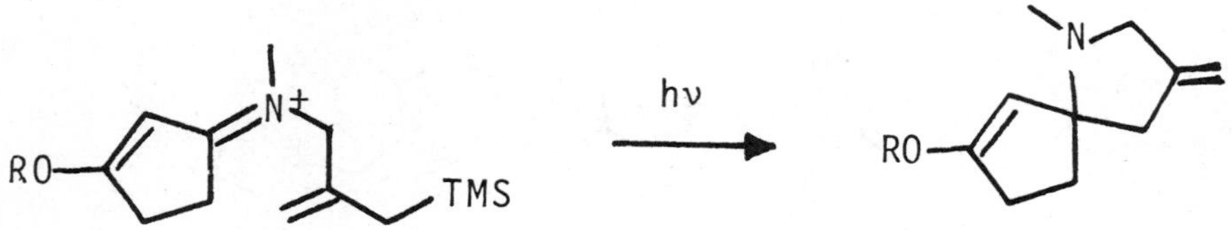

amine **67** served as a structure proof for the product of this cyclization. Similar cyclization of a vinylogous amide onto an aryl halide has been reported in the homoerythrina series [37].

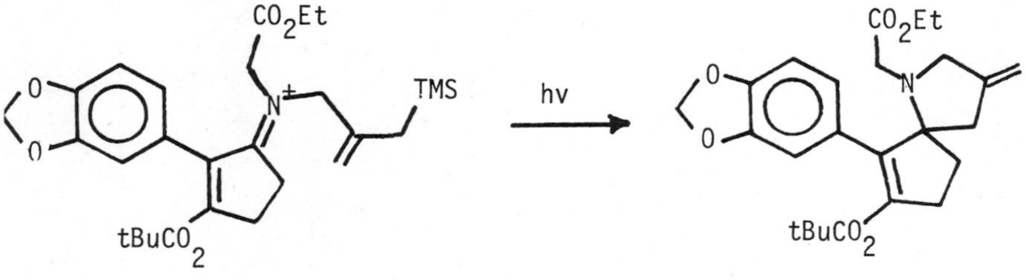

Several other attempts to prepare cephalotaxine have been reported and deserve mention. A model cyclization of an immonium salt **79** to amine **80** was investigated by Tiner-Harding et al. [38]. The ring closure occurred by electron-transfer-initiated photocyclization and holds promise in the construction of the nucleus of cephalotaxine by a strategy as yet untested.

Further development of this methodology resulted in the preparation of tetracyclic amine **82** via an immonium ion photocyclization of **81** [39]. This intermediate would, in principle, be convertible to the cephalotaxine nucleus by a Friedel–Crafts reaction of its side chain, but the important point of this strategy lies in the use of a hindered ester on the incipient E ring. Strategy of this type had not been attempted previously, and the initial success in model cyclization bodes well for a direct synthesis of cephalotaxine esters, provided the preparation of precursors can be shortened.

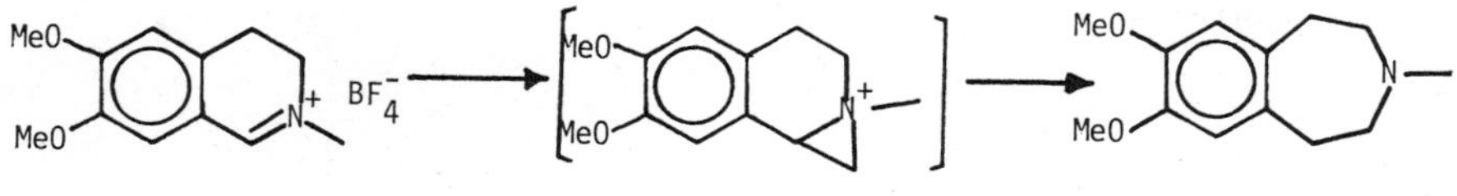

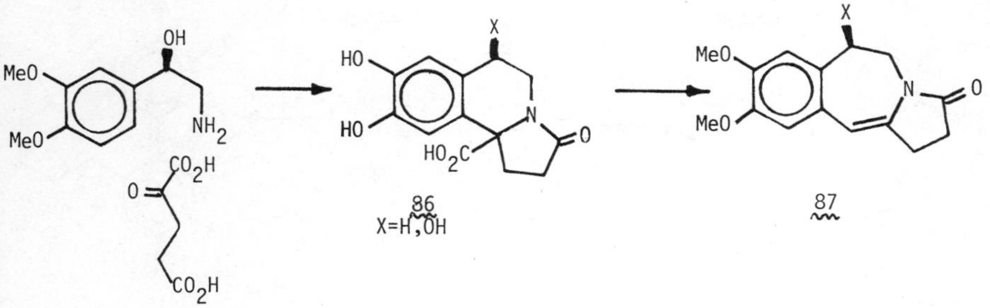

In an approach to cephalotaxine via ring expansion of an isoquinoline aziridinium ion of type **84**, Greene reported a successful transformation of **83** to benzazepine **85** as a prelude to the transformation of a more complex immonium salt, such as **86**, to cephalotaxine [40]. Similar transformations have been attempted by Lenz in the isoquinoline alkaloid series [41]. An attempt to construct cephalotaxine and 11-hydroxycephalotaxine via chiral precursors using an isoquinoline ring expansion of this type has been made. Condensation of biogenic amines with various pyruvic acids led to high yields of tetrahydroisoquinolinecarboxylic acids [42]. In this way acids **86** have been prepared and reduced to the corresponding alcohols, which were subjected to solvolyses under a variety of acidic conditions. Unoptimized yields of tricyclic enamines of type **87** have been obtained [43].

Biomimetic approaches to the synthesis of cephalotaxine have been attempted by several investigators. Kupchan et al. prepared tetrahydroisoquinoline **88** and subjected it to the conditions of oxidative coupling using vanadium oxide/trifluoroacetic acid/methylene chloride (Scheme 10) [44,45]. The resulting quinone **89** was cleaved to imine **90**, whose reduction, methylation, and debenzylation led to free amine **91**. Cyclization of **91** with potassium ferricyanide gave the assumed biogenic precursor to cephalotaxine, quinone **92** [45]. Marino and Samanen oxidized amine **91** with potassium hexacyanoferrate and obtained two dienones, **93** and **94** [46]. In a related experiment, McDonald and Suksamrarn obtained only the benzazepine dienone **93** [47].

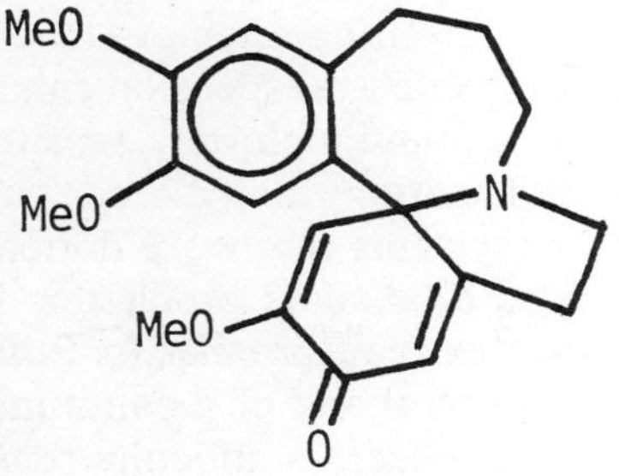

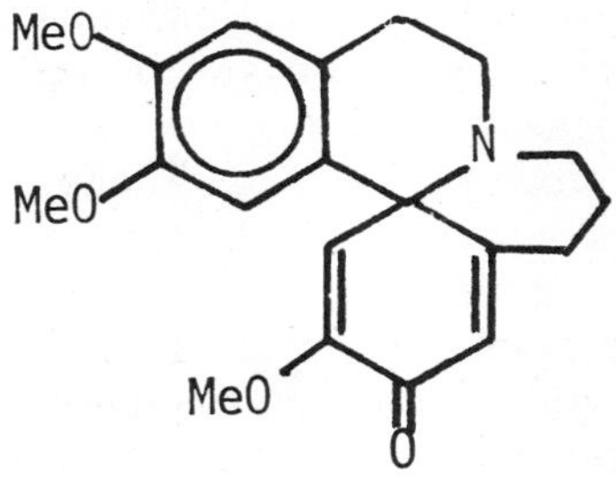

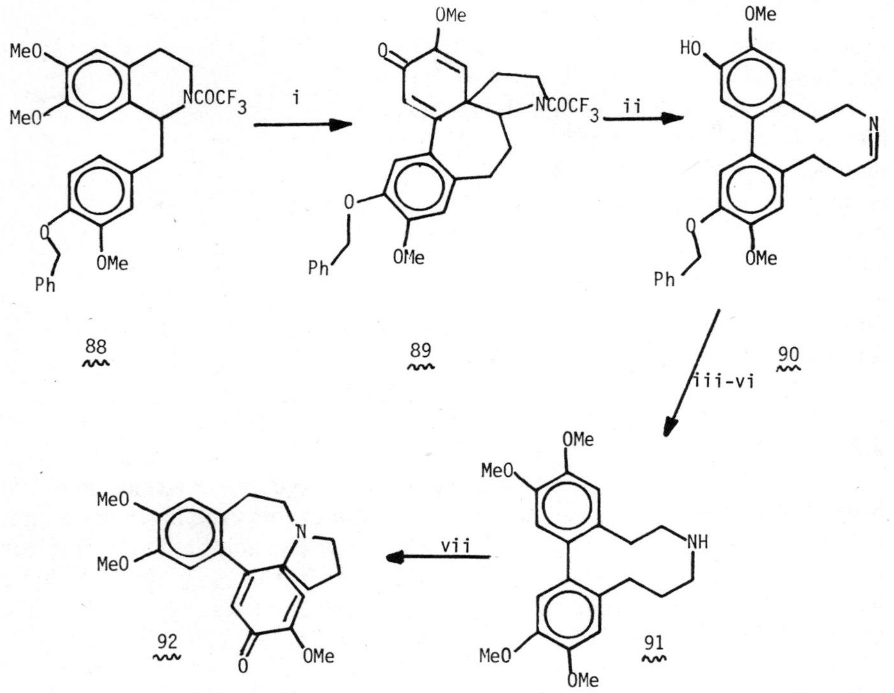

Scheme 10. Reagents: i, VO$_3$/CF$_3$CO$_2$H; ii, 1N NaOH; iii, NaBH$_4$; iv, (CF$_3$CO)$_2$O; v, CH$_2$N$_2$; vi, Pd(C), H$_2$; vii, K$_3$Fe(CN)$_6$, 10%.

Finally, serious attempts have been made by several groups to produce cephalotaxine and its esters by callus and suspension tissue cultures. Pioneering work in this novel approach was performed by Delfel and co-workers, who demonstrated that both cephalotaxine and its esters are produced by cultures [48–52]. Interestingly, the alkaloid profiles of several cultures were not identical, although the cultures originated from the same species, *C. harringtonia*. The ester content was highest during the early stages of development, whereas cephalotaxine content increased with the age of the cultures. Additional difficulties were encountered during attempts to produce homogeneous lines of callus cultures as a prelude to suspension culture propagation on a larger scale. The cultures were found uniformly sensitive to precise growth conditions and periods of illumination; they did not yield reliable results, perhaps because of enyzmatic systems that were dormant during different seasons. Nevertheless, the idea of alkaloid production by means of plant tissue culture is brilliant and holds extreme promise for future manufacture of the *Cephalotaxus* alkaloids in general and of its antitumor esters in particular. A precise study of the various factors affecting reproducibility of alkaloid profiles would no doubt cast additional light on the speculative nature of the arguments proposed for the biogenesis of these alkaloids.

In summary, it appears that the most efficient preparation of cephalotaxine to date is that of Auerbach and Weinreb. Although half the synthetic material must be sacrificed through resolution and although at present there is no asymmetric synthesis, this procedure is by far the simplest, and improvements of practical nature could be made only in the synthesis of enamine **36**. On the other hand, the tissue culture production of the alkaloids and perhaps metabolic alteration of the organism itself to maximize production of desired compounds would appear to be the most logical pursuit for the future.

4.2. Cephalotaxine Esters

Most of the synthetic approaches to cephalotaxine esters are concerned with the preparations of harringtonine, deoxyharringtonine, isoharringtonine, and homoharringtonine. This is hardly surprising in view of the antitumor activities of these alkaloids. Attempts to prepare these esters by direct esterification of a particular side-chain acid derivative and cephalotaxine have been uniformly unsuccessful [11,53]. This lack of success was attributed to two factors. First, the steric environment of the hydroxyl of cephalotaxine is extremely hindered, allowing the approach of an incoming acyl chain only from the direction perpendicular to the plane bisecting the pentacyclic skeleton. This phenomenon thus greatly reduces the probability of favorable configurations leading to bond formation in the transition state. Second, the esters are quite resistant to hydrolysis, attesting to the extreme steric hindrance of the carboxylate environment. Perhaps the most informative manuscript listing most of the unsuccessful attempts and advancing the probable reason for these failures is that of Kelly and the Peoria group. The descriptive information is contained in the footnotes of the paper dealing with the synthesis of harringtonine [54].

Harringtonine. The synthesis of the acyl portion of harringtonine was carried out in 1973 by Kelly et al. [55]. The protected propargylic alcohol **95** was condensed via its lithium acetylide with *t*-butyl ethyl oxalate to afford ester **96** (Scheme 11). Addition of methyl lithioacetate followed by treatment with trifluroacetic acid provided acid **97**, whose hydrogenation yielded the side-chain acid of harringtonine, **98**. This material lactonized spontaneously to **99**.

Two routes to harringtonine were reported almost simultaneously by Huang et al. [56] and Mikolajczak and Smith [57]. In the first of these, butyrolactone **100** (Scheme 12) was condensed with ethyl oxalate and the resulting pyruvate treated with acid to give hemiketal **101**. Acylation of this substance with cephalotaxine gave complex mixtures, and two solutions to this problem were put forth. First, the hydroxyl group was dehydrated and the acid converted to its acid chloride **102**. The esterification of cephalotaxine proceeded in very high yield, perhaps reflecting the unhindered surroundings of the sp^2-hybridized α carbon in **102**. Additional support for the ease of such esterifications, as opposed to those involving intermediates having a fully sp^3-hybridized α carbon, came from the later work of Hiranuma et al. [73]. Second, the unsaturated ester was hydrated to **104**, which could alternatively be obtained by esterification of the α-methoxy acid to produce **103**, whose hydration gave hemiketal **104**. The exposure of this material to the organozinc reagent derived from methyl bromoacetate and active zinc gave under the conditions of the Reformatsky reaction a mixture of harringtonine (**2**) and its epimer in a ratio of 1:1.1 [56].

The strategy used by the Peoria group is identical with the one just described except that acid chloride **102** was prepared from ethyl 4-methylpent-3-enoate via ester **105** [57]. Both groups reported diminished yields of the Reformatsky reaction, because cleavage of the ester under the reaction conditions gave back cephalotaxine. The yields of harringtonine were reported to be 20 and 10%, respectively.

Kelly et al. reported a conceptually different preparation of **2** starting from cyclohexene **106** (Scheme 13), produced as a major product in a Diels–Alder reaction of isoprene with the enol acetate of ethyl pyruvate [54]. This substance was protected as benzyl ether/benzyl ester **107** and the double bond cleaved oxidatively to provide keto acid **108**. Condensation with methylmagnesium bromide followed by lactonization and hydrogenation gave the

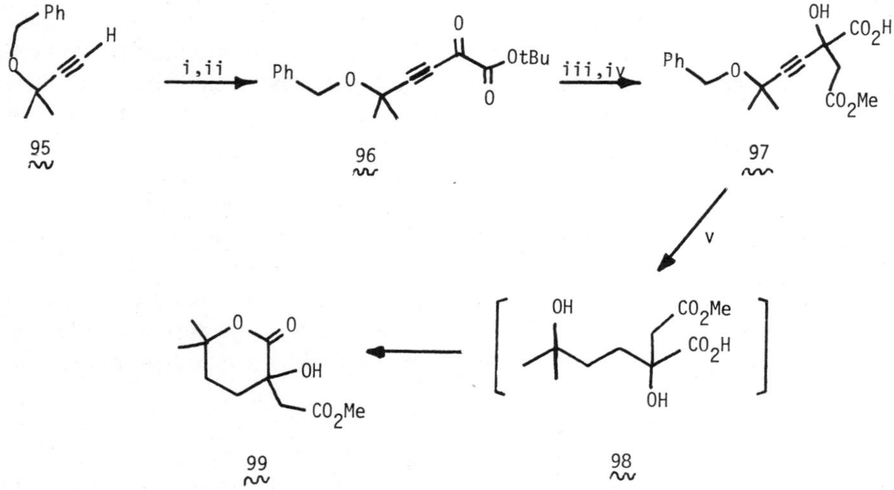

Scheme 11.   Reagents: i, *n*-BuLi; ii, EtO$_2$CCO$_2$*t*Bu; iii, LiCH$_2$CO$_2$Me; iv, CF$_3$CO$_2$H; v, Pd(C), H$_2$.

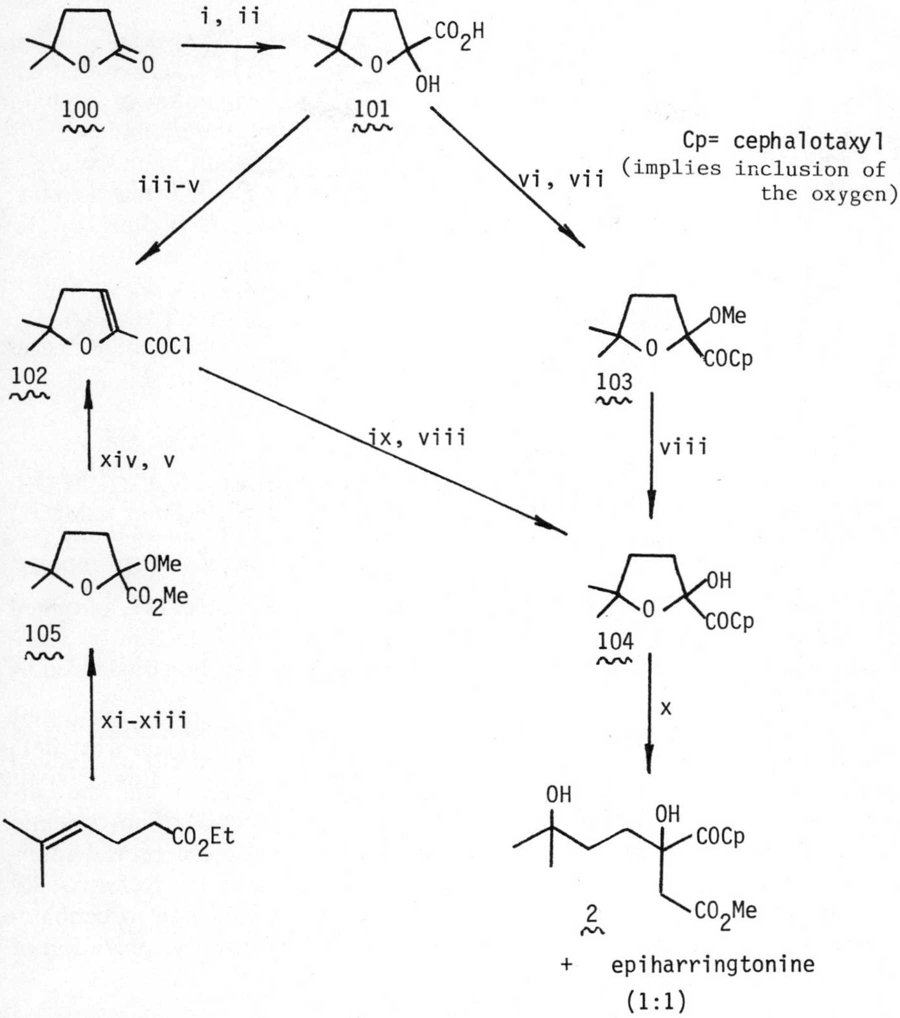

Scheme 12. Reagents: i, $(CO_2Et)_2$, OH^-; ii, H^+; iii, PhH, Δ; iv, Na_2CO_3; v, $(COCl)_2$; vi, MeOH; vii, DCC, cephalotaxine; viii, HOAc, HCl; ix, cephalotaxine; x, $BrCH_2CO_2Me$, $ZnCl_2$, K; xi, $(CO_2Et)_2$, NaH; xii, HCl, Δ; xiii, HCl, MeOH; xiv, NaOH.

lactone carboxylic acid **109**. Noteworthy in this preparation is the resistance of the benzyl ether moiety to undergo hydrogenolysis in the presence of a benzyl ester. The esterification of the acid chloride derived from **109** with cephalotaxine proceeded well to give **110**, which was subjected to methanolysis and hydrogenation to provide harringtonine and its epimer in 58% yield as a 1:1 mixture. The inertness of cephalotaxine enol ester to hydro-

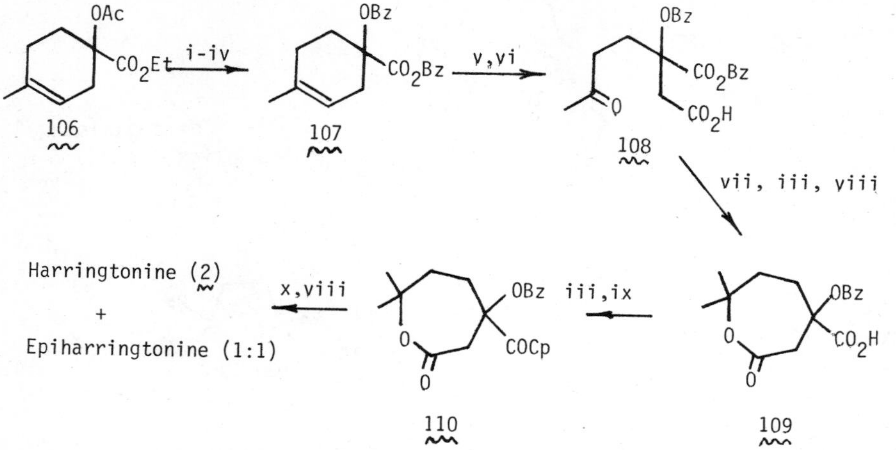

Scheme 13. Reagents: i, NaOH; ii, $KH/C_6H_5CH_2Br$; iii, $(COCl)_2$; iv, $C_6H_5CH_2OH$; v, OsO_4/IO_4^-; vi, CrO_3; vii, CH_3MgBr; viii, $H_2/Pd(C)$; ix, cephalotaxine; x, $MeO^-/MeOH$. Cp = cephalotaxyl.

genation was noted during these experiments. Because acid **111**, obtained by hydrolysis of **106**, has been prepared in proper configuration by resolution, the aforementioned synthesis of harringtonine can be considered as asymmetric [54].

A shortened version of the harringtonine synthesis appeared in 1975 as reported by the Tumor Research Group of the Chinese Academy of Medical Science [58]. Pyruvate **112** was prepared from cephalotaxine and the corresponding acid. Oxidation of this compound with mercuric trifluoroacetate followed by sodium borohydride reduction of the organomercurial intermediate gave hemiketal **104**, which was converted to **2** by the Reformatsky reaction. A notion was advanced that the excessive cleavage to cephalotaxine occurs during the Reformatsky reaction by internal lactonization of

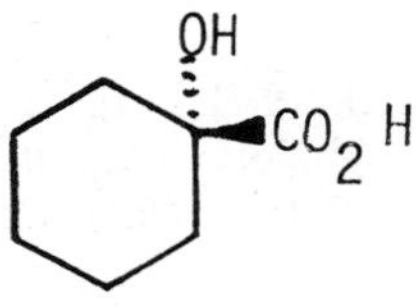

111

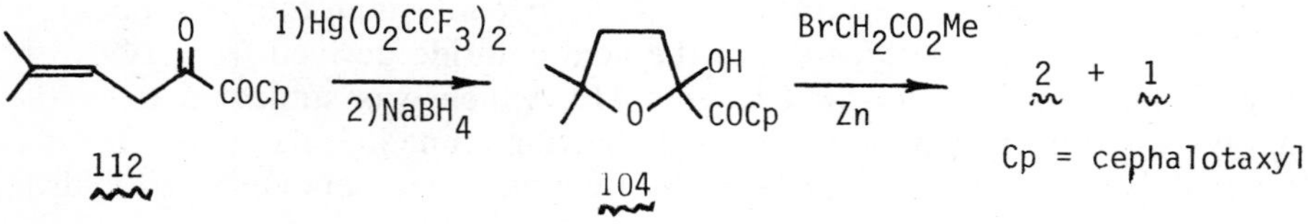

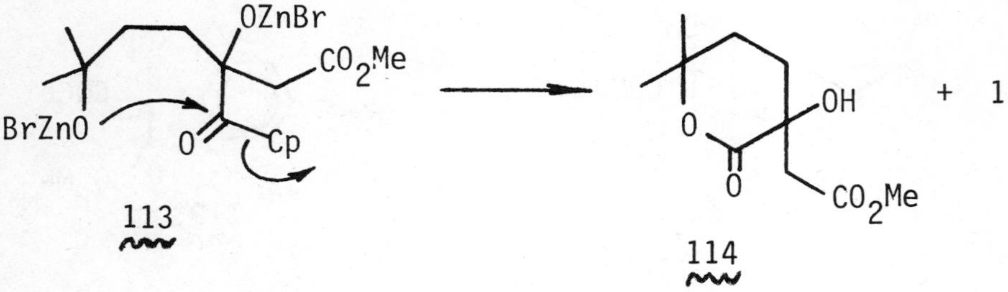

the zinc alkoxide (113) via a six-membered transition state to produce lactone 114 [58].

Deoxyharringtonine. The side-chain acid of this alkaloid ester was prepared by several groups independently and by similar strategies. Mikolajczak and co-workers reported in 1974 a partial esterification route to **3** by condensing methyl lithioacetate with cephalotaxyl pyruvate 117 (Scheme 14) [59].

The requisite acid 116 was available from the union of lithium 3-methylbutyne acetylide with *t*-butyl oxalate, followed by hydrogenation and hydrolysis. Deoxyharringtonine was prepared in 15% yield as a mixture of epimers. As in the harringtonine synthesis, large amounts of cephalotaxine were recovered from the reaction mixtures.

Investigators in the People's Republic of China reported two syntheses of deoxyharringtonine starting from the saturated 2-oxo-5-methylhexanoic acid (116). Huang used the Reformatsky reaction to introduce the acetate fragment and obtained a mixture of epimers of **3** with slight excess of deoxyharringtonine in 65% yield [60]. The isomers were separated by the crystallization of their picrates and subsequent recovery of the free bases.

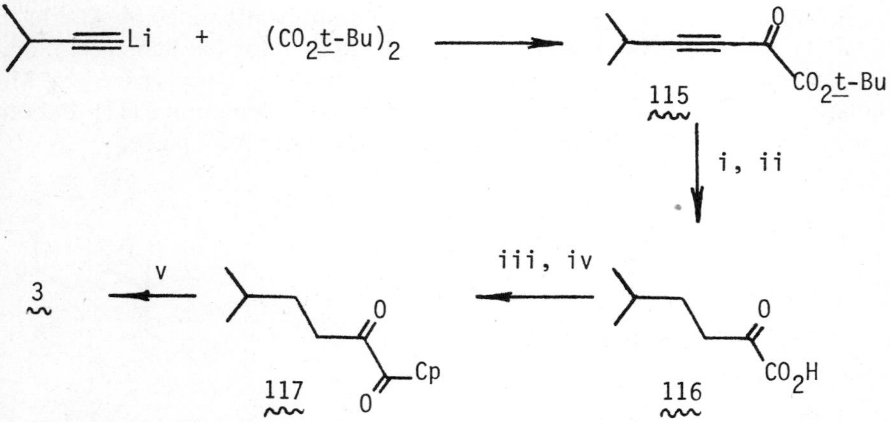

Scheme 14.   Reagents: i, H_2, Pd(C); ii, CF_3CO_2H; iii, $(COCl)_2$; iv, cephalotaxine; v, $LiCH_2CO_2Me$. Cp = cephalotaxyl.

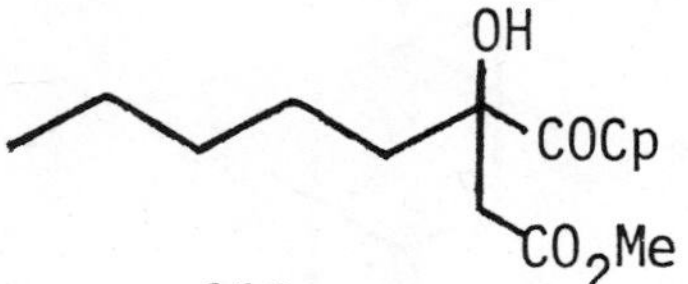

118

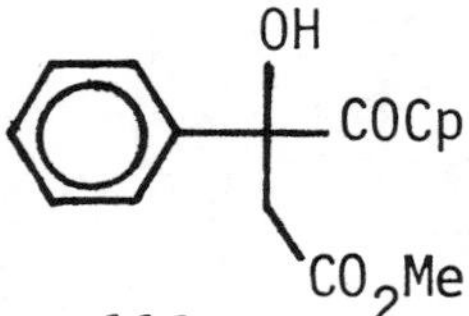

119

Cp = cephalotaxyl

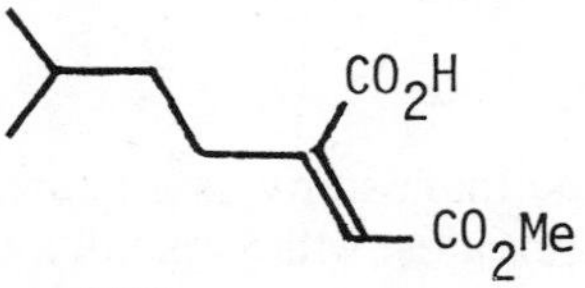

120 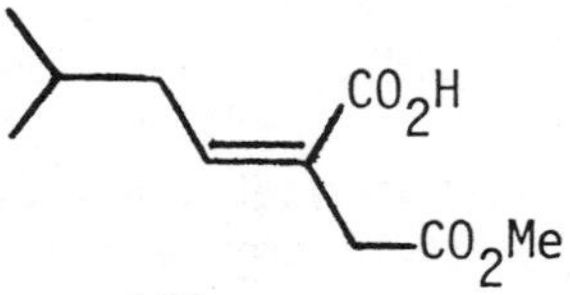   121

Li prepared deoxyharringtonine (3) from acid 116 as well as several unnatural esters of cephalotaxine by esterification of cephalotaxine with appropriate α-keto acids followed by the Reformatsky reaction [61]. Unnatural esters 118 and 119 have been prepared by this method.

Huang et al. reported an improved procedure of deoxyharringtonine synthesis by crystallizing the intermediate cephalotaxyl pyruvate, a compound of considerable lability [62].

Synthesis of deoxyharringtonine via partial esterification using acids 120 and 121 was attempted by Mikolajczak et al. [63] and Bates et al. [64]. The lack of success was attributed to both the steric bulk of acids and the sensitivity of the acrylate functionality.

A radically different approach was unsuccessfully attempted by Auerbach et al. The synthesis of acid 124 was performed as shown in Scheme 15, and the racemate was resolved [53]. The same acid was also prepared by Mikolajczak et al. during structural studies of deoxyharringtonine [11]. In both instances no esterification of 124 with cephalotaxine took place.

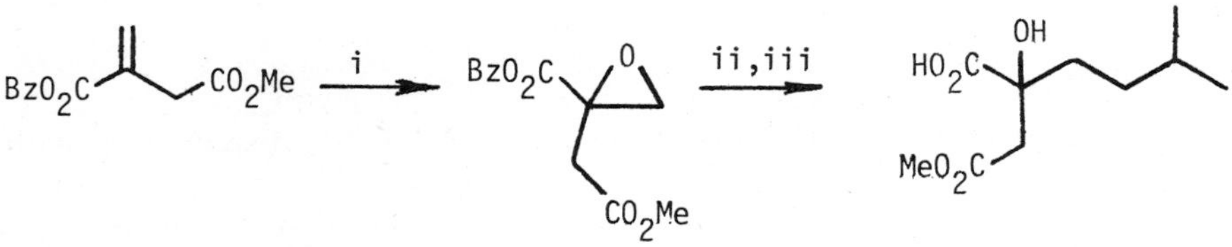

122   123   124

Scheme 15.   Reagents: i, CF_3CO_3H; ii, $(i\text{-}Bu)_2CuLi$; iii, H_2.

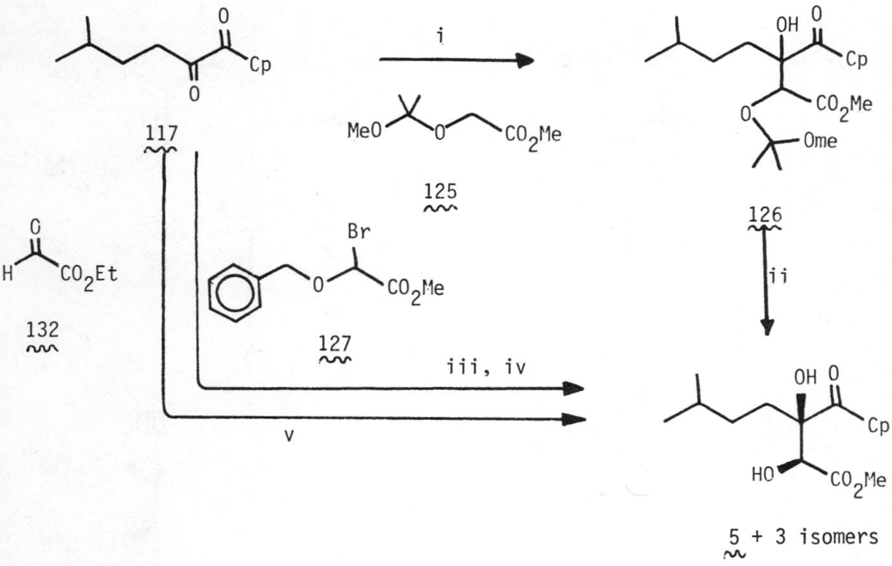

Scheme 16. Reagents: i, LDA/THF; ii, H⁺; iii, Zn/Et₂O; iv, H₂, Pd(C); v, (cyclopentadienyl)TiCl₃/LiAlH₄/THF. Cp = Cephalotaxyl.

Isoharringtonine. This ester has been synthesized in several laboratories by a strategy similar to the preparation of harringtonine and deoxyharringtonine. The handling of the reaction mixtures became more difficult because this alkaloid contains an additional chiral center, and mixtures of up to four isomers have been reported. Huang's synthesis reported in 1982 featured condensation of cephalotaxyl pyruvate **117**, the synthesis of which has been perfected to the tune of >92%, with mixed ketal **125** (Scheme 16) [65]. Lithium diisopropylamide was utilized to generate the anion of **125**, and the condensation product **126** was allowed to stand overnight at pH 3. Under these conditions the ketal was cleaved, and isoharringtonine was isolated from the resulting mixture in 15% yield and was shown to be a mixture of four stereoisomers.

The Beijing group reported a synthesis of **5** using benzyloxybromoacetate **127** under the conditions of the Reformatsky reaction. Although the yields of **5** were higher, no stereoselectivity was reported, and **5** was obtained as a mixture of four isomers [66]. The relative configuration of the isoharringtonine side-chain acid has been determined by Ipaktchi and Weinreb. They synthesized both fumarates **128** and **129** and performed hydroxylations with osmium tetroxide to produce diols **130** and **131**. Methanolysis of isoharringtonine provided diester **131** identical in all respects, except for optical activity, to erythro isomer **131**, thus confirming the relative configuration in **5** [67].

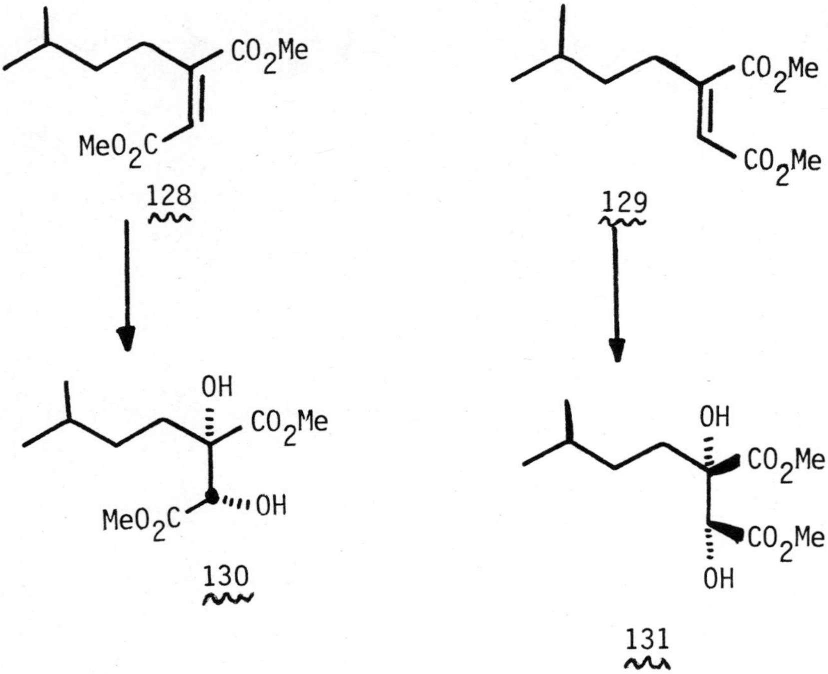

An improved yield of 51% was reported for the Reformatsky reaction of **127** with cephalotaxyl pyruvate **117**. All four isomers were produced, and some enrichment and separation were achieved by purification of the mixture before as well as after the removal of benzylic protecting groups [68].

A novel method of formation of **5** was reported in 1984 by Li et al. [69]. Pyruvate **117** was condensed with ethyl glyoxalate **132** by titanium-induced pinacol coupling to afford **5** directly as a mixture of isomers.

Homoharringtonine. The last of the four antileukemic esters to yield to synthesis was homoharringtonine (**4**). Three research groups reported independently its preparation by a strategy similar to those used in the syntheses of the other cephalotaxine esters. In the synthesis of Zhao et al., an attempt was made to extrapolate from the successful synthesis of harringtonine and to use pyranyl acid **134** prepared from 5,5-dimethyl-δ-valerolactone as shown in Scheme 17 [70]. Esterfication of the acid chloride of **134** with cephalotaxine proceeded well to give **135**; unfortunately this substance was inert to hydration, and hemiketal **136** was not obtained. The unsaturated acid **137**, also prepared from the δ-valerolactone, was esterified with cephalotaxine and subjected to the Reformatsky reaction to afford a mixture of epimers of dehydrohomoharringtonine (**139**). This unnatural alkaloid ester was prepared independently by Hiranuma and Hudlicky [72] from bromide **140**. In their hands the hydration of **139** to **4** would not be

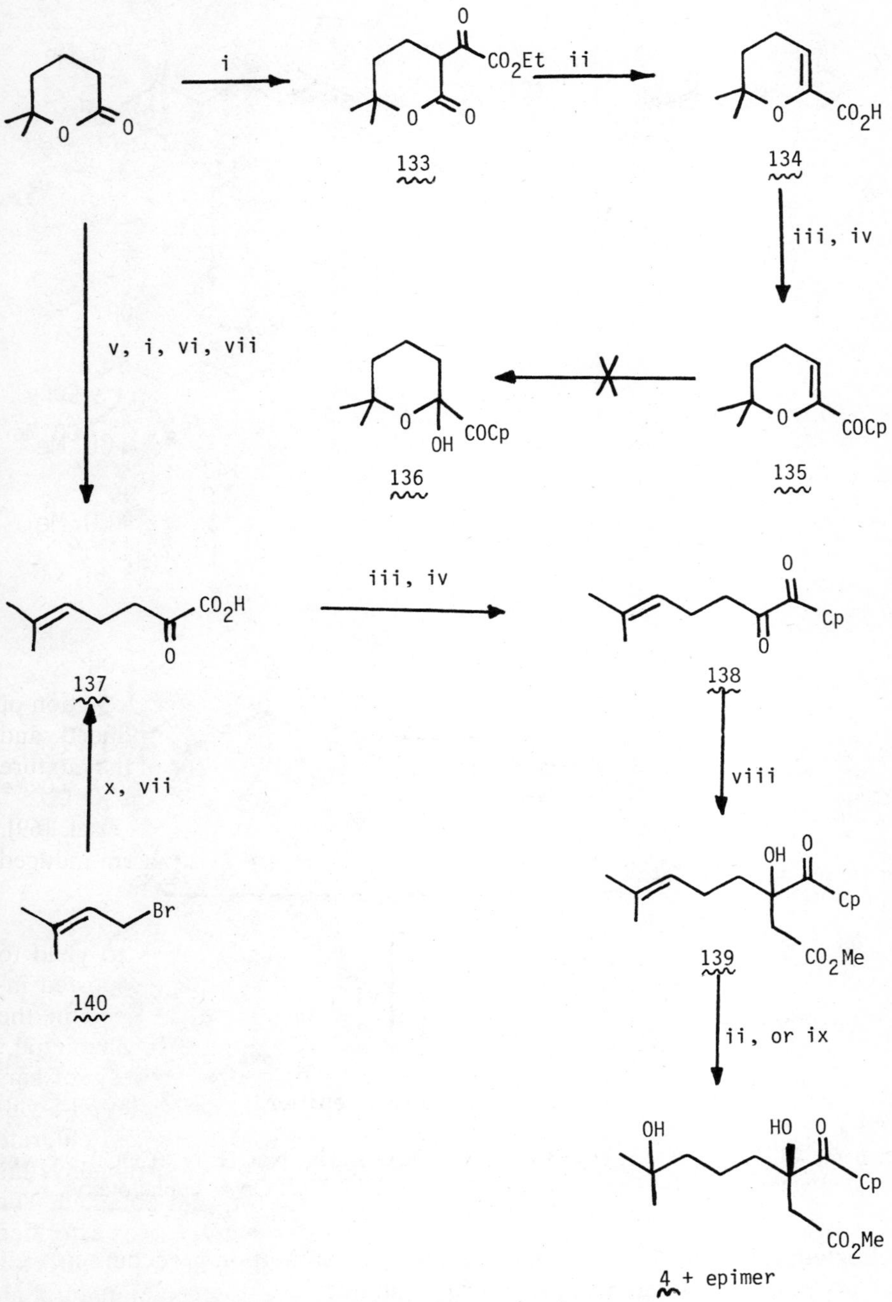

Scheme 17. Reagents: i, $(CO_2Et)_2/EtONa$; ii, H^+; iii, $(COCl)_2$; iv, cephalotaxine; v, HCl/EtOH; vi, DMSO/H_2O; vii, KOH/EtOH; viii, $BrCH_2CO_2Me/Zn$; ix, $Hg(CF_3CO)_2/NaBH_4$; x, Mg/$(CO_2Et)_2$. Cp = cephalotaxyl.

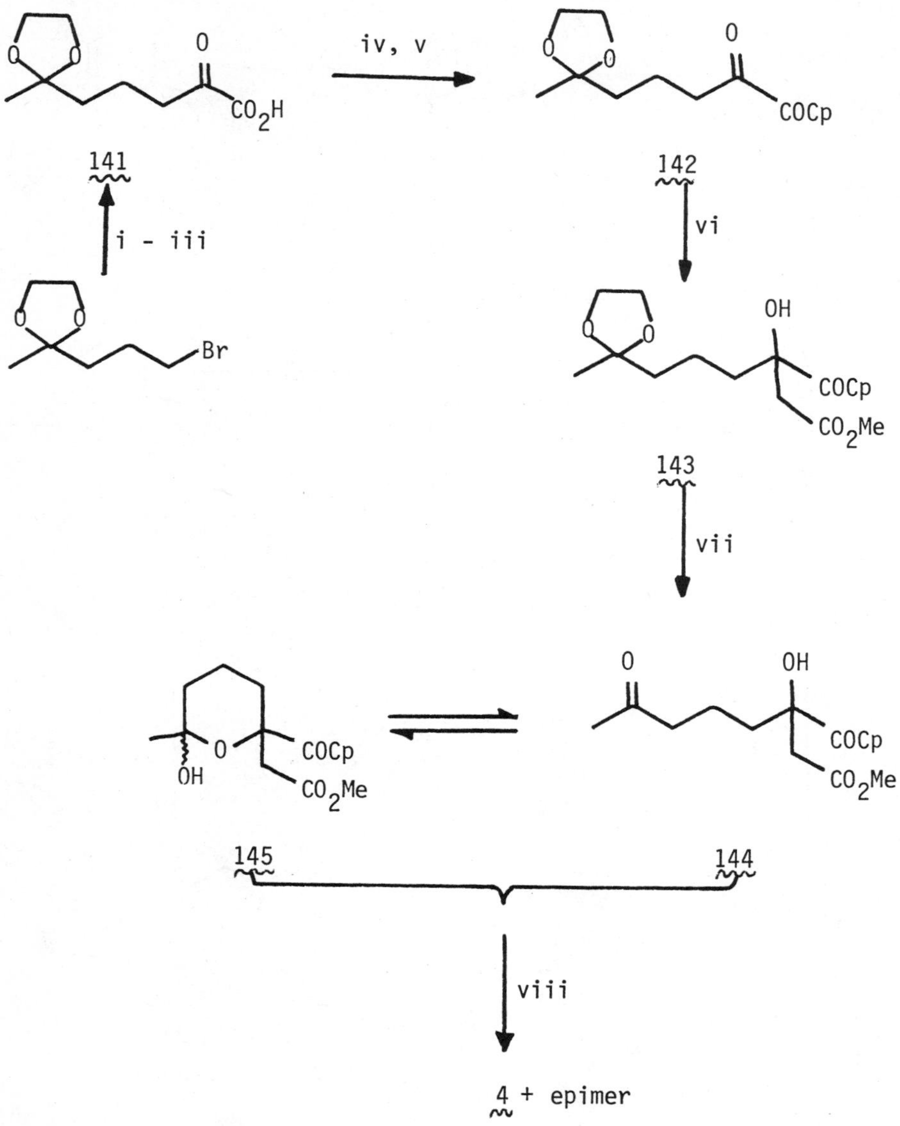

Scheme 18. Reagents: i, Mg/Et$_2$O; ii, ClCOCO$_2$Et; iii, NaOH; iv, (COCl)$_2$; v, cephalotaxine; vi, BrCH$_2$CO$_2$Me/Zn; vii, H$^+$; viii, MeMgI. Cp = cephalotaxyl.

effected; however, Zhao's group used an oxymercuration procedure as well as acid catalysis to arrive at homoharringtonine (**4**).

Wang reported an alternate synthesis of **4** starting with the protected keto acid **141** (Scheme 18) [71]. The preparation of cephalotaxyl ester **142** proceeded well, and Reformatsky reaction of this substance with methyl bromoacetate gave a mixture of diastereomeric ketals **143**. Mild hydrolysis of these compound gave a mixture of ketone **144** and its cyclic hemiketal **145**. Treatment of this mixture with methylmagnesium iodide gave homoharringtonine (**4**), its epimer, and large amounts of cephalotaxine produced by the cleavage of the ester linkage with a strong nucleophile.

Two routes to homoharringtonine were attempted by Hiranuma and Hudlicky [72]. The first of these was unsuccessful because the hydration of ester **139** did not produce homoharringtonine and because oxidative functionali-

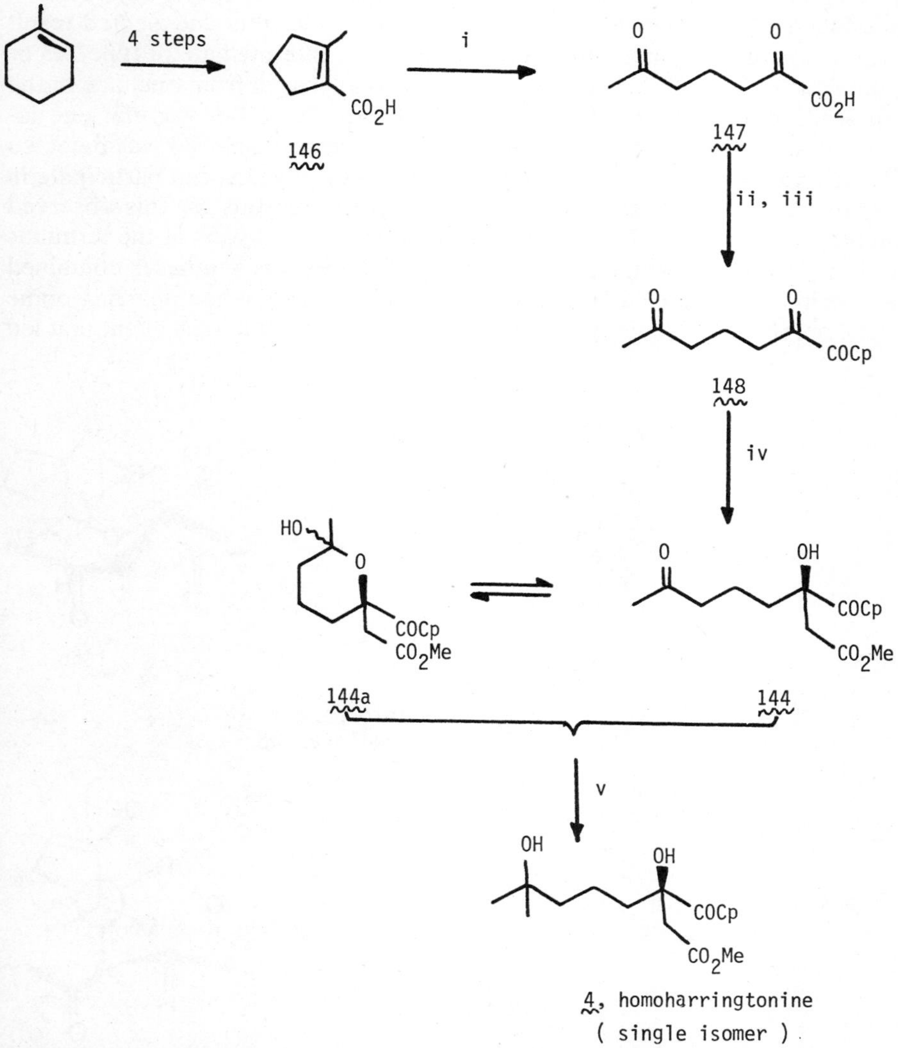

Scheme 19. Reagents: i, O_3/Me_2S; ii, $(COCl)_2$; iii, cephalotaxine; iv, $BrCH_2CO_2Me/Zn$; v, MeMgBr. Cp = cephalotaxyl.

zations of **139** led to complex mixtures. The second approach met with success and utilized the Reformatsky reaction as a means of constructing the side-chain chiral center. The synthesis was carried out first on racemic synthetic cephalotaxine and later on natural material. Ozonolysis of cyclopentenyl acid **146** produced diketo acid **147**, which was esterified with cephalotaxine in 70% yield to give **148** (Scheme 19) [72,73].

This compound was extremely labile and had to be used immediately in the next step. It decomposed in chlorinated solvents and during overnight attempts to record its ^{13}C NMR spectrum [73]. The Reformatsky reaction produced **144** as a single stereoisomer in 90% yield; this unexpected result was rationalized by invoking a crown-ether-like intermediate of type **145** or **146**, which allowed the organozinc reagent to approach from one face of the molecule only. Apparently the pro-(R) configuration **145** was the one involved in this reaction because none of the epimeric material was detected in the reaction mixtures. In **145** three of the four oxygens can participate in the formation of a cage-like chelate. One of the reasons for this observed stereospecificity may be the presence of a basic sp^2 oxygen at the terminus of the side chain. All the intermediates in the previous syntheses contained an olefin, a saturated carbon, or a ketal. The synthesis of homoharringtonine was completed by addition of methylmagnesium bromide—an event that led

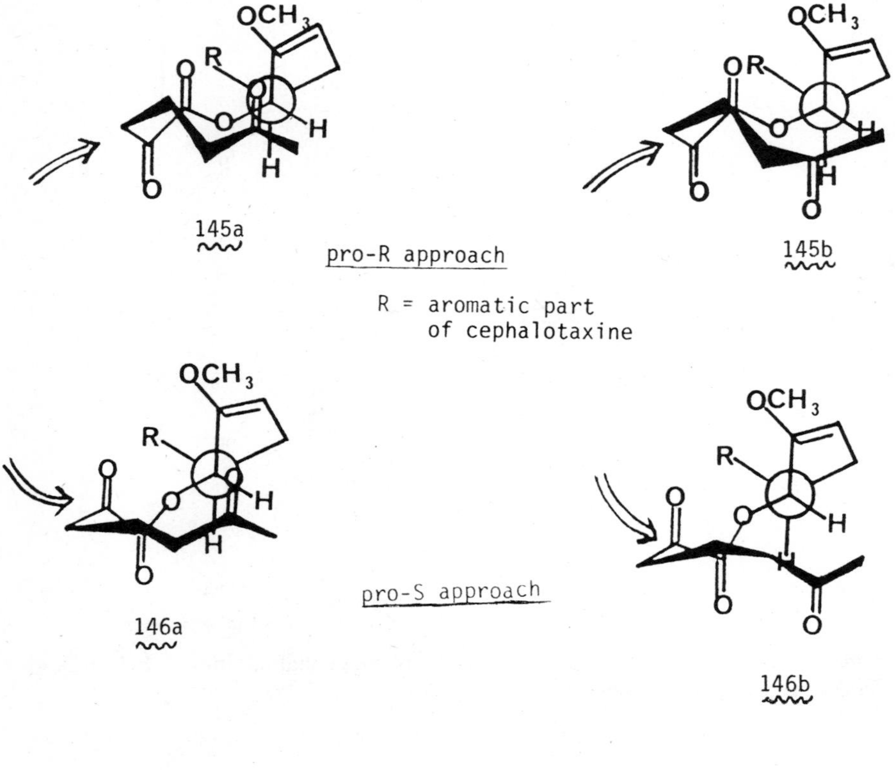

145a

pro-R approach

R = aromatic part
of cephalotaxine

145b

146a

pro-S approach

146b

to low yields of **4** through the aforementioned nucleophilic cleavage to cephalotaxine.

Additional investigation of the stereospecificity became available through the transformations of pyranyl acid **134**. Hiranuma et al. prepared **134** by formylation of **137** [73]. Depending on experimental conditions, either hydroxy acid **134a** or **134b** could be isolated in high yields (Scheme 20). Esterification of **134b** proceeded smoothly in greater than 90% yield to give **135**, which failed to hydrate under acidic conditions in accord with previously reported results of Zhao et al. [70]. However, exposure of **134b** to formic acid/perchloric acid did yield hydroxy acid **134a**. Surprisingly, the hydroxy acid **134a**, via its acid chloride, also formed cephalotaxine ester **136** in high yield. Apparently the hydroxyl group in **134a** is too hindered to participate in polymerization.

Exposure of **136** to two equivalents of methyl bromozinc acetate gave a mixture of homoharringtonine (**4**) and its epimer, in agreement with previously reported results [71]. Thus any sterospecificity in the Reformatsky reaction would appear to be the function of the terminal substituent of the side chain.

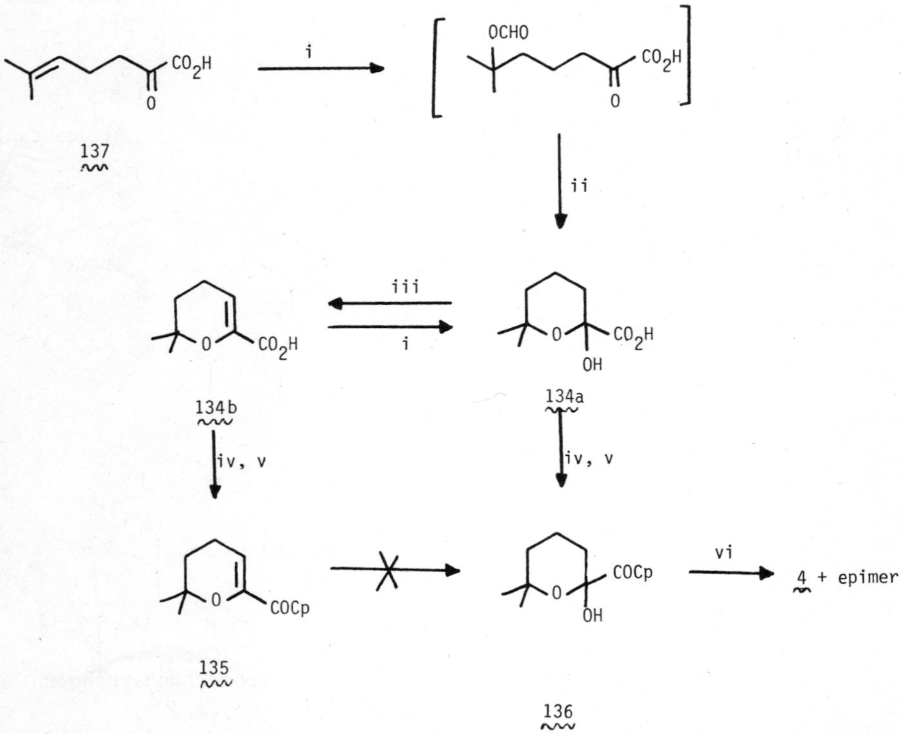

Scheme 20. Reagents: i, $HCO_2H/HClO_4$; ii, $NaOH/H_2O$; iii, HCl/H_2O; iv, $(COCl)_2$; v, cephalotaxine; vi, $BrCH_2CO_2Me/Zn$. Cp = cephalotaxyl.

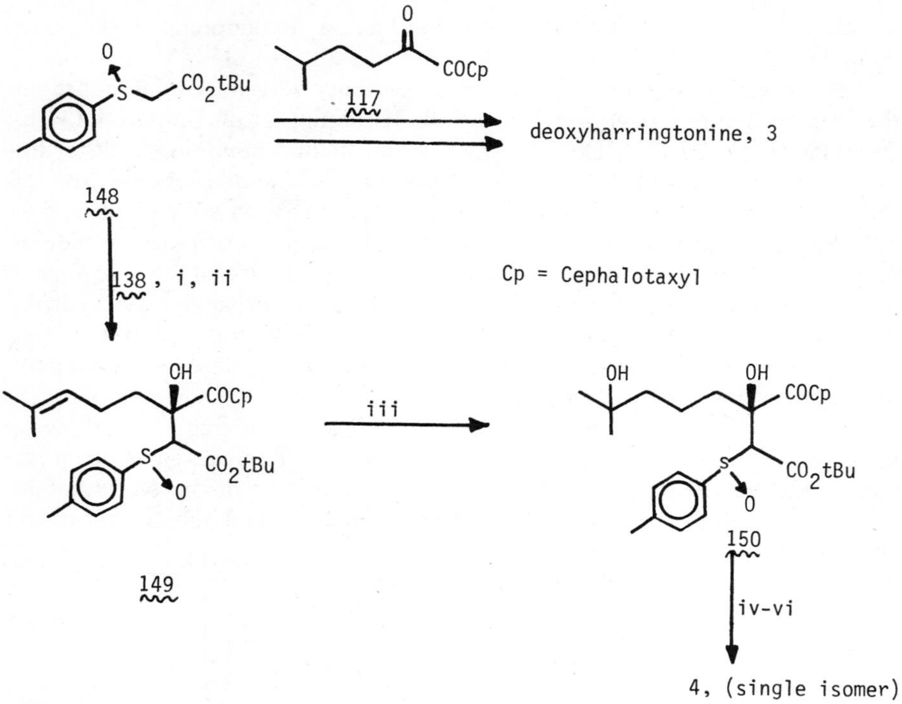

Scheme 21. Reagents: i, *t*BuMgBr/THF; ii, NH₄Cl/H₂O; iii, H₃O⁺; iv, Al amalgam; v, CF₃CO₂H; vi, CH₂N₂.

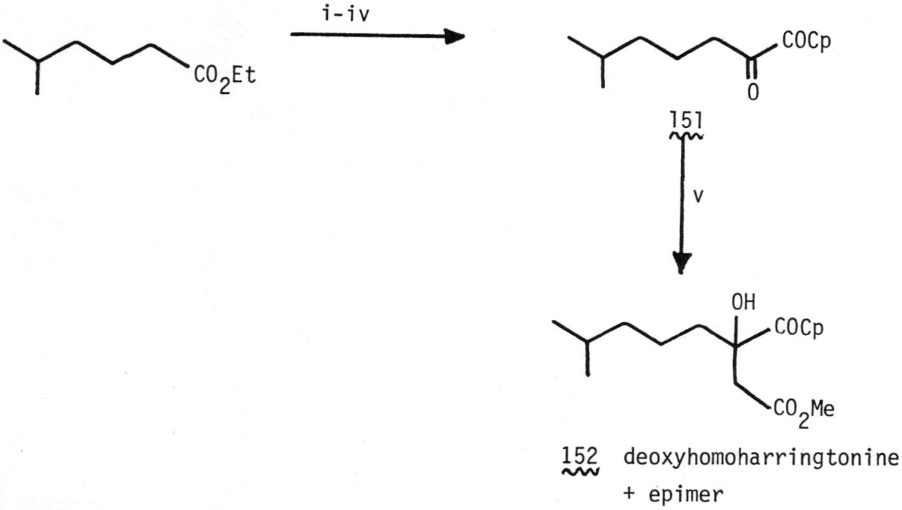

Scheme 22. Reagents: i, (CO₂Et)₂/NaH; ii, HBr/HOAc; iii, (COCl)₂; iv, cephalotaxine; v, BrCH₂CO₂Me/Zn. Cp = cephalotaxyl.

Cheng et al. reported an asymmetric synthesis of **4** in 1984 [74] and also prepared deoxyharringtonine by this method. Sulfinyl ester **148** in its proper chiral form was condensed with either of the cephalotaxyl esters **138** or **117**, leading to homoharringtonine and deoxyharringtonine, respectively (Scheme 21). Thus a single stereoisomer **149** was obtained and hydrated to the tertiary alcohol **150**. Desulfurization, ester hydrolysis, and methylation completed the asymmetric synthesis of homoharringtonine. Similarly, deoxyharringtonine was prepared via ester **117**.

A synthesis of deoxyhomoharringtonine was reported in 1982 by Wang et al. [75]. Ethyl 5-methylhexanoate was converted to cephalotaxyl pyruvate (**151**) as shown in Scheme 22. The Reformatsky reaction led to a mixture of deoxyhomoharringtonine (**152**) and its epimer.

4.3. Unnatural Derivatives of Cephalotaxine

Chemical testing of various alkaloids isolated from *Cephalotaxus* as well as the availability, through synthesis, of some of the esters epimeric with respect to side-chain stereocenters also prompted the investigation of unnatural esters of **1**.

All the new compounds were prepared by methods of esterification of natural or synthetic cephalotaxine, and some showed marginal biological activity. However, at the time of this writing no firm rationale exists for precise structure–activity relationship in cephalotaxine esters. Only seven of the new compounds showed activity, with harringtonine and homoharringtonine being the two most active compounds. Table 3 lists the side-chain residues of the unnatural esters [76–79].

Table 3.   Unnatural Esters and Derivatives of Cephalotaxine

2'-epiharringtonine
2'-epideoxyharringtonine
2'-epihomoharringtonine
2'-epideoxyhomoharringtonine
2',3'-epiisoharringtonines (4 isomers)

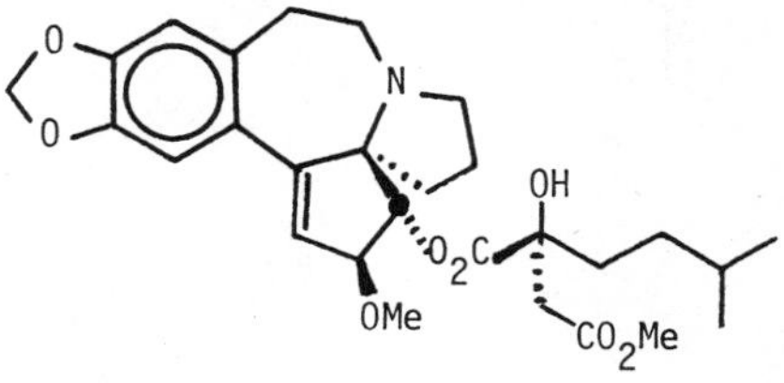

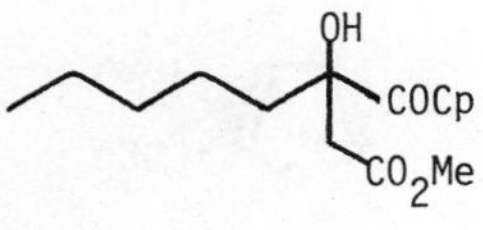

Table 3. (*continued*)

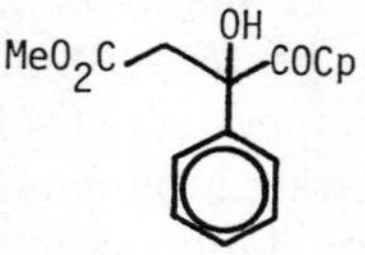

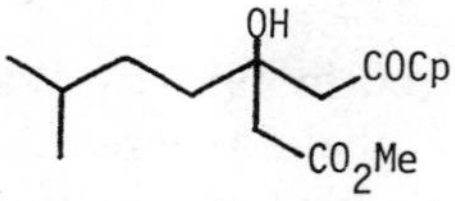

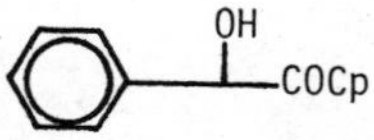

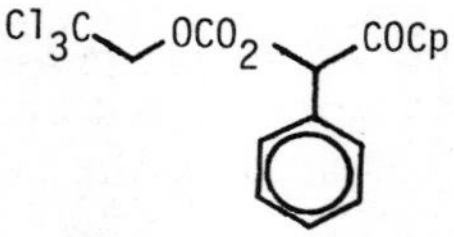

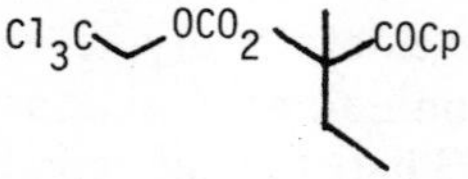

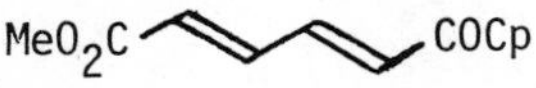

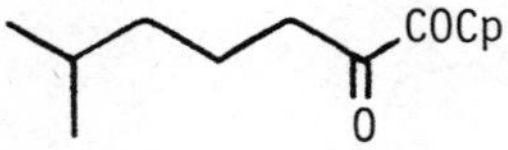

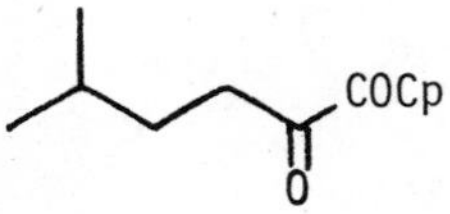

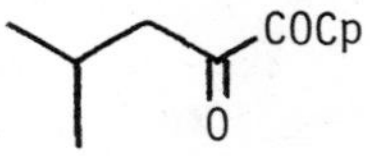

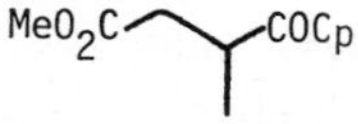

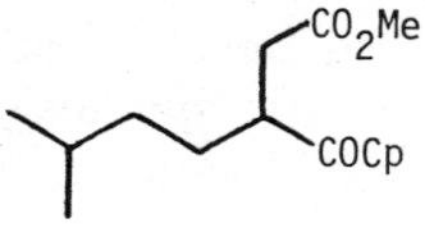

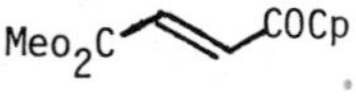

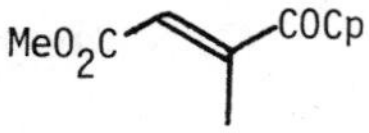

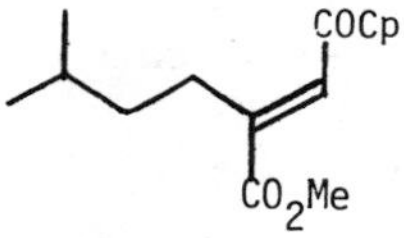

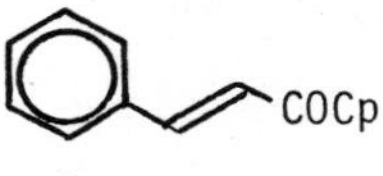

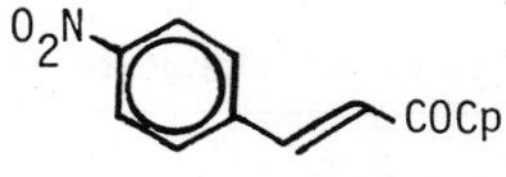

MeO₂C ... OH COCp

MeO₂C ... COCp

OH COCp CO₂Me

CO₂Me COCp

OH COCp

MeO₂C COCp *

Cl₃C OCO₂ COCp *

MeO₂C COCp

COCp CO₂Me *

Cl₃C OCO₂ COCp

MeO₂C COCp *

MeO₂C COCp

COCp

O₂N COCp

COCp O

C₆H₅COCp

COCp O

COCp N

COCp O

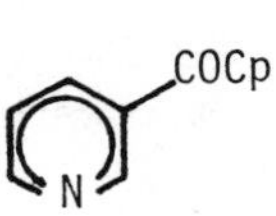

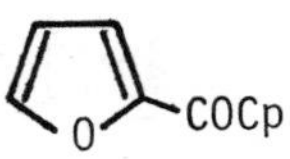

Table 3. (*continued*)

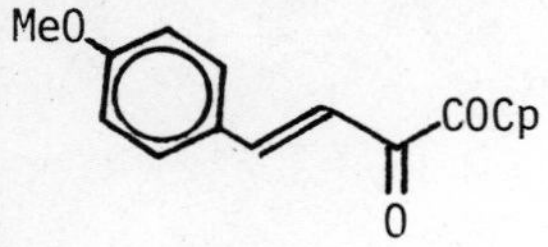

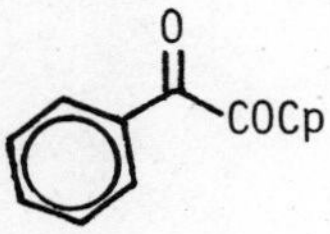

MeO — ⟨⟩ — CH=CH — COCp
                    ‖
                    O

⟨⟩ — CO — COCp

EtO₂CCOCp   *

CH₂=CH — COCp

CH₂=C(CH₃) — COCp

CH₃CH=CH — CH=CH — COCp   *

adamantane — COCp

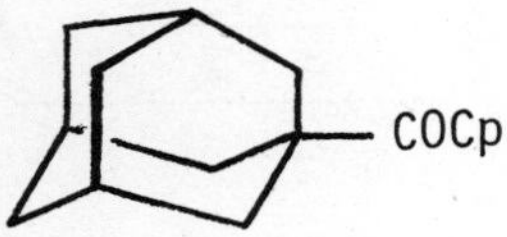

ClCH₂COCp

EtOCOCp

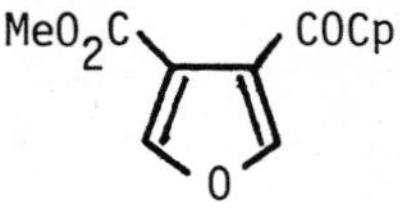

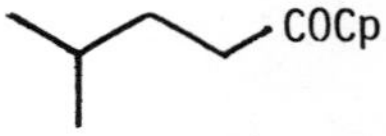

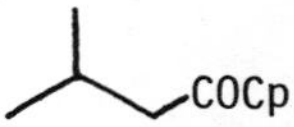

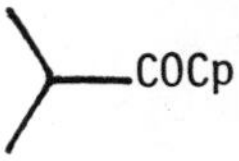

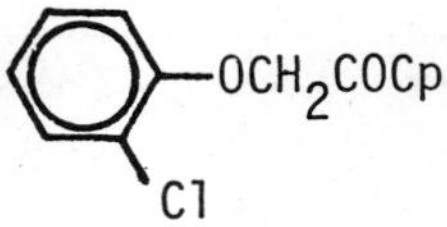

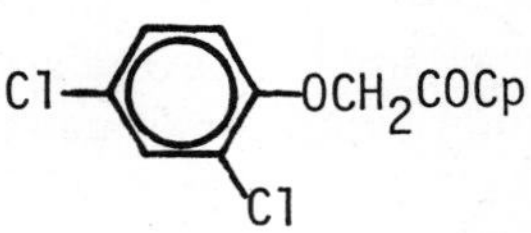

MeO₂C — [furan] — COCp

(CH₃)₂CH — CH₂ — CH₂ — COCp

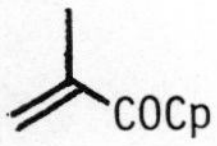

CH₃(CH₂)₅ — COCp

(CH₃)₂CH — CH₂ — COCp

CH₃CH₂CH₂ — COCp

(CH₃)₂CH — COCp

⟨⟩(Cl) — OCH₂COCp

Cl — ⟨⟩(Cl) — OCH₂COCp

Table 3. (*continued*)

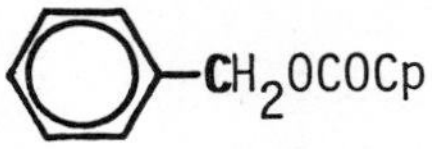

$C1_3CCH_2OCOCp$ *

H_3CSO_2Cp

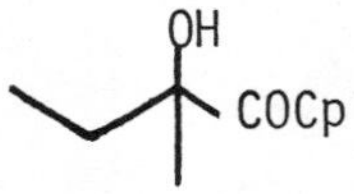

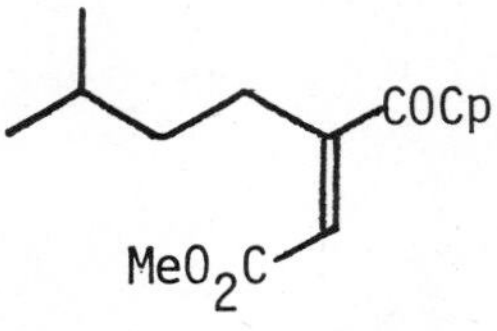

* Denotes derivatives possessing activity.
Cp = cephalotaxyl.

5. SUMMARY

The amount of experimental data available today in the field of synthesis and manipulation of *Cephalotaxus* alkaloids is formidable. The attainment of total synthesis of cephalotaxine and the successful approaches, through partial esterification, to cephalotaxine esters have provided organic chemists with the necessary departure point for providing these alkaloids in a reliable manner for medicinal screening. Their preparation is of course imperfect at this point, as it must be, because the "second-generation" studies of preparation of cephalotaxine esters have scarcely begun. Eventually, the chemical synthesis or the tissue culture preparations will compete successfully with isolation of these alkaloids.

The message left after nearly 15 years of intensive research in this field has informed synthetic chemists that the problem of esterification of cephalotaxine, the question of steric requirements of side-chain attachment, and the overall efficiency of the syntheses were all grossly misjudged at the onsets of their investigations. These problems were not simple and are but partially solved today only because scores of investigators devoted their time to examining every possible detail of a deceivingly elementary reaction in organic synthesis, namely, an esterification. The stage is now set for the implementation of improved procedures of both synthesis and separation of the compounds. Of special significance will be the design of syntheses with complete stereocontrol because the separation of isomers, especially the isomers differing in the stereochemistry of side-chain esters, proved extremely tedious and difficult [54,59,60]. Further improvements in isolation of harringtonine and homoharringtonine are being made [80–83], and some of the data have been summarized in an excellent recent review [3]. This same review offers also an excellent summary of spectral methods of analysis of *Cephalotaxus* alkaloids. The pharmacology and current status of clinical testing as well as possible reasons for activity of some of the *Cephalotaxus* alkaloids have been summarized recently [3,4]. The literature through August 1985 is covered in this chapter. Because this chapter cannot be exhaustive in all areas of research in *Cephalotaxus* alkaloids, all uncited references are listed at the end of this chapter. The references are complete with their titles and should make the retrieval of specific information an easy task.

Acknowledgments

The authors wish to express their heartfelt gratitude to Mr. Richard Powell of USDA, Peoria, and to Dr. Cecil Smith for their continuous support and assistance not only with the literature files for this review, but also with our research. We value the interest they showed in our endeavors. We appreciate the correspondence with Prof. Liang Huang, Inst. Mat. Med., Beijing, and the references she has sent us, as we appreciate the references provided to us by Prof. Pan (Lanzhou) and Prof. Ma (Sanghai). To Sayoko Hiranuma and Misako Shibata go our thanks for the experimental skills invested in the synthesis of homoharringtonine. Last but not least, the authors thank Ms. Sharon Matherly for the preparation of this manuscript.

REFERENCES

1. Y. H. Chen and G. Huang, *Zhongcaoyao Tongxun* **1977**, 254.
2. Authors' personal observations. *C. harringtonia* flourishes in the botanical gardens of Montevideo and Buenos Aires and is found growing wild in the pampas of the Plata region.

3. L. Huang and Z. Xue, in *The Alkaloids*, Vol. 23, F. M. Manske, Ed., Academic, New York, 1984, p. 157.

4. C. R. Smith, Jr., K. L. Mikolajczak, and R. G. Powell, *Antitumor Agents Based On Natural Product Models*, Academic, New York, 1980, Chapter 11.

5. J. A. Findlay, *Int. Rev. Sci.: Org. Chem. Ser. 2*, Vol. 9, D. H. Hey, Ed., Butterworth, London, 1976, p. 23.

5a. V. A. Snieckus, *Alkaloids* **5,** 176 (1975).

6. M. E. Wall, C. R. Eddy, T. T. Willaman, D. S. Correll, B. G. Schubert, and H. S. Gentry, *J. Am. Pharm. Assoc.* **43,** 503 (1954).

7. W. W. Paudler, G. I. Kerley, and J. McKay, *J. Org. Chem.* **28,** 2194 (1963); J. B. McKay, Ph.D. Thesis, Ohio University, Athens, OH, 1966, *Diss. Abstr. B* **27,** 763 (1966).

8. R. G. Powell, D. Weisleder, C. R. Smith, Jr., and I. A. Wolff, *Tetrahedron Lett.* **1969,** 4081.

9. D. J. Abraham, R. D. Rosenstein, and E. L. McGandy, *Tetrahedron Lett.* **1969,** 4085.

10. R. G. Powell, D. Weisleder, C. R. Smith, Jr., and W. K. Rohwedder, *Tetrahedron Lett.* **1970,** 815.

11. K. L. Mikolajczak, R. G. Powell, and C. R. Smith, Jr., *Tetrahedron* **28,** 1995 (1972).

12. S. K. Arora, R. B. Bates, R. A. Grady, and R. G. Powell, *J. Org. Chem.* **39,** 1269 (1974).

13. S. K. Arora, R. B. Bates, R. A. Grady, G. Germain, J. P. Declercq, and R. G. Powell, *J. Org. Chem.* **41,** 551 (1976).

14. S. Brandänge, S. Josephson, S. Vallen, and R. G. Powell, *Acta. Chem. Scand. B* **28,** 1237 (1974).

15. S. Brandänge, S. Josephson, and S. Vallen, *Acta. Chem. Scand. B* **28,** 153 (1974).

16. A. Gitterman, R. J. Parry, R. F. Dufresne, D. D. Sternbach, and M. D. Cabelli, *J. Am. Chem. Soc.* **102,** 2074 (1980); A. Gitterman, Ph.D. Thesis, Brandeis University, Waltham, MA, 1980, *Diss. Abstr. Int. B* **40,** 567 (1980).

17. R. J. Parry, "Biosynthesis of *Cephalotaxus* Alkaloids," in *Recent Advances in Phytochemistry*, T. Swaia and G. R. Wallen, Eds., 1979, Chapter 13, pp. 55–84.

18. R. J. Parry, M. N. T. Chang, J. M. Schwab, and B. M. Foxman, *J. Am. Chem. Soc.* **102,** 1099 (1980).

19. R. G. Powell, R. V. Madrigal, C. R. Smith, Jr., and K. L. Mikolajczak, *J. Org. Chem.* **39,** 676 (1974).

20. G. E. Ma, C. E. Lu, and G. J. Fan, *Chin. Pharm. Bull.* **17,** 205 (1982).

21. T. Hudlicky, unpublished observations. In our hands attempts at oxidative functionalization of cephalotaxine produced only fragmentation of the nucleus of **1.**

22. J. Auerbach and S. M. Weinreb, *J. Am. Chem. Soc.* **94,** 7172 (1972).

23. S. M. Weinreb and J. Auerbach, *J Am. Chem. Soc.* **97,** 2503 (1975).

24. J. VonBraun, G. Blessing, and R. S. Cahn, *Chem. Ber.* **57,** 908 (1924).

25. G. R. Proctor and R. A. Thompson, *J. Chem. Soc.* **1957,** 2502.

26. G. Harebourg and J. Gardent, *Compt. Rend. Acad. Sci. Fr.* **257,** 923 (1963).

27. L. J. Dolby, S. J. Nelson, and D. Senkovich, *J. Org. Chem.* **37,** 3691 (1972).

28. W. I. Taylor and M. M. Robinson, U.S. Patent 3,210,357 (1966); *CA* **65,** 2234e.

29. J. M. Schwab, R. J. Parry, and B. M. Foxman, *J. Chem. Soc., Chem. Commun.* **1975,** 906.

30. The authors had an occasion to reproduce the conditions of the Auerbach–Weinreb synthesis according to the updated experimentals used by Merck and kindly supplied to us by Prof. Weinreb. In our hands the overall yield was increased to ~17% by eliminating some of the purification steps.

31. M. F. Semmelhack, B. P. Chong, and L. D. Jones, *J. Am. Chem. Soc.* **94**, 8629 (1972).

32. M. F. Semmelhack, B. P. Chong, R. D. Stauffer, T. D. Rogerson, A. Chong, and L. D. Jones, *J. Am. Chem. Soc.* **97**, 2507 (1975).

33. M. F. Semmelhack, R. D. Stauffer, and T. D. Rogerson, *Tetrahedron Lett.* **1973**, 4519.

34. S. M. Weinreb and M. F. Semmelhack, *Acc. Chem. Res.* **8**, 158 (1975).

35. B. Weinstein and A. R. Craig, *J. Org. Chem.* **41**, 875 (1976).

36. I. Tse and V. Snieckus, *J. Chem. Soc., Chem. Commun.* **1976**, 505.

37. M. F. Semmelhack, B. P. Chong, and L. D. Jones, *J. Am. Chem. Soc.* **94**, 8679 (1972).

38. T. Tiner-Harding, J. W. Ulrich, F. T. Chiu, S. F. Chen, and P. S. Mariano, *J. Org. Chem.* **47**, 3360 (1982).

39. F. T. Chiu, J. W. Ulrich, and P. S. Mariano, *J. Org. Chem.* **49**, 228 (1984).

40. J. R. Greene, Ph.D. Thesis, University of Georgia, Athens, GA, 1983; *Diss. Abstr. Int. B* **44**, 3405 (1984).

41. G. Lenz, private communication.

42. T. Hudlicky, *J. Org. Chem.* **46**, 1738 (1981).

43. T. Hudlicky, unpublished observations.

44. S. M. Kupchan, O. P. Dhingra, and C. K. Kim, *J. Chem. Soc., Chem. Commun.* **1977**, 847.

45. S. M. Kupchan, O. P. Dhingra, and C. K. Kim, *J. Org. Chem.* **43**, 4464 (1978).

46. J. P. Marino and J. M. Samanen, *J. Org. Chem.* **41**, 179 (1976).

47. E. McDonald and A. Suksamrarn, *Tetrahedron Lett.* **1975**, 4425.

48. N. E. Delfel, *Plant. Physiol.* **59**, Suppl. 62 (1977).

49. N. E. Delfel, Abstr. Int. Plant Tissue Cell Cult. Conf., 4th, Calgary, Alberta, 1978, p. 124.

50. N. E. Delfel and J. A. Rothfus, *Phytochemistry* **16**, 1595 (1977).

51. N. E. Delfel, *Planta Med.* **39**, 168 (1980).

52. N. E. Delfel and L. J. Smith, *Planta Med.* **40**, 237 (1980).

53. J. Auerbach, T. Ipaktchi, and S. M. Weinreb, *Tetrahedron Lett.* **1973**, 4561.

54. T. R. Kelly, R. W. McNutt, Jr., M. Montury, N. P. Tosches, K. L. Mikolajczak, C. R. Smith, Jr., and D. Weisleder, *J. Org. Chem.* **44**, 63 (1979).

55. T. R. Kelly, J. C. McKenna, and P. A. Christensen, *Tetrahedron Lett.* **1973**, 3501.

56. L. Huang, Y. Xi, J. Guo, D. Liu, S. Xu, K. Wu, J. Cheng, Y. Jiang, Y. Gao, Z. Guo, L. Li, M. Zhang, and F. Chu, *Sci. Sin. (Engl. Ed.)* **22**, 1333 (1979); *CA* **92**, 147009t.

57. K. L. Mikolajczak and C. R. Smith, Jr., *J. Org. Chem.* **43**, 4762 (1978).

58. Tumor Research Group, *K'o Hsueh T'ung Pao*, **20**, 437 (1975); *CA* **84**, 105859z. *Ibid.* **21**, 512, 509 (1976); *CA* **86**, 171690.

59. K. L. Mikolajczak, C. R. Smith, Jr., D. Weisleder, T. R. Kelly, J. C. McKenna, and P. A. Christenson, *Tetrahedron Lett.* **1974**, 283.

60. W. Huang, Y. Li, and X. Pan, *Sci. Sin. (Engl. Ed.)* **23**, 835 (1980); *Lanzhou Daxue Xueboa, Ziran Kexueban* **1974**, 148.

61. S. Li and J. Dai, *Acta Chem. Sin.* **33**, 75 (1975); *CA* **84**, 150812q. See also *CA* **85**, 108857d.

62. W. Huang, Y. Li, and S. Pan, *K'o Hsueh Tung Pao* **21**, 178 (1976); *CA* **85**, 63208z. *Sci. Sin. (Engl. Ed.)* **23**, 835 (1980); *CA* **94**, 15928y.

63. K. L. Mikolajczak, C. R. Smith, Jr., and R. G. Powell, *J. Pharm. Sci.* **63**, 1280 (1974).

64. R. B. Bates, R. S. Cutter, and R. M. Freeman, *J. Org. Chem.* **42**, 4162 (1977).

65. X. Pan, Y. Li, S. Li, J. Tian, D. Zhang, Y. Cui, P. Cheng, Y. Wang, and W. Huang, *Kexue Tongbao* **27**, 1048 (1982); *CA* **98**, 4699f. *Kexue Tongbao (Engl. Ed.)* **28**, 1355 (1983); *CA* **100**, 139419v.

66. Y. Li, K. Wu, and L. Huang, *Acta Pharm. Sin.* **17**, 866 (1982); *CA* **98**, 126434m.

67. T. Ipaktchi and S. M. Weinreb, *Tetrahedron Lett.* **1973**, 3895.

68. Y. Li, K. Wu, and L. Huang, *Yaoxue Xuebao* **19**, 582 (1984); *CA* **102**, 132308r.

69. Y. Li, Y. Li, and W. Huang, *Kexue Tongbao* **11**, 703 (1984); *Engl. Ed.* **29**, 1131 (1984).

70. Z. Zhao, Y. Xi, H. Zhao, J. Hou, J. Zhang, and Z. Wang, *Acta Pharm. Sin.* **15**, 46 (1980); *CA* **94**, 103627e.

71. Y. Wang, Y. Li, X. Pan, S. Li, and W. Huang, *Kexue Tongbao* **25**, 576 (1980); *CA* **94**, 103628f. *Lan-chou Ta Hsueh Pao, Tzu Jan K'o Hsueh Pan* **1980**, 71; *CA* **95**, 220199y. *Huaxue Xuebao* **43**, 161 (1985); *CA* **103**, 71569y.

72. S. Hiranuma and T. Hudlicky, *Tetrahedron Lett.* **23**, 3431 (1982).

73. S. Hiranuma, M. Shibata, and T. Hudlicky, *J. Org. Chem.* **48**, 5321 (1983).

74. J. Cheng, J. Zhang, C. Zhang, J. Yang, and L. Huang, *Acta Pharm. Sin.* **19**, 178 (1984).

75. Y. Wang, Q. Mu, Y. Li, X. Pan, and W. Huang, *Kexue Tongbao* **27**, 856 (1982); *CA* **97**, 198426n. *Lanzhou Daxue Xuebao, Ziran Kexueban* **1981**, 153; *CA* **96**, 181475j. *Huaxue Xuebao* **41**, 648 (1983); *CA* **99**, 212773y.

76. K. L. Mikolajczak, C. R. Smith, Jr., and D. Weisleder, *J. Med. Chem.* **20**, 328 (1977).

77. K. L. Mikolajczak and C. R. Smith, Jr., U.S. Patent 4,152,333 (1979).

78. S. Li, H. Sun, X. Lu, S. Zhang, F. Lu, J. Dai, and Y. Xu, *Acta Pharm. Sin.* **16**, 821 (1981); *CA* **96**, 143126p.

79. K. L. Mikolajczak, R. G. Powell, and C. R. Smith, Jr., *J. Med. Chem.* **18**, 63 (1975).

80. Anonymous, *Kexue Tongbao* **23**, 696 (1978).

81. G. Zhang, Y. Zhou, Y. Guo, J. Guo, D. Liu, and F. Chu, *Fenxi Huaxue* **1981**, 291.

82. L. Huang, Q. Fang, J. Cheng, and Z. Jiang, *Fenxi Huaxue* **1983**, 158.

83. Y. Li, K. Wu, and L. Huang, unpublished work.

BIBLIOGRAPHY

Analysis

L. Tang and G. Weng, Application of orthogonal function spectrophotometry to the determination of harringtonine in liposomes, *Yaowu Fenxi Zazhi* **4**, 197 (1984); *CA* **101**, 177599g.

G. Zhang and H. Liu, Reversed-phase ion-pair HPLC analysis of epimers of partially synthesized homoharringtonine, *Yaoxue Xuebao* **19**, 697 (1984); *CA* **102**, 12497e.

G. Zhang and Z. Zhou, Analysis of *Cephalotaxus* alkaloids. III. High performance liquid chromatographic analysis of epimers of partially synthesized deoxyharringtonine and harringtonine, *Yaowu Fenxi Zazhi* **4**, 330 (1984); *CA* **102**, 84477w.

Z. Zhou and G. Zhang, Analysis of *Cephalotaxus* alkaloids as drugs. II. TLC separation and determination of two epimers of partially synthesized homoharringtonine, *Yaowu Fenxi Zazhi* **4**, 70 (1984); *CA* **101**, 177594b.

P. Cong, Mass spectroscopic study of cephalotaxine alkaloids, *Yaoxue Xuebao* **18**, 215 (1983); *CA* **99**, 54042c.

J. Greaves, J. Roboz, H. Jui, R. Suzuki, and J. Holland, Mass spectrometric studies on harringtonine, *Recent Dev. Mass Spectrom. Biochem. Med. Environ. Res.*, **8** (Anal. Chem. Symp. Ser., vol. 12), A. Frigerio, Ed., Elsevier, Amsterdam, 1983, pp. 135–142.

Q. Fang, Assay of harringtonine and epiharringtonine by liquid chromatography, *Yaowu Fenxi Zazhi* **3**, 346 (1983); *CA* **100**, 74034u.

L. Huang, Q. Fang, J. Cheng, and Z. Qiang, Semisynthetic diastereoisomers—separation and identification of harringtonine and epiharringtonine by preparative high-performance liquid chromatography, *Fenxi Huaxue* **11**, 158 (1983); *CA* **99**, 115256z.

Q. Mu, Y. Li, and W. Huang, Separation of the two epimers of three *Cephalotaxus* ester alkaloids by thin-layer chromatography, *Fenxi Huaxue,* **11**, 376 (1983); *CA* **99**, 122725e.

H. Jui and J. Roboz, Quantitation of harringtonine and homoharringtonine in serum by high-performance liquid chromatography, *J. Chromatogr.* **233**, 203 (1982).

S. Li, Y. Cui, Y. Li, X. Pan, Y. Wang, and W. Huang, Separation and determination of iso-harringtonine and its stereoisomers, *Lanzhou Daxue Xuebao, Ziran Kexueban* **18**, 179, 181 (1982); *CA* **99**, 71050w.

M. Ren and B. Wu, Estimation of the content of mixed harringtonine and homoharringtonine, *Ziran Zazhi* **5**, 637 (1982); *CA* **98**, 185637c.

J. Roboz, J. Greaves, H. Jui, and J. Holland, Quantification of homoharringtonine and har-ringtonine in serum by chemical ionization mass spectrometry, *Biomed. Mass Spectrom.* **9**, 510 (1982).

L. Li and S. Ding, Determination of labeling sites in [³H]-homoharringtonine, *Zhongyao Tong-bao* **8**, 30 (1983); *CA* **100**, 12500j.

X. Lu, S. Zhang, Z. Wang, and M. Wang, HPLC separation and determination of *Cephalotaxus* alkaloids, *Yaoxue Xuebao* **16**, 773 (1981); *CA* **96**, 91706f.

G. Zhang, Z. Zhou, Y. Gao, J. Guo, D. Liu, and F. Chu, Isolation and determination of the two epimers of partially synthesized harringtonine, *Fen Hsi Hua Hsueh* **9**, 291 (1981); *CA* **95**, 187494r.

K. Chang, C. Chou, F. Chu, Y. Kao, C. Kuo, and T. Liu, Preparation and quantitative de-termination of the epimers of partial synthetic harringtonine, *Yao Hsueh T'ung Pao* **15**, 45 (1980); *CA* **95**, 103355k.

D. Weisleder, R. Powell, and C. Smith, Carbon-13 nuclear magnetic resonance spectroscopy of *Cephalotaxus* alkaloids, *Org. Magn. Reson.* **13**, 114 (1980).

Chinese Academy of Medical Sciences, Institute of Materia Medica; Cancer Res., Separation and identification of harringtonine and epiharringtonine, *K'o Hsueh T'ung Pao* **23**, 696 (1978); *CA* **91**, 39699p.

G. Spencer, R. Plattner, and R. Powell, Quantitative gas chromotography and gas chromatog-raphy-mass spectrometry of *Cephalotaxus* alkaloids, *J. Chromatogr.* **120**, 335 (1976).

Biosynthesis

J. Guo, Present status of studies on the biosynthesis of *Cephalotaxus* alkaloids, *Zhongcaoyao* **15**, 367 (1984); *CA* **101**, 187915w.

M. Chang, Biosynthesis of the *Cephalotaxus* alkaloids, Ph.D. Thesis, Brandeis University, Waltham, MA, 1979, *Diss. Abstr. Int. B* **39**, 5922 (1979).

J. Schwab, M. Chang, and R. Parry, Biosynthesis of *Cephalotaxus* alkaloids. 3. Specific in-corporation of phenylalanine into cephalotaxine, *J. Am. Chem. Soc.* **99**, 2368 (1977).

R. Parry, D. Sternbach, and M. Cabelli, Biosynthesis of *Cephalotaxus* alkaloids. 2. Biosynthesis of the acyl portion of deoxyharringtonine, *J. Am. Chem. Soc.* **98**, 6380 (1976).

R. Parry and J. Schwab, Biosynthesis of *Cephalotaxus* alkaloids. I. Novel mode of tyrosine incorporation into cephalotaxine, *J. Am. Chem. Soc.* **97**, 2555 (1975).

Isolation and Structure Elucidation

W. Lin, R. Chen, and Z. Xue, Studies on the minor alkaloids of *Cephalotaxus fortunei* Hook. f, *Yaoxue Xuebao* **20**, 283 (1985); *CA* **103**, 85049r.

Y. Ma, M. Guo, T. Zhu, G. Ma, C. Lu, H. Huang, and Y. Yang, Determination of harringtonine and homoharringtone of antileukemic alkaloids in *Cephalotaxus fortunei* Hook. f. and *C. sinensis* (Rehd. et Wils.) Li, *Zhiwu Xuebao* **26**, 405 (1984); *CA* **101**, 207668y.

G. Ma, C. Lu, H. ElSohly, M. ElSohly, and C, Turner, Studies on the alkaloids of *Cephalotaxus*. Part 4. Fortuneine, a homoerythrina alkaloid from *C. fortunei, Phytochemistry* **22**, 251 (1983).

W. Pan, L. Mai, and Y. Li, Studies on the variation of harringtonine content in *Cephalotaxus oliveri* with seasons, parts and ages of the tree, *Zhongyao Tongbao* **8**, 15 (1983); *CA* **100**, 99907u.

G. Ma, G. Q. Sun, M. A. ElSohly, and C. E. Turner, Studies on the alkaloids of *Cephalotaxus*. III. 4-Hydroxycephalotaxine, a new alkaloid from *Cephalotaxus fortunei, J. Nat. Prod.* **45**, 585 (1982).

Z. Xue, N. Sun, D. Chen, L. Xu, W. Chen, L. Ren, X. Liang, and L. Huang, The studies of chemical constituents of *Cephalotaxus hainanensis* Li, *Chem. Nat. Prod., Proc. Sino-Am. Symp.*, Y. Wang, Ed., Gordon & Breach, New York, 1982, pp. 195–199.

L. Ren and Z. Xue, Studies on the antitumor constituents of *Cephalotaxus sinensis, Zhongcaoyao* **12**, 1 (1981); *CA* **96**, 3672p.

Z. Xue, L. Xu, D. Chen, and L. Huang, Studies on minor alkoids of *Cephalotaxus hainanensis* Li, *Yaoxue Xuebao* **16**, 752 (1981); *CA* **96**, 82690u.

Academia Sinica, Institute of Botany; Lab. Plant Chem., Studies on alkaloids of *Cephalotaxus sinensis* (Rehd. et Wils.) Li, *Chih Wu Hsueh Pao* **22**, 156 (1980); *CA* **93**, 235142g.

N. Delfel, Alkaloid distribution and catabolism in *Cephalotaxus harringtonia, Phytochemistry* **19**, 403 (1980).

R. Powell, R. Miller, and C. Smith, Cephalomannine, a new antitumor alkaloid from *Cephalotaxus mannii, J. Chem. Soc., Chem. Commun.* **1979**, 102.

G. Ma, L. Lin, T. Chao, and H. Fan, Studies on alkaloids of *Cephalotaxus*. II. Four minor alkaloids from *Cephalotaxus fortunei* Hook. f. and the structure of cephalofortuneine, *Hua Hsueh Hsueh Pao* **36**, 129 (1978); *CA* **89**, 193837x.

F. Zhang, Z. Wang, W. Pan, Y. Li, L. Mai, J. Sun, and G. Ma, A study on the antitumor plant *Cephalotaxus oliveri* Mast, *Chic Wu Hsueh Pao* **20**, 129 (1978); *CA* **89**, 190990f.

G. Ma, L. Lin, T. Chao, and H. Fan, Studies on the alkaloids of *Cephalotaxus*. I. Isolation and characterization of anticancer alkaloids and a new alkaloid (+)-acetylcephalotaxine from *Cephalotaxus fortunei, Hua Hsueh Hsueh Pao* **35**, 201 (1977); *CA* **90**, 19013m.

H. Furukawa, M. Itoigawa, M. Haruna, Y. Jinno, K. Ito, and S. Lu, Alkaloids of *Cephalotaxus wilsoniana* Hay, *Yakugaku Zasshi* **96**, 1373 (1976); *CA* **86**, 103043w.

R. Powell, and C. Smith, Harringtonine and isoharringtonine, U.S. Patent 3870727 (1975).

R. Powell, S. P. Rogovin, and C. Smith, Isolation of antitumor alkaloids from *Cephalotaxus harringtonia, Ind. Eng. Chem., Prod. Res. Develop.* **13**, 129 (1974).

S. Asada, Alkaloids isolated from the *Cephalotaxus durpacea, Yakugaku Zasshi* **93**, 916 (1973); *CA* **79**, 123699y.

W. Paudler and J. McKay, Structures of some of the minor alkaloids of *Cephalotaxus fortunei, J. Org. Chem.* **38**, 2110 (1973).

R. Powell, K. Mikolajczak, D. Weisleder, and C. Smith, Alkaloids of *Cephalotaxus wilsoniana, Phytochemistry* **11**, 3317 (1972).

R. Powell, Structures of homoerythrina alkaloids from *Cephalotaxus harringtonia, Phytochemistry* **11**, 1467 (1972).

Pharmacology

M. F. Lokhandwala, M. H. Sabouni, and B. S. Jandhyala, Cardiovascular actions of an experimental antitumor agent, homoharringtonine, in anesthetized dogs, *Drug Dev. Res.* **5**, 1573 (1985).

G. E. Umbach, V. Hug, G. Spitzer, B. Tomasovic, H. Thames, J. A. Ajani, and B. Drewinko, Survival of human bone marrow cells after *in vitro* treatment with 12 anticancer drugs and implications for tumor drug sensitivity assays, *J. Cancer Res. Clin. Oncol.* **109**, 130 (1985).

J. Yang, Z. Shen, Y. Sun, J. Han, and B. Xu, Cultured human hepatoma cell (BEL-7404) for anticancer drug screening. *Zhongguo Yaoli Xuebao* **6**, 144 (1985); *CA* **103**, 81176g.

J. Zhou, P. Zhou, Z. Dai, and B. Xu, Effect of homoharringtonine on the ultrastructure of murine leukemia P388 cells, *Zhongguo Yaoli Xuebao* **6**, 140 (1985); *CA* **103**, 64490n.

A. Boyd and J. Sullivan, Leukemic cell differentiation *in vivo* and *in vitro*: arrest of proliferation parallels the differentiation induced by the antileukemic drug harringtonine, *Blood* **63**, 384 (1984).

L. Cao, H. Huang, and C. Cao, Semisynthetic deoxyharringtonine for treating leukemia in mice and humans, *Int. J. Cell Cloning* **2**, 327 (1984).

R. Kato, S. Takamoto, M. Mizutani, M. Hato, and K. Ota, Antitumor activities and mechanisms of action of harringtonine and homoharringtonine, *Gan to Kagaku Ryoho* **11**, 2393 (1984); *CA* **102**, 55778q.

C. Qiu and G. Wu, Studies on the mechanism of inhibition of harringtonine on DNA synthesis. III. Kinetics of inhibition by harringtonine and comparison with its analogs, *Shengwu Huaxue Yu Shengwu Wuli Xuebao* **16**, 645 (1984); *CA* **103**, 16524e.

G. E. Umbach, V. Hug, G. Spitzer, H. Thames, and B. Drewinko, Responses of human bone marrow progenitor cells to fluoro-ara-AMP, homoharringtonine, and elliptinium, *Invest. New Drugs* **2**, 263 (1984).

J. M. Venditti, R. A. Wesley, and J. Plowman, Current NCI preclinical antitumor screening *in vivo*: results of tumor panel screening, 1976–1982, and future directions, *Adv. Pharmacol. Chemother.* **20**, 1 (1984).

N. Wang, W. Li, Y. Liang, and K. Wang, Effect of combination treatment with harringtonine and methotrexate on hematopoietic and immunological functions in mice, *Zhongliu Linchuang* **11**, 53 (1984); *CA* **101**, 143698y.

G. Wu, F. Fang, and J. Zuo, Preliminary studies on mechanism of inhibition of protein synthesis by harringtonine, *Yaoxue Xuebao* **19**, 167 (1984); *CA* **103**, 16415v.

S. Xue, P. Xu, S. Li, Y. Hu, J. Xiao, R. Han, and J. Feng, Effects of natural and semisynthetic harringtonine and homoharringtonine on the cell cycle of P388 murine leukemic cells as measured by flow cytometry, *Zhongguo Yixue Kexueyuan Xuebao* **6**, 28 (1984); *CA* **101**, 122659w.

Q. Zhou, Z. Pan, J. Feng, and R. Han, Comparison of effects of five antitumor drugs on the hemopoietic and P388 leukemic stem cells in mice, *Zhongguo Yaoli Xuebao* **5**, 66 (1984); *CA* **100**, 185410h.

R. Zhu, J. Jiang, J. Gong, and R. Cao, Effects of several natural alkaloids on the inhibition of elongation and breakdown of wheat and yeast cells and the synthesis of proteins and nucleic acids, *Zhiwu Shenglixue Tongxun* 22 (1984); *CA* **101**, 188156m.

Y. Cai, C. Xu, S. Li, and Y. Liu, Studies on sister chromatid exchanges induced by harringtonine, indirubin and pyquiton before or after activation with microsome enzyme, *Zhongguo Yixue Kexueyuan Xuebao* **5**, 161 (1983); *CA* **100**, 44999v.

T. Chou, F. Schmid, A. Feinberg, F. Philips, and J. Han, Uptake, initial effects, and chemotherapeutic efficacy of harringtonine in murine leukemic cells sensitive and resistant to vincristine and other chemotherapeutic agents, *Cancer Res.* **43**, 3074 (1983).

W. Cobb, A. Bogden, S. Reich, T. Griffin, D. Kelton, and D. LePage, Activity of two phase I drugs, homoharringtonine and tricyclic nucleotide, against surgical explants of human tumors in the 6-day subrenal capsule assay, *Cancer Treat. Rep.* **67**, 173 (1983).

C. C. Huang, C. S. Han, X. F. Yue, C. M. Shen, S. W. Wang, F. G. Wu, and B. Xu, Cytotoxicity and sister chromatid exchanges induced *in vitro* by six anticancer drugs developed in the People's Republic of China, *JNCI, J. Natl. Cancer Inst.* **71**, 841 (1983).

X. Ji, F. Zhang, and X. Dong, Studies on the antineoplastic effect and toxicity of semisynthetic harringtonine, *Yaoxue Xuebao* **18**, 299 (1983); *CA* **100**, 44972f.

T. L. Jiang, R. H. Liu, and S. E. Salmon, Comparative *in vitro* antitumor activity of homoharringtonine and harringtonine against clonogenic human tumor cells, *Invest. New Drugs* **1**, 21 (1983).

T. Jiang, S. Salmon, and R. Liu, Activity of camptothecin, harringtonin, cantharidin and curcumae in the human tumor stem cell assay, *Eur. J. Cancer Clin. Oncol.* **19**, 263 (1983).

K. Li, W. Chao, R. Han, and Z. Pan, Untrastructural and electron-microscope autoradiographic study on the mechanism of therapeutic effect of semisynthetic harringtonine and homoharringtonine on L1210 leukemic mouse cells, *Zhonghua Yixue Zazhi* **63**, 29 (1983); *CA* **98**, 172715b.

T. Okano, T. Ohnuma, J. Holland, H. Koeffler, and H. Jui, Effects of harringtonine in combination with acivicin, adriamycin, L-asparaginase, cytosine arabinoside, dexamethasone, fluorouracil or methotrexate on human acute myelogenous leukemia cell line KG-1, *Invest. New Drugs* **1**, 145 (1983).

Z. Li, Z. Sun, and R. Han, Effects of harringtonine and related alkaloids on the cAMP level of L615 and P388 leukemic cells, *Yaoxue Xuebao* **18**, 03 (1983); *CA* **100**, 17335u.

G. Wu, F. Fang, J. Zuo, R. Han, and Z. Sun, Epiharringtonine-induced enhancement of harringtonine inhibition on protein and DNA formation in tumor cells, *Zhongguo Yixue Kexueyuan Xuebao* **5**, 157 (1983); *CA* **100**, 44998u.

Q. Zhou, J. Feng, and R. Han, Comparison of the sensitivity of murine hemopoietic and P388 leukemic stem cells to five antitumor drugs, *Yaoxue Xuebao* **18**, 721 (1983); *CA* **100**, 44975j.

X. Ji, Y. Liu, H. Lin, and Z. Liu, Metabolism of homoharringtonine in rats and mice, *Yaoxue Xuebao* **17**, 881 (1982); *CA* **98**, 100715d.

R. Johnson and W. Howard, Development and cross-resistance characteristics of a subline of P388 leukemia resistant to 4'-(9-acridinylamino)methanesulfon-m-anisidide, *Eur. J. Cancer Clin. Oncol.* **18**, 479 (1982).

X. Kong, L. Yin, and L. Cheng, Transplantable lymphocytic mouse leukemia, L783. IV. Sensitivity test on antitumor drugs, *Shanghai Diyi Yixueyuan Xuebao* **9**, 443 (1982); *CA* **98**, 119037f.

R. Osieka and C. Schmidt, Primary and acquired resistance to alkylating agents in heterotransplants and human melanomas and colon carcinomas, *Proc. Int. Workshop Nude Mice, 3rd, 1979*, **2**, 675 (1982).

S. Takeda, N. Yajima, K. Kitazato, and N. Unemi, Antitumor activities of harringtonine and homoharringtonine, *Cephalotaxus* alkaloids which are active principles from plant by intraperitoneal and oral administration, *J. Pharmacobio-Dyn.* **5**, 841 (1982).

G. Wu, F. Fang, J. Zuo, R. Han, and Z. Sun, Studies on mechanism of inhibition of harringtonine on DNA synthesis. III. Relation between inhibition of harringtonine on DNA and on protein synthesis, *Zhongguo Yixue Kexueyuan Xuebao* **4**, 78 (1982); *CA* **97**, 174530r.

G. Wu, J. Nie, H. Yang, X. Chen, Q. Wei, H. Xiang, and Y. Zhang, Biochemical actions of some anticancer agents, *Beijing Shifan Daxue Xuebao, Ziran Kexueban*, **1982**, 57; *CA* **98**, 65182d.

G. Wu and C. Qiou, Studies on the mechanism of inhibition of harringtonine on DNA biosynthesis. II. Effect of harringtonine on DNA polymerase α, *Shengwu Huaxue Yu Shengwu Wuli Xuebao* **14**, 553 (1982); *CA* **98**, 191383h.

S. Yang, F. Fang, and G. Wu, Effects of harringtonine on nucleotide metabolism in tumor cells, *Yaoxue Xuebao* **17**, 721 (1982); *CA* **98**, 46556n.

R. Osieka, R. Becher, and C. Schmidt, Molecular pharmacology on human cancer xenografts, *Thymusaplastic Nude Mice Rats Clin. Oncol., Proc. Symp.*, 1979, G. Bastert, H. Fortmeyer, and H. Schmidt-Matthiesen, Eds., Fischer, Stuttgart, 1981, pp. 513–527.

Z. Wang, P. Chou, and B. Hus, Study on phytohemagglutinin-induced agglutination of rat erythrocytes: effect of ATP and other biologically active substances, *Seng Wu Hua Hsueh Yu Sheng Wu Wu Li Hsueh Pao* **13**, 121 (1981); *CA* **95**, 109459u.

G. Wu, F. Fang, R. Han, and Z. Sun, Studies on the mechanism of the inhibition of harringtonine on DNA synthesis. I. Effects of harringtonine on DNA template and metabolism, *Shengwu Huaxue Yu Shengwu Wuli Xuebao* **13**, 509 (1981); *CA* **97**, 16760m.

B. Xu, The influence of several anticancer agents on cell proliferation, differentiation and the cell cycle of murine erythroleukemia cells, *Am. J. Chin. Med.* **9**, 268 (1981).

B. Xu, Pharmacology of some natural products of China, *Trends Pharmacol. Sci.* **2**, 271 (1981).

Y. Xu and C. Du, Effect of harringtonine and its related alkaloids on DNA synthesis in mice bearing leukemias P388, L615 and normal mice, *Zhongguo Yaoli Xuebao* **2**, 252 (1981); *CA* **96**, 62700u.

Y. Xu, C. Du, F. Zhang, and X. Ji, Effect of harringtonine alkaloids on the protein biosynthesis in mouse leukemias L615 and P388 cells, *Yaoxue Xuebao* **16**, 661 (1981); *CA* **96**, 115592u.

P. Zhenkun, C. Zhongyi, W. Yongchao, L. Kun, S. Chiangyi, J. Xiujuan, X. Yuting, L. Zhanrong, and H. Jui, Studies on harringtonine as a new antitumor drug, *Cancer Res. People's Repub. China U.S.A.: Epidemiol., Causation New Approaches Ther., [Proc. Conf.], 1st,* 1980, P. A. Marks, Ed., Grune & Stratton, New York, 1981, pp. 261–267.

J. Zhou, G. He, L. Tao, and B. Xu, Effects of various antitumor agents on intracerebrally inoculated Ehrlich ascites carcinoma in mice, *Zhongguo Yaoli Xuebao* **2**, 256 (1981); *CA* **96**, 79559c.

X. Bin, J. Chen, J. Yang, S. Chang, H. Yueh, T. Wang, and C. Chou, New results in pharmacologic research of some anticancer agents, *Proc. U.S.–China Pharmacol. Symp.*, 1979, J. J. Burns and P. J. Tsuchitani, Eds., NAS, Washington, D.C., 1980, pp. 151–188.

S. Chen, C. Hua, and S. Hsi, Pharmacology of harringtonine and homoharringtonine, *K'o Hsueh T'ung Pao* **25**, 859 (1980); *CA* **94**, 234v.

A. Galsky, J. Wilsey, and R. Powell, Crown gall tumor disc bioassay. A possible aid in the detection of compounds with antitumor activity, *Plant Physiol.* **65**, 184 (1980).

A. Hobden and E. Cundliffe, Ribosomal resistance to the 12,13-epoxytrichothecene, *Biochem. J.* **190**, 765 (1980).

R. Miller, R. Powell, and C. Smith, Cephalomannine and its use in treating leukemic tumors, U.S. Patent 4205221 (1980).

Z. Pan, R. Han, and Y. Wang, The cytokinetic effects of harringtonine on leukemia L-1210 cells. I. Autoradiographic studies, *Sheng Wu Hua Hsueh Yu Sheng Wu Wu Li Hsueh Pao* **12**, 13 (1980); *CA* **93**, 88747c.

T. Ramabhadran and R. Thach, Specificity of protein synthesis inhibitors in the inhibition of encephalomyocarditis virus replication, *J. Virol.* **34**, 293 (1980).

Z. Wang and B. Hsu, Electron microscopic studies of the effect of homoharringtonine on hepatocyte nucleolus and myocardial cells, *Sheng Wu Hua Hsueh Yu Sheng Wu Wu Li Hsueh Pao* **12**, 231 (1980); *CA* **94**, 25139f.

P. Zhenkun, W. Yongchao, L. Kun, S. Chingyi, J. Xiujuan, X. Yuting, F. Yijun, L. Zhanrong, and H. Rui, Mechanisms of action of harringtonine, a new anticancer agent, *Proc. U.S.–*

China Pharmacol. Symp., 1979, J. J. Burns and P. J. Tsuchitani, Eds., NAS, Washington, D.C., 1980, pp. 69–92.

Chinese Academy of Medical Sciences; Inst. Mater Med. Chemical, pharmacological and clinical studies on the antitumor active principle of *Cephalotaxus hainanensis* Li, *Chung-hua Chung Liu Tsa Chih* **1**, 176 (1979); *CA* **92**, 121966c.

I. Fan and R. Han, Effect of harringtonine on L-1210 leukemia cells and normal bone marrow cells in mice, *Yao Hsueh Hsueh Pao* **14**, 467 (1979); *CA* **92**, 87872g.

M. Fresno and D. Vazquez, Initiation of protein synthesis in eukaryotic systems with native 40S ribosomal subunits: effects of translation inhibitors, *Methods Enzymol.* **60**, 566 (1979).

X. Ji, J. Liu, and Z. Liu, Metabolism of harringtonine, *Yao Hsueh Hsueh Pao* **14**, 234 (1979); *CA* **92**, 33731s.

J. Si, K. Son, W. Hu, S. Men, R. Han, Z. Pan, and Z. Li, Ultrastructural study on the effect of harringtonine on mouse leukemia L-1210 cells, *Sheng Wu Hua Hsueh Yu Sheng Wu Wu Li Hsueh Pao* **11**, 333 (1979); *CA* **93**, 19304u.

H. Wood, Development of natural products as antitumor drugs, *Med. Chem., Proc. Int. Symp., 6th,* 1978, M. A. Simkins, Ed., Cotswold Press, Oxford, 1979, pp. 265–280.

Y. Zhang, H. Yu, X. Luo, Y. Zheng, W. Li, X. Liu, and Y. Yuan, Toxicological studies of harringtonine and homoharringtonine, *Yao Hsueh Hsueh Pao* **14**, 135 (1979); *CA* **92**, 51963y.

A. Baez and D. Vazquez, Binding of [³H]narciclasine to eukaryotic ribosomes. A study on a structure-activity relationship, *Biochim. Biophys. Acta* **518**, 95 (1978).

R. Johnson, M. Chitnis, W. Embrey, and E. Gregory, *In vivo* characteristics of resistance and cross-resistance of an adriamycin-resistant subline of P388 leukemia, *Cancer Treat. Rep.* **62**, 1535 (1978).

D. Baaske and P. Heinstein, Cytotoxicity and cell cycle specificity of homoharringtonine, *Antimicrob. Agents Chemother.* **12**, 298 (1977).

Chinese Academy of Medical Sciences; Dep. Pharmacol., Inst. Mater. Med. The antitumor effects and pharmacologic actions of harringtonine, *Chin. Med. J. (Eng. Ed.)* **3**, 131 (1977).

T. Corbett, D. Griswold, B. Roberts, J. Peckham, and F. Schabel, Evaluation of single agents and combinations of chemotherapeutic agents in mouse colon carcinomas, *Cancer* **40**, 2660 (1977).

M. Fresno, A. Jimenez, and D. Vazquez, Inhibition of translation in eukaryotic systems by harringtonine, *Eur. J. Biochem.* **72**, 323 (1977).

D. Vazquez, A. Contreras, M. Fresno, and A. Jimenez, Interaction of translation inhibitors with ribosomes and polysomes, *Trans. Nat. Synth. Polynucleotides, [Proc. Int. Conf.]* A. B. Legocki, Ed., Univ. Agric. Poznan, Poznan, Poland, 1977, pp. 380–385.

D. Baaske, Preparation and analysis of protein conjugates of morphine, D-amphetamine and related compounds. I. The mode of action of homoharringtonine, a new potential antineoplastic alkaloid. II. Ph.D. Thesis, Purdue University, West Lafayette, IN, 1976, *Diss. Abstr. Int. B* **37**, 3972 (1977).

Chinese Academy of Medical Sciences; Chinese People's Liberation Army 187th Hospital; Inst. Mater. Med., Studies on the antitumor constituents of *Cephalotaxus hainanensis* Li, *Hua Hsueh Hsueh Pao* **34**, 283 (1976); *CA* **88**, 126256y.

M. Huang, Harringtonine, an inhibitor of initiation of protein biosynthesis, *Mol. Pharmacol.* **11**, 511 (1975).

J. Tscherne and S. Pestka, Inhibition of protein synthesis in intact HeLa cells, *Antimicrob. Agents Chemother.* **8**, 479 (1975).

I. Kline, Potentially useful combinations of chemotherapy detected in mouse tumor systems, *Cancer Chemother. Rep., Pt. 2,* **4**, 33 (1974).

R. Powell and C. Smith, Harringtonine and isoharringtonine for treating L1210 or P388 leukemic tumors in mice, U.S. Patent 3793454 (1974).

R. Powell, D. Weisleder, and C. Smith, Antitumor alkaloids from *Cephalotaxus harringtonia*. Structure and activity, *J. Pharm. Sci.* **61**, 1227 (1972).

Synthesis

D. J. Heacock, Part 1. Theoretical study of silicenium ions. Part 2. Syntheses of perhydrohistrionicotoxin. Part 3. Attempted synthesis of cephalotaxine, Ph.D. Thesis, University of Rochester, Rochester, NY, 1984, *Diss. Abst. Int.* B **45**, 2921 (1985).

R. J. Mattson, An approach to cephalotaxinone via the condensation of menolates with *O*-ethyl pyrrolidonium salts, Ph.D. Thesis, University of South Carolina, Columbia, SC, 1984, *Diss. Abstr. Int.* B **45**, 1472 (1984).

Iskura Sangyo Co., Ltd., Partial synthesis of harringtonine, Japanese Patent 58/32880 A2 [83/32880] (1983).

Y. Wang, Y. Li, Y. Li, X. Pan, S. Li, and W. Huang, Cephalotaxine dimethoxyacetate and its unexpected hydrolytic product—cephalotaxinone, *Lanzhou Daxue Xuebao, Ziran Kexueban* **19**, 17 (1983); *CA* **100**, 121418j.

W. Huang, Y. Li, X. Pan, S. Li, Q. Mu, and Y. Wan, Synthesis of *Cephalotaxus* ester alkaloids, *Chem. Nat. Prod., Proc. Sino-Am. Symp.*, Y. Wang, Ed., Gordon & Breach, New York, 1982, pp. 107–109.

T. Harding, Synthetic applications of photocyclization reactions. *N*-allyliminium salt photocyclizations and intramolecular photoarylation of *N*-haloarylethyl-β-enaminones, Ph.D. Thesis, Texas A. and M. University, College Station, TX, 1981, *Diss. Abstr. Int.* B **42**, 2836 (1982).

Kyowa Hakko Kogyo Co., Ltd., Cephalotaxine derivatives, Japanese Patent 57/64695 A2 [82/64695] (1982); *CA* **97**, 163308y.

Y. Li, S. Zhu, W. Wang, X. Pan, S. Li, Y. Wang, and W. Huang, New synthetic method for harringtonine, *Lanzhou Daxue Xuebao, Ziran Kexueban* **18**, 126 (1982); *CA* **98**, 16906j.

X. Pan, Y. Li, S. Li, Y. Wang, and W. Huang, Improved synthesis of cephalotaxinone, *Lanzhou Daxue Xuebao, Ziran Kexueban* **18**, 168 (1982); *CA* **97**, 182701n.

A. Pearson, P. Ham, and D. Rees, Organoiron complexes in organic syntheses. Part 14. Synthesis of azospirocyclic compounds via organoiron complexes. Potential synthetic routes to histrionicotoxin and *Cephalotaxus* alkaloids, *Tetrahedron Lett.* **21**, 4637 (1980).

P. Kaul, Studies on the synthesis of homoharringtonine, Ph.D. Thesis, Boston College, Chestnut Hill, MA, 1979, *Diss. Abstr. Int.* B **40**, 5273 (1980).

R. McNutt, Synthesis of harringtonine and 4,4-dimethylcyclobutenone, Ph.D. Thesis, Boston College, Chestnut Hill, MA, 1978, *Diss. Abstr. Int. B*, **38**, 5943 (1978).

C. Smith, Synthetic cephalotaxine esters having antileukemic activity, U.S. Patent Appl. 880000 (1978).

K. Mikolajczak and C. Smith, Synthesis of antitumor alkaloid deoxyharringtonine and its precursor, U.S. Patent 3959312 (1976).

A. Craig, Studies in the synthesis of peptide acyl activators. Studies in the synthesis of cephalotaxine, Ph.D. Thesis, University of Washington, Seattle, WA, 1975, *Diss. Abstr. Int.* B. **36**, 2805 (1975).

T. Utawanit, Part One: Stereospecific synthesis of trisubstituted olefins. Part Two: *Cephalotaxus* alkaloid esters, Ph.D. Thesis, University of Illinois, Urbana, IL, 1975, *Diss. Abstr. Int.* B **36**, 4499 (1976).

P. Christenson, Synthesis of harringtonine and cephalotaxine, Ph.D. Thesis, Boston College, Boston, MA, 1974, *Diss. Abst. Int. B.* **35**, 4830 (1975).

T. Rogerson, Reactions of 2-ethoxy-6,7-dihydro-5(H)-azepine. Methyl N-phenylacetimidate, new alkylating agent. Improved total synthesis of cephalotaxine, Ph.D. Thesis, Cornell University, Ithaca, NY, 1974, *Diss. Abst. Int. B*. **35**, 2663 (1974).

S. Nelson, Synthesis of the *Akuamma* alkaloids and cephalotaxine. Synthesis of the *Akuamma* alkaloids. Synthesis of cephalotaxine, Ph.D. Thesis, University of Oregon, Eugene, 1971, *Diss. Abst. Int. B*. **33**, 121 (1972).

Miscellaneous

M. Misawa, M. Hayashi, and S. Takayama, Production of antineoplastic agents by plant tissue cultures. I. Induction of callus tissues and detection of the agents in cultured cells, *Planta Med*. **49**, 115 (1983).

Kyowa Hakko Kogyo Co., Ltd., Cephalotaxines and their esters, Japanese Patent 57/102194 A2 [82/102194] (1982); *CA* **97**, 214276w.

M. Misawa, M. Hayashi, and S. Takayama, Production of antineoplastic agents by plant tissue cultures, *Plant Tissue Cult., Proc. Int. Congr. Plant Tissue Cell Cult., 5th,* A. Fujiwara, Ed., Maruzen: Tokyo, 1982, pp. 279–280.

Q. Yu, G. Ma, and Z. Huang, Studies on the fragmentation mechanisms of alkaloids isolated from *Cephalotaxus* under electron impact, *Huaxue Xuebao* **40**, 539 (1982); *CA* **97**, 182705s.

P. Conrad, Cephalotaxine support studies, Ph.D. Thesis, Purdue University, West Lafayette, IN, 1980, *Diss. Abstr. Int. B* **41**, 2178 (1980).

N. Delfel and J. Rothfus, Homodeoxyharringtonine and other cephalotaxine esters by tissue culture, U.S. Patent 4,152,214 (1979).

Honan Institute of Forestry, A preliminary study on the control of *Dendrolimus punctatus* with plant alkaloids, *K'un Ch'ung Hsueh Pao* **21**, 108 (1978); *CA* **89**, 1665v.

R. Powell and K. Mikolajczak, Desmethylcephalotaxinone and its correlation with cephalotaxine, *Phytochemistry* **12**, 2987 (1973).

Reviews

C. Smith, R. Powell, and K. Mikolajczak, The genus *Cephalotaxus*: source of homoharringtonine and related anticancer alkaloids, *Cancer Treat. Rep.* **60**, 1157 (1976).

A. S. Chawla and A. H. Jackson, *Erythrina* and related alkaloids, *Alkaloids (London)* **12**, 155 (1982).

Subject Index

691